HUMAN PHYSIOLOGY

HUMAN

The Benjamin/Cummings Series in the Life Sciences

F. J. Ayala
Population and Evolutionary Genetics: A Primer (1982)

F. J. Ayala and J. A. Kiger, Jr.
Modern Genetics (1980)

F. J. Ayala and J. W. Valentine
Evolving: The Theory and Processes of Organic Evolution (1979)

L. L. Cavalli-Sforza
Elements of Human Genetics, second edition (1977)

R. E. Dickerson and I. Geis
Hemoglobin (1983)

L. E. Hood, I. L. Weissman, W. B. Wood, and J. H. Wilson
Immunology, second edition (1984)

L. E. Hood, J. H. Wilson, and W. B. Wood
Molecular Biology of Eucaryotic Cells (1975)

J. B. Jenkins
Human Genetics (1983)

A. L. Lehninger
Bioenergetics: The Molecular Basis of Biological Energy Transformations, second edition (1971)

S. E. Luria, S. J. Gould, and S. Singer
A View of Life (1981)

E. Nicpon-Marieb
Human Anatomy and Physiology Laboratory Manual: Brief Edition (1983)

E. Nicpon-Marieb
Human Anatomy and Physiology Laboratory Manual: Cat and Fetal Pig Versions (1981)

E. B. Mason
Human Physiology (1983)

A. P. Spence
Basic Human Anatomy (1982)

A. P. Spence and E. B. Mason
Human Anatomy and Physiology, second edition (1983)

G. J. Tortora, B. R. Funke, and C. L. Case
Microbiology: An Introduction (1982)

J. D. Watson, N. Hopkins, J. Roberts, and J. Steitz
Molecular Biology of the Gene, fourth edition (1984)

I. L. Weissman, L. E. Hood, and W. B. Wood
Essential Concepts in Immunology (1978)

N. K. Wessells
Tissue Interactions and Development (1977)

W. B. Wood, J. H. Wilson, R. M. Benbow, and L. E. Hood
Biochemistry: A Problems Approach, second edition (1981)

PHYSIOLOGY

Elliott B. Mason

State University of New York College
at Cortland

Illustrations by Fran Milner

The Benjamin/Cummings Publishing Company, Inc.

Menlo Park, California • Reading, Massachusetts
London • Amsterdam • Don Mills, Ontario • Sydney

Dedication

To my wife Marsha and our children Jennifer, Julie, and Jessica. Their love, patience, encouragement, and understanding made this book possible.

Elliott B. Mason

Sponsoring Editor: *James W. Behnke*
Production Editor: *Karen Bierstedt*
Associate Editor: *Bonnie Garmus*
Copy Editor: *Carol Dondrea*
Developmental Editors: *John Hendry and Janet Wagner*
Cover and Book Designer: *John Edeen*
Artist: *Fran Milner*

About the cover: This scanning electron micrograph shows a number of nerve fibers leading to and crossing several striated muscle cells. © Lennart Nilsson.

Photo and art credits follow the glossary at the end of the text.

Library of Congress Cataloging in Publication Data
Mason, Elliott B., 1943–
    Human physiology.

    (The Benjamin/Cummings series in the life sciences)

    Includes index.
    1. Human physiology.   I. Title.   II. Series.
QP34.5.M463 1983   612   82-22805
ISBN 0-8053-6885-X

Second Printing

GHIJ–HA–89

The Benjamin/Cummings Publishing Company, Inc.
2727 Sand Hill Road
Menlo Park, California 94025

PREFACE

My major goal in writing this book is to provide a comprehensive treatment of human physiology accessible to students in a wide variety of programs, including allied health and nursing, physical education, the biological sciences, and the liberal arts. I have included enough of the relevant anatomy to make the inseparable relationship between structure and function meaningful. The book is written at a level that should make it readable by students who have had no prerequisite courses. Physical and chemical principles required to understand human physiology are presented within Chapter 2.

Sequence of Chapters

The chapters of this book are organized in the sequence most commonly used in the classroom. I recognize that there are other sequences in which the material might also be effectively presented so I have made each chapter as independent as possible. Cross-references have been included where they seem most useful.

Special Features

To enhance the text's usefulness as a teaching tool and to increase student interest, the following special features are included:

- An outstanding art program has been carefully integrated with the text. Clear and concise drawings by Fran Milner, complemented by many light micrographs and photographs, make the illustrations a valuable learning aid. When the book was designed, a special effort was made to ensure each illustration was positioned as closely as possible to the corresponding text. There are insets to help orient students to particular aspects of many of the figures.

- The introductory chapter begins with discussions of homeostasis and negative feedback mechanisms. These important concepts of human physiology are integrated throughout the remainder of the text and provide a unifying theme for the course.

- The specific immune responses and other mechanisms by which the body protects itself from infection and disease are topics of increasing importance. These topics are given special emphasis in Chapter 9, Defense Mechanisms of the Body.

- Many instructors include inheritance as a topic in their courses. Therefore, the material on reproduction in Chapter 20 includes a section presenting basic concepts of human genetics.

- Discussions of a number of diseases and dysfunctions and the effects of stress are included throughout the book. These serve to provide the student with a better understanding of the normal condition and emphasize the importance of maintaining homeostasis.

- Where relevant, brief discussions of the effects of aging are included. Aging is an area of active investigation, and students will gain some insight into its effects on homeostasis from the treatment it receives in this book.

Learning Aids

This book includes a number of learning aids to assist the diverse groups of students who take courses in human physiology.

Each chapter includes:

- learning objectives
- an introductory chapter outline
- a study outline (with page references back to the text)
- a self-quiz. (The answer key to the self-quiz is located in an appendix and includes references to the text pages on which the answers can be found.)

The appendices include:

- a guide to word roots, prefixes, suffixes, and combining forms
- a glossary, which contains over 1,000 definitions and provides a phonetic pronunciation of each term

Supplements

A comprehensive Instructor's Guide has been developed to accompany this book. The Instructor's Guide includes 40 chapter end quizzes set up for easy photocopying, plus two alternate final exams. Numerous references and resource bibliographies are also included.

Acknowledgments

In the preparation of this book, much was gained from the constructive comments and advice of a number of people who reviewed various portions and drafts of the manuscript (see list that follows). Their assistance and suggestions are gratefully acknowledged.

    Dr. Louis Gatto of the State University of New York College at Cortland generously provided many of the photomicrographs used in this book. Ms. Fayann Searfoss deserves special thanks for the many hours she spent typing the manuscript. Finally, a special thanks to the people at Benjamin/Cummings who were involved in this project. All of them provided valuable ideas, suggestions, and enthusiasm—particularly Jim Behnke, Executive Editor, Karen Bierstedt, Production Editor, and Bonnie Garmus, Associate Editor. The final form of this book owes much to the talents of John Hendry, the developmental editor, John Edeen, the book designer, and Fran Milner, the artist. Their contributions can never be adequately acknowledged.

List of Reviewers

Thomas Adams   Michigan State University
Robert M. Anthony   Triton College
Cynthia Carey   University of Colorado
Dwayne H. Curtis   California State University, Chico
Edward Donovan   Avila College
Steven A. Fink   West Los Angeles College
Steven Fisher   University of California, Santa Barbara
Lewis Greenwald   Ohio State University
John P. Harley   Eastern Kentucky University
August N. Jaussi   Brigham Young University
Ann Marie Kreuger   Boston State College
Stephen Langjahr   Antelope Valley College
Charles Leavell   Fullerton College
A. Kenneth Moore   Seattle Pacific University
David Saxon   Morehead State University
Tom Sourisseau   Cabrillo College

BRIEF CONTENTS

DETAILED CONTENTS

6 MUSCLE 120

7 BASIC FEATURES OF THE NERVOUS SYSTEM 150

13   BLOOD VESSELS AND THE LYMPHATIC SYSTEM   322

LEARNING OBJECTIVES

After completing this chapter, you should be able to:

- Define the term *physiology*.
- Cite three examples of the inter-relationship of structure and function.
- Describe the basic functions of the four types of tissues present in the body.
- Describe representative functions of the ten major body systems.
- Define homeostasis, and give examples of how the body maintains a homeostatic state.
- Explain the process of negative feedback, and describe how it is utilized to achieve a relatively stable, homeostatic control.
- Explain the process of positive feedback, and describe the response of a positive-feedback system.

CHAPTER CONTENTS

INTRODUCTION TO HUMAN PHYSIOLOGY

Physiology is the science that studies the functions of living organisms and their parts. It attempts to explain in physical and chemical terms the factors and processes involved in these functions. This book deals with human physiology—that is, with the specific aspect of physiology that examines the workings of the human body.

INTERRELATIONSHIP OF STRUCTURE AND FUNCTION

When studying human physiology, it is important to realize that structure and function are interrelated—indeed, they are inseparable. Just as the shapes and organization of the parts of a machine, such as an automobile, are appropriate for their functions, so the shapes and organization of the parts of the body are intimately associated with their functions. This interrelationship of structure and function is evident at all levels of body organization. At the gross-component level, for example, the structure of the human hand with its opposable thumb enables a person to grasp and manipulate objects efficiently. At the cellular level, the interrelationship of structure and function is exemplified by a nerve cell that has a long, thin process extending from its main portion. This process is well suited to the cell's function, which is to transmit information in the form of electrical signals from one region of the body to another. Even at the molecular level, the interrelationship of structure and function is apparent. For example, enzymes, which are protein molecules that speed up chemical reactions in the body, can act only if they have shapes that fit the shapes of the reacting molecules.

STRUCTURAL ORGANIZATION OF THE BODY

Because of the important relation between structure and function, the study of human physiology cannot be successfully undertaken without some familiarity with the structural organization of the body. In fact, it is largely through structure that function is accomplished.

There are four structural levels in the body: cells, tissues, organs, and systems.

CELLS

The body's basic unit of structural organization is the **cell,** and the body contains more than 75 trillion cells. Cells are living units that carry out those functions required for their own survival, such as acquiring and processing nutrients. Beyond this, however, the cells of the body tend to be specialized for the performance of specific activities that contribute to the well-being of the body as a whole. Muscle cells, for example, have a highly developed property of contractility (the ability to shorten or contract), which is of value in moving the body and its parts.

TISSUES

Within the body, groups of similar cells join together to form **tissues.** There are four principal types of tissues: epithelial, connective, muscle, and nervous.

Epithelial Tissue

Epithelial tissue is composed of cells that are very closely joined, with only a minimum of intercellular material between them. This tissue covers the surface of the body and lines body cavities, ducts, and vessels. It is also incorporated into various glands. Epithelial tissue forms a protective barrier between the body and the external environment, and it is involved in the absorption of materials into the body, in the secretion of special products within the body, and in the excretion of substances from the body.

Connective Tissue

Connective tissue is characterized by relatively few, widely spaced cells and an abundant intercellular substance. This tissue binds various body structures together and supports them in their proper locations.

Muscle Tissue

Muscle tissue is composed of specialized cells that are able to contract and thereby decrease in length. This tissue moves the bones of the skeleton, propels the blood throughout the body, and aids in digestion by moving food through the digestive tract.

Nervous Tissue

Nervous tissue contains many nerve cells, or neurons, that transmit information in the form of electrical signals from one point in the body to another. This tissue is found largely in the brain, the spinal cord, and nerves.

ORGANS

Organs consist of two or more different tissues combined in such a manner as to perform a specific function. For example, the stomach is a digestive organ. It is lined with epithelial tissue, which, in turn, is surrounded by muscle tissue. These tissues are held together by connective tissue and are supplied by nervous tissue.

SYSTEMS

Groups of organs that perform similar or related functions and that work together to accomplish a common purpose—for example, the digestion of food—make up a **system.** There are ten major systems in the human body: integumentary (the skin and associated structures), skeletal, muscular, nervous, endocrine, circulatory, respiratory, digestive, urinary, and reproductive. The functions of these systems are outlined in Table 1.1.

HOMEOSTASIS

The body's cells can survive and function efficiently only under relatively constant conditions of temperature, pressure, acidity, and so forth. However, the body as a whole is surrounded by an **external environment** in which there are rather wide fluctuations in temperature, humidity, and other factors. Therefore, if a person is to survive, his or her cells must be protected from the variability and extremes of this environment.

The body's cells are not directly exposed to the external environment. Instead, they exist in an aqueous **internal environment,** which is made up of the fluid portion of the blood and the interstitial fluid that continually bathes the cells (F1.1). The body maintains relatively constant chemical and physical conditions within this environment, and the existence of a relatively constant internal environment is referred to as **homeostasis.**

The fact that the internal environment is relatively constant does not imply that it is static or unchanging. Rather, a variety of occurrences continually tend to cause changes in this environment. For example, the activities of the body's cells remove materials such as glucose and oxygen from the internal environment and add materials such as urea and carbon dioxide. As a result, ho-

Table 1.1 *Major Body Systems*		
System	Major Components	Representative Functions
Integumentary	Skin and associated structures such as hair and nails	Protects internal body structures against injury and foreign substances; prevents fluid loss (dehydration); important in temperature regulation
Skeletal	Bones	Supports and protects soft tissues and organs
Muscular	Skeletal muscles	Moves body and its parts
Nervous	Brain, spinal cord, nerves, special sense organs	Controls and integrates body activities; responsible for "higher functions" such as thought and abstract reasoning
Endocrine	Hormone-secreting glands such as the pituitary, thyroid, parathyroids, adrenals, pancreas, and gonads	Controls and integrates body activities; function closely allied with that of the nervous system
Circulatory	Heart, blood vessels, blood	Links internal and external environments of body; transports materials between different cells and tissues
Respiratory	Nose, trachea, lungs	Transfers oxygen from atmosphere to blood, and carbon dioxide from blood to atmosphere
Digestive	Mouth, esophagus, stomach, small intestine, large intestine; accessory structures include salivary glands, pancreas, liver, gallbladder	Supplies body with substances (food materials) from which energy for activity is derived and from which components for the synthesis of required substances are obtained
Urinary	Kidneys, ureters, urinary bladder, urethra	Eliminates variety of metabolic end products such as urea; conserves or excretes water and other substances as required
Reproductive	Male: seminal vesicles, testes, prostate gland, bulbourethral glands, penis, associated ducts	Produces male gametes (sperm); provides method for introducing sperm into the female
	Female: ovaries, uterine tubes, uterus, vagina, mammary glands	Produces female gametes (ova); provides proper environment for development of fertilized ovum

meostasis can be maintained only if materials are added to the internal environment as rapidly as they are removed or removed as quickly as they are added. The internal environment, therefore, is not static, but rather exists in a dynamic steady state in which the input and output of materials are balanced.

Essentially, all the systems of the body contribute to the maintenance of homeostasis. For example, as the cells remove glucose and oxygen from the internal environment, the digestive and respiratory systems replace them. In addition, certain materials produced by the cells and added to the internal environment are removed by the

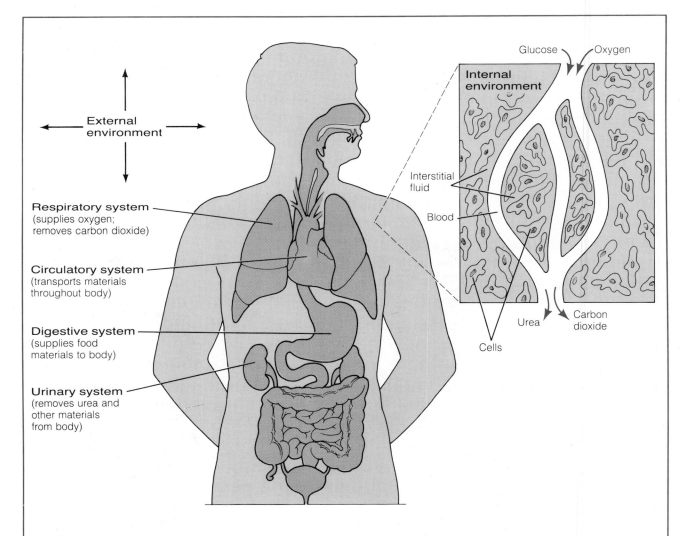

F1.1
The body's cells exist in an aqueous internal environment, which is made up of the fluid portion of the blood and the interstitial fluid that continually bathes the cells. Essentially all the systems of the body help maintain relatively constant conditions within this environment. The existence of a relatively constant internal environment is referred to as homeostasis.

urinary system. The circulatory system transports needed materials from areas such as the gastrointestinal tract or the lungs to the cells, and it also carries materials produced by the cells to organs such as the kidneys for removal from the body.

As long as the various systems function properly, the relative constancy of the internal environment is maintained, and the cells can survive and function efficiently. However, if the functioning of the systems is upset, the composition of the internal environment can change and become incompatible with survival.

HOMEOSTATIC MECHANISMS

In order to maintain homeostasis, the body must be able to sense changes in the internal environment. Moreover, it must be able to compensate for the changes, and this requires the ability to control the various systems concerned with maintaining the composition of the internal environment. The nervous and endocrine systems are the body's principal sensing and controlling systems.

NEGATIVE FEEDBACK

The body uses a regulatory principle known as **negative feedback** to maintain relatively constant, or stable, conditions in the internal environment.

Negative-feedback mechanisms have several components (F1.2a). One component, the *controlled system*, is a system whose activity is regulated in order to maintain the appropriate level of a particular variable—for example, temperature or oxygen. A second component, the *set point*, is a reference that calls for or indicates the level at which the variable is to be maintained. A third component, the *receptor*, monitors the variable and transmits information—referred to as feedback—to a fourth component, the *processing center*. The processing center compares and integrates information from the receptor about the actual level of the variable with information from the set point about the called-for level. If necessary, the processing center increases or decreases the activity of the controlled system in order to bring the actual level of the variable to the level called for by the set point.

A common example of a negative-feedback mechanism is the operation of a thermostatically controlled heater that keeps the temperature in a room comfortable and relatively constant when the temperature outside the room is low (F1.2b). The heater is the controlled system whose activity—heat production—is regulated in order to maintain the appropriate level of the variable, the room temperature. The set-point level of the variable is the temperature called for by setting the thermostat at a particular value. The thermostat contains a receptor that provides information about the actual room temperature and a processing center that compares this information with information about the set-point temperature. If the actual temperature differs from the set-point temperature, the thermostat turns the heater on or off as necessary to bring the actual room temperature to the set-point temperature.

Negative-feedback mechanisms are called negative because the feedback tends to cause the level of a variable to change in a direction opposite to that of an initial change. In the case of a thermostatically controlled heater, for example, an increase in the room temperature above the set-point level leads to increased feedback, which acts in an inhibitory fashion to turn the heater off and thereby allow the room temperature to decrease toward the set-point level. Conversely, when the room temperature falls below the set-point level, the inhibitory feedback decreases, the heater turns on, and the heat produced raises the room temperature toward the set-point level. Thus, negative-feedback mechanisms minimize the difference between the actual and the set-point level of a variable. Consequently, they tend to maintain relatively constant and stable conditions under circumstances that would otherwise cause the conditions to change.

Many negative-feedback mechanisms are present in the body. For example, the body's temperature control mechanism is believed to include a set point that calls for the maintenance of a particular body temperature. The body contains re-

and transmit feedback information to the brain, which contains a processing center. The processing center compares and integrates the information from the receptors about the actual body temperature with information from the set point about the called-for temperature, and, if necessary, initiates appropriate action to bring the two into line. For example, if the actual body temperature is below the set-point temperature, shivering may be initiated. The muscular contractions of shivering produce heat that tends to raise the body temperature to the set-point level. Conversely, if the body temperature is above the set-point temperature, sweating may be stimulated. The evaporation of the water in the sweat cools the body surface and tends to lower the body temperature to the set-point level.

POSITIVE FEEDBACK

Positive-feedback mechanisms are organized in a manner similar to that of negative-feedback mechanisms (see F1.2a). However, in contrast to negative-feedback mechanisms, positive-feedback mechanisms maximize, rather than minimize, the difference between the actual and the set-point level of a variable. They are called positive because the feedback tends to cause the level of the variable to change in the same direction as an initial change. With a positive-feedback mech-

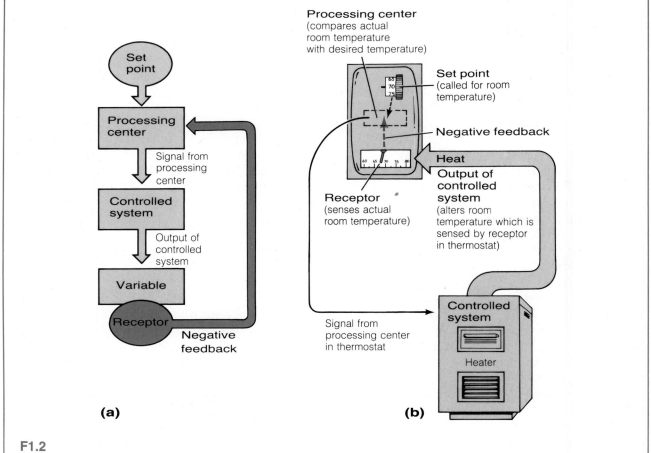

F1.2
(a) Schematic representation of the components of a negative-feedback mechanism.

(b) A thermostatically controlled heater is an example of a negative-feedback mechanism.

anism, an increase in the level of a particular variable above the set-point level leads to increased feedback that acts in a stimulatory fashion to increase the system's activity even more and thereby further increase the level of the variable. The increased level of the variable leads to still greater feedback, which in turn leads to a further increase in the activity of the system, and so on. Thus, positive feedback produces a cyclical effect in which a change in the level of a variable leads to further change in the same direction. In a positive-feedback situation, however, the level of a variable does not necessarily continue to change indefinitely. For example, the level to which a variable can rise may ultimately be limited by the controlled system's maximum level of activity or by the amount of energy or raw materials available to the system.

Positive feedback does not lead to the maintenance of stable, homeostatic conditions, and consequently it does not occur in the body as frequently as negative feedback. However, positive-feedback responses are occasionally evident. For example, during the birth of a baby, the pressure of the baby's head against the area around the opening of the mother's uterus stimulates the contraction of the uterine muscles (F1.3). The contractions, in turn, increase the pressure of the head against the area around the uterine opening, which further stimulates the contraction of the uterine muscles. In this instance, a positive-feedback response is clearly useful in promoting the expulsion of the baby from the uterus.

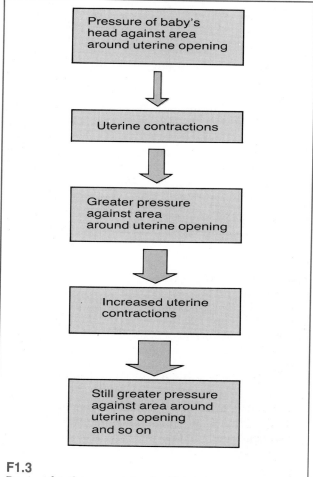

F1.3
During birth, a positive-feedback response stimulates increased uterine contractions.

STUDY OUTLINE

INTERRELATIONSHIP OF STRUCTURE AND FUNCTION Structure of body components is related to their function—for example, hand, nerve cells, enzymes. **p. 3**

STRUCTURAL ORGANIZATION OF THE BODY It is largely through structure that function is accomplished. **pp. 3–4**

Cells Living units that are the body's basic unit of structural organization.

Tissues Groups of similar cells.

EPITHELIAL TISSUE Covers body surface, lines body cavities, ducts, vessels; functions in protection, absorption, secretion, and excretion.

CONNECTIVE TISSUE Used for support, and to attach body structures to one another.

MUSCLE TISSUE Moves skeleton, pumps blood, moves food through digestive tract.

NERVOUS TISSUE Found largely in brain, spinal cord, nerves; neurons transmit electrical signals called nerve impulses throughout body.

Organs   Two or more different tissues combined in such a manner as to perform a specific function.

Systems   Groups of organs that perform similar or related functions and that work together to accomplish a common purpose; major systems: integumentary, skeletal, muscular, nervous, endocrine, circulatory, respiratory, digestive, urinary, reproductive.

HOMEOSTASIS   Refers to the existence of a relatively constant internal environment of the body. Required for normal cellular function.   **pp. 4–7**

HOMEOSTATIC MECHANISMS   Continued maintenance of homeostatis requires control and regulation of body systems.   **pp. 7–8**

Negative Feedback   Operates to minimize the difference between the actual and the called-for level of a variable. Provides means of relatively stable, homeostatic control.

POSITIVE FEEDBACK   Operates to maximize the difference between the actual and the called-for level of a variable; produces cyclical effects in which a change in the level of a variable leads to further change in the same direction.   **pp. 8–9**

SELF-QUIZ

1. Physiology is the science that studies the functions of living organisms and their parts. True or False?
2. Structure and function are interrelated at the cellular level of body organization but not at the molecular level. True or False?
3. Which tissue type has a highly developed property of contractility? (a) muscle tissue; (b) connective tissue; (c) nervous tissue; (d) epithelial tissue.
4. The stomach is a(n): (a) cell; (b) tissue; (c) organ; (d) system.
5. The body's internal environment is made up of the fluid portion of the blood and the interstitial fluid that continually bathes the cells. True or False?
6. The existence of a relatively constant internal environment of the body is referred to as: (a) corpostatin; (b) static maintenance; (c) homeostasis.
7. Which system is largely concerned with controlling various body activities? (a) endocrine; (b) reproductive; (c) digestive; (d) circulatory.
8. In negative feedback the effect of the feedback is to maximize the difference between the actual and the called-for level of a variable. True or False?
9. Negative feedback tends to result in: (a) a continually rising level of a particular variable; (b) a continually falling level of a particular variable; (c) stable, homeostatic control.

LEARNING OBJECTIVES

After completing this chapter, you should be able to:

- Define the term *chemical element*, and describe the three principal particles that make up an atom.
- Distinguish between polar and nonpolar covalent bonds, and give an example of each.
- Describe the composition of carbohydrates, lipids, proteins, and nucleic acids, and give an example of each.
- Describe how enzymes catalyze metabolic reactions, and cite two examples of the regulation of enzymatic activity.
- Distinguish between solutions, colloids, and suspensions.
- Explain three different ways in which the concentrations of solutions are expressed.
- Define an acid and a base.
- Explain the use of pH units in the measurement of the hydrogen ion concentration of a solution.
- Distinguish between diffusion and osmosis, and cite an example of each.
- Explain the process of dialysis, and describe its application in the artificial kidney.
- Distinguish between filtration and bulk flow, and cite an example of each.

CHAPTER CONTENTS

THE CHEMICAL AND PHYSICAL BASIS OF LIFE

Although the organs and tissues of the human body differ from one another in both form and function, they are composed of the same basic materials. The body is made up of chemical elements that interact with one another to form the anatomical structures and carry out the physiological processes characteristic of a living organism.

CHEMICAL ELEMENTS AND ATOMS

A **chemical element** is a substance that cannot be broken down into simpler material by chemical means. Each chemical element has its own name and a one- or two-letter symbol. For example, the symbol for the element oxygen is O, and the symbol for the element sodium is Na. At the present time, approximately 109 chemical elements are recognized, but only about 24 of these are normally found in the body (Table 2.1).

A chemical element is made up of extremely small units of matter called **atoms,** which are themselves composed of even smaller particles (F2.1). Positively charged particles called **protons** are located in the central area or nucleus of an atom. Uncharged particles called **neutrons,** if present, are also located in the nucleus. Negatively charged particles called **electrons** are in constant motion around the nucleus.

The number of protons and the number of electrons in an atom—and therefore, the number of positive charges and the number of negative charges—are equal. Consequently, an atom has no overall electrical charge and is electrically neutral.

Protons, neutrons, and electrons, like the atoms they compose, are matter, and, like all matter, they occupy space and possess mass. Each proton or neutron, however, has over 1800 times the mass of an electron. Thus, most of the mass of an atom is concentrated in the nucleus.

ELECTRON ENERGY LEVELS

Energy is the capacity to do work, and electrons possess different amounts of energy. In fact, the present view of atomic structure suggests that the electrons of an atom should be assigned to particular energy levels, which are numbered, starting with the lowest as

Table 2.1   *Chemical Elements of the Human Body*

Element	Symbol	Representative Functions
Carbon	C	A primary constituent of organic molecules such as carbohydrates, lipids, and proteins.
Hydrogen	H	A component of organic molecules and water. As an ion (H^+), it affects the pH of body fluids.
Nitrogen	N	A component of amino acids, proteins, and nucleic acids.
Oxygen	O	A component of many molecules, including water. As a gas, it is important in cellular respiration.
Calcium	Ca	A component of bones and teeth. It is required for proper muscle activity and blood clotting.
Chlorine	Cl	Ionic chlorine (Cl^-) is one of the major anions of the body.
Iodine	I	A constituent of the thyroid hormones thyroxine and triiodothyronine.
Iron	Fe	A constituent of the hemoglobin molecule. It is also a component of a number of respiratory enzymes.
Magnesium	Mg	Found in bone. It is also an important coenzyme in a number of reactions.
Phosphorus	P	A component of bones, teeth, many proteins, and nucleic acids. It is also a constituent of energy compounds such as adenosine triphosphate (ATP) and creatine phosphate (CP).
Potassium	K	As an ion, potassium (K^+) is the major intracellular cation. Potassium is important in the conduction of nerve impulses and in muscle contraction.
Sodium	Na	As an ion, sodium (Na^+) is the major extracellular cation. It is important in water balance and in the conduction of nerve impulses.
Sulfur	S	A component of many proteins, particularly the contractile protein of muscle.
Chromium	Cr	
Cobalt	Co	
Copper	Cu	
Fluorine	F	
Manganese	Mn	These substances are required by the body in very small amounts. They are referred to as trace elements.
Molybdenum	Mo	
Selenium	Se	
Silicon	Si	
Tin	Sn	
Vanadium	V	
Zinc	Zn	

one. The numbers are called **principal quantum numbers (n).** The lowest energy level ($n = 1$) can contain a maximum of 2 electrons, the second energy level ($n = 2$) can contain 8 electrons, the third energy level ($n = 3$) can contain 18 electrons, and the fourth energy level ($n = 4$) can contain 32 electrons (F2.2). Above the fourth energy level are still higher energy levels. (The electron energy levels are sometimes referred to as shells, with the K shell being equivalent to the $n = 1$ energy level, the L shell to the $n = 2$ energy level, the M shell to the $n = 3$ level, and so on.)

Electrons do not follow fixed pathways as they move around the nucleus of an atom, and it is impossible to determine the precise position of a specific electron at any one moment. However, electrons occupying higher energy levels are generally located farther from the nucleus than electrons occupying lower energy levels.

ATOMIC NUMBER, MASS NUMBER, AND ATOMIC WEIGHT

The number of protons in an atom is given by the atom's **atomic number.** The combined number of protons and neutrons is given by the atom's **mass number.** The number of neutrons in an atom is equal to the difference between the atom's atomic number and mass number. The atomic number of an atom is often indicated by a subscript preceding the symbol of the atom's chemical element, and the mass number is indicated by a superscript preceding the symbol. For example, an atom of oxygen with an atomic number of 8 and a mass number of 16 is written as $^{16}_{8}O$.

The total mass of an atom is called its **atomic weight.** The atomic weight of an atom is almost but not exactly equal to the sum of the masses of its constituent protons, neutrons, and electrons. The discrepancy is due to the fact that when protons, neutrons, and electrons combine to form an atom, some of their mass is converted to energy and is given off. The atomic numbers, mass numbers, and atomic weights of atoms of the chemical elements present in the body are given in Table 2.2.

ISOTOPES

All the atoms of a particular chemical element have the same number of protons. However, they can have different numbers of neutrons. These dif-

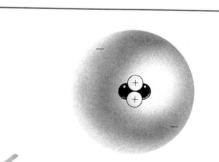

F2.1
An atom of the element helium. There are two protons (plus signs) and two neutrons (black circles) in the nucleus and two electrons (minus signs) in constant motion around the nucleus. The electrons do not follow fixed pathways as they move around the nucleus, but the color area indicates the region where they are located most of the time.

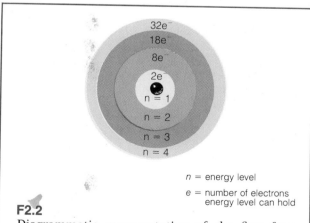

n = energy level

e = number of electrons energy level can hold

F2.2
Diagrammatic representation of the first four electron energy levels ($n = 1$, $n = 2$, $n = 3$, and $n = 4$). The circle in the center is the nucleus. The maximum number of electrons that can occupy each energy level is given.

ferent forms of a given element are called **isotopes.** Isotopes have the same atomic number (number of protons) but different mass numbers (due to different numbers of neutrons). For example, the most common isotope of the element oxygen has eight protons and eight neutrons in its nucleus ($^{16}_{8}O$). However, less common isotopes of oxygen containing eight protons and nine neutrons ($^{17}_{8}O$)

Table 2.2   *Atomic Numbers, Mass Numbers, and Atomic Weights of Atoms of Body Elements**

Element	Symbol	Atomic Number	Mass Number	Atomic Weight
Calcium	Ca	20	40	40.08
Carbon	C	6	12	12.011
Chlorine	Cl	17	35	35.453
Chromium	Cr	24	52	51.996
Cobalt	Co	27	59	58.933
Copper	Cu	29	63	63.546
Fluorine	F	9	19	18.998
Hydrogen	H	1	1	1.008
Iodine	I	53	127	126.905
Iron	Fe	26	56	55.847
Magnesium	Mg	12	24	24.305
Manganese	Mn	25	55	54.938
Molybdenum	Mo	42	98	95.94
Nitrogen	N	7	14	14.007
Oxygen	O	8	16	15.999
Phosphorus	P	15	31	30.974
Potassium	K	19	39	39.098
Selenium	Se	34	80	78.96
Silicon	Si	14	28	28.086
Sodium	Na	11	23	22.99
Sulfur	S	16	32	32.064
Tin	Sn	50	120	118.69
Vanadium	V	23	51	50.942
Zinc	Zn	30	64	65.38

*The mass numbers are those of the most common isotope of the element. The atomic weights (expressed in atomic mass units) are weighted averages of the atomic weights of the naturally occurring isotopes of the element.

and eight protons and ten neutrons ($^{18}_{8}O$) also exist. In nature, oxygen occurs as a mixture of these isotopes (99.76 percent $^{16}_{8}O$; 0.039 percent $^{17}_{8}O$; and 0.20 percent $^{18}_{8}O$).

RADIATION

A number of isotopes are radioactive; that is, they emit various kinds of radiation. Radioactive emissions can be either particlelike or they can take the form of electromagnetic rays.

PARTICULATE RADIATION

Alpha Radiation

An alpha particle is essentially a helium nucleus, composed of two protons and two neutrons, that is positively charged. Alpha particles are emitted by some radioactive isotopes, such as uranium ($^{238}_{92}U$).

Beta Radiation

A beta particle has essentially the same mass as an electron, but it can be either positively or negatively charged. Positive beta particles, which are called positrons, are emitted by such isotopes as $^{30}_{15}P$. Negative beta particles, which are called negatrons, are emitted by $^{14}_{6}C$ and $^{136}_{53}I$.

ELECTROMAGNETIC RADIATION: GAMMA RADIATION

A gamma ray can be regarded as a bundle of energy that is similar to a high-energy x-ray. It has no detectable mass or electrical charge. Gamma ray emissions frequently accompany positive or negative beta emissions.

TRANSMUTATION

When a radioactive element emits radiation, it is itself altered. The radioactive element disappears and a new element appears. Thus, atoms of one element spontaneously change into atoms of another element. This change of one element into another is called **transmutation** (radioactive decay). For example, the emission of an alpha particle ($^{4}_{2}He$) by uranium ($^{238}_{92}U$) forms thorium ($^{234}_{90}Th$) as follows:

$$^{238}_{92}U \rightarrow {}^{234}_{90}Th + {}^{4}_{2}He$$

BIOLOGICAL USES OF RADIOACTIVE ISOTOPES

Radioactive isotopes have proven very useful in studies of living organisms and the chemical reactions that occur within them. The radioactive emissions of different isotopes can be measured by a variety of means. Consequently, radioactive carbon and other radioactive isotopes are used as tracers that can be introduced into the body by ingestion, inhalation, or injection and followed through a variety of physiological processes. For example, the thyroid gland normally removes iodine from the blood and uses it in the formation of thyroid hormones. Thus, radioactive iodine, which is also taken up by the thyroid gland, can be used to study thyroid function.

Since radioactive emissions can damage tissues, high levels of radiation are used to treat certain disease conditions. In some forms of cancer, radiation treatments are used to destroy actively dividing cancer cells.

CHEMICAL BONDS

When atoms are close enough to one another, the outer electrons of one atom may interact with those of others. As a result, attractive forces can develop between atoms that are strong enough to hold the atoms together. These attractive forces are called **chemical bonds.**

In general, many atoms in the human body appear to be particularly stable when their highest electron energy levels are either filled or contain eight electrons, and much chemical bonding results in atoms that have either filled highest electron energy levels or highest levels that contain eight electrons.

COVALENT BONDS AND MOLECULES

In many cases, chemical bonds result from atoms sharing electrons. Bonds based on electron sharing are called **covalent bonds,** and two or more atoms held together as a unit by covalent bonds are known as a **molecule.** Many covalent bonds are *single covalent bonds*, in which two atoms share a pair of electrons, with one electron being provided by each atom. However, *double covalent bonds,* in which two atoms share two pairs of electrons, and *triple covalent bonds,* in which two atoms share three pairs of electrons, also occur.

Nonpolar Covalent Bonds

A covalent bond in which there is an equal sharing of electrons between two atoms, and in which one atom does not attract the shared electrons more strongly than the other atom, is called a **nonpolar covalent bond.** For example, the hydrogen atom possesses one proton in its nucleus and one electron in its $n = 1$ energy level. When two hydrogen atoms combine to form a hydrogen molecule, their single electrons are shared equally between the two nuclei so that each electron spends equal time in the vicinity of each nucleus (F2.3). Although neither atom gains complete possession of the other's electron, this sharing allows both to fill their highest ($n = 1$) electron energy levels.

Polar Covalent Bonds

A covalent bond in which there is an unequal sharing of electrons between two atoms, and in which one atom attracts the shared electrons more strongly than the other atom, is called a **polar covalent bond.** There can be varying degrees of polarity of such bonds, depending on how strongly one atom is able to attract shared electrons from another atom.

The unequal electron sharing that occurs in polar covalent bonds gives rise to polar molecules that have both positive and negative areas. For example, when two hydrogen atoms each bond covalently to an oxygen atom to form water, the shared electrons are more strongly attracted to the oxygen of the water molecule than to the hydrogens (F2.4). As a result, the shared electrons spend more time in the vicinity of the oxygen nucleus than in the vicinities of the hydrogen nuclei. Since electrons are negatively charged, the oxygen portion of the molecule becomes somewhat negative, and the hydrogen portions, with the positively charged protons of their nuclei less balanced by the presence of electrons, become somewhat positive.

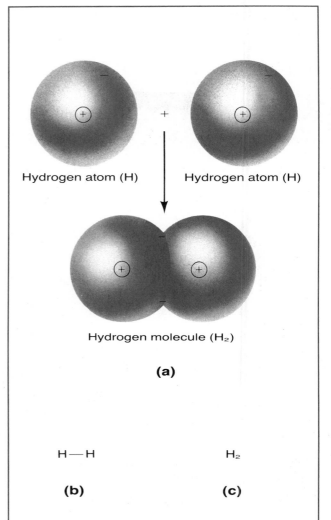

F2.3

(a) Covalent bonding of two hydrogen atoms to form a hydrogen molecule. In this situation, the electrons (minus signs) are shared equally between the two nuclei (plus signs), resulting in a nonpolar covalent bond. (b, c) Alternative methods of representing the composition of a molecule (in this case a hydrogen molecule) that will be used in future illustrations. In molecular diagrams (b) a shared pair of electrons (a single covalent bond) is represented by a straight line connecting the two bonded atoms. In molecular formulas (c) the number of atoms of each element that are present in a molecule is indicated, but there is little or no information as to how the atoms are connected.

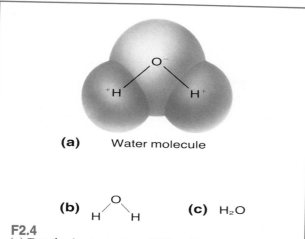

(a) Water molecule

(b) O / H H   **(c)** H₂O

F2.4
(a) Two hydrogen atoms (H) bonded covalently to an oxygen atom (O) to form a molecule of water. The shared electrons in this situation are attracted more strongly to the oxygen nucleus than to the hydrogen nuclei, resulting in polar covalent bonds. (b) A molecular diagram of a water molecule. (c) The molecular formula of a water molecule.

IONIC BONDS AND IONS

Often, the attraction of one atom for the electrons of another is so strong that electrons are not shared but are actually transferred from one atom to another; that is, they spend essentially all of their time in the vicinity of one nucleus and none in the vicinity of the other. This leaves one atom negatively charged (the one that gains electrons) and one atom positively charged (the one that loses electrons). Such charged atoms (or aggregates of atoms) are called **ions.** Positively charged ions are called **cations** and negatively charged ions are called **anions.**

Opposite charges attract one another, and oppositely charged ions can be held together by this attraction to form electrically neutral substances known as **salts.** Such attractions are called **ionic attractions** or **ionic bonds.** For example, in the reaction between sodium and chlorine, each sodium atom loses the single electron in its highest ($n = 3$) electron energy level to a chlorine atom (F2.5a). This produces positively charged sodium ions, which have their highest ($n = 2$) electron en-

ergy levels filled, and negatively charged chloride ions, which have a stable eight electrons in their highest ($n = 3$) electron energy levels. The positively charged sodium ions and negatively charged chloride ions attract one another, forming crystals of solid sodium chloride (table salt). In a sodium chloride crystal, sodium ions and chloride ions are packed into a three-dimensional lattice in such a way that each positive sodium ion is surrounded on four sides and top and bottom by negative chloride ions, and each chloride ion is similarly surrounded by six sodium ions (F2.5b). This is a particularly stable arrangement of positive and negative charges, and it occurs in many salts. Strictly speaking, there are no molecules in salts. There are only ordered arrays of ions in which no one positively charged ion belongs to any one negatively charged ion.

VAN DER WAALS ATTRACTIONS

Another type of chemical bond between atoms (and molecules) is known as **van der Waals attractions.** Van der Waals attractions are due mainly to momentary fluctuations in electron distributions around atoms. At any instant, the electrons of a particular atom may exist in an unsymmetrical distribution so that the atom has more of its electrons in one region than in another (F2.6). Thus, for a fraction of a second, the atom becomes polarized, with the region that has more electrons being slightly negative in relation to the region that has fewer electrons.

In this state, the atom behaves like a tiny electrical dipole (a dipole consists of a positive and an equal negative charge separated by some distance), and it can induce neighboring atoms to polarize. For example, in Figure 2.7, an instantaneous fluctuation in the electron distribution of the atom on the left causes the atom to have more of its electrons at the left than at the right. Consequently, the atom behaves as a dipole with a negative left side and a positive right side. The positive right side attracts electrons of the atom on the right in the figure and induces the formation of a similarly oriented dipole in this atom. The two atoms attract one another because the positive side of the left atom is close to the nega-

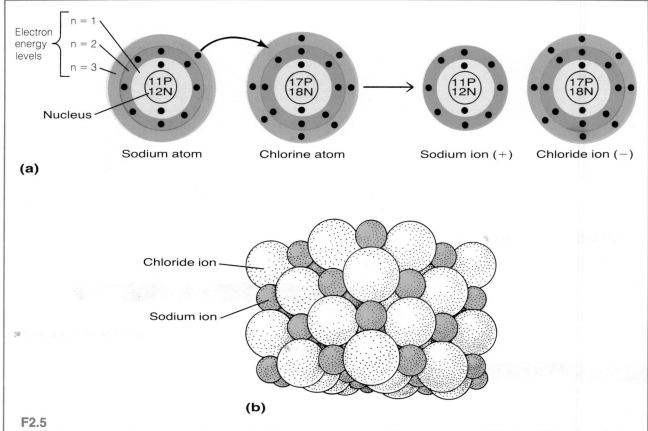

(a)

F2.5

(a) In the reaction between sodium and chlorine, each sodium atom loses the single electron (black dot) in its highest ($n = 3$) electron energy level to a chlorine atom. This produces positively charged sodium ions, which have their highest ($n = 2$) electron energy levels filled, and negatively charged chloride ions, which have a stable eight electrons in their highest ($n = 3$) electron energy levels. The number of protons (P) and neutrons (N) in the different nuclei is indicated. (b) Positively charged sodium ions and negatively charged chloride ions attract one another, forming crystals of solid sodium chloride (table salt). In a sodium chloride crystal, each sodium ion is surrounded by six chloride ions, and each chloride ion is surrounded by six sodium ions.

tive side of the right atom. This situation persists only for an extremely short time because the electrons are in motion. However, as electrons move from the left side to the right side of the atom on the left, electrons of the atom on the right move in a similar fashion. As a consequence, van der Waals attractions can be thought of as arising when electrons of neighboring atoms synchronize their motion in order to avoid one another as much as possible. This results in fluctuating electron distributions ("fluctuating dipoles") that give rise to extremely small but important instantaneous attractions between atoms.

HYDROGEN BONDS

Oppositely charged regions of polar molecules can attract one another, and an attraction of this sort that involves hydrogen and certain other atoms such as oxygen or nitrogen is called a **hydrogen bond.** For example, as noted previously, in the polar water molecule, the hydrogen portions of the molecule are somewhat positive, and the oxygen portion is somewhat negative. Consequently, the hydrogen portions of a water molecule can attract the oxygen portion of nearby water molecules, forming hydrogen bonds (F2.8). Hydrogen bonds

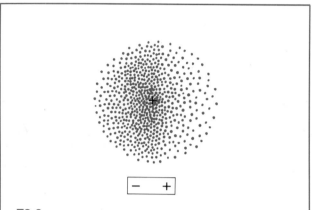

F2.6
Schematic illustration of the distribution of electrons around an atom. The greater the intensity of the stipling, the greater the probability that the electrons of the atom are located in that area. The + sign indicates the nucleus. For the fraction of a second indicated by this figure, the electron distribution is unsymmetrical, and more electrons are at the left than at the right. Consequently the atom is polarized as indicated.

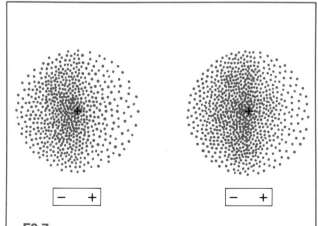

F2.7
The instantaneous polarization of one atom (left) can induce neighboring atoms (right) to polarize, and the polarized atoms can attract one another.

also occur in proteins and other large molecules found in the body.

CARBON CHEMISTRY

A tremendous number of molecules, especially those found in living organisms, contain the element carbon. A carbon atom has four electrons in its highest ($n = 2$) electron energy level, which it can share with other atoms. Carbon atoms, therefore, can each form four covalent bonds, and they commonly bond with hydrogen atoms or with other carbon atoms. Molecules composed only of carbon and hydrogen are called **hydrocarbons** (F2.9). Molecules that contain carbon, hydrogen, and additional elements are called **hydrocarbon derivatives.** Together, hydrocarbons and hydrocarbon derivatives constitute the class of molecules known as **organic molecules.** Organic molecules are so named because at one time they were thought to be made only by living organisms. All other molecules are classed as **inorganic molecules.**

Although both organic and inorganic molecules are essential to life, much of the body's chemistry is organic chemistry. Several major groups of organic molecules are of particular importance to the body.

CARBOHYDRATES

Carbohydrates, which include sugars and starch, are a major energy source for the body. In general, carbohydrates are composed of carbon, hydrogen, and oxygen, and the hydrogen and oxygen are frequently present in the same $2 : 1$ ratio that they are in water. An important group of carbohydrates is the **monosaccharides,** or single-unit sugars, such as glucose ($C_6H_{12}O_6$) (F2.10). Monosaccharide molecules can be linked together into larger molecules by synthetic reactions that generally involve the removal of a molecule of water (dehydration) at each linkage. The combination of two monosaccharide units produces a molecule called a **disaccharide.** Sucrose, or table sugar, is a disaccharide formed by bonding a glucose molecule to another monosaccharide called fructose. Much larger carbohydrate molecules can be formed by

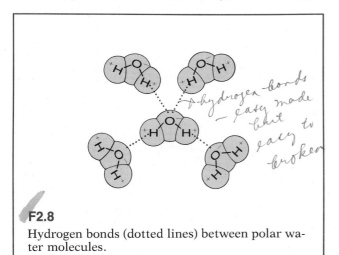

F2.8

Hydrogen bonds (dotted lines) between polar water molecules.

linking together many monosaccharide units. These molecules are called **polysaccharides.** Glycogen, a storage form of body carbohydrates, is a polysaccharide formed from thousands of glucose units bonded together into a single large molecule.

LIPIDS

Lipids are stored by the body as energy reserves, and they are utilized as structural components. Lipids are almost insoluble in water but very soluble in organic solvents such as benzene, ether, and chloroform. The major lipids are the **fatty acids** and **fats** which, like the carbohydrates, are composed of the elements carbon, hydrogen, and oxygen. However, fatty acids and fats contain only small amounts of oxygen.

Fatty acids consist of chains of carbon atoms with an acid carboxyl group (COOH) at one end of the chain (F2.11a). The most important of the fatty acids are: stearic ($C_{17}H_{35}COOH$); palmitic ($C_{15}H_{31}COOH$); oleic ($C_{17}H_{33}COOH$); linolenic ($C_{17}H_{29}COOH$); linoleic ($C_{17}H_{31}COOH$); and arachidonic ($C_{19}H_{31}COOH$).

Fats are formed from the union of fatty acids with the alcohol glycerol (F2.11b). The combination occurs at the carboxyl group end of the fatty acid by the removal of a molecule of water, which commonly occurs when the organic molecules of the body join with one another. If a single fatty acid molecule attaches to the glycerol molecule,

the product is called a **monoacylglycerol (monoglyceride).** If two fatty acids attach to the glycerol molecule, the product is a **diacylglycerol (diglyceride),** and if three fatty acids attach, a **triacylglycerol (triglyceride)** is formed. Triacylglycerols are stored by the body as energy reserves.

Another group of lipids, the **phospholipids (phosphoglycerides),** are important components of cellular membranes. A phospholipid consists of a glycerol molecule to which two fatty acids are attached (F2.12). In addition, a third molecule, containing a phosphate group and, usually, nitrogen is attached to the glycerol. The phosphate and nitrogen can ionize. As a result a phospholipid has a polar region at the end of the molecule where the glycerol and ionized phosphate and nitrogen are located, and a nonpolar region where the fatty acid chains of carbon atoms extend from the glycerol.

A molecule such as a phospholipid molecule that has polar or ionized groups at one end and a nonpolar region at the opposite end is called **amphipathic.** In an aqueous (water) environment, amphipathic molecules form spherical clusters known as **micelles** (F2.13a). The polar regions of the molecules are located at the surface of the micelle, where they associate (in part by hydrogen bonds) with water molecules. The nonpolar regions are oriented toward the center of the micelle, where they are attracted to one another by van der Waals attractions. In cellular membranes, amphipathic phospholipid molecules are organized in a similar manner, with the polar regions of the molecules located at the membrane surfaces and the nonpolar regions oriented toward the center of the membrane (F2.13b).

$$H - \overset{\displaystyle H}{\underset{\displaystyle H}{C}} - H$$

F2.9

A carbon atom bonded covalently to four hydrogen atoms to form a molecule of methane (CH_4), which is an example of a hydrocarbon.

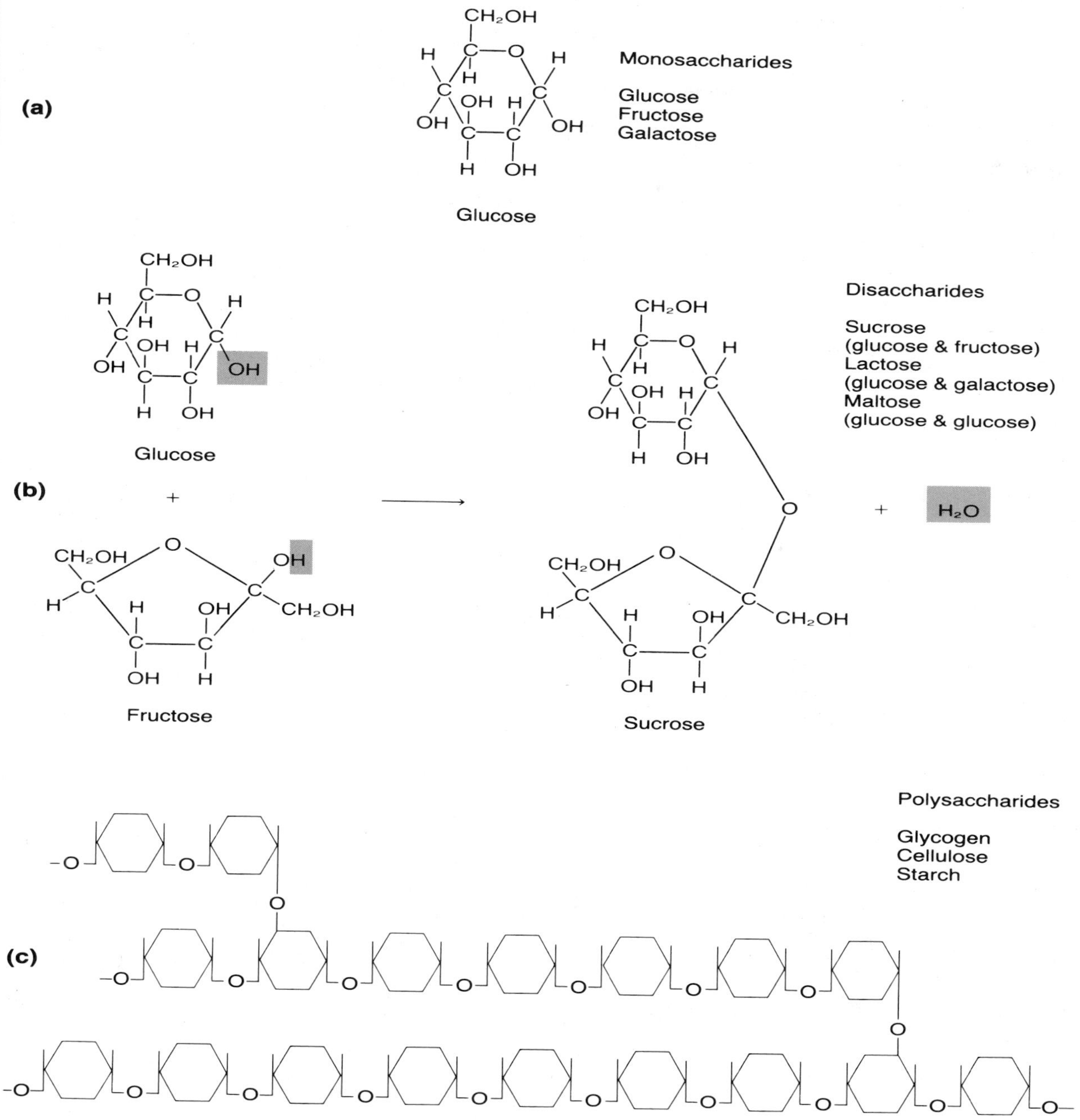

(a)

Monosaccharides

Glucose
Fructose
Galactose

Glucose

(b)

Glucose

+

Fructose

Disaccharides

Sucrose
(glucose & fructose)
Lactose
(glucose & galactose)
Maltose
(glucose & glucose)

Sucrose

+ H_2O

(c)

Polysaccharides

Glycogen
Cellulose
Starch

F2.10
Carbohydrates. (a) The monosaccharide glucose. (b) Monosaccharides are joined by bonds in a reaction that generally involves the removal of a molecule of water (a dehydration synthesis reaction, as shown here by the color areas). (c) A portion of the polysaccharide glycogen; the ring structures represent glucose molecules.

An additional group of lipids are the steroids. **Steroids** basically consist of four interconnected rings of carbon atoms that have few polar groups attached (F2.14). Cholesterol and some hormones—for example, the sex hormones—are steroids that are important in the body.

PROTEINS

Proteins are components of many body structures, and certain proteins—the enzymes—play critical roles in the chemical reactions that occur within the body. **Proteins** are large, complex molecules that are formed from smaller molecules called **amino acids.** Generally, amino acids have a central or alpha carbon to which is attached a hydrogen atom (H), an acid carboxyl group (COOH), an amino group (NH_2), and a fourth group that differs from amino acid to amino acid and is often indicated by the letter R (F2.15). Thus, amino acids, and consequently proteins, contain nitrogen (from the amino group) in addition to carbon, hydrogen, and oxygen. They may also contain other elements

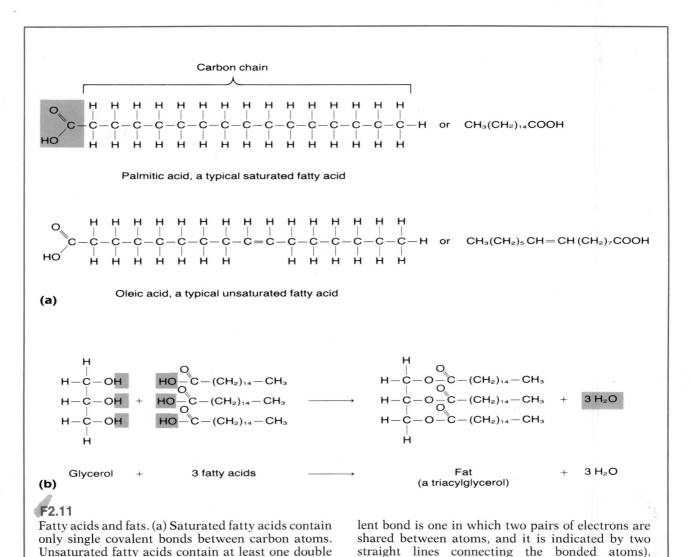

Palmitic acid, a typical saturated fatty acid

Oleic acid, a typical unsaturated fatty acid

(a)

Glycerol   +   3 fatty acids $\longrightarrow$ Fat (a triacylglycerol)   + $3 H_2O$

(b)

F2.11

Fatty acids and fats. (a) Saturated fatty acids contain only single covalent bonds between carbon atoms. Unsaturated fatty acids contain at least one double covalent bond between carbon atoms (a double cova- lent bond is one in which two pairs of electrons are shared between atoms, and it is indicated by two straight lines connecting the bonded atoms). (b) Glycerol and fatty acids combine to form fat.

such as sulfur, depending on the constitution of the individual R groups. Approximately 20 different amino acids (different because they possess different R groups) are commonly found in the proteins of the human body (Table 2.3). However, no one protein necessarily has all these different amino acids in its structure.

Proteins are formed from amino acids by reactions that bond the amino group of one amino acid to the acid carboxyl group of another with the loss of a molecule of water (F2.16). This bond is called a **peptide bond.** Two amino acids joined together by a peptide bond form a **dipeptide.** Approximately ten or more amino acids linked into a chain by peptide bonds form a **polypeptide.** A **protein** is a chain of approximately 100 or more amino acids linked by peptide bonds.

The sequence of amino acids in a polypeptide chain or protein constitutes what is called the **primary structure** of the molecule. Hydrogen bonds that occur principally between the constituents of the different peptide bonds of the linked amino

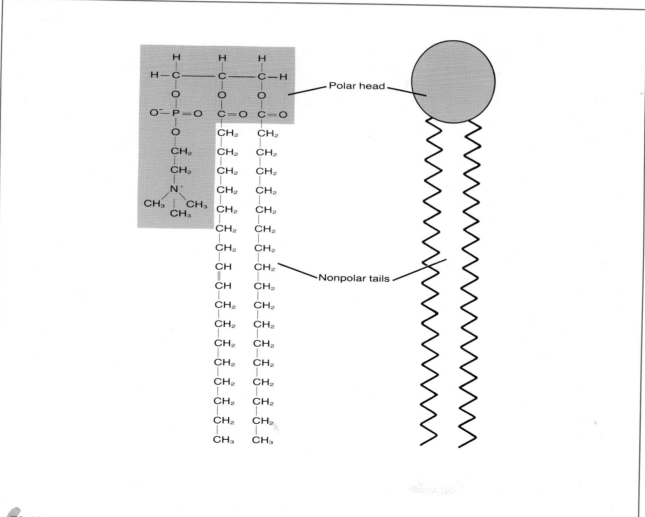

F2.12

(a) A phospholipid molecule. (b) Diagrammatic representation of a phospholipid molecule.

acids provide the polypeptide or protein with a **secondary structure.** For example, the hydrogen bonds cause some amino acid chains to form a coiled, helical structure called the alpha helix (F2.17). Interactions between atoms of the R groups of different amino acids of an amino acid chain also occur. These interactions cause the amino acid chain (which may be in a helical configuration) to fold into a particular three-dimensional configuration. This folding provides the polypeptide or protein with a **tertiary structure** (F2.18). Protein molecules can consist of a single amino acid chain, or several chains may link together through R-group interactions to form a multichain protein molecule. The interactions between the different amino acid chains of a multichain protein molecule provide still another level of structural organization to protein molecules—the **quaternary structure** (F2.19).

NUCLEIC ACIDS

Nucleic acids store and transmit information that is needed to synthesize the particular polypeptides and proteins present in the body's cells. Nucleic acids are complex molecules composed of structures known as purine and pyrimidine bases, five-carbon sugars (pentoses), and phosphate groups (which contain phosphorus and oxygen). A single base–sugar–phosphate unit is called a **nucleotide** (F2.20a). Individual nucleotides are linked together into a polynucleotide chain by bonds between the phosphate group of one nucleotide and the sugar of the next (F2.20b). If the nucleotides in the polynucleotide chain contain the sugar ribose, the chain is called **ribonucleic acid** or **RNA.** If the sugar is deoxyribose, the chain constitutes one portion of the two-chain molecule, **deoxyribonucleic acid** or **DNA.** A complete DNA molecule consists of two polynucleotide chains

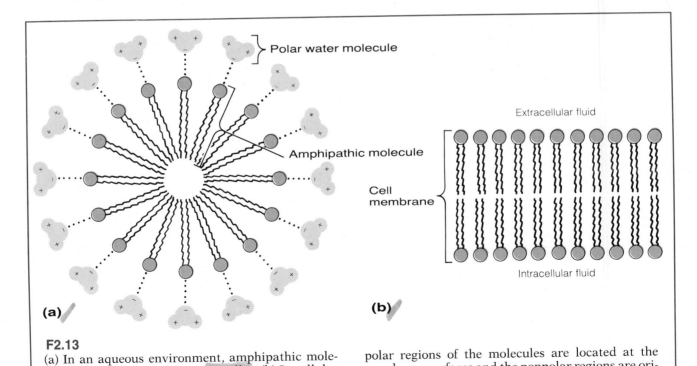

F2.13

(a) In an aqueous environment, amphipathic molecules become organized into micelles. (b) In cellular membranes, amphipathic phospholipid molecules are organized into a bimolecular layer in which the polar regions of the molecules are located at the membrane surfaces and the nonpolar regions are oriented toward the center of the membrane.

that run in opposite directions to one another. The purine and pyrimidine bases that are opposite one another in each polynucleotide chain link together by hydrogen bonds, and the two linked chains form a double spiral coil known as a *double helix* (F2.21). The complete, two-chain DNA structure is commonly called a DNA molecule even though the two polynucleotide chains are held together by hydrogen bonds rather than by covalent bonds. The purine bases of DNA are adenine and guanine, and the pyrimidine bases are cytosine and thymine. Because of structural and bonding considerations, when the two polynucleotide chains of DNA link with one another, an adenine of one

chain always bonds with a thymine of the other chain and vice versa, and cytosine always bonds with guanine and vice versa (see F2.21). This is called **complementary base pairing.** As a result of complementary base pairing, if the base sequence of one chain is known, the base sequence of the other can be predicted. The same bases are also found in RNA, with the exception that the base uracil substitutes for thymine. DNA is the genetic material of the cell, and it makes up a major portion of structures called chromosomes. DNA contains coded messages within its base sequence that instruct cells to synthesize particular polypeptides or proteins. RNA is involved in the trans-

F2.14

Steroids. Testosterone is a male sex hormone. Estradiol and progesterone are female sex hormones. As can be seen from the rings of carbon atoms, carbon atoms and the hydrogen atoms bound to them are often not specifically indicated by the letters C or

H, respectively, in the representation of the structure of a carbon-containing molecule. This method of representing molecular structure will be encountered in future figures.

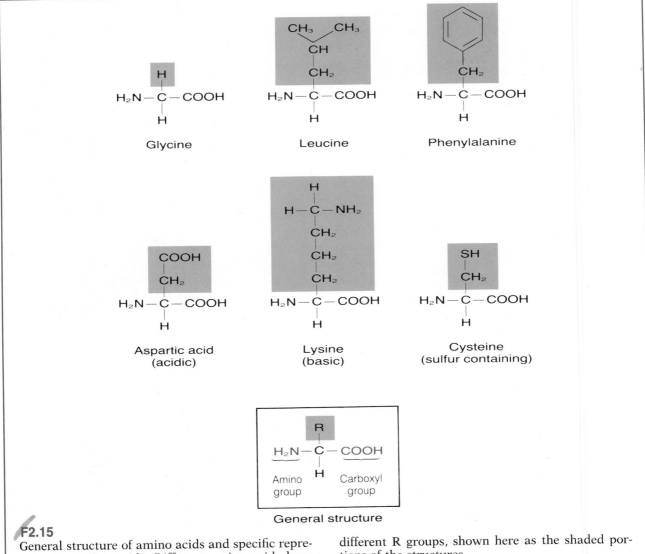

F2.15
General structure of amino acids and specific representative amino acids. Different amino acids have different R groups, shown here as the shaded portions of the structures.

F2.16
Peptide bonds link the amino group of one amino acid to the acid carboxyl group of another.

mission of the information of DNA to the active synthetic areas of the cells. These molecules are discussed further in Chapter 3.

ADENOSINE TRIPHOSPHATE

A substance called **adenosine triphosphate (ATP)** is the immediate source of energy for organismal activity—for example, muscle contraction. ATP is composed of the nitrogenous base adenine, the five-carbon sugar ribose, and three phosphate groups (F2.22). The phosphate groups are linked by high-energy chemical bonds that, when broken, provide energy to support the activities of the body. For example, when the terminal phosphate group is split away from a molecule of ATP, a molecule of adenosine diphosphate (ADP) is produced, and energy is released.

$$ATP \rightleftharpoons ADP + Phosphate + Energy$$

Once ADP has been formed, it can be resynthesized into ATP provided energy is available. As will be considered later (see Chapter 17), the breakdown of various food materials by chemical reactions that occur in the body releases energy that is utilized in ATP synthesis. In this way, energy contained within the chemical bonds of food materials is made available to the body in a useable form as ATP.

ENZYMES AND METABOLIC REACTIONS

The chemical reactions that constantly occur within the body are lumped together under the classification of **metabolism.** Metabolic reactions, in turn, are subdivided into anabolic, or synthesis, reactions that build up body structure, and catabolic, or decomposition, reactions that break down materials for various purposes such as the supply of energy. Metabolic reactions produce heat, and this heat can be of value because humans must maintain a constant body temperature. When humans are exposed to cold, metabolic rates increase and the heat generated is important in maintaining body temperature.

At normal body temperature, most metabolic reactions do not occur fast enough to benefit the body. Consequently, special catalysts—that is, substances that accelerate chemical reactions without undergoing any net chemical change during the reactions—are utilized to increase the rates of metabolic reactions to levels that can meet the body's needs. The biological catalysts include a group of proteins collectively termed **enzymes.**

Table 2.3 *The Twenty Amino Acids Found in Proteins*

Name	Three-Letter Abbreviation	One-Letter Abbreviation
Alanine	Ala	A
Arginine	Arg	R
Asparagine	Asn	N
Aspartic acid	Asp	D
Cysteine	Cys	C
Glutamic acid	Glu	E
Glutamine	Gln	Q
Glycine	Gly	G
Histidine	His	H
Isoleucine	Ile	I
Leucine	Leu	L
Lysine	Lys	K
Methionine	Met	M
Phenylalanine	Phe	F
Proline	Pro	P
Serine	Ser	S
Threonine	Thr	T
Tryptophan	Trp	W
Tyrosine	Tyr	Y
Valine	Val	V

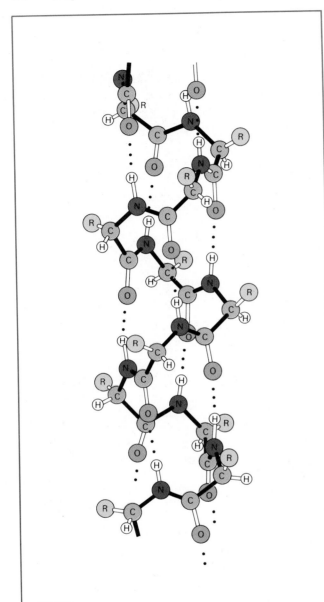

F2.17
Secondary alpha helical structure of proteins. Dotted lines indicate hydrogen bonds.

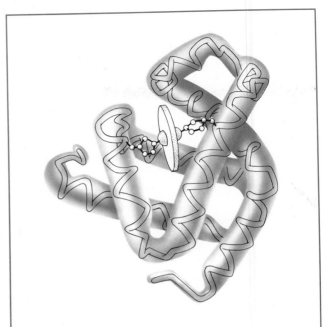

F2.18
Tertiary structure of a molecule of the protein myoglobin. Areas of helical secondary structure are also evident. The tube (color) that appears to enclose the secondary helical structure has been drawn here to make it easier to visualize the molecule's tertiary structure.

ACTION OF ENZYMES

Enzymes increase reaction rates by lowering the **activation energy** required for metabolic reactions to occur (F2.23). Under normal body conditions, few of the atoms and molecules that participate in a particular metabolic reaction have the necessary amount of energy to react with one another. In the presence of the proper enzyme, however, more of these atoms and molecules have the necessary energy to react because the enzyme lowers the amount of energy required. As a result, reactions that would otherwise proceed very slowly occur rapidly enough to be useful to the body.

Enzymes act by forming a temporary union with the reacting molecules, which are called **substrates.** This union is called an **enzyme–substrate (ES) complex.** The particular portion of an enzyme molecule with which a substrate combines is called the **active site** of the enzyme. Enzymes are very specific, and each catalyzes only individual reactions or limited classes of reactions. Only certain substrates can fit the specific three-dimensional structure of a given enzyme; therefore, only these substrates can form enzyme–substrate complexes, and react (F2.24). Thus, enzymes and substrates must fit together much like a key in a lock if a reaction is to be catalyzed successfully.

Since the specific three-dimensional configuration, or shape, of a protein that acts as an enzyme is essential to its ability to form an enzyme–substrate complex, factors that disrupt the chemical bonds determining that shape can denature the enzyme—that is, can alter its shape and thereby destroy its catalytic ability. Among the factors that can denature enzymatic proteins are variations in such internal environmental conditions of the body as temperature and acidity. The internal environment of the body, therefore, must be relatively constant—that is, homeostasis must be maintained—if the chemical reactions required for survival are to proceed in a stable fashion. Generally, enzymes are not destroyed during the course of the reactions they catalyze, and they appear at the conclusion of the reactions in the same states as when they entered. Many of the body's enzymes are initially produced in inactive forms (precursors) that must be activated before they will be effective catalysts.

REGULATION OF ENZYMATIC ACTIVITY

Because enzymes increase the rates of metabolic reactions, the amounts and types of different enzymes present in the body at any given moment play an important role in determining how rapidly different reactions can occur. Therefore, the regulation of enzymatic activity provides a method of regulating metabolic reactions.

Control of Enzyme Production and Destruction

One means of regulating enzymatic activity is to control the rates of production and destruction of particular enzymes, and thereby control the total amount of those enzymes present. If a particular enzyme is present in relatively large amounts, the reaction it catalyzes may proceed at a rapid rate, and much product may form. If only relatively small amounts of the enzyme are present, the re-

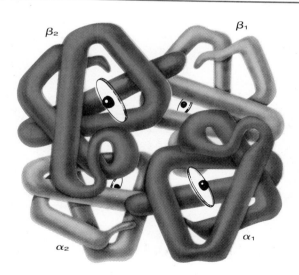

F2.19
Quaternary structure of the protein hemoglobin. A single hemoglobin molecule is composed of four polypeptide chains linked to one another. The particular configuration of the linked chains is called the quaternary structure of the protein.

action may proceed very slowly and only little product may form.

Control of Enzyme Activity

A second means of controlling enzymatic activity is to inhibit or enhance the activities of particular enzymes. In one type of inhibition, called **com-** petitive inhibition, an inhibitor molecule rather than a substrate molecule attaches reversibly to the active site of an enzyme molecule. Although the inhibitor molecule is generally similar in structure to the substrate molecule, it does not react to form a product as the substrate molecule does. Because both the inhibitor molecule and the

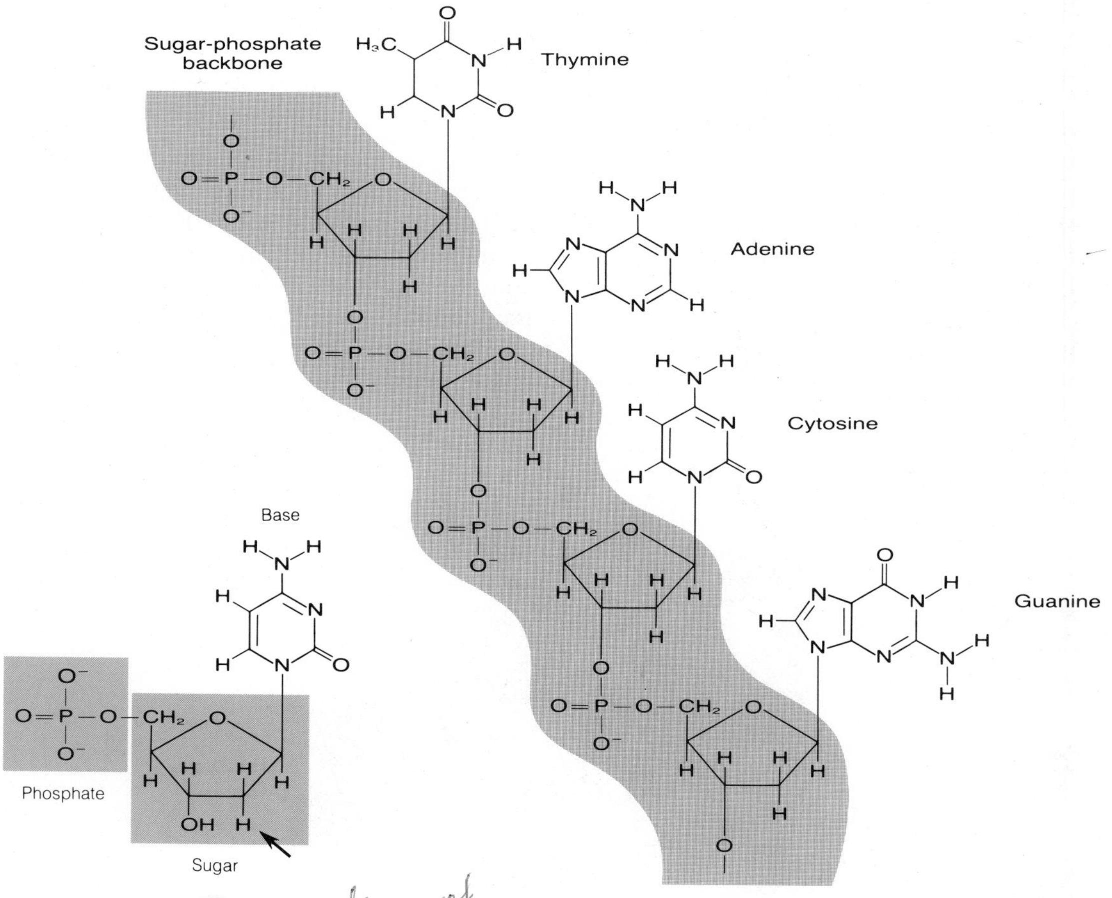

F2.20

(a) nucleotide   ↔ made up of phosphate & sugar bond

(b) Polynucleotide chain

(a) The nucleotide pictured here contains the base cytosine and the sugar deoxyribose. The sugar ribose has the same structure with the exception that the hydrogen indicated by the arrow is replaced by a hydroxyl (OH) group. (b) Four nucleotides linked in a polynucleotide chain. Adenine and guanine are purine bases and cytosine and thymine are pyrimidine bases.

substrate molecule can combine reversibly with the active site of the enzyme molecule there is a competition between them for the active site. If a great number of inhibitor molecules are present, many of them occupy the active sites of the enzyme molecules, and relatively few substrate molecules are able to form enzyme–substrate complexes, and react.

Another type of enzyme inhibition, called **noncompetitive inhibition,** can take several forms. In some cases, a noncompetitive inhibitor molecule combines irreversibly with the active site of an enzyme molecule. As a result, substrate molecules cannot combine with the active site. In other cases, a noncompetitive inhibitor molecule combines with an enzyme molecule at a site other than the active site. As a result of this combination, the structure of the enzyme molecule is altered so that it is less able to form an effective enzyme–substrate complex with its substrate.

Enzymes subject to regulation by small molecules have special binding sites called allosteric effector sites to which regulatory molecules attach by weak bonds. The combination of a regulatory molecule with an allosteric site alters the structure of the enzyme molecule and either activates or inhibits the enzyme. Often, in what is essentially a negative-feedback response, the final product of a series of enzymatically catalyzed reactions allosterically inhibits the first enzyme in the series. Thus, as the amount of product increases, the activity of the system that produces the product declines. Allosteric inhibition is a type of noncompetitive inhibition.

Some enzyme molecules are activated or inhibited by the chemical addition or removal of a phosphate (or other) group. The regulation of enzyme activity by phosphorylation or dephosphorylation differs from allosteric regulation in that it involves changes in covalent bonds in enzyme molecules, whereas allosteric regulation involves only patterns of weak bonds. The addition or removal of covalently bound groups requires the intervention of still other enzymes: kinase enzymes add phosphate groups, whereas phosphatase enzymes remove them. Since kinase and phosphatase enzymes are themselves subject to regulation—frequently by feedback mechanisms—enzyme regulation can involve a series of interacting events.

COFACTORS

Often, enzymes require the presence of nonprotein structures called **cofactors** to actively catalyze reactions. A cofactor may be either a metal ion or a complex organic molecule called a **coenzyme.** Many vitamins, for example, act as coenzymes.

SOLUTIONS

Water is the medium in which all living processes occur, and life as we know it would be inconceivable in the absence of this molecule. In fact, the chemical reactions that occur continuously within the body involve, for the most part, reactants that are in aqueous (water) solutions. A **solution** is a homogeneous mixture of two or more components that can be gases, solids, or liquids. The components of a true solution cannot be distinguished in the mixture, and they do not settle out at an appreciable rate. If a beam of light is passed through a true solution, the light path will not be visible. The particles dispersed within a true solution are very small—generally in the atomic and molecular size range. For example, sodium chloride dissolved in water forms a true solution. When dealing with solutions, the material present in the greatest amount is generally called the **solvent,** whereas substances present in smaller amounts are generally called **solutes.**

WATER AS A SOLVENT

In the body, water is the principal solvent, and the solutions most commonly encountered in living organisms result from dissolving gases, liquids, or solids in water. Water is an ideal solvent for living organisms for several reasons:

1. Water is a polar liquid whose chemical properties are such that many different materials can dissolve in it. For example, many salts dissolve easily in water because the positive and negative charges on the

polar water molecules can substitute for the positive and negative charges on the ions that compose the crystal lattices of the salts. When a salt crystal dissolves in water, each positively charged ion becomes surrounded by water molecules that have their negative oxygen portions turned toward the ion, and each negatively charged ion becomes surrounded by water molecules that have their positively charged hydrogen portions closest (F2.25). In this condition the ions from the salt crystal are said to be **hydrated.** Thus, when a salt crystal dissolves in water, it does not simply come apart into ions, but it is taken apart

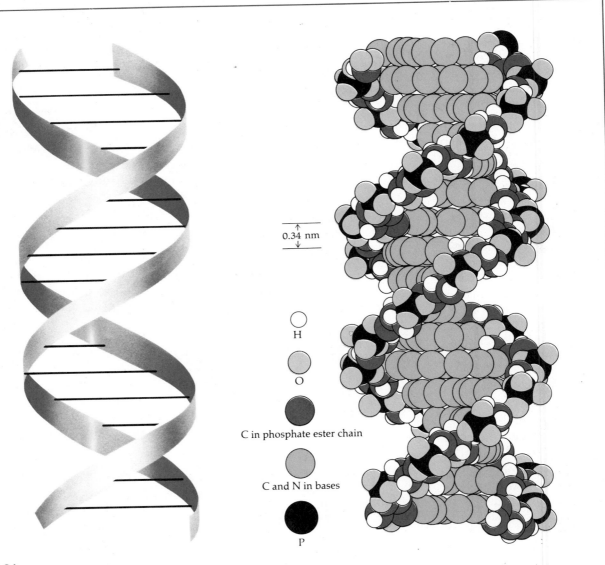

0.34 nm

H

O

C in phosphate ester chain

C and N in bases

P

F2.21
(a) The DNA double helix. The ribbons of the model on the left and the strings of dark and colored atoms in the space-filling model on the right represent the sugar–phosphate "backbones" of the two poly-nucleotide chains. The bases are stacked in the center of the molecule between the two backbones. The bases are 0.34 nm apart (nm = nanometer; 1 nm = 10^{-9}m).

by the water molecules. Some physiologically important salts and the ions into which they separate or dissociate when they dissolve in water are listed in Table 2.4.

2. Water has a high specific heat. This means that, compared with other liquids, water requires a good deal of heat to raise its temperature. Thus, the heat produced by metabolism does not affect body temperature as much as if some other solvent were present.

3. Water has a high latent heat of vaporization. This means that, compared with other liquids, water re-

(b) A portion of a two-chain DNA molecule. Complementary base pairs (A-T; G-C) are held together by hydrogen bonds (dotted lines). The sugar–phosphate "backbone" of each chain is in the colored region. The phosphate groups are shown in ionized form. Note that the two chains run in opposite directions—that is, they are antiparallel.

F2.22
Adenosine triphosphate (ATP). High-energy bonds are indicated by ~ .

quires a good deal of heat to change it from the liquid to the vapor state. Thus, the evaporation of water from body surfaces carries away large amounts of heat and provides the body with an effective cooling mechanism.

METHODS OF EXPRESSING THE CONCENTRATION OF A SOLUTION

It is often necessary to know the concentration of a solution, and concentrations are expressed in a number of ways. One way is simply to indicate the percent of solute in a solution by weight (wt./wt.), by volume (v./v.), or by a combination of the two (wt./v.). For example, a 10% solution by weight can be made by dissolving 10 grams of a solute (such as glucose) in enough solvent (such as water) to make 100 grams of solution. Similarly, a 10% solution by volume can be made by mixing 10 milliliters of a solute (such as ethyl alcohol) with enough solvent (water, for example) to make 100 milliliters of solution. Alternatively, a 10% solution by weight/volume can be made by dissolving 10 grams of a solute (such as glucose) in enough solvent (such as water) to make 100 milliliters of solution. When the concentration of a solution is expressed in percent, the measurements employed—wt./wt., v./v., or wt./v.—should always be indicated.

A second way of expressing the concentration of a solution is in terms of its **molarity.** In this method the amount of solute is not expressed in terms of weight or volume but in terms of moles. A **mole** of a substance is the amount of the substance in grams that is equal to the molecular weight of the substance. The **molecular weight** of a substance can be determined by adding together the atomic weights of the atoms that make up a molecule of the substance. For example, a molecule of glucose has a chemical formula of $C_6H_{12}O_6$, and is composed of 6 carbon atoms, 12 hydrogen atoms, and 6 oxygen atoms. The atomic weight of carbon is 12.011, the atomic weight of hydrogen is 1.008, and the atomic weight of oxygen is 15.999. A mole of glucose would therefore weigh:

$$
\begin{aligned}
6 \times 12.011 &= 72.066 \text{ (for carbon)} \\
12 \times 1.008 &= 12.096 \text{ (for hydrogen)} \\
6 \times 15.999 &= \underline{95.994} \text{ (for oxygen)} \\
&\quad\ 180.156 \text{ grams}
\end{aligned}
$$

A mole of any substance contains the same number of molecules as a mole of any other substance. This number is called *Avogadro's number,* and it is 6.022×10^{23}.

Expressing the concentration of a solution in terms of molarity indicates the number of moles of solute in a liter of solution. For example, a one molar (mol/L) solution can be made by adding one mole of a solute to enough solvent to make one liter (1000 milliliters) of solution. If the molecules of the solute remain intact in the solution, the solution will contain 6.022×10^{23} molecules of the solute.

The term *mole* can be applied to atoms and ions as well as to molecules. For example, a mole of potassium is equal to the atomic weight of potassium in grams (39.098 grams). A mole (39.098 grams) of potassium contains 6.022×10^{23} potassium atoms. In the case of salts, which are made up of ions, the chemical formula of a salt can be used to calculate the number of grams in a mole of the salt. Sodium chloride, for example, has a chemical formula of NaCl, and a mole of sodium chloride is 58.443 grams of sodium chloride (because the atomic weights of sodium and chlorine are 22.99 and 34.453, respectively). A mole of sodium chloride contains 6.022×10^{23} sodium ions (one mole of sodium ions) and 6.022×10^{23} chloride ions (one mole of chloride ions). Similarly, a mole of the salt calcium chloride ($CaCl_2$) contains 6.022×10^{23} calcium ions (one mole of calcium

ions) and $2 \times 6.022 \times 10^{23}$ or 12.044×10^{23} chloride ions (2 moles of chloride ions).

Frequently, the amounts of ionized inorganic constituents in the body fluids are expressed in terms of **equivalents (Eq)** per liter of solution rather than in terms of moles per liter of solution. For present purposes, the number of grams in 1 equivalent of a particular ion can be said to be equal to the sum of the atomic weights of the atoms that compose the ion, divided by the charge of the ion, without regard for the sign (+ or −) of the charge (Table 2.5). For example, 1 equivalent of phosphate ions ($PO_4^{\equiv}$) is equal to:

$$1 \times 30.974 = 30.974 \text{ (for phosphorus)}$$
$$4 \times 15.999 = \underline{63.996} \text{ (for oxygen)}$$
$$94.970 \div 3 = 31.657 \text{ grams of phosphate ions}$$

Similarly, 1 equivalent of calcium ions (Ca^{++}) is equal to 40.08/2 or 20.04 grams of calcium ions.

The solutions encountered in the body generally have low concentrations, and the amounts of solute present are often expressed in terms of millimoles (1 millimole = 1/1000th of a mole) or mil-

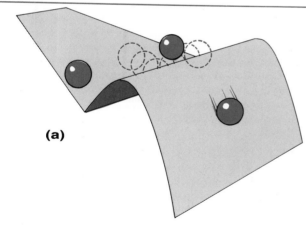

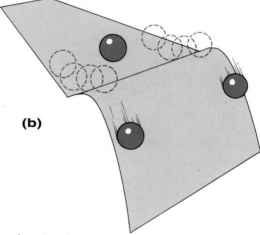

(a)

(b)

F2.23
Diagrammatic representation of activation energy. (a) Just as the balls must have sufficient energy to roll up the small slope before they can roll down the large one, the atoms and molecules of the body must have sufficient energy before they can react with one another. Under normal body conditions, very few at- oms and molecules have the required energy. (b) In the presence of an enzyme the required amount of energy is lowered (the small uphill slope gets smaller), and more atoms and molecules will have this lowered required amount of energy to react with one another.

liequivalents (1 milliequivalent = 1/1000th of an equivalent), rather than in terms of moles or equivalents. The concentrations of solutions considered in this manner are then expressed as millimoles per liter (millimolar; mmol/L) or milliequivalents per liter (mEq/L).

SUSPENSIONS

Other types of mixtures besides true solutions are possible. Among these are **suspensions.** In a suspension, the dispersed particles are so large that they can be kept dispersed only by constant agitation. The components of the suspension remain distinct from one another, thus creating a heterogeneous mixture. If left to stand, the dispersed particles settle. A mixture of sand in water, for example, is an obvious suspension, and in the

body blood cells are suspended in the fluid portion of the blood.

COLLOIDS

The transition from the homogeneity of true solutions to the heterogeneity of obvious suspensions is not a sudden one, and there are many gradations in between. **Colloids** are intermediate between true solutions and obvious suspensions. Colloids consist of particles that are dispersed in a medium, much like the composition of obvious suspensions. However, the particles are small enough that they do not readily settle out if left to stand. Colloidal systems can be distinguished from true solutions by passing a beam of light through them. If the system is a colloid, the beam will be scattered by the dispersed particles, and

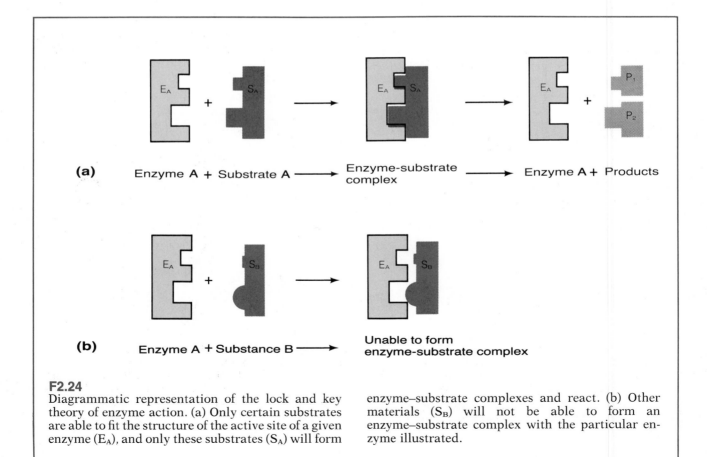

(a)   Enzyme A + Substrate A ⟶ Enzyme-substrate complex ⟶ Enzyme A + Products

(b)   Enzyme A + Substance B ⟶ Unable to form enzyme-substrate complex

F2.24
Diagrammatic representation of the lock and key theory of enzyme action. (a) Only certain substrates are able to fit the structure of the active site of a given enzyme (E_A), and only these substrates (S_A) will form enzyme–substrate complexes and react. (b) Other materials (S_B) will not be able to form an enzyme–substrate complex with the particular enzyme illustrated.

the light path will be visible when viewed from the side. In nature, colloids are very common. Milk is a colloid, and the protoplasm of living cells is considered to have colloidal properties.

ACIDS, BASES, AND pH

Water molecules exist mostly in an undissociated state, with the two hydrogens chemically bonded to the oxygen of the molecule. However, a very small percentage of water molecules (0.0000002%) dissociate into hydrogen ions (H^+) and hydroxide ions (OH^-). Although only about 1 in 500 million water molecules actually dissociates, this small amount of dissociation is one of the most important properties of water, and many metabolic reactions are critically dependent on the hydrogen ion concentration of the solution in which they occur.

Substances that alter the hydrogen ion concentration can be added to water. Substances that increase the hydrogen ion concentration are called **acids,** and substances that decrease the hydrogen ion concentration are called **bases.** Alternatively, acids may be defined as *proton donors* (a hydrogen ion is equivalent to a proton), whereas bases are *proton acceptors.* For example, hydrochloric acid (HCl) is an acid because it can dissociate into hydrogen ions (H^+) and chloride ions (Cl^-), and thereby serve as a hydrogen ion (proton) donor which can increase the hydrogen ion concentration of a solution.

$$HCl \rightleftharpoons H^+ + Cl^-$$

Conversely, ammonia (NH_3) is a base because it can accept hydrogen ions (protons) to form ammonium ions (NH_4^+), and thereby decrease the hydrogen ion concentration of a solution.

$$NH_3 + H^+ \rightleftharpoons NH_4^+$$

The hydrogen ion concentration of the body fluids must by maintained within narrow limits, and it is often necessary to know the hydrogen ion concentration of a solution. In pure water the small dissociation of water molecules results in a hydrogen ion concentration of 0.0000001

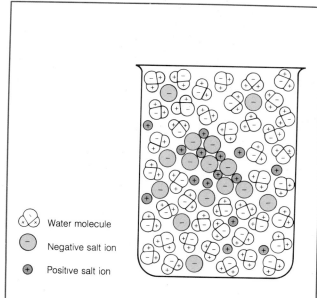

F2.25
Breakup of a salt crystal by water molecules, with hydration of ions. Each salt ion in solution is surrounded by polar water molecules with the opposite charge to that of the ion turned toward it.

Water molecule

Negative salt ion

Positive salt ion

Table 2.4	*Some Salts of Importance to the Body**
Salt	Ions
Sodium chloride (NaCl) $\rightarrow$	Sodium (Na^+) and chloride (Cl^-)
Potassium chloride (KCl) $\rightarrow$	Potassium (K^+) and chloride (Cl^-)
Calcium chloride ($CaCl_2$) $\rightarrow$	Calcium (Ca^{+2}) and chloride ($2\,Cl^-$)
Magnesium chloride ($MgCl_2$) $\rightarrow$	Magnesium (Mg^{+2}) and chloride ($2\,Cl^-$)
Calcium carbonate ($CaCO_3$) $\rightarrow$	Calcium (Ca^{+2}) and carbonate (CO_3^{-2})
Calcium phosphate ($Ca_3[PO_4]_2$) $\rightarrow$	Calcium ($3\,Ca^{+2}$) and phosphate ($2\,PO_4^{-3}$)
Sodium sulfate (Na_2SO_4) $\rightarrow$	Sodium ($2\,Na^+$) and sulfate (SO_4^{-2})

*The arrows indicate the ions into which these salts dissociate when they are dissolved in water.

Table 2.5 *Charges of Common Body Ions*	
Ion	Charge
Bicarbonate (HCO_3^-)	−1
Calcium (Ca^{++})	+2
Chloride (Cl^-)	−1
Hydrogen (H^+)	+1
Magnesium (Mg^{++})	+2
Phosphate ($PO_4^{\equiv}$)	−3
Potassium (K^+)	+1
Sodium (Na^+)	+1
Sulfate ($SO_4^=$)	−2

(1×10^{-7}) moles per liter. When expressed in this manner, the hydrogen ion concentration is such a small number that it is difficult to work with. Consequently, the hydrogen ion concentration of a solution is commonly expressed as a logarithmic value called the **pH,** which is determined according to the following relationship:

$$pH = \log_{10} \frac{1}{[H^+]}$$

where $[H^+]$ is the molar concentration of hydrogen ions in the solution. The pH of pure water is 7, and a solution with a pH of 7 is considered to be a neutral solution. As the hydrogen ion concentration of a solution increases, the pH value drops. Each drop of one unit on the pH scale indicates a tenfold increase in the hydrogen ion concentration. A solution with a pH of 6, therefore, has ten times the hydrogen ion concentration of a solution with a pH of 7, and a solution with a pH of 5 has 100 ($10 \times 10 = 100$) times the hydrogen ion concentration of a solution with a pH of 7. The lower the pH, therefore, the more acidic the solution (F2.26). Similarly, pH values above 7 indicate progressively more basic, less acidic solutions.

DIFFUSION

Atoms, molecules, and ions are constantly in motion. Thus, they possess kinetic energy, or energy

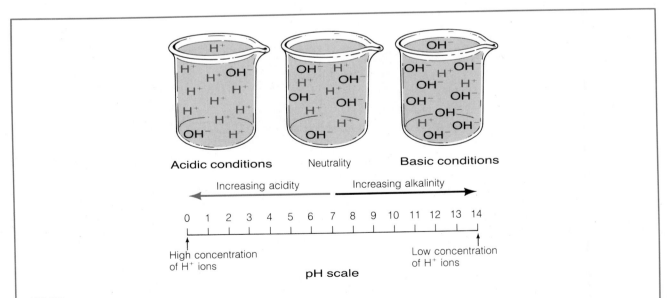

F2.26
The pH scale. pH values below 7 indicate acidic conditions, and values above 7 indicate basic or alkaline conditions.

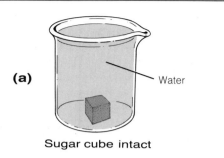

(a)

Water

Sugar cube intact

(b)

Sugar molecules

Sugar cube partially dissolved

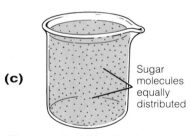

(c)

Sugar molecules equally distributed

Sugar cube dissolved showing state of equilibrium

F2.27

Diffusion. (a) Initially all the sugar molecules are within the sugar cube. (b) As the cube dissolves, sugar molecules disperse in the water. Note that their concentration is highest near the sugar cube, where they are entering solution. (c) When all of the sugar has dissolved and the system is at equilibrium, sugar molecules are dispersed evenly throughout the water as a result of their random movement.

of motion. The velocity at which an atom, molecule, or ion moves is a function of temperature. The higher the temperature, the greater the velocity of movement.

Diffusion is the movement of atoms, molecules, or ions from one location to another as a consequence of their thermal motion. For example, if a cube of sugar is placed in a beaker of water, the sugar will dissolve, and the sugar molecules will eventually diffuse throughout the water (F2.27). In the diffusion process, individual sugar molecules move in a random fashion. However, since initially there are more sugar molecules in the area that surrounds the sugar cube (where the sugar molecules are entering solution) than in areas of the water farther from the cube, it is probable that more sugar molecules will move away from the area around the sugar cube than will move toward the area. Thus, although individual sugar molecules move at random, the net movement of sugar molecules by diffusion (that is, the net diffusion) is from regions of high concentrations of sugar molecules to regions of low concentrations of sugar molecules (provided the temperature and pressure throughout the system are constant). **Net diffusion,** then, is the movement of a substance from a region of higher to a region of lower concentration as a consequence of the thermal motion of the atoms, molecules, or ions of the substance when the temperature and pressure throughout the system are constant.

In the preceding example, the movement of sugar molecules by diffusion will eventually result in a uniform distribution of sugar molecules throughout the water, and no differences in concentration will exist. When this occurs, the system is said to be in equilibrium. A state of equilibrium, however, does not imply a state where there is no longer any movement. Rather, it means that as many atoms, molecules, or ions of a substance—in this case, sugar molecules—enter a particular area at any one time as leave it. Thus, although the same atoms, molecules, or ions of a substance may not be in a given area, the same number of atoms, molecules, or ions of the substance will be. As a result, there is no net change in concentration.

Water molecules can also diffuse, and they can exhibit net diffusion from regions of higher to re-

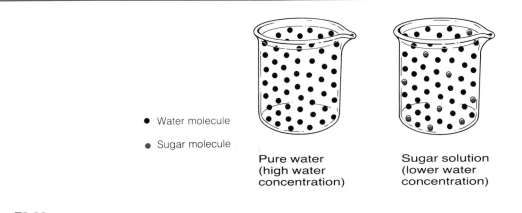

- ● Water molecule
- ● Sugar molecule

**Pure water
(high water
concentration)**

**Sugar solution
(lower water
concentration)**

F2.28
When solute molecules (in this case sugar) are added to water, the concentration of water molecules in a given volume decreases. There can be different concentrations of water (solvent), just as there can be different concentrations of solute in a solution.

gions of lower concentration. But how can there be different concentrations of water? Consider the following. If one beaker is filled with pure water, there will be 100% water in the beaker, and the entire volume of the beaker will be occupied by water molecules. If an identical second beaker is filled with sugar solution, some of the volume will be occupied by water molecules and some by sugar molecules. Therefore, there will not be 100% water molecules in this second beaker, and the concentration of water in this beaker will be less than the concentration of water in the first beaker (F2.28).

If, instead of keeping the pure water and the sugar solution in separate beakers, they are placed on separate sides of a single container that has a removable barrier between the sides, the following will occur when the barrier is removed (F2.29). The sugar molecules will show a net diffusion from their region of higher concentration (the sugar solution side) to their region of lower concentration (the pure water side). Likewise, water molecules will exhibit a net diffusion from their region of higher concentration (the pure water side) to their region of lower concentration (the sugar solution side). Both processes will continue until equilibrium is reached—that is, until both sugar molecules and water molecules are equally distributed throughout the system. At this point, no regions of different concentration exist and thus no further net diffusion of either sugar molecules or water molecules will occur.

OSMOSIS

Consider another situation. Suppose pure water placed on one side of a container is separated by a membrane from a sugar solution placed on the other side of the container. Suppose further that the membrane is a **semipermeable membrane** that allows the passage of solvent (water molecules) but not solutes (sugar molecules) (F2.30). In this situation, water will exhibit a net movement from its region of higher concentration through the membrane to its region of lower concentration. Sugar molecules, however, will not be able to move through the membrane. As a result, there will be a net movement of water into the sugar solution, but no movement of sugar into the pure water. The movement of water that takes place across the semipermeable membrane in this instance is an example of osmosis. More generally, **osmosis** is the movement of solvent through any membrane in response to a concentration difference (a concentration gradient) across the membrane.

In the preceding example, the occurrence of osmosis will cause the volume of pure water on the one side of the membrane to decrease gradually, and the volume of sugar solution on the other side to increase gradually. However, there will always be a higher concentration of water on the pure water side of the membrane than on the sugar solution side. Nevertheless, the net movement of water into the sugar solution does not go on indefinitely. Eventually, the pressure of the additional volume of fluid on the sugar solution side (the *hydrostatic pressure*) rises to a point at which it is able to balance the force tending to move water into the sugar solution by osmosis, and no further net movement of water into the sugar solution occurs. At this point, the hydrostatic pressure that is exerted against the membrane on the sugar solution side is great enough to force water molecules across the membrane from the sugar solution into the pure water as fast as they move from the pure water into the sugar solution. The result is an equilibrium in which there are not equal concentrations of sugar molecules and water molecules in all parts of the system, but one in which opposing forces prevent the net movement of water molecules and a membrane prevents any movement of sugar molecules.

The pressure required to prevent the net movement of pure water into a solution when the water is separated from the solution by a semipermeable membrane is a measure of the solution's **osmotic pressure.** The osmotic pressure of a solution, which is expressed in units called osmoles or milliosmoles, indicates the tendency of water to move by osmosis into the solution. The osmotic pressure of a solution depends basically on the number of solute particles present and not on their nature. The greater the number of solute particles in a given volume of solution, the greater the osmotic pressure of the solution. If two solutions have the same osmotic pressure, they are said to be **isosmotic** (*iso* = same). If two solutions have different osmotic pressures, the one with the higher osmotic pressure is said to be **hyperosmotic** to the one with the lower osmotic pressure, and the solution with the lower osmotic pressure is said to be **hypoosmotic** to the solution with the higher os-

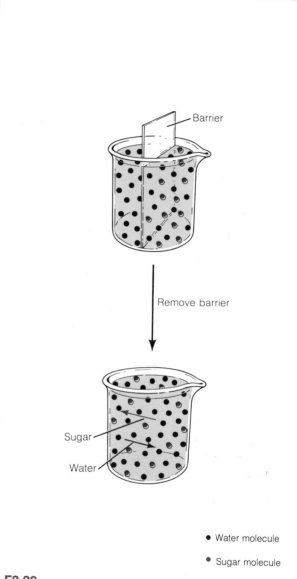

F2.29
Redistribution of water and sugar molecules when a barrier to molecular movement is removed. Sugar molecules show a net diffusion from their region of high concentration to their region of low concentration. Likewise, water molecules show a net diffusion from their region of high concentration to their region of low concentration. At equilibrium, both sugar and water molecules are randomly and equally distributed throughout the container.

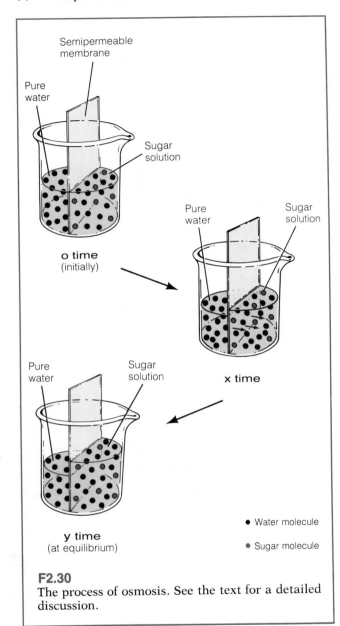

Semipermeable membrane

Pure water

Sugar solution

o time
(initially)

Pure water

Sugar solution

x time

Pure water

Sugar solution

y time
(at equilibrium)

● Water molecule

● Sugar molecule

F2.30
The process of osmosis. See the text for a detailed discussion.

least one solute across itself. Thus, a selectively permeable membrane is not equally permeable to all solute particles present. Osmosis can occur across a selectively permeable membrane in response to different concentrations of water on either side of the membrane.

DIALYSIS

Some membranes are permeable to water molecules and small particles—for example, sodium ions and chloride ions—that may be present in the water, but they are not permeable to large particles such as protein molecules. If such a membrane is placed between a sodium chloride–protein solution on one side and pure water on the other, water molecules, sodium ions, and chloride ions will be able to pass through the membrane but protein molecules will not. As a result, there will be a net diffusion of sodium ions and chloride ions from the solution into the pure water, and water will exhibit a net movement into the sodium chloride–protein solution. If the pure water side is constantly drained away and replenished so that no equilibrium is established, the sodium chloride can be removed from the sodium chloride–protein solution. This process of selectively separating substances in a liquid by taking advantage of their differing diffusibilities through porous membranes is called **dialysis.**

Dialysis achieves its most dramatic application in the artificial kidney. In this device, blood is passed through a membranous tube immersed in a bathing medium of known composition. The membranous tube is permeable to water molecules and other small particles, but it is not permeable to large particles such as protein molecules. Since the atoms, molecules, or ions of most wastes are small particles, these can be removed from the blood while vital protein molecules are retained. Substances required by the body that are also composed of small particles can be retained by having them present in the bathing fluid in the same concentration as they are in the blood. Thus, there will be no concentration difference of these substances on either side of the membrane and, therefore, no net diffusion.

motic pressure (*hyper* = above; *hypo* = below).

The membrane that surrounds living cells is not a simple semipermeable membrane but behaves like a **selectively permeable membrane.** A selectively permeable membrane is a membrane that does not permit the free unhampered movement of all solutes present, but maintains a differential concentration (a concentration gradient) of at

BULK FLOW

Bulk flow is the movement of atoms, molecules, ions, or other particles as a unit in one direction as the result of forces that push them from one point to another. For example, when a sodium chloride solution is pumped through a pipe, the ions and molecules of the solution travel in one direction as a unit.

The bulk flow of a liquid or gas depends on an inequality of pressure that acts on the liquid or gas. If different regions of a liquid or gas are subjected to unequal pressures, the liquid or gas will flow from the region of higher pressure to the region of lower pressure. The flow of blood within the blood vessels and the movement of air into and out of the lungs during respiration are examples of bulk flow that occur in the body.

FILTRATION

Filtration is a process that separates one or more components of a mixture from the others. In this process, the mixture is forced through a porous filter or membrane by mechanical forces such as hydrostatic pressure. The direction of movement is from a region of higher pressure to a region of lower pressure, and large particles that are present in the mixture may not be able to pass through the filter or membrane. For example, consider a glucose solution to which sand has been added. If this mixture is forced through a membrane containing pores that are too small for the sand to pass, water molecules and glucose molecules will move through the membrane, but sand particles will remain behind.

In the body, a filtration process that allows the passage of only small particles commonly occurs. The blood pressure within blood vessels called capillaries acts to force the fluid portion of the blood, including small dissolved particles such as glucose molecules or the ions of salts, across the capillary walls and out of the blood vessels. However, blood cells and protein molecules that are too large to leave the vessels remain behind in the blood.

STUDY OUTLINE

CHEMICAL ELEMENTS AND ATOMS

Chemical Element A substance that cannot be broken down into simpler material by chemical means.

Atoms Make up chemical elements and are composed of: protons (in nucleus; have positive charge); neutrons (in nucleus; electrically neutral); electrons (move around nucleus; have negative charge). **p. 13**

ELECTRON ENERGY LEVELS Present view of atomic structure suggests that electrons should be assigned to particular energy levels. The statistical distribution of electrons is such that electrons occupying higher energy levels will generally be found farther from the nucleus of an atom than electrons occupying lower energy levels. **p. 13–15**

ATOMIC NUMBER, MASS NUMBER, AND ATOMIC WEIGHT **p. 15**

Atomic Number Number of protons in an atom.

Mass Number Combined number of protons and neutrons in an atom.

Atomic Weight Total mass of an atom.

ISOTOPES Atoms of an element may differ from one another in numbers of neutrons; these different forms of given elements are called isotopes. Isotopes have same number of protons (atomic number) but different number of neutrons (mass numbers differ). **p. 15–17**

RADIATION A number of isotopes emit radiation in form of particles or electromagnetic rays. **p. 17**

Particulate Radiation

ALPHA RADIATION Particle is essentially a helium nucleus (two protons, two neutrons); has positive charge.

BETA RADIATION Particle has mass of electron but may have positive or negative charge.

Electromagnetic Radiation: Gamma Radiation Gamma ray can be regarded as a bundle of energy similar to a high-energy x-ray.

Transmutation As radioactive element emits radiation, the element disappears and a new element appears.

BIOLOGICAL USES OF RADIOACTIVE ISOTOPES Radioactive isotopes used as tracers can be followed through physiological processes. Radiation is used to treat diseases such as cancer. **p. 17**

CHEMICAL BONDS Attractive forces between atoms that are strong enough to hold atoms together. **p. 17–21**

Covalent Bonds and Molecules Bonds based on electron sharing are called covalent bonds, and two or more atoms held together by covalent bonds are known as a molecule.

NONPOLAR COVALENT BONDS Involve equal sharing of electrons between atoms.

POLAR COVALENT BONDS Involve unequal sharing of electrons between atoms; condition gives rise to polar molecules that have both positive and negative areas.

Ionic Bonds and Ions Transfer of electrons from one atom to another creates charged atoms (or aggregates of atoms) called ions; ionic bonds are attractions between oppositely charged ions that can hold ions together to form electrically neutral substances known as salts.

Van der Waals Attractions Chemical bond between atoms (and molecules) based on momentary fluctuations in electron distributions around atoms.

Hydrogen Bonds Attractions between oppositely charged regions of polar molecules that involve hydrogen and certain other atoms such as oxygen or nitrogen.

CARBON CHEMISTRY Molecules composed only of carbon and hydrogen are called hydrocarbons; molecules containing carbon, hydrogen, and other elements are called hydrocarbon derivatives; hydrocarbons and hydrocarbon derivatives constitute organic molecules; all other molecules called inorganic. **p. 21–29**

Carbohydrates Are composed of carbon, hydrogen, oxygen, with hydrogen and oxygen frequently in a $2:1$ ratio.

MONOSACCHARIDES Single-unit sugars, such as glucose.

DISACCHARIDES Two monosaccharide units, for example, sucrose.

POLYSACCHARIDES Many monosaccharide units, for example, glycogen.

Lipids Several major types.

FATTY ACIDS Chains of carbon atoms with acid carboxyl group (COOH) at one end.

FATS Union of fatty acids with glycerol.

PHOSPHOLIPIDS Glycerol united with two fatty acid molecules and a third molecule containing a phosphate group and, usually, nitrogen; important components of cellular membranes.

STEROIDS Basically consist of four interconnected rings of carbon atoms that have few polar groups attached; examples are cholesterol and sex hormones.

Proteins Large, complex molecules formed from amino acids. About 20 different amino acids form human proteins. Proteins are formed from amino acids by linking the amino acids with peptide bonds.

Nucleic Acids Composed of purine and pyrimidine bases, five-carbon sugars, and phosphate groups; single base–sugar–phosphate unit called a nucleotide.

RNA Formed when the sugar in a polynucleotide chain is ribose.

DNA Formed when the sugar in two polynucleotide chains is deoxyribose; makes up major portion of chromosomes.

Adenosine Triphosphate Composed of adenine, ribose, and three phosphate groups; serves as immediate source of energy for organismal activity.

ENZYMES AND METABOLIC REACTIONS
Sum total of chemical reactions in body is called metabolism. Metabolic reactions are subdivided into anabolic, or synthesis, reactions and catabolic, or decomposition, reactions. Enzymes are biological catalysts that speed up metabolic chemical reactions by lowering activation energies. **p. 29–33**

Action of Enzymes Enzyme forms enzyme–substrate complex with substrate at enzyme active site.

Regulation of Enyzmatic Activity Provides a method of regulating metabolic reactions.

CONTROL OF ENZYME PRODUCTION AND DESTRUCTION Controls amounts of particular enzymes present.

CONTROL OF ENZYME ACTIVITY
1. Competitive inhibition—inhibitor molecule binds reversibly to active site of enzyme.
2. Noncompetitive inhibition—inhibitor molecule binds irreversibly to active site or to site other than active site of enzyme.

Cofactors Substances often needed by enzymes to catalyze reactions.

SOLUTIONS
Homogeneous mixtures of two or more components that can be gases, solids, liquids; material present in a solution in greatest amount is generally called the solvent; materials present in lesser amounts are generally called solutes. **p. 33–38**

Water as a Solvent Water is ideal solvent for living organisms because:
1. Many different materials dissolve in water.
2. Water has a high specific heat.
3. Water has a high latent heat of vaporization.

Methods of Expressing the Concentration of a Solution
1. As a percent of a solute by weight, by volume, or by a combination of the two.
2. As molarity (moles of a solute per liter of solution).
3. As equivalents per liter of solution (for ionized inorganic constituents of body fluids).

SUSPENSIONS
Mixtures in which the dispersed particles are so large that they tend to settle out. **p. 38**

COLLOIDS
Between true solutions and obvious suspensions; dispersed particles small enough that they do not readily settle out. **p. 38–39**

ACIDS, BASES, AND pH
In pure water there is a balance between hydrogen ions and hydroxide ions, since dissociation of a water molecule produces one ion of each type. **p. 39–40**

Acids Substances that increase hydrogen ion concentration of an aqueous solution (proton donors).

Bases Substances that decrease the hydrogen ion concentration of an aqueous solution (proton acceptors).

pH Scale A logarithmic mathematical scale used to indicate the hydrogen ion concentration of a solution.

DIFFUSION   The movement of atoms, molecules, or ions from one location to another as a consequence of their thermal motion. Net diffusion is the movement of a substance from a region of higher concentration to a region of lower concentration as a consequence of the thermal motion of the atoms, molecules, or ions of the substance.   **p. 40–42**

OSMOSIS   The movement of solvent through any membrane in response to a concentration difference across the membrane.   **p. 42–44**

Osmotic Pressure   The tendency of water molecules to enter a solution in response to a concentration gradient.

Isosmotic   Describes two solutions having the same osmotic concentration.

Hyperosmotic and Hypoosmotic   Describe two solutions of different osmotic concentrations where the more concentrated one is hyperosmotic to the less concentrated, and the less concentrated is hypoosmotic to the more concentrated.

DIALYSIS   A process that selectively separates substances in a liquid by taking advantage of their differing diffusibilities through porous membranes.   **p. 44**

Application in Artificial Kidney   Blood is passed through a membranous tube immersed in bathing medium of known composition; wastes that are small particles are removed; vital protein molecules are retained; essential small particles are also retained by having them at same concentration in bathing medium as in blood, hence no net diffusion across membrane.

BULK FLOW   The movement of atoms, molecules, ions, or other particles as a unit in one direction as the result of forces that push them from one point to another.   **p. 45**

FILTRATION   A process that separates one or more components of a mixture from the others. In this process, the mixture is forced through a porous filter or membrane by mechanical forces such as hydrostatic pressure.   **p. 45**

SELF-QUIZ

1. At the present time, approximately 109 elements are recognized but only about 24 of these are normally found in the body. True or False?
2. Most of the mass of an atom is concentrated in the atom's: (a) outer electron energy levels; (b) ionic bonds; (c) nucleus.
3. A neutral atom contains the same number of electrons as it does: (a) protons; (b) neutrons; (c) energy levels.
4. The number of neutrons contained in $^{17}_{8}O$ is: (a) 8; (b) 9; (c) 17.
5. Salts are made up of positive and negative ions that form nonpolar covalent bonds with one another. True or False?
6. Interactions between atoms that are due mainly to momentary fluctuations in electron distributions around atoms are called: (a) polar covalent bonds; (b) ionic bonds; (c) van der Waals attractions.
7. Molecules composed of only carbon and hydrogen are called: (a) hydrocarbons; (b) carbohydrates; (c) inorganic molecules.
8. Fats are formed from the union of fatty acids with the alcohol glycerol. True or False?
9. Proteins are formed from amino acids by linking the amino group of one amino acid to the acid carboxyl group of another with: (a) an ionic bond; (b) a peptide bond; (c) a disulfide bond.

10. Match the terms with the appropriate lettered descriptions.

Proteins	Primary structure
Amino acids	Secondary structure
Peptide bond	Tertiary structure
Dipeptide	Quaternary structure
Polypeptide	

  (a) The sequence of amino acids in a polypeptide chain.
  (b) The means by which an amino group of one amino acid is joined to the acid carboxyl group of another amino acid.
  (c) Composed of 100 or more amino acids linked into a chain.
  (d) The interactions of different amino acid chains of multichain protein molecules.
  (e) Two amino acids bonded chemically.
  (f) Result of hydrogen bonds that occur principally between the constituents of a protein's peptide bonds.
  (g) These substances are the basic units of proteins.
  (h) The folding of an amino acid chain into a particular three-dimensional configuration.
  (i) Ten amino acids linked into a chain.

11. RNA makes up a major portion of chromosomes, which contain hereditary genetic information for directing the activities of the body's cells. True or False?

12. Enzymes are very specific, and each will catalyze only individual reactions or limited classes of reactions. True or False?

13. Generally, enzymes are destroyed during the course of the reactions they catalyze. True or False?

14. The type of enzyme inhibition in which an inhibitor molecule rather than a substrate molecule combines reversibly with an enzyme molecule at the active site of the enzyme is called: (a) noncompetitive inhibition; (b) competitive inhibition; (c) allosteric inhibition.

15. Match the terms with the appropriate lettered descriptions.

   Solvent   Suspension
   Solution   Colloid
   Solute

   (a) In a solution, the substance present in the smaller amount is generally this substance.
   (b) The dispersed particles tend to settle out in this mixture.

(c) In the body, water is this substance.
(d) A homogeneous mixture of two or more components: gases, solids, or liquids.
(e) In this mixture dispersed particles tend not to settle out, and a beam of light passing through the mixture is scattered by the particles.

16. Substances that increase the hydrogen ion concentration of a solution are called: (a) salts; (b) bases; (c) acids.

17. Net diffusion can be viewed as the movement of a substance from a region of lower concentration to a region of higher concentration. True or False?

18. The movement of solvent through any membrane in response to a concentration difference across the membrane is termed: (a) diffusion; (b) osmosis; (c) dialysis.

19. If two solutions have the same osmotic concentration, they are said to be: (a) isosmotic; (b) hyperosmotic; (c) hypoosmotic.

20. The movement of atoms, molecules, ions, or other particles as a unit in one direction as the result of forces that push them from one point to another is called: (a) osmosis; (b) diffusion; (c) bulk flow.

THE CELL 3

The cell is the basic structural and functional unit of the human body, and the cellular nature of the body's tissues, organs, and systems is evident upon microscopic examination. Consequently, a knowledge of cellular function is essential to an understanding of human physiology.

CELL COMPONENTS

Cells are highly organized units made up of many different components (F3.1). They contain a variety of structures, collectively called **organelles** (little organs), that are responsible for specific cellular functions. Cells also contain chemical substances, such as glycogen granules or lipid droplets, that are collectively called **inclusions.**

A cellular component called the **nucleus** is the control center of the cell. The nucleus is surrounded by a region known as the **cytoplasm.** Most of the cell organelles are in the cytoplasm, suspended in the intracellular fluid known as the **cytosol.** A membrane called the **plasma membrane** (or **cell membrane**) surrounds the cytoplasm and forms the limiting boundary of the cell. Membranes also surround many organelles. Membranes are selectively permeable barriers that regulate the movement of different materials into and out of cells, as well as into and out of different organelles. They also provide points of attachment for various cell components. For example, the contractile elements of muscle cells are attached to the plasma membrane, and the enzymes that mediate certain chemical reactions are bound to the membranes of particular organelles. Although some differences are evident, the basic organization and fundamental properties of the various cell membranes are believed to be similar to those of the plasma membrane.

PLASMA MEMBRANE

The **plasma membrane** is composed of a bimolecular layer of lipid (particularly phospholipid and cholesterol), and various proteins are either attached to or embedded in the lipid in an asymmetrical pattern (F3.2). Carbohydrates are often covalently linked to lipid and protein molecules at the extracellular surface of the membrane; and in some areas, groups of proteins extend through the membrane, forming aqueous channels or pores that connect the interior of the cell with the external environment.

Movement of Materials Across the Plasma Membrane

All materials that enter or leave a cell must pass either through the plasma membrane or through

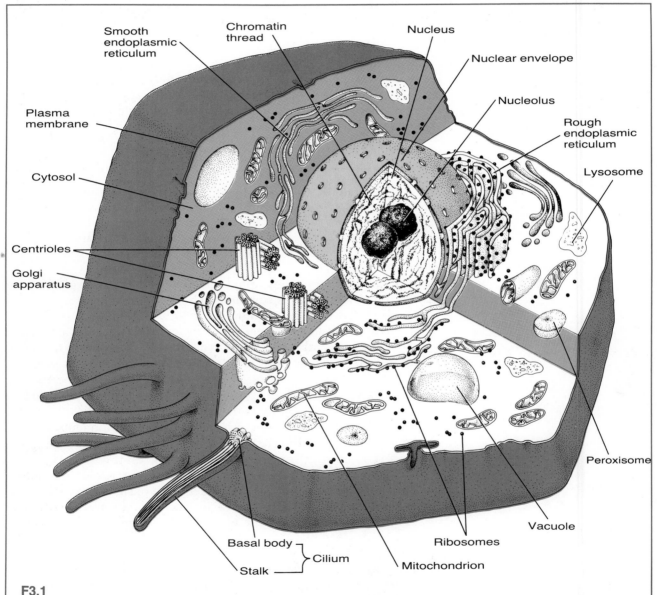

F3.1
A "typical" cell showing subcellular organelles. There is probably no actual cell that can be considered "typical" in all respects.

channels in the membrane. Thus, the properties of the membrane (as well as the properties of the penetrating materials) are important in determining the ease with which substances enter or leave cells.

Diffusion Through the Lipid Portion of the Membrane.

Many nonpolar substances that are soluble in lipids move easily through the plasma membrane by simple diffusion. However, water-soluble polar substances that are not very lipid-soluble diffuse through the membrane only with difficulty, if at all. Thus, the lipid portion of the membrane is a selectively permeable barrier to substances entering or leaving cells.

Movement Through Membrane Channels.

The membrane channels provide an alternative route through the membrane for polar molecules and ions. However, this route is limited to substances small enough to fit through the channels, which are estimated to be about 0.8 nanometer (nm) in diameter (1 nm = 1×10^{-6}mm). In addition, the movement of ions through the channels is highly specific, and certain ions pass most easily through certain channels. For example, sodium ions pass easily through channels called sodium channels, and potassium ions pass easily through channels called potassium channels.

Various signals can open or close membrane channels, and thereby alter membrane permeability. The signals, which are frequently chemical or electrical in nature, are believed to open or close channels by causing changes in the conformations (shapes) of the proteins forming the channels.

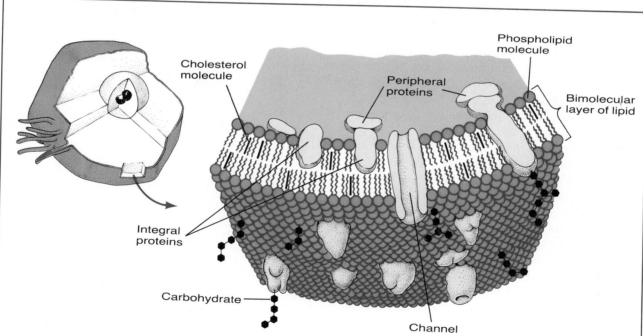

F3.2

The plasma membrane is composed of a bimolecular layer of lipid (particularly phospholipid and cholesterol). Various proteins are either attached to or embedded in the lipid in an asymmetrical pattern: integral proteins are embedded in the lipid, and peripheral proteins are loosely bound to the membrane surface. Carbohydrate molecules are often covalently linked to lipid and protein molecules at the extracellular surface of the membrane.

Mediated Transport. Many polar molecules larger than 0.8 nanometer in diameter enter cells readily even though they are quite insoluble in lipids and are too large to fit through membrane channels. Various forms of mediated transport account for the ability of these substances to cross the plasma membrane. **Mediated transport** makes use of molecules called *carrier molecules,* which are part of the membrane itself. Substances to be transported across the membrane attach to specific binding sites on the carrier molecules. This attachment enables the transported substance to cross the membrane and, depending on the direction of transport, either enter or leave the cell.

The exact mechanisms by which carrier molecules enable substances to pass through the membrane are not known. However, one theory proposes that carrier molecules are fixed in place within the structure of the membrane (F3.3). Each carrier possesses a binding site that is oriented so that a substance to be transported that is located on one side of the membrane can attach reversibly to the site. Once the attachment occurs, the carrier molecule undergoes a conformational change that moves the binding site and the attached substance to the other side of the membrane where the substance is released.

Mediated transport systems exhibit the characteristics of specificity, saturation, and competition (F3.4). **Specificity** means that only certain substances are carried by mediated transport systems. Each carrier molecule binds only with a select group of substances, and only substances that can attach to the various carrier molecules can cross the membrane by mediated transport. **Saturation** means there is a limit to the amount of a substance that can cross the membrane by mediated transport in a given time. When all the carrier molecules for a given substance are being utilized, the addition of more of the substance will not increase its rate of transport. **Competition** means that different substances can compete for the services of the same carrier. Suppose, for example, that substance A and substance B are both transported across the membrane by attaching to the same binding site on carrier X. If a substantial

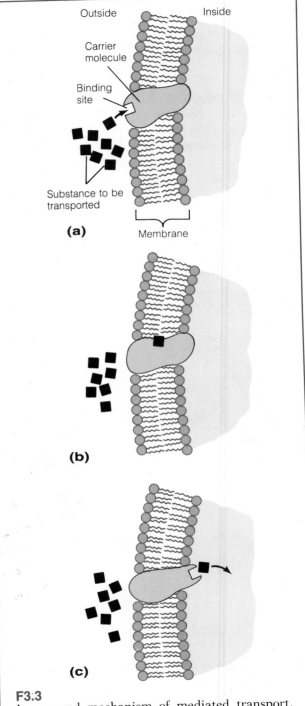

F3.3
A proposed mechanism of mediated transport. See the text for details.

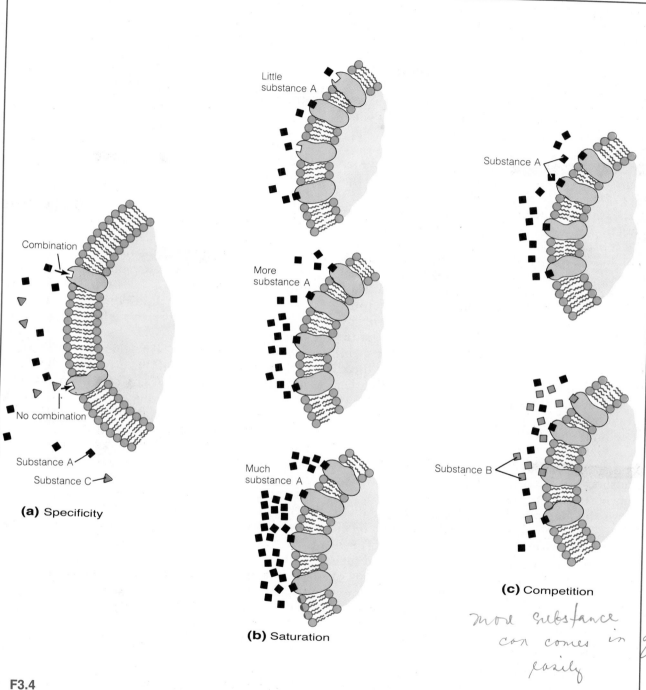

Combination

No combination

Substance A

Substance C

(a) Specificity

Little substance A

More substance A

Much substance A

(b) Saturation

Substance A

Substance B

(c) Competition

[handwritten: more substance w/the con comes in easily gradient]

F3.4

Mediated transport systems exhibit characteristics of (a) specificity, (b) saturation, and (c) competition. See the text for details.

amount of substance A but no substance B is present, then all the carrier X molecules will be utilized in transporting substance A. If substance B is added, it will compete with substance A for the services of the carrier X molecules. In this situation, some substance A and some substance B will be transported, and the transport rate of substance A will decrease compared to its rate when it was the only substance present.

Facilitated Diffusion. One type of mediated transport is called **facilitated diffusion.** In this type of transport, carrier molecules move materials across the membrane equally well in either direction, into or out of the cell. However, if more of a substance to be transported is outside the cell than inside, more transfer will occur from outside to inside than in the opposite direction, resulting in a net movement of the substance inward. Thus, the net movement in facilitated diffusion is from a region of high concentration to a region of low concentration. When equal concentrations of the substance are reached on either side of the membrane, there will be equal transport in both directions and, therefore, no net movement in either direction. Facilitated diffusion is basically a passive process that does not require the expenditure of cellular energy.

Active Transport. Another type of mediated transport is called **active transport.** In this type of transport, carrier molecules move a substance across the membrane from one side to the other regardless of the substance's concentration on either side of the membrane. Active transport systems can move materials across the membrane from regions of low concentration to regions of high concentration, against normal concentration or diffusion gradients. Consequently, they can accumulate material on one side of the membrane at concentrations that are many times those of the material on the opposite side of the membrane. Energy must be expended to move substances uphill against concentration or diffusion gradients, and active transport depends on the metabolic processes of the cell to supply the needed energy.

If a cell is unable to supply the required amount of energy, active transport cannot occur.

Endocytosis and Exocytosis

Some cells form vesicles that enclose a small volume of extracellular material (F3.5). In this process, a cell surrounds the material with a section of its plasma membrane. This section of the membrane then separates from the plasma membrane and moves into the interior of the cell. The general term for this process is **endocytosis.** If the contents of the vesicle are large particles suspended in the aqueous medium of the extracellular fluid, the process of vesicle formation is called **phagocytosis** (cell eating). If the contents are materials in solution or very small particles in suspension, the process is called **pinocytosis** (cell drinking). The mechanisms that regulate endocytosis are not fully understood, but in at least some cases, endocytosis is stimulated by the binding of specific particles to sites on the surface of the plasma membrane.

Substances such as cell products and secretions that are contained within membrane-bounded vesicles leave cells by **exocytosis** (F3.6). In this process, a membrane-bounded vesicle within a cell fuses with the plasma membrane, and the contents of the vesicle are released from the cell. In addition to releasing material from the cell, exocytosis can generate additional plasma membrane when the membranous vesicles fuse with the plasma membrane. Exocytosis can be triggered by specific stimuli that cause an increase in the concentration of free calcium ions within cells. The increased concentration of calcium ions leads to the fusion of intracellular vesicles with the plasma membrane, perhaps by activating contractile proteins in the membrane and surrounding cytoplasm, which act upon the vesicles. Both endocytosis and exocytosis are active, energy-requiring processes.

When considering the different ways in which materials enter or leave cells, it is important to remember that a given substance may enter or leave a cell by several different routes. For example, sodium ions and potassium ions can move through membrane channels, and they are also actively transported.

TONICITY

The plasma membrane is permeable to water. When a cell is immersed in an aqueous solution, the movement of water across the plasma membrane can influence the state of tension or tone of the cell, and the cell may swell due to water entry or shrink due to water loss. Consequently, a solution can be described in terms of its **tonicity**—that is, in terms of its ability to influence the state of tension or tone of cells immersed in it, as a result of the movement of water into or out of the cells. An **isotonic** solution is a solution in which cells maintain their normal tone and in which there is no net movement of water into or out of the cells. A **hypotonic** solution is a solution that produces a change in the tone of cells immersed in it as a result of the net movement of water from the solution into the cells. A **hypertonic** solution is a solution that produces a change in the tone of cells

immersed in it as a result of the net movement of water out of the cells and into the solution.

The properties of the plasma membrane are important in determining the tonicity of a solution. Consequently, a solution that is isosmotic to particular cells may or may not be isotonic. For example, the plasma membrane of red blood cells is permeable to water and urea. However, the membrane prevents the net entry of sodium chloride into the cells, and it prevents the net exit of osmotically active solute particles from the cells. It is possible to make solutions of urea and solutions of sodium chloride that are isosmotic to one another and also to red blood cells. That is, all three have the same osmotic concentration. If red blood cells are placed in the sodium chloride solution, they neither shrink due to water loss nor swell due to water entry because no osmotic difference on either side of the plasma membrane exists, and the membrane prevents the net exit of osmotically active solute particles from the cells as well as the net entry of sodium and chloride ions from the

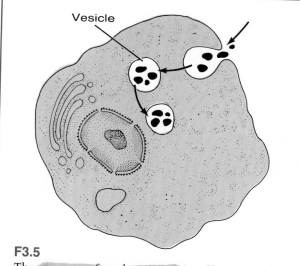

F3.5
The process of endocytosis. A cell surrounds a small volume of extracellular material with a section of its plasma membrane. This section of the membrane then separates from the plasma membrane, giving rise to a vesicle that moves to the interior of the cell.

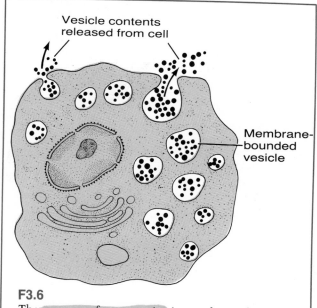

F3.6
The process of exocytosis. A membrane-bounded vesicle fuses with the plasma membrane, and the contents of the vesicle are released from the cell.

sodium chloride solution. Thus, the red blood cells maintain their normal tone, and the sodium chloride solution is both isosmotic and isotonic to the cells.

If red blood cells are placed in the urea solution, a different result is observed. Since the plasma membrane of red blood cells is permeable to urea and since a greater concentration of urea is outside the cells than inside them, there will be a net movement of urea into the cells. This increases the

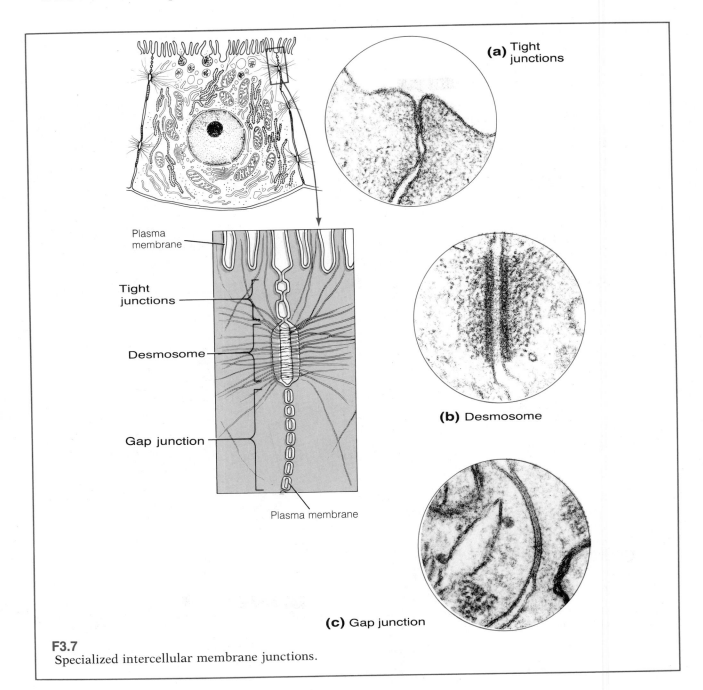

Plasma membrane

Tight junctions

Desmosome

Gap junction

Plasma membrane

(a) Tight junctions

(b) Desmosome

(c) Gap junction

F3.7
Specialized intercellular membrane junctions.

number of osmotically active solute particles within the cells (and decreases the concentration of water within the cells). It also decreases the number of osmotically active solute particles outside the cells (and increases the concentration of water outside the cells). As a result, there will be a net movement of water into the cells by osmosis and the cells will swell and perhaps burst. Thus, in spite of the fact that initially the urea solution had the same osmotic concentration as red blood cells, it was not isotonic to the cells since it did not maintain the normal tone of the cells.

SPECIALIZED INTERCELLULAR MEMBRANE JUNCTIONS

A number of specialized intercellular membrane junctions occur between cells.

Tight Junctions

At a **tight junction** the plasma membranes of adjacent cells essentially fuse with one another, eliminating any extracellular space between the cells (F3.7a). Tight junctions commonly occur between epithelial cells, where they serve to prevent or limit the extracellular passage of substances from one side of a sheet of epithelial cells to the other.

Desmosomes

At most **desmosomes,** called spot desmosomes, the plasma membranes of adjacent cells are separated by an extracellular space about 25 to 35 nanometers wide that contains filamentous material (F3.7b). Along the inner surfaces of the membranes are disc-shaped plaques onto which filamentous structures converge. Spot desmosomes are analagous to rivets or spot welds. They attach one cell to another, and they are important in the maintenance of cell-to-cell adhesion. Other desmosomes, called belt desmosomes, are also involved in cell-to-cell adhesion, and desmosomes called hemidesmosomes attach cells to an underlying matrix of connective tissue.

Gap Junctions

At a **gap junction** the plasma membranes of adjacent cells are separated by a narrow extracellular space about 2 to 4 nanometers wide (F3.7c). Small tubular channels about 2 nanometers in diameter extend across the space and directly link the cytoplasm of the adjacent cells. Gap junctions are sites at which small molecules and ions can pass from one cell to another, and they play an important role in the transmission of electrical activity between cells.

THE NUCLEUS

The **nucleus** is a large organelle that is separated from the rest of the cell by a pore-containing, double membrane called the **nuclear envelope** (F3.8). The nucleus contains the genetic material of the cell in the form of chromatin threads composed of DNA combined with protein and a small amount of RNA. DNA contains coded messages that instruct the cell to synthesize particular polypeptides or proteins. Once the polypeptides or proteins are synthesized, they act as enzymes, as structural proteins, or in other ways to accomplish the work of the cell.

All cells except the reproductive cells contain a person's full genetic complement of DNA. However, all of the possible polypeptides and proteins the DNA can instruct a cell to synthesize are not found at all times in all cells. Various control mechanisms regulate the production of polypeptides and proteins by cells, and a particular polypeptide or protein may be produced by one type of cell but not by another. This is important because a cell's functions depend in large measure on the proteins it produces. For example, if a cell is to carry out a specific sequence of metabolic reactions, it must produce the particular enzymatic proteins required for the reactions. Thus, thyroid gland cells must produce the enzymes needed to synthesize thyroid hormones, and adrenal gland cells must produce the enzymes required to manufacture adrenal hormones. Because it provides the instructions for synthesizing particular polypeptides or proteins, DNA plays a central role in determining cell function.

RNA Synthesis (Transcription)

The DNA in the nucleus is unable to pass through the nuclear envelope into the cytoplasm; yet it is in the cytoplasm that polypeptides and proteins are synthesized. As a result of this situation, DNA must transfer its instructions for synthesizing particular polypeptides or proteins to molecules that carry the instructions from the nucleus to the cytoplasm. It does this by a process called **transcription,** in which DNA serves as a template for the assemblage of molecules of RNA.

As discussed in Chapter 2, both DNA and RNA are made up of base–sugar–phosphate units called *nucleotides,* and a DNA molecule is composed of two polynucleotide chains. The two chains are joined by hydrogen bonds between the bases of the different chains, and the bases always pair with one another in a predictable, complementary fashion. Adenine always links with thymine and vice versa, whereas cytosine always links with guanine and vice versa.

When DNA serves as a template for RNA synthesis, the two linked chains of DNA separate from one another for some distance, exposing part of the base sequence of the DNA (F3.9). An enzyme called *RNA polymerase* links RNA nucleotides into an RNA chain in the order dictated by the sequence of bases in the template DNA. This occurs according to the complementary base-pairing

rules—with the exception that in RNA the base uracil substitutes for thymine (and the sugar unit of RNA nucleotides is ribose rather than deoxyribose as in DNA). Thus, RNA polymerase incorporates an adenine-containing nucleotide into RNA wherever the template chain of DNA has thymine, a guanine-containing nucleotide where the DNA has cytosine, a uracil-containing nucleotide where the DNA has adenine, and a cytosine-containing nucleotide where the DNA has guanine. Following this, the newly formed RNA chain separates from the DNA template as an independent RNA molecule.

Usually, only one of the two chains of any given segment of DNA serves as a template for RNA synthesis. However, one DNA chain may serve as the template at some sites, and the other DNA chain may be the template at other sites.

Messenger RNA

The RNA molecules that carry instructions from DNA in the nucleus to the cytoplasmic sites of polypeptide and protein synthesis are called **mes-**

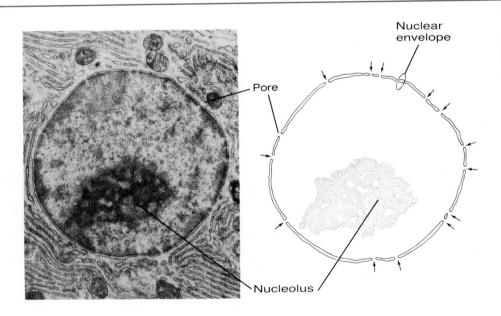

F3.8

An electron micrograph of the nucleus of a pancreatic cell. Note the pores (indicated by arrows) in the nuclear envelope that surrounds the nucleus. The dark area within the nucleus is the nucleolus.

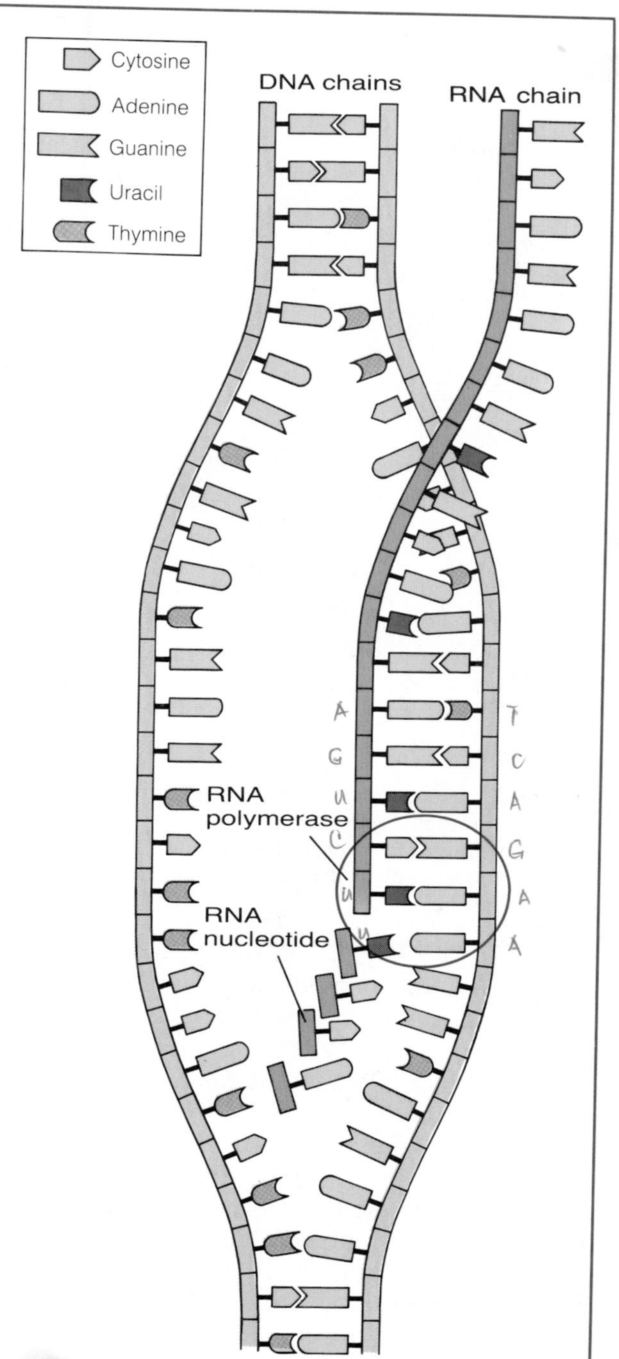

Functional Aspects

Cytosine
Adenine
Guanine
Uracil
Thymine

DNA chains   RNA chain

RNA
polymerase

RNA
nucleotide

F3.9

The enzyme RNA polymerase catalyzes the addition of nucleotides to the growing RNA chain in the sequence dictated by the order of bases in the DNA template.

senger **RNA** (**m-RNA**) molecules. Each m-RNA molecule contains a coded message derived from DNA that specifies the sequence in which different amino acids are to be joined to form a specific polypeptide or protein. The message is contained within the base sequence of the m-RNA, and this base sequence is determined by the base sequence of the DNA that serves as a template for RNA synthesis. Each sequence of three bases in DNA or m-RNA constitutes a "word" of the message. For example, the three-base sequence cytosine–guanine–adenine along a chain of a DNA molecule is a code word that specifies the amino acid alanine of a polypeptide or protein. When m-RNA is synthesized, the DNA code word cytosine–guanine–adenine appears in the m-RNA as guanine–cytosine–uracil, in accordance with the rules of complementary base pairing. Thus, the "words" of m-RNA, which are called *codons*, are complementary to the DNA code words, and the codon guanine–cytosine–uracil also specifies the amino acid alanine (F3.10).

The four principal bases of DNA—adenine (A), cytosine (C), guanine (G), and thymine (T)—can be arranged to form 64 different three-base code words; the bases of m-RNA can form an equal number of complementary codons. Since only about 20 amino acids are found in polypeptides and proteins, this system has more than enough capacity to specify each amino acid. In fact several different "words" can specify the same amino acid. For example, the DNA code words CGA, CGG, CGT, and CGC all specify the amino acid alanine. In addition, some base sequences serve as "punctuation," indicating starting and stopping points for reading the DNA or m-RNA messages.

Ribosomal RNA

There are other types of RNA besides m-RNA. One type, called **ribosomal RNA** (**r-RNA**), is closely associated with a structure known as the **nucleolus,** which is contained within the nucleus of the cell (F3.8). The nucleolus is the site of synthesis (on DNA templates) of a precursor RNA that is ultimately processed into ribosomal RNA and, in the nucleolar area, r-RNA combines with protein. Ribosomal RNA–protein complexes migrate out of the nucleus and into the cytoplasm where they

Enzymes are also found in nucleoplasm.

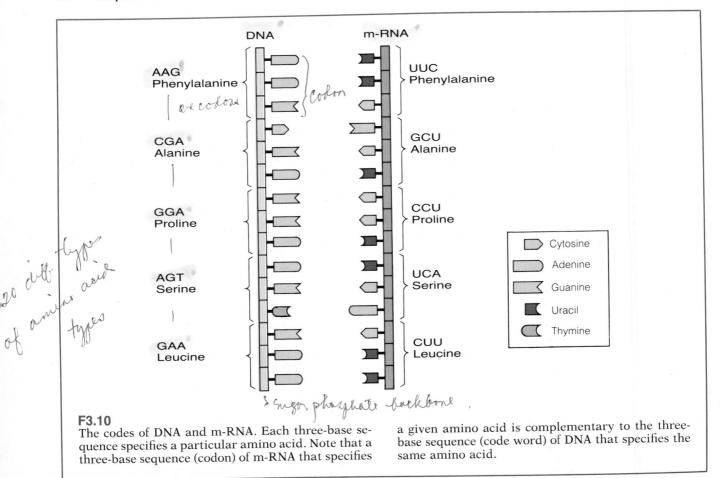

F3.10
The codes of DNA and m-RNA. Each three-base sequence specifies a particular amino acid. Note that a three-base sequence (codon) of m-RNA that specifies a given amino acid is complementary to the three-base sequence (code word) of DNA that specifies the same amino acid.

become organized into structures called *ribosomes*.

RIBOSOMES

Ribosomes are small cytoplasmic organelles. A single ribosome consists of two subunits: a larger 60S subunit and a smaller 40S subunit (60S and 40S refer to the relative rates of sedimentation of the subunits in a centrifugal field). Each subunit is composed of ribosomal RNA and protein.

Protein Synthesis (Translation)

It is at a ribosome that amino acids are linked together in the sequence specified by messenger RNA to form a polypeptide or protein (F3.11). In this process, which is called **translation**, an m-RNA strand that carries coded instructions for synthesizing a specific polypeptide or protein becomes attached to a 40S ribosomal subunit, which then binds to a 60S subunit to form a fully functional ribosome. The ribosome "reads" the m-RNA message and assembles the particular polypeptide or protein specified by the m-RNA. The m-RNA is always "read" in the same direction from the starting point, and a polypeptide or protein is synthesized from the end with the free amino group to the end with the free acid carboxyl group. Frequently, a number of ribosomes attach to the same strand of m-RNA, forming what is known as a polyribosome or polysome. Each ribosome assembles a polypeptide or protein according to the instructions of the m-RNA.

The amino acids that are incorporated into a

growing polypeptide chain are brought to a ribosome from the cytoplasm by a type of RNA called **transfer RNA (t-RNA)**, and there is at least one different t-RNA molecule for each different amino acid. A specific t-RNA molecule contains within its base sequence a region called the **anticodon,** which can attach to a particular m-RNA codon. This allows the amino acid the t-RNA carries to be inserted into a growing polypeptide at the position specified by m-RNA. For example, the DNA code word CGA, which specifies the amino acid alanine, appears in m-RNA as the codon GCU. A t-RNA molecule that carries the amino acid alanine to a ribosome has as its anticodon a base sequence that can pair with the GCU codon of m-RNA. When a ribosome reaches a GCU codon, a t-RNA molecule that carries alanine attaches to the ribosome, and an enzyme that is bound to the ribosome joins the alanine to the growing polypeptide. Another t-RNA molecule then attaches according to the next codon of the m-RNA message, and the amino acid it carries is joined to the alanine of the polypeptide. The t-RNA that carries alanine is then released.

When a ribosome has "read" an entire m-RNA message, the newly formed polypeptide or protein is released. The ribosome dissociates into its 60S and 40S subunits, and the m-RNA is released. The ribosomal subunits can re-form ribosomes that "read" other m-RNA molecules and assemble new polypeptides or proteins. The m-RNA can be utilized by other ribosomes.

ENDOPLASMIC RETICULUM

A membranous network of tubular or saclike channels called the **endoplasmic reticulum** (*endo* = within; *plasm* = cytoplasm; *reticulum* = network) extends throughout much of the cytoplasm of the cell and is almost always interconnected with the nuclear envelope (F3.12). The walls of the endoplasmic reticulum contain enzymes that play a role in fatty acid and steroid synthesis, and an extensive endoplasmic reticulum is found in cells that secrete steroid hormones. In addition, the endoplasmic reticulum is believed to be the site of synthesis of cellular membranes.

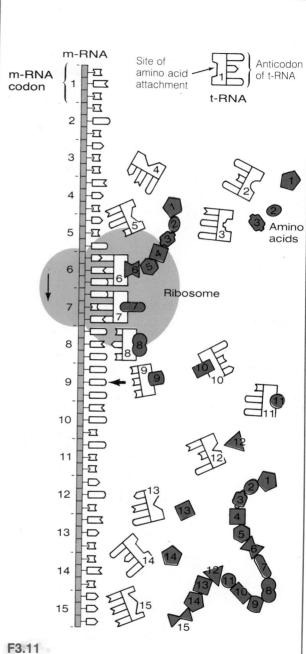

F3.11

A ribosome "reading" the code of m-RNA and assembling a polypeptide or protein chain of amino acids. Molecules of t-RNA bring amino acids to the ribosome for incorporation into the growing chain.

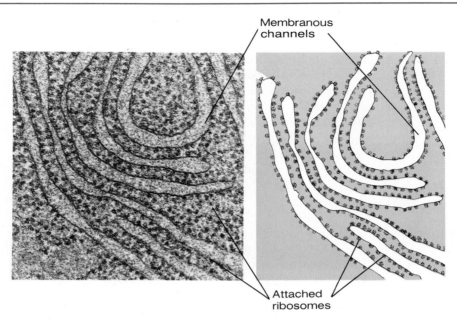

F3.12

An electron micrograph of rough endoplasmic reticulum. The granules attached to the reticulum are ribosomes.

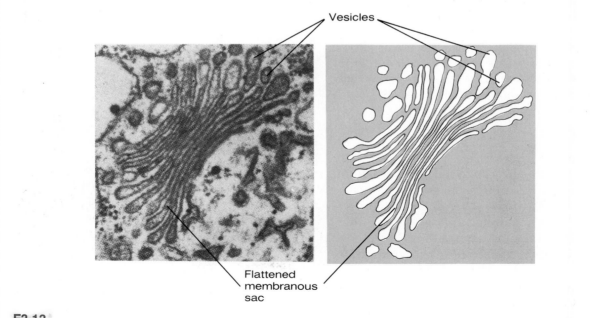

F3.13

An electron micrograph of the Golgi apparatus.

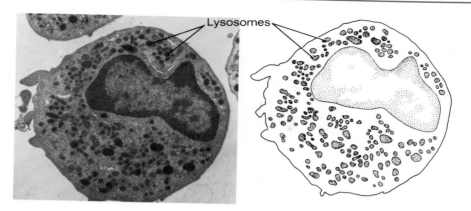

F3.14
An electron micrograph of lysosomes in a white blood cell. The large structure is the nucleus.

Ribosomes are often attached to the endoplasmic reticulum. Endoplasmic reticulum that has ribosomes attached is called **rough** endoplasmic reticulum, and endoplasmic reticulum that does not have attached ribosomes is called **smooth** endoplasmic reticulum. Proteins synthesized by the ribosomes of rough endoplasmic reticulum can pass into the channels of the reticulum, and portions of the reticulum can pinch off or break away, giving rise to membrane-bounded vesicles that contain protein.

GOLGI APPARATUS

The **Golgi apparatus** is an organelle that consists of a series of flattened, membranous sacs that are frequently located near the nucleus of the cell (F3.13). Protein-containing vesicles from the endoplasmic reticulum can fuse with the membranes of the Golgi apparatus, generating additional Golgi membranes. Within the Golgi apparatus, the proteins are processed and modified, and they are believed to be sorted for delivery to different destinations. Ultimately, vesicles containing processed protein pinch off from the Golgi membranes, and the contents of the vesicles become progressively more concentrated as a result of fluid removal. In secretory cells, small, membrane-bounded vesicles called *secretory vesicles* or *secretory granules* migrate from the Golgi

region to the plasma membrane. The vesicles fuse with the membrane, and their contents are released from the cell by exocytosis.

LYSOSOMES

Lysosomes (*lyso* = dissolution; *soma* = body) are membrane-bounded cytoplasmic organelles that appear granular during inactivity but assume the appearance of vesicles when active (F3.14). Lysosomes originate from the Golgi apparatus, and they contain strong digestive enzymes that are capable of breaking down proteins, lipids, certain carbohydrates, DNA, and RNA. Lysosomes form digestive vacuoles by attaching to vesicles that are present within the cell as the result of phagocytosis (F3.15). After attachment, the lysosomes release their enzymes into the vacuoles, and the enzymes digest the phagocytized material, often bacteria, converting the material to products that can enter the cytoplasm and be utilized by the cell. Undigested material remains within the vacuoles, which are then called *residual bodies*. Lysosomes are particularly abundant in certain white blood cells whose principal activity is the phagocytosis of foreign materials in the body.

Parts of a cell itself are sometimes broken down within lysosomal vacuoles that are called autophagic vacuoles. Such an event can occur during starvation, allowing the cell to use a part of its own substance for fuel without doing itself irreparable harm. After a cell has been severely injured or has died, lysosomal membranes may rupture. The enzymes released then digest the material of the cell itself.

PEROXISOMES

Peroxisomes are membrane-bounded cytoplasmic organelles that are very similar to lysosomes in microscopic appearance, although they may actually be larger, irregularly shaped structures. Peroxisomes differ from lysosomes in that lysosomes contain digestive enzymes, whereas peroxisomes contain a variety of powerful oxidative enzymes as well as enzymes that catalyze the oxidative decomposition of hydrogen peroxide.

MITOCHONDRIA

Mitochondria (*mito* = thread; *chondros* = granule) are cytoplasmic organelles that are bounded by a double membrane (F3.16). The outer membrane is smooth and surrounds the mitochondrion itself. The inner membrane folds at intervals into the central portion of the mitochondrion, forming partitions known as *cristae*. A semisolid system called the *matrix* is found within the mitochondrion and is surrounded by the inner membrane. Mitochondria are involved in the generation of metabolic energy for cellular activities, and the specific metabolic reactions that take place within them are discussed in Chapter 17.

MICROTUBULES

The cytoplasm of many cells contains an array of very small, hollow, cylindrical, unbranched tubules called **microtubules.** Microtubules are composed of subunits of a protein called tubulin as well as various proteins known as microtubule-associated proteins. Microtubules are involved in cell movement processes such as the translocation of organelles from one place to another. Microtubules also play a structural role in the development and maintenance of cell shape.

CILIA, FLAGELLA, AND BASAL BODIES

Many cells have one or more thin, cylindrical structures that are motile (capable of movement)

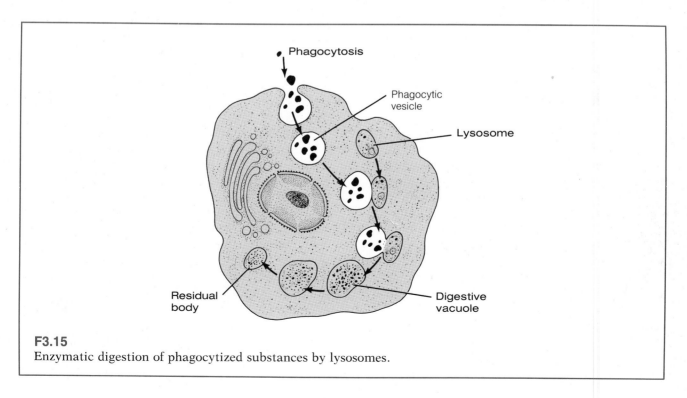

F3.15
Enzymatic digestion of phagocytized substances by lysosomes.

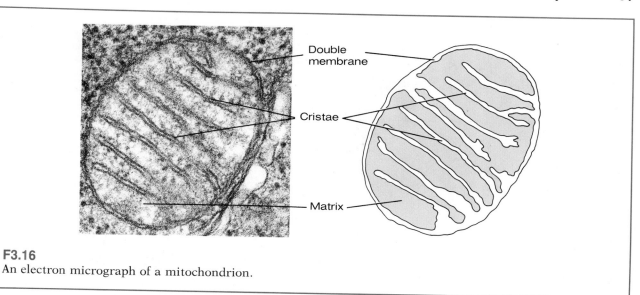

F3.16
An electron micrograph of a mitochondrion.

projecting from their surfaces. These structures can move substances over the cell surface, or they can move an entire cell through a liquid medium. If the structures are short and numerous, they are called **cilia.** If they are longer and fewer, they are called **flagella.** Both cilia and flagella have the same basic organizational pattern, and both originate from cytoplasmic structures called **basal bodies.** Cilia and flagella consist of a membranous sheath that encloses a series of microtubules. The sheath is continuous with the plasma membrane, and the microtubules are arranged in a characteristic circular pattern of nine groups of tubules with two tubules per group (F3.17*a*). Two additional microtubules are found in the center of this circular pattern. Basal bodies also have a characteristic circular pattern of groups of microtubules, but basal bodies have three tubules per group and they do not have the two central tubules (F3.17*b*).

CENTRIOLES

Structures called **centrioles,** which are involved in cell division, are found near the nucleus of the cell in a region called the *centrosome* or *centrosphere.* Centrioles normally occur in pairs, with each member of the pair oriented at right angles to the other. Centrioles are cylindrical in shape, and they contain a series of microtubules that are arranged in the same pattern as the microtubules of basal

bodies (F3.17*b* and *c*). In fact, basal bodies and centrioles are believed to have common origins, and they may even be identical structures.

MICROFILAMENTS

The cytoplasm of some cells contains very small fibrils called **microfilaments.** Microfilaments generally occur in bundles or other groupings rather than singly. Microfilaments are associated with contractile activities involved in cell movement, and muscle cells possess an extensive array of microfilaments.

INCLUSIONS

Cells contain a wide variety of chemical substances that are collectively called **inclusions.** The hemoglobin molecules of red blood cells, which transport oxygen and carbon dioxide, are inclusions. So are particles of the pigment melanin, which is found in some cells of the eyes, the skin, and the hair. Several metabolically important substances are also found in cells as inclusions. For example, the polysaccharide glycogen, a storage form of carbohydrate, is particularly evident in liver and muscle cells, and fats are stored in adipose tissue cells. When a cellular inclusion is a liquid that could mix with the cytoplasm of a cell, the inclusion is often surrounded by a membrane, forming a structure called a *vacuole.*

EXTRACELLULAR MATERIALS

Many body substances are found outside of cells rather than within them. These are collectively called **extracellular materials.** Extracellular materials include the body fluids and the extracellular framework in which many cells are embedded. Many extracellular materials are products of the cells themselves. Among these are chondroitin sulfate, which is a jellylike substance found in bone,

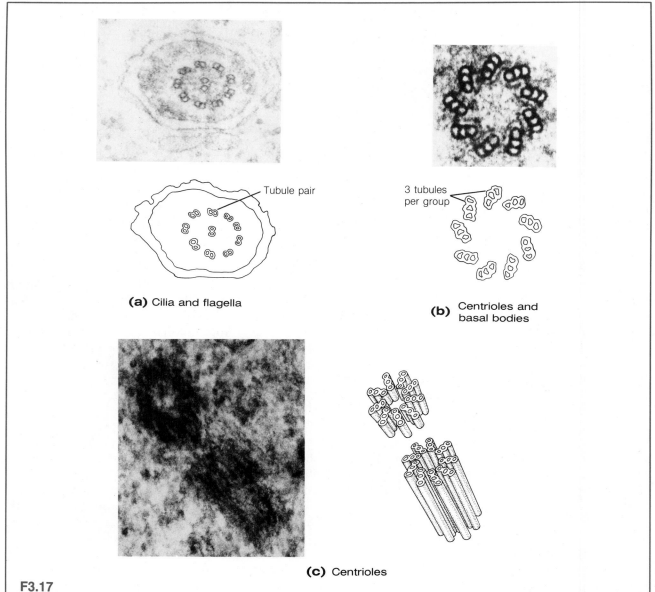

(a) Cilia and flagella

(b) Centrioles and basal bodies

(c) Centrioles

F3.17

(a) Cross-sectional view of a typical 9 + 2 arrangement of microtubules in cilia and flagella. (b) Cross-sectional view of a typical arrangement of tubules in basal bodies and centrioles. Note that there are three tubules per group and no tubules in the center. (c) Centrioles.

cartilage, and heart valves; and hyaluronic acid, which is a viscous, fluidlike substance present in a number of tissues. A variety of fibrous materials such as the proteins collagen and elastin also occur extracellularly. Connective tissue is particularly rich in extracellular materials.

CELL DIVISION

Some highly specialized cells such as muscle and nerve cells do not undergo cell division once they have developed fully. Other cells, however, such as liver, intestinal, bone marrow, and epidermal cells are able to divide and reproduce themselves. The processes by which such cells divide encompass several basic events. One is the replication of the genetic material within the nucleus of the cell. A second is the redistribution of the genetic material into two new nuclei. The process of redistributing the genetic material into two new nuclei that each contain the same genetic information as the original nucleus is called **mitosis.** A third event is the division of the cell's cytoplasm to produce two new cells (called daughter cells), each with its own nucleus. This process is called **cytokinesis.** Usually mitosis and cytokinesis occur simultaneously.

INTERPHASE: REPLICATION OF DNA

The period between active cell divisions is called **interphase.** During interphase, the DNA molecules that comprise the genetic material of the cell (and the protein and RNA associated with the DNA) appear only as indistinct chromatin threads within the nucleus. During this period, DNA molecules serve as templates for the replication of additional DNA molecules (F3.18). In this process, the two linked chains of a DNA molecule separate from one another for some distance, and each chain acts as a template that specifies the order in which individual DNA nucleotides are to be incorporated into a new DNA chain. DNA polymerase and polynucleotide ligase enzymes link the nucleotides into the new DNA chain according to the complementary base-pairing pattern. That is, an adenine-containing nucleotide is incorporated into the new DNA chain wherever the template

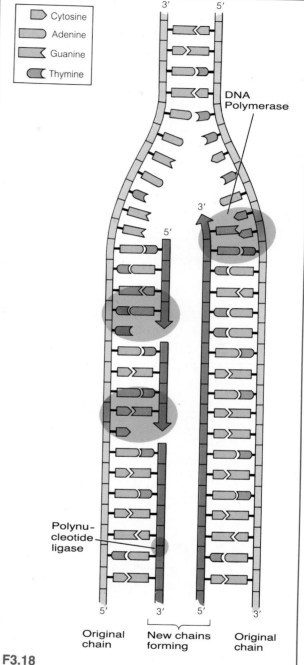

F3.18

Replication of DNA. The original two chains of a DNA molecule separate, each chain then serving as the template for the assembly of a new complementary chain. This results in two DNA molecules, each exactly like the original.

chain has thymine; a guanine-containing nucleotide is incorporated where the template DNA has cytosine; a thymine-containing nucleotide, where the template has adenine; and a cytosine-containing nucleotide, where the template has guanine. The end result is a new DNA chain that is the complement of the original template DNA chain. The newly replicated chain of DNA remains

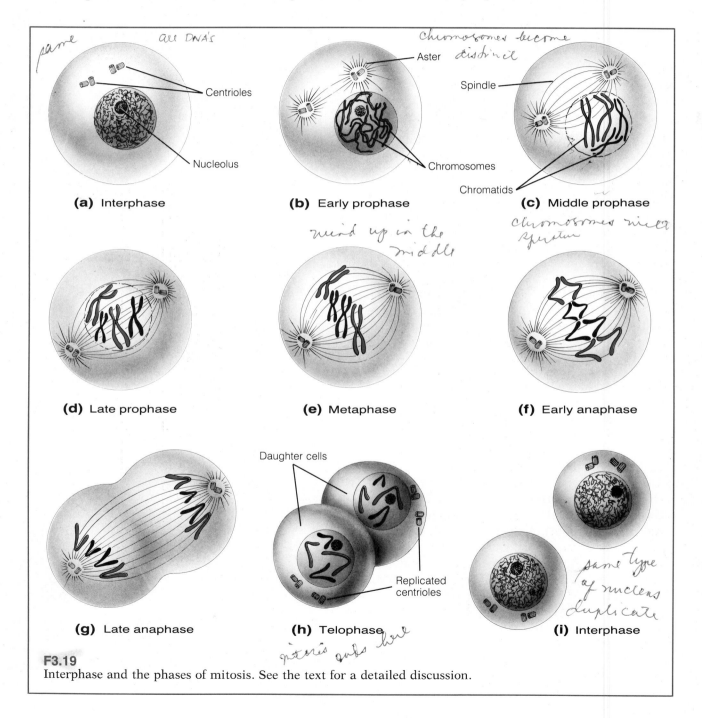

(handwritten annotations: "same", "all DNA's", "Chromosomes become distinct", "wind up in the middle", "Chromosomes wind spirtn", "same type of nucleus duplicate", "mitosis ends here")

(a) Interphase — Centrioles, Nucleolus

(b) Early prophase — Aster, Chromosomes

(c) Middle prophase — Spindle, Chromatids

(d) Late prophase

(e) Metaphase

(f) Early anaphase

(g) Late anaphase

(h) Telophase — Daughter cells, Replicated centrioles

(i) Interphase

F3.19
Interphase and the phases of mitosis. See the text for a detailed discussion.

attached to the original template DNA, thereby forming a complete, two-chain DNA molecule exactly like the original. Thus, each new two-chain DNA consists of one chain from the original DNA, which acted as a template, and one newly synthesized chain.

MITOSIS

Several phases in the overall process of mitosis (nuclear division) are recognized (F3.19). However, it must be emphasized that mitosis itself is a continuous event and not a series of discrete steps.

Prophase

The initial phase of mitosis is called **prophase** (F3.19*b–d*). Early in prophase, two pairs of centrioles are present. As prophase proceeds (and in some cells beginning in interphase), one pair of centrioles moves toward one end, or pole, of the cell, and the other pair moves toward the opposite end. Fibers (microtubules) that project in all directions from the regions of the centrioles become visible, the nucleolus disappears, and the nuclear envelope begins to disintegrate. Also during prophase, the chromatin threads of DNA, protein, and RNA become tightly coiled and visible as structures called **chromosomes** (F3.20). Each chromosome is made up of two separate strands called **chromatids**, and at one point along its length each chromatid has a special region called a **centromere.** The two chromatids are held together at their centromere regions by as yet unidentified forces. One chromatid includes a new, two-chain DNA molecule that was replicated during interphase using one chain of an original DNA molecule as a template. The second chromatid includes the new DNA molecule that was formed using the other chain of the original DNA as a template.

By the end of prophase, the centrioles have nearly reached opposite poles of the cell, and the chromosomes have moved toward a position at the middle or equator of the cell halfway between the two centriole pairs. Some of the fibers that radiate from the centriole regions end blindly. These are known as *astral fibrils,* and they form an aster about each centriole pair. Other fibers, called *spindle fibers,* form an organized array known as a **spindle** between one centriole pair and

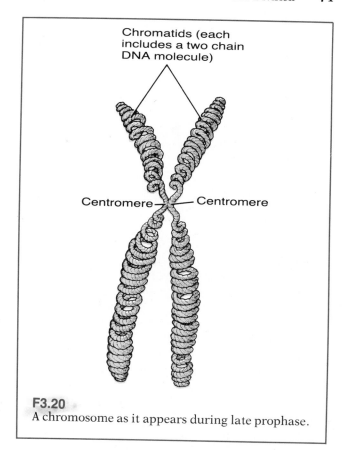

F3.20
A chromosome as it appears during late prophase.

the other. The spindle is composed of at least two kinds of spindle fibers. One kind extends from one pole of the cell to the center of the spindle, where it overlaps with similar fibers coming from the opposite pole. A second kind of spindle fiber also extends from a pole of the cell toward the center of the spindle, but it eventually attaches to a chromosome.

METAPHASE

By the beginning of **metaphase** the nuclear envelope has completely disappeared (F3.19*e*). As metaphase proceeds, the chromatids of each chromosome can be seen to be attached by their centromeres to spindle fibers along the central or equatorial plate of the spindle. At the conclusion of this stage, the chromatids of each chromosome uncouple, and each of the former chromatids becomes an independent, single-stranded chromosome.

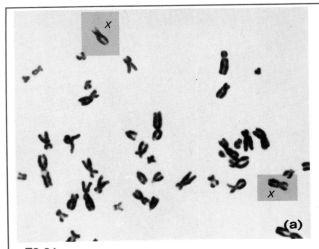
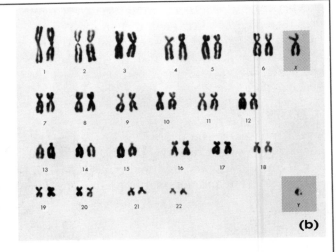

F3.21
Human chromosomes. (a) Chromosomes of a female with X sex chromosomes indicated. (b) Chromosomes of a male with homologous chromosomes arranged in pairs. X and Y are sex chromosomes. Note X and Y chromosomes in color boxes.

ANAPHASE

During **anaphase,** the single-stranded chromosomes separate and move toward opposite poles of the cell (F3.19*f* and *g*). By the end of anaphase, the single-stranded chromosomes (each of which includes a complete, two-chain DNA molecule) have reached the poles of the cell. When mitosis and cytokinesis occur together, the beginning of cytokinesis is generally evident during anaphase as an inward pinching of the cell membrane in the equatorial region.

TELOPHASE

In **telophase,** a new nuclear envelope forms, presumably from endoplasmic reticulum, and the nucleolus reappears (F3.19*h*). If mitosis and cytokinesis are occurring together, cytokinesis is completed when the continued inward pinching of the cell membrane separates the cell into two daughter cells. Also during telophase, the spindle disappears, and the chromosomes become less distinct, gradually assuming their interphase appearance of chromatin threads. In many cells, the centrioles are replicated during this phase, but in other cells they are replicated at other times. By the end of telophase, the two daughter cells have assumed the interphase appearance, and the division cycle is complete.

MEIOSIS

There are 46 chromosomes in each of the human somatic cells (all cells except the reproductive cells). Two of these are **sex chromosomes** (two X chromosomes in females, and one X and one Y chromosome in males). The remaining 44 chromosomes are called **autosomes.** The 44 autosomes consist of 22 pairs of similar-appearing chromosomes (F3.21). One member of each pair contains genetic information derived from the individual's father, and the other member of each pair contains information derived from the individual's mother. Each pair makes up a set of **homologous chromosomes.** The two sex chromosomes of the female (XX) are also homologous, but the two sex chromosomes of the male (X and Y) are not.

The 46 chromosomes of human somatic cells actually consist of two 23-chromosome sets (22 autosomes and 1 sex chromosome per set), with one set having been derived from the individual's father and one from the individual's mother. Thus the **gametes,** or reproductive cells, of the male (the sperm from the testes) and female (the ova from the ovaries) each contain only 23 chromosomes. When a sperm cell fertilizes an ovum, each contributes its 23 chromosomes, thereby establishing the full 46 chromosomes of the new individual.

Cells with two complete sets of chromosomes (46 chromosomes) are known as **diploid cells.** The formation of gametes, however, must result in the formation of cells that have only one set of chromosomes (23 chromosomes rather than 46 chromosomes). Such cells are called **haploid cells,** and the normal processes of mitosis do not produce such cells. A second type of cell division, known as **meiosis,** is responsible for the production of haploid reproductive cells.

Two successive division sequences occur in meiosis (F3.22). In the first sequence, prophase occurs essentially as in mitosis with the exception that all 46 chromosomes do not move separately toward the spindle fibers. Rather, homologous chromosomes pair (synapse) with one another (as do the nonhomologous sex chromosomes of the male). The paired chromosomes move toward the spindle fibers as two chromosome units. During metaphase, the chromatids of the synapsed chromosomes do not uncouple. At anaphase, the paired homologous chromosomes simply move apart, with one member of the pair moving to one pole of the cell and the other member of the pair moving to the opposite pole. Thus, 23 double-stranded chromosomes are moved to each pole during the first division sequence of meiosis, rather than the 46 single-stranded chromosomes that were moved to each pole in mitosis. The remaining events of first anaphase and first telophase occur as in mitosis (with cytokinesis), resulting in two daughter cells that each contain only 23 chromosomes, but the chromosomes are double-stranded rather than single-stranded. Following the first division sequence of meiosis, a short period called **interkinesis** occurs. During this period, which is similar to the interphase period between mitotic divisions, the 23 double-stranded chromosomes of the daughter cells do not duplicate themselves.

Following the interkinesis period, the second division sequence of meiosis occurs. Each of the two 23-chromosome daughter cells undergoes a typical mitotic division with cytokinesis. The result of this second division sequence is four haploid cells, each of which contains 23 single-stranded chromosomes.

Meiosis allows a great deal of genetic diversity in the makeup of reproductive cells (sperm and ova). The chromosomes that originally came from the individual's male parent, for example, do not necessarily all line up toward one centriole pair, nor do those from the individual's female parent all line up toward the other centriole pair during the synapsis of chromosomes that occurs in the first division sequence of meiosis (F3.23). Rather, a mixture of positions occurs, so that the resulting daughter cells each receive, in random assortment, some chromosomes derived from the individual's male parent and some from the female parent.

Further genetic diversity can result from the process of **crossing over,** which takes place occasionally during the first stage of meiosis, while the chromosomes are synapsed (F3.24). In this process a chromatid of one chromosome of a synapsed pair breaks, and a corresponding break occurs in a chromatid of the other chromosome of the pair. The two free chromatid fragments then exchange places so that when the breaks are repaired, the fragment originating from the chromatid of the one synapsed chromosome is attached to the chromatid of the other chromosome and vice versa. When the synapsed chromosomes move apart and daughter cells are ultimately produced, the daughter cells can have a different genetic composition than they would have if crossing over had not occurred. Thus, the gametes that are ultimately produced from different cells differ from one another genetically depending on the particular chromosomal redistribution that occurs during meiosis.

CANCER

In its essence, cancer is simply an uncontrolled division of cells that gives rise to abnormal growths called malignant tumors. Sometimes the growths are localized and, if discovered early enough, can be removed surgically. Cancer cells,

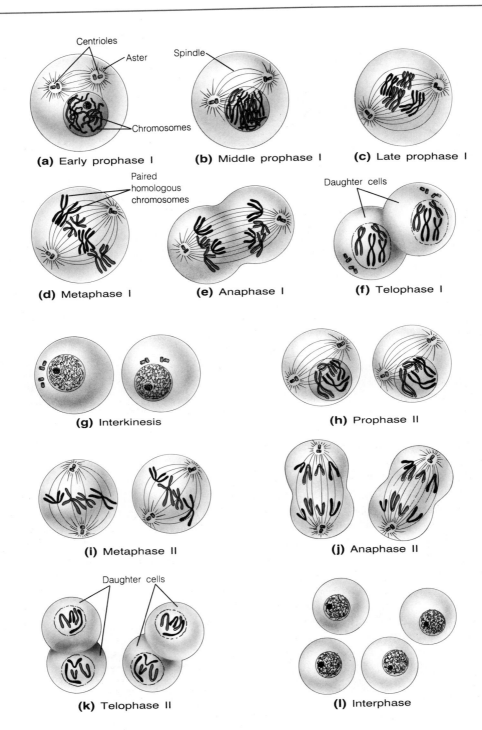

F3.22
The phases of meiosis. See the text for a detailed discussion.

however, are not very adhesive, and they may spread (metastasize) throughout the body and lodge in various organs and tissues. Such an occurrence renders the cancer inoperable, and other means of treatment must be employed. Radiation therapy is sometimes used in such cases because rapidly dividing cells are very susceptible to radiation. Various drugs and chemotherapeutic agents that attack rapidly growing cells are also employed. One problem with radiation therapy and chemotherapy, however, is that cells such as white blood cells, which protect the body from foreign organisms, also come from dividing cells that may be injured by these therapies, leaving the individual less able to ward off other illnesses and infections.

Although a number of carcinogens—cancer-causing agents—have been discovered, the exact

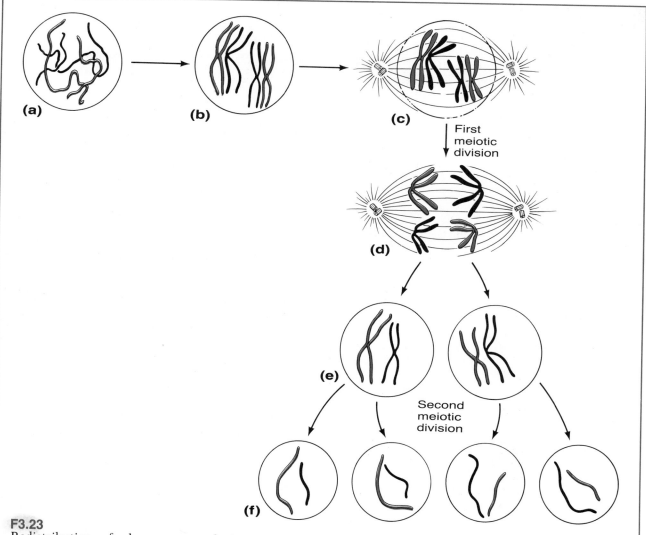

F3.23
Redistribution of chromosomes during meiosis. Chromosomes derived originally from the male parent (color) and from the female parent (black) do not necessarily remain together. The resulting cells have a mixture of chromosomes from both parents.

nature of the disease remains a mystery. Cancer may be due to changes that remove cells from whatever controls exist to regulate their division. Certain viruses have been shown to cause cancer in some animal species and there is some evidence that viruses can also be carcinogenic in humans.

GENETIC DISORDERS

Many diseases and abnormalities can be traced to genetic defects. Occasionally, incorrect bases are incorporated into DNA chains, and the three-base code words containing the incorrect bases specify amino acids that are different from those that would normally have been called for. This results in the production of incorrect m-RNA molecules. When ribosomes "read" these erroneous m-RNA molecules, incorrect amino acids are inserted into the proteins being formed. These incorrect proteins may not function properly and disease can be one result.

Such is the case in the genetic disease sickle-cell anemia, which occurs when an incorrect amino acid becomes incorporated into one of the polypeptide chains of the oxygen-transporting molecule hemoglobin. Red blood cells that contain the abnormal hemoglobin molecules are fragile and misshapen. They rupture within, and block, blood vessels, leading to an impaired delivery of oxygen to the body's cells.

Also, since many proteins function as enzymes that catalyze cellular chemical reactions, errors in their construction can make it impossible for certain reactions to occur. The genetic disease phenylketonuria results from the lack of a functional enzyme required to convert the amino acid phenylalanine to tyrosine. The inability of the cells to carry out this reaction efficiently can ultimately produce mental retardation.

Genetic disorders range from those in which a single amino acid is incorrectly incorporated into a polypeptide chain to those in which entire polypeptides or proteins are not manufactured at all. Alterations in the genetic material of the cell that lead to such occurrences are called **mutations.** In most instances, mutations are found to be harmful, or, at best, neutral to cell function. Occasionally, however, they may lead to an improved condition as far as the organism is concerned. Mutations that occur in reproductive cells can be preserved and passed from parent to offspring. Genetic disorders are discussed further in Chapter 20.

AGING

From the moment of fertilization, the human organism grows and develops until full maturity is reached. After this period, the individual gradually ages. The aging processes themselves are currently attracting a great deal of interest. Although theories abound, little is known about the exact nature of these processes. The following are some current thoughts on the matter.

In the mature individual, some highly specialized cells such as muscle and nerve cells do not

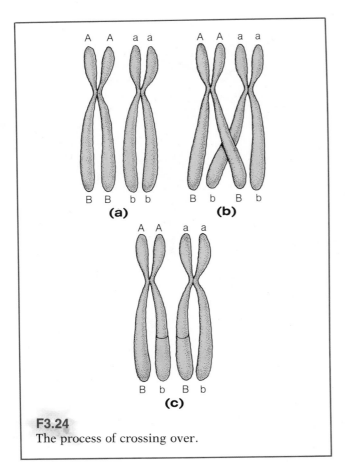

F3.24
The process of crossing over.

undergo cell division. When these cells die, as a result of disease or injury, they are generally not replaced, and remaining muscle or nerve cells must assume their functions. This places an additional stress on the remaining cells and may hasten their death, thus contributing to aging. Other cells of the body, such as those of the liver and pancreas, retain the ability to divide, and it appears that these organs age more slowly than muscle or nerve tissues.

Several theories have been proposed to explain what actually causes the death of cells and, thus, aging. These range from ideas that external events such as x-rays or cosmic radiation—or the more nebulous "stresses of life"—gradually take their toll on cells, to ideas that are more internally oriented, such as those that propose that aging changes are built into the genetic apparatus of the cell.

STUDY OUTLINE

CELL COMPONENTS   Nucleus surrounded by cytoplasm, which contains numerous structures called organelles. Limiting boundary of cell is plasma membrane.   **p. 51**

Plasma Membrane   A highly selective barrier that affects the movement of substances into or out of cells. Membrane is lipid bilayer with protein interspersed; channels in membrane connect cell interior with outside. Materials that enter or leave cell must pass through membrane or channels.

MOVEMENT OF MATERIALS ACROSS THE PLASMA MEMBRANE

Diffusion Through the Lipid Portion of the Membrane Many nonpolar substances soluble in lipids move easily through plasma membrane by simple diffusion.

Movement Through Membrane Channels   Small polar molecules and ions move through channels.

Mediated Transport   Carrier molecules in plasma membrane transport substances across membrane. Mediated transport systems exhibit characteristics of specificity, saturation, and competition.

Facilitated Diffusion   Carrier molecules move material across the membrane equally well in either direction. Net movement is from region of high concentration to region of low concentration; does not require cellular energy.

Active Transport   Carrier molecules move materials across the membrane from regions of low concentration to regions of high concentration, against normal concentration or diffusion gradients; requires cellular energy.

Endocytosis and Exocytosis   Some cells take in substances by forming vesicles that enclose small portions of the external environment, a process called endocytosis. Substances leave cell by the fusion of membrane-bounded vesicles with plasma membrane, a process called exocytosis.

Tonicity   Refers to the tendency of a solution to influence the state of tension or tone of cells placed in it due to water movement into or out of the cells. The tonicity of a solution is influenced by the properties of a cell's plasma membrane.

SPECIALIZED INTERCELLULAR MEMBRANE JUNCTIONS

Tight Junctions   Plasma membranes of adjacent cells essentially fuse with one another, limiting the extracellular passage of materials between cells.

Desmosomes   At spot desmosomes, the plasma membranes of adjacent cells are separated by an extracellular space that contains filamentous material. Desmosomes are important in the attachment of one cell to another.

Gap Junctions   Plasma membranes of adjacent cells are separated by a relatively narrow extracellular space; small channels extend across this space and link the cytoplasm of adjacent cells; the transfer of materials from cell to cell occurs through channels.

The Nucleus
1. Contains the genetic material of the cell in the form of chromatin threads (DNA combined with protein and RNA).
2. DNA instructs cells to synthesize particular polypeptides or proteins.
3. Polypeptides or proteins synthesized according to DNA specifications act as enzymes, as structural proteins, or in other ways to accomplish the work of the cell.

RNA SYNTHESIS (TRANSCRIPTION)   DNA serves as the template for RNA synthesis.

MESSENGER RNA
1. M-RNA contains coded information, derived from DNA, that is used to direct protein synthesis.
2. The principal bases of DNA—adenine, cytosine, guanine, thymine—can be arranged to form 64 different three-base code words; bases of m-RNA can form an equal number of complementary three-base codons.
3. Code word and codon sequence specifies the amino acid sequence of a polypeptide or protein.

RIBOSOMAL RNA
1. Nucleoli are found in the nucleus and are associated with r-RNA production.

2. R-RNA becomes a part of ribosomes.

Ribosomes Cytoplasmic organelles that consist of r-RNA and protein.

PROTEIN SYNTHESIS (TRANSLATION) At ribosomes, amino acids are linked together in the sequence specified by m-RNA to form polypeptides and proteins.

Endoplasmic Reticulum Membranous network of tubular or saclike channels; walls contain enzymes that play a role in fatty acid and steroid synthesis; believed to be sites of synthesis of cellular membranes. Proteins synthesized by attached ribosomes can enter channels, and protein-containing vesicles can break away.

Golgi Apparatus Flattened membranous sacs with which vesicles from endoplasmic reticulum can fuse. Proteins are processed, modified, and sorted in Golgi apparatus; vesicles containing processed proteins pinch off from Golgi membranes.

Lysosomes Membrane-bounded cytoplasmic organelles that contain digestive enzymes that act on proteins, lipids, certain carbohydrates, DNA, and RNA.

Peroxisomes Contain a variety of powerful oxidative enzymes, as well as enzymes that catalyze the oxidative decomposition of hydrogen peroxide.

Mitochondria Involved in the generation of metabolic energy for cell activity.

Microtubules Small, hollow, cylindrical, unbranched tubules involved in cell movement processes and in the development and maintenance of cell shape.

Cilia, Flagella, and Basal Bodies Cilia and flagella are motile projections from the plasma membrane that can move substances over cell surfaces, or move entire cells about. Both are believed to arise from basal bodies. Cilia, flagella, and basal bodies contain microtubules.

Centrioles Similar in structure to basal bodies; involved in cell division.

Microfilaments Occur in bundles or other groupings; associated with contractile activities involved in cell movement.

Inclusions Many chemical substances in cells, such as hemoglobin.

EXTRACELLULAR MATERIALS Include body fluids and substances in which many cells are embedded. Connective tissues are particularly rich in extracellular materials. **p. 68**

CELL DIVISION Several basic events are involved, including replication of genetic material in nucleus, redistribution of this material into two new nuclei, and the division of cytoplasm into two new cells, each with its own nucleus. **p. 69**

Interphase Period between active cell divisions. DNA replication occurs.

Mitosis A continuous event, although four phases are observable.

PROPHASE Chromosomes become visible; centrioles move toward opposite poles of cell; spindle forms.

METAPHASE Chromatids attached by centromeres to spindle fibers; chromatids uncouple and each becomes a separate, single-stranded chromosome.

ANAPHASE Single-stranded chromosomes separate and move to opposite poles of cell.

TELOPHASE Spindle disappears and chromosomes assume interphase appearance as indistinct chromatin threads.

Meiosis Cell division process that produces haploid cells; involves two successive division sequences.

FIRST SEQUENCE

Prophase Essentially as in mitosis, except that homologous chromosomes pair and move toward spindle fibers as two chromosome units.

Metaphase Chromatids of synapsed chromosomes do not uncouple.

Anaphase Paired chromosomes move apart, one member of the pair moves to one pole of cell and the other member to the opposite pole.

Telophase Cytokinesis produces two new cells, each with 23 double-stranded chromosomes.

SECOND SEQUENCE Each of the 23 chromosome daughter cells undergoes mitosis with cytokinesis. Result is four haploid cells, each with 23 single-stranded chromosomes.

CANCER   Uncontrolled division of cells. May be treated surgically if discovered early enough; radiation therapy and chemotherapy are other forms of treatment.   **p. 73**

GENETIC DISORDERS   Many diseases and abnormalities are traceable to genetic defects. Genetic disorders can cause incorrect protein synthesis; erroneous protein might then lead to disease. Alterations in the genetic material of the cell are called mutations; in most cases mutations are harmful.   **p. 76**

AGING   X-rays, cosmic radiation may damage or kill cells that are not replaced, and thus cause aging. Aging changes may be programmed into genetic apparatus of cells.   **p. 76**

SELF-QUIZ

1. Many nonpolar substances that are soluble in lipids move relatively easily through the plasma membrane by: (a) exocytosis; (b) active transport; (c) simple diffusion.
2. Match the following terms associated with the plasma membrane with the appropriate descriptions.

   Specificity   Facilitated diffusion
   Saturation   Active transport
   Competition

   (a) Carrier molecules move material across the membrane equally well in either direction; net movement is from region of high concentration to region of low concentration.
   (b) Each carrier molecule binds with only a select group of substances.
   (c) Carrier molecules move materials across the plasma membrane against a concentration or diffusion gradient.
   (d) The amount of a substance that can cross the plasma membrane by mediated transport in a given time is limited.
   (e) A particular binding site on a carrier molecule can bind with more than one substance.
3. The term for the process by which a cell takes in material from the external environment by means of membrane-bounded vesicles is: (a) exocytosis; (b) endocytosis; (c) cytokinesis.
4. A solution that produces a change in the tone of cells immersed in it as a result of the net movement of water from the solution into the cells is a(an): (a) isotonic solution; (b) hypotonic solution; (c) hypertonic solution.
5. Specialized intercellular membrane junctions at which the transfer of materials from cell to cell occurs are called: (a) gap junctions; (b) tight junctions; (c) desmosomes.
6. DNA contains coded messages that instruct a cell to synthesize particular polypeptides or proteins. True or False?
7. A three-base sequence in m-RNA that specifies a particular amino acid of a polypeptide or protein is called a(an): (a) code word; (b) codon; (c) anticodon.
8. Ribosomes are formed from r-RNA–protein complexes that migrate out of the nucleus and into the cytoplasm. True or False?
9. Enzymes that play a role in fatty acid and steroid synthesis are associated with: (a) endoplasmic reticulum; (b) ribosomes; (c) lysosomes.
10. Enzymes capable of digesting proteins, lipids, DNA, and RNA are associated with: (a) ribosomes; (b) microtubules; (c) lysosomes.
11. Organelles involved in the generation of metabolic energy for cellular activities are: (a) peroxisomes; (b) mitochondria; (c) lysosomes.
12. Thin, short, and numerous projections that can move substances over the surface of a cell are known as: (a) cilia; (b) flagella; (c) basal bodies.
13. Cilia, flagella, basal bodies, and centrioles all contain: (a) mitochondria; (b) microtubules; (c) peroxisomes.
14. Match the following cytoplasmic structures with their related functions.

    Ribosomes   Basal bodies
    Golgi apparatus   Microfilaments
    Peroxisomes

    (a) Organelles that contain powerful oxidative enzymes
    (b) Structures from which cilia and flagella arise
    (c) Structures that are associated with contractile activities involved in cell movement
    (d) The site of polypeptide and protein synthesis
    (e) A membranous organelle that adds carbohydrate to many proteins

15. Mitosis and cytokinesis always occur simultaneously. True or False?
16. The DNA molecules that comprise the genetic material of the cell appear only as indistinct chromatin threads within the nucleus during: (a) metaphase; (b) anaphase; (c) interphase.
17. During both RNA production and DNA replication, nucleotides attach to an exposed template DNA chain according to a complementary base-pairing pattern. True or False?
18. Match the following items associated with meiosis with the appropriate lettered descriptions.

Autosomes   23
XY   Haploid
Homologous chromosomes   Interkinesis
46   Crossing over
Diploid

(a) XX

(b) The period between the first and second meiotic divisions
(c) Cells with two complete sets of chromosomes are said to be this type.
(d) This process contributes to genetic diversity.
(e) This cell type is characteristic of gametes.
(f) The total number of chromosomes in a human somatic cell
(g) Male sex chromosomes
(h) Chromosome classification that does not include sex chromosomes
(i) The number of chromosomes contained in each of the human male and female gametes

19. A doctor would be likely to use surgery but not radiation or chemotherapeutic agents to treat a patient in whom cancer cells had metastasized. True or False?
20. In most instances, mutations are found to be: (a) neutral; (b) helpful; (c) harmful.

LEARNING OBJECTIVES

After completing this chapter, you should be able to:

- Compare and contrast the functions of mucous membranes and serous membranes.
- Cite the two main components of bone that provide it with its strength.
- Describe five functions of bone.
- Describe the functions of osteoblasts and osteoclasts.
- Explain the function of the reticuloendothelial system.
- Compare and contrast the functions of sweat glands and sebaceous glands.
- Describe four functions of the skin.
- Discuss the healing of cut skin.
- Explain the classification of burns.
- Describe the physiological consequences of a third-degree burn.
- Describe the process of split skin grafting.

CHAPTER CONTENTS

TISSUES AND SKIN

4

Groups of cells that are similar in structure, function, and embryonic origin, and that are bound together with varying amounts of intercellular material, are referred to as **tissues.** Tissues join together to form the organs of the body, and an understanding of tissue functions contributes to an understanding of the functions of the organs. There are four principal types of tissues in the body: epithelial, connective, muscle, and nervous.

EPITHELIAL TISSUE

Epithelial tissue is composed of cells that are very closely joined, with only a minimum of intercellular material between them (F4.1). Epithelial tissue covers most of the free surfaces of the body, both internal and external. It forms the outer layer of the skin (the epidermis), the lining of the digestive tube, the linings of the thoracic (chest) and abdominopelvic (peritoneal) cavities, and the lining of the blood vessels (the endothelium). In addition, epithelial tissue is incorporated into various glands, and it forms a number of glandular ducts and tubules.

FUNCTIONS OF EPITHELIAL TISSUE

Epithelial tissue, either alone or in combination with other tissues, is important in a variety of body functions. The epithelium of the skin is a protective barrier between the body and the external environment. Other epithelia are involved in the absorption of materials into the body, the secretion of special products within the body, and the excretion of waste products from the body. The functions of several body components that are composed primarily of epithelial tissue are particularly worthy of examination.

Mucous Membranes

Mucous membranes line the digestive, respiratory, urinary, and reproductive tracts, all of which open to the exterior of the body. The activities of the epithelial cells of mucous membranes include both absorption and secretion. For example, food substances are absorbed into the body through many of the epithelial cells of the mucous membranes of the digestive tract. Other epithelial cells, called goblet cells, secrete *mucus*, a viscous fluid that keeps the free surfaces of the membranes moist and lubricates food material as it passes through the tract.

Serous Membranes

Serous membranes line the thoracic and abdominopelvic cavities and also cover many of the organs within these cavities. The epithelial cells of serous membranes secrete a clear, watery *serous fluid*. The serous fluid keeps the membranes moist and reduces friction when the organs rub against one another or against the walls of the cavities.

Glands

A **gland** is a group of cells that are specialized for secretion. Most of the glands of the body are composed of epithelial cells that produce specific secretions such as sweat, milk, hormones, or enzymes.

CONNECTIVE TISSUE

In contrast to epithelial tissue, which consists of closely packed cells and relatively little inter-cellular material, **connective tissue** is characterized by relatively few, widely spaced cells and an abundant intercellular substance (F4.2). The intercellular substance of connective tissue includes various types of fibers, such as tough, inelastic collagenous fibers that are composed of the protein collagen.

Several types of cells are associated with connective tissue, but *fibroblasts* and *macrophages* are the most common. Fibroblasts produce the intercellular substance of many types of connective tissue. Macrophages are active *phagocytes* (*phago* = to eat). Macrophages move through the loose types of connective tissue and engulf foreign matter as well as dead or dying cells. Macrophages are active in inflammatory conditions and are part of the reticuloendothelial system, which is discussed later in this chapter.

FUNCTIONS OF CONNECTIVE TISSUE

Connective tissue performs a variety of functions. It serves as the framework upon which epithelial

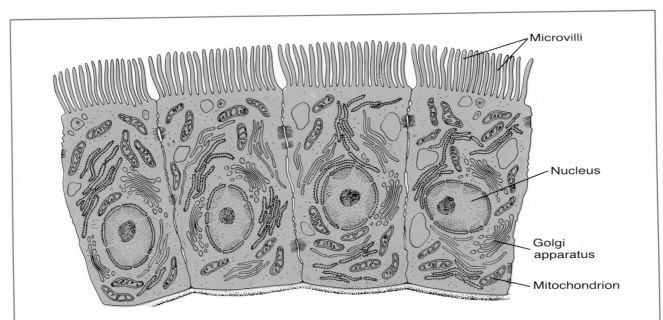

F4.1
Epithelial tissue. The epithelial cells shown here have their free surfaces folded into tiny projections called microvilli that greatly increase the free surface area. Other epithelial cells have free surfaces that are smooth, and still others have free surfaces that are ciliated.

cells cluster; it binds various tissues and organs together, supporting them in their proper locations; it serves as a storage site for food materials; and it forms the rigid framework of the body (the skeleton). The functions of several specialized types of connective tissue are especially noteworthy.

Adipose Tissue

Adipose tissue is composed of fat cells that are dispersed within loose connective tissue (F4.3). Each fat cell contains a large droplet of fat that squeezes and flattens the nucleus and forces the cytoplasm into a thin ring around the cell's periphery. Adipose tissue serves as a storage area for fat. It also pads and protects certain regions of the body.

Cartilage

Cartilage is a specialized fibrous connective tissue characterized by a relatively firm intercellular substance that includes numerous collagenous fibers. Because of the firmness of its intercellular substance, cartilage functions as a structural support. However, it also possesses a certain amount of flexibility.

Bone

Bone is another type of connective tissue (F4.4). The intercellular substance of bone includes collagenous fibers and inorganic salts that contain substantial amounts of calcium and phosphate. The collagenous fibers provide bone with great tensile strength—that is, great resistance to stretching and twisting. The salts enable bone to withstand compression. This combination of fibers and salts makes bone exceptionally strong without being brittle.

Bone performs several important functions:

1. *Support* Bone forms a framework—the skeleton—that supports the body's soft tissues.

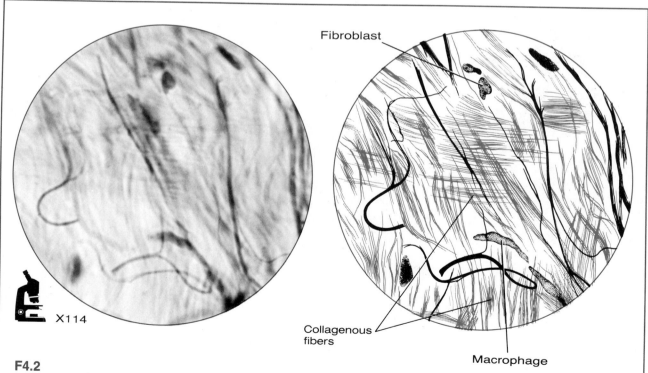

F4.2
Connective tissue. The type of connective tissue shown here is loose connective tissue.

X114

Fibroblast

Collagenous fibers

Macrophage

2. *Movement* Many of the body's muscles attach to bone, and many bones articulate with one another at movable joints. Consequently, bone plays an important role in determining the kind and extent of movement the body is capable of.

3. *Protection* Many vital organs are protected from injury by bone. The brain is surrounded by the bones of the skull, the spinal cord is within the canal formed by the vertebrae, the thoracic organs are protected by the rib cage, and the urinary bladder and internal reproductive organs are protected by the bony pelvis.

4. *Mineral reservoir* Bone contains large deposits of calcium, phosphorus, sodium, potassium, and other minerals. These minerals can be mobilized as required and distributed by the circulatory system to other regions of the body.

5. *Hemopoiesis (blood-cell formation)* Many bones contain spaces that are filled with a material called *red bone marrow*. Cells within the red bone marrow produce the blood cells found within the circulatory system.

Bone Formation and Breakdown.

Bone is a living tissue that is continually being formed and broken down. Cells called *osteoblasts* are involved in bone formation. Osteoblasts secrete the organic portion of the intercellular substance of bone, which includes collagenous fibers and glycoproteins such as chondroitin sulfate. The inorganic salts are deposited in the organic framework. It is believed that, initially, salts such as calcium phosphate $[Ca_3(PO_4)_2]$ are formed; then, through a process of substitution and addition of ions, these salts are converted to the inorganic material of bone, which is composed of highly insoluble crystals of hydroxyapatite $[3Ca_3(PO_4)_2 \cdot Ca(OH)_2]$.

Although the events of mineralization are not

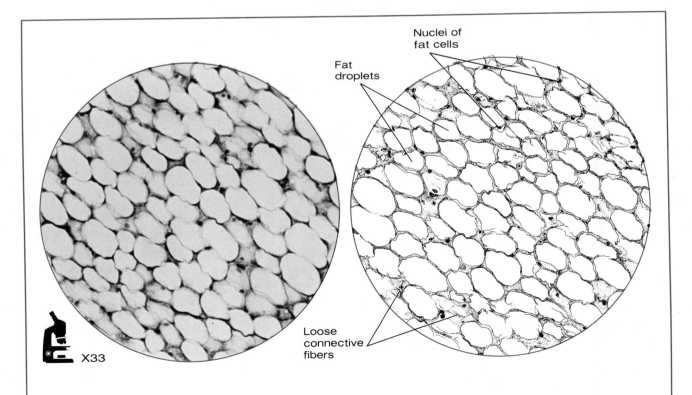

Nuclei of fat cells

Fat droplets

Loose connective fibers

X33

F4.3
Adipose tissue. Adipose tissue is composed of fat cells that are dispersed within loose connective tissue.

fully known, one theory proposes that bone cells concentrate large quantities of calcium and phosphate and that the cells subsequently release calcium phosphate compounds into the extracellular fluid. The small areas of calcium phosphate salts that result serve as sites for further salt deposition. An alternative theory suggests that the initial formation of bone mineral takes place in membrane-bounded vesicles called *matrix vesicles*, which are synthesized and secreted into the extracellular fluid by bone cells.

In contrast to osteoblasts, cells called *osteoclasts* break down bone. The osteoclasts secrete enzymes that digest the protein portion of the intercellular substance. The osteoclasts are also believed to secrete acids, such as citric acid and

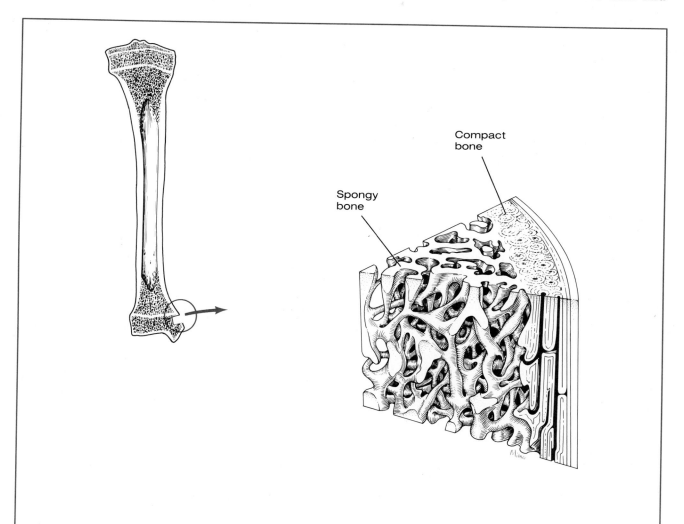

F4.4

Bone. Two forms of bone are visible. Spongy bone consists of a three-dimensional network of interconnecting pieces of bone with many spaces between the pieces. The spaces often contain a substance called red bone marrow. Compact bone is a relatively solid, continuous mass of bone in which the only spaces are microscopic.

lactic acid, that lower the pH of the surrounding environment and cause bone minerals to dissolve.

Bone Strength. The relationship between bone formation and bone breakdown determines the strength of a bone at any moment, and various physical, nutritional, and hormonal factors influence this relationship. For example, if a bone is subjected to repeated strain, increased amounts of collagenous fibers and inorganic salts are deposited, strengthening the bone. Thus, regular exercise can alter bone, and the bones of an athlete tend to be thicker and stronger than the bones of an untrained person. Conversely, if a bone is not subjected to strain, salts are withdrawn from the bone. Thus, when a limb is paralyzed or immobilized in a cast, the bones of the limb tend to waste away and weaken.

One interesting property of bone is that when it is deformed, a voltage can be recorded between two opposite surfaces. When a bone is bent, the convex surface becomes negatively charged, and new bone deposition occurs at this negative surface, where it strengthens the bone most effectively. In fact, weak electrical currents are sometimes used to promote the healing of broken bones that do not mend properly.

Reticuloendothelial System

The **reticuloendothelial system** is often considered to be a type of connective tissue, even though the cells of this system do not really form a distinct tissue. The cells of the reticuloendothelial system, however, share the common function of phagocytosis. The reticuloendothelial system is also called the *macrophage system* because most of the phagocytic cells are macrophages. Macrophages are found in many places throughout the body, including loose connective tissue, lymphatic tissue, mesenteries of the digestive tract, bone marrow, the spleen, the adrenal glands, and the pituitary gland. In some locations, they are given special names—such as the Kupffer cells that line the blood sinusoids of the liver, the dust cells of the lungs, and the microglia of the central nervous system. Although, in general, macrophages are phagocytic, those in specific locations often are selective as to what they ingest. Macrophages in the spleen, for example, are particularly active in breaking down aging red blood cells.

Blood

Blood, which consists of cells dispersed in an intercellular substance, is usually considered to be a specialized connective tissue. The intercellular substance that surrounds the blood cells is an extracellular fluid called *plasma.* Blood is considered further in Chapter 11.

MUSCLE TISSUE

Muscle tissue is composed of long, thin cells called *fibers.* It is important to realize that muscle fibers are living cells, and are in no way similar to the fibers of connective tissue. Muscle fibers have a highly developed property of contractility, which enables muscle tissue to move the skeleton, propel the blood through the body, and aid digestion by moving food through the digestive tract. Muscle tissue is considered further in Chapters 6 and 12.

NERVOUS TISSUE

Nervous tissue contains many highly specialized nerve cells, or *neurons,* that transmit information in the form of electrical signals from one part of the body to another. This tissue is found largely in the brain, the spinal cord, and nerves, and the activity of neurons is important in the control and regulation of many body functions. Nervous tissue is studied in greater detail in Chapters 5, 7, and 8.

SKIN

The **skin** forms the entire external covering of the body. It is continuous with, but differs structurally from, the mucous membranes that line the respiratory, digestive, urinary, and reproductive tracts. The skin is not a tissue. Rather, it is a combination of tissues that form a simple organ. Although the skin is not often viewed as an organ, it is, in fact, the largest organ of the body.

The skin is composed of two main layers: (1) the *epidermis,* a surface layer of closely packed epithelial cells, and (2) the *dermis,* a deeper layer of dense irregular connective tissue (F4.5). The dermis is well supplied with blood vessels, lymph vessels, nerves, and sensory receptors. It also contains sweat glands and glands called sebaceous glands.

The skin is connected to underlying structures by a layer of loose connective tissue, the *hypodermis*, or *subcutaneous tissue*. In many cases, fat cells are deposited in the loose connective tissue, forming adipose tissue, and in some regions such as over the abdomen and the buttocks, much fat can accumulate.

NOURISHMENT OF THE SKIN

In contrast to the well-vascularized dermis, there are no blood vessels within the epidermis. As a result, the cells of the epidermis must obtain nutrients by diffusion from the vessels of the dermis. This activity is able to meet the needs of cells lo-

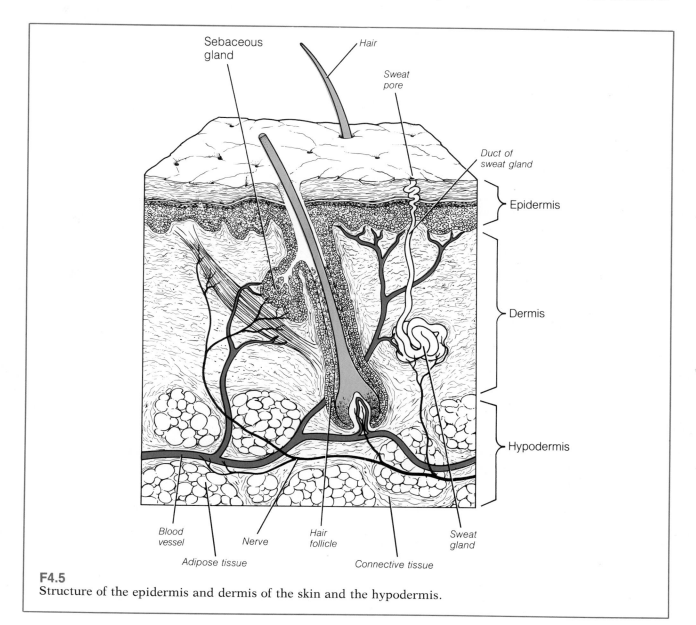

F4.5
Structure of the epidermis and dermis of the skin and the hypodermis.

cated deep within the epidermis, close to the dermis. However, these cells are continually undergoing mitosis, forming epidermal cells that are forced toward the body surface. As the cells move toward the surface and farther away from the source of nourishment, they die and their cytoplasm is replaced by a fibrous protein called *keratin*. The outermost layers of dead epidermal cells are constantly being lost due to abrasion—for example, as a result of friction with clothing. These cells are replaced, however, by cells moving outward from the deep regions of the epidermis.

GLANDS OF THE SKIN

Two types of glands—sweat glands and sebaceous glands—are widely distributed in the skin.

Sweat Glands

Sweat (sudoriferous) glands are located over most of the body surface, except the lips, the nipples, and portions of the skin of the genital organs (F4.5). The sweat glands secrete a watery solution called *sweat* that contains sodium chloride and traces of urea, sulfates, and phosphates. The sweat glands are stimulated by nerves that lead to them, and the amount of sweat secreted depends on such factors as the environmental temperature and humidity, the amount of muscular activity, and various conditions that cause stress.

Sebaceous Glands

A few **sebaceous glands** empty their secretions directly onto the surface of the skin. However, most of the glands empty their secretions into hair follicles, which are epithelial-lined structures from which hairs grow (F4.5). The secretion of the sebaceous glands, which is called *sebum*, is an oily substance rich in lipids. Sebum travels along the shafts of the hairs to the surface of the skin. The sebum oils the skin and the hairs, preventing them from drying. It also contains substances that are toxic to certain bacteria. The sebaceous glands, which are stimulated by the sex hormones (mainly testosterone), are particularly active during adolescence. If sebum accumulates within the duct of a gland, it forms a white pimple. If this blocked sebum becomes oxidized, it darkens, forming a "blackhead."

FUNCTIONS OF THE SKIN

The skin performs a number of functions that are important in maintaining homeostasis.

Protection

The skin forms a physical barrier that prevents microorganisms and other substances (including water) from invading the body. It protects body structures against excessive ultraviolet radiation, and it greatly reduces the loss of body water to the environment. The surface of the skin is coated with a thin liquid film that tends to be acidic (pH 4–6.8). This film acts as an antiseptic layer that retards the growth of microorganisms on the surface of the skin. When subjected to repeated trauma such as rubbing or scraping, the skin becomes thickened, forming calluses in the more severe instances.

Body Temperature Regulation

The skin plays an important role in maintaining a constant body temperature. For example, if the body temperature begins to rise, blood vessels in the dermis dilate, increasing the flow of warm blood to the body surface, where heat is lost to the environment. In addition, increased secretory activity by the sweat glands causes the body surface to become wet with sweat, which evaporates and further facilitates the loss of body heat. Conversely, if the body temperature begins to fall, blood vessels in the dermis constrict, decreasing the flow of blood to the body surface. This response diminishes the loss of heat to the environment and conserves body heat.

Sensation

Because it contains nerve endings and specialized sensory receptors, the skin provides the body with much information about the external environment. Events such as temperature changes, light touch, pressure, and trauma that causes pain all stimulate receptors located in the skin. These receptors, in turn, alert the central nervous system to the particular event, thus enabling the person to take appropriate action. This action might be simple and automatic, such as the withdrawal of a hand from a hot dish, or it may require a more complicated act, such as deciding that a heavier coat is needed because of a temperature change.

Vitamin D Production

In the presence of sunlight or other ultraviolet radiation, a steroid substance in the skin (7-dehydrocholesterol) is altered in such a way that it forms vitamin D_3 (cholecalciferol). After being metabolically transformed in the liver and kidneys, vitamin D_3 promotes the absorption of calcium from the digestive tract. Vitamin D_3, therefore, is important in maintaining body calcium at optimum levels.

HEALING OF CUT SKIN

When the skin is cut and the edges of the wound are drawn together by sutures or other means, a V-shaped slit extends from the surface to the subcutaneous tissue (F4.6*a*). Soon after the skin is cut, the process of blood coagulation causes a small amount of a fibrous substance called *fibrin* to form near the bottom of the slit. With time, epidermis from the surface grows down into the slit, adhering to sound tissue on either side. By about one week, the epidermis extends down into the dermis (F4.6*b*). The epidermis continues its downward growth, adhering to healthy dermis and passing beneath any fibrin that is present. By about two weeks, epidermal continuity is restored when the epidermis growing down one side of the slit meets the epidermis growing down the opposite side, near the bottom of the slit (F4.6*c*).

While the epidermis is growing down the sides of the slit, fibroblasts and blood vessels from the subcutaneous tissue are engaged in repairing the connective tissue of the skin. At the junction of the dermis and the subcutaneous tissue in the region of the wound, an abundant growth of fibroblasts and small blood vessels (capillaries) forms a ridge of new tissue (F4.6*c*). As this tissue grows, it bulges up at the bottom of the epidermis-lined slit and pushes the slit upward until it becomes level with the surface (F4.6*d*). Thus, the area previously occupied by the slit becomes occupied by new connective tissue that is derived chiefly from subcutaneous tissue and is covered by a thin epidermis. This epidermis remains thin for some time.

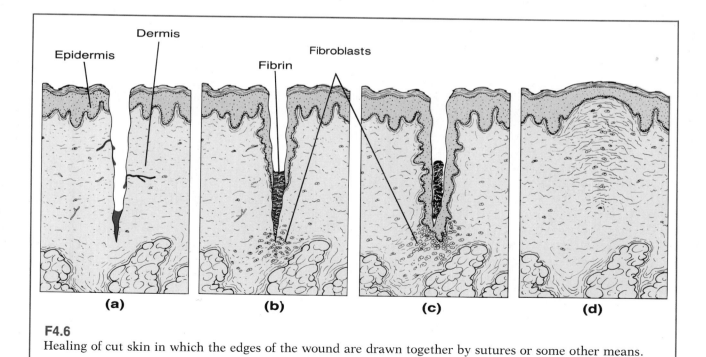

F4.6
Healing of cut skin in which the edges of the wound are drawn together by sutures or some other means.

BURNS AND SKIN GRAFTS

Burns can destroy the skin and cause serious problems that clearly demonstrate the importance of the skin to the body. When the skin is destroyed, there is a large loss of interstitial fluid and blood plasma. Lost along with these fluids are plasma proteins and mineral salts. If unchecked, these losses upset the osmotic equilibrium of the body and cause dehydration, kidney malfunction, and shock. In addition, with the protection of the skin gone, it is very easy for infectious agents to invade the body.

Burns are classified according to their severity. First-degree burns are those in which only the epidermis is damaged. The symptoms of a first-degree burn include localized pain, redness, and swelling. Sunburn is usually a first-degree burn. Second-degree burns are those in which both the epidermis and dermis are damaged. However, the damage is not severe, and the skin quickly regenerates. Third-degree burns are those in which both the epidermis and the dermis are so severely damaged that they can only regenerate from the edges of the wound. If the burned area is extensive, this is a slow process during which body fluids are constantly being lost from the damaged area, and the possibility of infection is high. In addition, such wounds can result in the formation of extensive scar tissue, which is not only disfiguring, but can also restrict the movement of the damaged part.

Serious damage to the skin, such as can result from burns, may necessitate the use of **skin grafts** to hasten healing, reduce the loss of body fluids, and minimize the formation of scar tissue. In one method of skin grafting—called split skin grafting—skin taken from one part of a person's

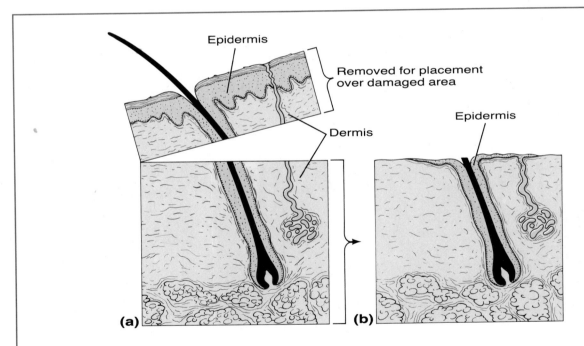

F4.7
Split skin grafting. (a) The outer portion (the epidermis and perhaps half the dermis) is removed from an area of healthy skin and placed over a region where serious skin damage has occurred. (b) The exposed, nonepidermal surface that remains becomes covered with new epidermis that grows over the surface from remaining hair follicles and the ducts of sweat glands.

body is used to cover another part. Split skin grafting is useful because it eventually increases the total amount of skin on the body surface. With this method of skin grafting, the outer portion (the epidermis and perhaps half the dermis) of the skin is removed from an undamaged area and placed over a region where serious skin damage has occurred (F4.7a). The cells of the graft are nourished by interstitial fluid from the damaged surface. Gradually, connective-tissue cells form new intercellular substance that attaches the graft in place, and eventually the graft becomes vascularized.

The exposed, nonepidermal surface in the region from which the graft was taken becomes covered with new epidermis that grows over the surface from remaining hair follicles and the ducts of sweat glands, which originate embryologically from the same tissue that gives rise to the epidermis (F4.7b).

EFFECTS OF AGING ON THE SKIN

With aging, the skin tends to become thinned, somewhat wrinkled, dry, and occasionally scaly.

The skin can also become more permeable and allow substances to pass through it more readily. In addition, with aging, collagen fibers within the dermis become thicker, and there is a gradual decrease in the underlying subcutaneous fat. There is also a decrease in the number and activity of hair follicles, sweat glands, and sebaceous glands. Consequently, aging is often accompanied by a loss of hair, reduced sweating, and decreased oil (sebum) production.

Skin that has been exposed to sunlight over a lifetime shows changes that are more severe than those due to the intrinsic aging process alone. For example, aging skin that has had excessive exposure to sunlight shows more marked wrinkling and furrowing, and may develop nodules of an abnormal type of collagen. In addition, some cutaneous cancers tend to occur most often on areas of the body that have had long exposure to the sun, particularly in fair-skinned people.

STUDY OUTLINE

EPITHELIAL TISSUE Composed of cells that are very closely joined, with only a minimum of intercellular material between the cells. **p. 83**

Functions of Epithelial Tissue Protection, absorption, excretion, secretion.

MUCOUS MEMBRANES Epithelial cells of mucous membranes are both absorptive and secretory; goblet cells secrete mucus.

SEROUS MEMBRANES Epithelial cells of serous membranes secrete a watery fluid called serous fluid, which keeps membranes moist and reduces friction when organs rub against one another or against walls of body cavities.

GLANDS Epithelial cells of glands produce specific secretions such as sweat, milk, hormones, or enzymes.

CONNECTIVE TISSUE Characterized by relatively few widely spaced cells and an abundant intercellular substance. Intercellular substance includes fibers such

as collagenous fibers. Fibroblasts produce the intercellular substance of many types of connective tissue. Macrophages are phagocytic cells found in connective tissue **pp. 84–88**

Functions of Connective Tissue Framework, binding, support, storage.

ADIPOSE TISSUE Contains fat cells that store fat; pads and protects certain body regions.

CARTILAGE Relatively firm intercellular substance and numerous collagenous fibers; serves as structural support.

BONE Intercellular substance includes collagen fibers and inorganic salts; functions: support, movement, protection, mineral reservoir, hemopoiesis.

Bone Formation and Breakdown Osteoblasts secrete organic portion of intercellular substance; inorganic salts are deposited in organic framework. Osteoclasts secrete enzymes that digest protein portion of intercellular substance; also believed to secrete acids that cause bone minerals to dissolve.

Bone Strength   Depends on relationship between bone formation and breakdown; influenced by physical, nutritional, and hormonal factors. Repeated strain leads to increased deposition of collagenous fibers and inorganic salts, strengthening bone. No strain leads to withdrawal of salts and weakening of bone. When bone is bent, convex surface becomes negatively charged and new bone deposition occurs at this surface.

RETICULOENDOTHELIAL SYSTEM   Cells do not form distinct tissue, but all are phagocytic; most are macrophages; certain cells are often selective as to what they ingest.

BLOOD   Intercellular substance that surrounds blood cells is extracellular fluid called plasma.

MUSCLE TISSUE   Composed of contractile cells; moves skeleton, propels blood through body, moves food through digestive tract.   **p. 88**

NERVOUS TISSUE   Contains cells called neurons that transmit signals called nerve impulses; nerve impulse has nature of electrical pulse that travels along a neuron; activity important in control and regulation of many body functions.   **p. 88**

SKIN   Forms external covering of body; not a tissue, but a combination of tissues that form a simple organ.
1. Epidermis—surface layer of closely packed epithelial cells.
2. Dermis—deeper layer of dense irregular connective tissue well supplied with blood vessels, lymph vessels, nerves, and sensory receptors. Contains glands called sweat glands and sebaceous glands.
Skin is connected to underlying structures by a layer of loose connective tissue called the hypodermis or subcutaneous tissue.   **pp. 88–91**

Nourishment of Skin   Epidermis has no blood vessels, and epidermal cells must obtain nourishment by diffusion from vessels of dermis. This activity is able to meet the needs of the deepest epidermal cells. However, these cells continually undergo mitosis, forming cells that are forced toward the body surface. As the cells move toward the surface and farther away from the source of nourishment, they die, and their cytoplasm is replaced by a fibrous protein called keratin.

Glands of the Skin

SWEAT GLANDS   Secrete a watery solution called sweat that contains sodium chloride and traces of urea, sulfates, and phosphates. Nerves to sweat glands stimulate secretion of sweat.

SEBACEOUS GLANDS   Secrete sebum, which is an oily substance rich in lipids. Sebum oils skin and hair and contains substances toxic to certain bacteria. Sebaceous glands are stimulated by sex hormones.

Functions of Skin

PROTECTION   Skin forms physical barrier that prevents microorganisms and other substances from invading body. It protects against excessive ultraviolet radiation, reduces loss of body water to environment, and is coated with liquid film that acts as antiseptic.

BODY TEMPERATURE REGULATION   Dilation of blood vessels in dermis and sweating increase heat loss from body; constriction of vessels in dermis conserves body heat.

SENSATION   Nerve endings and specialized sensory receptors in skin provide the body with much information about the external environment.

VITAMIN D PRODUCTION   Sunlight or other ultraviolet radiation converts a steroid in skin into vitamin D_3. After being metabolically transformed, vitamin D_3 promotes absorption of calcium from digestive tract.

Healing of Cut Skin   When skin is cut and edges of wound are drawn together:
1. Epidermis grows down sides of cut and meets at bottom.
2. Fibroblasts and capillaries derived chiefly from subcutaneous tissue form ridge of new tissue at bottom of cut. As tissue grows, it pushes new epidermis upward until it is level with surface.

Burns and Skin Grafts   Burns can destroy skin with loss of interstitial fluid, blood plasma, plasma proteins, and mineral salts, leading to upset osmotic equilibrium, dehydration, kidney malfunction, and shock. Burns are classified according to severity.
1. First-degree—damage only to epidermis.
2. Second-degree—damage to both epidermis and dermis, but skin quickly regenerates.

3. Third-degree—damage to epidermis and dermis is so severe that they can regenerate only from edges of wound.

Split skin grafting can be used to repair serious damage to skin. Outer portion (epidermis and perhaps half the dermis) of healthy skin is removed and placed over damaged area and gradually grows into place. Area from which skin was taken becomes covered with new epidermis that grows from hair follicles and ducts of sweat glands.

Effects of Aging on the Skin   With aging, skin tends to become thinned, somewhat wrinkled, dry, and occasionally scaly. Skin can become more permeable, and aging is often accompanied by loss of hair, reduced sweating, and decreased oil production.

SELF-QUIZ

1. The respiratory tract is lined with: (a) mucous membrane; (b) serous membrane; (c) endothelium.
2. The thoracic and abdominopelvic cavities are lined with: (a) mucous membrane; (b) serous membrane; (c) epidermis.
3. Certain epithelial cells produce: (a) mucus; (b) hydroxyapatite; (c) collagenous fibers.
4. Which tissue is characterized by relatively few, widely spaced cells and an abundant intercellular substance? (a) epithelial tissue; (b) connective tissue; (c) muscle tissue.
5. Which cells are primarily phagocytic? (a) fibroblasts; (b) osteoblasts; (c) macrophages.
6. Cartilage is a specialized fibrous connective tissue that is characterized by a relatively firm intercellular substance that does not include any collagenous fibers. True or False?
7. A combination of collagenous fibers and inorganic salts makes bone exceptionally strong without being brittle. True or False?
8. Osteoclasts: (a) secrete the organic framework of bone; (b) release enzymes that break down the proteins of the intercellular substance of bone; (c) cause bone mineralization by secreting hydroxyapatite crystals.
9. When a bone is subjected to repeated strain: (a) salts are withdrawn from the bone; (b) osteoblasts break down the intercellular substance of the bone; (c) the bone often becomes thicker and stronger.
10. When bone is bent, the convex surface becomes positively charged and new bone deposition occurs at this surface. True or False?
11. Although, in general, macrophages are phagocytic, those in specific locations are often selective as to what they ingest. True or False?
12. The epidermis of the skin contains: (a) dead cells whose cytoplasm has been replaced by keratin; (b) sensory receptors; (c) sweat glands.
13. Sweat glands are stimulated to secrete sweat by: (a) nerves; (b) sex hormones; (c) vitamin D_3.
14. Sebum is a substance that is rich in: (a) sulfates; (b) urea; (c) lipids.
15. The surface of the skin tends to highly alkaline, with a pH of 6.8. True or False?
16. The loss of body heat to the environment is diminished by: (a) an increased dilation of blood vessels of the epidermis; (b) an increased secretion of sweat; (c) a decreased dilation of blood vessels of the dermis.
17. Vitamin D_3: (a) increases the excretion of potassium by the kidneys; (b) promotes the absorption of calcium from the digestive tract; (c) is formed in the skin from the breakdown of collagenous fibers.
18. Sunburn is usually a first-degree burn. True or False?
19. A burn victim who has experienced damage to both the epidermis and dermis, but whose lost tissue will quickly regenerate is suffering from a: (a) first-degree burn; (b) second-degree burn; (c) third-degree burn.
20. With aging, collagen fibers within the dermis of the skin become thicker, and there is a gradual increase in the underlying subcutaneous fat. True or False?

LEARNING OBJECTIVES

After completing this chapter, you should be able to:

- Describe the function of a dendrite and an axon.
- Describe the movements of potassium, sodium, and chloride ions across an unstimulated neuronal membrane.
- Describe the role of active transport mechanisms in maintaining the resting membrane potential of a neuron.
- Describe the movements of sodium and potassium ions during an action potential.
- Explain the all-or-none nature of an action potential and a nerve impulse.
- Cite two factors that influence the velocity at which an axon conducts a nerve impulse.
- Discuss the events involved in the transfer of information from one neuron to another at a synapse.
- Distinguish between an excitatory postsynaptic potential and an inhibitory postsynaptic potential.
- Distinguish between temporal summation and spatial summation.
- Compare and contrast a generator potential and a receptor potential.
- Describe the process of adaptation.

CHAPTER CONTENTS

NEURONS
Structure of a Neuron
Dendrites
Axons
The Formation of Myelinated Axons

BASIC ELECTRICAL CONCEPTS

MEMBRANE POTENTIALS
Resting Membrane Potential
Development of the Resting Membrane Potential
Movement of Ions across the Unstimulated Neuronal Membrane
Potassium Ions
Sodium Ions
Chloride Ions
Role of Active Transport Mechanisms in Maintaining the Resting Membrane Potential
Local Potential
Action Potential
Depolarization and Polarity Reversal
Return to the Unstimulated Condition
The Nerve Impulse
Refractory Periods
All-or-None Response
Direction of Nerve Impulse Conduction
Conduction Velocities

SYNAPSES
Electrical Synapses
Chemical Synapses
Excitatory Synapses
Inhibitory Synapses
Synaptic Delay
Neurotransmitters

NEURAL INTEGRATION
Divergence and Convergence
Summation
Temporal Summation
Spatial Summation
Facilitation
Determination of Postsynaptic Neuron Activity
Modification of Neuronal Activity
Presynaptic Inhibition
Neuromodulators

RECEPTORS
Generator Potentials
Receptor Potentials
Discrimination of Differing Stimulus Intensities
Adaptation

EFFECTORS

NEURONS, SYNAPSES, AND RECEPTORS 5

The nervous system provides the body with a rapid means of internal communication that is critically important in regulating and coordinating the activities of the billions of body cells. As a result, neural activity is essential to the body's ability to maintain homeostasis. This chapter focuses on the physiology of the *neuron*, which is the basic structural and functional component of the nervous system. It describes how a neuron transmits information from one point to another in the form of electrical signals, and it examines the ways in which one neuron communicates with another at specialized junctions known as *synapses*. It also considers the function of *receptors*, which are structures that convert information about conditions in the body's internal and external environments into neural signals.

NEURONS

Neurons are specialized cells that can respond to stimulation by initiating and conducting electrical signals. Certain other cells, including many muscle cells, can also initiate and conduct electrical signals, but the unique shapes and arrangements of neurons make them especially well suited for transmitting information from one point to another.

STRUCTURE OF A NEURON

A neuron is composed of a **cell body** and one or more thin, cytoplasm-containing processes that extend from the cell body (F5.1). The cell body contains a large nucleus and many organelles that are involved in maintaining the metabolism of the neuron.

Two types of processes—dendrites and axons—extend from the cell bodies of neurons. According to the classical usage of the terms, a dendrite is a process that conducts electrical signals toward a cell body, and an axon is a process that conducts electrical signals away from a cell body. It is becoming more and more common, however, to differentiate dendrites and axons as follows.

Dendrites

The **dendritic zone** is the receptive portion of a neuron in which an electrical signal originates. The dendritic zone can include the cell body as well as branching processes, or **dendrites,** that are either direct extensions of the cell body or more remote branchings separated from the cell body by a length of axon (F5.2). The number, length, and extent of branching of the dendrites varies in different neurons.

Axons

The **axon (nerve fiber)** is the conductive process of a neuron, and a specific type of electrical signal called a *nerve impulse* travels along it. A neuron has only one axon, but the axon generally has several branches called *collaterals*. The cell body is often located between the axon and the dendrites, but it may also be located along the axon or to one side of the axon (F5.2). Axons vary in length, and some axons extend a meter or more, from the spinal cord in the lower back to the feet and toes.

The Formation of Myelinated Axons

Most axons are covered by a thick layer of the fatty substance **myelin,** which is formed by two different types of cells. In the peripheral nervous system—that is, outside the brain and spinal cord—a myelinated axon is surrounded by *Schwann cells,* which are arranged sequentially along the axon (F5.1). Each Schwann cell wraps itself around the axon several times, pushing the cytoplasm and nucleus of the Schwann cell peripherally (F5.3). The layers of lipids and proteins of the Schwann cell membrane make up the myelin covering the axon. The myelin is interrupted at regular intervals by **nodes of Ranvier,** which are junctions between successive Schwann cells. In the central nervous system—that is, within the brain and spinal cord—cells known as *oligodendrocytes* send out processes that spiral around axons, forming myelin coverings.

BASIC ELECTRICAL CONCEPTS

As previously indicated, a neuron transmits information from one point to another in the form of electrical signals. Therefore, in order to understand the function of a neuron, it is necessary to understand the basic principles of electricity.

There are two types of electrical charge: positive and negative. The total amount of positive

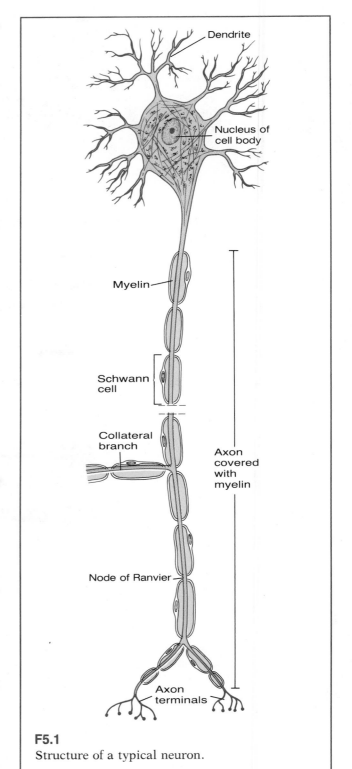

F5.1
Structure of a typical neuron.

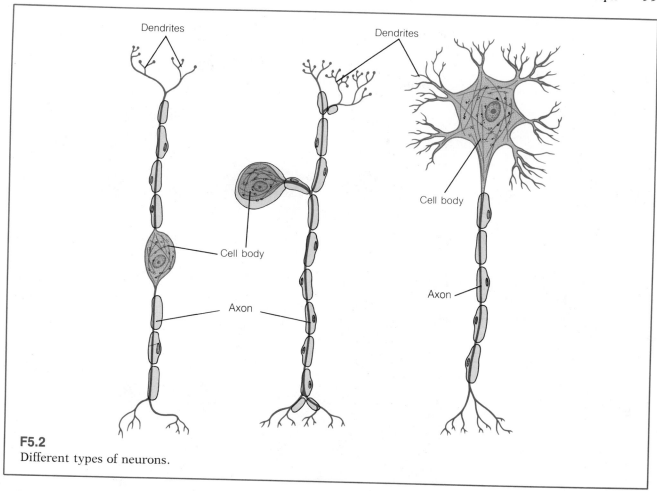

F5.2
Different types of neurons.

charge in the universe is believed to equal the total amount of negative charge, and consequently the universe is electrically neutral. However, limited areas within the universe—and within the human body as well—can have more positive than negative charge or vice versa. These areas are said to be either positively or negatively charged, depending on which charge predominates.

Like charges repel one another; that is, positive charge repels positive charge, and negative charge repels negative charge. However, an electrical force occurs between opposite—that is, positive and negative—charges that attracts them to one another and tends to draw them together. This force increases as the amount of charge increases and as the distance between the charges decreases.

Because of the attractive force between positive and negative charges, energy must be expended and work must be done to separate them. Conversely, if positive and negative charges are allowed to come together, energy is liberated, and this energy can be used to perform work. Thus, when positive and negative charges are separated from one another, they have the *potential* to perform work. The measure of this potential is known as **voltage,** and the units of measurement are volts or millivolts (one millivolt = 0.001 volt). Voltage is always measured between two points in a system, and it is often referred to as the potential difference, or potential, between the points.

The actual movement, or flow, of electric charge from one point to another is known as **cur-**

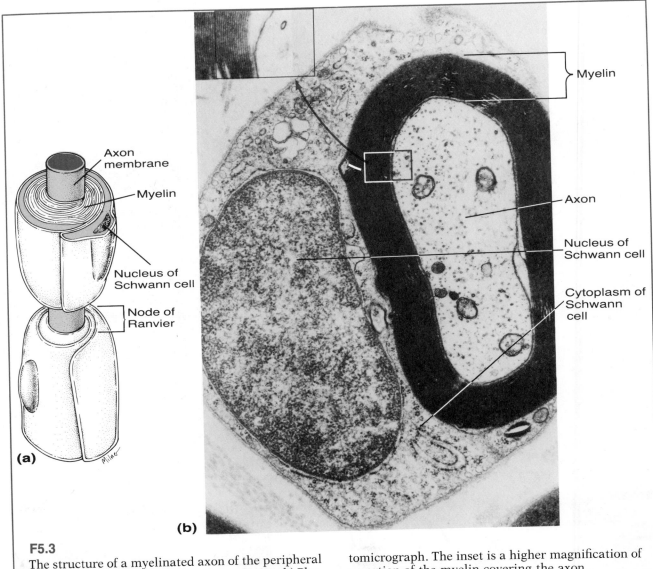

(a)

- Axon membrane
- Myelin
- Nucleus of Schwann cell
- Node of Ranvier

- Myelin
- Axon
- Nucleus of Schwann cell
- Cytoplasm of Schwann cell

(b)

F5.3

The structure of a myelinated axon of the peripheral nervous system. a) Schematic representation. b) Pho-tomicrograph. The inset is a higher magnification of a portion of the myelin covering the axon.

rent. The amount of charge that moves between two points depends on the voltage and on the hindrance to the movement of charge offered by the material between the points. This hindrance is called **resistance.**

The relation between voltage (E), current (I), and resistance (R) is expressed by Ohm's law:

$$I = E/R$$

Thus, when the resistance is constant, the current flow between two points increases when the voltage between the points increases; and, when the voltage is constant, the current flow decreases when the resistance increases.

Ions are electrically charged particles, and if positively charged ions are separated from negatively charged ions, a voltage develops. Moreover, the movement of ions from one point to another

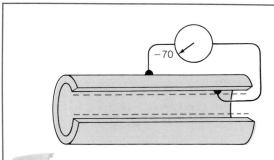

F5.4
Like all body cells, an unstimulated neuron exists in an electrically polarized state. Unstimulated neurons exhibit voltage differences across their cell membranes of about −70 to −85 millivolts (the inside of the cell is relatively negative compared to the outside).

produces a current. Many ions are present in the body—particularly in the intracellular and extracellular fluids—and the distribution and movements of ions give rise to electrical phenomena that are important in the function of neurons and other cells.

MEMBRANE POTENTIALS

All body cells are electrically polarized, with the inside of the cells relatively negative compared to the outside. This polarity can be measured as a difference in electrical potential or voltage between the inside and the outside of a cell, and this voltage is known as the **membrane potential.** Changes in the membrane potential convey important information to cells, and a change called an *action potential* is the basis of the transmission of nerve impulses by neurons.

RESTING MEMBRANE POTENTIAL
The membrane potential of an unstimulated neuron, which is called the **resting membrane potential,** is about −70 to −85 millivolts (F5.4). (The negative sign indicates that the inside of the neuron is relatively negative compared to the outside.) The resting membrane potential is a result of differences in the ionic compositions of the intracellular and extracellular fluids (see, for exam-

ple, Table 5.1). These differences are due to the characteristics and function of the plasma membrane of the neuron, which include the following:

1. The plasma membrane contains energy-utilizing, active transport mechanisms that "pump" certain ions into cells and other ions out of cells. For example, the neuronal membrane contains a *sodium-potassium pump* that transports sodium ions outward and potassium ions inward. As a result, the intracellular fluid of an unstimulated neuron contains a higher concentration of potassium ions and a lower concentration of sodium ions than the extracellular fluid.
2. The plasma membrane is not equally permeable to all ions. For example, the plasma membrane of a neuron contains a number of permanently open channels that are quite permeable to potassium ions but relatively impermeable to sodium ions. Consequently the unstimulated neuronal membrane is 50 to 100 times more permeable to potassium ions than to sodium ions.

As a consequence of the differences in the ionic compositions of the intracellular and extracellular fluids that result from the characteristics and function of the plasma membrane, a very slight excess of positively charged ions accumulates immediately outside the membrane, and an equal number of negatively charged ions accumulates immediately inside the membrane. This sepa-

Table 5.1 *Representative Concentrations of Selected Ions in the Extracellular Fluid and the Intracellular Fluid of an Unstimulated Neuron.* *		
Ion	Extracellular Fluid	Intracellular Fluid
Potassium (K^+)	5	150
Sodium (Na^+)	150	15
Chloride (Cl^-)	125	10

*Values are in milliequivalents per liter. Many other ions are also present in the extracellular and intracellular fluids but potassium, sodium, and chloride ions are particularly important to neural function.

ration of positive and negative charges across the membrane gives rise to the resting membrane potential. The development of this potential involves only a minute quantity of the total ions present, and it does not significantly affect the overall electrical neutrality of the intracellular or extracellular fluids.

Development of the Resting Membrane Potential

Of particular importance in establishing the resting membrane potential are the facts that: (1) the intracellular fluid of an unstimulated neuron contains a higher concentration of potassium ions and a lower concentration of sodium ions than the extracellular fluid, and (2) the unstimulated neuron is 50 to 100 times more permeable to potassium ions than to sodium ions. As a result, potassium ions can diffuse relatively easily through the neuronal membrane from the intracellular fluid, where they are in high concentration, to the extracellular fluid, where they are in lower concentration. However, sodium ions have difficulty diffusing from the extracellular fluid, where they are in high concentration, to the intracellular fluid, where they are in lower concentration. Since both potassium ions and sodium ions are positively charged, the overall result is that, initially, a somewhat greater number of positively charged ions moves out of the neuron than into it, leaving behind a slight excess of negatively charged ions. This contributes to the separation of positive and negative charges across the neuronal membrane that gives rise to the resting membrane potential. However, like charges repel and unlike charges attract one another. Thus, as the resting membrane potential develops and the inside of the neuron becomes relatively negative compared to the outside, an electrical force begins to oppose the outward movement of positively charged ions like potassium ions and tends to attract positively charged ions into the neuron. Ultimately, a steady state is reached at which the forces tending to move potassium ions—and also sodium and other ions—out of the neuron are balanced by forces tending to move the ions into the neuron. In this state, which is the normal state of

unstimulated neurons in the body, the neuron is polarized at its resting membrane potential.

Movement of Ions across the Unstimulated Neuronal Membrane

The movements of potassium, sodium, and chloride ions across the plasma membrane of an unstimulated neuron that is in a steady state and polarized at its resting membrane potential are influenced by both concentration and electrical forces as well as by active transport mechanisms.

Potassium Ions.   The plasma membrane of an unstimulated neuron is relatively permeable to potassium ions, and a concentration force favors a net diffusion of potassium ions from the intracellular fluid, where they are in high concentration, to the extracellular fluid, where they are in lower concentration (Table 5.1). However, in an unstimulated neuron that is polarized at its resting membrane potential, the inside of the neuron is relatively negative compared to the outside. Consequently, an electrical force opposes a net outward movement of positively charged ions like potassium ions and tends to attract positively charged ions into the neuron. This electrical force is not sufficient by itself to prevent a net outward diffusion of potassium ions, but the concentration force tending to move potassium ions outward is balanced by a combination of this electrical force and the active transport of potassium ions from the extracellular fluid to the intracellular fluid by the sodium-potassium pump (F5.5). Thus, equal numbers of potassium ions move across the membrane in each direction, and there is no net movement in either direction.

Sodium Ions.   In an unstimulated neuron that is polarized at its resting membrane potential, a concentration force favors a net diffusion of sodium ions from the extracellular fluid, where they are in high concentration, to the intracellular fluid, where they are in lower concentration (Table 5.1). In addition, the electrical force resulting from the polarity of the neuron also favors a net inward movement of positively charged sodium ions (F5.5). Nevertheless, there is no net move-

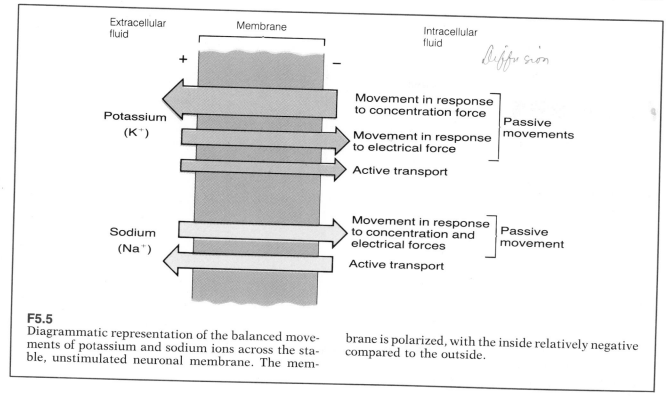

F5.5
Diagrammatic representation of the balanced movements of potassium and sodium ions across the stable, unstimulated neuronal membrane. The membrane is polarized, with the inside relatively negative compared to the outside.

ment of sodium ions across the neuronal membrane because the membrane is relatively impermeable to sodium ions, and the sodium-potassium pump within the membrane actively transports any sodium ions that enter the neuron back into the extracellular fluid.

Chloride Ions. The plasma membrane of an unstimulated neuron that is polarized at its resting membrane potential is permeable to chloride ions, and a concentration force favors a net diffusion of chloride ions from the extracellular fluid, where they are in high concentration, to the intracellular fluid, where they are in lower concentration (Table 5.1). This force is opposed by the electrical force due to the polarity of the neuron, which favors a net outward movement of negatively charged ions. Until recently it was thought that the concentration and electrical forces balanced one another, and chloride ions were passively distributed—that is, in electrochemical equilibrium—across the membrane. Now it is

known that in many neurons the electrical force is not sufficient by itself to prevent a net entry of chloride ions. Consequently, an active transport mechanism called a chloride pump, which moves chloride ions outward, is believed to be important in preventing a net entry of chloride ions into at least some unstimulated neurons.

Role of Active Transport Mechanisms in Maintaining the Resting Membrane Potential

As previously indicated, the resting membrane potential is a result of differences in the ionic compositions of the extracellular and intracellular fluids, and these differences are maintained by active transport mechanisms. For example, if the sodium-potassium pump stops operating, a net movement of potassium ions out of the neuron occurs in response to a concentration force that is not balanced by an opposing electrical force, and a net movement of sodium ions into the neuron occurs in response to both concentration and electrical forces. These movements alter the ionic compositions of the extracellular and intracellular fluids and this, in turn, alters the resting membrane potential.

LOCAL POTENTIAL

The plasma membrane of an axon contains selectively permeable *voltage-gated channels*, which open and close in response to alterations in membrane polarity. When an axon is stimulated so that it undergoes some degree of depolarization—that is, its resting membrane potential changes toward zero—a number of voltage-gated channels that are selectively permeable to sodium ions open. Consequently, in the stimulated area, the permeability of the membrane to sodium ions increases, and there is a net movement of sodium ions across the membrane from the extracellular fluid to the intracellular fluid. (Recall that both a concentration force and an electrical force favor this movement.) Since the axon is polarized, with the inside relatively negative compared to the outside, and since sodium ions are positively charged, the net movement of sodium ions into the axon in the stimulated area tends to enhance the depolarization due to the stimulus. However, if the depolarization resulting from the stimulus is slight, the increased permeability of the membrane to sodium ions is only slight, and the net movement of sodium ions into the axon is quickly balanced by a net outward movement of potassium ions. (Note that the electrical force opposing the net outward movement of potassium ions in the unstimulated axon is diminished by the depolarization of the axon.) In this situation, only a transient local depolarization called a *local potential* occurs in the area of stimulation, and the axon returns to its unstimulated, fully polarized state when the stimulus is removed. The local potential is a graded potential; that is, its magnitude increases with increasing strengths of stimuli (F5.6).

ACTION POTENTIAL

If a stimulus of sufficient intensity is applied to an axon, the local depolarization in the area of stimulation reaches a critical level called **threshold.** (Generally, threshold is reached when an axon de-

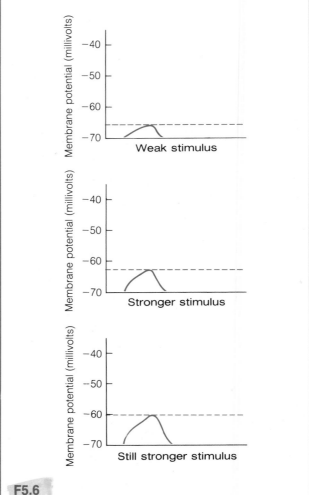

F5.6

Record of the response of the axon of a nerve cell to locally applied stimulation. Note that the response is graded: with increased strengths of stimuli, greater local potentials occur. These potentials are transient, and the axon returns to its unstimulated state when the stimulus is removed.

polarizes by 15 to 20 millivolts from its resting level.) Once threshold is reached, the axon continues to depolarize without further stimulation. The membrane potential quickly decreases to zero and then reverses so the inside of the axon is relatively positive compared to the outside (F5.7). This rapid depolarization and polarity reversal is called an **action potential,** and it is unique to nerve and muscle cells. The next two sections describe how an action potential develops and sub-sides in one area of the plasma membrane of an axon. The section following these explains how an action potential travels along the membrane as a nerve impulse.

Depolarization and Polarity Reversal

As a local depolarization of an axon approaches threshold, more and more voltage-gated sodium channels open, and the plasma membrane in the depolarized area becomes increasingly permeable to sodium ions. Consequently, there is an in-creasing net movement of sodium ions into the axon. At threshold, the opening of voltage-gated sodium channels has increased the sodium perme-ability of the membrane to the point where the net movement of sodium ions into the axon is too great to be balanced by an equal net movement of potassium ions outward. As a result, the net in-ward movement of positively charged sodium ions increases the depolarization of the axon, which in turn causes more voltage-gated sodium channels to open. The opening of these channels further in-creases the permeability of the membrane to sodi-um ions and leads to a net movement of even more sodium ions into the axon. The net inward move-ment of more sodium ions further depolarizes the axon, causing still more voltage-gated sodium channels to open, and so on and so on. Thus, when an axon is depolarized to threshold, a regen-erative, positive-feedback cycle occurs that no longer requires the participation of the stimulus. The changing permeability of the membrane to sodium ions and the increased net movement of sodium ions into the axon during an action poten-tial continue the depolarization of the axon inde-pendently of the stimulus. As a result, the original polarity of the axon in the region of depolarization decreases to zero and then reverses so that the inside of the axon becomes relatively positive compared to the outside.

Return to the Unstimulated Condition

The voltage-gated sodium channels remain open for only a brief interval, and the actual time of increased permeability to sodium ions at any one point on the axonal membrane during an action

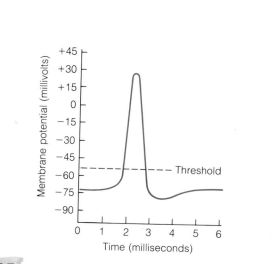

F5.7

Record of an action potential in an axon. When a local depolarization reaches threshold, the mem-brane potential quickly decreases to zero and then reverses so the inside of the axon is relatively pos-itive compared to the outside. Following this, the axon returns to its unstimulated state.

potential is quite short—on the order of a milli-second (F5.8). Soon after the voltage-gated sodium channels open, voltage-gated channels that are se-lectively permeable to potassium ions open. These channels remain open somewhat longer than the voltage-gated sodium channels, and while they are open, the permeability of the membrane to potassium ions is increased. As a result, there is a significant net movement of positively charged potassium ions out of the axon. (Note that both concentration and electrical forces favor this movement when the polarity reverses during an action potential.) The net outward movement of potassium ions causes the inside of the axon to become less positive and then more negative com-pared to the outside. The rapid decrease in mem-brane permeability to sodium ions and the increase in permeability to potassium ions rees-tablishes the original unstimulated polarity of the axon, with the inside being relatively negative in relation to the outside. Following these events, the sodium-potassium pump moves potassium ions back into the cell and sodium ions out of the cell.

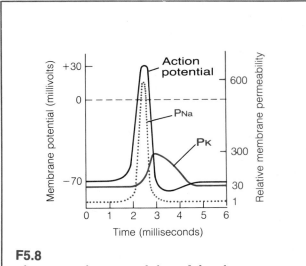

F5.8

Changes in the permeability of the plasma membrane of an axon to sodium (P_{Na}) and potassium (P_K) ions that are associated with an action potential.

Thus, not only is the normal unstimulated polarity of the neuron reestablished, but so is the normal distribution of ions.

THE NERVE IMPULSE

When one area of the axonal membrane reverses its polarity during an action potential, the voltage difference between that area and the immediately adjacent area of the membrane leads to a local current flow between the areas (F5.9). The current is carried by ions and, by convention, the direction of current flow is the direction of movement of positively charged ions. Along the inside of the membrane, positively charged ions are repelled from the area of polarity reversal, and negatively charged ions are attracted toward the area. Along the outside of the membrane, positively charged ions are attracted toward the area of polarity reversal, and negatively charged ions are repelled from the area. Consequently, current flows away from the area of polarity reversal along the inside of the membrane and toward the area of polarity reversal along the outside of the membrane. The current flow depolarizes the area of the membrane immediately adjacent to the area of polarity reversal, and this area becomes more permeable to

sodium ions. When the depolarization of the area reaches threshold, an action potential occurs. A local current flow between this area and the next adjacent area of the membrane depolarizes the next adjacent area, and an action potential occurs there. This activity continues along the length of the membrane, producing a **propagated action potential**, or **nerve impulse,** that moves along the axon.

REFRACTORY PERIODS

The movements of sodium and potassium ions across any given region of the axonal membrane during an action potential involve only a few ions and require only a few milliseconds. During a short period that essentially corresponds to the period of sodium permeability changes, no additional stimulus applied to a region of the membrane that is undergoing these changes can evoke another action potential, regardless of how strong that stimulus is. This period is called the **absolute refractory period,** and its length limits the number of action potentials—and therefore the number of nerve impulses—that can be produced in a given period of time. Intact nerve cells can generally produce action potentials at frequencies between 0 and 500 per second, although some cells are capable of much higher frequencies for brief periods of time.

During a period that roughly corresponds to the period of increased potassium permeability, a stimulus stronger than that normally required to reach threshold may be able to initiate another action potential. This period is known as the **relative refractory period.**

ALL-OR-NONE RESPONSE

As is evident from the events involved in the generation of a propagated action potential, the conduction of a nerve impulse depends on the properties of the nerve-cell membrane and its changing permeability to sodium and potassium ions. Once an axon is depolarized to threshold and a nerve impulse is triggered, the impulse travels along the axon at a rate and with a magnitude that is characteristic of the particular neuron being stimulated, and independent of the strength of the initial stimulus. Thus, a nerve impulse is not a

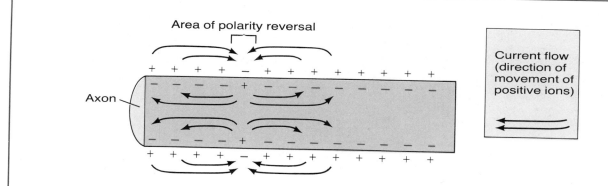

F5.9
Current flow between an area of polarity reversal and adjacent areas of the axonal membrane during an action potential.

graded response, but instead has the same magnitude (under similar physiological conditions) whether it is triggered by a weak stimulus that just causes the axon to reach threshold or by a much stronger stimulus. This type of response is called an *all-or-none response*.

DIRECTION OF NERVE IMPULSE CONDUCTION

If nerve impulses are triggered experimentally in the middle of an axon, they travel away from their site of origin in both directions toward the two ends of the axon. However, in the body, the initial action potentials and polarity reversals that give rise to nerve impulses normally occur at one end of an axon. Consequently, nerve impulses travel in one direction toward the other end of the axon. Note, however, that this undirectional transmission of nerve impulses is due to their site of origin and not to any inability of an axon to conduct impulses in the opposite direction.

CONDUCTION VELOCITIES

The axons of different neurons have different thresholds and conduct nerve impulses at different velocities. Generally, the larger the axon diameter, the lower the threshold and the greater the conduction velocity.

Myelinated axons display a type of nerve impulse conduction called **saltatory** (jumping) **con-**duction (F5.10). Because myelin is a good insulator with a high resistance to current flow, in a myelinated axon the local current that gives rise to a propagated action potential flows from one node of Ranvier, where the myelin is interrupted, to the next. Thus, action potentials do not occur all along the axon, but only at the nodes of Ranvier. As a result, a nerve impulse in a myelinated axon "jumps" quickly along the axon from node to node. Because of this saltatory, or "jumping," conduction, a nerve impulse travels faster in a myelinated axon than in an unmyelinated axon.

The saltatory conduction of nerve impulses in myelinated axons is energetically more efficient than the conduction of impulses in unmyelinated axons. In a myelinated axon, where action potentials occur only at the nodes of Ranvier, fewer sodium ions enter the axon and fewer potassium ions leave during the conduction of a nerve impulse than in an unmyelinated axon, where action potentials occur all along the axon. Consequently, less metabolic energy is required to restore the resting distribution of ions in a myelinated axon than in an unmyelinated axon.

SYNAPSES

For information to be transmitted throughout the nervous system, not just one neuron but chains of

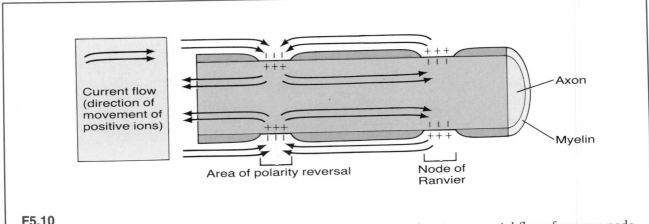

F5.10
In a myelinated axon, the local current that gives rise to a propagated action potential flows from one node of Ranvier to the next.

them must be traversed. This activity requires a means of passing information from neuron to neuron as well as the ability to transmit information along the axon of a single neuron in the form of nerve impulses.

Information is transferred from one neuron to another at neuronal junctions called **synapses.** A neuron that transmits information toward a synapse is called a **presynaptic neuron,** and a neuron that transmits information away from a synapse is called a **postsynaptic neuron.** A single neuron can be a presynaptic neuron at one synapse and a postsynaptic neuron at another (F5.11).

Near its end, an axon branches into numerous **axon terminals,** and most synapses occur between an axon terminal of a presynaptic neuron and a dendrite or cell body of a postsynaptic neuron. However, in some cases, synapses occur between two axons, two dendrites, or a dendrite and a cell body.

There are two types of synapses: electrical and chemical.

ELECTRICAL SYNAPSES

At an **electrical synapse,** a presynaptic and postsynaptic neuron are connected by a gap junction (see Chapter 3). This connection allows local electrical currents resulting from action potentials in the presynaptic neuron to pass directly to the postsynaptic neuron and influence its activity.

CHEMICAL SYNAPSES

At a **chemical synapse,** an axon terminal of a presynaptic neuron closely approaches a dendrite or cell body of a postsynaptic neuron (F5.12). The two neurons, however, remain separated by a small **synaptic cleft,** which prevents a nerve impulse in the presynaptic neuron from crossing directly to the postsynaptic neuron. Instead, when a nerve impulse arrives at an axon terminal of a presynaptic neuron, it increases the permeability of the terminal to calcium ions, and there is a net movement of calcium ions into the cell. This triggers the release of a chemical **neurotransmitter** substance from the axon terminal. The molecules of the neurotransmitter substance diffuse across the synaptic cleft and attach to receptors on the membrane of the postsynaptic neuron. This attachment alters the three-dimensional shape of the receptors, and initiates a series of events that influence the activity of the postsynaptic neuron.

In the human nervous system, chemical synapses occur much more often than electrical synapses, and the following sections deal with the function of chemical synapses.

A chemical synapse can be either excitatory or inhibitory, depending on the effect the neurotransmitter released by the presynaptic neuron has on the postsynaptic neuron.

Excitatory Synapses

At an excitatory synapse, the binding of neurotransmitter molecules to receptors on the membrane of the postsynaptic neuron increases the likelihood that the neuron will transmit a nerve impulse. In general, the combination of neurotransmitter molecules with receptors at an excitatory synapse produces changes in the permeability of the membrane of the postsynaptic neuron that lead to a depolarization of the neuron called an **excitatory postsynaptic potential (EPSP).** At many excitatory synapses, the neurotransmitter-receptor combination triggers the opening of specific *chemically gated channels* in the membrane of the postsynaptic neuron that increases the permeability of the membrane to potassium ions and sodium ions. This increased permeability results in a net outward movement of some potassium ions and a net inward movement of a greater amount of sodium ions, which depolarizes the cell. At other excitatory synapses, permeability changes to ions such as calcium ions may contribute to the depolarization of the postsynaptic neuron and the production of an excitatory postsynaptic potential.

An excitatory postsynaptic potential is a graded potential that varies in magnitude with the degree of stimulation of the postsynaptic neuron by neurotransmitter molecules. If the stimulation is sufficiently intense, the axon of the postsynaptic neuron depolarizes to threshold, and a nerve impulse is triggered.

Inhibitory Synapses

At an inhibitory synapse, the binding of neurotransmitter molecules to receptors on the membrane of a postsynaptic neuron decreases the likelihood that the neuron will transmit a nerve impulse. In general, the combination of neurotransmitter molecules with receptors at an inhibitory synapse triggers the opening of chemically gated channels in the membrane of the postsynaptic neuron that increases the permeability of the membrane to potassium ions and/or

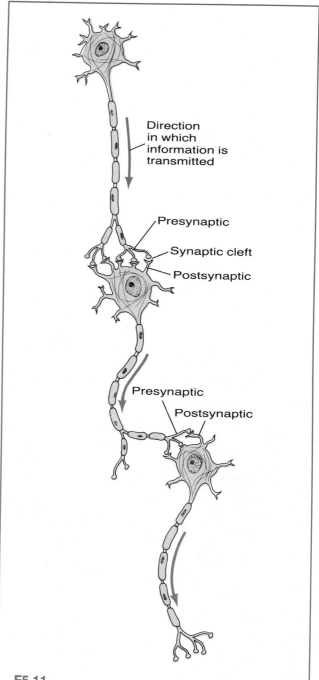

Direction in which information is transmitted

Presynaptic

Synaptic cleft

Postsynaptic

Presynaptic

Postsynaptic

F5.11
A single neuron can be a presynaptic neuron at one synapse and a postsynaptic neuron at another.

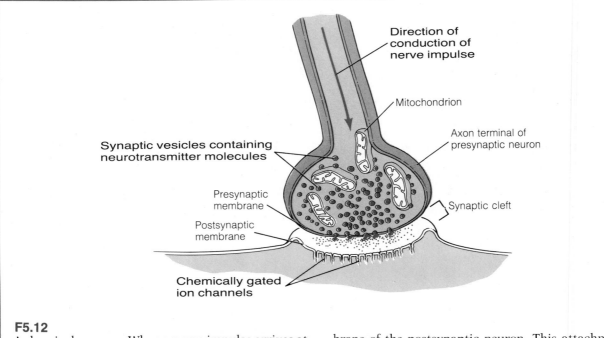

F5.12

A chemical synapse. When a nerve impulse arrives at an axon terminal, chemical neurotransmitter molecules are released. The molecules diffuse across the synaptic cleft and attach to receptors on the membrane of the postsynaptic neuron. This attachment alters the three-dimensional shapes of the receptors and initiates a series of events that influence the activity of the postsynaptic neuron.

chloride ions. This increased permeability usually results in a net outward movement of potassium ions and/or a net inward movement of chloride ions, which causes an increase in the resting polarity of the neuron called an **inhibitory postsynaptic potential (IPSP).** In some cases, however, the permeability change leads to a stabilization of the membrane potential without a hyperpolarization. In any event, a stronger than normal excitatory stimulus is required to trigger a nerve impulse in the postsynaptic neuron.

Synaptic Delay

A neuron can conduct a nerve impulse quite rapidly (speeds up to 250 miles per hour have been recorded), but the speed of transmission at chemical synapses is much slower. The time required to cross a chemical synapse is called the **synaptic delay.** This delay—which is approximately 0.5 to 1 millisecond—is primarily due to the time required for the release of neurotransmitter mole-

cules from the presynaptic ending upon the arrival of a nerve impulse.

Neurotransmitters

Neurotransmitter substances are manufactured by the neurons that release them. After their manufacture, neurotransmitter molecules are stored in small membrane-bounded sacs called **synaptic vesicles.** When a nerve impulse arrives at an axon terminal and the permeability of the terminal to calcium ions increases, some of the synaptic vesicles fuse with the axon terminal membrane, and neurotransmitter molecules are released into the synaptic cleft (see F5.12). Usually only one type of neurotransmitter is released from the terminals of a particular neuron.

When neurotransmitter molecules combine with receptors on the membrane of a postsynaptic neuron, the permeability changes and postsynaptic potentials that result frequently last only a few milliseconds. However, in some cases, they last hundreds of milliseconds and perhaps longer. Depending on the particular synapse involved, the neurotransmitter molecules themselves are ultimately removed from the region of the receptors by enzymatic inactivation, by diffusion away from the region, or by being taken up by the presynaptic neuron.

At the present time, some 30 different substances are known or suspected to serve as neurotransmitters. One of these, *acetylcholine,* is released by neurons at their junctions with skeletal muscle cells. In the brain, the transmitter *norepinephrine* is thought to be involved in the maintenance of arousal, in the brain system of reward, in dreaming sleep, and in the regulation of mood. The chemical substance *dopamine* has been implicated in the regulation of emotional responses and the control of complex movements, and *serotonin* is believed to play a role in temperature regulation, sensory perception, and the onset of sleep.

The amino acid *gamma-aminobutyric acid (GABA)* is a common inhibitory transmitter in the brain, and it is thought that a number of other amino acids also act as neurotransmitters. For example, *glutamic acid* and *aspartic acid* powerfully excite many neurons, and *glycine* has an inhibitory effect in the spinal cord.

Chains of amino acids called *neuropeptides* are also believed to act as neurotransmitters. A neuropeptide called *substance P* is released from the axon terminals of some sensory neurons at their synapses with spinal cord neurons. Substance P appears to excite spinal neurons that conduct nerve impulses associated with painful stimuli. Neuropeptides called *enkephalins* and *endorphins* also seem to be involved in the perception and integration of pain as well as in emotional experiences.

NEURAL INTEGRATION

The nervous system integrates information from many different sources in order to produce useful, coordinated responses by the body to a wide variety of conditions. At the cellular level, this integrative ability depends on such occurrences as divergence, convergence, summation, and facilitation.

DIVERGENCE AND CONVERGENCE

The axon of a presynaptic neuron may branch many times and, thus, may synapse with many postsynaptic neurons (F5.13). This phenomenon, which is called **divergence,** permits a nerve impulse in a single presynaptic neuron to affect many postsynaptic neurons. Conversely, axon terminals of many different presynaptic neurons may all synapse with a single postsynaptic cell. This phenomenon is called **convergence.** Divergence and convergence allow for a wide variety of neuronal interactions and information transfers within the nervous system.

SUMMATION

The postsynaptic potentials produced by the release of neurotransmitter molecules at chemical synapses can add together, or **summate,** to influence the activity of a postsynaptic neuron. In fact, summation is normally required to trigger a nerve impulse in a postsynaptic neuron because the excitatory postsynaptic potential (EPSP) produced by the neurotransmitter released in response to the arrival of a single nerve impulse at a single excitatory synapse is rarely, if ever, large enough to depolarize the axon of the neuron to threshold.

There are two types of summation: temporal and spatial.

Temporal Summation

Temporal summation is the summation that occurs when many nerve impulses arrive at a single synapse within a short period of time. For example, the arrival of one nerve impulse at an excitatory synapse leads to the release of some neurotransmitter molecules, which produce a relatively small EPSP in the postsynaptic neuron. However, if a second nerve impulse arrives and a second EPSP is produced before the initial EPSP dies away, the two EPSPs can add together, or summate. This summation produces a greater depolarization of the postsynaptic neuron than would result from either EPSP alone. If enough nerve

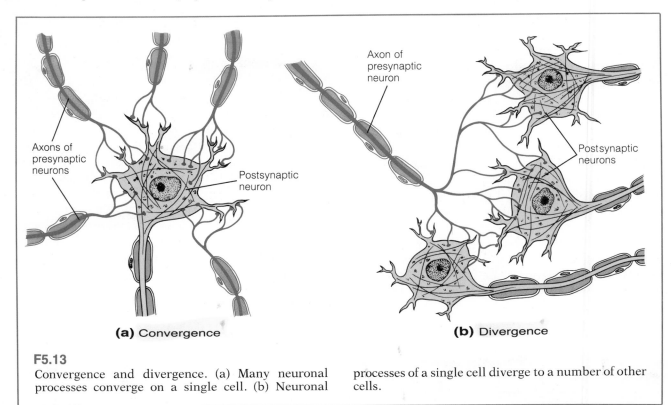

Axons of presynaptic neurons

Postsynaptic neuron

(a) Convergence

Axon of presynaptic neuron

Postsynaptic neurons

(b) Divergence

F5.13
Convergence and divergence. (a) Many neuronal processes converge on a single cell. (b) Neuronal processes of a single cell diverge to a number of other cells.

impulses arrive at the synapse close enough together in time, the summation of the EPSPs that are produced can depolarize the axon of the postsynaptic neuron to threshold and trigger a nerve impulse in it.

Spatial Summation

Spatial summation is the summation that occurs when nerve impulses arrive very close together in time at a number of synapses between different presynaptic axon terminals and the same postsynaptic neuron. For example, as previously indicated, the EPSP produced by the neurotransmitter released in response to the arrival of a single nerve impulse at a single excitatory synapse is not normally very great. However, if nerve impulses arrive at many excitatory synapses at about the same time, the EPSPs produced at each synapse can summate and depolarize the axon of the postsynaptic neuron to threshold.

FACILITATION

Quite often, as a result of spatial summation, the axon of a postsynaptic neuron is depolarized toward threshold by EPSPs produced at some excitatory synapses. Consequently, less depolarization is required for the axon to reach threshold when additional EPSPs occur. This phenomenon is called **facilitation** because the initial EPSPs that depolarize the axon toward threshold facilitate the attainment of threshold and the generation of a nerve impulse in the axon when additional EPSPs occur.

DETERMINATION OF POSTSYNAPTIC NEURON ACTIVITY

As is the case for EPSPs, inhibitory postsynaptic potentials (IPSPs) produced by neurotransmitter molecules released at inhibitory synapses can also undergo temporal and spatial summation. Moreover, a postsynaptic neuron often receives thousands of presynaptic inputs, some of which may form excitatory synapses with the neuron and some of which may form inhibitory synapses.

Whether or not the axon of a postsynaptic neuron is depolarized to threshold depends on the relationship between the excitatory and inhibitory events influencing the neuron at any moment. If excitatory events are sufficiently dominant, the axon may reach threshold and transmit a nerve impulse. If inhibitory events predominate, the axon will be unlikely to transmit a nerve impulse. Thus, the activity of a postsynaptic neuron is determined by the integrated activities of many presynaptic inputs from various sources. Integrative processes of this sort, involving many neurons in multineuronal pathways, allow the nervous system to generate appropriate responses to many different circumstances.

MODIFICATION OF NEURONAL ACTIVITY

The transmission of information from one neuron to another at synapses can be altered by a phenomenon known as presynaptic inhibition and also by the influence of chemical substances known as neuromodulators.

Presynaptic Inhibition

In some cases, the amount of chemical neurotransmitter released when a nerve impulse arrives at an excitatory synapse is reduced by the process of **presynaptic inhibition.** In this process, an inhibitory neuronal ending makes synaptic contact with the ending of a presynaptic neuron at an excitatory synapse (F5.14). When the inhibitory neuronal ending is activated, less neurotransmitter is released from the ending of the presynaptic neuron when a nerve impulse arrives than would otherwise be the case. As a consequence, the stimulation of the postsynaptic neuron is reduced. Presynaptic inhibition provides the nervous system with a means of regulating the influence that nerve impulses in a particular pre-

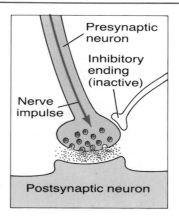

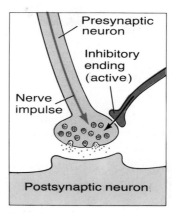

F5.14
Presynaptic inhibition. An inhibitory neuronal ending makes synaptic contact with the ending of the presynaptic neuron at an excitatory synapse. When the inhibitory neuronal ending is activated, less neurotransmitter is released from the ending of the presynaptic neuron when a nerve impulse arrives than would otherwise be the case. As a consequence, the stimulation of the postsynaptic neuron is reduced.

synaptic neuron will have on a postsynaptic neuron.

Neuromodulators — *amplifying or dumping*

Neuromodulators are chemical substances that alter neuronal activity by altering neurons directly or by influencing the effectiveness of a neurotransmitter. For example, a neuromodulator

Stimulate

may alter the synthesis or release of a neurotransmitter from a presynaptic neuron or it may increase or decrease the sensitivity of a postsynaptic neuron to a neurotransmitter. A number of hormones are believed to function as neuromodulators, and many substances currently believed to be neuromodulators also exert nonneural effects. Moreover, it is possible that a particular substance may act as a neurotransmitter in one region of the nervous system and as a neuromodulator in another. However, it is not known if a neuron can secrete both a neurotransmitter and a neuromodulator.

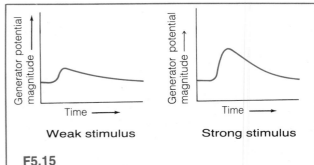

F5.15
Adequate stimuli applied to neuronal endings that act as receptors produce generator potentials. Generator potentials are graded potentials—that is, the stronger the stimulus, the greater the generator potential produced by the stimulus.

RECEPTORS

Receptors are structures that convert information about conditions in the body's internal or external environments into neural signals. Some receptors, such as those in the nose for smell, are specialized endings of neurons. Others, such as those of the ears for sound, are separate cells that are synaptically connected to neurons. Regardless of their nature, however, receptors act as *transducers*—that is, they convert various forms of energy (light, heat, pressure, sound, and so on) into the ionic-electrical phenomena that ultimately result in the generation of neural signals.

The environmental factors that activate receptors are called *stimuli*. However, not all receptors respond to all stimuli. The receptors of the eyes, for example, are normally activated by light but not by sound, and the receptors of the ears are usually activated by sound but not by light. The particular stimuli that activate a given receptor are referred to as *adequate stimuli* for that receptor.

GENERATOR POTENTIALS

When an adequate stimulus is applied to a receptor that is a specialized neuronal ending, it responds with a general increase in permeability to all small ions that produces a depolarization called a **generator potential** (F5.15). A generator potential is a graded potential. The stronger the stimulus applied to a receptor ending, the greater

the magnitude of the generator potential. In addition, the rate of application or removal of a stimulus can affect the magnitude of a generator potential. Thus, a stimulus that builds or diminishes rapidly in intensity usually produces a greater generator potential than a stimulus that builds or diminishes slowly in intensity. A generator potential generally lasts longer than the 1 or 2 milliseconds of an action potential, and it does not exhibit a refractory period. Consequently, if a second stimulus is applied before a generator potential resulting from an initial stimulus disappears, a second generator potential can add to the first, producing an even greater depolarization. Thus, a *summation* of generator potentials can occur. A generator potential of sufficient magnitude depolarizes the axon of the neuron to threshold and triggers a nerve impulse in the axon.

RECEPTOR POTENTIALS

When an adequate stimulus is applied to a receptor that is a separate cell synaptically connected to a neuron, it responds with a depolarization that is called a **receptor potential.** A receptor potential is a graded potential that exhibits characteristics similar to those of a generator potential. The receptor potential causes the release of a chemical neurotransmitter from the receptor cell, and the neurotransmitter alters the permeability and polarity of the associated neuron. If the axon of the associated neuron is depolarized to threshold, a nerve impulse is triggered.

DISCRIMINATION OF DIFFERING STIMULUS INTENSITIES

The fact that a nerve impulse is conducted in an all-or-none fashion regardless of the intensity of the stimulus that triggers it raises some question about neural function. If the magnitude of a nerve impulse does not vary with the intensity of the stimulus, how does the nervous system transmit information about stimulus intensity? For example, how can a light touch on the hand be distinguished from a much stronger touch? One way is to have more neurons activated by strong stimuli than by weak stimuli. Another, less obvious, way is to have different frequencies of nerve impulses generated in a single neuron in response to varying intensities of stimuli. The frequency of nerve impulses in a neuron associated with a receptor is related to the magnitude of the generator or receptor potential produced by a stimulus. A strong stimulus produces a large generator or receptor potential, which, in turn, triggers a high frequency of nerve impulses in the neuron, and a weak stimulus produces a smaller generator or receptor potential, which, in turn, triggers a lower frequency of nerve impulses in the neuron. In this way, nerve impulse frequency provides the nervous system with a means of distinguishing stimulus intensity.

ADAPTATION

When a stimulus is continuously applied, the frequency of nerve impulses in a neuron may diminish with time even though the intensity of the applied stimulus remains the same. This phenomenon is called **adaptation**. In some cases, adaptation is due, at least in part, to the fact that the responsiveness of the receptor membrane diminishes with time, causing the magnitude of the generator or receptor potential evoked by the stimulus to diminish also. Consequently, a lower frequency of nerve impulses is triggered in the neuron.

The occurrence of adaptation may help explain the observation that many sensations—smell, for instance—often diminish substantially in intensity when stimuli are applied for long periods. However, some sensations—pain, for one—frequently diminish only slightly or not at all in intensity when stimuli are applied for long periods. A pain sensation that does not diminish greatly in intensity with time can be beneficial to a person because it may warn of a potentially harmful situation. Unfortunately, the person cannot always control the situation—or the pain.

EFFECTORS

Ultimately, neural signals are transmitted to structures called **effectors** to bring about responses to the various stimuli received by the nervous system. Effectors capable of responding to neural signals include muscle cells and the secretory cells of glands and organs. A neuron–effector junction is similar to a chemical synapse between two neurons, and information is transmitted from neurons to effectors by chemical neurotransmitters. When a nerve impulse arrives at the ending of a neuron at a neuron–effector junction, chemical neurotransmitter molecules are released. The neurotransmitter molecules diffuse to the effector cell and alter its activity. The influences of neural signals on the activity of specific effectors are discussed when the effectors are considered in later chapters.

STUDY OUTLINE

NEURONS Basic structural and functional components of the nervous system; transmit electrical signals. **pp. 97–98**

Structure of a Neuron Cell body and one or more thin cytoplasm-containing processes.

DENDRITES Dendritic zone is receptive portion of

a neuron, can include cell body and branching processes, or dendrites.

AXONS Conductive process of a neuron along which electrical signals called nerve impulses travel.

THE FORMATION OF MYELINATED AXONS Schwann cells spiral around axons to form myelin covering of many peripheral nervous system axons. Processes of oligodendrocytes form myelin of central nervous system axons.

BASIC ELECTRICAL CONCEPTS pp. 98–101

1. There are two types of electrical charge: positive and negative.
2. Like charges repel one another, and unlike charges attract one another.
3. Voltage is a measure of the potential of separated positive and negative charges to perform work.
4. Current is the actual movement of electric charge from one point to another.
5. Resistance is the hindrance to the movement of charge.
6. Ions are electrically charged particles, and their distribution and movements in the body give rise to electrical phenomena that are important in the function of neurons and other cells.

MEMBRANE POTENTIALS All body cells are electrically polarized, and the voltage between the inside and outside of a cell is the membrane potential. pp. 101–107

Resting Membrane Potential Result of differences in ionic composition of intracellular and extracellular fluid due to the characteristics and function of the plasma membrane of a neuron.
1. Plasma membrane contains active transport mechanisms that pump ions in and out of cells.
2. Plasma membrane is not equally permeable to all ions.
3. Differences in ionic compositions of intracellular and extracellular fluid result in accumulation of positively charged ions outside membrane and negatively charged ions inside membrane that gives rise to resting membrane potential.

DEVELOPMENT OF THE RESTING MEMBRANE POTENTIAL Of particular importance are the facts that
1. the intracellular fluid of an unstimulated neuron contains a higher concentration of potassium ions and a lower concentration of sodium ions than the extracellular fluid

2. the unstimulated neuron is 50 to 100 times more permeable to potassium ions than to sodium ions.

MOVEMENT OF IONS ACROSS THE UNSTIMULATED NEURONAL MEMBRANE The movements of potassium, sodium, and chloride ions across the plasma membrane of an unstimulated neuron that is in a steady state and polarized at its resting membrane potential are influenced by both concentration and electrical forces as well as by active transport mechanisms.

Potassium Ions Concentration force favoring net outward movement of potassium ions is balanced by electrical force favoring net inward movement of positively charged ions and by inward pumping of potassium ions.

Sodium Ions Electrical and concentration forces favoring net inward movement of sodium ions are balanced by outward pumping of sodium ions. Membrane is not very permeable to sodium ions.

Chloride Ions Concentration force that favors a net inward movement of chloride ions is balanced by electrical force favoring net outward movement of negatively charged ions and in at least some neurons by outward pumping of chloride ions.

ROLE OF ACTIVE TRANSPORT MECHANISMS IN MAINTAINING THE RESTING MEMBRANE POTENTIAL Active transport mechanisms maintain the differences in the ionic compositions of the extracellular and intracellular fluids that give rise to the resting membrane potential.

Local Potential Axon of nerve cell stimulated so that it undergoes some degree of depolarization. Membrane becomes more permeable to positively charged sodium ions, and sodium ions enter cell, which tends to enhance depolarization due to stimulus. If depolarization resulting from stimulus is slight, increased sodium permeability is only slight and net inward movement of sodium ions is balanced by net outward movement of potassium ions. Cell returns to unstimulated, fully polarized state when stimulus is removed. Local potential is a graded potential—that is, its magnitude increases with increasing strengths of stimuli.

Action Potential When an axon reaches threshold, the membrane potential quickly decreases to zero and then reverses so the inside of the axon is relatively positive compared to the outside. This rapid depolarization and polarity reversal is called an action potential.

DEPOLARIZATION AND POLARITY REVERSAL Due to entry of sodium ions into the axon through voltage-gated channels.

RETURN TO THE UNSTIMULATED CONDITION Due to decreased entry of sodium ions into the axon and increased movement of potassium ions out of the axon. Sodium-potassium pump restores normal distribution of ions.

Nerve Impulse Axonal membrane adjacent to area of initial action potential is depolarized and generates an action potential. Action potentials continue to be generated along the axon, producing a propagated action potential, or nerve impulse.

Refractory Periods
1. ABSOLUTE REFRACTORY PERIOD Time after an action potential when no additional stimulus can evoke another action potential; length essentially corresponds to period of sodium permeability changes.
2. RELATIVE REFRACTORY PERIOD Period when a stimulus greater than that normally required can initiate an action potential; length roughly corresponds to period of increased potassium permeability.

All-Or-None Response Under similar physiological conditions, nerve impulse has same magnitude regardless of stimulus strength.

Direction of Nerve Impulse Conduction In the body a nerve impulse normally originates at the end of an axon and is conducted in one direction toward the other end of the axon.

Conduction Velocities
1. LARGER-DIAMETER AXONS Lower threshold; conduct nerve impulses at greater velocity than smaller-diameter axons.
2. MYELINATED AXONS Conduct nerve impulses faster than unmyelinated axons because of saltatory conduction.

SYNAPSES Junctions between neurons at which information is transferred from one neuron to another. **pp. 107–111**

Electrical Synapses Presynaptic and postsynaptic neurons are connected by a gap junction. Electrical currents in the presynaptic neuron can pass directly to the postsynaptic neuron.

Chemical Synapses Axon terminal of a presynaptic neuron is separated from a postsynaptic neuron by a synaptic cleft. Axon terminal releases neurotransmitter molecules that diffuse across the synaptic cleft and attach to receptors on postsynaptic neuron.

EXCITATORY SYNAPSES The combining of neurotransmitter molecules with receptors on the plasma membrane of the postsynaptic neuron produces changes in the permeability of the membrane that lead to a depolarization of the neuron called an excitatory postsynaptic potential.

INHIBITORY SYNAPSES The combining of neurotransmitter molecules with receptors on the plasma membrane of the postsynaptic neuron produces changes in the permeability of the membrane that usually cause an increase in the resting polarity of the neuron called an inhibitory postsynaptic potential. Sometimes the permeability changes lead to a stabilization of the membrane potential without a hyperpolarization.

SYNAPTIC DELAY Time required to cross a chemical synapse.

NEUROTRANSMITTERS Manufactured by neurons; stored in synaptic vesicles. Some 30 different substances are presently known or suspected to be chemical transmitters.
1. *Acetylcholine* Released by neurons at their junctions with skeletal muscle cells.
2. *Norepinephrine* In the brain it is thought to be involved in the maintenance of arousal, brain system of reward, dreaming sleep, regulation of mood.
3. *Dopamine* Implicated in regulation of emotional responses, and in control of complex movements.
4. *Serotonin* Believed to play a role in temperature regulation, sensory perception, and onset of sleep.
5. *Gamma-aminobutyric acid* Common inhibitory transmitter in brain.
6. *Glutamic acid* and *aspartic acid* Exert powerful excitatory effects on many neurons.
7. *Glycine* Inhibitory transmitter in spinal cord.
8. *Substance P* Believed to be involved in transmission of pain-related information.
9. *Enkephalins* and *endorphins* Seem to be in-

volved in perception and integration of pain as well as in emotional experiences.

NEURAL INTEGRATION
The nervous system integrates information from many different sources in order to produce useful, coordinated responses by the body to a wide variety of conditions. **pp. 111–114**

Divergence and Convergence
1. Divergence: presynaptic neuron processes branch many times to synapse with many postsynaptic neurons; permits nerve impulse in presynaptic neuron to affect many postsynaptic neurons.
2. Convergence: processes of many presynaptic neurons synapse with a single postsynaptic cell.

Summation The postsynaptic potentials produced by the release of neurotransmitter molecules at chemical synapses can add together, or summate, to influence the activity of a postsynaptic neuron.

TEMPORAL SUMMATION Summation that occurs when many nerve impulses arrive at a single synapse within a short period of time.

SPATIAL SUMMATION Summation that occurs when nerve impulses arrive very close together in time at a number of synapses between different presynaptic axon terminals and the same postsynaptic neuron.

Facilitation An initial depolarization of the axon of a postsynaptic neuron by excitatory postsynaptic potentials produced at some excitatory synapses; facilitates the attainment of threshold and the generation of a nerve impulse when additional excitatory postsynaptic potentials occur.

Determination of Postsynaptic Neuron Activity The activity of a postsynaptic neuron is determined by the integrated activities of many presynaptic inputs from various sources.

Modification of Neural Activity Presynaptic inhibition and neuromodulators can alter the transmission of information from one neuron to another at synapses.

PRESYNAPTIC INHIBITION An inhibitory neuronal ending makes synaptic contact with the ending of a presynaptic neuron at an exictatory synapse. When the inhibitory neuronal ending is activated, less neurotransmitter is released from the ending of the presynaptic neuron when a nerve impulse arrives than would otherwise be the case. As a consequence, the stimulation of the postsynaptic neuron is reduced.

NEUROMODULATORS Chemical substances that alter neuronal activity by altering neurons directly or by influencing the effectiveness of a neurotransmitter.

RECEPTORS
Receptors act as transducers that convert various forms of environmental energy into ionic-electrical phenomena that result in the generation of neural signals. **pp. 114–115**

Generator Potentials Depolarization in response to adequate stimulation of a neuronal ending that acts as a receptor; a graded potential with a longer duration than an action potential and no refractory period.

Receptor Potentials Depolarization in response to an adequate stimulus of a separate receptor cell synaptically connected to a neuron. Similar to a generator potential but receptor cell releases chemical transmitter.

Discrimination of Differing Stimulus Intensities Strong stimuli may activate more neurons than weak stimuli; different frequencies of nerve impulses in a single neuron in response to varying intensities of stimuli.

Adaptation A phenomenon that occurs when a stimulus is continuously applied, but frequency of nerve impulses diminishes with time even though intensity of applied stimulus is constant.

EFFECTORS
Muscle cells and secretory cells of glands and organs. **p. 115**

SELF-QUIZ

1. In an unstimulated neuron: (a) the inside is relatively negative compared to the outside; (b) the inside is relatively positive compared to the outside; (c) the sodium-potassium pump does not operate.

2. The resting, unstimulated neuron has: (a) a greater concentration of sodium inside than outside; (b) a greater concentration of potassium inside than outside; (c) an equal concentration of both sodium and potassium inside and outside.

3. In an unstimulated neuron, a concentration force favors the net movement of: (a) potassium ions into the neuron; (b) sodium ions out of the neuron; (c) chloride ions into the neuron.

4. In an unstimulated neuron, an electrical force favors the movement of: (a) potassium ions out of the neuron; (b) sodium ions into the neuron; (c) chloride ions into the neuron.

5. The membrane of an unstimulated neuron is: (a) essentially impermeable to chloride ions; (b) quite permeable to potassium ions; (c) extremely permeable to sodium ions.

6. During a local potential in an axon there is: (a) a net movement of potassium ions into the axon; (b) an increased permeability of the axonal membrane to sodium ions; (c) a net movement of sodium ions out of the axon.

7. When the axon of a neuron reaches threshold and an action potential is generated: (a) potassium ions rapidly enter the axon; (b) the membrane permeability to both sodium and potassium ions decreases substantially; (c) sodium ions rapidly enter the axon.

8. The actual time of increased permeability to the entrance of sodium ions at any one point on the axonal membrane at the onset of an action potential is quite long and generally lasts more than a millisecond. True or False?

9. Which of the following is an all-or-none response? (a) action potential; (b) excitatory postsynaptic potential; (c) generator potential.

10. The axons of different neurons all transmit nerve impulses at the same velocity. True or False?

11. Saltatory, or "jumping," conduction of nerve impulses occurs in: (a) large-diameter myelinated axons; (b) large-diameter unmyelinated axons; (c) small-diameter unmyelinated axons.

12. At a chemical synapse the plasma membranes of the presynaptic and postsynaptic neurons are: (a) fused with one another; (b) separated by a synaptic cleft; (c) connected by a gap junction.

13. At an excitatory synapse, the combination of neurotransmitter molecules with receptors: (a) depolarizes the postsynaptic neuron; (b) increases the resting polarity of the postsynaptic neuron; (c) stabilizes the polarity of the postsynaptic neuron at the normal resting membrane potential.

14. Gamma-aminobutyric acid is a stimulatory chemical transmitter substance released by neurons at their junctions with effectors. True or False?

15. A transmitter substance believed to be involved in the transmission of neural signals associated with painful stimuli is: (a) dopamine; (b) substance P; (c) serotonin.

16. The postsynaptic potentials produced by the release of neurotransmitter molecules at chemical synapses can add together, or summate, to influence the activity of a postsynaptic neuron. True or False?

17. Generator potentials: (a) can summate; (b) have a long absolute refractory period; (c) generally are of shorter duration than action potentials.

18. A generator potential of sufficient magnitude in a neuronal ending that acts as a receptor depolarizes the axon of the neuron to threshold and triggers nerve impulses. True or False?

19. Differing stimulus intensities may lead to different: (a) magnitudes of nerve impulses in a given neuron; (b) velocities of conduction of nerve impulses in a given neuron; (c) frequencies of nerve impulses in a given neuron.

20. Muscle cells are effectors that are unable to respond to neural signals. True or False?

MUSCLE PHYSIOLOGY 6

Muscle is composed of contractile cells that actively develop tension and shorten. As a result, muscle is important in the movement of body parts, the alteration of the diameters of tubes within the body, the propulsion of materials within the body, and the excretion of substances from the body. In addition, muscle contractions produce significant amounts of heat that can be used to maintain normal body temperature. Because of its many functions, muscle contributes importantly to the maintenance of homeostasis.

MUSCLE TYPES

The body contains three types of muscle: skeletal muscle, smooth muscle, and cardiac muscle.

SKELETAL MUSCLE

As the name implies, most **skeletal muscle** attaches to the bones of the skeleton. The contractions of skeletal muscle exert force on the bones and move them. Consequently, skeletal muscle is responsible for activities such as walking and manipulating objects in the external environment.

Skeletal muscle is *voluntary* muscle—that is, its contractions are normally controlled by the conscious desires of the individual. However, under many conditions, skeletal muscle contractions do not require conscious thought. For example, a person does not usually have to think about contracting the skeletal muscles involved in maintaining posture. Skeletal muscle contractions are regulated by signals transmitted to the muscle by a portion of the nervous system known as the somatic nervous system.

When viewed microscopically, skeletal muscle cells exhibit alternating transverse light and dark bands that give them a striped, or *striated*, appearance.

SMOOTH MUSCLE

Smooth muscle is so named because its cells lack the striations evident in skeletal muscle cells. Smooth muscle is found in the walls of hollow organs and tubes such as the stomach, intestines, and blood vessels, and its contractions govern the movement of materials through these structures.

Smooth muscle is *involuntary* muscle—that is, its contractions are not normally controlled by the conscious desires of the individual. However, under appropriate circumstances (see Chapter 7), a person can gain some voluntary control over smooth muscle. Smooth muscle contractions are regulated by factors intrinsic to the muscle itself, by hormones, and by signals transmitted to the muscle by a portion of the nervous system known as the autonomic nervous system.

CARDIAC MUSCLE

Cardiac muscle is a specialized type of muscle that forms the wall of the heart. Cardiac muscle is involuntary muscle, and its contractions are regulated by intrinsic factors, hormones, and the autonomic nervous system. Cardiac muscle cells are striated. The physiology of cardiac muscle is considered in Chapter 12.

STRUCTURE OF SKELETAL MUSCLE

A skeletal muscle is composed of many individual muscle cells called **muscle fibers.** The muscle fibers are held together by thin sheets of fibrous connective tissue known as *fascia.* The fascia covers the entire muscle and also penetrates it, dividing the muscle fibers into bundles and enveloping each fiber (F6.1). Blood vessels and nerves pass into the muscle with the fascia to supply the muscle fibers.

A skeletal muscle is usually attached to a bone by a strong, fibrous *tendon,* which is formed by the connective tissue of the muscle as it extends beyond the end of the muscle. Some tendons are quite short, whereas others are more than a foot in length.

STRUCTURE OF SKELETAL MUSCLE FIBERS

Skeletal muscle fibers are multinucleate cells approximately 10 to 100 microns in diameter that are frequently many centimeters long. Each fiber contains several hundred to several thousand regularly ordered, threadlike **myofibrils** that extend lengthwise throughout the cell (F6.2). When high-

ly magnified, a myofibril exhibits alternating transverse light and dark bands that are responsible for the striated appearance of the cell. The light bands are named **isotropic** or **I bands,** and the dark bands are named **anisotropic** or **A bands.** Crossing the center of each I band is a dense, fibrous **Z line.** The Z lines divide the myofibrils into a series of repeating units known as **sarcomeres.** In the center of a sarcomere and, therefore, in the center of an A band, is a somewhat less dense region, the **H zone.** A thin, dark **M line** crosses the center of the H zone.

A sarcomere contains two distinct types of longitudinally oriented **myofilaments:** thick filaments and thin filaments (F6.3). **Thick filaments** occupy the A band, and the H zone contains only thick filaments. The M line is formed by linkages between the thick filaments that hold them in a parallel arrangement. **Thin filaments** occupy the I band and part of the A band. The thin filaments attach to the Z lines. In the region of the A band where thick and thin filaments overlap, there is a hexagonal arrangement of thin filaments around each thick filament (F6.4).

COMPOSITION OF THE MYOFILAMENTS

The thick filaments consist mainly of the protein **myosin.** A myosin molecule is made up of two identical subunits, each shaped something like a golf club. The two subunits are tightly wound around each other so that a complete myosin molecule has two rather bulbous heads protruding from one end of a straight shaft. A thick filament contains approximately 200 myosin molecules arranged in such a way that the shafts of the molecules are bundled together, with the heads of the molecules (called **cross bridges**) facing outward (F6.5a). The myosin molecules face in opposite directions on either side of the center of a thick filament, with the shafts of the molecules directed toward the center. Because of this arrangement, the central area of the filament contains the shaft portions of myosin molecules but no heads (F6.5c).

The thin filaments consist mainly of the proteins **actin, tropomyosin,** and **troponin** (F6.5b). The actin portion of a thin filament consists of spherical subunits of globular (G) actin that are

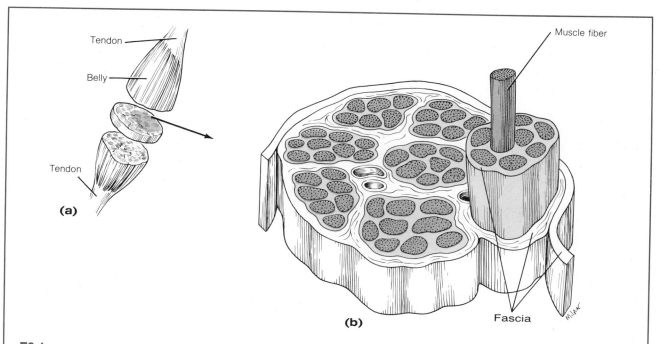

F6.1
Structure of a skeletal muscle. (a) Entire muscle with the belly sectioned. (b) Enlargement of a cross section of the belly.

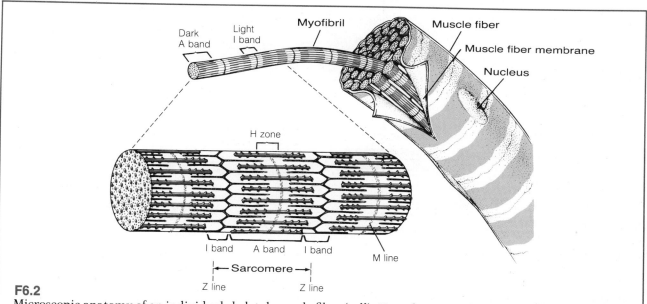

F6.2
Microscopic anatomy of an individual skeletal muscle fiber (cell). Note the striated (striped) appearance of the muscle fiber and the myofibrils.

organized into a double chain of fibrous (F) actin. The F-actin structure resembles two strings of pearls that have been twisted around one another in a spiral, with each pearl being equivalent to a G-actin subunit. Although the G-actin subunits are spherical, they have a definite polarity, and they are linked to one another from front to back. Associated with each chain of G-actin subunits are threadlike molecules of tropomyosin. The tropomyosin molecules lie end to end along the surfaces of the actin chains, and each tropomyosin molecule covers approximately seven G-actin subunits. Attached to each tropomyosin molecule and also to actin is a smaller molecule of the protein troponin. The arrangement of thick and thin filaments in a sarcomere is illustrated in F6.5c.

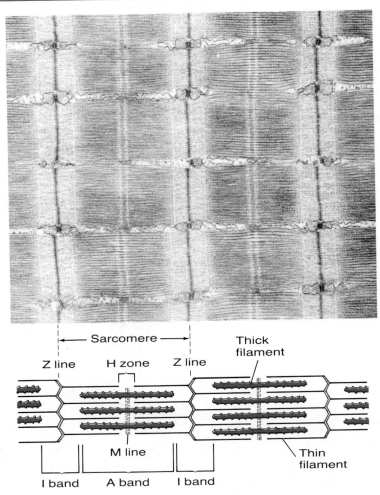

F6.3

Longitudinal view of the structure of sarcomeres. The I band consists only of thin filaments and is divided by the Z line. The A band consists of thick filaments that overlap at either end with the thin filaments. The region where only thick filaments occur is the H zone. A single sarcomere extends from one Z line to the next. Thus, half of each I band is associated with one sarcomere and half with the neighboring sarcomere.

TRANSVERSE TUBULES AND SARCOPLASMIC RETICULUM

Tubules known as **transverse tubules** (*t* **tubules**) pass deep into a skeletal muscle fiber from the plasma membrane (F6.6). In addition, a membranous network, the **sarcoplasmic reticulum,** extends throughout the fiber and surrounds each myofibril. The sarcoplasmic reticulum is in some respects similar to the endoplasmic reticulum of other cells. Elements of the sarcoplasmic reticulum and *t*-tubule system lie close to one another at the junctions of the A and the I bands of the myofibrils. At these locations, structures consisting of three tubules (triads) are formed.

CONTRACTION OF SKELETAL MUSCLE

Experimentally, two basic types of muscle contraction—isometric contraction and isotonic contraction—are frequently employed to study muscle function. An **isometric** (*iso* = equal; *metric* = measure) contraction is a contraction during which the length of a muscle remains constant (F6.7a). In this type of contraction, the muscle actively develops tension and exerts force on an object, but does not shorten. In the body, isometric types of muscle contraction occur when a person supports an object in a fixed position or attempts to lift an object that is too heavy to move. An **isotonic** (*iso* = equal; *tonic* = tension) contraction is a contraction during which a muscle shortens while under a constant load (F6.7b). In this type of contraction, the force against which the muscle shortens remains constant even though the length of the muscle changes considerably. In the body, where most skeletal muscles attach to bones and more than one muscle is usually involved in a movement, pure isotonic contractions seldom occur because the load on a muscle frequently changes as the muscle shortens. Nevertheless, contractions during which a muscle shortens, such as the contractions that move the

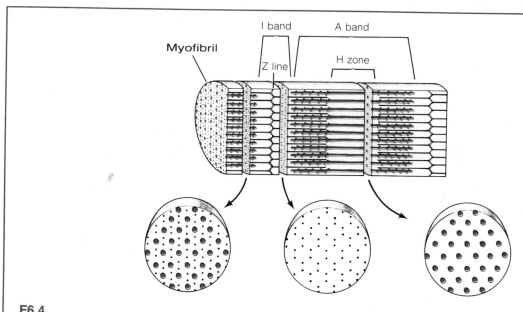

F6.4

Cross sections through different regions of sarcomeres of a myofibril. Note that in the regions of the A bands where thick and thin filaments overlap, each thick filament is surrounded by six thin filaments, and each thin filament is surrounded by three thick filaments.

legs during walking or the arms when lifting an object, are often referred to as isotonic contractions. Regardless of whether or not the load on a muscle changes, before a muscle can shorten, it must first develop sufficient tension to equal and overcome the resistance of the load against which it contracts—only then can shortening occur (F6.8).

CONTRACTION OF A SKELETAL MUSCLE FIBER

At the cellular level, the same basic events occur during both isometric and isotonic skeletal mus-

cle contractions. A skeletal muscle contracts when stimuli from the nervous system excite the individual muscle fibers. This excitation initiates a series of events that lead to interactions between the thick and thin filaments of the sarcomeres of the fibers. These interactions are responsible for the development of tension and the shortening of the fibers.

The Neuromuscular Junction

Nerve cells known as **motor neurons** supply the neural stimulation that skeletal muscle fibers require in order to contract. These neurons form

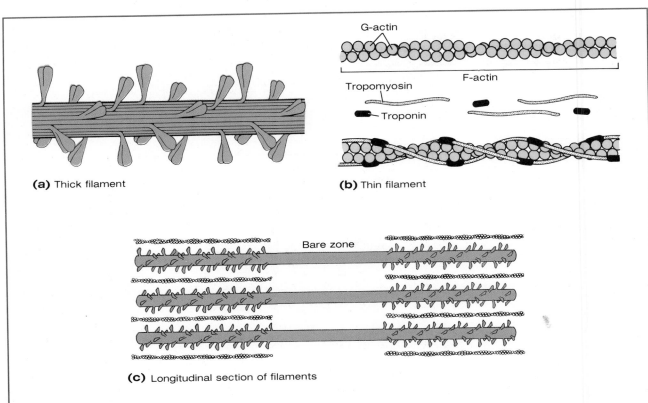

(a) Thick filament

(b) Thin filament

(c) Longitudinal section of filaments

F6.5

Composition of the myofilaments. (a) The thick filament consists mainly of golf-club-shaped molecules of myosin bundled together so that the heads of the molecules project from the shaft of the filament and spiral around the shaft. (b) The thin filament. Individual G-actin subunits link into a double chain of F-actin. In a single thin filament, two chains of G-actin subunits twist around one another. Along the surface of each chain lie threadlike tropomyosin molecules that each cover seven G-actin subunits. Each tropomyosin molecule is attached to a molecule of troponin. Troponin is also attached to actin. (c) Longitudinal view of thick and thin filaments as arranged in a sarcomere. Note that the myosin molecules project in opposite directions on either side of the bare zone.

specialized junctions called **neuromuscular junctions** with skeletal muscle fibers (F6.9a). At a neuromuscular junction, an ending of a motor neuron closely approaches a specialized point along the plasma membrane of a skeletal muscle fiber, but it does not directly contact the membrane. Instead, a small gap separates the motor neuron ending from the muscle fiber membrane. Most skeletal muscle fibers have only one neuromuscular junction.

Excitation of a Skeletal Muscle Fiber

A nerve impulse in a motor neuron does not directly stimulate a skeletal muscle fiber because it cannot cross the gap that separates a motor neuron ending from the muscle fiber plasma membrane. Instead a nerve impulse indirectly stimulates a skeletal muscle fiber in the following way. When a nerve impulse arrives at a neuromuscular junction, the chemical neurotransmitter acetylcholine is released from the terminal endings of the neuron (F6.9b). The acetylcholine diffuses to the muscle fiber plasma membrane where it attaches to acetylcholine receptors. This attachment leads to an increase in the permeability of the membrane to both sodium and potassium ions, and the fiber depolarizes in the region of the junction. This depolarization, which is called an *end-plate potential*,

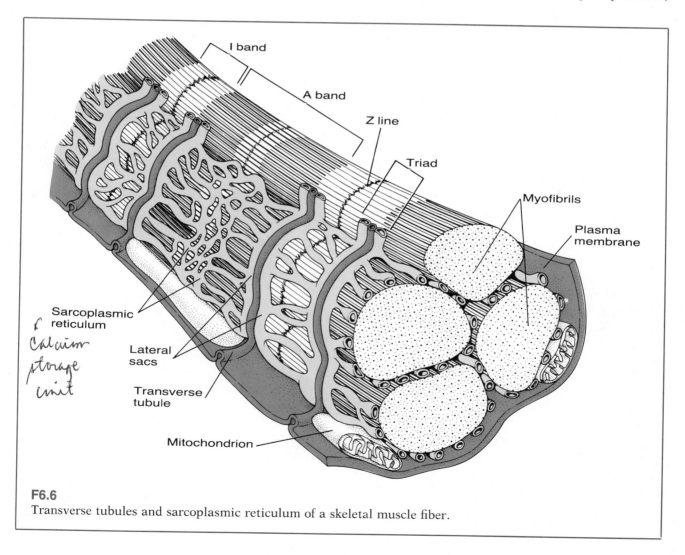

F6.6
Transverse tubules and sarcoplasmic reticulum of a skeletal muscle fiber.

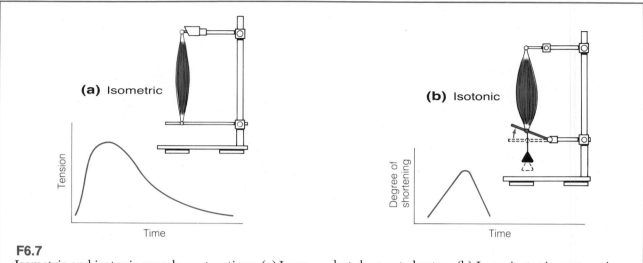

F6.7

Isometric and isotonic muscle contractions. (a) In an isometric contraction, the muscle develops tension but does not shorten. (b) In an isotonic contraction, the muscle shortens while under a constant load.

is sufficient to produce a propagated action potential that travels along the membrane. The propagated action potential triggers a series of intracellular events that culminate in interactions between the thick and thin filaments of the sarcomeres and the contraction of the fiber.

The effective combination of acetylcholine with the receptors of the muscle fiber plasma membrane lasts only a few milliseconds. Following its release from the terminal endings of a motor neuron at a neuromuscular junction, acetylcholine is rapidly inactivated by an enzyme called *acetylcholine esterase*, which is located on the muscle fiber membrane very close to the acetylcholine receptor sites. The inactivation of acetylcholine terminates its action on the muscle fiber.

Excitation-Contraction Coupling

The series of events by which a propagated action potential in the plasma membrane of a skeletal muscle fiber causes interactions between the thick and thin filaments of the sarcomeres and the contraction of the fiber are grouped together under the heading of **excitation-contraction coupling.** From the plasma membrane, the propagated action potential passes along the *t* tubules into the central areas of the fiber. As it moves along the *t*

tubules, it triggers the release of calcium ions from sites in the sarcoplasmic reticulum called lateral sacs (F6.6). The calcium ions bind to troponin molecules of the thin filaments and, as is explained in the following section, this leads to

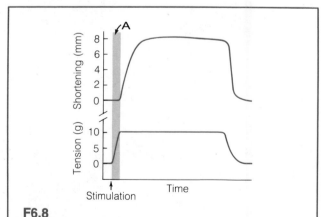

F6.8

Record of a contraction during which a muscle shortens and lifts a 10-gram load. Immediately following its stimulation, there is a period (A) during which the muscle develops sufficient tension to equal and overcome the resistance of the load but does not shorten. Following this the muscle shortens and lifts the load.

interactions between the thick and thin filaments of the sarcomeres. These interactions are directly responsible for muscle contraction.

Mechanism of Contraction

Muscle contraction requires energy, which is supplied by adenosine triphosphate (ATP). In a muscle fiber, an ATP molecule occupies a binding site on a club-shaped head of a myosin molecule of a thick filament (F6.10). The myosin molecule possesses enzymatic activity, and it splits the ATP into adenosine diphosphate (ADP) and phosphate. This reaction releases energy that is transferred to the myosin, producing a high-energy form of myosin.

In addition to an ATP binding site, a myosin head contains a binding site that can attach to a complementary site on an actin subunit of a thin filament, and a high-energy form of myosin has a strong tendency to bind to actin. In a relaxed, unstimulated muscle fiber, this binding is prevented by tropomyosin, which lies along the surface of the actin and physically blocks interactions between high-energy myosins and actin subunits. However, when a muscle fiber is stimulated and calcium ions are released from the sarcoplasmic reticulum, the calcium ions bind to troponin molecules, which are linked to both actin and tropomyosin. The binding of calcium ions to troponin weakens the linkage between troponin and actin. The weakening of this linkage allows tropomyosin to move away from its blocking position and deeper into the groove formed by the twisting of the two actin chains around one another. With the tropomyosin out of the way, high-energy myosins can link with G-actin subunits.

When a high-energy myosin combines with an actin subunit, the energy stored within the myosin

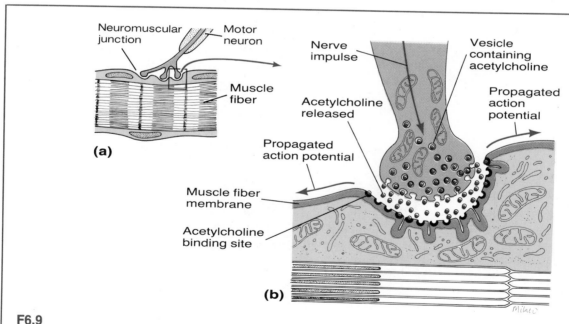

F6.9

A neuromuscular junction. (a) Diagrammatic representation of the general structure of a neuromuscular junction. (b) A more highly magnified section of the junctional area, illustrating the events that occur at a neuromuscular junction. When a nerve impulse arrives at a terminal ending of a motor neuron, acetylcholine is released. The acetylcholine binds with receptors on the muscle fiber plasma membrane at the junction. This binding leads to a change in the membrane's permeability to sodium and potassium ions, and produces a propagated action potential that travels along the membrane.

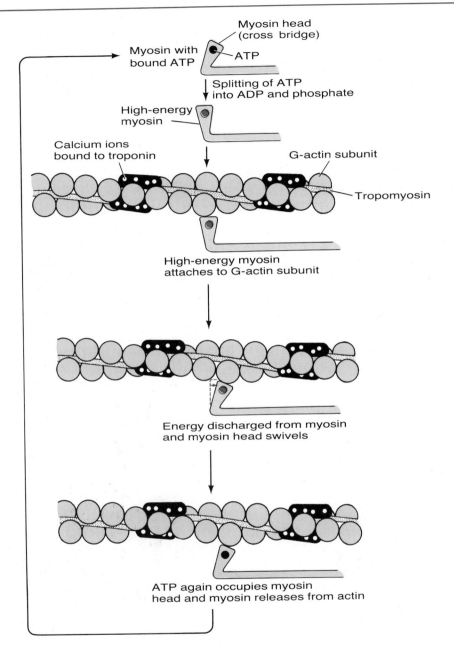

Myosin head (cross bridge)

Myosin with bound ATP

ATP

Splitting of ATP into ADP and phosphate

High-energy myosin

Calcium ions bound to troponin

G-actin subunit

Tropomyosin

High-energy myosin attaches to G-actin subunit

Energy discharged from myosin and myosin head swivels

ATP again occupies myosin head and myosin releases from actin

F6.10

Mechanism of muscle contraction. A myosin head of a thick filament has ATP bound to it. The ATP is split, producing a high-energy form of myosin. The high-energy myosin binds to an actin subunit of the thin filament. The energy of the high-energy myosin is discharged and the myosin head swivels, pulling on the actin-containing thin filament. The myosin re- mains bound to the actin until ATP again occupies the myosin head. When this occurs, the ATP is split and the cycle is repeated. For this contractile cycle to occur, troponin must bind calcium. This binding re- sults in the removal of tropomyosin from a position blocking the myosin-actin interaction.

is discharged, with the resultant production of a force that causes the myosin head (cross bridge) to move. In essence, the myosin head swivels toward the center of the sarcomere, pulling on the actin-containing thin filament. The myosin remains attached to the actin until an ATP molecule again occupies the myosin head. When this occurs, the myosin releases from the actin. It again splits ATP into ADP and phosphate, producing a high-energy form of myosin that attaches to an actin subunit, and the cycle is repeated. At any instant during the contraction of a skeletal muscle fiber, approximately 50% of the myosin heads are attached to actin subunits, and the rest are at intermediate stages of the activity cycle.

The force of the myosin heads of the thick filaments pulling on the actin-containing thin filaments is transmitted to the plasma membrane of the muscle fiber and, ultimately, to the load. If enough force is developed by the fiber—and by other fibers involved in the muscle contraction—to overcome the resistance of the load, the repeated cycling of the myosin heads pulls the thin filaments past the thick filaments toward the centers of the sarcomeres (F6.11). This draws the Z lines closer together, and the muscle fiber shortens.

Regulation of the Contractile Process

Once a nerve impulse stimulates a skeletal muscle fiber and calcium ions are released from the sarcoplasmic reticulum, why doesn't the formation of attachments between thick and thin filaments and, therefore, the contractile process continue indefinitely? In fact, the calcium ions released in response to a single nerve impulse are free for only a short time. Following their release, an active transport mechanism quickly pumps the calcium ions back into the sarcoplasmic reticulum. With the return of the calcium ions to the sarcoplasmic reticulum, troponin strengthens its connection with actin, pulling the tropomyosin back into its blocking position. When tropomyosin returns to its blocking position, no further interactions between high-energy myosins and actin subunits are possible. Consequently, the contractile process ceases, and the muscle fiber relaxes. When another nerve impulse stimulates the muscle fiber and calcium ions are again released from the sarcoplasmic reticulum, the contractile process occurs once more. If many nerve impulses arrive at a skeletal muscle fiber in rapid succession so that calcium ions continue to be available to bind with troponin, the fiber does not relax between successive impulses, and the contractile process con-

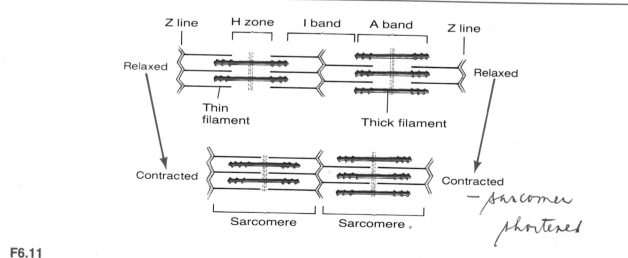

F6.11

A muscle fiber shortens when the thin filaments move past the thick filaments toward the centers of the sarcomeres, and the Z lines are drawn closer together.

tinues until the impulses cease and the calcium ions are returned to the sarcoplasmic reticulum.

The sequence of events involved in the excitation and contraction of a skeletal muscle fiber is outlined in Table 6.1.

Table 6.1 *Sequence of Events Involved in the Excitation and Contraction of a Skeletal Muscle Fiber*

1. A nerve impulse arrives at a neuromuscular junction. Acetylcholine is released from the motor neuron, and it binds to receptors on the muscle fiber plasma membrane.

2. A propagated action potential travels over the plasma membrane of the muscle fiber and along the transverse tubules into the interior of the cell.

3. The propagated action potential triggers the release of calcium ions from lateral sacs of the sarcoplasmic reticulum.

4. Calcium ions bind with troponin.

5. Tropomyosin moves away from its blocking position, and this allows myosin and actin to interact with one another.

6. High-energy myosins bind with actin subunits of the thin filaments.

7. Energy stored in the high-energy myosins is discharged, and the myosin heads swivel, pulling on the thin filaments.

8. ATP binds with myosin heads, which release from actin subunits.

9. ATP is split into ADP and phosphate, again producing high-energy myosins, and the cycle (steps 6–9) is repeated over and over as long as calcium ions are bound to troponin.

10. Calcium ions are returned to the lateral sacs of the sarcoplasmic reticulum.

11. Tropomyosin moves back into its blocking position, preventing further interaction between high-energy myosins and actin subunits.

12. Contraction ceases, and the muscle fiber relaxes.

SOURCES OF ATP FOR MUSCLE CONTRACTION

The metabolic breakdown of fatty acids or glucose by skeletal muscle fibers produces ATP for muscle contraction. Under conditions of mild to moderate exercise, these substances are broken down and ATP is produced by processes that utilize oxygen (**aerobic processes**). However, during periods of intense activity, oxygen cannot be supplied to many muscle fibers fast enough, and oxidative metabolism cannot produce all the ATP required for contraction (F6.12). During such periods, non-oxygen-requiring metabolic processes (**anaerobic processes**) provide additional ATP. These processes break down glucose and stored glycogen to lactic acid, which diffuses out of the muscle fibers and into the blood.

Immediately following the initiation of muscular activity, many actively contracting muscle fibers utilize ATP faster than the metabolic reactions just discussed can supply it. However, muscle fibers contain a substance called **creatine phosphate** that provides them with a means of rapidly forming ATP. Creatine phosphate contains energy and phosphate that are transferred to ADP to produce ATP:

$$\text{Creatine phosphate} + \text{ADP} \rightleftharpoons \text{ATP} + \text{Creatine}$$

Following the initiation of muscular activity as ATP is being utilized, the net direction of this reaction is from right to left, forming ATP. During inactive periods when metabolic reactions are producing ATP not immediately required for contraction, the overall direction of the reaction is from right to left, regenerating creatine phosphate.

Oxygen Debt

As previously indicated, during intense activity, oxidative metabolism cannot supply all the ATP that muscles require. Therefore, anaerobic processes and creatine phosphate are used to provide additional ATP. The use of anaerobic processes and creatine phosphate allows the muscles to maintain a high level of activity for a longer period of time than would be possible if only oxidative processes supplied the ATP for contrac-

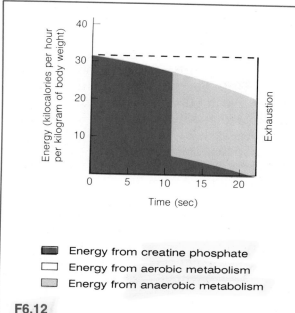

F6.12

Energy sources for muscular contraction during strenuous exercise (running at 18 km/hr on a treadmill inclined upward at an angle of 15°). Energy from creatine phosphate is the initial and major energy source during superexertion. Energy from aerobic metabolism increases exponentially from the onset of exercise, but the mechanism is sluggish and accounts for only a small amount of the energy used during the first few seconds. The remaining energy is acquired from anaerobic metabolism.

Key:
- ■ Energy from creatine phosphate
- □ Energy from aerobic metabolism
- ▨ Energy from anaerobic metabolism

tion. However, when intense muscular activity ceases, aerobic processes must provide ATP for the resynthesis of creatine phosphate, and the lactic acid that is produced during anaerobic metabolism can be converted back to glucose and glycogen in the liver by oxygen-requiring processes. These oxygen-requiring aerobic events account for the deep, rapid breathing that continues after muscular activity has ceased. In essence, an **oxygen debt** is built up during periods of intense muscular activity when nonoxidative sources of ATP (anaerobic metabolism and creatine phosphate) are used to support muscle contraction. This debt is paid back by an elevated rate of respiration during the period that follows activity. The elevated rate of respiration provides the oxygen required to produce ATP for the resynthesis of creatine phosphate and to convert lactic acid back to glucose and glycogen.

Muscle Fatigue

Intense skeletal muscle activity cannot continue indefinitely. Eventually **fatigue** sets in, and muscles no longer respond. Muscle fatigue is a complex phenomenon. However, it can apparently result from the depletion of ATP due either to a failure to produce it rapidly enough to keep up with the demand, or to the depletion of metabolic reserves of materials such as glycogen. In addition, the buildup of metabolic by-products such as lactic acid can cause the pH of muscles to become more acidic, thus reducing enzyme activity and contributing to the development of fatigue. (Note that "psychological fatigue" can cause a person to stop exercising even though the muscles themselves are still able to contract—that is, before any of the factors just mentioned create an actual physiological fatigue.)

THE MOTOR UNIT

There are more muscle fibers in a skeletal muscle than there are neurons to supply the muscle. Consequently, each neuron must branch to supply several muscle fibers, and a nerve impulse transmitted along a particular neuron reaches all the fibers supplied by the neuron.

A single neuron and all the muscle fibers it supplies make up a **motor unit** (F6.13). In the body, the motor unit and not the individual muscle fiber can be thought of as the functional unit of muscle activity because a nerve impulse in a single neuron stimulates all the muscle fibers supplied by the neuron. Muscles used for fine movements over which great control is exercised, such as the muscles of the hands, generally contain motor units in which each neuron supplies a relatively small number of muscle fibers, perhaps only a dozen or so. Muscles whose contractions are less precise, such as the muscles of the back or calf, have motor units in which a single neuron may supply several hundred muscle fibers.

SKELETAL MUSCLE RESPONSES

The contractions of the muscle fibers of the motor units of a muscle combine to produce contractions

The Muscle Twitch

If the fibers of a skeletal muscle are stimulated with a single brief stimulus, the muscle will contract once rapidly and then relax. This response, which is called a **muscle twitch**, shows three distinct phases (F6.14). Immediately following the arrival of the stimulus at the muscle, there is a short *latent period* during which no response is seen. During this period, the processes associated with excitation-contraction coupling occur. Following the latent period, a *period of contraction* occurs. During this period, the muscle actively develops tension, and, if enough tension develops to overcome the resistance of the load, the muscle shortens. The final phase of the muscle twitch is the *period of relaxation*. During this period, the tension that was actively developed by the muscle diminishes, and, if the muscle shortened, it returns to its original unstimulated length.

Graded Muscular Contractions

When a skeletal muscle fiber is stimulated and a propagated action potential travels over the plasma membrane and along the *t* tubules, enough calcium ions are released from the sarcoplasmic reticulum to completely activate the fiber. Consequently, when a nerve impulse triggers a propagated action potential in the membrane of a skeletal muscle fiber, the fiber contracts to the maximum extent possible for the existing conditions. (Note that this does not mean the contraction of a skeletal muscle fiber is exactly the same at all times. The physiological condition of a fiber at the time it is stimulated affects its contraction, and different conditions can exist at different times. For example, a fiber may or may not have contracted previously, or it may contain greater or lesser amounts of the by-products of contraction such as lactic acid.)

Even though individual skeletal muscle fibers contract to the maximum extent possible for the existing conditions, a muscle as a whole responds in a graded fashion to meet the demands of the tasks at hand. For example, the biceps brachii muscle of the arm does not contract the same when a person lifts a feather as it does when the person lifts a bowling ball. A smooth, graded response by a muscle depends on such factors as the number of motor units activated at any particular

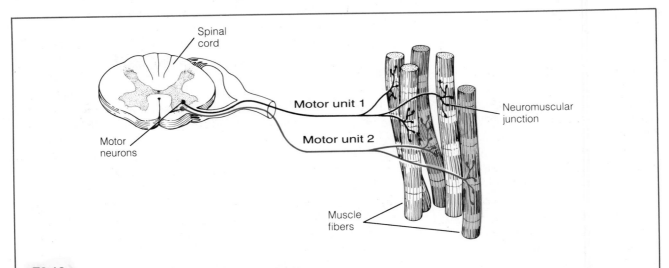

F6.13

Two motor units of a muscle. Each motor unit consists of a motor neuron and all the muscle fibers supplied by the neuron.

time, the frequency of nerve impulses, and the asynchronous activation of different motor units.

Multiple Motor Unit Summation.

The term **multiple motor unit summation** (also called spatial summation or recruitment) refers to the ability of the individual motor units of a muscle to combine their simultaneous activities to influence the degree of contraction of the entire muscle. If only a few motor units are active at any one time, the muscle contraction will be relatively weak. If many motor units are active, the muscle contraction will be relatively strong.

Wave Summation and Tetanus.

Following the stimulation of a skeletal muscle fiber and the triggering of a propagated action potential that leads to the contraction of the fiber, there is a brief period of time during which a second stimulus will not produce another propagated action potential no matter how strong the stimulus is. This period is called the *absolute refractory period*. Following the absolute refractory period there is another brief interval during which a stronger than normal stimulus is required to produce another prop-

agated action potential. This period is called the *relative refractory period*. The absolute and relative refractory periods usually last only 1 or 2 milliseconds, and they end well before the actual contraction of the muscle fiber (which can take 20 to 100 milliseconds) is completed. Consequently, a skeletal muscle fiber can be stimulated to contract a second time before it has completely relaxed from an initial contraction, a third time before it has completely relaxed from the second contraction, and so on.

In the body, the neurons that supply the muscle fibers of the motor units of a skeletal muscle do not normally transmit just a single stimulatory impulse that would produce only a twitch-type response. Instead, the neurons transmit volleys of impulses to the fibers, in which one impulse closely follows another. Consequently, the muscle fibers are stimulated a second time before they have completely relaxed from their initial contractions, a third time before they have relaxed from their second contractions, and so on. This pattern of stimulation produces a summation of individual contractions, or twitches, called **wave summation** (temporal summation) that can result in a state of more or less sustained contraction

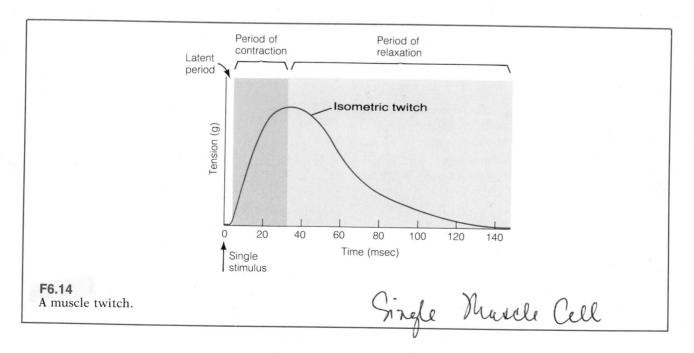

F6.14
A muscle twitch.

Single Muscle Cell

called **tetanus** (F6.15). If the successive stimuli arrive far enough apart in time that the muscle is able to relax partially between stimuli, a condition of *incomplete tetanus* is seen. If the stimuli arrive so rapidly that no relaxation of the muscle occurs between stimuli, a condition of *complete tetanus* exists.

Asynchronous Motor Unit Summation. A single muscle, such as the biceps brachii of the arm, consists of many motor units, and a sustained, coordinated contraction of the muscle is due in part to the asynchronous activation of different motor units (F6.16). Some motor units are activated initially, then they relax while other motor units are activated; then these relax and still other motor units become active. The result is a smooth, sustained contraction of the muscle as a whole.

FACTORS INFLUENCING THE DEVELOPMENT OF MUSCLE TENSION

The amount of force, or tension, developed by a skeletal muscle is influenced by the composition of the muscle and by the length of the muscle at the time it contracts.

Contractile and Series Elastic Elements

A muscle contains both contractile elements and series elastic elements. The **contractile elements** are those structures actively involved in contraction, such as the thick and thin filaments of the muscle fibers. The **series elastic elements** are structures that resist stretch (but can be stretched), which are located between the contractile elements and the load. They include connective tissue and some parts of the muscle fibers themselves. When a muscle contracts, the force generated by the contractile elements stretches the series elastic elements which, in turn, exert force on the load.

Influence of Series Elastic Elements on Muscle Tension. Because of the presence of series elastic elements, the tension developed by the contractile elements during a muscle contraction—that is, the internal tension, or active state, of the muscle fibers—is not always equivalent to the external tension exerted on the load. When a muscle is stimulated, the formation of attachments between the thick and thin filaments and the development of internal tension by the contractile elements of the muscle occur rather rapidly. However, following the formation of attachments between the thick and thin filaments and the development of internal tension, some time and effort are required to stretch or take up any slack in the series elastic elements of the muscle. During a single muscle twitch, the internal tension reaches its peak and decreases before the series elastic ele-

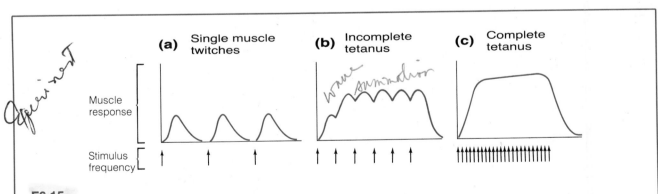

F6.15
Record of response of an isolated skeletal muscle stimulated with increasing frequencies of stimuli of sufficient intensity to produce a maximal response from the muscle. (a) Indicates single muscle twitches; (b) illustrates partial or incomplete tetanus (summation of twitches); (c) shows complete tetanus.

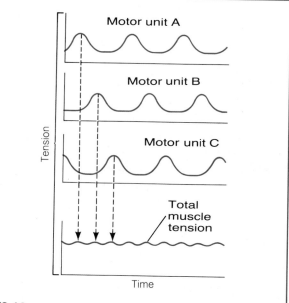

F6.16

The maintenance of a nearly constant tension in an entire muscle is a result of the asynchronous activity of individual motor units of the muscle.

ments are stretched to a tension equal to the maximum internal tension. As a result, less than the full internal tension is transmitted to the load (F6.17*a*). During a tetanic contraction, however, the repeated stimulation of the muscle maintains the internal tension long enough for the series elastic elements to be stretched to a tension similar to the internal tension (F6.17*b*). As a result, more of the internal tension is transmitted to the load during a tetanic contraction than during a single muscle twitch, and the response of a muscle during a tetanic contraction is greater than that during a single muscle twitch (see F6.15).

Active and Passive Tension. Because a muscle contains both contractile and elastic elements, it can develop two types of tension: active tension and passive tension. **Active tension** is the tension due to the activity of the contractile elements when a muscle is stimulated and contracts. **Passive tension,** in contrast, is a consequence of a muscle's elasticity, and a muscle does not have to contract to exert passive tension. When a skeletal muscle is stretched so that it lengthens, it behaves

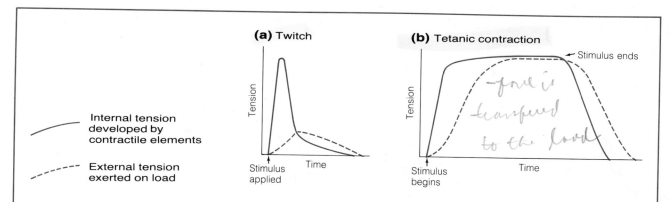

F6.17

(a) During a single muscle twitch, the internal tension developed by the contractile elements (the active state of the muscle fibers) reaches its peak and decreases before the series elastic elements are stretched to a tension equal to the maximum internal tension. As a result, less than the full internal tension is transmitted to the load. (b) During a tetanic contraction, the repeated stimulation of the muscle maintains the internal tension long enough for the series elastic elements to be stretched to a tension similar to the internal tension. As a result, more of the internal tension is transmitted to the load during a tetanic contraction than during a single muscle twitch, and the response of a muscle during a tetanic contraction is greater than that during a single muscle twitch.

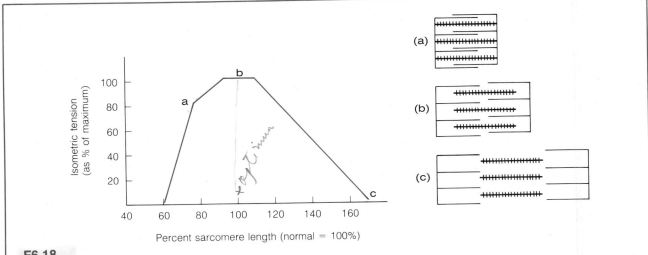

F6.18
Active tension developed by a skeletal muscle fiber at different initial fiber lengths, expressed in terms of percent sarcomere length.

much like a spring or rubber band. Within limits, the more it is stretched, the greater the passive tension it exerts.

Influence of Length on the Development of Muscle Tension

The amount of active tension developed by a skeletal muscle fiber when it contracts varies with the length of the fiber at the time of contraction (F6.18). This situation involves the fact that the fiber contracts as a result of the formation of attachments between the thick and thin filaments of its sarcomeres. When the thin filaments completely overlap the portions of the thick filaments that possess cross bridges, more attachments can form than when the muscle fiber is stretched so there is less overlap. If a muscle fiber is stretched to the point where no overlap occurs, no attachments between thick and thin filaments can form, and the fiber cannot contract. Conversely, when a muscle fiber is compressed, the thin filaments overlap and interfere with one another, and the thick filaments are forced against the Z lines of the sarcomeres. As a result, the active tension the muscle fiber can develop when it contracts is less than maximal.

A whole skeletal muscle exhibits a similar relationship between its length and the amount of active tension it develops when it contracts, and each skeletal muscle has an optimal length from which it can develop the maximum active tension (F6.19). In addition, as a muscle is lengthened—that is, stretched—it exerts increasing amounts of passive tension, and the total tension exerted by a

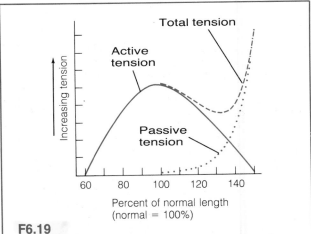

F6.19
Graph showing the influence of muscle length on the tension developed by a skeletal muscle.

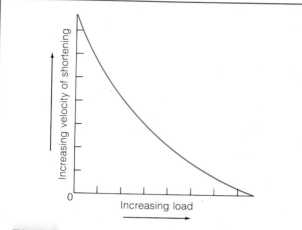

F6.20
Relation of load to velocity of shortening of a skeletal muscle. As the load increases the velocity of shortening decreases.

muscle when it contracts is equal to the sum of its active and passive tensions. In the body, the lengths of most relaxed, unstimulated muscles are such that they allow the development of maximal active tension when the muscles are stimulated.

RELATION OF LOAD TO VELOCITY OF SHORTENING

The greater the load on a muscle, the slower the velocity of shortening (F6.20). With only a small load, the velocity of shortening is relatively fast. As the load increases, the velocity of shortening decreases until a point is reached at which the load is too great for the muscle to move. At this point, the velocity of shortening is zero, and the contraction of the muscle is isometric.

RELATIONSHIP BETWEEN LEVERS AND MUSCLE ACTIONS

The movements brought about by the actions of most skeletal muscles involve the use of levers. A **lever** is a rigid structure that is capable of moving around a pivot point, called a **fulcrum,** when a force is applied. In the body, the bones of the skeleton function as levers, the joints serve as fulcrums, and skeletal muscles provide the force to move the bones.

CLASSES OF LEVERS

There are three classes of levers.

Class I Levers

In *Class I levers*, the fulcrum is located between the point at which a force is applied and the weight that is to be moved (F6.21*a*). A seesaw is a common example of a Class I lever. In the body, an example of a Class I lever is raising the face to look up. The joint between the bottom of the skull and the first vertebra serves as the fulcrum, the facial portion of the skull is the weight, and the force (pull) is applied to the back of the skull by the contraction of muscles located along the back of the neck.

Class II Levers

In *Class II levers*, the weight to be moved is between the fulcrum and the point at which the force is applied (F6.21*b*). This type of lever is used in the wheelbarrow. The best example in the body is raising the body on the toes. In this case, the base of the toes serves as the fulcrum, the toes support the weight, and the contraction of muscles of the calf causes a force (pull) to be exerted at the heel of the foot.

Class III Levers

In *Class III levers*, the weight is at one end, the fulcrum at the other, and the force is applied between them (F6.21*c*). Lifting a shovel uses this type of leverage. There are many examples of Class III levers in the body, since it is the most common lever system used. One example is bending, or flexing, the elbow (decreasing the angle between adjoining bones). The weight is at the hand, the fulcrum is the elbow joint, and the force (pull) is exerted by the contraction of muscles that attach to the forearm between the fulcrum and the weight.

EFFECTS OF LEVERS ON MOVEMENTS

The portion of a lever located between the fulcrum and the point where a force is applied is called the *power arm;* the portion between the fulcrum and the weight is the *weight arm*. When the weight arm is long in relation to the power arm, a weight can be moved rapidly over a considerable distance, but a strong force is required. Conversely, when the weight arm is short in relation to the power arm, the same weight can be moved with less force, but both the speed of movement and the distance of movement are reduced. Thus, depending on the arrangement of particular muscles and bones, the levers of the body may enable muscles to move loads faster over a greater distance than would otherwise be possible, or they may enable muscles to move heavier loads than they otherwise could.

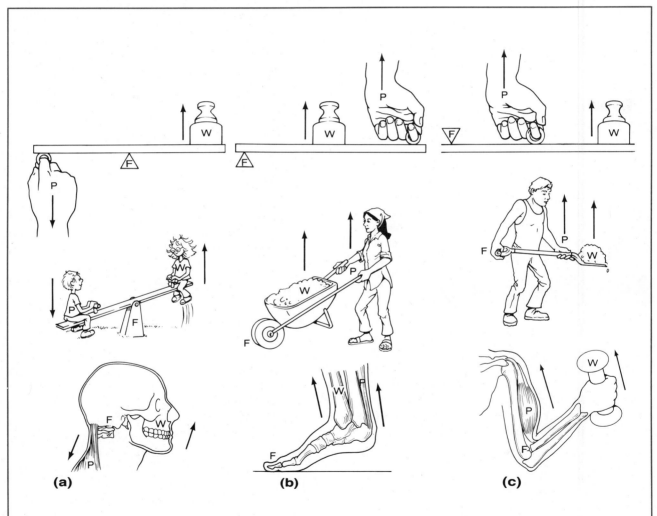

(a)   **(b)**   **(c)**

F6.21

(*a*) Class I lever. The fulcrum (F) is located between the weight (W), or resistance, and the force, or pull (P). Arrows indicate the direction of movement. (*b*) Class II lever. The weight (W), or resistance, is located between the fulcrum (F) and the point of force, or pull (P). Arrows indicate the direction of movement. (*c*) Class III lever. The force, or pull (P), is applied between the fulcrum (F) and the weight (W), or resistance. Arrows indicate the direction of movement.

TYPES OF SKELETAL MUSCLE FIBERS

Not all skeletal muscle fibers are identical. For example, the myosin molecules of some fibers split ATP more rapidly than the myosin molecules of other fibers. Because of this, cross-bridge cycling occurs more rapidly in some fibers than others, and some fibers contract faster than others. In addition, the metabolic processes by which muscle fibers produce the ATP necessary for contraction differ from one fiber to another, and the ability of a fiber to produce ATP influences its resistance to fatigue.

Based on their contraction speed and fatigue resistance, three types of skeletal muscle fibers can be distinguished.

SLOW TWITCH, FATIGUE-RESISTANT FIBERS

Slow twitch, fatigue-resistant fibers contain myosin molecules that split ATP at a slow rate. Consequently, cross-bridge cycling occurs slowly, and the fibers contract at a slow speed. Slow twitch, fatigue-resistant fibers have a highly developed capacity for oxidative metabolism. They are surrounded by many blood vessels and they contain large amounts of an oxygen-binding protein known as **myoglobin.** Myoglobin increases the rate of oxygen diffusion into the fibers, and it provides them with small stores of oxygen. In general, the oxidative metabolic processes of slow twitch, fatigue-resistant fibers can meet their needs for ATP, and these fibers are extremely resistant to fatigue.

FAST TWITCH, FATIGUE-RESISTANT FIBERS

Fast twitch, fatigue-resistant fibers are also well supplied by blood vessels, contain large amounts of myoglobin, and have a highly developed capacity for oxidative metabolism. However, they contain myosin molecules that split ATP at a rapid rate. Consequently, cross-bridge cycling occurs rapidly, and they contract at a fast speed. The highly developed oxidative processes of these fibers can supply most of their needs for ATP, and

the fibers are quite resistant to fatigue (although less so than slow twitch, fatigue-resistant fibers).

FAST TWITCH, FATIGABLE FIBERS

Fast twitch, fatigable fibers contain myosin molecules that split ATP rapidly, and they contract at a fast speed. However, these fibers are not well supplied by blood vessels, and they do not contain large amounts of myoglobin. Instead, they contain large stores of glycogen, and they are especially geared to the utilization of anaerobic metabolic processes. These processes are unable to supply the fibers continually with the amount of ATP they require, and the fibers fatigue easily.

UTILIZATION OF DIFFERENT FIBER TYPES

In some skeletal muscles, one type of muscle fiber predominates. However, most skeletal muscles contain all three fiber types in proportions that vary from muscle to muscle. Consequently, muscles show a range of contraction speeds and fatigue resistances.

The presence of varying proportions of the three fiber types in different skeletal muscles is consistent with the following facts:

1. Not all skeletal muscles perform the same functions, and different functions often require different types of muscle contractions. For example, the postural muscles of the back and legs that support the body against the force of gravity must be able to undergo sustained contractions without fatigue, and the muscles that move the arms must be able to develop large amounts of tension quickly in order to allow the rapid lifting of heavy objects.
2. The same skeletal muscle often performs different functions at different times. For example, certain muscles of the legs may be involved in supporting the body against the force of gravity at one time, and they may participate in moving the legs during walking or running at another.

Even though most skeletal muscles contain all three fiber types, all the fibers of any one motor unit are the same type. Moreover, the different fiber types tend to be used in characteristic fashions when a muscle contracts. For example, during activities of short duration, if only a weak contraction is required from a muscle, just the slow twitch, fatigue-resistant fibers are activated. If a stronger contraction is required, the fast twitch, fatigue-resistant fibers are also activated, and if a still stronger contraction is necessary, the fast twitch, fatigable fibers are added.

EFFECTS OF EXERCISE ON SKELETAL MUSCLE

When skeletal muscles are not used, either because the nerves supplying them no longer function, or because a limb is immobilized as the result of a broken bone, the muscle fibers atrophy (diminish in size). Conversely, regular exercise can produce increases in muscle size, endurance, and strength.

Different types of exercise cause different kinds of changes in the fibers of a skeletal muscle. Endurance types of exercise such as running are associated with a transformation of fast twitch, fatigable fibers into fast twitch, fatigue-resistant fibers with relatively small increases in the diameters of the fibers or their strength. High-intensity, short-duration types of exercise such as weight lifting, however, cause fast twitch, fatigable fibers to hypertrophy. The fibers increase in diameter and there is an increased synthesis of actin and myosin filaments, which results in a large increase in the strength of the fibers.

The increases in strength that occur with exercise do not always seem to be accounted for simply by increased muscle size. It has been suggested that, in addition to increased muscle size, more nerve pathways that activate more motor units may be utilized. Normally, all the motor units of a muscle are not activated simultaneously; if they were, the muscle's tendons could be torn loose from their attachments. As exercise develops a muscle, however, perhaps more motor units can be activated simultaneously, thereby increasing strength.

Endurance exercises produce cardiovascular and respiratory changes that cause muscles to be better supplied with materials such as oxygen and carbohydrates. Thus, the improved muscular performance that results from exercise is not solely the result of muscular changes but involves other systems as well.

SKELETAL MUSCLE DYSFUNCTIONS

CRAMPS

Cramps are painful, involuntary muscle contractions that are slow to relax. They can occur during exercise or at rest, and their precise cause is not known. Cramps may be caused by conditions within the muscle itself—for example, a low oxygen supply—or by stimulation from the nervous system. There is some evidence that cramps that occur during heavy exercise are caused by low blood levels of sodium and chloride ions, as the result of the loss of these ions by sweating. However, it is not clear whether the depletion of sodium chloride acts on the muscles or on the nervous system.

MUSCULAR DYSTROPHY

The term *muscular dystrophy* refers to a group of diseases characterized by progressive muscular weakness. The weakness is the result of the degeneration of muscle fibers, an increase in connective tissue within the muscles, and in some forms, the replacement of muscle fibers by fatty tissue. The muscular dystrophies are genetically transmitted. Some forms of muscular dystrophy are fatal, but other forms have a more favorable outlook.

MYASTHENIA GRAVIS

Myasthenia gravis is a chronic condition characterized by extreme muscle weakness. It is caused by an abnormal response of the body's immune system that disrupts acetylcholine receptors on the muscle cell membranes at the neuromuscular junctions. This disruption decreases the responsiveness of the muscle fibers to acetylcholine released from motor neuron endings. About 10% of the people who have this disease die from it. However, if an afflicted individual survives the first three years, there is a good chance that the condition will stabilize with some degree of recovery.

EFFECTS OF AGING ON SKELETAL MUSCLE

When a person reaches the middle twenties, a progressive and continuous loss of skeletal muscle mass begins, and much of the loss is replaced by fat. Since fat weighs less than muscle, the normal body weight at age 50 is less than that at age 20. As the muscle mass decreases, so does the maximal strength, which declines by about 50% between 20 and 80 years of age.

SMOOTH MUSCLE

Smooth muscle fibers are uninucleate, spindle-shaped cells that are considerably smaller than skeletal muscle fibers. Smooth muscle fibers have a well-developed capacity for anaerobic metabolism, but their overall metabolic machinery is not as highly developed as that of skeletal muscle fibers.

Smooth muscle fibers possess thick filaments that contain myosin and thin filaments that contain actin and tropomyosin. However, the filaments are not organized into regularly ordered sarcomeres, and smooth muscle fibers are not striated. Nevertheless, smooth muscle uses cross-bridge movements between myosin and actin to generate force, and it is believed that smooth muscle contraction occurs by a sliding filament mechanism similar to that in skeletal muscle.

Smooth muscle contraction is triggered by calcium ions obtained from two sources—the sarcoplasmic reticulum and the extracellular fluid—and most smooth muscle fibers rely upon both sources to some extent. When a smooth muscle fiber is stimulated, calcium ions are released from the sarcoplasmic reticulum. Moreover the plasma membrane becomes more permeable to calcium ions, which enter the fiber from the extracellular fluid. The calcium ions bind to the myosin-containing thick filaments rather than to the thin filaments as in skeletal muscle. Just how the binding allows cross-bridge activity to proceed is presently unclear. Smooth muscle relaxation occurs when the calcium ions are pumped back into the sarcoplasmic reticulum or across the plasma membrane and out of the fiber.

STRESS-RELAXATION RESPONSE

Smooth muscle can be stretched much more than skeletal muscle before exhibiting any marked changes in tension. When a smooth muscle is suddenly stretched, it initially displays an increased tension. However, this tension begins to decrease almost immediately, and within a few minutes it has returned to its original level. This phenomenon, which is called **stress-relaxation,** is important because many smooth muscles are located in the walls of hollow structures such as the stomach, intestines, and urinary bladder. As these structures become filled, they enlarge, stretching their walls and the smooth muscles associated with them. Within limits, however, the stress-relaxation response of smooth muscle allows these hollow structures to enlarge with little appreciable change in the pressures exerted on their contents.

ABILITY TO CONTRACT WHEN STRETCHED

Smooth muscle is able to contract and actively generate tension under greater stretching than skeletal muscle. This ability is at least partly explained by the fact that the filaments of smooth muscle fibers are not organized into any regular pattern of sarcomeres but rather exist in a somewhat irregular array. Because of this, the problem of stretching smooth muscle fibers to the point where filaments no longer overlap one another is less serious than in skeletal muscle cells. At many degrees of stretch, some filaments still overlap one another and the fibers can still contract.

The ability to contract when stretched is important to smooth muscles that are associated with hollow structures. During the filling and distension of these structures, the smooth muscles associated with them undergo considerable stretching. Despite this stretching, however, the muscles maintain the ability to contract and actively generate tension.

DEGREE OF SHORTENING DURING CONTRACTION

When smooth muscles contract, they can shorten far more as a percentage of their length than can skeletal muscles. The useful distance of contraction for a skeletal muscle is about 25% to 35% of the length of the muscle, whereas a smooth muscle can contract from twice its normal length to one-half its normal length. The ability of smooth muscles to shorten considerably when they contract allows structures such as the stomach, intestines, or urinary bladder to vary the diameters of their lumens (cavities) from considerably large values to practically zero.

SMOOTH MUSCLE TONE

Compared with skeletal muscle, smooth muscle contracts rather slowly. Smooth muscle, however, is quite resistant to fatigue, and many smooth muscles undergo sustained, long-term (tonic) contractions that consume relatively small amounts of energy. The tonic contraction of smooth muscle (*smooth muscle tone*) is important in the circulatory system, where the smooth muscles of many blood vessels must contract for long periods. Also, the tonic contraction of smooth muscles of the intestine exerts a steady pressure on the intestinal contents.

TYPES OF SMOOTH MUSCLE

Two basic arrangements of smooth muscle occur in the body: single-unit and multiunit. Although some intergrading between the two types is evident, most smooth muscles adhere basically to one pattern or the other.

Single-Unit Smooth Muscle

The cells of *single-unit smooth muscle* are connected by gap junctions at which electrical impulses can spread from cell to cell. As a result, many cells respond as a unit to stimulation. Single-unit smooth muscle is self-excitable, and it contracts without external stimulation. Occa-

sionally, one cell of the unit becomes spontaneously stimulated, and this stimulus passes to other cells causing the entire unit to contract. This type of muscle is sensitive to stretch, and a quick stretch of the muscle can produce a contraction. Single-unit smooth muscle is found in small arteries and veins, the intestine, and the uterus.

Multiunit Smooth Muscle

The individual cells of *multiunit smooth muscle* are usually sufficiently separated from one another so that a stimulus to one cell is not transferred to neighboring cells. As a result, each cell responds independently to stimulation. Multiunit smooth muscle is generally not self-excitable, and it requires external stimulation to initiate a contraction. This type of smooth muscle is not responsive to stretch. Multiunit smooth muscle is found in large arteries and the large airways to the lungs.

INFLUENCE OF EXTERNAL FACTORS ON SMOOTH MUSCLE CONTRACTION

Smooth muscle contraction is influenced by external factors that include neural activity and chemical substances such as hormones. As mentioned in the previous section, external factors are generally required to initiate the contraction of multiunit smooth muscle. Moreover, external factors modulate—that is, enhance or inhibit—the frequency and intensity of the spontaneous contractions of single-unit smooth muscle. The responses of different smooth muscles to particular factors vary greatly, and a factor that stimulates one smooth muscle may inhibit another. As particular smooth muscles are discussed in later sections of the text, their responses to specific substances will be considered.

STUDY OUTLINE

MUSCLE TYPES

Skeletal Muscle   Most attaches to skeleton; striated cells; voluntary control; regulated by somatic nervous system.

Smooth Muscle   In walls of hollow organs and tubes; cells lack striations; involuntary control; regulated by factors intrinsic to muscle itself, hormones, autonomic nervous system.

Cardiac Muscle Forms wall of heart, striated cells, involuntary control; regulated by factors intrinsic to muscle itself, hormones, autonomic nervous system.

STRUCTURE OF SKELETAL MUSCLE
A skeletal muscle is composed of muscle cells called muscle fibers; bound together by fascia; usually attached to bones by tendons. **p. 122**

Structure of Skeletal Muscle Fibers
Fibers (cells) contain:
1. *Myofibrils* Longitudinal threadlike arrays of proteins; cross-striated by alternating light and dark bands.
   Anisotropic (A, dark) bands have less dense H zone in center; H zone is crossed by M line.
   Isotropic (I, light) bands have dense Z line crossing center; Z lines divide myofibrils into sarcomeres.
2. *Sarcomeres* Repeating units of myofibrils; contain filamentous structures called myofilaments.
   Thick filaments Only in A band; H zone has only thick filaments; each thick filament is surrounded by six thin filaments.
   Thin filaments I band; part of A band; attach to Z lines.

Composition of the Myofilaments
1. *Thick filaments* consist mainly of the protein myosin.
2. *Thin filaments* consist mainly of the proteins actin, tropomyosin, and troponin.

Transverse Tubules and Sarcoplasmic Reticulum
1. *Transverse tubules (t tubules)* Run deep into muscle cell from the plasma membrane.
2. *Sarcoplasmic reticulum* Membranous network that runs throughout cell and surrounds myofibrils.

CONTRACTION OF SKELETAL MUSCLE p. 125
1. *Isometric* muscle actively develops tension but does not shorten.
2. *Isotonic* muscle actively develops tension and shortens.

Contraction of a Skeletal Muscle Fiber
At the cellular level, the same basic events occur during both isometric and isotonic skeletal muscle contraction.

THE NEUROMUSCULAR JUNCTION
Ending of a motor neuron approaches specialized point along the plasma membrane of a skeletal muscle fiber, forming a neuromuscular junction.

EXCITATION OF A SKELETAL MUSCLE FIBER
Nerve impulse reaches neuromuscular junction; acetylcholine released; permeability of muscle fiber plasma membrane at junction changes; propagated action potential is produced.

EXCITATION-CONTRACTION COUPLING
Propagated action potential moves along *t* tubules to fiber interior; calcium ions released from recticular sites called lateral sacs; calcium ions bind to troponin molecules of thin filaments of sarcomeres; binding leads to interactions between thick and thin filaments and muscle contraction.

MECHANISM OF CONTRACTION
1. High-energy myosin formed when ATP is split.
2. When calcium ions bind to troponin molecules, tropomyosin molecules move aside.
3. High-energy myosin links with G-actin subunit.
4. Energy discharged from high-energy myosin; myosin head swivels and pulls on actin-containing filament.
5. When ATP again occupies myosin head, the cycle repeats.

REGULATION OF THE CONTRACTILE PROCESS
Active transport mechanism returns calcium ions to sarcoplasmic reticulum; when calcium ions return to sarcoplasmic reticulum, tropomyosin returns to blocking position.

Sources of ATP for Muscle Contraction
1. Oxidative metabolism provides ATP (aerobic process).
2. ATP and lactic acid produced during intense activity by nonoxidative (anaerobic) metabolism.
3. Creatine phosphate rapidly produces ATP.

OXYGEN DEBT
Built up during periods of intense muscular activity; oxygen debt repaid by elevated respiration rate after activity.

MUSCLE FATIGUE
Can apparently result from ATP depletion; lactic acid buildup may cause pH change in muscle.

The Motor Unit
A single neuron and all the muscle cells it supplies.

Skeletal Muscle Responses
Contractions of muscle fibers of motor units combine to produce contractions of whole muscle.

THE MUSCLE TWITCH
1. *Latent period* Follows arrival of stimulus at muscle.
2. *Period of contraction* Active development of tension and possible shortening.
3. *Period of relaxation.*

GRADED MUSCULAR CONTRACTIONS

When an action potential occurs in the membrane of a skeletal muscle fiber, the fiber contracts to the maximum extent possible for conditions existing at the time. However, an entire muscle responds in graded fashion to meet demands of particular tasks.

Multiple Motor Unit Summation   Strength of a muscle contraction can be varied by varying the number of motor units active at any one time.

Wave Summation and Tetanus   Repeated stimulations produce summation of individual twitches; tetanus—state of sustained contraction.

Asynchronous Motor Unit Summation   All the motor units of a muscle may not be active at the same time; some motor units may be active initially and then others, producing a smooth continuous contraction of the muscle as a whole.

Factors Influencing the Development of Muscle Tension   Include composition of a muscle and length of a muscle at the time it contracts.

CONTRACTILE AND SERIES ELASTIC ELEMENTS

1. Contractile elements are those structures actively involved in contraction, such as the thick and thin filaments of the muscle fibers.
2. Series elastic elements are structures that resist stretch (but can be stretched), which are located between the contractile elements and the load.

Influence of Series Elastic Elements on Muscle Tension   Because of the presence of series elastic elements, the tension developed by the contractile elements during a muscle contraction—that is, the internal tension, or active state, of the muscle fibers—is not always equivalent to the external tension exerted on the load.

Active and Passive Tension
1. Active tension is tension due to the activity of the contractile elements when a muscle is stimulated and contracts.
2. Passive tension is a consequence of a muscle's elasticity; a muscle does not have to contract to exert passive tension. When a muscle is stretched, it behaves much like a spring or rubber band and exerts passive tension.

INFLUENCE OF LENGTH ON THE DEVELOPMENT OF MUSCLE TENSION   Greatest active tension develops at optimal muscle length; decreases when muscle is stretched or compressed. Within limits, passive tension increases with stretch of muscle.

Relation of Load to Velocity of Shortening   The greater the load on a muscle, the slower the velocity of shortening.

RELATIONSHIP BETWEEN LEVERS AND MUSCLE ACTIONS   Movements brought about by actions of most skeletal muscles involve the use of levers. p. 139

Classes of Levers

CLASS I LEVERS   Example is raising the face to look up.

CLASS II LEVERS   Example is raising the body on the toes.

CLASS III LEVERS   Example is flexing the elbow.

Effects of Levers on Movements   Depending on the arrangement of particular muscles and bones, the levers of the body may enable muscles to move loads faster over a greater distance than would otherwise be possible, or they may enable muscles to move heavier loads than they otherwise could.

TYPES OF SKELETAL MUSCLE FIBERS   Three types, based on contraction speed and fatigue resistance. p. 141

Slow Twitch, Fatigue-Resistant Fibers   Myosin splits ATP at slow rate; slow speed of contraction; many blood vessels; much myoglobin; well-developed oxidative metabolic processes; very fatigue-resistant.

Fast Twitch, Fatigue-Resistant Fibers   Myosin splits ATP at rapid rate; fast speed of contraction; many blood vessels; much myoglobin; well-developed oxidative metabolic processes; quite fatigue-resistant.

Fast Twitch, Fatigable Fibers   Myosin splits ATP at rapid rate; fast speed of contraction; few blood vessels; little myoglobin; geared to anaerobic metabolic processes; fatigue easily.

UTILIZATION OF DIFFERENT FIBER TYPES
Most skeletal muscles contain all three fiber types and each type tends to contribute in a characteristic fashion to a muscle contraction; all the fibers of any one motor unit are the same type. p. 141

EFFECTS OF EXERCISE ON SKELETAL MUSCLE
1. Muscle fiber atrophy results from disuse of skeletal muscles.
2. Muscle size and strength can increase with exercise.

3. Endurance-type exercise causes transformation of fast twitch, fatigable fibers into fast twitch, fatigue-resistant fibers with small increase in fiber mass.
4. High-intensity, short-duration exercise causes hypertrophy of fast twitch, fatigable fibers, with large increase in fiber mass and strength.   **p. 142**

SKELETAL MUSCLE DYSFUNCTIONS   p. 142

Cramps Involuntary painful muscle contractions, slow to relax; may be caused by low O_2 supply in muscles; may be caused by nervous system stimulations; those that occur during heavy exercise may be due to low levels of sodium and chloride ions in the blood.

Muscular Dystrophy Genetically transmitted progressive muscular weakness that results from muscle cell degeneration; increase in connective tissue or replacement of muscle cells by fatty tissue.

Myasthenia Gravis Rare chronic condition of extreme muscle weakness; related to inability of muscle fiber plasma membrane at neuromuscular junction to respond to acetylcholine.

EFFECTS OF AGING ON SKELETAL MUSCLE
Progressive and continuous loss of skeletal muscle mass occurs with aging.   **p. 143**

SMOOTH MUSCLE   p. 143
1. Differs from skeletal muscle.
2. Smaller cells, no regular cross striations.
3. No regularly ordered myofibrils; but does possess filaments that contain actin and tropomyosin, and filaments that contain myosin.
4. Calcium source extracellular as well as from sarcoplasmic reticulum; calcium enters, is then pumped out of cell.

Stress-Relaxation Response Does not increase tension greatly when stretched.

Ability to Contract When Stretched Can contract and actively develop tension when stretched.

Degree of Shortening During Contraction Can shorten relatively large amounts.

Smooth Muscle Tone Generally slower response than skeletal muscle; can undergo long-term, sustained contractions, smooth muscle tone.

Types of Smooth Muscle

SINGLE-UNIT SMOOTH MUSCLE
1. Cells connected by gap junctions; impulses spread from cell to cell.
2. Self-excitable; spontaneous contractions.
3. Small arteries, veins, intestines, uterus.

MULTIUNIT SMOOTH MUSCLE
1. Cells separated; stimulus to one not transferred to other cells; cells generally require external stimulation.
2. Large arteries; large airways to lungs.

Influence of External Factors on Smooth Muscle Contraction Neural activity and chemical factors such as hormones can influence contraction.

SELF-QUIZ

1. Skeletal muscles are voluntary muscles. True or False?
2. Skeletal muscle fibers: (a) exhibit cross striations; (b) are innervated by the autonomic nervous system; (c) are uninucleate.
3. The thick filament of a sarcomere of a skeletal muscle cell contains: (a) actin; (b) troponin; (c) myosin.
4. The threadlike molecule that lies along the surface of actin is: (a) myosin; (b) tropomyosin; (c) troponin.
5. Structures that run from the plasma membrane into the interior of a skeletal muscle fiber form the: (a) *t* tubule network; (b) sarcoplasmic reticulum; (c) myofibers.
6. Acetylcholine causes a change in the permeability of the plasma membrane of a skeletal muscle fiber at the neuromuscular junction. True or False?
7. The membrane system of skeletal muscle fibers that contains calcium ions necessary for contraction is the: (a) sarcomere; (b) sarcoplasmic reticulum; (c) myofibril.
8. During a skeletal muscle contraction, ATP occupies: (a) actin; (b) myosin; (c) troponin.
9. Actin acts as an enzyme to split ATP into ADP and phosphate. True or False?
10. In an unstimulated skeletal muscle fiber, the interaction of actin and myosin is believed to be directly blocked by: (a) calcium ions; (b) *t* tubules; (c) tropomyosin.
11. During a skeletal muscle contraction, calcium ions

bind with: (a) tropomyosin; (b) myosin; (c) troponin.

12. During a contraction in which a skeletal muscle fiber shortens: (a) troponin shortens; (b) the Z lines are drawn closer together; (c) calcium binds with myosin.

13. During sustained, intense, rapid exercise, a skeletal muscle produces: (a) oxygen; (b) glycogen; (c) lactic acid.

14. Which substance provides skeletal muscle fibers with a means of rapidly forming ATP immediately after the initiation of muscular activity? (a) fatty acids; (b) creatine phosphate; (c) kinase.

15. Within a skeletal muscle: (a) there are more neurons than muscle fibers; (b) there are no neuromuscular junctions between nerve endings and muscle fibers; (c) each neuron branches to supply a number of muscle fibers.

16. An individual neuron, together with all of the skeletal muscle fibers it supplies, makes up a: (a) sarcomere; (b) motor unit; (c) myofibril.

17. When a nerve impulse triggers a propagated action potential in the membrane of a skeletal muscle fiber, the fiber contracts to the maximum extent possible for the existing conditions. True or False?

18. Fast twitch, fatigable skeletal muscle fibers are: (a) well supplied by blood vessels; (b) myoglobin rich; (c) geared for the utilization of anaerobic metabolic processes.

19. With training of a high-intensity, short-duration type: (a) skeletal muscle strength decreases; (b) fast twitch, fatigable fibers hypertrophy; (c) slow twitch, fatigue-resistant fibers are converted to fast twitch, fatigable fibers.

20. Smooth muscle fibers are: (a) multinucleate; (b) found in the walls of internal or visceral organs; (c) principally under voluntary control.

21. Which type of muscle is likely to become spontaneously stimulated? (a) single-unit smooth muscle; (b) skeletal muscle; (c) multiunit smooth muscle.

22. The muscular dystrophies are generally considered to be genetically transmitted. True or False?

BASIC FEATURES OF THE NERVOUS SYSTEM 7

The nervous system can respond to stimuli that activate receptors by stimulating effectors. In this activity **afferent (sensory) neurons** carry nerve impulses from receptors to the brain and spinal cord. Within the brain and spinal cord, **interneurons** conduct nerve impulses from one neuron to another. Ultimately, **efferent (motor) neurons** conduct nerve impulses away from the brain and spinal cord to effectors (F7.1). In addition to stimulating effectors, the nervous system can also integrate and store information it receives from receptors, providing for such capacities as abstract reasoning, conceptualization, imagination, and memory.

Although there is only one nervous system, it is frequently thought of as being separated into several divisions (F7.2). Bear in mind, however, that these divisions—which are themselves called nervous systems—are all integral parts of a single nervous system.

ORGANIZATION OF THE NERVOUS SYSTEM

The nervous system is usually divided into two major parts: the central nervous system and the peripheral nervous system (F7.3). The **central nervous system (CNS)**, which consists of the brain and spinal cord, is the integrative and control center of the nervous system. It receives sensory input from the peripheral nervous system and formulates responses to this input. The **peripheral nervous system (PNS)** connects receptors and effectors in outlying parts of the body with the central nervous system. It consists of **nerves**, which are the processes of many neurons held together by connective tissue, and **ganglia**, which are groups of nerve cell bodies associated with nerves.

THE CENTRAL NERVOUS SYSTEM

THE BRAIN

The brain is the major integrative center of the nervous system, and specific brain areas are associated with specific functions.

Cerebrum

The **cerebrum** is the largest portion of the brain. It is divided into right and left **cerebral hemispheres,** and each hemisphere is divided into **frontal, parietal, temporal,** and **occipital** lobes (F7.4). On the ventral surface of each hemisphere is a small **olfactory bulb** and its associated **olfactory tract** (F7.5).

The cerebrum has an outer surface of **gray matter,** the **cerebral cortex,** which is composed of nerve cell bodies and unmyelinated nerve fibers (F7.6). Additional gray matter structures, the **basal ganglia,** are located deep inside each cerebral hemisphere. The gray matter of the cortex is separated from the basal ganglia by **white matter,** which is composed of bundles, or **tracts,** of myelinated nerve fibers.

Cerebral Cortex.   The cerebral cortex is popularly thought of as the site of the mind and intellect, but physiologically it is an integrating area that brings together basic sensory information into complex perceptual images, and refines signals sent from the brain to other areas of the body.

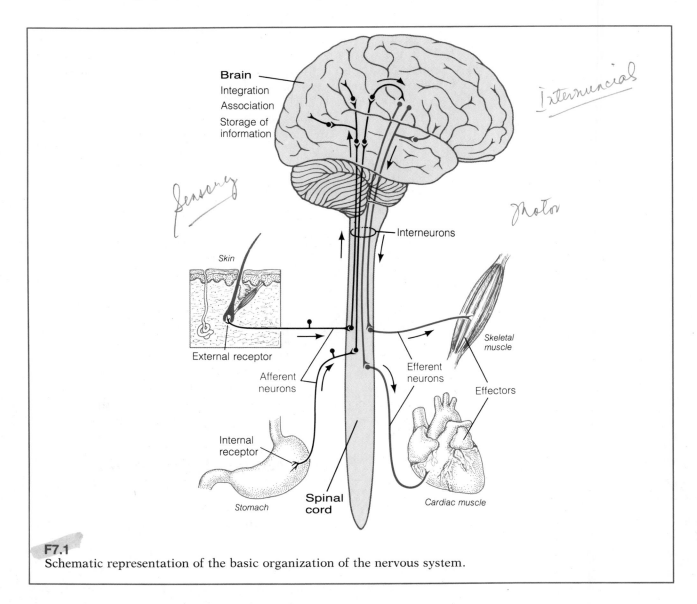

F7.1
Schematic representation of the basic organization of the nervous system.

Certain areas of the cortex are related to specific functions. These areas have been precisely mapped, but for our purposes it is sufficient to consider only the general locations of some of the major functional areas (F7.7). It should be kept in mind, however, that Figure 7.7 represents an oversimplification of a very complex structure. Because of extensive interconnections between various cortical areas, any function attributed to a specific area most probably involves several areas.

The **primary motor area** is located at the rear of the frontal lobe (F7.7). This area is a stimulatory area involved in controlling the voluntary contractions of skeletal muscles. The neurons of this area are organized sequentially as shown in Figure 7.8. Most of the neural pathways that transmit impulses from the primary motor area to the skeletal muscles cross from one side of the nervous system to the other. Thus, the primary motor area

of the right cerebral hemisphere largely influences the skeletal muscles on the left side of the body, and the primary motor area of the left cerebral hemisphere largely influences the skeletal muscles on the right side of the body.

The **primary sensory area** (somatic sensory area; somesthetic area) is located at the front of the parietal lobe (F7.7). Within this area are the terminations of the pathways that transmit general sensory information concerning temperature, touch, pressure, and pain from body receptors to the cerebral cortex. These pathways cross from one side of the nervous system to the other as they ascend to the cortex. Thus, the primary sensory area of the right cerebral hemisphere receives information from the left side of the body, and the primary sensory area of the left cerebral hemisphere receives information from the right side of the body. The neurons of the primary sensory area

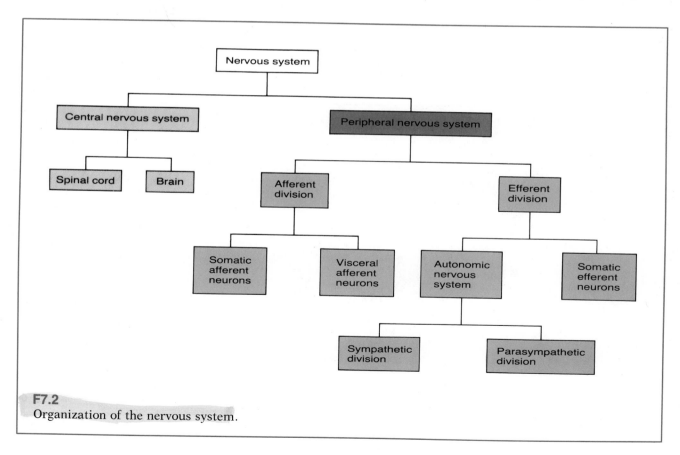

F7.2
Organization of the nervous system.

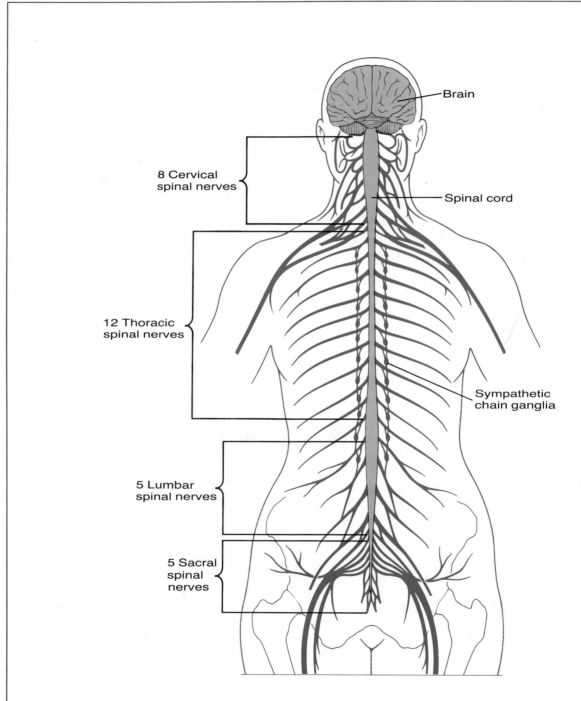

F7.3

The central nervous system and the proximal portions of the peripheral nervous system.

are organized sequentially in a manner similar to the neurons of the primary motor area (F7.8).

The **visual area** is in the posterior portion of the occipital lobe. The **auditory area,** which receives nerve impulses associated with hearing, is located along the upper margin of the temporal lobe. The area concerned with the sense of smell, the **olfactory area,** is located on the medial surface of the temporal lobe. The **taste area** is located in the parietal lobe.

When dealing with sensory areas, it is important to realize that only nerve impulses, not sensations, are transmitted from receptors to the brain. It is in the brain that nerve impulses are consciously interpreted as particular sensations (pain, touch, sound, taste, and so on). The brain then projects the sensations—usually with considerable accuracy—to the locations of the stimuli that are activating the receptors.

Several **association areas** surround the sensory and motor areas of the cerebral cortex. These areas are involved in processing, integrating, and interpreting sensory information and in formulating patterns of motor responses.

Basal Ganglia. Like the primary motor area, the **basal ganglia (basal nuclei)** are involved in controlling skeletal muscle activity. Consequently, the contractions of skeletal muscles are controlled both by neurons of the cerebral cortex and by neurons located elsewhere in the brain, including the basal ganglia. However, in contrast to the stimulatory effects of the cortical neurons, neurons of the basal ganglia act in part to inhibit muscle contraction.

Hemispheric Specialization. Each cerebral hemisphere is somewhat specialized for carrying out

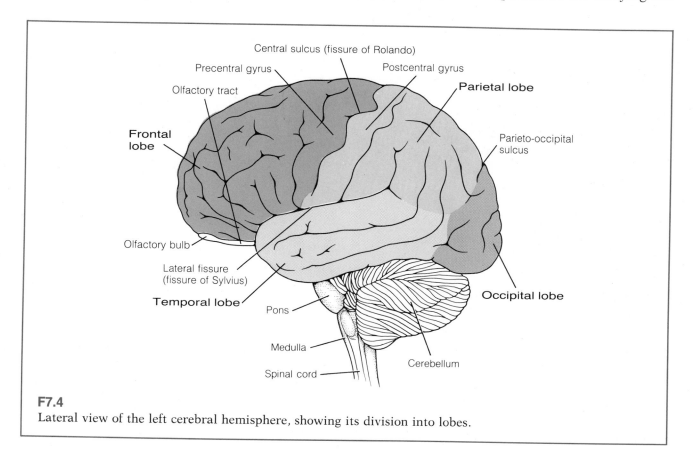

F7.4
Lateral view of the left cerebral hemisphere, showing its division into lobes.

certain kinds of mental processes. For example, language ability tends to be localized in the left cerebral hemisphere. In general, the left hemisphere is concerned with verbal and sequential processes and behaviors such as writing business letters and solving simple equations. This hemisphere excels at performing rational, linear, verbally oriented tasks, and it seems to process information in a fragmentary or analytical way. In contrast, the right cerebral hemisphere is primarily concerned with the recognition of complex visual patterns or with mentally picturing objects in three-dimensional space. The right hemisphere is also more involved in the expression and recognition of emotion and in certain musical or artistic abilities, such as identifying a theme in an unfamiliar piece of music. The information processing in the right hemisphere tends to be holistic and unitary rather than fragmentary and analytical.

The ability to recognize faces is well localized within the brain without being a function of pri-

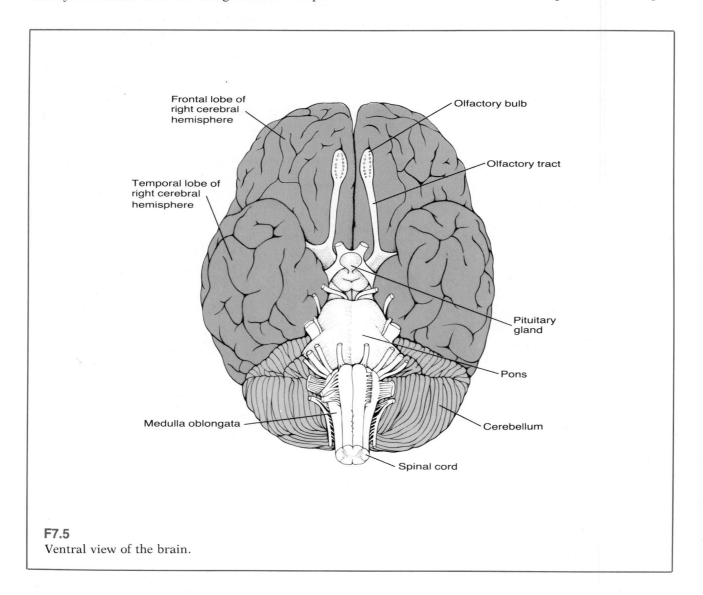

F7.5
Ventral view of the brain.

marily one cerebral hemisphere or the other. Damage to the medial undersides of both occipital lobes and the ventromedial surfaces of the temporal lobes produces a failure to recognize a person by sight. People with brain damage in these areas can usually correctly name objects but cannot identify faces. Such individuals may even fail to recognize their parents, spouses, or children by sight. Nevertheless, when a familiar but unrecognized person speaks, the person with brain damage can immediately recognize the other person's voice and can then name him or her.

Thalamus

The **thalamus,** which consists of two oval masses of gray matter containing over 20 functionally separate groups of nerve cell bodies, or **nuclei,** is an important relay and integrating center of the brain (F7.6; F7.9). All the pathways that transmit nerve impulses from receptors to the cerebral cortex, except those associated with olfaction, synapse within the thalamus. From the thalamus, nerve impulses are relayed to specific regions of the cortex as well as to subcortical areas such as the basal ganglia and hypothalamus.

Hypothalamus

The **hypothalamus,** which lies below the thalamus, also contains a number of nuclei (F7.9). It is involved in regulating body temperature, water balance, appetite, gastrointestinal activity, sexual activity, and emotions such as fear and rage. The hypothalamus also regulates the release of hormones from the pituitary gland (see Chapter 10).

Limbic System

Although emotions are strongly influenced by the hypothalamus, emotional responses involve a

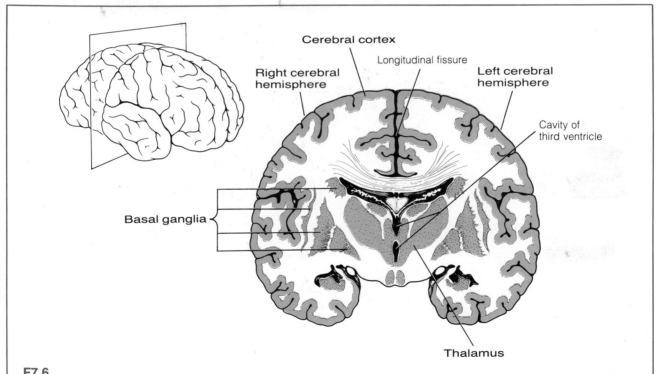

F7.6
Frontal section of the brain, showing the cerebral cortex (gray matter) surrounding the white matter, and the basal ganglia deep within the white matter.

complex interaction of structures in several different regions of the brain, including the cerebrum. A group of structures collectively referred to as the limbic system are particularly important in emotional responses (F7.10).

In animal experiments, it is possible to electrically stimulate various centers within the limbic system. When certain centers are stimulated, the animal responds to the stimulation as if it were pleasant ("pleasure centers"). When other centers are stimulated, the animal responds to the stimulation as if it were unpleasant ("punishment centers"). A wide variety of emotional behavior patterns can be caused by stimulating or removing specific regions of the limbic system. It is assumed, therefore, that the structures of the lim-

bic system play an important role in regulating emotional behavior.

Cerebellum

The **cerebellum** projects from the dorsal surface of the brain beneath the cerebrum (F7.4; 7.5; 7.9). It makes extensive interconnections with other regions of the central nervous system, and has widespread input and output capabilities. The cerebellum is an important coordinator of skeletal muscle activity. It receives information from receptors for position and movement, equilibrium, balance, touch, vision, and sound. It also receives information about signals sent to the muscles by other areas of the brain. The cerebellum integrates the information it receives and, in turn,

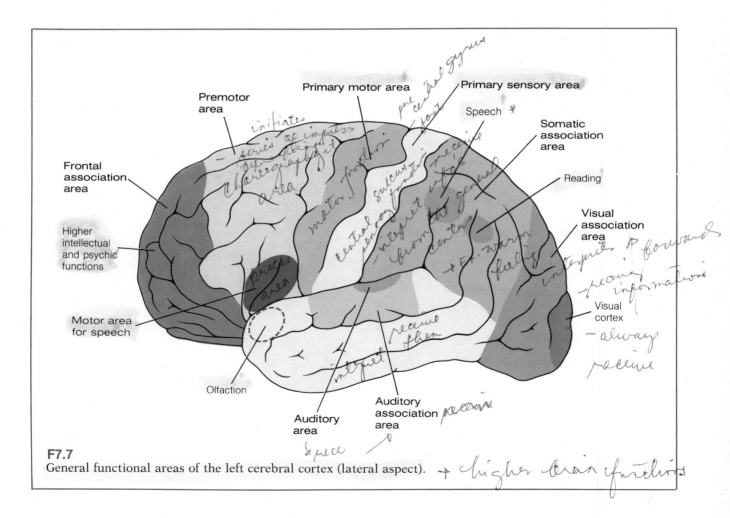

F7.7
General functional areas of the left cerebral cortex (lateral aspect).

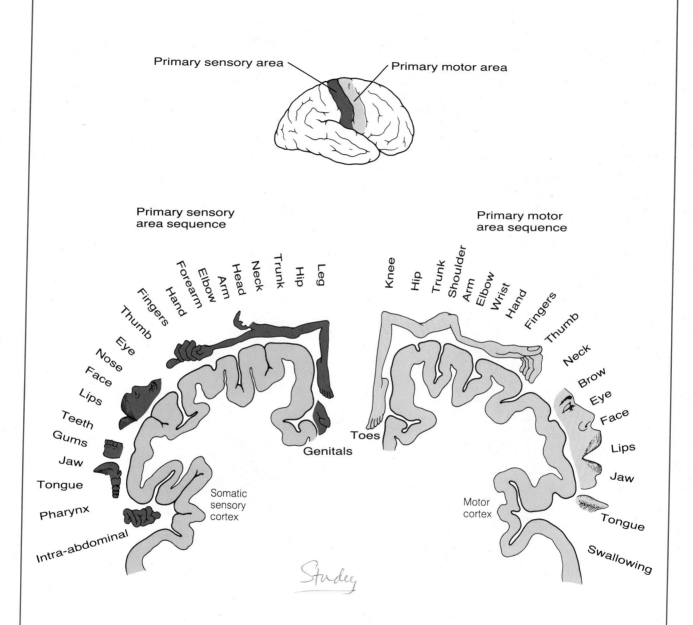

F7.8

Frontal section of the cerebrum. Left half: through the primary sensory area, showing the locations of regions of the cerebral cortex that receive sensory nerve impulses from specific body structures. Right half: through the primary motor area, showing the locations of neurons within the cerebral cortex involved in the control of voluntary movement of specific body structures.

sends impulses to both higher brain centers and the spinal cord to coordinate movements.

Pons

The **pons** connects the cerebellum with the brain stem, and provides connections between upper and lower levels of the central nervous system (F7.5; F7.9). In addition, the stimulation of certain areas within the pons affects the rate of respiration (see Chapter 15).

Medulla Oblongata

The **medulla oblongata** is the most inferior division of the brain (F7.5; F7.9). The medulla contains neurons involved in the control of several vital functions, including heart rate, respiration, dilation and constriction of blood vessels, coughing, swallowing, and vomiting.

Brain Stem and Reticular Formation

The medulla, the pons, and the area known as the midbrain together form the **brain stem** (F7.9). Extending throughout the brain stem and up into the thalamus and hypothalamus is a region of gray matter containing a network of interlacing nerve fibers called the **reticular formation** (F7.11). The reticular formation receives nerve impulses from receptors located in peripheral areas of the body, as well as from the basal ganglia, the cerebellum, and other brain areas. Selected impulses that pass

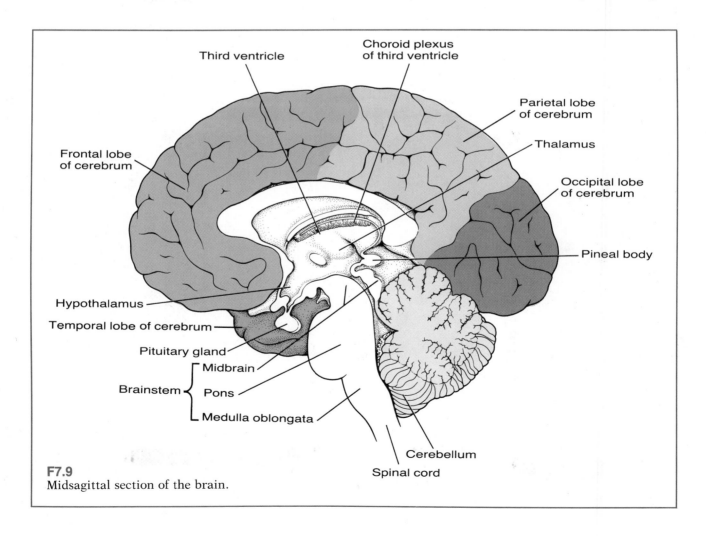

F7.9
Midsagittal section of the brain.

through the reticular formation are relayed to the cerebral cortex, thus activating the cortex. Because it exerts this control over the cortex, the reticular formation is considered to include an activating or arousal system, the **reticular activating system**, that is essential in maintaining wakefulness.

THE SPINAL CORD

Below the medulla, the central nervous system continues as the **spinal cord** (F7.12). The spinal cord contains centrally located gray matter that is composed of nerve cell bodies and unmyelinated nerve fibers (F7.13). The gray matter is surrounded by white matter that is composed of bun-

dles, or tracts, of myelinated nerve fibers. **Ascending tracts** transmit impulses up the spinal cord to higher levels within the cord or brain, and **descending tracts** transmit impulses down the spinal cord from the brain or higher levels of the cord. The dorsal and ventral roots of 31 pairs of spinal nerves arise from the spinal cord, and each portion of the cord that gives rise to a pair of spinal nerves is called a spinal segment.

The Spinal Reflex Arc

Some afferent neurons that carry nerve impulses to the spinal cord synapse directly or through interneurons with efferent neurons in the gray matter of the cord at the same level at which they

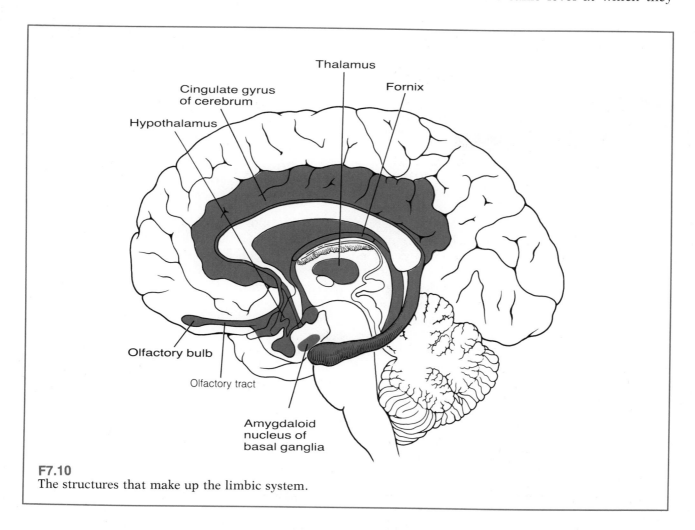

F7.10
The structures that make up the limbic system.

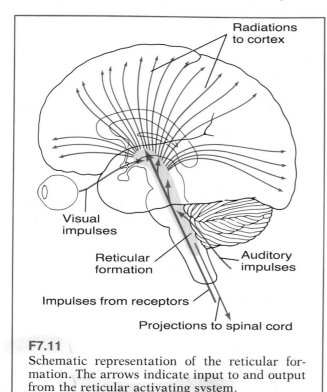

Visual impulses

Radiations to cortex

Reticular formation

Auditory impulses

Impulses from receptors

Projections to spinal cord

F7.11
Schematic representation of the reticular formation. The arrows indicate input to and output from the reticular activating system.

enter the cord. Other afferent neurons travel only a short distance up or down the cord before synapsing either directly or indirectly with efferent neurons. Thus, the spinal cord contains neural pathways by which nerve impulses from receptors can reach effectors without traveling to the brain. These pathways are called **spinal reflex arcs.**

The presence of spinal reflex arcs makes possible automatic, stereotyped reactions to stimuli in which a particular stimulus always elicits a particular response. Such reactions are called **reflexes.** Since spinal reflexes occur at the level of the spinal cord, without involving the brain, they are involuntary responses even though they often involve skeletal muscles. While a reflex is occurring, however, information about the stimulus that initiated the reflex may also be transmitted to the brain where a conscious sensation is elicited. For example, when a person's hand touches something hot, the person becomes aware of it through impulses transmitted to the brain. However, by the time a painful, burning sensation is felt, the

person has already withdrawn the hand from the hot object as a result of a reflex action. If the spinal cord is severed in the neck region, thus making it impossible for any impulses to reach the brain from the spinal cord, the hand would still be withdrawn as a result of reflex activity that occurs within the spinal cord itself. However, there would be no sensation of pain.

A spinal reflex arc has at least five components (F7.14):

1. A **receptor,** which can be a peripheral ending of an afferent neuron or a specialized cell associated with a peripheral ending of an afferent neuron.
2. An **afferent neuron,** which transmits nerve impulses from the receptor to the spinal cord.
3. A **synapse** within the spinal cord between an afferent neuron and an efferent neuron. If an afferent neuron synapses directly with an efferent neuron, the spinal reflex arc is a *monosynaptic reflex arc.* If there are one or more interneurons between an afferent and an efferent neuron, requiring more than one synapse, it is a *polysynaptic reflex arc.* Almost all central nervous system reflexes are polysynaptic.
4. An **efferent neuron,** which transmits nerve impulses from the spinal cord to an effector.
5. An **effector,** which responds to the efferent nerve impulses. Muscle tissue (skeletal, smooth, or cardiac) and glands serve as effectors.

Two important spinal reflexes—the stretch reflex and the tendon reflex—influence the contraction of skeletal muscles.

Stretch Reflex. The muscle **stretch reflex** is initiated by skeletal muscle receptors called **muscle spindles** (F7.15). The muscle spindles provide information about the lengths of skeletal muscles, and they respond to both the rate and the magnitude of a length change. Within a muscle spindle are three to ten specialized muscle cells called *intrafusal fibers* that differ from the cells (*extrafusal fibers*) of the muscle as a whole. The intrafusal fibers are contractile only at their ends, and the ends of the fibers are supplied by efferent neurons called *gamma efferent neurons.* The noncontractile, central portions of the intrafusal fibers are supplied by afferent neurons that spiral around the intrafusal fibers. The frequency of nerve impulses in the afferent neurons increases in re-

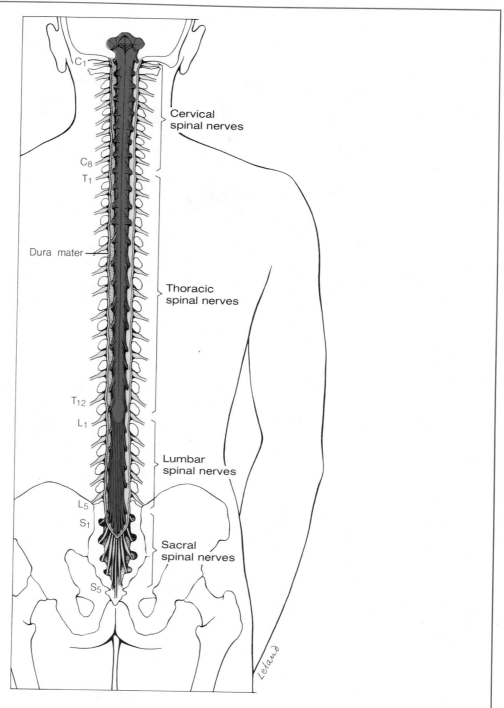

F7.12

The spinal cord and the proximal portions of the spinal nerves. The letters indicate specific groups of spinal nerves. C = cervical; T = thoracic; L = lumbar; S = sacral.

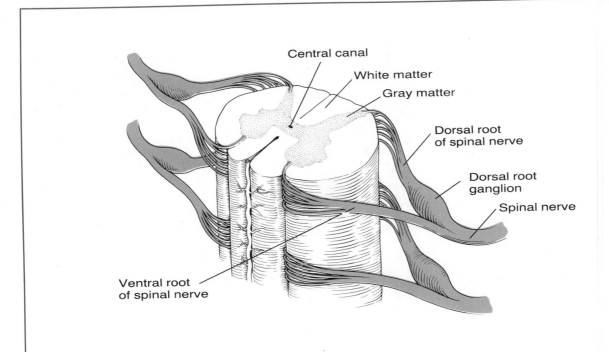

F7.13
A cross section of the spinal cord. The dorsal and ventral roots of the spinal nerves are evident.

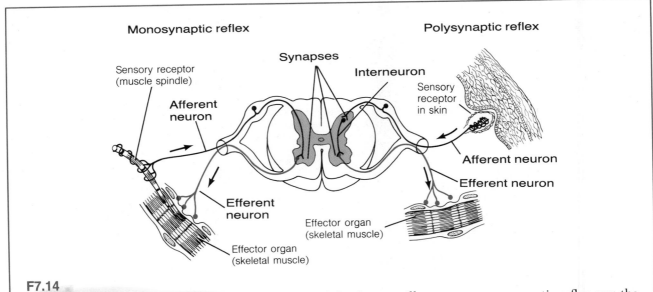

F7.14
The components of a spinal reflex arc. The left side of the diagram illustrates a monosynaptic reflex arc; the right side shows a polysynaptic reflex arc.

sponse to both the sudden and maintained stretch of the central areas of the intrafusal fibers and decreases when the central areas of the fibers are compressed. A muscle spindle is attached in parallel with the extrafusal fibers of a skeletal muscle so that stretching the muscle stretches the muscle spindle (and its intrafusal fibers), and contracting the muscle compresses the muscle spindle.

In the stretch reflex, the stretching of a muscle results in an increased frequency of nerve impulses in the afferent neurons associated with the muscle spindles (F7.16). Within the spinal cord, the afferent neurons synapse with efferent neurons called *alpha motor neurons* that supply the extrafusal fibers of the muscle. As a result, the increased frequency of nerve impulses in the afferent neurons increases the stimulation of the alpha motor neurons, causing the muscle to contract and resist the stretch. The afferent neurons also synapse with inhibitory neurons that, in turn, synapse with efferent neurons controlling muscles whose activities oppose the contraction of the stretched muscle (that is, antagonistic muscles). Thus, when the stretch reflex stimulates the stretched muscle to contract, antagonistic muscles that oppose the contraction are inhibited. This occurrence is called **reciprocal inhibition** and the neuronal mechanism that causes it is called reciprocal innervation.

The *patellar reflex* (knee jerk) is a common example of a stretch reflex. Striking the patellar tendon at the knee pulls on the tendon and thus stretches the quadriceps muscles of the anterior thigh. If all the components of the reflex arc are intact, the muscle spindles within the quadriceps muscles initiate a stretch reflex that causes the quadriceps muscles to contract and swing the leg forward.

In addition to their synapses with neurons involved in the stretch reflex, the afferent neurons from muscle spindles also synapse with neurons that relay impulses to the brain. As a result, nerve impulses from muscle spindles provide information about the state of stretch or contraction (that is, length) of skeletal muscles that is important in maintaining body posture and in coordinating muscular activity.

The activity and sensitivity of a muscle spindle is influenced by nerve impulses carried to the in-

trafusal fibers of the spindle by the gamma efferent neurons. The contraction of the ends of the intrafusal fibers due to gamma efferent stimulation stretches the central, noncontractile portions of the fibers. This stretching increases the activity of the afferent neurons and enhances the response of the spindle to further stretch. Thus, the excitability of the muscle spindle and the amount of muscle stretch necessary to initiate a stretch reflex can be altered by altering the degree of contraction of the intrafusal fibers.

Nerve impulses to a skeletal muscle from high-

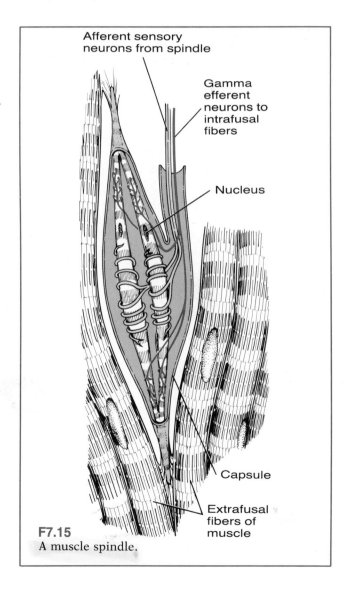

F7.15
A muscle spindle.

Afferent sensory neurons from spindle

Gamma efferent neurons to intrafusal fibers

Nucleus

Capsule

Extrafusal fibers of muscle

er centers often simultaneously stimulate both the extrafusal fibers of the muscle (by way of the alpha motor neurons) and the intrafusal fibers of the spindles (by way of the gamma efferent neurons). This simultaneous stimulation maintains the responsiveness of the spindles when the muscle shortens, and it may allow the basic stretch reflex mechanism to assist in attaining the proper degree of muscular contraction. For example, suppose picking up an object requires a certain degree of contraction by a muscle. If both the extrafusal fibers of the muscle and the intrafusal fibers of the muscle spindles are stimulated, and if the contraction of the muscle is equal to the contraction of the intrafusal fibers of the spindles, the degree of stimulation of the muscle spindles will not change as a result of the contraction of the muscle. However, if for some reason (for example, fatigue) the stimulation of the extrafusal fibers results in a muscle contraction that is less than the contraction of the intrafusal fibers of the spindles, the central portions of the intrafusal fibers will be stretched. As a result, the activity of the afferent

neurons from the spindles increases. This increased activity further stimulates the alpha motor neurons to the extrafusal fibers, and thereby helps achieve the needed degree of muscular contraction (perhaps by activating additional motor units). In addition, nerve impulses from the muscle spindles are relayed to higher brain centers, and these centers may also alter their output to the alpha motor neurons in order to attain the proper degree of muscular contraction.

Tendon Reflex.   The **tendon reflex** helps protect tendons and their associated muscles from damage that could result from excessive tension. The receptors for this reflex are **Golgi tendon organs,** which are encapsulated structures located within tendons near the junction of a tendon with a muscle. Unlike muscle spindles, which are sensitive to muscle length, Golgi tendon organs are sensitive to tension. Within a Golgi tendon organ are afferent neuron endings that are wrapped around small bundles of collagen fibers. When the tension applied to a tendon increases, so does the fre-

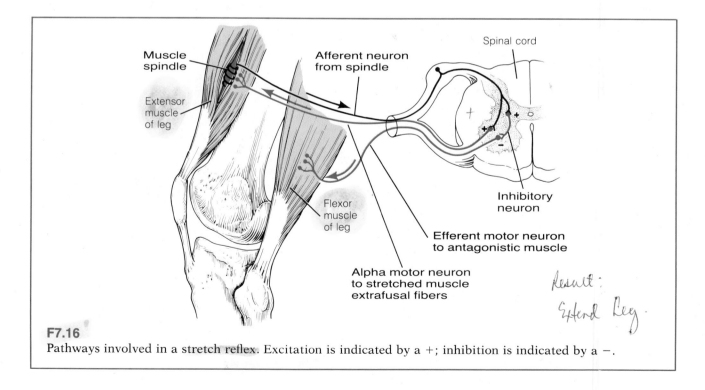

F7.16
Pathways involved in a stretch reflex. Excitation is indicated by a +; inhibition is indicated by a −.

quency of nerve impulses in the afferent neurons; when the tension decreases, so does the afferent impulse frequency. Because of their locations within tendons, the Golgi tendon organs are in series with the muscles themselves, and they respond to changes in tension that result from either passive stretch or muscular contraction.

In the tendon reflex, an increase in the tension applied to a tendon—most commonly as a result of muscular contraction—increases the frequency of nerve impulses in the afferent neurons associated with the tendon organs (F7.17). Within the spinal cord, the afferent neurons synapse with inhibitory neurons that, in turn, synapse with alpha motor neurons that supply the muscle associated with the tendon. As a result, the increased frequency of nerve impulses in the afferent neurons from the tendon organs ultimately diminishes the stimulation of the alpha motor neurons to the muscle, and may even completely inhibit muscular contraction. Thus, the tendon reflex protects the muscle and tendon from excessive tension. The afferent neurons from the Golgi tendon or-

gans also synapse with stimulatory neurons that, in turn, synapse with motor neurons controlling antagonistic muscles. As a result, the tendon reflex is generally accompanied by a reciprocal stimulation of the antagonistic muscles.

In addition to synapsing with neurons involved in the tendon reflex, the afferent neurons from the Golgi tendon organs also synapse with neurons that relay impulses to the brain. Thus, just as nerve impulses from muscle spindles provide the brain with information about the length of a skeletal muscle, nerve impulses from Golgi tendon organs provide information about the tension developed by the muscle.

NEURON POOLS

Within the central nervous system, neurons are organized into functional groups called **neuron pools,** and information transmitted from a receptor is often received by a particular neuron pool. Within the pool, this information is processed and integrated with information received from other sources. It is then transmitted from the pool to

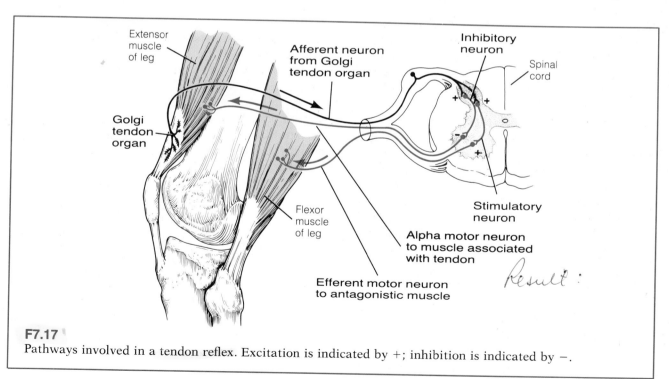

F7.17
Pathways involved in a tendon reflex. Excitation is indicated by +; inhibition is indicated by −.

various destinations. Within a neuron pool, the processing of information involves integrative activities such as divergence, convergence, summation, and facilitation (see Chapter 5).

Neuron pools occur at all levels of the central nervous system, including the cerebral cortex, and there are numerous interconnections among the various pools. Although the neural pathways involved in particular activities are commonly depicted as simple neuron-to-neuron chains, the performance of even the simplest actions generally requires the participation of many neurons whose activities are coordinated in complex circuits in neuron pools. For example, the alpha motor neurons to a skeletal muscle are influenced by nerve impulses arriving from muscle spindles, Golgi tendon organs, and higher brain centers. As a result, the degree of activity of the alpha motor neurons at any one moment is determined by a combination of influences.

MENINGES, VENTRICLES, AND CEREBROSPINAL FLUID

The entire central nervous system is covered by three layers of connective tissue known as the **meninges** (singular: *meninx*). The innermost meninx, the **pia mater,** is a delicate, vascular membrane that adheres closely to the brain and spinal cord (F7.18). The middle meninx, the **arachnoid,** is a delicate structure that is attached to the pia mater by filamentous strands. Between the arachnoid and the pia mater is a subarachnoid space. The outermost meninx, the **dura mater,** is a strong, fibrous connective tissue membrane. Around the brain it is a double-layered structure whose outer layer adheres closely to the skull. In certain areas the two layers of the dura mater are separated, forming blood-filled venous sinuses through which most of the blood supply of the brain flows on its way to the heart. Thin projections of the arachnoid, the **arachnoid villi,** extend into the venous sinuses.

Within the brain are four interconnected cavities or **ventricles,** which are continuous with a narrow **central canal** that extends the length of the spinal cord (F7.19).

The subarachnoid space, the ventricles of the brain, and the central canal of the spinal cord contain a clear, watery **cerebrospinal fluid,** which cushions the central nervous system, protecting it from jolts and blows. Cerebrospinal fluid is secreted into the ventricles by highly vascular tissues, the **choroid plexuses.** Because it is a selective secretion, the composition of cerebrospinal fluid differs from that of the blood plasma. For example, proteins, potassium, and calcium are present in lower concentration in the cerebrospinal fluid than in the plasma, and sodium and chloride are present in higher concentration.

There is normally a slight pressure within the ventricles, and the cerebrospinal fluid circulates slowly through them. In the brain-stem region, the fluid passes through small holes into the subarachnoid space. Within the subarachnoid space, the cerebrospinal fluid circulates slowly around the brain and spinal cord. It is ultimately reabsorbed into the blood through the arachnoid villi.

If cerebrospinal fluid were to accumulate, it would exert pressure that would be damaging to the brain. Normally, however, cerebrospinal fluid is reabsorbed into the blood at the same rate that it is formed.

Substances such as nutrients and the end products of metabolism can diffuse between the neural tissue and the cerebrospinal fluid, and an examination of cerebrospinal fluid provides a means of determining whether infectious organisms are present in the central nervous system. Samples of cerebrospinal fluid are withdrawn by inserting a hollow needle between the third and fourth lumbar vertebrae into the subarachnoid space. This procedure is called a lumbar puncture (F7.20*a* and *b*). Spinal anesthesias (spinal blocks) are sometimes administered in a similar manner. In an attempt to identify damaged intervertebral discs, a contrast medium that is observable on an x-ray is sometimes injected into the subarachnoid space. The x-ray taken after the injection of the contrast medium is referred to as a lumbar myelogram (F7.20*c*). Any obstruction of the contrast medium indicates possible disc pathology.

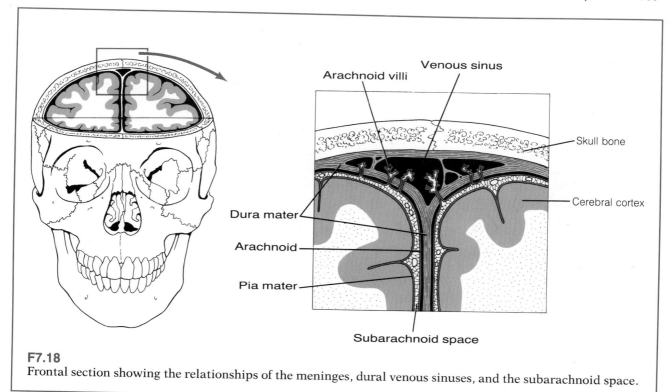

F7.18
Frontal section showing the relationships of the meninges, dural venous sinuses, and the subarachnoid space.

DISORDERS OF THE CENTRAL NERVOUS SYSTEM

Because the brain and spinal cord are such complex structures, innumerable abnormalities can occcur within the central nervous system as a result of injury or disease. Central nervous system disorders are particularly serious because the neurons within the central nervous system do not effectively regenerate once they have been damaged. Consequently, recovery from central nervous system disorders is often incomplete.

Central nervous system disorders that interfere with the transmission of nerve impulses to skeletal muscles can cause muscular paralysis. Paralysis may be either flaccid, in which the affected muscles lose their normal slight contraction (tonus) and reflexes are absent, or it may be spastic, with increased tonus and reflexes. Central nervous system disorders that interfere with the transmission of nerve impulses from receptors to the brain can cause the loss of particular sensations in certain areas of the body.

Parkinsonism

Parkinsonism is a disorder of the basal ganglia. Its onset is gradual, and it usually affects people over 50. A number of factors are thought to produce Parkinson's syndrome, all of which result in abnormalities of cells within the basal ganglia. As previously indicated, neurons within the basal ganglia regulate skeletal muscle activity in part by inhibiting contractions. Consequently, parkinsonism results in spastic movements that are beyond the control of the individual. This condition is characterized by useless contractions that cause muscular tremor and rigidity, as well as by a decrease in muscular movements such as swinging the arms while walking and changing facial expressions related to emotions. In fact, many sufferers exhibit a seemingly emotionless, masklike facial expression. There is considerable evidence that parkinsonism is related to the abnormal metabolism of the neurotransmitter dopamine by the basal ganglia. In victims of this disease, the level of dopamine within the basal ganglia is substantially below normal.

Poliomyelitis

Poliomyelitis is caused by a virus that destroys the cell bodies of efferent neurons to skeletal muscles, especially in the cervical and lumbar areas of the spinal cord. This condition is characterized by fever, severe headache, and a flaccid paralysis of the muscles that are supplied by the neurons. After several weeks of paralysis the muscles begin to atrophy, and eventually the muscle tissue may be almost completely replaced by connective tissue and adipose tissue. Death can result from respira-

tory failure if the virus invades nerve cells within the regulatory centers of the medulla.

Multiple Sclerosis

Multiple sclerosis is a chronic condition that results in widespread destruction of the myelin sheaths of neurons in the brain and spinal cord. The destroyed myelin sheaths are replaced by hardened plaques that interfere with the normal transmission of nerve impulses. The cause of the destruction is not known, but a virus is suspected by some investigators. Multiple sclerosis causes a great variety of symptoms, depending on the areas of the central nervous system where plaque formation occurs. Common symptoms include abnormal sensations, spastic paralysis, and exaggerated reflexes. Although the disease is chronic and progressive, it is not unusual for the symp-

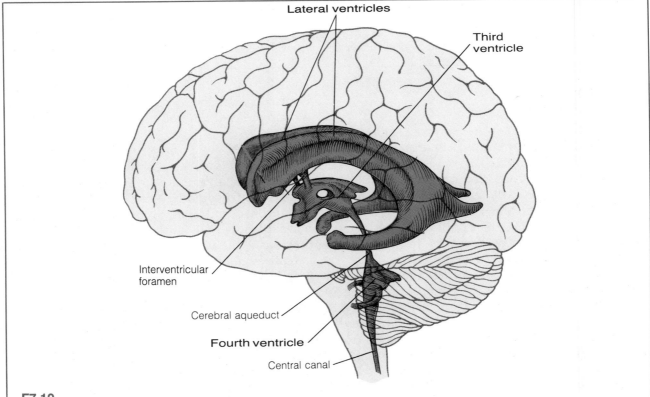

F7.19
Ventricles of the brain, viewed as if they could be seen from the surface of the brain.

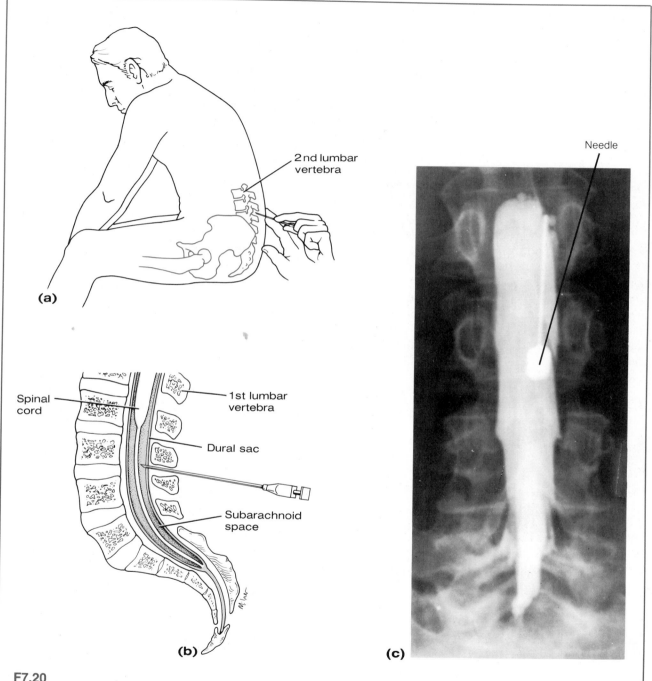

F7.20

(a) The technique of a lumbar puncture. (b) Position of the needle within the subarachnoid space below the termination of the spinal cord. (c) Lumbar myelo-gram. The white material is contrast medium that was injected through the needle.

toms of the disease to disappear for years at a time. During such periods the disease is said to be in remission.

Encephalitis and Myelitis

The invasion of nervous tissue by bacteria, viruses, fungi, and other agents can produce inflammation of the central nervous system. Inflammation of the brain is called *encephalitis.* Inflammation of the spinal cord is known as *myelitis.* In both conditions there are a variety of possible motor and sensory symptoms including paralysis, coma, and death.

Meningitis

Meningitis is an inflammation of the meninges. It is caused by various microorganisms, the most common being the meningococci, streptococci, pneumococci, and tubercle bacilli. The organisms are thought to enter the body through the nose and throat. Meningitis causes a high fever with a severe headache and stiffness of the neck. In the more serious cases, meningitis can cause coma and death.

Tumors of the Central Nervous System

Tumors within the brain and the spinal cord can produce a variety of dysfunctions, depending on their locations. The symptoms of a tumor can include headaches, convulsions, a change in behavior patterns, pain, and paralysis. These dysfunctions are the result of the destruction of nervous tissue by the tumor, increased pressure within the skull, or edema of the nervous tissue. Tumors may be treated by surgical removal or by chemical and radiation therapy.

THE PERIPHERAL NERVOUS SYSTEM

The **peripheral nervous system (PNS)** consists of (1) an afferent division whose neurons transmit impulses from receptors to the central nervous system, and (2) an efferent division whose fibers transmit impulses from the central nervous system to effectors. Included in the peripheral ner-

vous system are 12 pairs of cranial nerves and 31 pairs of spinal nerves. The cranial nerves arise from the brain and, with the exception of the vagus nerves, they supply structures located in the head and neck regions (F7.21). Table 7.1 is a summary of the cranial nerves and their functions. The spinal nerves are formed from the unions of dorsal and ventral roots that arise from the spinal cord (F7.13). The dorsal roots contain the processes of afferent neurons that pass from the spinal nerves into the spinal cord. The cell bodies of these neurons are located outside the spinal cord within **dorsal root ganglia.** The ventral roots contain the processes of efferent neurons that pass from the spinal cord into the spinal nerves. The cell bodies of these neurons are located within the spinal cord.

AFFERENT DIVISION

The afferent division of the peripheral nervous system is composed of **somatic afferent neurons** and **visceral afferent neurons.** The somatic afferent neurons transmit impulses to the central nervous system from receptors in the skin, fascia, and joints. These impulses frequently reach the conscious level. The visceral afferent neurons transmit impulses to the central nervous system from receptors in the internal organs. Impulses from these receptors generally do not reach the conscious level.

EFFERENT DIVISION

The efferent division of the peripheral nervous system is composed of **somatic efferent neurons** and **visceral efferent neurons.** The somatic efferent neurons transmit nerve impulses from the central nervous system to skeletal muscles. The visceral efferent neurons make up the autonomic nervous system.

The Autonomic Nervous System

The **autonomic nervous system** is composed entirely of visceral efferent neurons that innervate and regulate the activities of cardiac muscle, smooth muscle, and glands (F7.22). This system is normally an involuntary system that operates below the conscious level.

In contrast to somatic efferent pathways, in which a single neuron extends from the central nervous system to the structure innervated, the autonomic pathways that extend from the central nervous system to the effectors are composed of two neurons. One of these neurons, called a **preganglionic** (presynaptic) **neuron,** has its cell body located within the central nervous system. The axon of a preganglionic neuron travels to a ganglion located outside the central nervous system, where it synapses with second neurons, called **postganglionic** (postsynaptic) **neurons.** The axons of postganglionic neurons travel to the various effectors.

Sympathetic and Parasympathetic Divisions. The autonomic nervous system can be separated both structurally and functionally into two divisions, the **sympathetic division** and the **parasympathetic division** (F7.22). The cell bodies of the preganglionic neurons of the sympathetic division of the autonomic nervous system are located within the gray matter of the spinal cord from the first thoracic through the second lumbar segments. For

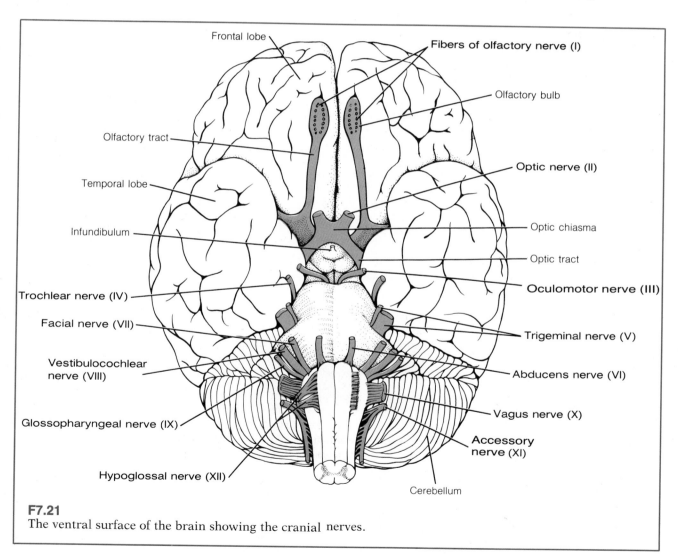

F7.21
The ventral surface of the brain showing the cranial nerves.

Table 7.1 *Summary of the Cranial Nerves*	
Nerves	Basic Functions*
I Olfactory	*Sensory* olfaction (sense of smell)
II Optic	*Sensory* vision
III Oculomotor	*Motor* levator palpebrae superioris and the external eye muscles, except superior oblique and lateral rectus *Proprioception* from the innervated muscles *Parasympathetic* ciliary muscle of the lens and the sphincter of pupil
IV Trochlear	*Motor* superior oblique muscle of the eye *Proprioception* from the superior oblique muscle
V Trigeminal	

Ophthalmic division
    Maxillary division
    Mandibular division
	Sensory cornea, skin of nose, forehead, and scalp
	Sensory nasal cavity, palate, upper teeth, skin of cheek, and upper lip
	Sensory tongue, lower teeth, skin of chin, lower jaw, and temporal regions
	Motor muscles of mastication
	Proprioception from muscles of mastication

VI Abducens	*Motor* lateral rectus muscle of the eye *Proprioception* from lateral rectus muscle
VII Facial	*Motor* muscles of facial expression *Proprioception* from muscles of facial expression *Sensory* taste from the anterior two-thirds of tongue *Parasympathetic* sublingual and submandibular salivary glands, and lacrimal glands
VIII Vestibulocochlear	
Vestibular division Cochlear division	*Sensory* head position and movement *Sensory* hearing
IX Glossopharyngeal	*Motor* pharyngeal muscles *Proprioception* from innervated muscles *Parasympathetic* parotid salivary glands *Sensory* taste from the posterior one-third of the tongue; pharynx; middle-ear cavity; carotid sinus
X Vagus	*Motor* muscles of the pharynx and the larynx *Proprioception* from innervated muscles *Sensory* skin of the external ear; taste from the rear of the tongue; visceral sensory from the thoracic and the abdominal organs *Parasympathetic* organs of the thoracic and abdominal cavities
XI Accessory	*Motor* muscles of the neck *Proprioception* from innervated muscles
XII Hypoglossal	*Motor* intrinsic and extrinsic muscles of the tongue *Proprioception* from innervated muscles

*Proprioception, as used here, refers to the transmission of information from muscle spindles.

this reason, the sympathetic division is also called the *thoracolumbar division*. The ganglia of the sympathetic division are generally located close to the spinal cord (the sympathetic chain ganglia) or halfway between the spinal cord and the innervated organ. The cell bodies of the preganglionic neurons of the parasympathetic division of the autonomic nervous system are located either within the brain or within the gray matter of the spinal cord in the second, third, and fourth sacral segments. For this reason, the parasympathetic division is also called the *craniosacral division*. The ganglia of the parasympathetic division are generally located close to the innervated organs (terminal ganglia).

Parasympathetic preganglionic and postganglionic fibers, as well as sympathetic preganglionic fibers, release acetylcholine—the same transmitter secreted by somatic efferent neurons (F7.23). Therefore, these fibers are called *cholinergic fibers*. In contrast, most sympathetic postganglionic fibers secrete norepinephrine (noradrenaline). Consequently, these fibers are called *adrenergic fibers*. Some postganglionic sympathetic fibers, however, secrete acetylcholine. For example, the sympathetic postganglionic fibers that innervate the sweat glands are cholinergic.

Norepinephrine (as well as the closely related substance epinephrine) is also secreted by the adrenal medulla, an endocrine gland. The adrenal medulla possesses similar physiological and biochemical properties to those of the sympathetic nervous system, and it is innervated by cholinergic preganglionic sympathetic neurons that do not synapse before reaching the gland. Consequently, the adrenal medulla is often considered to be a modified sympathetic ganglion. Norepinephrine has the same effects, whether released into the bloodstream by the adrenal medulla or secreted directly onto an organ by a sympathetic fiber. However, when it is released into the bloodstream, it is carried to all parts of the body, and thus its effects may be more widespread.

Many body organs (for example, the heart, the intestine, and the lungs) are supplied by both the sympathetic and parasympathetic divisions of the autonomic nervous system. In such cases, the two

divisions frequently (but not always) cause opposite responses. Thus, if one division increases the activity of an organ, the other may decrease it. Although most organs are predominantly controlled by one division or the other, the dual innervation of an organ by both divisions of the autonomic nervous system contributes to the precise control of the organ's activity. The effects of sympathetic and parasympathetic stimulation on a number of body organs are summarized in Table 7.2.

There is no generalization that will explain whether sympathetic or parasympathetic stimulation will excite or inhibit a particular organ. However, when viewed in broad terms, parasympathetic stimulation tends to produce responses that are primarily concerned with maintaining bodily functions under relatively quiet conditions. For example, parasympathetic stimulation decreases the heart rate and promotes digestive activities. Sympathetic stimulation, on the other hand, tends to produce responses that prepare a person for strenuous physical activity, as may be required in emergency situations or in situations that lead to aggressive or defensive behavior. In fact, emotional states such as rage or fear are generally accompanied by a widespread activation of the sympathetic division of the autonomic nervous system. This broad sympathetic activity produces a group of responses such as an increased heart rate and a dilation of the bronchii of the lungs that increase the capability of the body to perform vigorous muscular activity. These responses are particularly beneficial to a person who must defend against or flee from a physical threat or challenge and, consequently, they are frequently called "fight or flight" responses.

BIOFEEDBACK

The fact that the autonomic nervous system normally functions below the conscious level implies that a person has no control over the activities governed by this system. However, this is not entirely true.

Normally, a person receives only limited information at the conscious level about what is occurring within the body. For example, blood pressure may fluctuate or brain-wave patterns may change

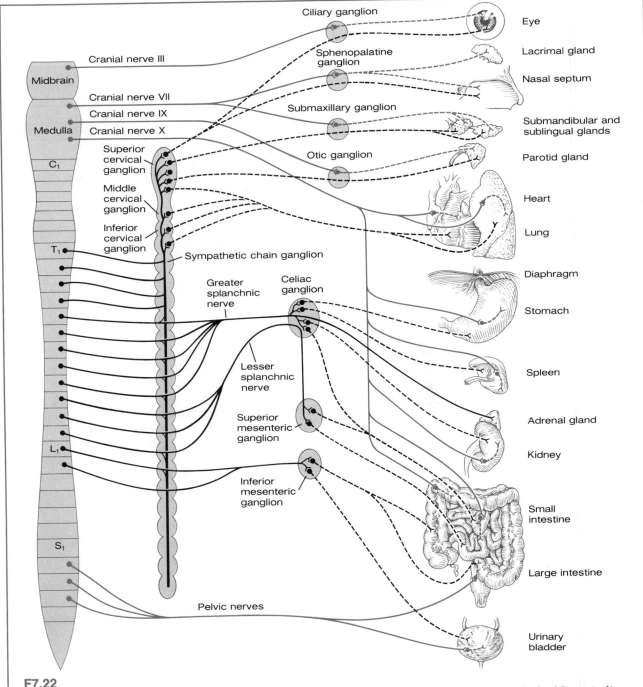

F7.22

The autonomic nervous system. The parasympathetic division is shown in red; the sympathetic division is shown in black. The solid lines indicate preganglionic nerve fibers; the dashed lines indicate postganglionic nerve fibers.

without the person becoming aware of the changes. Consequently, no conscious effort is made to react to or to control such changes. However, the technique of **biofeedback** provides a person with conscious information about body events that usually go unnoticed. Biofeedback utilizes electronic instruments to monitor some of the normally subconscious activities that occur within the body and raise them to the conscious level. The instruments provide information about such events as temperature changes, changes in heart rate, nerve impulse patterns, and so forth. With this conscious knowledge of body activities that had previously been subconscious, it has become possible in some cases for people to learn to control certain of these activities. For example, using biofeedback techniques, people have learned to lower their heart rates, lower their blood pressures, increase the circulation of blood through their limbs, relieve migraine headaches by reducing the blood pressure within the vessels of the head, and control epileptic seizures. Biofeedback, then, shows considerable promise as a self-administered therapeutic technique with broad applications.

DISORDERS OF THE PERIPHERAL NERVOUS SYSTEM

The most common disorders involving neurons of the peripheral nervous system are those associated with damage or inflammation.

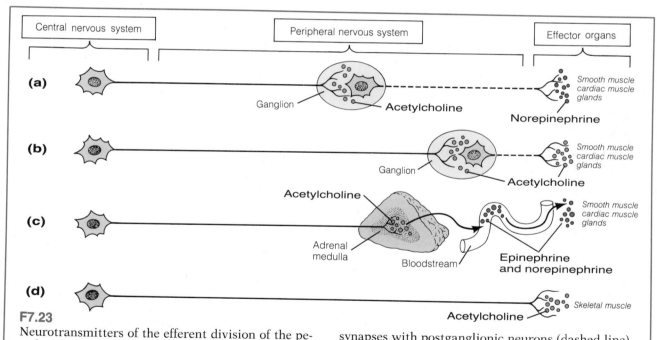

F7.23

Neurotransmitters of the efferent division of the peripheral nervous system. (a) Preganglionic neurons (solid line) of the sympathetic division of the autonomic nervous system release acetylcholine at their synapses with postganglionic neurons (dashed line). Although exceptions occur, the postganglionic neurons release mainly norepinephrine at their junctions with effectors. (b) Preganglionic neurons (solid line) of the parasympathetic division of the autonomic nervous system release acetylcholine at their synapses with postganglionic neurons (dashed line), and the postganglionic neurons also release acetylcholine at their junctions with effectors. (c) The adrenal medulla is supplied by preganglionic sympathetic neurons that release acetylcholine. The adrenal medulla releases epinephrine and norepinephrine into the bloodstream, which carries the secretions to effectors. (d) Somatic efferent neurons release acetylcholine at their junctions with effectors.

Table 7.2 *Effects of the Autonomic Nervous System*

Structure	Effects of Sympathetic Stimulation	Effects of Parasympathetic Stimulation
Heart	Increase rate	Decrease rate
Lungs		
Bronchioles	Dilation	Constriction
Bronchial glands	Possible inhibition of secretion	Stimulation of secretion
Salivary Glands	Secretion of viscous fluid	Secretion of watery fluid
Stomach		
Motility	Decreased	Increased
Secretion	Possible inhibition	Stimulation
Intestine		
Motility	Decreased peristalsis	Increased peristalsis
Secretion	Possible inhibition	Stimulation
Pancreas (Exocrine Portion)		Stimulation of secretion
Liver	Increased release of glucose	
Eye		
Iris	Dilation of pupil (contraction of radial muscles)	Constriction of pupil (contraction of sphincter muscles)
Ciliary muscle	Slight relaxation	Contraction (accommodates for near vision)
Sweat Glands	Stimulation of secretion (cholinergic)	
Adrenal Medulla	Stimulation of secretion (cholinergic preganglionic neurons)	
Urinary Bladder	Relaxation	Contraction
Blood vessels of:		
Skin	Constriction	
Salivary glands	Constriction	
Abdominal viscera	Constriction	
External genitalia	Constriction	Dilation

Injury and Regeneration of Peripheral Nerves

When a peripheral nerve is severely damaged or severed, the portion of the nerve beyond the injury undergoes degenerative changes (F7.24). Within a few days following the injury, the neuronal processes and the myelin sheaths that surround them are broken down by macrophages.

Following the injury, Schwann cells proliferate, forming cords within the connective-tissue tubes that normally surround the nerve fibers. In a matter of days, sprouts form on the stumps of the damaged neurons. Some of these sprouts grow into the connective-tissue tubes, and if there are no obstructions (for example, scar tissue) within the tubes, the neurons may again grow out to the periphery and eventually innervate the structures that were separated from their nerve supply. As the new neuronal processes grow along the connective-tissue tubes, they are surrounded by Schwann cells.

Because neurons regenerate at a rate of from 1 to 4 millimeters per day, it is possible to estimate the length of time that will be required for the nerve supply to return to the denervated structure. The first indication that a peripheral nerve has reached the vicinity of a denervated structure is an improved blood supply to the area. This occurrence is followed by the return of sensory function to the structure. The motor function, in the case of a paralyzed skeletal muscle, is the last to return.

Neuritis

The term *neuritis* means "inflammation of a nerve," but many of the conditions that are referred to as neuritis are more degenerative than inflammatory. Neuritis may be characterized by a range of sensations, from mild tingling to sharp, stabbing pains. It can result from a number of conditions, including mechanical damage to the involved nerves, prolonged pressure on the nerves, vascular disorders that involve the nerves, or invasion of the nerves by pathological organisms.

Neuralgia

Neuralgia refers to attacks of severe pain along the path of a peripheral nerve. There are many vari-

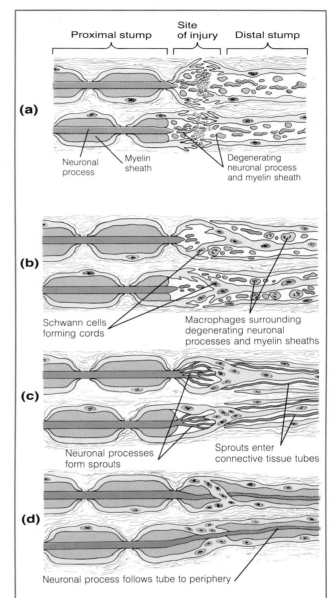

F7.24
Regeneration of a peripheral nerve. (a) Nerve fibers degenerate distal (and a short distance proximal) to the site of the injury. (b) Schwann cells become active, forming cords within the connective-tissue tubes. (c) Neuronal processes then form sprouts that grow into the tubes alongside the Schwann cell cords. (d) Neuronal processes follow the Schwann cell cords. The Schwann cells then wrap around the processes, forming myelin sheaths around the new processes.

eties of neuralgia, most of which have unknown causes. In some types of neuralgia, such as trigeminal neuralgia (tic douloureux), the attacks of pain are short at first, and they may be followed by periods of several weeks with no pain. With time, however, the attacks generally become more frequent and of longer duration.

STUDY OUTLINE

ORGANIZATION OF THE NERVOUS SYSTEM

Central nervous system—brain and spinal cord—is integrative and control center; peripheral nervous system—nerves and ganglia—connects receptors and effectors in outlying parts of body with central nervous system. **p. 151**

THE CENTRAL NERVOUS SYSTEM pp. 151–172

The Brain Major integrative center of the nervous system.

CEREBRUM Largest portion of brain; includes cerebral cortex, basal ganglia, olfactory bulbs and tracts.

Cerebral Cortex Primary motor area involved in controlling voluntary contractions of skeletal muscles; primary sensory area is termination area of pathways carrying sensory information for temperature, touch, pressure, pain. Cortex contains visual, auditory, olfactory, taste, and association areas.

Basal Ganglia Involved in controlling skeletal muscle activity; neurons act in part to inhibit muscle contractions.

Hemispheric Specialization Each cerebral hemisphere is somewhat specialized for carrying out certain kinds of mental processes.

THALAMUS Important relay and integrating center of brain.

HYPOTHALAMUS Involved in regulating body temperature, water balance, appetite, gastrointestinal activity, sexual activity, emotions such as fear and rage, release of pituitary hormones.

LIMBIC SYSTEM A group of brain structures that are particularly important in regulating emotional behavior.

CEREBELLUM Coordinates activities of skeletal muscles.

PONS Connects cerebellum with brain stem; connects upper and lower levels of central nervous system; influences respiration.

MEDULLA OBLONGATA Contains neurons involved in control of several vital functions, including heart rate, respiration, dilation and constriction of blood vessels, coughing, swallowing, and vomiting.

BRAIN STEM AND RETICULAR FORMATION Medulla, pons, and midbrain form brain stem. Reticular formation extends throughout the brain stem and into the thalamus and hypothalamus; it includes a reticular activating system that activates the cerebral cortex.

The Spinal Cord Contains ascending and descending tracts.

THE SPINAL REFLEX ARC Neural pathway by which nerve impulses from receptors reach effectors without traveling to the brain. A spinal reflex arc is composed of at least five components: receptor, afferent neuron, synapse, efferent neuron, effector.

Stretch Reflex Initiated by muscle spindles that respond to stretch; stretch of spindle increases activity of afferent neurons from spindle, which increases stimulation of alpha motor neurons to muscle (for example, the patellar reflex). Muscle spindles provide information about lengths of skeletal muscles that is important in coordination of muscular activity. Activity of muscle spindles may assist in attaining proper degree of muscular contraction.

Tendon Reflex Golgi tendon organs respond to increased tension with increased afferent neuron activity that ultimately inhibits alpha motor neurons to muscle associated with tendon. Golgi tendon organs provide information about tension developed by muscle.

Neuron Pools Functional groups of neurons within central nervous system; divergence, convergence, summation, and facilitation occur within pools; even simple actions require the participation of many neurons whose activities are coordinated in complex circuits in neuron pools.

Meninges, Ventricles, and Cerebrospinal Fluid

Central nervous system is covered by three connective-tissue meninges: pia mater, arachnoid, dura mater; the brain contains four interconnected ventricles that are continuous with central canal of spinal cord. The subarachnoid space, the ventricles, and the central canal of spinal cord contain cerebrospinal fluid secreted by choroid plexuses. Cerebrospinal fluid cushions central nervous system.

Disorders of the Central Nervous System   Disorders that interfere with transmission of nerve impulses to skeletal muscles can cause muscular paralysis; disorders that interfere with impulse transmission from receptors to the brain can cause loss of sensations.

PARKINSONISM   Disorder of basal ganglia characterized by movements that are beyond the control of the individual; related in some way to abnormal dopamine metabolism.

POLIOMYELITIS   Virus destroys cell bodies of somatic efferent neurons to skeletal muscles, causing flaccid paralysis.

MULTIPLE SCLEROSIS   Chronic widespread destruction of myelin sheaths of brain and spinal cord neurons interferes with nerve impulse conduction.

ENCEPHALITIS AND MYELITIS   Encephalitis is brain inflammation; myelitis is spinal cord inflammation.

MENINGITIS   Inflammation of meninges.

TUMORS OF THE CENTRAL NERVOUS SYSTEM
Symptoms include headaches, convulsions, behavior changes, pain, paralysis; tumors may be treated by surgical removal or by chemical and radiation therapy.

THE PERIPHERAL NERVOUS SYSTEM   Includes 12 pairs of cranial nerves and 31 pairs of spinal nerves.
pp. 172–180

Afferent Division   Includes somatic afferent neurons from skin, fascia, and joints, and visceral afferent neurons from internal organs.

Efferent Division   Includes somatic efferent neurons to skeletal muscles, and visceral efferent neurons, which make up autonomic nervous system.

THE AUTONOMIC NERVOUS SYSTEM   Composed entirely of visceral efferent neurons to cardiac muscle, smooth muscle, glands; normally an involuntary system that functions below the conscious level; two neuron (preganglionic and postganglionic) pathways.

Sympathetic and Parasympathetic Divisions   Cell bodies of sympathetic preganglionic neurons are located from first thoracic through second lumbar segments of spinal cord. Cell bodies of parasympathetic preganglionic fibers within brain or within second, third, and fourth sacral segments of spinal cord. Most postganglionic sympathetic fibers secrete norepinephrine, postganglionic parasympathetic fibers secrete acetylcholine. Many organs are innervated by both sympathetic and parasympathetic fibers, which generally cause opposite responses; in general, parasympathetic stimulation tends to produce responses that are primarily concerned with maintaining bodily functions under relatively quiet conditions, and sympathetic stimulation tends to produce responses that prepare a person for strenuous physical activity.

Biofeedback   Technique by which a person made aware of normally subconscious activities can exert some voluntary control over them.

Disorders of the Peripheral Nervous System

INJURY AND REGENERATION OF PERIPHERAL NERVES   Macrophages break down neuronal processes and myelin sheaths distal to injury site; Schwann cells proliferate, forming cords; sprouts form on stumps of neurons; sprouts grow toward periphery; Schwann cells surround new neuronal processes.

NEURITIS   Degenerative or inflammatory condition that causes mild tingling to sharp, stabbing pains.

NEURALGIA   Spasms of severe pain along path of a peripheral nerve.

SELF-QUIZ

1. The white matter of the brain and spinal cord is composed of: (a) unmyelinated nerve fibers; (b) myelinated nerve fibers; (c) nerve cell bodies.
2. Neurons involved in controlling the voluntary contractions of skeletal muscles are located in the: (a) medulla oblongata; (b) hypothalamus; (c) cerebrum.
3. The primary sensory area receives impulses from receptors concerned with: (a) vision; (b) temperature; (c) olfaction.
4. The taste area is located in the parietal lobe. True or False?
5. The hypothalamus of the brain is involved in: (a) the regulation of body temperature; (b) the initiation of voluntary skeletal muscle contractions; (c) the interpretation of nerve impulses from the auditory receptors of the ears.
6. Emotions such as fear and rage are strongly influenced by the: (a) medulla oblongata; (b) hypothalamus; (c) cerebellum.
7. "Pleasure centers" and "punishment centers" of the brain are associated with the: (a) reticular activating system; (b) limbic system; (c) cerebellum.
8. The part of the brain that is an important coordinator of skeletal muscle activity is the: (a) hypothalamus; (b) cerebellum; (c) medulla oblongata.
9. The control of heart rate, coughing, and swallowing is a function of the: (a) thalamus; (b) pons; (c) medulla oblongata.
10. Match the terms associated with the brain with their appropriate descriptions.

    Basal ganglia   Pons
    Cerebral hemisphere   Hypothalamus
    Primary motor area   Ventricles
    Visual area   Reticular formation
    Olfactory area

    (a) This structure is divided into frontal, parietal, temporal, and occipital lobes
    (b) Located in the posterior portion of the occipital lobe
    (c) Connects the cerebellum with the brain stem
    (d) Controls many vital processes, including sexual activity and appetite
    (e) A fluid-filled system of cavities within the brain
    (f) A network of interlacing nerve fibers that activates the cerebral cortex
    (g) Located on the medial surface of the temporal lobe
    (h) Structures involved in controlling skeletal muscle activity in part by inhibiting muscle contraction
    (i) A stimulatory area involved in the voluntary control of skeletal muscle contractions

11. Ascending tracts of the spinal cord transmit impulses up the spinal cord to higher levels within the cord or brain. True or False?
12. Increasing the tension applied to a Golgi tendon organ within a tendon: (a) decreases the activity of the afferent neurons from the tendon organ; (b) decreases the stimulation of alpha motor neurons to the muscle associated with the tendon; (c) increases the contraction of the intrafusal fibers of the tendon organ.
13. Cushioning the central nervous system against jolts and blows is a function of the: (a) pia mater; (b) cerebrospinal fluid; (c) choroid plexuses.
14. Parkinsonism is a dysfunction of the: (a) basal ganglia; (b) brain stem; (c) cerebellum.
15. Nerve impulses are transmitted to the spinal cord through the: (a) dorsal roots of spinal nerves; (b) ventral roots of spinal nerves; (c) neurons of the autonomic nervous system.
16. The autonomic nervous system normally is a voluntary system that functions at the conscious level. True or False?
17. The autonomic nervous system has both motor and sensory functions. True or False?
18. The efferent pathways of the autonomic nervous system, which run from the central nervous system to the effectors are composed of how many neurons? (a) 1; (b) 2; (c) 4.
19. Most sympathetic postganglionic fibers secrete: (a) norepinephrine; (b) acetylcholine; (c) serotonin.
20. The adrenal medulla causes responses that are similar to those caused by the parasympathetic division of the autonomic nervous system. True or False?
21. It is possible for a peripheral nerve fiber that has been severed to regenerate and innervate the same structures that it supplied before it was damaged. True or False?

THE SPECIAL SENSES 8

The body possesses specialized receptors for vision (sight), hearing (audition), head position and movement (the labyrinthine or vestibular sensations), smell (olfaction), and taste (gustation).

VISION—THE EYE

The **eye** is the sense organ for vision. It is housed in a bony skull cavity called the orbit, which protects much of the eye from physical injury.

MAJOR STRUCTURES OF THE EYE

The eye is essentially a spherical structure composed of three basic coats, or layers: the fibrous tunic, the vascular tunic, and the internal tunic, or retina (F8.1).

Fibrous Tunic

The **fibrous tunic** is the outermost layer of the eye. The posterior five-sixths of the fibrous tunic, which is called the **sclera,** is composed of white colored, dense connective tissue. The sclera protects the inner structures of the eye and helps maintain the shape of the eye. The anterior one-sixth of the fibrous tunic, which is called the **cornea,** is clear and has a greater curvature than the sclera. As light enters the eye, it passes through the cornea.

Vascular Tunic

The **vascular tunic** is located beneath the fibrous tunic. The vascular tunic is composed of the **choroid,** the **ciliary body,** and the **iris.** The choroid, which lines most of the internal region of the sclera, is darkly pigmented and contains many blood vessels. Around the edge of the cornea the choroid forms the ciliary body, which contains smooth muscles called ciliary muscles. The anterior portion of the vascular tunic is the iris. The iris, which is basically a continuation of the choroid, is a thin, pigmented diaphragm that contains smooth muscles arranged in both circular and radial patterns. The pigmentation of the iris is responsible for eye color. In the center of the iris is a rounded opening, the **pupil,** through which light enters the interior regions of the eye.

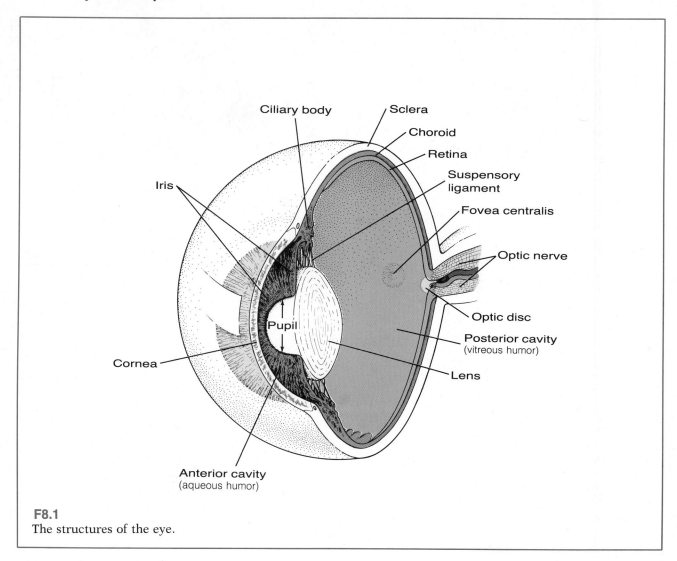

F8.1
The structures of the eye.

Retina

The internal tunic, or **retina,** is the innermost layer of the eye. The retina consists of an outer **pigmented layer** and inner **nervous-tissue layers.** The pigmented layer is composed of a single layer of epithelial cells that lie in contact with the choroid. Both the pigmented layer of the retina and the choroid contain the brown-black pigment melanin. The dark pigmentation of these structures reduces the reflection of light that enters the eye.

The nervous-tissue layers of the retina contain the actual receptors for light—photoreceptors called **rods** and **cones**—as well as numerous neural interconnections. After several synapses within the retina, nerve fibers converge and exit through the rear of the eye as the **optic nerve.** At the point of exit, no photoreceptors are present; hence, this area is called the **blind spot** (optic disc) (F8.2). Near the blind spot but closer to the posterior pole of the eye is a depression called the **fovea centralis** (that is, central depression). The fovea, which is the portion of the retina where visual acuity is greatest, does not contain any rods, but it does contain very densely packed cones. When a person looks directly at an object, the image of the object is generally focused on the fovea.

F8.2

Demonstration of the blind spot. Close your left eye, and focus your right eye on the X. Move the page toward or away from you until the black spot disap- | pears. When this occurs, the image of the spot has fallen on the blind spot (the optic disc) of the retina.

Lens

Behind the pupil is a clear **lens,** which is held in position by a fibrous **suspensory ligament** that attaches to the ciliary body. The lens is a relatively elastic, biconvex structure that can undergo changes of shape. The lens aids in focusing light on the retina.

Cavities and Humors

The lens separates the interior of the eye into two cavities. In front of the lens is an **anterior cavity,** and behind the lens is a **posterior cavity.** The anterior cavity contains a clear fluid, the **aqueous humor,** and the posterior cavity contains a transparent, semifluid, jellylike substance, the **vitreous humor.** The aqueous humor is produced at a rate of about 5 to 6 milliliters per day by folds called ciliary processes, which project from the ciliary body. The aqueous humor drains into a canal near the junction of the cornea and iris and eventually reaches the bloodstream. The rates of production and removal of aqueous humor are such that a relatively constant pressure of about 15 millimeters of mercury (the intraocular pressure) is maintained within the eye.

LIGHT

Light is the small portion of the electromagnetic spectrum to which the receptors of the eye are sensitive (F8.3). Light possesses both particlelike and wavelike properties. From a particlelike viewpoint, light is propagated in small, discrete packets of energy called photons. The travel of light, however, is often described in terms of wavelike properties such as wavelengths and frequencies (F8.4).

The eye is stimulated by those wavelengths of electromagnetic radiation that range from about 400 to 700 nanometers (1 nanometer = one-billionth of a meter) (F8.5). Within this range, the eye is able to distinguish an immense number of different wavelengths that are interpreted by the brain as different colors. The shorter wavelengths are interpreted as blues and violet, and the longer wavelengths are sensed as oranges and reds. A mixture of all the wavelengths of light is interpreted as white, and the absence of light is black.

OPTICS

Light rays traveling in a straight line through a uniform medium of a given optical density (such

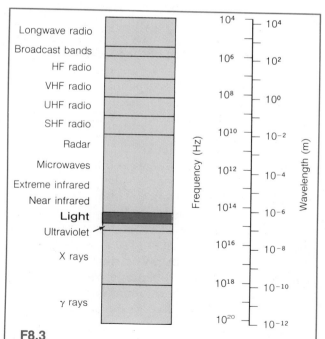

F8.3

The electromagnetic spectrum. The eye is sensitive to the limited range of wavelengths called light. Frequency is measured in cycles per second (Hz), and wavelength is given in meters (m).

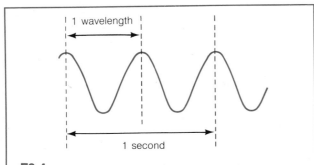

F8.4
Wavelike properties. The distance between successive peaks is the wavelength. The number of peaks that pass a fixed point in a given time—usually 1 second—is the frequency.

as air) travel at a uniform speed. If the rays enter a second medium with a different optical density (such as glass), their speed is altered. Moreover, if they strike the surface of the second medium at an angle other than perpendicular to the medium's surface, they are bent (refracted) (F8.6). The greater the deviation from the perpendicular at which the rays strike the second medium, the greater the bending. If the optical density of the second medium is greater than that of the first (for example, when light rays traveling in air strike glass), the rays are bent toward the perpendicular. If the optical density of the second medium is less than that of the first (for example, when light rays traveling through glass strike air), the rays are bent away from the perpendicular.

Suppose that light rays traveling through air from a point of light strike a spherical, biconvex glass lens (F8.7). Rays from the point of light that strike the lens at angles other than the perpendicular are bent as they enter and leave the lens. Moreover, light rays striking the periphery of the lens are bent more than rays striking the central areas. This difference in bending occurs because rays striking the peripheral areas enter and leave the lens at angles more divergent from the perpendicular than rays striking the central areas. As a result of this differential refraction, the rays converge and come to a focus at a single point behind the lens called the focal point. The distance from the lens to the focal point increases as the curva-

ture of the lens diminishes, and also as the distance from the light source to the lens diminishes.

If the light source is an object instead of a point, an image of the object forms behind the lens. The image is inverted (upside down), and its right and left are reversed. Usually, the image is reduced in size, and the farther away the object, the smaller its image (F8.8).

FOCUSING OF IMAGES ON THE RETINA

When light rays strike the cornea, they are bent (refracted) in much the same way as light rays striking the lens in the previous example. In fact, when light rays enter the eye, the greatest bending occurs when the rays pass into the cornea. Additional bending occurs as the light rays pass from the cornea to the aqueous humor, and also when the rays enter and leave the lens. The lens carries out the focusing adjustments required to ensure that images are formed sharply and clearly on the retina.

Emmetropia

The intraocular pressure exerted by the fluid within the eye forces the walls of the eye apart. Because

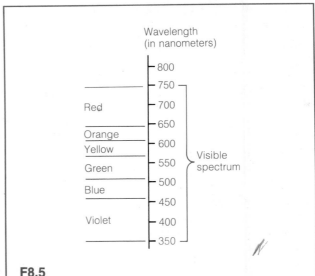

F8.5
The visible portion of the electromagnetic spectrum.

the lens is attached to the walls by way of the suspensory ligament that connects to the ciliary body, it is under tension. Since the lens is elastic, the tension flattens it somewhat, giving it a less

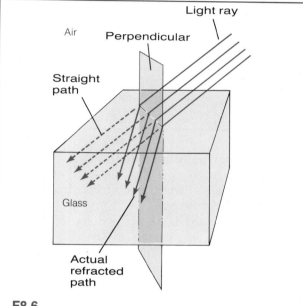

F8.6
Bending of light rays as they travel from air into glass. Note that in this case the light rays are bent toward the perpendicular because the optical density of the glass is greater than that of the air.

pronounced curvature than it might otherwise assume. In this condition, the normal or **emmetropic** eye focuses sharply on the retina parallel light rays from distant points (light rays from points farther than 20 feet away are considered parallel). Divergent light rays that enter the eye from points closer than 20 feet are focused behind the retina.

Accommodation

The focusing system of the eye must adjust in order to focus light rays from objects closer than 20 feet on the retina. The adjustment process is referred to as **accommodation,** and it is accomplished by the ciliary muscles of the ciliary body, which contract in response to parasympathetic stimulation. The contraction of the ciliary muscles pulls the ciliary body slightly forward and inward, narrowing the ring of the ciliary body. This action lessens the tension on the suspensory ligament and permits the elastic lens to assume a more pronounced curvature. The greater curvature results in a greater bending of light rays that strike the lens at angles other than the perpendicular, and it allows the focusing on the retina of divergent light rays from points closer than 20 feet. Under such conditions, light rays from distant points are focused in front of the retina.

Information that reaches the visual association areas of the occipital cortex from the retina ap-

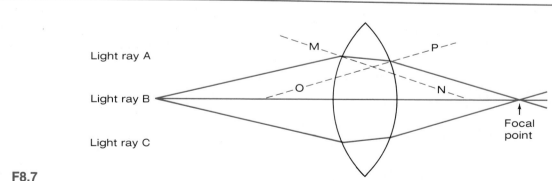

F8.7
Bending of light rays from a point of light as they strike a biconvex lens. Line M-N is the perpendicular for ray A as it enters the lens. Line O-P is the perpendicular for ray A as it leaves the lens. Note that ray B, which strikes the very center of the lens perpendicular to its surface, is not bent.

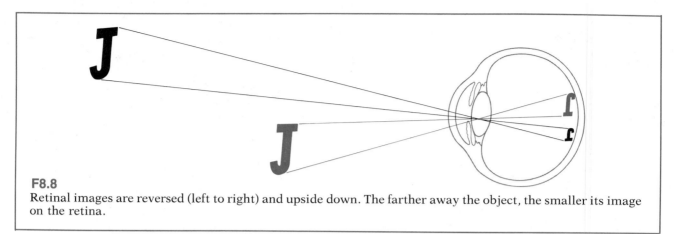

F8.8
Retinal images are reversed (left to right) and upside down. The farther away the object, the smaller its image on the retina.

pears to be important in accommodation. When the image on the retina is out of focus, the visual association areas transmit signals to the ciliary muscles to alter the curvature of the lens and focus the image on the retina. These focusing adjustments usually require less than a second.

During accommodation, the pupil constricts, eliminating the most divergent light rays that would otherwise pass through the most peripheral portions of the lens. The elimination of these rays is beneficial because even a good lens may not be perfect, and light rays that pass through the peripheral regions of the lens do not always focus at exactly the same point as those that pass through the central areas. Thus, the constriction of the pupil aids in the formation of a sharp image on the retina (and also reduces the amount of light that enters the eye).

Another event associated with accommodation is the convergence of the eyes. That is, the eyes turn inward when viewing close objects so that the image falls on the fovea of each retina, which is the portion of the retina where visual acuity is greatest.

Near and Far Points of Vision

The closer an object is to the eye, the more divergent the light rays from the object, and the greater the accommodation required to focus the light rays on the retina. However, the ability of the eye to accommodate for viewing near objects is limited, and the distance from the eye to the nearest point whose image can be focused on the retina is called the **near point of vision.** The near point of vision varies with age. It is close to the eye in youth and recedes farther and farther from the eye as a person gets older.

The **far point of vision** is the distance from the eye to a point whose image can be focused on the retina without accommodation. The far point of vision for the normal emmetropic eye is infinity (any point beyond 20 feet).

CONTROL OF EYE MOVEMENTS

Six straplike skeletal muscles, the extrinsic eye muscles, extend from the orbit to the connective tissue of the eye (F8.9). The extrinsic eye muscles are responsible for eye movements, and they are among the most rapid acting and precisely controlled skeletal muscles in the body. The motor units of the extrinsic eye muscles each contain only a few muscle fibers, and relatively high frequencies of stimuli are required to produce tetanus (350 stimuli per second versus 100 stimuli per second for some other skeletal muscles).

The movements of the eyes are synchronized with one another so that both eyes move in a coordinated manner when viewing an object. When a person wants to look at something, he or she moves the eyes voluntarily to find the object and fix upon it. These *voluntary fixation movements* are controlled by a small area of the cerebral cortex located bilaterally in the premotor regions of the frontal lobes. Once the object is found, *involuntary fixation movements* keep the image of the object fixed on the foveal portion of the retina. The in-

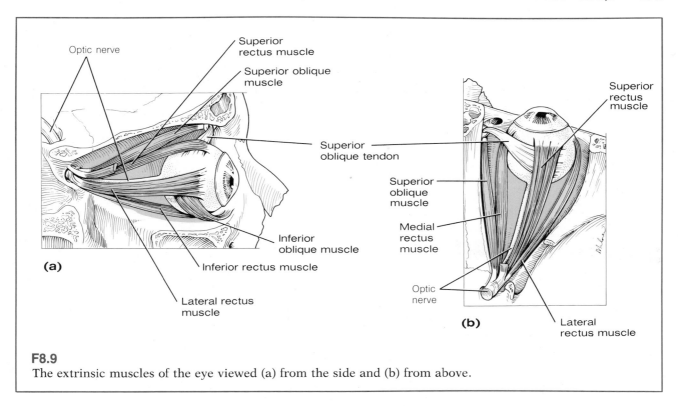

F8.9
The extrinsic muscles of the eye viewed (a) from the side and (b) from above.

voluntary fixation movements are controlled by the occipital region of the cortex, particularly the visual association areas. The visual fixation mechanisms depend on feedback from the retina to keep the image of an object on the fovea. For example, each time the image drifts to the edge of the fovea, sudden flicking movements of the eye occur. These movements, which are automatic reflex responses controlled by the involuntary fixation mechanisms, prevent the image from leaving the fovea.

BINOCULAR VISION AND DEPTH PERCEPTION

Humans have **binocular vision**—that is, the eyes view portions of the external world that overlap considerably with one another. However, the visual field of the left eye is not identical with that of the right eye (F8.10). Moreover, the two eyes do not form exactly identical images of an object because they occupy slightly different locations. This *retinal disparity,* as it is called, contributes to a person's ability to judge relative distances when

objects are nearby (that is, closer than about 70 meters). With only one eye, depth perception depends partly on the use of learned cues such as the fact that the sizes of objects in the environment appear to diminish with distance.

Diplopia (Double Vision)

Light rays that come from an object strike the retinas of both eyes. However, a person normally perceives only one object and not two. This is because the retinas possess *corresponding points* that, when stimulated, result in the perception of a single object. Normally, when an individual looks at an object, the eyes move in such a manner that light rays from the object fall on corresponding points of the foveas of the retinas. However, if the extrinsic eye muscles are weak or paralyzed, or if the muscles of the two eyes act in an unequal or uncoordinated fashion, light rays from an object may not fall on corresponding points of the retinas. As a result, the individual may develop **diplopia** (double vision) and perceive two objects instead of one. Transient diplopia can develop in acute alcoholic intoxication due to the partial paralysis of the extrinsic eye muscles.

Strabismus

In the condition known as **strabismus** (squint, cross-eyedness) the movements of the two eyes are not properly coordinated with one another when a person looks at an object. In some cases of strabismus, the eyes alternate in fixing on an object. In others, the same eye is used all the time. In such cases, the image formed by the unused eye can eventually become suppressed and, if the condition is not corrected, the repressed eye may become functionally blind.

PHOTORECEPTORS OF THE RETINA

The outermost portion of the nervous-tissue layers of the retina (the portion closest to the pigmented layer of the retina and the choroid) contains the light-sensitive photoreceptors: the **rods** and **cones** (so named because of their microscopic appearance) (F8.11). The rods and cones contain photopigments whose configurations are altered when light strikes them and they absorb the light. These alterations lead to changes in the polarity of the photoreceptors, and they ultimately result in the transmission of neural signals from the retina to the brain that are interpreted as visual events.

There are four different photopigments, each consisting of a protein called an *opsin* to which a chromophore molecule called *retinal* is attached. The opsins differ from pigment to pigment and confer specific light-sensitive properties on each pigment. Retinal is produced from vitamin A_1, and retinal and vitamin A_1 can be interconverted. The pigmented layer of the retina contains stores of vitamin A_1.

When light of the proper wavelength strikes a photopigment, the chemical configuration of the

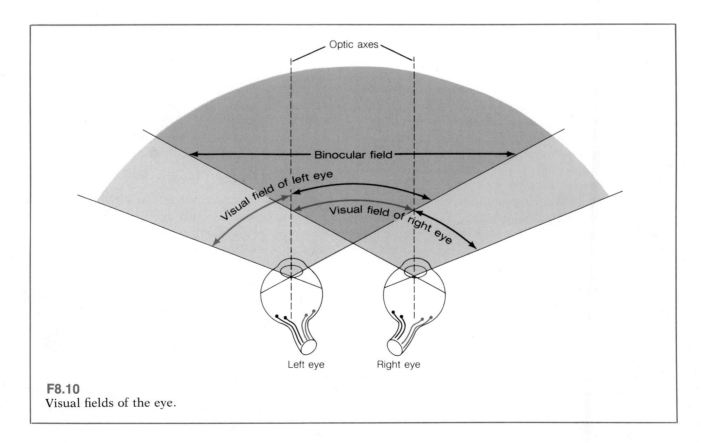

F8.10
Visual fields of the eye.

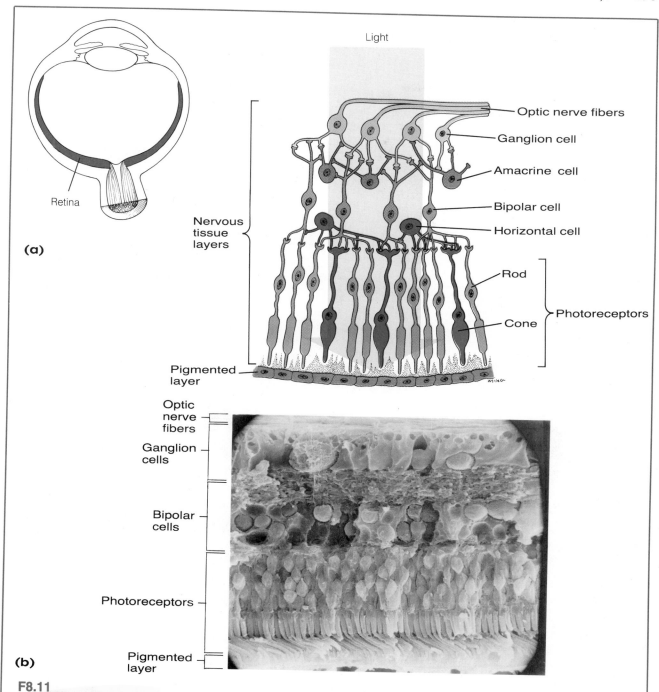

Light

Optic nerve fibers

Ganglion cell

Amacrine cell

Bipolar cell

Horizontal cell

Rod

Cone

Photoreceptors

Nervous tissue layers

Pigmented layer

Retina

(a)

Optic nerve fibers

Ganglion cells

Bipolar cells

Photoreceptors

Pigmented layer

(b)

F8.11
The structure of the retina: (a) Schematic view. (b) Scanning electron micrograph. (From *Tissues and Organs: A Text Atlas of Scanning Electron Microscopy* by Richard G. Kessel and Randy H. Kardon, W. H. Freeman and Company, Copyright © 1979).

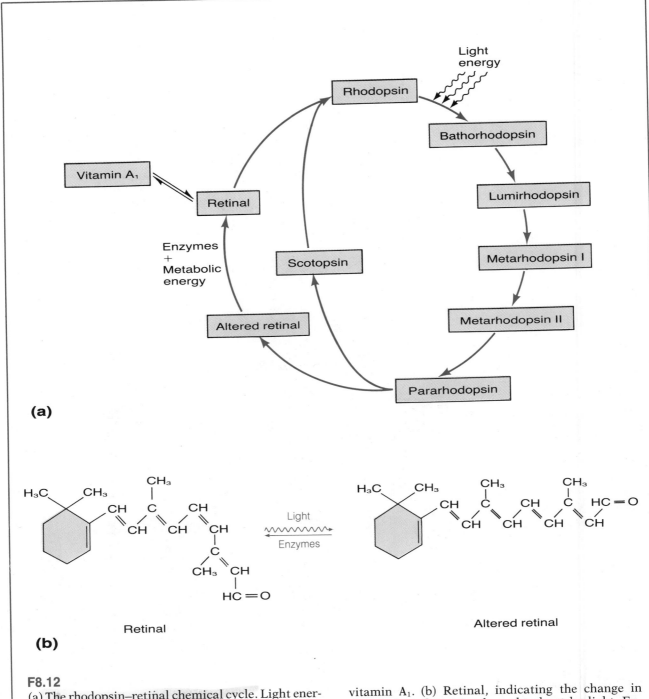

(a)

(b)

Retinal

Altered retinal

F8.12

(a) The rhodopsin–retinal chemical cycle. Light energy alters the chemical configuration of the retinal of rhodopsin, leading to the formation of bathorhodopsin. Note that retinal can be formed from vitamin A₁. (b) Retinal, indicating the change in chemical configuration brought about by light. Enzymes and metabolic energy are required to convert altered retinal back to retinal.

retinal is altered, and the retinal breaks away from the opsin. After such a light-induced breakdown of the photopigment, the altered retinal molecule is rearranged and rejoined to the opsin to restore the photopigment.

Rods

The photopigment contained within the rods is called *rhodopsin*. Rhodopsin is composed of retinal and the opsin scotopsin. Vitamin A_1 absorbed by the rods is converted to retinal, which combines with scotopsin to form rhodopsin. If the eyes are not exposed to light, the concentration of rhodopsin can build up to a very high level.

When light strikes the rods, the chemical configuration of the retinal is altered, and the altered retinal begins to pull away from the scotopsin, forming a substance called bathorhodopsin (prelumirhodopsin) (F8.12). Bathorhodopsin is very unstable; it quickly decays into lumirhodopsin, which in turn rapidly decomposes into metarhodopsin I. Metarhodopsin I spontaneously forms metarhodopsin II, which in turn becomes pararhodopsin. Pararhodopsin then splits into altered retinal and scotopsin. Following these events, the altered retinal is rearranged and rejoined to scotopsin to form rhodopsin. The rearrangement of the altered retinal is enzymatically mediated and requires metabolic energy.

The rods are very light-sensitive, and they can respond to low levels of illumination such as those present at night or in dimly lit areas. The responses of the rods indicate degrees of brightness but they do not indicate color, and rod responses are interpreted only in shades of gray.

Cones

The chemical events that occur when light strikes the cones are similar to those that take place when light strikes the rods. However, there are three different types of cones. Each type contains a different photopigment and each is selectively sensitive to particular wavelengths of light (F8.13). Red cones respond more intensely than the others to those wavelengths of light that the brain interprets as red. Green cones respond more intensely than the others to those wavelengths of light that the brain interprets as green. Blue cones respond

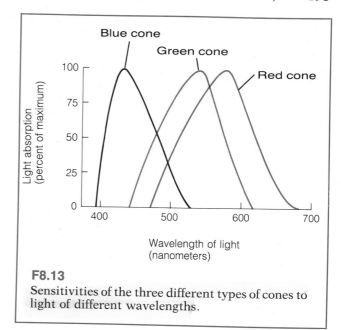

F8.13
Sensitivities of the three different types of cones to light of different wavelengths.

most intensely to those wavelengths of light that the brain interprets as blue or violet.

The cones are responsible for color vision. The ability to perceive many different colors rather than only three colors (red, green, and blue) is due to the fact that different wavelengths of light striking the retina evoke different ratios of response from the three cone types. These varied responses are interpreted by the brain as a tremendous variety of different colors. For example, when light with a wavelength of 580 nanometers strikes the retina, the red cones respond more intensely than the green cones, and the blue cones do not respond at all. The brain interprets this pattern of responses as the color yellow.

The cones operate only at relatively high levels of illumination, and they are the principal photoreceptors during daylight or in brightly lit areas. The cones are concentrated in the center of the retina and most highly in the fovea, whereas the rods are more numerous in the peripheral retina.

NEURAL ELEMENTS OF THE RETINA

The portions of the nervous-tissue layers of the retina closest to the interior of the eye are composed of neural elements (F8.11). The middle portion of the nervous-tissue layers contains **bipolar cells,** and the innermost portion contains **ganglion cells.** The rods and cones synapse with bipolar cells, which in turn synapse with ganglion cells.

In general, the neural pathways that transmit the responses of the rods display greater convergence than the pathways that transmit the responses of the cones. For example, in the peripheral regions of the retina many rods synapse with a single bipolar cell, and many bipolar cells synapse with a single ganglion cell. In contrast, within the fovea there are approximately equal numbers of cones, bipolar cells, and ganglion cells, and relatively little convergence occurs. Consequently, the cones in the fovea provide more precise information about the area of the retina stimulated than do the rods in the peripheral retina. As a result, vision is more precise in the fovea than in the peripheral retina. On the other hand, the convergence that occurs along the visual pathways of the rods provides numerous opportunities for spatial summation, and this, together with the differences in light sensitivity of the rods and cones themselves, contributes to the fact that rod vision is more effective in dim light than cone vision.

In addition to bipolar cells and ganglion cells, the retina contains neural cells called **horizontal cells** and **amacrine cells** (F8.11). These cells are important in lateral interactions that occur within the retina. For example, when rods or cones respond to light, their responses are transmitted to horizontal cells as well as to bipolar cells. The horizontal cells act in an inhibitory fashion on bipolar cells adjacent to the bipolar cells directly affected by the rods and cones. This *lateral inhibition* enhances the signal contrast between strongly stimulated regions of the retina and adjacent, more weakly stimulated regions.

VISUAL PATHWAYS

The axons of the retinal ganglion cells come together and leave the eye as the **optic nerve.** The two optic nerves (one from each eye) meet at a structure called the **optic chiasma,** located just anterior to the pituitary gland (F8.14). Within the optic chiasma, ganglion cell axons from the medi-

al half of each retina cross to the opposite side, and those from the lateral half of each retina remain on the same side. From the optic chiasma, the axons continue as the **optic tracts.** Thus, the left optic tract consists of ganglion cell axons from the lateral half of the retina of the left eye and the medial half of the retina of the right eye (the right portion of the visual field of each eye), and the right optic tract consists of axons from the lateral half of the retina of the right eye and the medial half of the retina of the left eye (the left portion of the visual field of each eye).

Most of the ganglion cell axons within the optic tracts travel to the lateral geniculate bodies of the thalamus where they synapse with neurons that form pathways called optic radiations, which terminate in the visual cortex of the occipital lobes. Some of the axons within the optic tracts travel to midbrain nuclei called pretectal nuclei where they are involved in a reflex called the pupillary light reflex (see page 199).

PROCESSING OF VISUAL SIGNALS

When discussing the processing of visual signals, the response of a visual pathway cell is commonly described in terms of its receptive field. The *receptive field* of a cell is the area of the retina that, when illuminated, influences the activity of the cell. In essence, it is an area of photoreceptors that provide input to the cell. This input is often provided indirectly by way of other cells, and it can be inhibitory as well as stimulatory. For example, a photoreceptor may provide input to a bipolar cell by way of an inhibitory horizontal cell.

Because of such occurrences as convergence and lateral interactions, the processing of visual signals begins in the retina, and the ganglion cells do not transmit to the brain a simple mosaic pattern of the image on the retina. The receptive field of a ganglion cell is circular and is composed of two parts: a small central area and a ringlike peripheral area (F8.15). In the dark, a ganglion cell continually transmits nerve impulses to the brain. However, when the receptive field of a ganglion cell is illuminated, the frequency of nerve impulse transmission changes. In one type of ganglion cell, called an on-center cell, positioning a spot of light on the central area of a cell's receptive field increases the frequency of nerve impulse transmission. Conversely, illuminating the peripheral area decreases the frequency of nerve impulse transmission. In a second type of ganglion cell, called an off-center cell, positioning a spot of light on the central area of a cell's receptive field decreases the frequency of nerve impulse transmission, and illuminating the peripheral area increases the frequency of nerve impulse transmission. In either cell type, nerve impulse frequency varies with the degree of illumination of the central area of a cell's receptive field compared to the degree of illumination of the peripheral area. That is, when both areas are illuminated simultaneously, the activity of the ganglion cell is a reflection of both stimulatory and inhibitory influences. Consequently, the signals transmitted to the brain by a ganglion cell indicate such things

as the light level in one small area of the visual scene compared to the average illumination of the immediately surrounding region.

The neurons of the lateral geniculate bodies of the thalamus that transmit nerve impulses to the visual cortex have receptive fields that are similar to those of ganglion cells—that is, their receptive fields are circular with either an excitatory central area and an inhibitory peripheral region or vice versa.

In the visual cortex there is a hierarchy of cells with receptive fields that vary widely in organization. Some cortical cells have circularly symmetrical receptive fields, and their responses resemble those of ganglion cells and lateral geniculate neurons. However, the receptive fields of other cortical cells are organized so the cells respond best not to spots of light but to specifically oriented line segments (such as narrow slits of

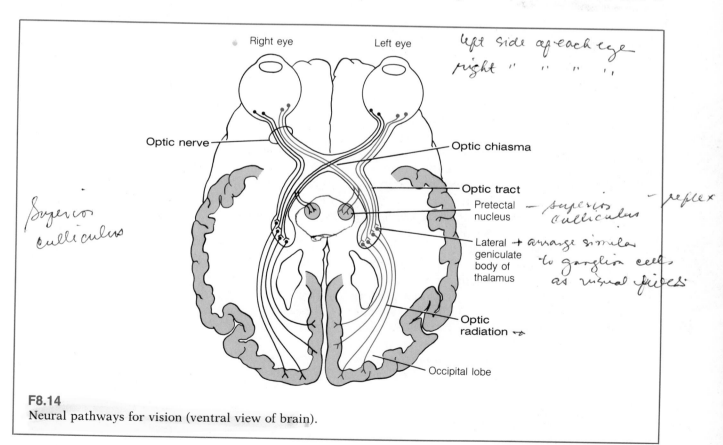

F8.14
Neural pathways for vision (ventral view of brain).

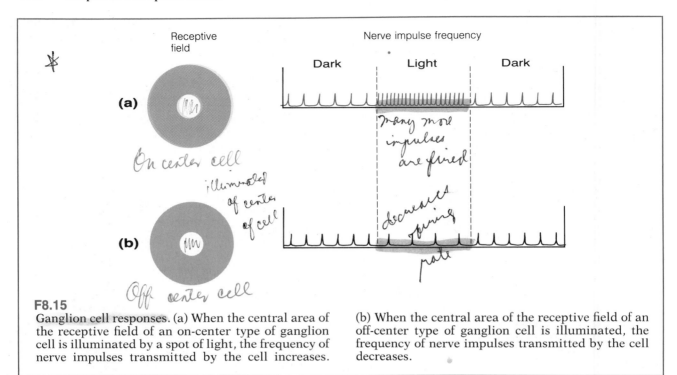

F8.15

Ganglion cell responses. (a) When the central area of the receptive field of an on-center type of ganglion cell is illuminated by a spot of light, the frequency of nerve impulses transmitted by the cell increases.

(b) When the central area of the receptive field of an off-center type of ganglion cell is illuminated, the frequency of nerve impulses transmitted by the cell decreases.

light) that are positioned on the retina. For example, cells called simple cells respond only when the line is oriented at a precise angle in a particular portion of a cell's receptive field (F8.16a). The responses of these cells provide information about lines and borders. Other cells, called complex cells, respond when the line is oriented at a precise angle regardless of its position in a cell's receptive field (F8.16b). The responses of these cells help indicate the movement of a visual stimulus. Thus, the processing of visual signals that begins in the retina continues in other parts of the visual pathway. As a result, the components of the pathway transmit a coded message about certain aspects of the image on the retina rather than a mosaic pattern of the image. The message contains information about characteristics of the image such as contrast, movement, and color, and it is this information that the brain utilizes in developing a visual representation of the image.

LIGHT AND DARK ADAPTATION

When a person moves from a dimly lit area to a brightly lit area, the person's eyes often feel un-comfortable, and vision is poor for several minutes. With time, however, **light adaptation** occurs, and vision improves. Similarly, when a person moves from a well-lit area to a dimly lit area, the person cannot see well initially. However, as time passes, **dark adaptation** occurs, and vision gets better. Several mechanisms contribute to light and dark adaptation.

Photopigment Concentration

Relatively small changes in the concentration of rhodopsin within the rods can greatly influence their sensitivity to light. For example, with a decrease in rhodopsin content of only 0.6% from the maximum, rod sensitivity declines about 3000 times. A similar situation is believed to exist with respect to the cones and their photopigments.

When a person moves from a dimly lit area to a brightly lit area, significant amounts of photopigments (particularly rhodopsin) are broken down by the light. This occurrence, which requires about five to ten minutes for completion, decreases the sensitivity of the eyes and raises the visual threshold so that a greater intensity of light is required for effective vision.

When a person moves from a well-lit area to a dimly lit or dark area, the photopigments of the rods and cones that were broken down in bright light are regenerated. The regeneration of the photopigments lowers the visual threshold and increases the sensitivity of the eyes to light. Although this process can continue for hours, a considerable increase in sensitivity occurs within 20 to 40 minutes. The increase in visual sensitivity that occurs during the first ten minutes or so is relatively slight, and has been attributed primarily to the cones. This increase is followed by a much greater increase in sensitivity that is due to the more slowly adapting rods, which are primarily responsible for vision in dim light.

Pupillary Light Reflex

The **pupillary light reflex** constricts (reduces the diameter of) the pupils in bright light and helps regulate the amount of light entering the eyes. In this reflex, light striking the retina initiates neural signals that are sent to the pretectal nuclei of the midbrain. These signals ultimately lead to an increased stimulation of the circular smooth muscles of the iris by neurons of the parasympathetic division of the autonomic nervous system. The contraction of the smooth muscles constricts the pupils. In dim light, the stimulation of the circular smooth muscles decreases, and the pupils dilate (increase their diameter).

Neural Adaptation

Neurons of the visual pathways in the retinas undergo rapid neural adaptation. When a person moves from a dimly lit area to a brightly lit area, the signals transmitted by the bipolar cells, horizontal cells, amacrine cells, and ganglion cells are initially very intense. However, within a fraction of a second, the intensity of the signals declines severalfold.

VISUAL ACUITY

Visual acuity is the ability of the eye to distinguish detail. It is related to resolving power, which is the ability to distinguish two closely spaced points as separate. For two closely spaced points to be distinguished as separate, it is believed that light rays entering the eye from one of the two points must stimulate a different receptor unit in the retina than light rays from the other point, and that there must be at least one unstimulated or only weakly stimulated unit between the two receptor units (F8.17). Whether or not this pattern of

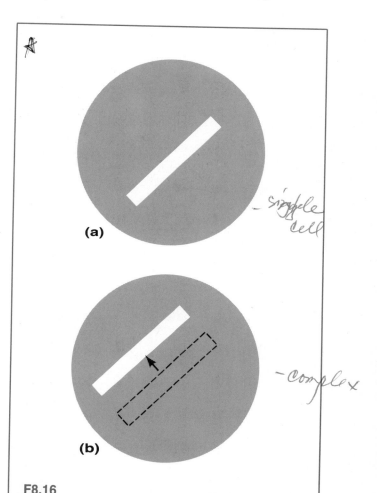

F8.16
Responses of cells of the visual cortex. (a) A simple cell responds when a line is oriented at a precise angle in a particular portion of the cell's receptive field. (b) A complex cell responds when the line is oriented at a precise angle regardless of its position in the cell's receptive field. Thus, the cell continues to respond as a properly oriented line moves across its receptive field. (Some complex cells respond better when the line moves in one direction than in the other.)

stimulation occurs depends on both the focusing system of the eye and on the area of the retina that the light rays strike. Light rays that enter the eye from two points have the greatest likelihood of stimulating different receptor units in the fovea where the cones are packed closely together, and each cone has a relatively direct path to the brain. Moreover, in the fovea the inner layers of the retina are generally displaced to one side, allowing light to pass relatively unimpeded to the cones. Therefore, visual acuity is greatest in the region of the fovea.

Visual acuity is often determined with the aid of eye charts, which frequently consist of a series of letters of varying sizes. A person with normal visual acuity should be able to read the leters on the 20-foot line of the eye chart at a distance of 20 feet. If a person can read only the 40-foot line at 20 feet, the person is said to have 20/40 vision (meaning that the person's eyes cannot distinguish beyond 20 feet what normal eyes could distinguish at 40 feet).

VISUAL DISORDERS

Many disorders can affect the eyes. Among the most common are focusing problems.

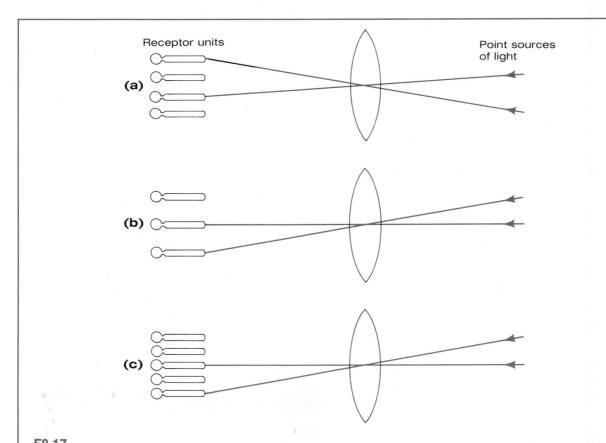

F8.17
Resolving power of the eye. In order for two point sources of light to be discriminated as separate, light rays from one point must stimulate a different receptor unit of the retina than light rays from the second point, with at least one unstimulated or only weakly stimulated unit between the two. The more closely the receptor units are packed, the more likely this will occur. In (a), two widely spaced points may be detected as separate. In (b), the two points will not be discriminated as separate. In (c), two closely spaced points may be detected as separate.

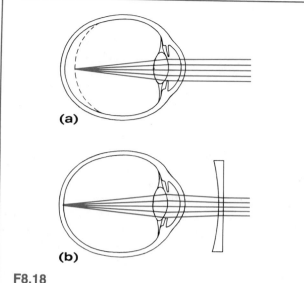

F8.18
Myopia. (a) In myopia, light rays from distant objects focus in front of the retina. (b) Most cases of myopia can be corrected by a concave lens, which causes light rays to diverge as they enter the eye.

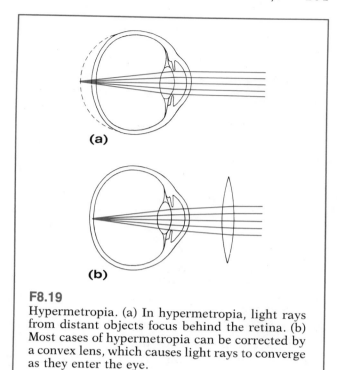

F8.19
Hypermetropia. (a) In hypermetropia, light rays from distant objects focus behind the retina. (b) Most cases of hypermetropia can be corrected by a convex lens, which causes light rays to converge as they enter the eye.

Myopia

Myopia is nearsightedness. In this condition, light rays from distant objects are focused in front of the retina, and only light rays from close objects can be focused accurately on the retina. Myopia is caused by a focusing system that has too great a refractive power with respect to the position occupied by the retina. Most commonly, this is due to an elongated eyeball, which can result from a weakness of the coats of the eye (F8.18). In myopia, the near point of vision is closer than normal and the far point of vision is closer than 20 feet. In all but the severest cases, myopia can be corrected by eyeglasses with concave lenses. These lenses cause light rays to diverge slightly as they enter the eye, helping the refractive system of the eye focus them on the retina.

Hypermetropia

Hypermetropia (hyperopia) is farsightedness. In this condition, light rays from distant objects are focused behind the retina when the eye is at rest, and accommodation is necessary in order to focus the rays on the retina. Since the eyes can accommodate only so much, a hypermetropic individual has trouble viewing close objects. Moreover, accommodation is required to view any object close or far. Thus, in hypermetropia there is no far point of vision, and the near point lies farther away than normal. Hypermetropia is most commonly due to a shortened eyeball (F8.19). The condition can usually be corrected with eyeglasses having convex lenses. These lenses cause light rays to converge as they enter the eye, and thus assist the eye's refractive system in focusing the rays on the retina.

Astigmatism

Previous considerations of the refractive system of the eye have assumed that all elements of the system possess uniformly curved surfaces (much like a marble). Often, however, the surface of one or more elements of the system is not uniformly curved in all planes. For example, the surface of the cornea may have a different curvature in the horizontal plane than in the vertical plane (much

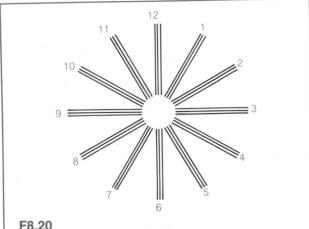

F8.20
Astigmatism may be detected by viewing a series of radiating lines. To astigmatic individuals, some lines appear more distinct than others.

like a chicken's egg). As a result, light rays that enter the eye in different planes are focused at different points, causing an out-of-focus image. The condition that results from unequal curvatures of portions of the eye's refractive system is called *astigmatism*. If an astigmatic individual examines a figure like Figure 8.20, in which straight lines radiate from a central point, some lines are sharply focused on the retina and seen clearly, while others are focused in front of or behind the retina and seen indistinctly. All but very severe cases of astigmatism can be corrected by a lens (eyeglasses) that has a greater curvature in one plane than another. The lens is oriented in front of the eye so that the differential bending of light rays passing through the lens compensates for the differential bending of light rays by the eye.

Cataract

In some cases, the lens of the eye or a portion of it becomes cloudy or opaque so that vision is impaired. This cloudiness or opacity is known as a *cataract* (F8.21). Often lenses with cataracts are removed surgically, and effective vision is restored by the use of special eyeglasses that compensate for the loss of the lens.

Glaucoma

Glaucoma is an abnormal elevation of the intraocular pressure, often resulting from deficient drainage of the aqueous humor. When severe, the high intraocular pressure of glaucoma squeezes shut blood vessels that supply the eye, leading to the degeneration of the retina and blindness.

Color Blindness

Color blindness is a deficiency in color perception that ranges from an inability to distinguish certain shades of color to a complete lack of color perception. The condition is caused by deficiency or absence of one or more of the different cone types involved in color vision. Difficulty in distinguishing reds and greens (red-green color blindness) is the most common form of color blindness. There is a strong hereditary component in color blindness, and about 8% of males but less than 1.5% of females are red-green color blind.

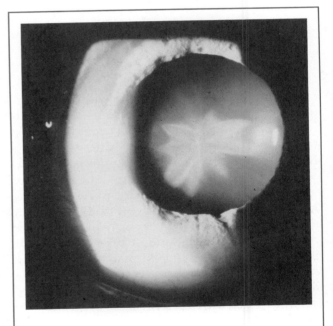

F8.21
Rose petal anterior cataract.

EFFECTS OF AGING ON THE EYE

With aging, the lens of the eye becomes yellowed due to the effects of ultraviolet rays from such sources as sunlight. In addition, the lens is one of the few body structures that exhibits increased cellular growth with age, and this may contribute to the development of cataracts.

As an individual ages, the pupil of the eye is no longer able to dilate fully, and the amount of light that reaches the photoreceptors of the retina by age 70 may be only 50% of the amount that reaches the retina during youth. The inability of the pupil to dilate fully may also contribute to poor drainage of the aqueous humor, resulting in an increased intraocular pressure and a greater likelihood of glaucoma.

The continuous exposure of the rods and cones to light causes damage to their membranes and generates cellular debris. The debris can be removed by cells of the pigmented layer of the retina, and, as long as these cells function properly, no problems occur. However, with aging, these cells may become congested and debris from the rods and cones may accumulate, thus contributing to a loss of visual acuity. This loss is particularly striking if it occurs in the fovea.

With aging, the lens gradually loses its elasticity, and thereby loses some of its ability to change shape during accommodation for the viewing of near objects. As a result, the near point of vision recedes farther and farther from the eye. Although the loss of lens elasticity begins early in life, the greatest loss occurs after about the age of 40 (F8.22). As the ability to accommodate for the viewing of near objects diminishes, it becomes necessary to hold reading materials farther and farther from the eye in order to focus the printed letters on the retina. Ultimately, books and papers may have to be held so far from the eye that the images of the letters on the retina are too small to be recognized. This condition, called *presbyopia*, can be corrected by the use of eyeglasses with convex lenses. The convex lens increases the convergence of the light rays so that the refractive system of the eye can focus them on the retina

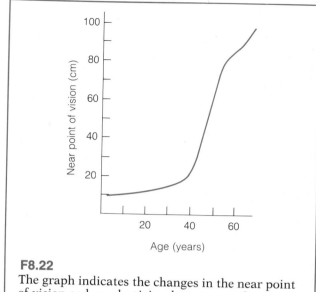

F8.22
The graph indicates the changes in the near point of vision as lens elasticity decreases with age.

when the printed matter is held reasonably close to the eye.

HEARING AND HEAD POSITION AND MOVEMENT— THE EAR

The ear contains receptors for sound as well as receptors that detect head position and movement.

STRUCTURE OF THE EAR

The ear is divided into external, middle, and inner regions (F8.23).

External Ear

The **auricle,** or **pinna,** is the most prominent portion of the **external ear** (F8.23). It consists of an irregularly shaped framework of elastic cartilage that is covered with skin. The auricle directs sound waves into the external auditory meatus.

The **external auditory meatus** (canal) is a curved passageway approximately 2.5 centimeters long that extends from the auricle to the eardrum. The meatus is lined with skin, and near its entrance are fine hairs and sebaceous glands. It also contains modified sweat glands called *ceruminous glands* that secrete cerumen (earwax). The hairs and the cerumen help prevent small foreign objects from reaching the eardrum.

Middle Ear

The **middle ear** is a small, air-filled chamber within the bone of the skull (F8.23). It is separated from the external auditory meatus by the **tympanic membrane** (eardrum), and from the inner ear by a bony wall in which there are two small membrane-covered openings—the **oval window** and the **round window.**

Three **ossicles,** or middle-ear bones, form an articulated (jointed) bridge across the middle-ear chamber between the tympanic membrane and the oval window. The handle-shaped portion of the ossicle called the **malleus** (hammer) attaches to the inner surface of the tympanic membrane. The base of the ossicle called the **stapes** (stirrup) fits against the oval window. The third ossicle, the **incus** (anvil), lies between the malleus and the stapes and articulates with them. The articulations between the ossicles are freely movable joints. When a sound wave (or, more precisely, an air-pressure wave) strikes the tympanic membrane and causes it to vibrate, the lever system formed by the ossicles picks up the vibrations and transmits them to the oval window. Two small muscles attach to the ossicles. The tensor tympani muscle attaches to the handle of the malleus, and the stapedius muscle attaches to the stapes. The tensor tympani muscle (as well as ligaments) pulls the handle of the malleus inward, keeping the tympanic membrane taut.

An **auditory tube** connects the middle-ear chamber with the nasopharynx. This tube, which is closed most of the time in adults, can be opened

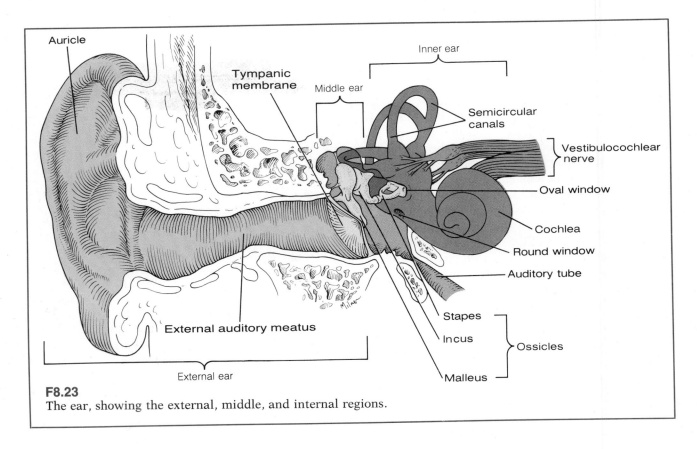

F8.23
The ear, showing the external, middle, and internal regions.

by swallowing or yawning, and it provides a passageway by which the air pressure within the middle-ear chamber is equalized with the atmospheric pressure. If the two pressures were not equal and, for example, the pressure within the middle ear were higher than the atmospheric pressure, the tympanic membrane would bulge outward. Similarly, if the pressure within the middle ear were lower than the atmospheric pressure, the tympanic membrane would bulge inward. Either occurrence would not only be painful but would impair hearing by interfering with the vibration of the tympanic membrane.

Inner Ear

The **inner ear** is located medially to the middle ear, within the bone of the skull. It consists of a series of canals called the **osseous labyrinth,** which are hollowed out of the bone (F8.24). Within the osseous labyrinth, and following its course, is

a **membranous labyrinth.** The membranous labyrinth is a continuous series of ducts that are filled with a fluid called **endolymph.** In some regions, the membranous labyrinth is attached to the osseous labyrinth. However, in most regions, the membranous labyrinth is separated from the osseous labyrinth by a fluid called **perilymph,** which fills all the space in the osseous labyrinth that is not occupied by the membranous labyrinth. The osseous labyrinth is divided into three areas: the vestibule, the semicircular canals, and the cochlea.

The Vestibule. The **vestibule** is a chamber just medial to the middle-ear chamber. The oval window forms a membranous partition between the middle-ear chamber and the vestibule. Vibrations of the oval window induced by the stapes are transmitted to the perilymph of the vestibule. Within the vestibule are two enlargements of the

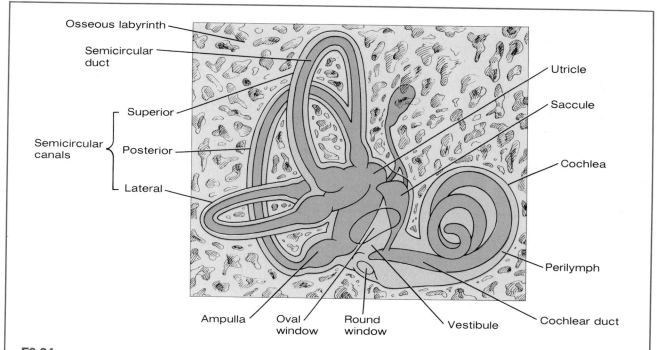

F8.24
Structures of the inner ear. The membranous labyrinth (shown in darker shade of color) is filled with endolymph and is separated from the osseous labyrinth by perilymph.

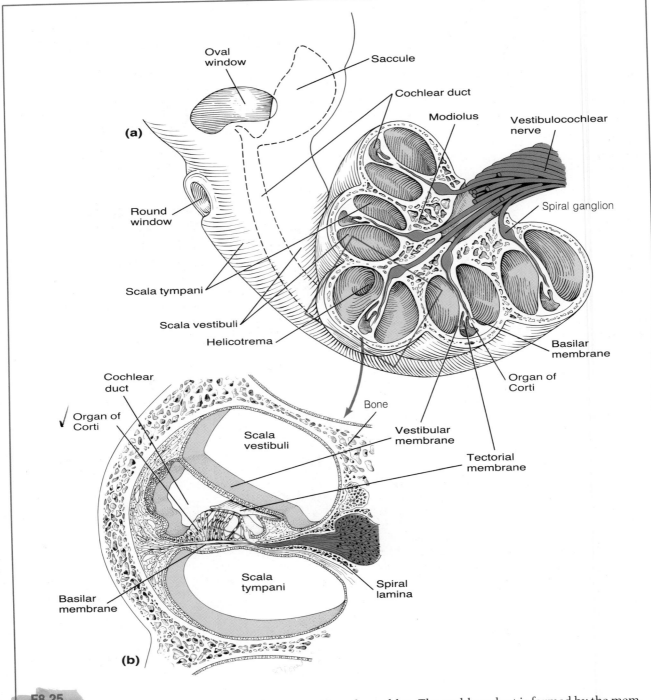

(a) Section through the cochlea. The scala tympani and scala vestibuli are continuous through the helicotrema. (b) Magnified cross section of one turn of the cochlea. The cochlear duct is formed by the membranous labyrinth and contains endolymph. The scala tympani and scala vestibuli contain perilymph.

membranous labyrinth, the **utricle**, and the **saccule.** The utricle (and perhaps the saccule) contains receptor cells that detect the position and movement of the head.

The Semicircular Canals.

Within the inner ear are three bony **semicircular canals** that are arranged at right angles to one another. Each contains a membranous **semicircular duct.** An enlargement of each duct, called an **ampulla,** contains receptor cells that detect certain movements of the head.

The Cochlea.

The portion of the inner ear concerned with hearing is the **cochlea.** It resembles a snail shell, spiraling 2.5 turns around a central bony core called the modiolus (F8.25*a*). A bony shelf called the spiral lamina extends into the cochlea from the modiolus (F8.25*b*). Two membranes, the **vestibular membrane** and the **basilar membrane,** extend across the cochlea from the spiral lamina, dividing the cochlea into three longitudinal tunnels. The central tunnel between the vestibular and basilar membranes is the **cochlear duct.** The cochlear duct is the lumen of the membranous labyrinth of the cochlea and is filled with endolymph. The upper tunnel, the **scala vestibuli,** is separated from the cochlear duct by the vestibular membrane. The lower tunnel, the **scala tympani,** is separated from the cochlear duct by the basilar membrane. The scala vestibuli and scala tympani both contain perilymph, and are continuous with one another at the apex of the cochlea through an opening called the **helicotrema.** The scala vestibuli opens into the vestibule, and its perilymph is continuous with the perilymph of the vestibule that bathes the oval window. The scala tympani ends at the round window. The relationships of different portions of the cochlea are more easily understood if the cochlea is visualized in an uncoiled condition, as shown in Figure 8.26.

The **organ of Corti,** which contains the receptors for hearing, is located on the basilar membrane (F8.26). It consists of a series of receptor cells, called hair cells, and supporting cells (F8.27). The hair cells are innervated by afferent fibers from the cochlear division of the ves-

tibulocochlear nerve (cranial nerve VIII). Overhanging the organ of Corti is a flexible **tectorial membrane,** which is anchored to the spiral lamina. Thin processes (referred to as hairs) that extend from the hair cells of the organ of Corti are in contact with the tectorial membrane.

HEARING

Hearing is the perception of sound.

Sound

Sound travels in the form of pressure waves that are produced when molecules of air are alternate-

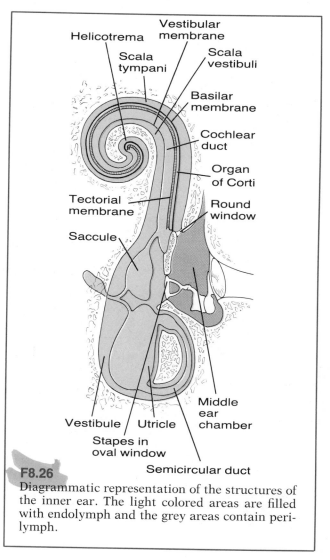

F8.26 Diagrammatic representation of the structures of the inner ear. The light colored areas are filled with endolymph and the grey areas contain perilymph.

ly compressed and rarefied (F8.28). For example, when the arm of a vibrating tuning fork moves in one direction, air molecules ahead of the arm are pushed together, or compressed, and the pressure in the area increases. In turn, air molecules in the area of compression bump into other air molecules ahead of them, pushing these air molecules together and creating a new region of compression. Through many repetitions of this process, the initial compression of air molecules by the arm of the tuning fork can be transmitted a considerable distance, even though individual air molecules move only a short distance. Similarly, when the arm of the tuning fork moves back in the opposite direction, the compression of the air molecules and the pressure are decreased even below normal. These changes, too, are transmitted from air molecule to air molecule. Thus, a sound wave consists of regions of compression, in which the air molecules are close together and the pressure is relatively high, alternating with areas of rar-efaction, where the molecules are farther apart and the pressure is lower.

The intensity or loudness of a sound is related to the amplitude of the sound wave—that is, to the pressure difference between a zone of compression and a zone of rarefaction. The pitch of a sound is related to the frequency of the sound wave (usually expressed in cycles per second or Hertz, Hz). In general, the greater the amplitude of a particular sound wave, the louder the sound, and the higher the frequency of a sound wave, the higher the pitch of the sound.

The human ear is capable of detecting sound waves with frequencies between about 20 and 20,000 cycles per second. However, the ear is most sensitive to sound waves with frequencies between 1000 and 4000 cycles per second. Sound waves with frequencies outside this range must have greater amplitudes than sound waves with frequencies within the range if they are to be detected.

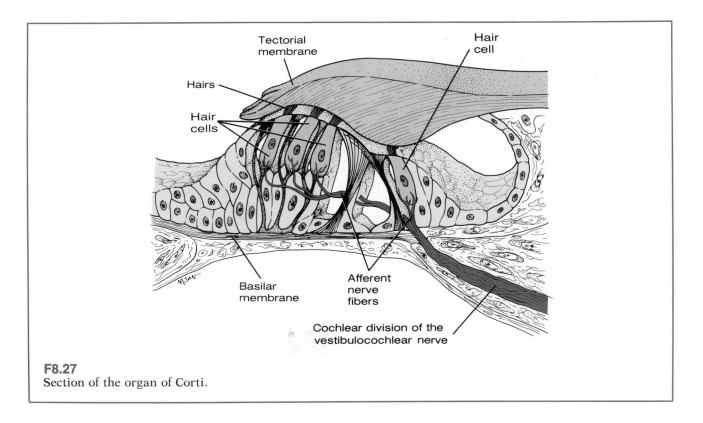

F8.27
Section of the organ of Corti.

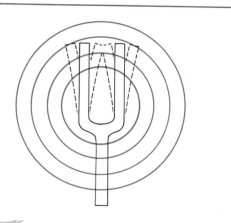

F8.28

Production of a sound wave. As a sound source—here a tuning fork—vibrates, it creates alternating areas of compression (higher pressure), in which the air molecules are close together, and rarefaction (lower pressure), where the molecules are farther apart. This activity produces a pressure wave that radiates outward from the sound source. This pressure wave is called a sound wave.

Transmission of Sound Waves to the Inner Ear — *amplify sound*

If sounds are to be perceived—that is, if a person is to hear a sound—the vibratory movements of sound waves traveling in air must be transferred to the fluids within the inner ear. However, fluids have much greater inertia than air, and greater pressures are required to cause movements of fluids than are required to cause movements of air. The pressures necessary to move the fluids of the inner ear result in part from the arrangement of the ossicles of the middle ear and in part from the difference in area between the tympanic membrane and the oval window.

A sound wave enters the external auditory meatus and strikes the tympanic membrane. The alternating higher and lower pressure regions of the sound wave cause the tympanic membrane to move forward and backward (vibrate) at the same frequency as the sound wave. The vibratory movements of the tympanic membrane are transmitted across the middle-ear cavity by the malleus, incus, and stapes to the oval window of the inner ear.

Because of the arrangement of the lever system of the ossicles, the force exerted on the oval window by the stapes is about 1.3 times greater than the force exerted on the malleus at the tympanic membrane (although the distance of movement of the stapes at the oval window is only about 75% as great as the distance of movement of the malleus at the tympanic membrane). Moreover, the area of the tympanic membrane that the sound wave can strike (about 55 mm²) is approximately 17 times greater than the area of the oval window (about 3.2 mm²). Thus, the pressure applied to the perilymph within the cochlea by the movement of the stapes against the oval window is about 22 times (1.3×17) greater than the pressure that would be exerted if the sound wave were to strike the oval window directly. It is this higher pressure that acts on the perilymph to cause its movement.

A reflex called the *tympanic* or *sound attenuation reflex* occurs in response to loud sounds. After a latent period of about 40 milliseconds following the occurrence of a loud sound, there is a reflexive contraction of the tensor tympani and stapedius muscles. The simultaneous contraction of these muscles dampens the vibrations of the ossicles and decreases the transmission of sounds (particularly sounds with frequencies below 1000 cycles per second) through the middle ear to the cochlea. This reflex protects the auditory receptors of the cochlea from damage by excessively loud sounds, and it may also mask low-frequency sounds in loud environments. Because of the latent period, however, the reflex provides little protection from sudden, explosive types of loud sounds.

Function of the Cochlea — *amplify the intensity of the ray*

If the forward movement of the stapes at the oval window is very slow, the pressure exerted on the perilymph pushes the perilymph along the scala vestibuli, through the helicotrema, and into the scala tympani, causing the round window to bulge outward into the middle ear (F8.29a). When the stapes moves backward and exerts less pressure, the perilymph moves back from the scala tympani into the scala vestibuli. Thus, very low frequency pressure waves have little effect on the basilar membrane. Because of the inertia of the fluids of the inner ear, however, higher-frequency pressure waves of the sort associated with sound perception do not follow this pathway. Instead, these pressure waves are transmitted from the perilymph of the scala vestibuli near the oval window, through the flexible vestibular membrane, to the endolymph of the cochlear duct. From the endolymph, the waves are transmitted through the basilar membrane to the perilymph of the scala tympani, and finally to the round window. The transmission of the pressure waves through the basilar membrane at the base of the cochlea near the oval and round windows causes this area of the membrane to vibrate (move downward and upward). The vibration of the basilar membrane near the base of the cochlea initiates a wave that

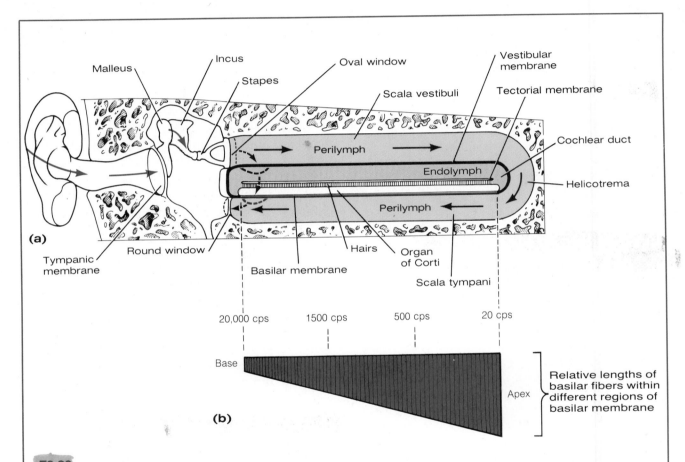

F8.29
(a) Schematic representation of the transmission of pressure waves in the cochlea. Solid black arrows indicate the transmission pathway when the movement of the stapes is slow. Dashed black arrows indicate the transmission pathway of higher-frequency pressure waves of the sort associated with sound perception.
(b) Graphic representation of the basilar membrane showing that the end of the membrane toward the middle ear contains shorter basilar fibers than the end toward the helicotrema. Maximum vibrations occur in the region of the basilar membrane that has the same resonant frequency as the sound wave that causes the pressure wave. Low-frequency sound waves cause maximum vibration toward the apex of the basilar membrane. High-frequency waves have their maximal effect toward the base of the basilar membrane.

travels along the basilar membrane toward the helicotrema.

The basilar membrane contains about 25,000 basilar fibers that project from the spiral lamina toward the outer wall of the cochlea. The basilar fibers increase in length and decrease in thickness and rigidity from the base of the cochlea toward the helicotrema (F8.29b). Because of this, the portion of the basilar membrane near the base of the cochlea tends to vibrate at high frequencies, and the portion near the helicotrema tends to vibrate at lower frequencies. Moreover, when a portion of the basilar membrane vibrates, the fluid between the vibrating portion of the membrane and the oval and round windows must also move. Since less fluid mass must move (that is, less inertia must be overcome) when a region of the membrane near the base of the cochlea vibrates than when a portion of the membrane near the helicotrema vibrates, the fluid loading of the basilar membrane also favors high-frequency vibration at the base of the cochlea near the oval and round windows and low-frequency vibration at the apex of the cochlea near the helicotrema. Thus, the sympathetic vibration or resonance of the basilar membrane in response to high-frequency sound waves is greatest near the base of the cochlea, and the sympathetic vibration or resonance of the basilar membrane in response to low-frequency sound waves is greatest near the apex of the cochlea.

As a particular frequency wave travels along the basilar membrane from its point of initiation near the oval and round windows, the vibration of the basilar membrane increases in amplitude and is greatest at that portion of the membrane that has the same natural resonant frequency as the frequency of the wave (F8.30). At this point the energy of the wave is dissipated, and beyond this region the wave quickly dies away. Thus, high-frequency sound waves cause maximal vibration of the basilar membrane near the base of the coch-

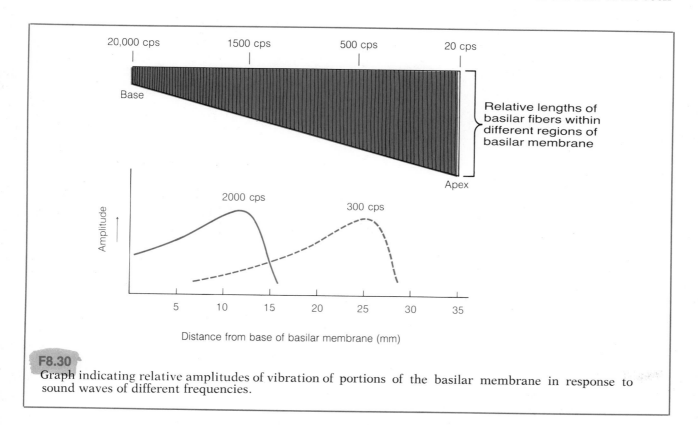

F8.30 Graph indicating relative amplitudes of vibration of portions of the basilar membrane in response to sound waves of different frequencies.

lea near the oval and round windows, and low-frequency sound waves cause maximal vibration of the basilar membrane near the apex of the cochlea near the helicotrema.

The hairs of the receptor cells of the organ of Corti, which is located on the basilar membrane, are in contact with or are embedded in the tectorial membrane that overhangs them (see F8.27). When the basilar membrane vibrates, the hairs are displaced, altering the polarity of the hair cells. When the basilar membrane moves upward toward the scala vestibuli, the hair cells of the affected region depolarize, and afferent nerve impulses are generated in neurons of the cochlear division of the vestibulocochlear nerve that are associated with the hair cells. When the basilar membrane moves downward, the affected hair cells hyperpolarize, and the generation of afferent impulses within the neurons of the cochlear division of the vestibulocochlear nerve decreases.

Auditory Pathways

The neural pathways between the organ of Corti and the auditory portion of the cerebral cortex include synapses in the medulla, the inferior colliculi of the midbrain, and the medial geniculate bodies of the thalamus (F8.31). Auditory signals from each ear are transmitted on both sides of the brain stem and cortex, and dysfunctions of the auditory pathways on one side do not greatly affect hearing in either ear.

Determination of Pitch

The ability to determine the pitch of most sounds is related to the fact that a sound wave of a particular frequency causes a specific region of the bas-

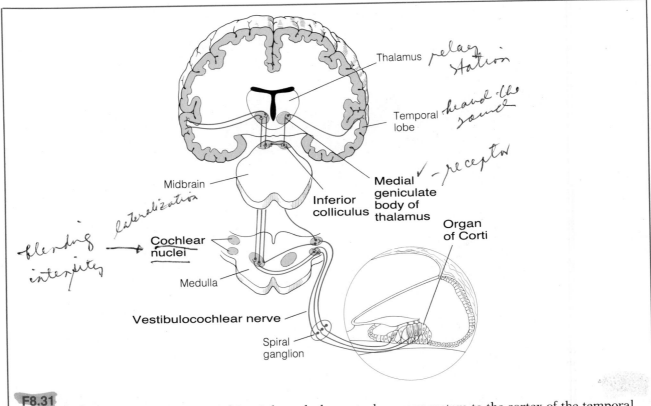

F8.31
Auditory pathways from the organ of Corti through the central nervous system to the cortex of the temporal lobe of the brain.

ilar membrane to vibrate more intensely than other regions, and the brain is able to detect which area of the basilar membrane is most stimulated. In fact, certain neurons within the auditory cortex respond only to a narrow range of sound frequencies. When the maximal vibration of the basilar membrane is near the base of the cochlea, the sounds perceived are interpreted as high-pitched; when the maximal vibration is in the middle of the cochlea, the sounds are interpreted as being of intermediate pitch; and when the maximal vibration is near the apex of the cochlea, the sounds are interpreted as low-pitched.

Determination of Loudness

High-intensity (high-amplitude) sound waves cause a greater amplitude of vibration of the basilar membrane than low-intensity (low-amplitude) waves. As the amplitude of vibration of the basilar membrane increases, hair cells of the organ of Corti are more strongly stimulated. In fact, certain hair cells apparently do not become strongly stimulated until the amplitude of vibration of the basilar membrane becomes quite high. The stronger stimulation of the hair cells activates more neurons and increases the frequency of transmission of afferent nerve impulses to the brain. The brain interprets this increased neural activity as an increase in the loudness of the sound.

Sound Localization

Several mechanisms are involved in localizing the source of a sound. The auditory system is able to detect differences in the time of arrival of a sound wave at each ear, and this ability provides one sound localization mechanism. For example, a sound wave coming from a point directly in front of a person reaches both ears at the same time, but a sound wave coming from a point to the person's right reaches the right ear slightly before it reaches the left ear. This aspect of sound localization is most effective at low frequencies (that is, frequencies less than 3000 cycles per second).

The auditory system is also able to detect differences in the intensity of a sound wave at one ear compared to the other, and this ability provides a second sound localization mechanism. The head acts as a sound barrier, particularly for higher-frequency sound waves. Thus, if a sound wave comes from a point directly in front of a person, both ears will receive the same intensity of stimulation. However, if the sound wave comes from a point to the person's right, the right ear will be more intensely stimulated than the left ear.

Impairment of Hearing

Middle-Ear Infections. Infections of the mucous membrane of the throat can travel through the auditory tube and cause an inflammation of the mucous membrane that lines the middle-ear cavity, including the inner surface of the tympanic membrane. The inflammation may cause fluid to collect within the middle ear, temporarily interfering with the ability to hear.

Deafness. Diseases of the ear or injuries to the ear can cause partial or total deafness. Such hearing losses can be classified as conduction deafness or nerve deafness.

In conduction deafness, there is interference with the transmission of sound waves through the external or middle ear. The interference can be caused by a physical blockage of the external auditory meatus by a foreign object or by cerumen (earwax), inflammation of the eardrum, adhesions between the ossicles, or thickening of the oval window. There is no damage to the receptor cells of the organ of Corti or to the nerve pathways in conduction deafness. Therefore, hearing aids that transmit sound waves to the inner ear through the bone of the skull rather than through the middle ear can aid a person suffering from this condition.

In nerve deafness, the loss of hearing results from disorders that affect the sound receptors in the inner ear, the vestibulocochlear nerve, or nerve pathways or centers within the central nervous system. Such disorders can be caused by a number of factors including infections, tumors, and trauma. Hearing aids may not be helpful in cases of nerve deafness in which the hearing loss is due to destroyed receptor cells or nerve pathways.

HEAD POSITION AND MOVEMENT

In addition to its role in hearing, the inner ear provides information about the position and movement of the head. This information is utilized in the coordination of movements that maintain body equilibrium and balance (see Chapter 9). Moreover, signals from the inner ear are involved in controlling the extrinsic eye muscles so that the eyes can remain fixed on the same point despite changes in the position of the head. The portion of the inner ear involved in these activities is called the **vestibular apparatus.** Receptor cells of the vestibular apparatus are located within the utricle (and perhaps the saccule) and within the ampullae of the semicircular ducts.

Function of the Utricle

Receptor cells of the utricle provide information about the position of the head with respect to gravity and about the linear acceleration or deceleration of the head (such as would occur when a person starts to run forward in a straight line).

The receptor structure within the utricle is called a **macula.** The macula is composed of groups of hair cells whose protruding hairs are embedded in a gelatinous substance that contains tiny particles of calcium carbonate called *otoliths.* The otoliths make the gelatinous substance heavier than the surrounding endolymph and, under the influence of gravity, the gelatinous substance exerts force on the hair cells. Even when the head is upright, neurons associated with the hair cells transmit a continuous series of impulses to the brain by way of the vestibular division of the vestibulocochlear nerve (F8.32). When the position of the head changes, the direction of the force of the gelatinous substance on the hair cells also changes, causing the hairs to be displaced from their normal position. The displacement of the hairs alters the stimulation of the neurons associated with the hair cells, and a different pattern of nerve impulses is transmitted to the brain. For example, when the head is tilted to the left, the transmission of nerve impulses from the left utricle increases and the transmission of impulses from the right utricle decreases. When the head is

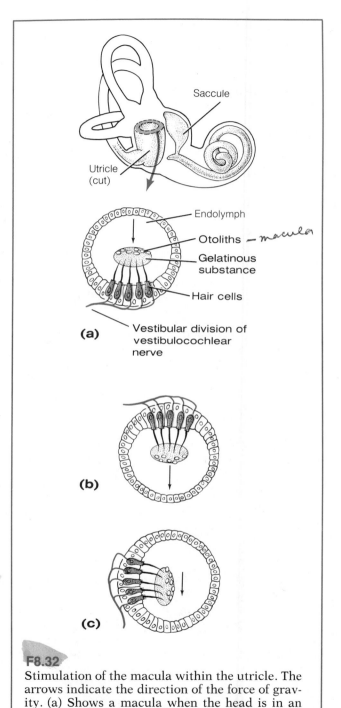

F8.32

Stimulation of the macula within the utricle. The arrows indicate the direction of the force of gravity. (a) Shows a macula when the head is in an upright position. (b) Illustrates a macula when the head is inverted. (c) Shows a macula when the head is in a horizontal position.

upright, it is possible to detect a change in head position of as little as one-half degree.

The hair cells of the macula also detect linear acceleration or deceleration. When the head undergoes linear acceleration, the inertia of the gelatinous substance causes it to lag behind the movement of the head. The lag of the gelatinous substance displaces the hairs of the hair cells, and an altered pattern of nerve impulses is transmitted to the brain. For example, the lag of the gelatinous substance that occurs when a person starts to run forward displaces the hairs of the hair cells to the rear. This is similar to the displacement that would occur if the head were tilted backward so that the face was oriented upward. If the linear movement of the head continues at a constant velocity, the inertia of the gelatinous substance is overcome, and it eventually moves at the same rate that the head is moving. During linear deceleration, the momentum of the gelatinous substance causes it to continue to move as the head slows and stops. The continued movement of the gelatinous substance also displaces the hairs of the hair cells and alters the pattern of nerve impulses that is transmitted to the brain. Thus, the hair cells of the macula of the utricle detect linear acceleration or deceleration, but they provide little information during linear motion at a constant velocity.

Function of the Saccule

The role of the saccule in humans is not entirely clear. Its receptor cells are organized into a macula like that of the utricle, and the saccule is generally considered to function in a similar manner as a component of the vestibular apparatus. However, some investigators believe the saccule to be part of the auditory system, at least in certain lower vertebrates.

Function of the Semicircular Ducts

The semicircular ducts detect rotational (angular) acceleration or deceleration of the head (such as would occur when the head is suddenly turned to one side). The semicircular ducts are filled with endolymph, and the ampulla of each duct contains a group of hair cells called a crista (F8.33).

The hairs are embedded in a gelatinous mass called the cupula. In general, the hairs of the hair cells within at least one semicircular duct are displaced by rotational acceleration or deceleration in any given plane.

When the head undergoes rotational acceleration, the inertia of the endolymph causes it to lag behind the movement of the semicircular duct. In

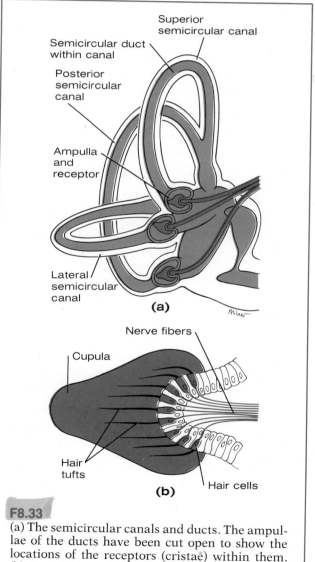

F8.33

(a) The semicircular canals and ducts. The ampullae of the ducts have been cut open to show the locations of the receptors (cristae) within them. (b) An enlargement of a crista.

effect, this response is equivalent to a flow of endolymph in a direction opposite to the movement of the duct—that is, opposite to the direction of the head's rotation. The lag of the endolymph alters the force exerted on the cupula within the ampulla of the duct and displaces the cupula and the hairs of the hair cells from their normal position. The displacement of the hairs alters the stimulation of neurons associated with the hair cells, and a different pattern of nerve impulses is transmitted to the brain. If the rotation of the head continues at a constant velocity, the inertia of the endolymph is overcome, and it eventually moves at the same rate that the semicircular duct is moving. Under these conditions, the cupula returns to its normal position and afferent nerve impulse transmission returns to the resting pattern within about 20 seconds. During rotational deceleration, the momentum of the endolymph causes it to continue to move as the semicircular duct slows and stops. In effect, this response is equivalent to a flow of endolymph within the semicircular duct in the direction of rotation. The continued movement of the endolymph also alters the force exerted on the cupula within the ampulla of the duct, and displaces the cupula and the hairs of the hair cells so that a different pattern of nerve impulses is transmitted to the brain. Thus, the hair cells of the cristae detect rotational acceleration or deceleration, but they provide little information during periods of rotation at a constant velocity.

Nystagmus

Nystagmus is a characteristic movement of the eyes that is associated with their ability to remain fixed on the same point despite changes in the position of the head. A type of nystagmus called *vestibular nystagmus* occurs even when the eyes are closed. Vestibular nystagmus is produced when the cupulae and the hairs of the hair cells within the ampullae of the semicircular ducts are displaced. For example, if the head undergoes horizontal rotational acceleration (for example, the person suddenly spins around and around), the eyes move relatively slowly in a direction opposite to the direction in which the head is turning (as

would occur if the eyes were fixed on some object in the environment). After the eyes have turned far to the side, they jump rapidly back to their forward position, and the movements are repeated. When the head undergoes horizontal deceleration (for example, the person suddenly stops spinning), nystagmus also occurs, but the directions of the eye movements are opposite to those that took place during acceleration (that is, the slow drift of the eyes occurs in the direction in which the rotation took place). These opposite directions of eye movements occur because, during deceleration, the cupulae and the hairs of the hair cells of the cristae are displaced in a direction opposite to that which took place during acceleration; therefore, a different pattern of nerve impulses is transmitted to the brain.

Another type of nystagmus, called *optokinetic nystagmus*, is produced when visual images move across the retinas of the eyes. For example, when the head moves horizontally in relation to the visual scene (such as when a person looks out of a window of a moving train), the eyes fix on an object and follow it, moving slowly until they have turned far to the side. The eyes then jump rapidly back to their forward position, fix upon a new object, and the movements are repeated. Optokinetic nystagmus does not occur when the eyes are closed.

SMELL (OLFACTION)

The receptors for smell are specialized neurons located in the epithelium of the nasal mucosa in the upper portion of the nasal cavity (F8.34). These neurons have two processes. One of the processes, together with similar processes from other olfactory receptor cells, travels to the olfactory bulbs in the brain as a component of the olfactory nerves. Within the olfactory bulbs, the processes of the receptor cells synapse with neurons that leave the bulbs as the olfactory tracts. From the olfactory tracts, nerve fibers travel to areas that include the amygdaloid nuclei and the primary olfactory cortex on the medial sides of the temporal lobes.

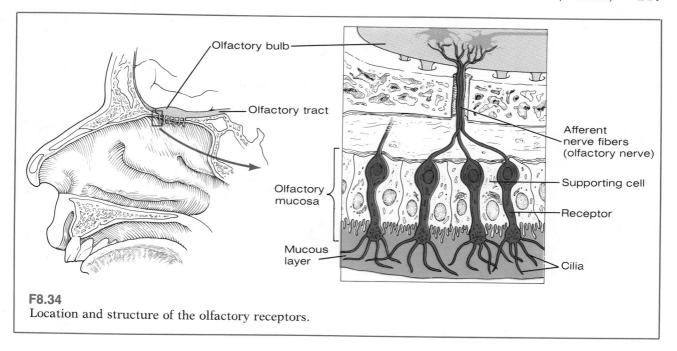

F8.34

Location and structure of the olfactory receptors.

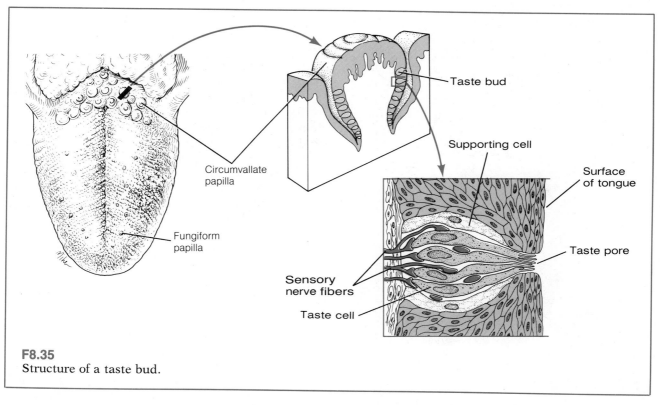

F8.35

Structure of a taste bud.

The second process of an olfactory receptor cell bears many fine projections called olfactory hairs or cilia that extend into the nasal cavity and are embedded in the mucous lining of the cavity.

For an odorous substance to be detected, it must reach the olfactory receptors. Normally, air moving through the nose does not pass through the upper portion of the nasal cavity where the olfactory receptors are located, so odorous substances must diffuse to the region. This process is aided by sniffing. Once in the region of the receptors, an odorous substance must dissolve in the mucous layer that covers the receptors and interact in some way with the receptors. This interaction is believed to depolarize the receptor, resulting in a generator potential that can initiate nerve impulses that are conveyed back to the brain.

Tens of thousands of different odors can be discriminated, but the basis for this discrimination is largely unknown. One theory is that a receptor cell membrane possesses 20 to 30 different types of receptor sites where the molecules of various odorous substances can interact and stimulate the depolarization of the cell. Although any one site is thought to be able to interact with a wide variety of odorous molecules, it responds best to molecules of particular sizes, polarities, and shapes. Thus, a particular molecule may interact well at some receptor sites and not as well at others. The better the interaction of the odorous molecule with a specific receptor site, the greater will be the depolarization resulting from the interaction. The depolarizations that occur at the different receptor sites of a single receptor cell are believed to be additive, resulting in a generator potential that determines the firing rate of the receptor cell. Different receptor cells are believed to have different combinations and concentrations of specific types of receptor sites. Thus, a given odorous substance may cause a simultaneous but differential stimulation of different receptor cells. The system is thus provided with the ability to discriminate many different odors.

TASTE (GUSTATION)

The receptors for taste are called **taste buds.** Most of them are located on the tongue, but some are found on the roof of the mouth, and in the pharynx and larynx. A taste bud consists of a series of receptor cells and supporting cells arranged much like the sections of a grapefruit. The upper portions of the receptor cells extend into a small pore at the surface of the taste bud where they are bathed by the fluids of the mouth (F8.35).

Afferent nerve fibers synapse with the receptor cells of the taste buds. A single nerve fiber may synapse with several different receptor cells, and a single receptor cell may have several different nerve fibers synapsing with it. Impulses concerned with taste are ultimately conveyed to the taste area of the cerebral cortex in the parietal lobe.

If a substance is to evoke a taste sensation, it must dissolve in the fluids that bathe the tongue and interact with the receptor cells of the taste

F8.36
Areas of the tongue that are most sensitive to particular taste sensations.

buds. Four primary taste sensations have traditionally been identified—sweet, salt, bitter, and sour—with each taste being detected best in specific regions of the tongue (F8.36). It appears, however, that there is no corresponding specificity of taste receptor cell types. A given taste receptor cell can interact with and respond to a variety of different substances that belong to more than one of the specific taste categories.

Taste receptor cells have been postulated to possess several different types of receptor sites that can form loose combinations with different types of molecules. The formation of these combinations depolarizes the receptor cell, leading to the generation of nerve impulses in the sensory neurons that contact the receptor cell. Thus, the sensation of a specific taste may be a complex phenomenon that involves the relative activities and firing patterns of a number of different sensory neurons from a variety of different taste receptor cells.

Much of what is commonly considered to be "taste" actually involves the stimulation of olfactory receptors. This fact is particularly evident when a person has a cold that produces congestion of the nasal mucosa that blocks olfactory stimulation. Under such conditions, the person often reports an inability to "taste" food, even though his or her actual taste system continues to function normally.

STUDY OUTLINE

VISION—THE EYE pp. 185–203

Major Structures of the Eye

FIBROUS TUNIC Outermost layer.

Sclera Posterior; white; serves as protection and maintains eye shape.

Cornea Anterior; clear; allows light passage.

VASCULAR TUNIC Beneath fibrous tunic.

Choroid Posterior; darkly pigmented; contains many blood vessels; reduces reflection of light that enters the eye.

Ciliary body Around edge of cornea; contains smooth ciliary muscles.

Iris Anterior; thin muscular diaphragm; pigmentation responsible for eye color.

Pupil Opening in center of iris through which light passes.

RETINA Innermost layer of eye (internal tunic) consists of outer pigmented layer and inner nervous-tissue layers.

Pigmented layer Composed of epithelial cells in contact with choroid; reduces reflection of light that enters the eye.

Nervous-tissue layers Contain photoreceptors: rods and cones.

Optic nerve Area of exit from eye is blind spot.

Fovea centralis Has a high concentration of cone receptors.

LENS Aids in focusing light on retina; relatively elastic, biconvex structure.

CAVITIES AND HUMORS Lens separates eye into two cavities: anterior cavity—contains aqueous humor; posterior cavity—contains vitreous humor.

Light Portion of electromagnetic spectrum to which eye receptors are sensitive.

Optics Light rays are bent (refracted) when striking a biconvex lens; light rays converge at focal point behind lens.

Focusing of Images on the Retina

EMMETROPIA Lens under tension appears relatively flat; eye focuses parallel light rays from distant points (farther than 20 feet) sharply upon retina.

ACCOMMODATION Focusing adjustments for viewing objects closer than 20 feet; ciliary muscles contract and permit greater lens curvature; pupil constriction eliminates divergent rays; convergence of eyeballs.

NEAR AND FAR POINTS OF VISION Near point

of vision varies with age—close to the eye when young and recedes with age; far point of vision: any object beyond 20 feet for normal eye can be focused on retina without accommodation.

Control of Eye Movements
The extrinsic eye muscles are among the most rapid acting and precisely controlled skeletal muscles.

Voluntary fixation movements   Controlled by a small area of the premotor regions of the frontal lobes of the cerebral cortex; move the eyes voluntarily to find an object and fix on it.

Involuntary fixation movements   Controlled by the occipital region of the cortex; keep the image of the object focused on the foveal portion of the retina.

Binocular Vision and Depth Perception
Retinal disparity enhances a person's ability to judge relative distances when objects are nearby.

DIPLOPIA (DOUBLE VISION)   May occur when light rays from an environmental object do not fall on corresponding points on retinas.

STRABISMUS   Improper coordination of eye movements. Eyes may alternate in fixing on environmental object, or one eye may be used all the time and the other eye may become functionally blind.

Photoreceptors of the Retina
Rods and cones contain photopigments. A photopigment consists of a protein called an opsin to which a chromophore molecule called retinal is attached. The opsins differ from pigment to pigment. Light alters the chemical configuration of retinal and the retinal breaks away from the opsin. This occurrence leads to changes in the polarity of the receptor cells that contain the photopigment.

RODS   Responses indicate degrees of brightness, but do not indicate color; contain rhodopsin; numerous in peripheral retina.

CONES   Responses indicate color; three different cone types: red, green, blue; cones are concentrated in center of retina and fovea.

Neural Elements of the Retina
1. Rods and cones synapse with bipolar cells, which synapse with ganglion cells.
2. Horizontal cells and amacrine cells are involved in lateral interactions that occur within the retina.
3. Neural pathways within the retina are partially responsible for differences in rod vision and cone vision.

Visual Pathways
Ganglion cell axons from medial half of each retina cross over at optic chiasma; optic tracts to thalamus; optic radiations to visual cortex.

Processing of Visual Signals
Begins in retina and continues throughout visual pathway. Components of pathway transmit coded message about certain aspects of image on retina.

Light and Dark Adaptation
Due to several mechanisms.

PHOTOPIGMENT CONCENTRATION   Photopigments are broken down in light, regenerated in darkness.

PUPILLARY LIGHT REFLEX   In bright light pupils constrict, and in dim light they dilate.

NEURAL ADAPTATION   When a person moves from a dimly lit area to a brightly lit area, the signals transmitted by the bipolar cells, horizontal cells, amacrine cells, and ganglion cells are initially very intense. However, within a fraction of a second, the intensity of the signals declines severalfold.

Visual Acuity
Ability of eye to distinguish detail—greatest in fovea; light rays stimulate different receptor units with an unstimulated or weakly stimulated unit between them.

Visual Disorders

MYOPIA   Nearsightedness.
1. Light rays from distant objects focused in front of retina.
2. Near point of vision closer than normal; far point of vision closer than 20 feet.
3. Corrected by concave lenses.

HYPERMETROPIA   Farsightedness.
1. No far point of vision; near point farther away than normal.
2. Increase refraction with convex lenses.

ASTIGMATISM   Unequal curvatures of portions of the refractive system of eye. Light rays that enter the eye in one plane focus at a point different from that of light rays that enter the eye in another plane.

CATARACT   Cloudiness or opaqueness of lens of eye or a portion of it.

GLAUCOMA   Abnormal elevation of intraocular pressure.

COLOR BLINDNESS   Absence or inadequate representation of certain cones.

Effects of Aging on the Eye
1. Lens becomes yellow due to exposure to ultraviolet rays.
2. Potential development of cataracts with age.
3. Pupil is less able to dilate, which results in a reduction of light that reaches photoreceptors.
4. Poor drainage of aqueous humor increases likelihood of glaucoma.
5. Debris from rods and cones may cause loss of visual acuity.
6. Lens loses elasticity, leading to presbyopia.

HEARING AND HEAD POSITION AND MOVEMENT—THE EAR   The ear contains sound receptors and receptors that detect head position and movement.   **pp. 203–216**

Structure of the Ear   Divided into external, middle, and inner regions.

EXTERNAL EAR
Auricle (pinna) directs sound waves into external auditory meatus.
External auditory meatus (canal) contains hairs and cerumen; serves as resonator.

MIDDLE EAR   Air chamber in temporal bone; extends from tympanic membrane to oval window.
Ossicles (malleus, incus, stapes) transmit vibrations of tympanic membrane to oval window.
Auditory tube regulates pressure.

INNER EAR   Series of fluid-filled (perilymph) canals called osseous labyrinth that contain membranous labyrinth, which is filled with endolymph.

The Vestibule   Utricle detects head position and movement. Saccule may function like utricle.

The Semicircular Canals   Detect certain head movements.

The Cochlea   Portion of inner ear for hearing; spiral organ of Corti.

Hearing

SOUND   Travels in the form of waves that consist of regions of compression, in which air molecules are close together and the pressure is relatively high, alternating with areas of rarefaction, where the molecules are farther apart and the pressure is lower. The ear can detect sound waves with frequencies between about 20 and 20,000 cycles per second but is most sensitive to frequencies between 1000 and 4000 cycles per second.

TRANSMISSION OF SOUND WAVES TO THE INNER EAR   A sound wave enters the external auditory meatus and causes the tympanic membrane to vibrate. Vibrations of the tympanic membrane are transmitted across the middle-ear cavity by the malleus, incus, and stapes to the oval window of the inner ear.

FUNCTION OF THE COCHLEA   Pressure waves generated by the movement of the stapes against the oval window cause the basilar membrane to vibrate near the base of the cochlea, initiating a wave that travels along the basilar membrane toward the helicotrema. The vibration of the basilar membrane alters the polarity of the hair cells of the spiral organ of Corti, which is located on the membrane, leading to the generation of nerve impulses that are transmitted to the brain and interpreted as sounds. High-frequency sound waves cause maximal vibration of the basilar membrane near the base of the cochlea, and low-frequency sound waves cause maximal vibration near the apex of the cochlea.

AUDITORY PATHWAYS   Auditory signals from each ear are transmitted on both sides of the brain stem and cortex, and dysfunctions of the auditory pathways on one side do not greatly affect hearing in either ear.

DETERMINATION OF PITCH   A sound wave of particular frequency causes a particular region of the basilar membrane to vibrate maximally. The brain can detect which area of the basilar membrane is most stimulated and interpret this as a particular pitch.

DETERMINATION OF LOUDNESS   High-intensity sound waves cause greater vibration of the basilar membrane than low-intensity sound waves. The brain detects this difference and interprets the greater vibration as a louder sound.

SOUND LOCALIZATION
1. The auditory system is able to detect differences in the time of arrival of a sound wave at each ear.
2. The auditory system is able to detect differences in the intensity of a sound wave at each ear.

IMPAIRMENT OF HEARING

Middle-Ear Infections Inflammation of mucous membrane of middle ear.

Deafness

Conduction deafness Interference with sound-wave transmission by physical blockage, inflammation of eardrum, ossicle adhesions, or oval window thickening.

Nerve deafness Loss of hearing that involves damage to sensory cells or nerve pathways.

Head Position and Movement Vestibular apparatus of inner ear provides information about head position and movement. Information is utilized in the coordination of movements that maintain body equilibrium and balance, and in controlling the extrinsic eye muscles so that the eyes can remain fixed on the same point despite a change in the position of the head.

FUNCTION OF THE UTRICLE Provides information about the position of the head with respect to gravity and about the linear acceleration or deceleration of the head.

Macula Receptor structure composed of hair cells with hairs embedded in gelatinous substance containing calcium carbonate. Hairs are displaced and the stimulation of the hair cells is altered by changing the position of the head or by linear acceleration or deceleration.

FUNCTION OF THE SACCULE Role not clear. May function like the utricle, but some investigators believe the saccule to be a part of the auditory system—at least in certain lower vertebrates.

FUNCTION OF THE SEMICIRCULAR DUCTS

Detect rotational acceleration or deceleration. Crista–receptor structure composed of hair cells, with hairs embedded in a gelatinous mass called the cupula. Hairs are displaced and the stimulation of the hair cells is altered by rotational acceleration or deceleration.

NYSTAGMUS A characteristic movement of the eyes that is associated with the ability of the eyes to remain fixed on the same point despite changes in the position of the head.

Vestibular nystagmus Produced when the cupulae and the hairs of the hair cells within the ampullae of the semicircular ducts are displaced.

Optokinetic nystagmus Produced when visual images move across the retinas of the eyes.

SMELL (OLFACTION)   p. 216

1. Receptors are neurons in epithelium of nasal mucosa in upper nasal cavity.
2. Odorous substances must dissolve in mucous layer that covers receptors, and must interact with receptors.
3. Different receptors have different combinations and concentrations of specific types of receptor sites.

TASTE (GUSTATION)   p. 218

1. Receptors are taste buds on tongue, roof of mouth, pharynx, larynx.
2. Much of what is commonly considered to be "taste" actually involves the stimulation of olfactory receptors.

SELF-QUIZ

1. The pigmented layer of the retina: (a) helps reduce the reflection of light that enters the eye; (b) contracts in bright light; (c) is responsible for eye color.
2. When a person looks directly at an object, the image of the object is generally focused on the optic disc. True or False?
3. The aqueous humor is produced by: (a) the cornea; (b) ciliary processes that project from the ciliary body; (c) the vitreous body.
4. The greater the deviation from the perpendicular at which light rays enter a block of glass: (a) the greater the wavelength of the light rays; (b) the less the bending of the light rays; (c) the greater the bending of the light rays.
5. During accommodation for viewing near objects: (a) the pupils constrict; (b) the ciliary muscles relax; (c) the eyes diverge.
6. The extrinsic eye muscles are the least rapid acting but among the most precisely controlled skeletal muscles in the body. True or False?
7. The different photopigments of the cones: (a) are composed of the same opsins; (b) are altered when light strikes them and they absorb the light; (c) do not contain retinal as a part of their structure.

8. Those eye structures that enable a person to perceive color are: (a) cones; (b) rods; (c) ossicles.

9. The processing of visual signals begins in the thalamus, and the ganglion cells of the retina transmit a simple mosaic pattern of the image on the retina to the brain. True or False?

10. When a person spends a period of time in bright light, significant amounts of photopigments (particularly rhodopsin) are broken down. This photopigment breakdown decreases the sensitivity of the eyes and raises the visual threshold so that a greater intensity of light is required for effective vision. True or False?

11. The constriction and dilation of the pupil occurs primarily as a voluntary response and serves in part to regulate the amount of light that enters the eye. True or False?

12. Visual acuity is greatest in the: (a) fovea centralis; (b) peripheral retina; (c) ciliary body.

13. An abnormal elevation of the intraocular pressure is: (a) strabismus; (b) astigmatism; (c) glaucoma.

14. The loss of lens elasticity that occurs with aging is called: (a) presbyopia; (b) myopia; (c) hyperopia.

15. The cochlea is the portion of the inner ear that is associated with hearing. True or False?

16. The organ of Corti is located on the: (a) tympanic membrane; (b) basilar membrane; (c) vestibular membrane.

17. The loudness of a sound is determined primarily by which aspect of a sound wave? (a) amplitude; (b) frequency; (c) pitch.

18. Vibratory movements of the tympanic membrane are transmitted across the middle-ear cavity by the malleus, incus, and stapes to the oval window of the inner ear. True or False?

19. The tympanic or sound attenuation reflex is particularly useful in protecting against sudden explosive types of loud sounds. True or False?

20. High-frequency sound waves cause maximal vibration of the basilar membrane near the helicotrema at the apex of the cochlea. True or False?

21. Conduction deafness: (a) may be caused by impacted cerumen; (b) cannot be corrected by hearing aids; (c) involves damage to the sensory cells of the organ of Corti.

22. When the head is in an upright position: (a) neurons from the utricle do not transmit impulses to the brain; (b) the gelatinous substance within the utricle exerts force on the hair cells; (c) it is possible to detect a change in head position only if the change exceeds ten degrees.

23. The hairs of the hair cells within at least one semicircular duct are generally displaced by rotational acceleration or deceleration in any given plane. True or False?

24. Receptor cells of the semicircular ducts of the inner ear are best able to detect: (a) linear motion of the head; (b) rotational motion of the head at a constant velocity; (c) rotational acceleration.

25. During a sudden rotational acceleration of the head: (a) the endolymph moves before the semicircular duct moves; (b) the cupula is displaced from its normal position; (c) the afferent pattern of nerve impulses transmitted to the brain does not change.

26. For an odorous substance to be detected, it must dissolve in the mucous layer that covers the olfactory receptors and interact with the receptors. True or False?

27. There is a specific receptor cell type for each of the four traditional primary taste sensations: sweet, salt, bitter, and sour. True or False?

LEARNING OBJECTIVES

After completing this chapter, you should be able to:

- Describe the flexor reflex and the associated crossed extensor reflex.
- Explain how general sensory input plays an important role in maintaining the arousal of the brain.
- Distinguish between slow-wave sleep and paradoxical sleep.
- Describe possible mechanisms involved in the sleep-wakefulness cycle.
- Distinguish between short-term and long-term memory, and cite the possible physical-chemical mechanisms of each.
- Define and cite an example of a conditioned response.
- Describe and cite an example of instrumental learning.
- Describe three neural mechanisms involved in supporting the body against the force of gravity.
- Explain the role of the cerebellum in coordinating movements.
- Describe the processes by which skilled movements may be learned and performed.

CHAPTER CONTENTS

INTEGRATIVE SPINAL REFLEXES

MENTAL PROCESSES
The Electroencephalogram
Consciousness
  The Aroused Brain
  Sleep
  Sleep-Wakefulness Cycle
Attention
Emotions and Behavior
Pain
  Referred Pain
  Phantom Pain
  Anesthesia
Memory
  Short-Term Memory
  Long-Term Memory
Learning

CONTROL OF BODY MOVEMENTS
  Support of the Body Against Gravity
    Reflexes
    Higher Centers
  Maintenance of Equilibrium
  Locomotion
  The Cerebellum and the Coordination of Movement
  Skilled Movements

LANGUAGE

INTEGRATIVE FUNCTIONS OF THE NERVOUS SYSTEM

9

The nervous system brings together information from many different sources, and it coordinates and regulates numerous interrelated activities that take place within the body. Consequently, the nervous system is an integrative system that is vital in maintaining homeostasis.

INTEGRATIVE SPINAL REFLEXES

Even basic nervous system activities such as reflex responses exemplify the integrative function of the nervous system. This integrative function is evident in the **flexor** or **withdrawal reflex** and the associated **crossed extensor reflex.** The flexor reflex results in the withdrawal of a limb, most commonly in response to some noxious stimulus. For example, if a person's foot is pricked by a pin, nerve impulses resulting from the stimulus (the prick) are transmitted to the spinal cord by afferent neurons (F9.1). Within the spinal cord, the afferent neurons stimulate interneurons that, in turn, stimulate efferent neurons supplying the flexor muscles of the leg on the same side of the body as the site of stimulation (the ipsilateral side). In addition, the afferent neurons stimulate interneurons that inhibit efferent neurons supplying the extensor muscles of the ipsilateral leg. Thus, the flexor muscles of the ipsilateral leg are activated, and the extensor muscles are inhibited. As a result, the leg is flexed, and the foot is withdrawn.

The crossed extensor reflex, which is initiated by the same stimulus responsible for the flexor reflex, results in the extension of the corresponding limb on the side of the body opposite the site of stimulation (the contralateral side). For example, when a person's foot is pricked with a pin, the afferent neurons not only stimulate neurons involved in the flexor reflex, but they also stimulate commissural interneurons that cross to the opposite side of the spinal cord. There, the commissural interneurons stimulate interneurons that, in turn, stimulate efferent neurons supplying the extensor muscles of the contralateral leg. The commissural interneurons also stimulate interneurons that inhibit efferent neurons supplying the flexor muscles of the contralateral leg. Thus, the extensor muscles of the contralateral leg are activated and the flexor muscles are inhibited. As a result, the leg is extended.

Even though the flexor reflex and the associated crossed extensor reflex occur at the spinal cord level, they nevertheless result in a coordinated, adaptive integration of separate muscular activities. The withdrawal of the stimulated limb removes it from a potential source of injury, and the extension of the corresponding contralateral limb can help the individual maintain balance.

Although some integrative activities occur at the spinal cord level, the brain is the major integrative center of the nervous system. For example, a stimulus such as a pin prick not only elicits reflex responses, but nerve impulses are transmitted to the brain where further integrative activities occur. Pain may be sensed, and conscious responses to the pin prick initiated. Ultimately, the event may be committed to memory, and learning may take place. This learning may take the form of avoiding in the future conditions under which similar injury may occur.

MENTAL PROCESSES

The brain is an extremely complex organ. It has been studied intensively for many years, but researchers have so far achieved only enticing bits of

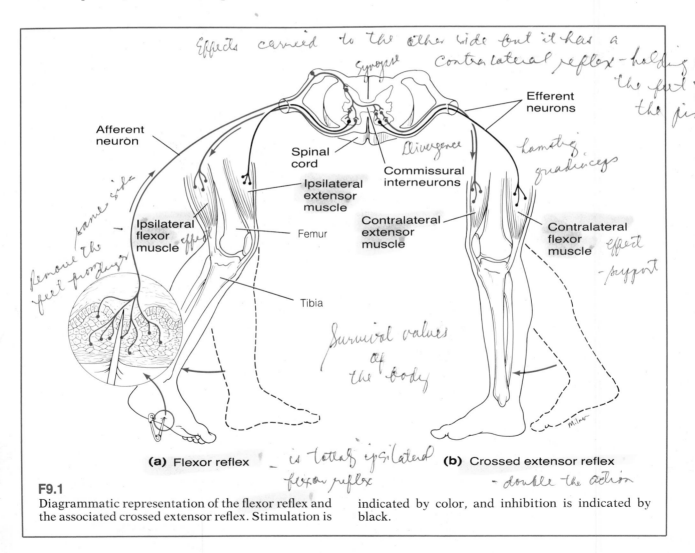

(a) Flexor reflex   **(b)** Crossed extensor reflex

F9.1
Diagrammatic representation of the flexor reflex and the associated crossed extensor reflex. Stimulation is indicated by color, and inhibition is indicated by black.

understanding about its function. The processes involved in memory, learning, concept formation, and abstract reasoning have largely eluded investigators. Bear in mind, then, that the following discussions of mental processes—particularly those dealing with complex phenomena such as memory and learning—are largely theoretical. As more information becomes available, some of these ideas may have to be abandoned in favor of new ones.

No area or structure of the brain acts entirely on its own. The removal of portions of the brain or the severing of tracts within the brain—experimentally in animals, or as the result of trauma in humans—reveal that the most complex of the so-called higher functions of the brain (such as memory and learning) are generally whole-brain functions. That is, the performance of complex functions involves more than one brain area, and each area is probably involved in many functions. Even those functions that occur automatically below the conscious level generally entail input from several different sources.

THE ELECTROENCEPHALOGRAM

The degree and pattern of electrical activity that occurs within the brain can be detected by electrodes placed on the head (F9.2). The record of the electrical waves produced during brain activity is called an **electroencephalogram (EEG).** The electroencephalogram is essentially a record of waves of different amplitudes (that is, heights) and frequencies that arise from the cerebral cortex, perhaps at times as a result of the influence of subcortical centers such as the thalamus. These brain waves, as they are called, are always present, even during unconsciousness, indicating that the brain is constantly active as long as it is alive.

There is some correlation between brain waves and body activity (F9.3). For example, when a person is awake, brain-wave recordings show low-amplitude, high-frequency waves. When a person falls asleep, the frequency of the brain waves slows, but the amplitude increases. A person who is anesthetized also exhibits slow waves.

CONSCIOUSNESS

In order to perform most of its higher functions, the brain must be in a state of awareness, or **consciousness.** Consciousness, however, is difficult to define, and there are gradations of awareness be-

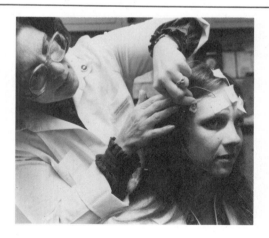

F9.2
Recording the degree and pattern of electrical activity occurring within the brain by means of electrodes placed on the head.

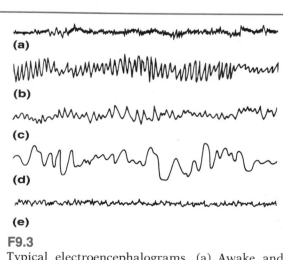

F9.3
Typical electroencephalograms. (a) Awake and alert. (b) Relaxed with eyes closed. (c) Drowsy. (d) Asleep, slow-wave sleep. (e) Asleep, paradoxical sleep.

tween complete unconsciousness and alert wakefulness. Consciousness is sometimes considered to be the state in which thoughts or instants of awareness occur. Although this concept makes a thought seem like a simple episode, each thought probably requires the activity of several portions of the brain. In other words, a thought is the result of a pattern of neural activities that occur simultaneously in several locations, such as the cerebral cortex, the thalamus, the limbic system, and the reticular formation.

If a single thought requires the activity of several brain centers, it seems reasonable to suppose that a state of consciousness requires the activation or arousal of widespread areas of the brain. Experimental evidence suggests that this is the case.

The Aroused Brain

When the brain is aroused or awake, it is in a state of readiness and is able to react consciously to stimuli. The attainment of this state seems to depend in large part on nerve impulses sent throughout the brain by the reticular activating system, or RAS (see page 161). Stimulation of the RAS in a sleeping animal produces an EEG like that of an aroused brain and causes the animal to awaken. Conversely, destruction of the RAS produces a permanent coma, with an EEG characteristic of the sleeping state.

The activity of the RAS is influenced by nerve impulses from cutaneous, visual, auditory, muscular, and visceral receptors. These impulses, which are called *arousal signals*, stimulate the RAS to send impulses throughout the brain to arouse it. Consequently, general sensory input plays an important role in maintaining the arousal of the brain.

Sleep

Sleep can be defined as a state of altered consciousness or partial unconsciousness from which a person can be aroused by appropriate stimuli. In contrast, a coma is a state of unconsciousness from which a person cannot be aroused.

As a person becomes drowsy and falls asleep, there is a gradual shift in the EEG toward higher-amplitude, slower (lower-frequency) waves. The state of sleep characterized by high-amplitude, low-frequency EEG waves is called **slow-wave sleep.** During deep slow-wave sleep, the respiratory cycle is regular and deep, the heart rate is rhythmic and slow, and the blood pressure is below waking levels.

Approximately once every 90 minutes during a normal night of sleep, the EEG of slow-wave sleep is interrupted by episodes, lasting from 5 to 20 minutes, during which the EEG resembles that from an aroused brain. This state of sleep is called **paradoxical sleep.** Because the eyes move rapidly behind the closed eyelids during paradoxical sleep, it is also referred to as **rapid-eye-movement** (*REM*) sleep. Paradoxical sleep occurs in conjunction with slow-wave sleep, and the first paradoxical sleep episode normally takes place 80 to 100 minutes after a person falls asleep. A specific region of the pons called the locus caeruleus appears to induce paradoxical sleep, and damage to this region prevents its occurrence. Neurons of this area contain an abundance of norepinephrine, suggesting that its release leads to paradoxical sleep.

During paradoxical sleep, muscle tone throughout the body is greatly depressed, although periodic twitching of the facial muscles and limbs occurs. In addition, the respiration and heart rate are irregular, and the blood pressure may rise or fall. If a person is awakened every time a period of paradoxical sleep begins so that a paradoxical sleep deficit develops, the person spends a greater proportion of the total sleep time in paradoxical sleep when allowed to sleep undisturbed. Most dreaming occurs during paradoxical sleep (F9.4).

Sleep-Wakefulness Cycle

The normal sleep-wakefulness cycle is one of the most obvious human rhythms, and both mental and physical benefits are derived from alternating periods of sleep and wakefulness. Although the processes that underlie this cycle are not fully understood, awakening appears to occur when signals from the RAS reach a sufficient intensity to arouse the brain. Conversely, sleep seems to result when the activity of the RAS declines to a level no longer adequate to maintain arousal.

The RAS not only transmits nerve impulses throughout the brain but, in what is basically a positive-feedback response, brain areas such as the cerebral cortex send stimulatory signals to the RAS (F9.5). In addition, a second positive-feedback response occurs in which the RAS not only receives nerve impulses from receptors in the peripheral musculature, but it also transmits signals to the muscles. Thus, signals from the RAS that enhance the activity of the brain or the muscles lead to the transmission of stimulatory impulses from these structures to the RAS. These positive-feedback responses are believed to be part of a neural wakefulness circuit that operates as follows.

When the relatively dormant RAS of a sleeping individual is stimulated by arousal signals from various receptors, the RAS sends more intense signals to the brain and muscles. When this occurs, the positive-feedback responses just described further stimulate the RAS. The RAS, in turn, further stimulates the brain, and so on and so on. When

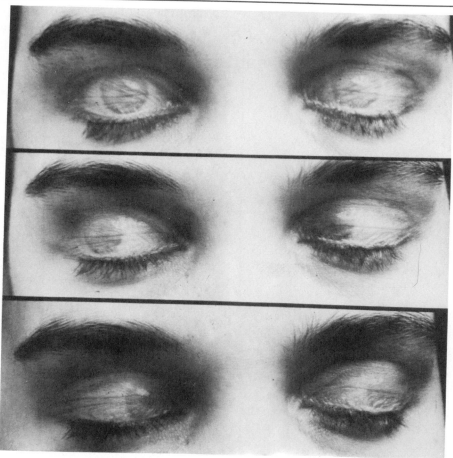

F9.4

These photographs were taken using time-lapse photography during a period of paradoxical, or rapid-eye-movement, sleep. Note how the eyes move from side to side behind the eyelids. Although the reason for this eye movement is not certain, some researchers hypothesize that the sleeper is surveying dream imagery, perhaps giving the imagery logic and direction by supplying movement and depth to the characters and objects in the dream.

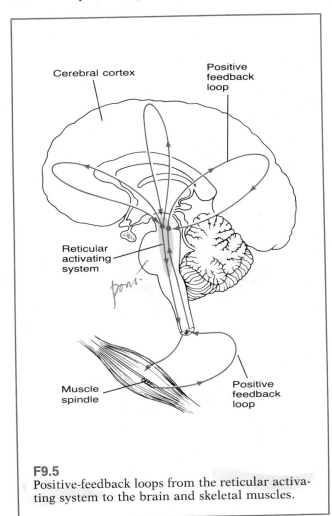

F9.5
Positive-feedback loops from the reticular activating system to the brain and skeletal muscles.

ties of the neurons involved to transmit nerve impulses. Therefore, once particular loops are activated, their level of activity does not continue to increase indefinitely but finally stabilizes.

With the passage of time, the excitability of the wakefulness circuit is believed to diminish, and consequently the intensity of the signals from the RAS to the brain declines. As a result, the individual becomes drowsy and falls asleep. During sleep, the excitability of the wakefulness circuit increases, and arousal signals again set in motion the positive-feedback events that lead to awakening. Because of the cyclical variation in the excitability of the wakefulness circuit, arousal signals that are not strong enough to produce wakefulness early in the sleep period may be sufficient to awaken the individual later in the period. Moreover, the system exerts selectivity as to the types of signals that produce arousal. For example, a mother may quickly awaken if her child cries, but she may not be awakened by the sounds of traffic, trains, or airplanes.

Researchers have proposed a number of theories to explain the cyclical variation in the excitability of the wakefulness circuit. One suggestion is that with continued activity during the waking period, neurons of the positive-feedback loops become depleted of neurotransmitter substances. Consequently, the feedback diminishes, the activity of the RAS declines, and the person becomes drowsy and falls asleep. During the sleep period, the neurotransmitter substances are replenished, and the feedback loops recover their excitability.

Another suggestion is that the accumulation or removal of various chemical substances influences the excitability of the wakefulness circuit. For example, inhibitory chemical substances could gradually build up during the waking period and, through their influence on the RAS, lead to drowsiness and sleep. During the sleep period, the substances are removed, leading to an increased excitability of the RAS and the arousal of the brain. There is some evidence that supports this possibility. If dialyzed blood or whole cerebrospinal fluid from animals that have been kept awake for several days is injected into the brain ventricle

the signals from the RAS to the brain become sufficiently intense, the brain is aroused and the individual awakens.

The wakefulness circuit contains numerous individual positive-feedback loops, and the degree of wakefulness at any moment depends on the number of active loops and on their level of activity. If many loops are activated simultaneously in a sleeping person, the positive-feedback aspect of the wakefulness circuit produces a rapidly increasing response that could explain the often sudden transition between the sleeping and the waking states. The ultimate level of activity of the positive-feedback loops is limited by the capaci-

systems of other animals, it can put them to sleep. Moreover, neurons within a region of the brain stem called the median raphe secrete serotonin into the RAS, and damage to these neurons induces a state of relative sleeplessness. Perhaps, then, sleep is caused by the secretion of serotonin from these neurons.

Researchers have proposed that a feedback relationship exists between the serotonin-secreting brain-stem neurons that may cause sleep and the neurons of the pons that appear to induce paradoxical sleep. According to this proposal, the brain-stem neurons facilitate the neurons of the pons, whose activity leads to paradoxical sleep. These neurons, in turn, stimulate the brain-stem neurons to take up serotonin that was previously secreted. This uptake decreases the extracellular serotonin concentration and lessens the inhibition of the RAS, permitting a return to the waking state. As a result, periods of paradoxical sleep would be important because they would remove an inhibitory chemical substance from the system.

ATTENTION

The term **attention** refers to the ability of an individual to maintain a selective awareness of some aspect of the environment, as well as to the ability to respond selectively to a particular stimulus. A person does not always respond to every stimulus. Instead, a selection process occurs so that the person pays attention to only certain stimuli. The person's previous experience is important in determining which stimuli warrant attention at any particular moment, and interest as well as actual biological urgency influence the priority given to particular stimuli.

A person's general level of attentiveness is influenced by the activity of the reticular activating system mechanisms involved in wakefulness and sleep. In addition, certain thalamic areas of the RAS can selectively activate specific areas of the cerebral cortex, and thus direct a person's attention to particular aspects of the environment. Moreover, other central nervous system mechanisms can selectively inhibit or enhance afferent signal input to the brain, and thus focus a person's attention on a particular aspect of the environment. For example, the auditory portion of the cerebral cortex can inhibit or facilitate signals from the cochleas of the ears, and the visual cortex can influence the intensities of signals from the retinas of the eyes.

EMOTIONS AND BEHAVIOR

The integrative functions of the limbic system (see page 157) influence emotions such as fear and anxiety as well as emotional behavior patterns such as rage. In addition, the limbic system is important in determining whether a particular activity is pleasant and rewarding or painful and punishing. In animal experiments, the electrical stimulation of certain areas of the limbic system appears to please or satisfy the animal. Conversely, the stimulation of other areas appears to cause pain, fear, or other elements of punishment, and the animal may exhibit rage or escape behaviors. Similarly, the stimulation of certain limbic areas in humans during neurosurgery results in pleasurable sensations, and the stimulation of other areas leads to vague feelings of fear or anxiety.

The reward and punishment regions of the brain have a great influence on a person's behavior. In fact, much of what a person does depends on reward and punishment. If an activity is pleasurable or rewarding, a person will generally continue the activity. If an activity is unpleasant or punishing, the person will generally not continue the activity. Moreover, reward and punishment have much to do with learning (see page 236).

Neurotransmitter substances such as the monoamines norepinephrine, dopamine, and serotonin appear to be involved in certain emotional and behavioral states and in learning. For example, drugs that enhance the effect of norepinephrine increase rage, and drugs that block its effect diminish rage behavior. Moreover, increased norepinephrine is associated with elevated mood, and decreased norepinephrine with depression. In this regard, it is believed that some antidepressant drugs act by increasing the brain concentration of norepinephrine.

PAIN

Stimuli strong enough to cause tissue damage commonly give rise to a sensation of pain. Stimuli such as excessive mechanical stress, extremes of heat or cold, and various chemical substances stimulate pain receptors. Nerve impulses from the receptors are transmitted along pain fibers that enter the spinal cord, ascend or descend one or two segments within the cord, and terminate on neurons of the gray matter. From this point, the nerve impulses probably pass along one or more short neurons before reaching neurons of ascending tracts. The impulses are ultimately transmitted to brain areas that include the reticular formation, the thalamus, and the somatic sensory areas of the cerebral cortex. In addition to eliciting a sensation of pain, the transmission of pain signals to the brain leads to emotional reactions (for example, crying, anxiety, or fear) and behavioral responses (such as withdrawal or defensive responses).

Referred Pain

A person can often identify the location of a stimulus that produces pain because the brain usually projects the sensation of pain to the site of the stimulus. In some instances, however, impulses from pain receptors in internal organs are incorrectly interpreted by the brain as coming from areas quite distant from the actual sites of stimulation—particularly as coming from sites on the body surface (F9.6). This phenomenon is called **referred pain.** The locations of some referred pains are so consistent that they are used by physicians in diagnosing visceral dysfunction. For example, a heart attack frequently causes referred pain in the skin over the heart, the left shoulder, and down the medial surface of the left arm.

One attempt to explain the cause of referred pain suggests that afferent neurons that transmit pain signals from a particular area of the body surface and afferent neurons that transmit pain signals from an internal organ connect within the spinal cord with the same ascending neurons. Thus, these ascending neurons carry pain signals to the brain from both the particular body surface area and the internal organ. Because cutaneous pain is much more common than visceral pain, pain signals carried over the ascending neurons are interpreted by the brain as having originated in the skin rather than in the viscera, and the pain sensation is projected to the skin site.

Phantom Pain

Phantom pain is the phenomenon whereby a person who has undergone an amputation continues to feel pain that he or she perceives as coming from the amputated body part. Phantom pain, like referred pain, is a case of inaccurate projection of the pain sensation by the brain. The neurons that supplied the affected structure are, of course, severed as a result of the amputation. However, the remaining portions of the neurons may continue to send nerve impulses to the same area of the brain that they did previously. For some time the brain continues to interpret impulses from the severed neurons as originating from the same body region that they normally would. As a result, the sensations evoked in the brain are projected to that region. In this manner, pain (and other sensations) may still be "felt," for example, in the toes, even after the foot has been amputated.

Anesthesia

Various anesthetics are commonly used to reduce a person's sensitivity to painful stimuli. A number of local anesthetics (for example, procaine and tetracaine) exert their effects in circumscribed areas by decreasing the permeability of nerve-cell membranes to sodium ions. This decreased sodium permeability reduces the excitability of the neurons so that they do not transmit nerve impulses. General anesthetics reduce sensitivity to pain by rendering a person unconscious. This result depends at least partially on the ability of the general anesthetic to inhibit or depress conduction in the reticular activating system.

MEMORY

Memory refers to the ability to store experiences, thoughts, and sensations for later recall. There are many unanswered questions about the processes that make memory possible, but it appears that there are at least two basic types of memory: short-term and long-term.

Different physical-chemical mechanisms are believed to be responsible for the two types of memory. In short-term memory, a bioelectric process is thought to be involved, whereas in long-term memory, structural as well as biochemical changes in neurons or synapses are thought to occur.

Short-Term Memory

Short-term memory allows the recall of information for only short periods following its initial presentation. Theorists believe that short-term memory decays in a matter of seconds unless the information within the short-term memory is rehearsed. Rehearsal, which is essentially the process of keeping one's attention on the information to be remembered, holds information within the short-term memory for an indefinite period of time. Rehearsal is also believed to be important in the transfer of information from short-term to long-term memory.

Some researchers suggest that short-term memory depends on the activation of *reverberating (oscillating) circuits* of neurons within the brain by signals from various receptors (F9.7). The com-

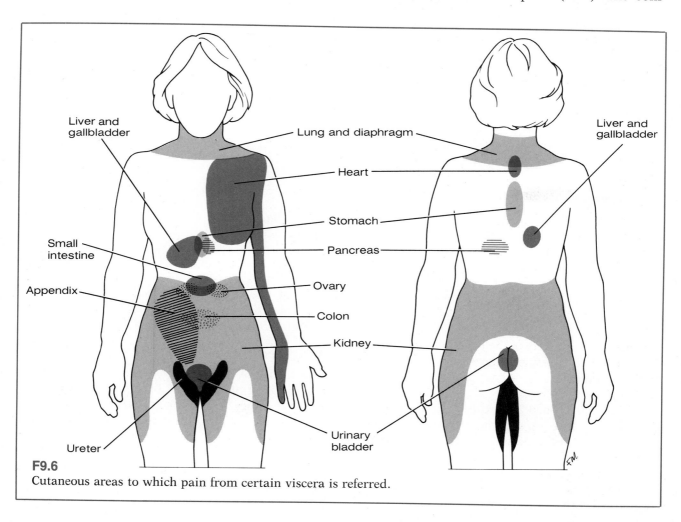

F9.6
Cutaneous areas to which pain from certain viscera is referred.

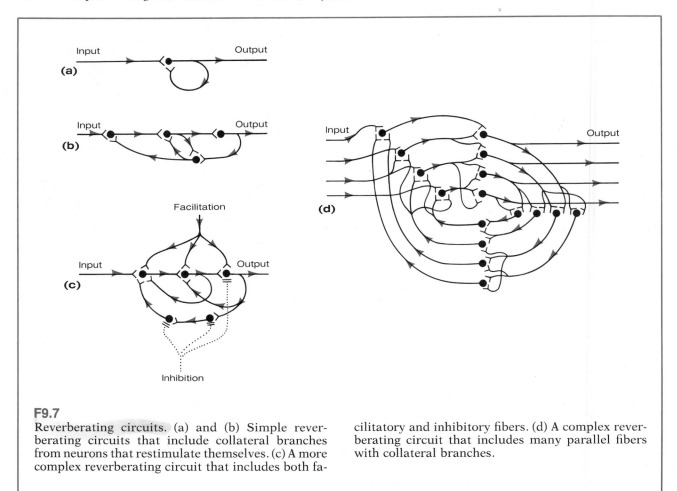

F9.7

Reverberating circuits. (a) and (b) Simple rever-berating circuits that include collateral branches from neurons that restimulate themselves. (c) A more complex reverberating circuit that includes both fa-cilitatory and inhibitory fibers. (d) A complex rever-berating circuit that includes many parallel fibers with collateral branches.

ponents of the reverberating circuits repeatedly reactivate each other for a period of time, thereby retaining information within the circuits. Without rehearsal, the activity of the reverberating circuits is believed to diminish, the short-term memory fades, and the person can no longer remember the information.

It has also been proposed that short-term memory depends on a temporary facilitation of nerve impulse transmission at the synapses of par-ticular neuronal circuits. This facilitation could be due either to an increase in the release of neu-rotransmitters at synapses following the repeated activation of the neurons of the circuit, or to facil-itatory changes in the resting membrane poten-tials of the neurons. In either case, without rehearsal, the facilitation—and the memory—are thought to decline.

Long-Term Memory

The ability of the nervous system to recall infor-mation long after it was first presented is **long-term memory.** The consolidation of a bit of information within the long-term memory ap-pears to take some time. During this time, various events can occur that interfere with the entrance of the information into the long-term memory. To some investigators, this possibility for inter-ference suggests that information must remain within the short-term memory while the activities necessary to incorporate it into the long-term memory take place.

Theorists do not believe that long-term memory is due to the same type of neural activity responsible for short-term memory. Rather, they believe that long-term memory is the result of changes in the structure or the biochemistry of neurons or synapses.

Researchers have found that rats trained to perform various activities or exposed to visually stimulating environments have heavier and thicker cerebral cortices than rats kept in isolation or maintained under monotonous conditions. They have also discovered that rats subjected to a stimulating environment have larger cortical neuron cell bodies and more small projections extending from the dendrites of certain cortical neurons than rats maintained in a monotonous environment. (Note that these are structural changes in the existing neurons—not changes in the total number of neurons within the nervous system. It is well established that neurons do not divide, except during embryological development.)

Differences such as those observed in the cerebral cortices of animals from different environments have led some investigators to suggest that long-term memory involves alterations in the functional contacts between neurons in the brain (for example, changes in the number or area of synaptic junctions between neurons). Other investigators have suggested that the excitability of synapses is altered by a change in the ability of presynaptic neurons to secrete transmitter substances, or by an alteration in the sensitivity of postsynaptic neurons. Moreover, some investigators have proposed that certain molecules—RNA, proteins, peptides, or glycoproteins—serve as "memory molecules" that store coded information and perhaps influence the structures or activities of neurons.

Alterations in the structure or biochemistry of neurons or synapses, such as those just discussed, are thought to provide for memory by producing long-term or permanent facilitation of nerve impulse transmission at the synapses of particular neuronal circuits, thus allowing the circuits to be easily reexcited by incoming signals at later dates.

Such hypothetical facilitated neuronal circuits are called *memory traces (memory engrams)*.

LEARNING

Learning, which is usually defined as a relatively permanent change in behavior as a result of experience, is a complex brain function that enables a person to adapt to a wide variety of circumstances and situations. Learning can occur very quickly—in the case of some adaptations, in a second or less. Learning is also flexible—learned behaviors can be unlearned and new behaviors learned in their place. Although a number of behavioral responses such as the integrative spinal reflexes discussed earlier are not learned, most of a person's behaviors are learned.

A simple type of learning called **classical conditioning** leads to the development of conditioned responses. A *conditioned response* is a response to a stimulus (called a conditioned stimulus) that did not previously cause the response as a consequence of the pairing of this stimulus with another stimulus (called an unconditioned stimulus) that naturally produces the response. A number of human behaviors can be influenced by classical conditioning, particularly those involving reflexes or emotional responses. For example, a puff of air blown into the eye is an unconditioned stimulus that naturally causes an eye-blink reflex. The sound of a bell is a conditioned stimulus that does not naturally cause this reflex. If a bell is rung just before a puff of air is blown into a person's eye, and the sequence is repeated several times, the person will eventually blink in response to the sound of the bell, even before the puff of air occurs. The blink in response to the sound of the bell is a conditioned response. In a similar manner, fear, which is an emotional response to the threat of pain, is quite easily conditioned. If a person who does not normally have a fear of dogs is bitten by a dog (an unconditioned stimulus), the person may, in the future, respond to the sight of the dog (a conditioned stimulus) with physical signs and subjective feelings of fear (a conditioned response). Moreover, the fear response may be generalized to all dogs, not just to the dog that bit the person.

Another type of learning is **instrumental learning,** or **operant conditioning.** In this process, the outcome or result of a particular behavior is important in determining whether the behavior will be strengthened (positively reinforced) or suppressed (negatively reinforced). If a particular behavior leads to a pleasant result or to a result that enables a person to escape from or avoid punishment, the behavior tends to be strengthened and becomes more likely to occur. If the behavior leads to the removal of a pleasant stimulus or to punishment, the behavior tends to be suppressed and becomes less likely to occur. There are many situations in which instrumental learning influences behavior, and parents often use instrumental learning techniques to teach children useful habits. For example, a parent may train a child to pick up his or her toys after play by smiling and praising the child each time the toys are picked up. In this situation, the picking-up behavior is positively reinforced by the pleasant result of parental approval. Consequently, it becomes more likely to occur.

A more complex type of learning, called **cognitive learning,** also occurs in humans. The human brain is capable of forming new thought patterns by recalling and integrating previous experiences. The brain can correlate information—gained from previous experiences and stored in the memory—into new concepts by abstract reasoning.

CONTROL OF BODY MOVEMENTS

The nervous system coordinates and controls the contractions of the many skeletal muscles involved in body movements.

SUPPORT OF THE BODY AGAINST GRAVITY

Spinal reflexes, such as the positive supportive reaction and muscle stretch reflexes, as well as higher centers within the brain, regulate the contractions of the skeletal muscles that support the body against the force of gravity and maintain an upright posture.

Reflexes

The **positive supportive reaction** is a complex reflex that is initiated when pressure is applied to the bottoms of the feet. Nerve impulses from pressure receptors are transmitted to the spinal cord, where the incoming signals are diverged through a complex pathway of interneurons, and stimulatory impulses are sent to the alpha motor neurons of the extensor muscles of the legs. The contractions of the extensor muscles stiffen the legs, helping to support the individual.

The skeletal muscles exist in a continual state of slight contraction known as **skeletal muscle tone (tonus)** that is important in supporting the body against gravity and in maintaining an upright posture. A basic neural mechanism involved in the maintenance of muscle tone is the stretch reflex (see Chapter 7). Consequently, the stretch reflex is important in maintaining an upright posture. For example, when the force of gravity causes the knees to buckle, the extensor muscles of the legs are stretched, initiating stretch reflexes that stimulate the alpha motor neurons of the extensor muscles. The contractions of the extensor muscles stiffen the legs, helping to support the individual.

Higher Centers

The excitability of the alpha motor neurons to skeletal muscles is influenced not only by nerve impulses from receptors involved in spinal reflexes, but also by nerve impulses from higher centers within the brain. Nerve impulses from these centers may be sent either to the alpha motor neurons themselves or to the intrafusal fibers of the muscle spindles. Impulses to the intrafusal fibers alter the sensitivity of the muscle spindles, and this, in turn, can alter the stimulation of the alpha motor neurons by way of the stretch reflex pathway.

The reticular formation and associated nuclei in the medulla-pons region (the vestibular nuclei) are intrinsically excitable, and they transmit stimulatory signals to motor areas of the spinal cord. Although the full activity of the reticular formation and vestibular nuclei is normally held in check by inhibitory signals from the basal ganglia, when a person is in a standing position continuous impulses are sent to the spinal cord. These impulses contribute to the excitation of the alpha motor neurons to the skeletal muscles, and they provide much of the intrinsic excitation required to maintain tone in the extensor muscles that support the body against gravity.

MAINTENANCE OF EQUILIBRIUM

Nerve impulses transmitted to the brain from receptors that provide information about the position and movement of the body or its parts strongly influence the activity of muscles involved in maintaining equilibrium or balance. Among these receptors are those of the eyes and the vestibular apparatus of the ears, as well as receptors in the skin, muscles, and joints.

It is difficult to assign varying degrees of importance to the different types of information employed in maintaining equilibrium. Visual information is important, but so are other types of information. For example, receptors in the neck provide information about the position of the head relative to the body. This information is important because the vestibular apparatus of the ears provides information only about the position or movement of the head itself. Thus, information from the neck receptors helps the system determine if signals from the vestibular apparatus are due to the movement of the head alone (such as bending the head to one side), in which case the equilibrium of the body may not be endangered, or to the movement of the whole body (such as tilting the entire body to one side), in which case the equilibrium of the body may be endangered, and adjustments may be necessary to prevent a fall (F9.8). Signals from the various receptors are integrated within the central nervous system to provide information about the position of the body and its parts in space, and this information helps determine the degree of activity in individual muscles that is required to maintain balance.

LOCOMOTION

The alternating flexion of one limb and extension of the other that occurs during locomotion (walking or running) depends on rhythmic patterns of activity occurring within the spinal cord. These patterns, it is thought, make use of reverberating circuits or interneurons to stimulate alternately the flexor and extensor muscles of each leg, while reciprocal inhibition mechanisms keep one leg extended when the other is flexed. These activity patterns are believed to be coordinated by the integrative actions of higher centers, including the cerebral cortex, the basal ganglia, the cerebellum, and the reticular formation. The higher centers control these activities according to the desires of the individual.

THE CEREBELLUM AND THE COORDINATION OF MOVEMENT

The cerebellum is an especially important controller of muscular activities, particularly very rapid activities such as typing or running. During movement, the cerebellum receives information about the signals being sent to the muscles from areas of the motor system such as the cerebral cortex, the basal ganglia, and the reticular formation. The cerebellum also receives information from muscle spindles, Golgi tendon organs, joint receptors, the vestibular apparatus, and the eyes about the position, rate of movement, inertia, momentum, and so forth of body parts. The cerebellum compares the information about the movement called for by the motor system with information about the actual status of the body parts. If necessary, it sends signals back to the motor system to bring the actual movement into line with the intended one.

The cerebellum appears to play an important role in providing the central nervous system with the ability to predict the future position of a body part in the next few hundredths of a second during a movement. Some time before a moving body part reaches its intended position, signals are sent from the cerebellum to slow the moving part and stop it at its intended point. This activity involves the excitation of antagonistic muscles near the end of a movement coupled with the inhibition of the muscles that started the movement.

The cerebellum is also important in maintaining body equilibrium. When the direction of movement changes, signals transmitted to the cerebellum from receptors such as the vestibular apparatus of the ears help a person anticipate an impending loss of equilibrium and take corrective action before the equilibrium loss actually occurs.

SKILLED MOVEMENTS

Almost all skilled movements (such as the hand movements involved in typing or playing a violin) are learned movements whose mastery requires repetition. Such movements are usually slow and awkward at first, but their speed and efficiency generally improve with practice. It has been proposed that as a skilled movement is learned, the pattern of neural activity necessary to perform the movement becomes stored within the brain, perhaps in the form of a facilitated neuronal circuit (that is, as a memory trace). When the learned, preprogrammed movement is to be repeated, this stored pattern is called forth, and the muscle contractions necessary to perform the movement are repeated in the orderly sequence dictated by the pattern. The precise parts of the brain in which such patterns may be stored are not known. However, the motor cortex, the sensory cortex, and deeper centers such as the basal ganglia and even the cerebellum may be involved. Of particular interest is a cortical area in front of the primary motor cortex called the premotor area. Electrical stimulation of this area sometimes produces skilled patterns of movement such as hand movements, and this area is thought to be particularly involved in the occurrence of learned movements that can be rapidly performed.

As a skilled movement is being learned, it is usually performed slowly and deliberately. Dur-

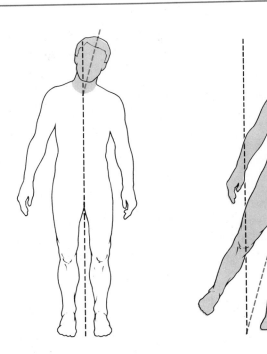

F9.8
Proprioceptive joint receptors in the neck (as well as receptors in other body areas) play an important role in maintaining equilibrium. They provide information to determine if signals from head region receptors such as the vestibular apparatus are due to the movement of the head alone or to the movement of the body as a whole.

ing this time, feedback from receptors in muscles and joints, from cutaneous receptors, and from the eyes is important in indicating the effectiveness and degree of success of the movement and in establishing a pattern of learned motor functions in the brain. Even after a precision movement has been mastered, feedback is important in monitoring and, if necessary, correcting the movement, if it is performed slowly. However, once they are learned, many skilled movements are performed so rapidly that there is no time for feedback control or correction to occur. Nevertheless, feedback is important in determining if the movement has been correctly performed and in correcting the skilled movement pattern if necessary so that the movement is executed properly the next time it is performed. The feedback signals are believed to be sent not only to the cerebellum but also to the sensory areas of the cerebral cortex which, in turn, relays signals to the motor cortex.

The primary motor area of the cerebral cortex was long believed to be the area principally responsible for initiating voluntary motor activity. However, recent studies have indicated that this is not the case. When monkeys were taught to per-

form simple hand movements in response to cues, it was found that neurons in the motor cortex, the cerebellum, and the basal ganglia all discharged impulses prior to the occurrence of the movement. Moreover, other studies indicated that cerebellar neurons become active before neurons of the motor cortex. These findings have led to the proposal that subcortical structures such as the cerebellum and basal ganglia are involved in initiating activity in motor cortex neurons. According to this view, signals related to the performance of learned, preprogrammed movements reach the motor cortex by way of the cerebellum and the basal ganglia. Moreover, it has been suggested that the cerebellum is particularly involved in the performance of rapid movements, the basal ganglia are important in the performance of slower movements, and the motor cortex is involved in the more precise integration of both rapid and slow movements.

LANGUAGE

Language is a complex form of communication in which spoken or written words represent objects or concepts. One person communicates with or transfers information to another primarily by language.

The areas of the cerebral cortex concerned with language are usually located in only one cerebral hemisphere—the left in 95% of people (F9.9). A portion of the left frontal lobe known as *Broca's area* is concerned with the motor processes of word formation. If Broca's area is damaged, the person's speech is slow and labored, and he or she has difficulty articulating words. The person can decide what he or she wants to say but cannot make the vocal system perform smoothly.

A portion of the left temporal lobe known as *Wernicke's area* is responsible for choosing appropriate words to express a person's thoughts and for attaching meanings to words. If Wernicke's area is damaged, the person can still articulate words, but the words are often inappropriate or even nonsense words. Moreover, a person with extensive damage to Wernicke's area loses all comprehension of spoken language.

F9.9
Brain areas associated with language.

Broca's area and Wernicke's area cooperate to produce spoken language. Wernicke's area determines the appropriate choice and sequence of words to express a person's thoughts, and transmits signals to Broca's area. In Broca's area the incoming signals are processed and a detailed motor program for the muscular movements of vocalization is generated. This program is transmitted to the primary motor cortex, which, in turn, sends nerve impulses to the speech muscles.

STUDY OUTLINE

INTEGRATIVE SPINAL REFLEXES A noxious stimulus can cause flexion of limb on ipsilateral side of body; commissural interneuron connections lead to extension of corresponding limb on contralateral side of body. **pp. 225–226**

MENTAL PROCESSES No area or structure of the brain acts entirely on its own. The control of particular functions involves more than one area of the brain, and each brain area is probably involved in many functions. **pp. 226–236**

The Electroencephalogram Record of electrical waves produced during brain activity; spontaneous waves present, even during unconsciousness; some correlation between brain waves and body activity.

Consciousness State of awareness; state in which thoughts or instants of awareness occur; result of activation or arousal of widespread areas of the brain.

THE AROUSED BRAIN Able to consciously react to stimuli; the reticular activating system is important in arousing the brain; general sensory input plays an important role in maintaining brain arousal.

SLEEP State of partial or complete unconsciousness from which a person can be aroused by appropriate stimuli; *slow-wave sleep*: high-amplitude, low-frequency EEG; *paradoxical sleep*: rapid-eye-movement sleep; dreams.

SLEEP-WAKEFULNESS CYCLE Awakening appears to occur when impulses to the cerebral cortex from the RAS reach a sufficient intensity to arouse the brain.
1. Positive-feedback responses involving brain areas such as the cerebral cortex as well as the peripheral musculature contribute to the activity of the RAS leading to wakefulness.
2. Over time, the excitability of the wakefulness circuit diminishes, the intensity of signals from RAS declines, and the individual becomes drowsy and falls asleep.
3. The depletion of chemical transmitter substances from positive-feedback loop neurons or the accumulation of inhibitory chemical substances may influence the excitability of the wakefulness circuit (for example, serotonin secretion may lead to sleep).

4. Brain-stem neurons that may contribute to slow-wave sleep by secreting serotonin may be involved in a feedback relationship with paradoxical sleep neurons of pons. The neurons of the pons may stimulate the serotonin-secreting neurons of the brain stem to take up serotonin, leading to wakefulness.

Attention
1. Ability of an individual to maintain selective awareness of some aspect of the environment; ability to respond selectively to a particular stimulus.
2. Interest and biological urgency influence priority given to particular stimuli.
3. Influenced by specific thalamic areas of reticular activating system.
4. Other central nervous system mechanisms can selectively inhibit or enhance afferent signal input to brain.

Emotions and Behavior Limbic system is important in emotions and emotional behavior patterns; it is also important in determining whether a particular activity is pleasant and rewarding or painful and punishing. Reward and punishment areas of brain have a great influence on a person's behavior. Monoamines such as norepinephrine, dopamine, and serotonin are involved in certain emotional and behavioral states and in learning.

Pain Signals to brain from pain receptors elicit sensation of pain and also lead to such activities as emotional reactions and behavioral responses.

REFERRED PAIN Inaccurate projection of pain sensation to site other than site of stimulus.

PHANTOM PAIN Person feels pain from amputated body part.

ANESTHESIA Anesthetics reduce sensitivity to painful stimuli. Local anesthetics can decrease permeability of nerve-cell membranes to sodium ions so neurons do not transmit impulses. General anesthetics render person unconscious, at least partially by inhibiting or depressing conduction in RAS.

Memory Ability to store experiences, thoughts, and sensations for later recall.

SHORT-TERM MEMORY Decays rapidly unless rehearsed; may depend on activation of reverberating circuits of neurons within the brain by signals from receptors.

LONG-TERM MEMORY May involve actual alterations in the structure or biochemistry of neurons or synapses; may involve changes in number or area of synaptic junctions between neurons; may involve facilitated neuronal circuits called memory traces (memory engrams).

Learning A relatively permanent change in behavior as a result of experience.
1. Classical conditioning: conditioned responses; produced by pairing an unconditioned stimulus with a conditioned stimulus.
2. Instrumental learning (operant conditioning): the outcome or result of a particular behavior is important in determining whether the behavior will be strengthened or suppressed.
3. Cognitive learning: recall and integrate previous sensory experiences; concept formation; abstract reasoning.

CONTROL OF BODY MOVEMENTS Nervous system coordinates and controls contractions of skeletal muscles. **pp. 236–239**

Support of the Body Against Gravity Involves reflexes and higher brain centers.

REFLEXES Extensor thrust reflex and muscle stretch reflexes are important in supporting the body against the force of gravity.

HIGHER CENTERS Nerve impulses from higher centers (such as the reticular formation and vestibular nuclei) contribute to excitation of alpha motor neurons, helping to maintain tone in extensor muscles that support the body against gravity.

Maintenance of Equilibrium Nerve impulses from various receptors provide information about the position and movement of the body or its parts. These signals are integrated within the central nervous system, and they help determine the degree of activity in individual muscles that is required to maintain balance.

Locomotion
1. Believed to depend on rhythmic patterns of activity occurring within the spinal cord that involve reverberating neuronal circuits and reciprocal inhibition.
2. The rhythmic activity patterns are coordinated to produce effective locomotion by the integrative actions of higher centers (such as the cerebral cortex, basal ganglia, cerebellum, reticular formation).
3. The higher centers control the activities according to desires of the individual.

The Cerebellum and the Coordination of Movement
1. During movement, the cerebellum receives signals from motor areas about the intended movement and from the periphery about the actual movement.
2. The cerebellum compares information and, if necessary, sends signals to motor areas to bring actual movement into line with intended movement.
3. The cerebellum appears to play an important role in providing the central nervous system with the ability to predict the future position of a body part during a movement.
4. The cerebellum is important in the maintenance of equilibrium.

Skilled Movements
1. Almost all skilled movements are learned movements that require repetition to master.
2. As a movement is learned, the pattern of neural activity necessary to perform it may become laid down in the brain.
3. When the movement is to be repeated, the pattern is called forth and the muscular contractions necessary to perform the movement are repeated.
4. Premotor area of the cortex may be important in pattern storage.
5. Sensory feedback is important in learning skilled movements and establishing patterns of learned motor function in brain; also involved in monitoring and perhaps correcting pattern of movement after movement is mastered.
6. Cerebellum and basal ganglia appear to be involved in initiating activity in neurons of the motor cortex during learned, preprogrammed movements; cerebellum is particularly involved in rapid movements, basal ganglia in slow movements.

LANGUAGE Complex form of communication in which spoken or written words represent objects or concepts; Broca's area is concerned with motor processes of word formation; Wernicke's area is responsible for choosing appropriate words to express thoughts and for attaching meanings to words. **p. 239**

SELF-QUIZ

1. During the flexor reflex and the associated crossed extensor reflex the: (a) ipsilateral flexor muscles are stimulated; (b) contralateral extensor muscles are inhibited; (c) ipsilateral extensor muscles are stimulated.
2. Electroencephalogram records during slow-wave sleep show waves of: (a) low amplitude; (b) low frequency; (c) no amplitude or frequency.
3. Electroencephalogram records of an alert or awake brain show waves of: (a) low amplitude and high frequency; (b) high amplitude and low frequency; (c) high amplitude and high frequency.
4. Destruction of the reticular activating system produces: (a) temporary loss of memory; (b) permanent coma; (c) an inability to sleep soundly.
5. General sensory input may play an important role in maintaining the arousal of the brain. True or False?
6. REM sleep is the same as: (a) slow-wave sleep; (b) drowsy sleep; (c) paradoxical sleep.
7. Neurons of the pons that may be involved in the occurrence of paradoxical sleep contain an abundance of: (a) norepinephrine; (b) serotonin; (c) substance P.
8. The previous experience of an individual appears to be important in determining which sensory stimuli may warrant the individual's attention at any particular moment. True or False?
9. Brain structures that are particularly important in relation to emotions form the: (a) reticular activating system; (b) limbic system; (c) sensorimotor system.
10. The reward and punishment regions of the brain have a great influence on a person's behavior. True or False?
11. Local anesthetics are believed to exert their effect primarily by inhibiting the reticular activating system. True or False?
12. In order to keep information in the short-term memory for an indefinite period of time: (a) an alteration in the structure of synapses must occur; (b) an alteration in the structure of neurons must occur; (c) rehearsal is necessary.
13. Rats that have been trained to perform various activities show a(an): (a) increase in the total number of cortical neurons; (b) increase in thickness of the cerebral cortex; (c) decrease in the size of the pons.
14. Learning in which the outcome of a particular behavior is important in determining whether the behavior will be strengthened or suppressed is called: (a) classical conditioning; (b) instrumental learning; (c) cognitive learning.
15. Impulses sent to motor areas of the spinal cord from the reticular formation and vestibular nuclei contribute to the excitation of the alpha motor neurons to the skeletal muscles. True or False?
16. The alternating flexion of one limb and extension of the other that occurs during locomotion is believed to depend on rhythmic patterns of activity occurring within the: (a) spinal cord; (b) pons; (c) cerebral peduncles.
17. The cerebellum appears to play an important role in: (a) coordinating body movements; (b) emotion; (c) attention.
18. Almost all skilled movements are learned movements that require varying amounts of repetition to master. True or False?
19. Which brain structure may be particularly involved in the performance of learned, preprogrammed slow movements? (a) hypothalamus; (b) basal ganglia; (c) pons.
20. The brain area concerned with the motor processes of word formation is: (a) Wernicke's area; (b) Penfield's area; (c) Broca's area.

LEARNING OBJECTIVES

After completing this chapter, you should be able to:

- Describe two ways that hormones exert their effects at the cellular level.
- List the hormones of the major endocrine glands and describe their effects.
- Explain how the release of pituitary hormones is influenced by the nervous system.
- Explain the application of feedback control to endocrine regulation.
- Describe the synthesis, storage, and release of the thyroid hormones.
- Describe the control of the adrenal medulla.
- Describe four factors that influence the release of aldosterone.
- Cite the symptoms of adrenal cortical hypofunction and hyperfunction, and relate them to the effects of the adrenal cortical hormones.
- Distinguish between the roles of glucagon and insulin in controlling blood-glucose levels.
- Cite the symptoms of diabetes mellitus, and relate them to the effects of the hormone insulin.

CHAPTER CONTENTS

THE ENDOCRINE SYSTEM

The endocrine system is a regulatory system that coordinates and integrates many body processes. Thus, its role is similar to that played by the nervous system. In fact, the endocrine and nervous systems are very closely related. Like the nervous system, the endocrine system contributes importantly to the maintenance of homeostasis. It helps maintain a relative constancy of nutrients, salts, fluid volume, tonicity, and temperature in the internal environment of the body. It also influences growth, maturation, reproduction, and other body functions.

This chapter provides an overview of the activities of the endocrine system and the major endocrine glands. The activities of individual endocrine structures are considered further in those chapters dealing with the specific body processes they influence.

BASIC ENDOCRINE FUNCTIONS

Although a precise definition of an endocrine gland is difficult to formulate, **endocrine glands** are generally ductless glands composed of epithelial cells that secrete their products into the extracellular space around the cells and from which the products enter the bloodstream (F10.1). (In contrast, exocrine glands typically secrete their products by way of ducts onto body surfaces—either external surfaces like the skin or internal surfaces like the lining of the digestive tract.) Besides the clearly recognized endocrine glands such as the thyroid gland and the adrenal glands, other structures—for example, the gastrointestinal tract—also exhibit endocrine activity.

The products of the endocrine structures—the **hormones**—are chemical messengers that are carried by the blood to the body's cells where they exert their effects. Hormones do not fall into any easily defined class of chemical substances. Some are steroids (for example, cortisol), others are proteins (for example, growth hormone), still others are polypeptides (for example, parathyroid hormone) or are closely related to amino acids (for example, epinephrine). The major endocrine glands and their hormones are summarized in Table 10.1.

TRANSPORT OF HORMONES

Within the bloodstream, many hormones, particularly the steroid hormones, are bound to specific carrier proteins, and the free and bound forms of a hormone exist in equilibrium with one another. The reversible association between a hormone and a carrier protein provides both a storage and buffer system for the hormone. Usually only a small fraction of the hormone is present in the free form, and the protein-bound portion is a readily available reserve. The binding of a hormone to a carrier protein can protect the body against excessive concentrations of the hormone, and it can delay the degradation of the hormone. The binding can also protect certain small hormone molecules from excretion by the kidneys.

MECHANISMS OF HORMONE ACTION

The bloodstream transports hormones throughout the body. However, a particular hormone does not necessarily affect all the body's cells. Only certain cells, called **target cells,** respond to any given hormone. The target cells of a particular hormone possess receptors to which molecules of the hormone can attach. When the hormone molecules attach to the receptors, a series of events is triggered that influences such cellular activities as the rates of certain reactions or the permeability of plasma membranes, leading ultimately to the observed effects of the hormone on body processes. Receptors for a number of hormones, particularly the protein, peptide, and catecholamine hormones, are located on the plasma membranes of their target cells. Receptors for some hormones, particularly the steroid hormones and possibly several nonsteroid hormones as well, are located within their target cells.

Different hormones affect the activities of their target cells by many different mechanisms. However, two important general mechanisms by which hormones act are by utilizing intracellular mediators and by activating genes within cells.

Utilization of Intracellular Mediators

Many hormones (and also other substances) that bind to receptors on plasma membranes influence cellular function by way of intracellular mediators. One intracellular mediator is a substance known as *3',5' cyclic adenosine monophosphate (cyclic AMP)*. Cyclic AMP is formed from adenosine triphosphate (ATP) by the enzyme adenylate cyclase, which is bound to the inner surface of the plasma membrane. When molecules of a hormone that utilizes cyclic AMP as an intracellular mediator attach to receptors on the plasma membrane, the activity of adenylate cyclase is altered, perhaps by way of intermediate molecules within the membrane called transducers (F10.2). The alteration of adenylate cyclase activity leads to changes in the level of cyclic AMP within the cell,

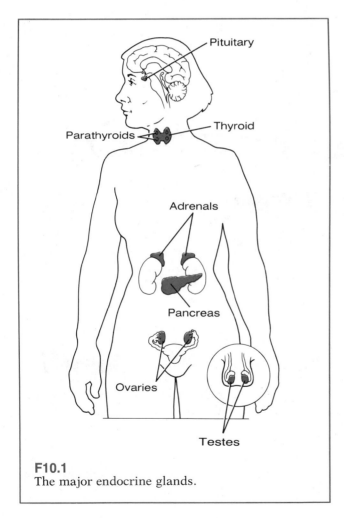

F10.1
The major endocrine glands.

Table 10.1 *Major Endocrine Glands and Their Hormones*

Gland	Hormone	Representative effects	Selected disorders
Pituitary F10.4 Neurohypophysis (hormones are actually manufactured in brain)	Antidiuretic Hormone (ADH)	Promotes reabsorption of water from urine-forming structures of kidneys	Undersecretion leads to diabetes insipidus.
	Oxytocin	Stimulates contraction of uterine smooth muscle and myoepithelial cells around alveoli of mammary glands. As such, it is involved in birth processes and milk "let down" during nursing.	
Adenohypophysis	Follicle-Stimulating Hormone (FSH) and Luteinizing Hormone (LH)	Stimulate gonads to produce gametes and sex hormones	Undersecretion causes gonadal inactivity in males, and menstrual failure in females.
	Thyrotropin (TSH)	Stimulates thyroid gland to secrete thyroid hormones	Undersecretion leads to symptoms of hypothyroidism.
	Adrenocorticotropin (ACTH)	Stimulates adrenal cortex to secrete glucocorticoids (such as cortisol)	Undersecretion leads to symptoms of adrenal cortical insufficiency. Oversecretion leads to symptoms of adrenal cortical hyperfunction.
	Growth Hormone (GH)	Stimulates growth in general, and growth of skeletal system in particular. It also affects metabolic functions.	Undersecretion produces pituitary dwarfs. Oversecretion causes gigantism or, in adults, acromegaly.
	Prolactin	Involved in milk secretion in females.	Undersecretion may cause failure to lactate after giving birth. Oversecretion may lead to lactation without recently having given birth.
Thyroid F10.10	Thyroxine (T_4, or tetraiodothyronine) and Triiodothyronine (T_3)	Increase oxygen consumption and heat production (calorigenic effect). Important for	Undersecretion leads to symptoms of hypothyroidism, possibly causing cretinism in children, or

Table 10.1 *Major Endocrine Glands and Their Hormones (continued)*

Gland	Hormone	Representative effects	Selected disorders
		normal growth and development; these hormones affect many metabolic processes.	myxedema in adults. Oversecretion leads to hyperthyroidism possibly causing Graves' disease.
	Calcitonin	Lowers blood calcium and phosphate levels	
Parathyroids F10.14	Parathyroid Hormone	Affects calcium and phosphate metabolism to raise plasma calcium levels and decrease plasma phosphate levels	Undersecretion leads to nervous excitability and tetanus. Oversecretion leads to bone decalcification, and calcification of soft tissues such as the kidneys may occur.
Adrenals F10.15 Medulla	Epinephrine	Affects carbohydrate metabolism generally leading to hyperglycemia. Constricts vessels in skin, mucous membranes, and kidneys.	
	Norepinephrine	Increases heart rate and force of contraction of cardiac muscle, constricts blood vessels in almost all areas of the body	
Cortex	Mineralocorticoids (such as aldosterone)	Promote reabsorption of sodium, and excretion of potassium from urine-forming structures of kidneys	Undersecretion may lead to decreased fluid volume and circulatory difficulties, and contribute to Addison's disease. Oversecretion may cause increased fluid volume, edema, and hypertension.
	Glucocorticoids (such as cortisol)	Affect many aspects of carbohydrate metabolism; generally leads to mobilization and hyperglycemia	Undersecretion contributes to Addison's disease. Oversecretion leads to Cushing's syndrome.

Table 10.1 *Major Endocrine Glands and Their Hormones (continued)*			
Gland	Hormone	Representative effects	Selected disorders
Pancreas F10.18	Insulin	Affects many aspects of carbohydrate metabolism; generally causes hypoglycemia	Relative deficiency of insulin leads to hyperglycemia and diabetes mellitus.
	Glucagon	Affects metabolism in fashion generally opposite of insulin; generally causes hyperglycemia	
Gonads		The gonadal hormones are involved in the processes of reproduction. Their functions are discussed in Chapter 20.	
Ovaries	Estrogens Progesterone		
Testes	Androgens (such as testosterone)		

which, in turn, can affect various cellular functions such as enzyme activities, secretory activities, and cell permeability. The specific effects of changes in cyclic AMP levels on particular cells depend on the characteristics of the cells. For example, when the pancreatic hormone glucagon binds to liver cells, elevated cyclic AMP concentrations promote the breakdown of glycogen into glucose, and when thyroid-stimulating hormone from the pituitary gland binds to cells of the thyroid gland, increased cyclic AMP leads to the release of the thyroid hormones.

One of the major functions of cyclic AMP is to activate a class of enzymes called *protein kinases.* The protein kinases attach phosphate groups to specific proteins, which are often enzymes themselves. This phosphorylation can activate or inhibit enzymatic proteins, leading to the stimulation or inhibition of various cellular reactions or processes. There are different protein kinases, and there is generally more than one protein kinase in a given type of cell. Each protein kinase phosphorylates a different protein, and therefore influences different cellular reactions or processes. Thus, the effects of cyclic AMP in a given cell depend at least in part on the particular protein kinases present.

Other substances besides cyclic AMP have been suggested as possible mediators of hormonal effects. Many of the effects attributed to cyclic AMP are dependent on, mimicked by, or reinforced by calcium ions, and it has been proposed that calcium ions are involved in the mediation of hormonal effects, in some cases regulating or being regulated by cyclic AMP. Researchers have also suggested that 3',5' cyclic guanosine monophosphate (cyclic GMP) is involved in the mediation of hormonal effects. Cyclic GMP is produced from guanosine triphosphate (GTP) by the enzyme guanylate cyclase, which occurs in the cytoplasm as well as in plasma membranes. Cyclic GMP often promotes responses that are opposite to those promoted by cyclic AMP, and it has been proposed that cyclic AMP and cyclic GMP mediate the actions of different hormones that produce opposing effects on target cells.

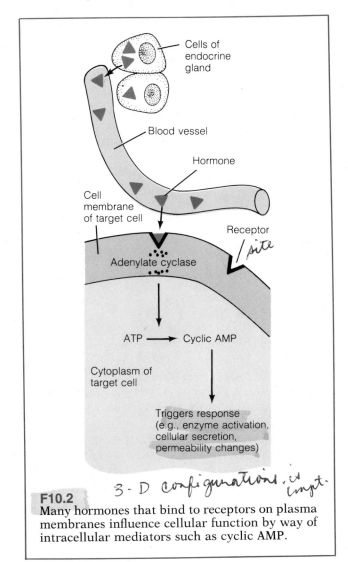

Cells of
endocrine
gland

Blood vessel

Hormone

Cell
membrane
of target cell

Receptor
/site

Adenylate cyclase

ATP → Cyclic AMP

Cytoplasm of
target cell

Triggers response
(e.g., enzyme activation,
cellular secretion,
permeability changes)

3-D configurations is impt.

F10.2
Many hormones that bind to receptors on plasma membranes influence cellular function by way of intracellular mediators such as cyclic AMP.

Activation of Genes

Some hormones, particularly the steroid hormones, exert at least some of their effects by activating genes within a cell (F10.3). In this activity, hormone molecules enter a cell and combine with receptors. The hormone-receptor complex interacts either directly or indirectly with the genetic material to activate certain genes. This gene activation leads to the synthesis of m-RNA and ultimately to the production of proteins (for example, enzymes) that influence cellular reactions or processes.

HORMONAL INTERRELATIONSHIPS

There is a strong interrelationship between different endocrine glands and hormones. In fact, almost every activity influenced by the endocrine system is affected by a balance among different hormones acting either together or in sequence. For example, hormones secreted by the pituitary gland, the thyroid gland, the adrenal glands, and the pancreas all influence carbohydrate metabolism.

In some cases, one hormone affects the activity of a second hormone. One way that this occurs is for the first hormone to increase or decrease the number or affinity of the target-cell receptors for the second hormone, and thereby increase or decrease the ability of the second hormone to influence its target cells.

A number of hormones influence the rates of production of other hormones. As discussed in the following sections, the pituitary gland produces several hormones that act in this fashion.

PITUITARY GLAND

The **pituitary gland (hypophysis),** which is located beneath the brain, has two major divisions: the neurohypophysis and the adenohypophysis (F10.4). The **neurohypophysis** consists of the **lobus nervosus** (neural lobe, or **pars nervosa**) and the **infundibulum.** The **adenohypophysis** includes the **pars distalis, pars tuberalis,** and **pars intermedia.** The pars intermedia, however, is virtually missing from the human pituitary, although it is present in most other vertebrates.

The pituitary gland can also be divided into an **anterior lobe,** which includes the pars distalis and pars tuberalis, and a **posterior lobe,** which includes the lobus nervosus and, when present, the pars intermedia.

There is a close relationship between the pituitary and the nervous system. The neurohypophysis is directly connected to the brain by way of the infundibulum. The adenohypophysis is also closely associated with the brain, although it is not directly connected to it. Within the lower hypothalamic region of the brain (that is, the median eminence) is a capillary bed known as the *primary capillary plexus* (F10.5). Blood in the primary capillary plexus flows by way of *hypothalamic-hypophyseal portal veins* directly to a *secondary capillary plexus* in the adenohypophysis. Thus, the brain and the adenohypophysis are linked by the circulatory system.

NEUROHYPOPHYSEAL HORMONES AND THEIR EFFECTS

The lobus nervosus releases two peptide hormones (Table 10.1).

Antidiuretic Hormone

Antidiuretic hormone (ADH), also called **vasopressin,** promotes the reabsorption of water from the urine-forming structures of the kidneys and thus helps retain fluid within the body. At moderate to high concentrations, ADH constricts blood vessels called arterioles.

Oxytocin

Oxytocin stimulates the smooth muscles of the uterus. It also promotes the contraction of myoepithelial cells surrounding the saclike alveoli of the mammary glands, resulting in the "let down" of milk during lactation. Oxytocin has no known function in the male.

RELEASE OF NEUROHYPOPHYSEAL HORMONES

Although both ADH and oxytocin are released from the lobus nervosus, neither hormone is manufactured there. Rather, both are manufactured in certain hypothalamic nuclei by specialized neural cells called **neurosecretory cells.** These cells possess long processes (axons) that extend from their cell bodies in the hypothalamic nuclei along the infundibulum to the lobus nervosus (F10.6). ADH and oxytocin synthesized in the brain by neurosecretory cells move down the cell processes to the lobus nervosus, from which they are released. Both hormones are transported from the brain to the lobus nervosus attached to molecules of a carrier protein called *neurophysin.*

In addition to synthesizing and transporting ADH and oxytocin, the specialized neurosecretory

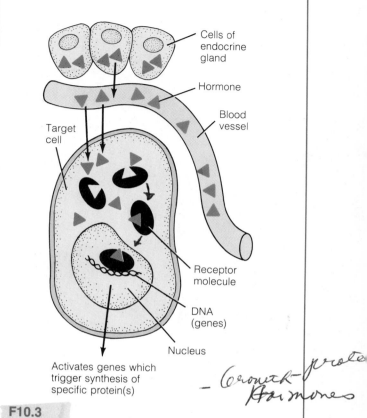

Activates genes which trigger synthesis of specific protein(s)

F10.3

Mechanism by which some hormones, particularly the steroid hormones, activate genes within a cell.

— Growth — protein Hormones
↳ are all lipid

cells also conduct nerve impulses. This activity is involved in controlling the release of the neurohypophyseal hormones.

Because ADH and oxytocin are synthesized in the hypothalamus by specialized neurosecretory cells, the synthesis and release of these hormones can be influenced by neural inputs to the brain. Information from appropriate receptors is neurally relayed to the brain to alter the production and release of ADH or oxytocin in accordance with the regulatory requirements of the body.

ADENOHYPOPHYSEAL HORMONES AND THEIR EFFECTS

The adenohypophysis of the pituitary releases several hormones (Table 10.1).

Insulin — limited to cell membrane — cause of activity of a substance — glucose

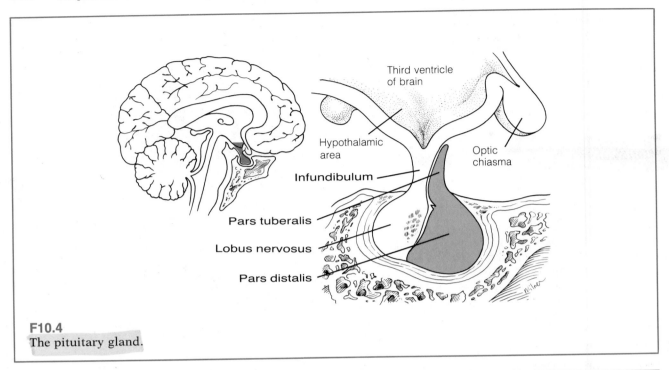

F10.4
The pituitary gland.

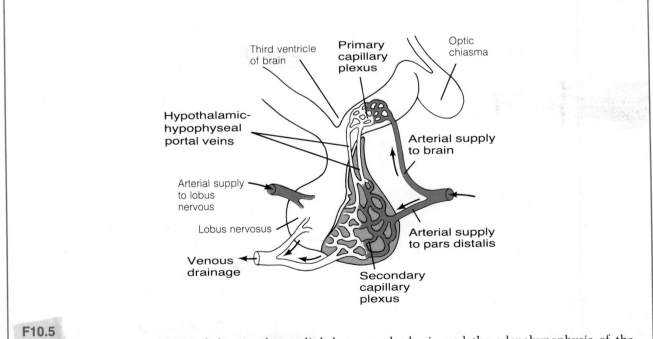

F10.5
Diagrammatic representation of the circulatory link between the brain and the adenohypophysis of the pituitary gland.

Gonadotropins

Two of the adenohypophyseal hormones are called **gonadotropins** because they particularly affect the gonads (ovaries and testes). In females, **follicle-stimulating hormone (FSH)** stimulates the development of structures called follicles within the ovaries and the secretion of the estrogenic female sex hormones. In this activity, FSH works in conjunction with the other pars distalis gonadotropin, **luteinizing hormone (LH).** Surging levels of LH together with FSH lead to ovulation and the formation of a structure called the corpus luteum from an ovarian follicle. The corpus luteum produces estrogenic hormones, as well as the female sex hormone progesterone.

Both FSH and LH are also present in males. FSH is involved in the development and maturation of sperm. LH, which is sometimes called **interstitial-cell–stimulating hormone (ICSH)** in males, stimulates the interstitial cells of Leydig of the testes to produce the male sex hormones (that is, androgens such as testosterone). Since the male sex hormones are themselves involved in sperm production, LH can be regarded as important in this process as well, and FSH may also play some role in androgen production. Both FSH and LH are glycoproteins (proteins combined with carbohydrate molecules).

Thyrotropin

Thyrotropin, also called **thyroid-stimulating hormone (TSH),** stimulates the synthesis and release of the thyroid hormones from the thyroid gland. Like FSH and LH, thyrotropin is a glycoprotein.

Adrenocorticotropin

Adrenocorticotropin, or **adrenocorticotropic hormone (ACTH),** stimulates the release of hormones from the cortical region of the adrenal gland, particularly cortisol and other glucocorticoid hormones that are active in carbohydrate metabolism. ACTH is a polypeptide hormone composed of 39 amino acids.

Growth Hormone

Growth hormone (GH), also called **somatotropin,** stimulates growth in general, and the growth of

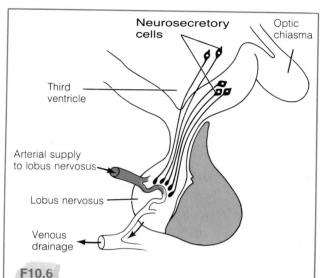

F10.6

Neurosecretory cells in the hypothalamic region of the brain. These cells manufacture antidiuretic hormone (ADH) and oxytocin, which are transported along the axons of the neurosecretory cells to the lobus nervosus of the pituitary. From the lobus nervosus, the hormones are released into the circulation.

the skeletal system in particular. GH enhances the entrance of amino acids into cells and favors their incorporation into protein. In addition, it increases the release of fatty acids from adipose tissue into the blood. The fatty acids can be taken up and used as energy sources by most cells. Growth hormone also promotes glucose formation from liver glycogen. Although GH may initially increase the rate of glucose uptake by adipose tissue, heart, and skeletal muscle cells, after some hours glucose uptake and utilization are reduced, an effect that tends to contribute to high blood-glucose levels (hyperglycemia). In many instances, growth hormone acts synergistically with other hormones to enhance their effects. GH is a protein hormone consisting of 191 amino acids.

A number of growth hormone effects are brought about through the mediation of a group of growth-promoting peptides found in the blood plasma that are collectively called **somatomedins.** For example, the ability of growth hormone to stimulate cartilage and bone growth occurs, at least in part, by way of the somatomedins, which directly stimulate cartilage and bone. Growth hormone stimulates the production of somatomedins by the liver and perhaps by the muscles and kidneys as well.

The release of growth hormone increases in response to declining concentrations of blood glucose and also in response to elevated blood levels of certain amino acids, particularly arginine. An increased secretion of growth hormone is associated with fasting, hypoglycemia (low blood-glucose levels), exercise, and certain kinds of stress.

Prolactin

Prolactin is involved in the initiation and maintenance of milk secretion in females. Like oxytocin, it plays no clearly recognized role in males. The structure of prolactin, which is a protein hormone, resembles that of GH.

OTHER ADENOHYPOPHYSEAL SUBSTANCES

In addition to the hormones just discussed, the adenohypophysis contains other substances that may have hormonal effects. One of these is a substance called **beta-lipotropin (β-LPH)**, which is a polypeptide 91 amino acids long. There is evidence that β-LPH stimulates the release of the hormone aldosterone, which is involved in sodium and potassium metabolism, from the cortical regions of the adrenal glands. There is also evidence that it causes lipid breakdown in fat cells.

The adenohypophysis also contains several forms of a substance called **melanocyte-stimulating hormone (MSH)**. A form of MSH called alpha-MSH causes changes in the skin color of some lower vertebrates such as frogs, but the roles of the MSHs in higher vertebrates such as humans is unclear. When a form of MSH called gamma-MSH is injected into a cerebral ventricle of an experimental animal, the animal's body temperature decreases. Moreover, there is evidence that certain cells of the hypothalamus manufacture gamma-MSH and use it as a neurotransmitter. Thus, gamma-MSH (and other substances) may have neural as well as hormonal effects.

PRO-OPIOMELANOCORTIN AND THE PRODUCTION OF ADENOHYPOPHYSEAL HORMONES

A number of the peptide and protein hormones of the adenohypophysis, as well as other peptide and protein substances, are believed to be produced initially as parts of large precursor molecules that are subsequently split by enzymes into fragments of precise length and biological activity. For example, cells in the adenohypophysis and also in the hypothalamus of the brain synthesize a protein molecule called *pro-opiomelanocortin* that contains the amino acid sequences of ACTH, β-LPH, and several MSHs, as well as the sequences of an enkephalin and an endorphin (F10.7). Thus, pro-opiomelanocortin is apparently a precursor molecule from which ACTH, β-LPH, and other substances can be formed.

It should be noted that not all cells containing a precursor like pro-opiomelanocortin necessarily secrete the same hormones or other substances. In some cells a precursor may be split to produce one substance, and in others it may be split to form another substance. It is also possible that the same cell may produce more than one substance from a precursor molecule, or that it may form one substance at one time and another substance at some other time. In any event, the release of each hormone or other substance is precisely controlled, and there is no massive, undifferentiated secretion of multiple materials. The mechanisms that control the release of the adenohypophyseal hormones are discussed in the following sections.

RELEASE OF ADENOHYPOPHYSEAL HORMONES

The brain exercises a good deal of control over the synthesis and release of the adenohypophyseal hormones even though the adenohypophysis is not directly connected to the brain.

Hormone Releasing and Inhibiting Substances

Within the brain, specialized neurosecretory cells manufacture a group of molecules known collectively as **releasing** and **inhibiting substances.** If the chemical identity of a particular releasing or inhibiting substance is known, the substance is referred to as a hormone. If its chemical composition is unknown, it is referred to as a releasing or inhibiting factor. The neurosecretory cells in the brain that manufacture releasing or inhibiting substances secrete their products in the region of the primary capillary plexus (F10.8.) The releasing or inhibiting substances travel through the circulation by way of the hypothalamic-hypophyseal portal veins to the adenohypophysis, where they influence the release of adenohypo-

physeal hormones.

The following releasing or inhibiting substances have been either postulated or identified: **gonadotropin-releasing hormone (GnRH),** which stimulates the release of both FSH and LH; **thyrotropin-releasing hormone (TRH); corticotropin-releasing factor (CRF); prolactin-releasing factor (PRF); and growth-hormone–releasing factor (GH-RF).** In addition, there appears to be a **prolactin-inhibiting factor (PIF),** and there is an inhibiting factor for growth hormone called **somatostatin.**

F10.7

The amino acid sequence of pro-opiomelanocortin (as determined in cattle). This molecule incorporates the amino acid sequences of ACTH (positions 1 through 39), beta-lipotropin (42-132), an enkephalin (102-106), and an endorphin (102-132). It also contains several forms of melanocyte-stimulating hormone, or MSH (alpha-MSH: 1-13; beta-MSH: 82-99; and gamma-MSH: -44 through -55).

Nervous System Effects on Adenohypophyseal Hormone Release

Regardless of whether a particular factor or hormone promotes or inhibits the release of certain adenohypophyseal hormones, the adenohypophyseal hormones are under the influence of the nervous system. For example, information neurally relayed to the brain can alter the secretion of various releasing or inhibiting substances, and an altered secretion of these substances, in turn, influences the release of adenohypophyseal hormones. In addition, since many adenohypophyseal hormones stimulate the release of other hormones (for example, those of the thyroid, adrenals, and gonads), a pathway exists by which neural activity can ultimately influence a wide variety of endocrine functions. Even brain activities of the sort that are responsible for emotions can affect hormonally controlled events. For instance, emotional trauma can upset the menstrual cycle or nursing.

PITUITARY DISORDERS

Pituitary malfunctions can be caused by disorders of the gland itself or by difficulties involving the releasing or inhibiting substances from the brain. Pituitary disorders can involve the functioning of major segments of the gland, or they may be limited to disorders in the release of a single pituitary hormone.

Individual hormone disorders produce a number of observable conditions. Deficiencies of the gonadotropins upset normal gonadal function, and serious deficiencies can lead to gonadal inactivity in males and to menstrual failure in females. TSH deficiency results in underactivity of the thyroid (hypothyroidism). Deficiencies of prolactin may result in a failure of lactation after delivery, whereas prolactin overproduction can lead to lactation in a woman who has not recently given birth to a child. ACTH deficiency leads to adrenal cortical insufficiency, whereas excess ACTH produces symptoms of adrenal cortical hyperfunction. A deficiency of growth hormone in the young can result in impaired growth and may produce pituitary dwarfs. An excess of growth hormone before the growth in length of the long bones is complete results in great height and gigantism. After this time, an excess secretion of growth hormone leads to acromegaly. In acromegaly, height does not increase as in gigantism, but bones thicken and soft tissues increase in size. Structures such as the jaw, hands, feet, and tongue enlarge, and there is a coarsening of the facial features. A deficiency of antidiuretic hormone results in diabetes insipidus, which is characterized by copious urination.

F10.8 Neurosecretory cells in the hypothalamic region of the brain produce releasing substances that travel through the circulation to the pars distalis of the pituitary.

FEEDBACK CONTROL OF HORMONE RELEASE

Feedback control, particularly negative feedback (Chapter 1), is an important regulator of endocrine function. For example, a stimulus may promote an increased secretion of a releasing substance from the hypothalamus (F10.9). The increased secretion of the releasing substance causes an augmented release of a pituitary hormone, which, in turn, stimulates a target endocrine gland to release more of its hormone. The increased amount of hormone released by the target endocrine gland, in addition to exerting its normal physiological effects, may provide negative feedback to either the hypothalamus to inhibit the secretion of the releasing substance, or to the pituitary gland to inhibit the release of the pituitary hormone. Either occurrence leads to a diminished release of hormone from the target endocrine gland. This particular type of feedback pathway (from target gland to pituitary or brain) is called a *long feedback loop*.

Additionally, the pituitary hormones themselves may feed back to the brain to influence the secretion of releasing or inhibiting substances. This feedback pathway is called a *short feedback loop*. Both the long and short feedback loops provide mechanisms by which endocrine function can influence the activity of the endocrine system itself.

THRYOID GLAND

The **thyroid gland** is located around the trachea just below the larynx (F10.10). It is well vascularized and is composed of hollow balls of cells called follicles that are bound together with connective tissue (F10.11). The central region of each follicle contains a protein substance, the *colloid*, which is a stored form of thyroid hormones.

THE THYROID HORMONES: TRIIODOTHYRONINE AND THYROXINE

The thyroid hormones contain iodine. Most of the body's iodine is obtained from the diet and appears in the blood as ionic iodide (I^-), which is actively taken up by the thyroid follicles (F10.12). Within the follicles, the iodide is converted to an active form of iodine, which is attached to the amino acid tyrosine to form monoiodotyrosine

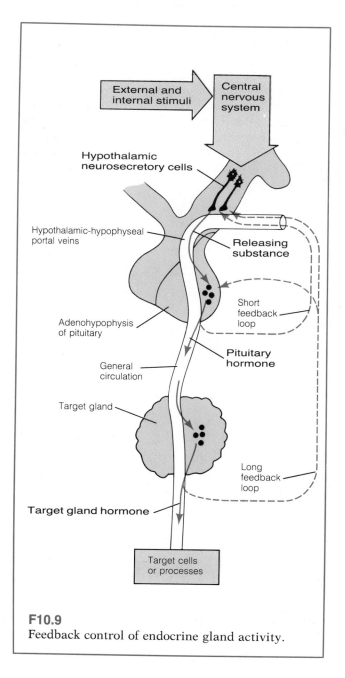

F10.9
Feedback control of endocrine gland activity.

(MIT) (F10.13). A second iodine can also be attached to the tyrosine to form diiodotyrosine (DIT). Two diiodotyrosines are then coupled to form **tetraiodothyronine (T_4, or thyroxine).** Alternatively, a monoiodotyrosine and a diiodotyrosine can combine to form **triiodothyronine (T_3).** These synthetic reactions do not occur with the tyrosines in the free state. Rather, the tyrosines are part of a glycoprotein molecule known as *thyroglobulin*, which is manufactured by the follicle cells and secreted into the colloid area. The iodination of the tyrosines appears to occur at the cell-colloid interface as the thyroglobulin is secreted, and the iodinated tyrosines are then coupled to one another. As a result, thyroglobulin molecules containing MIT, DIT, T_3, and T_4 are produced, and these are stored in the colloid areas of the follicles.

When the thyroid is actively secreting, thyroglobulin molecules are taken into the follicle cells by endocytosis. Within the cells, lysosomes fuse with the endocytic vesicles, and lysosomal enzymes break down the thyroglobulin molecules, freeing MIT, DIT, T_3, and T_4. The MIT and DIT are deiodinated, and the iodine can be recycled for use in other iodinations. The T_3 and T_4, which are collectively referred to as the thyroid hormones, pass out of the follicle cells and enter the bloodstream. Within the bloodstream almost all of the thyroid hormones are bound to plasma proteins such as *thyroid-binding globulin (TBG)*. The thyroid normally produces about 10% T_3 and 90% T_4. However, in the tissues, much of the T_4 is converted to T_3, and there is increasing evidence that T_3 is the major active form of the thyroid hormones at the cellular level.

Among the most evident effects of the thyroid hormones are those associated with metabolism. The administration of thyroid hormones increases the body's oxygen consumption and heat production (the calorigenic effect). Although most body tissues are responsive to this influence, some tissues—including the spleen, retina, brain, uterus, and testes—are not. When thyroid hormones are administered, the calorigenic effect normally does not become evident for some time, but it may last several days.

The administration of moderate doses of thyroxine to thyroidectomized rats stimulates protein synthesis, but large doses inhibit protein synthesis and increase the free amino acids in the plasma. In adult humans, the calorigenic effect produced by the administration of thyroxine is accompanied by an increased breakdown of body protein. In children, the thyroid hormones are essential for normal growth and maturation, and growth is retarded by hypothyroidism. The administration of small doses of thyroxine to hypothyroid children increases protein synthesis and favors growth, but the administration of large doses of thyroxine to either normal or hypothyroid children produces protein breakdown, as it does in adults.

The thyroid hormones affect almost all aspects of carbohydrate metabolism, and many of their influences depend on or are modified by other hormones, particularly the catecholamines and insulin. The thyroid hormones apparently influence

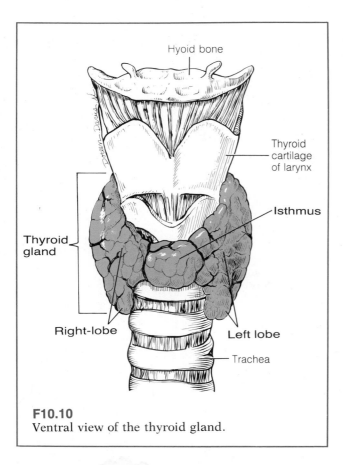

F10.10
Ventral view of the thyroid gland.

Hyoid bone

Thyroid cartilage of larynx

Isthmus

Thyroid gland

Right-lobe

Left lobe

Trachea

the actions of the catecholamine epinephrine in breaking down glycogen (glycogenolysis) and elevating blood glucose (hyperglycemia). They also appear to potentiate the glycogen-synthesizing and glucose-utilizing effects of insulin. In rats, small doses of T_4 enhance glycogen synthesis in the presence of insulin, whereas large doses increase liver glycogenolysis, thus causing glycogen depletion. The rate of intestinal absorption of glucose is enhanced by the thyroid hormones as is the uptake of glucose by muscle and adipose tissue. Generally, the thyroid hormones increase the oxidation of carbohydrates and accelerate the liberation of energy.

The thyroid hormones stimulate many phases of lipid metabolism including synthesis, mobilization, and degradation. Generally, degradation effects predominate, and excess thyroid hormones cause a decrease in lipid stores and in the plasma concentrations of triacylglycerols, phospholipids, and cholesterol.

RELEASE OF THE THYROID HORMONES

Thyrotropin (TSH) from the pars distalis of the pituitary controls the synthesis and release of the thyroid hormones, and TSH, in turn, is regulated by thyrotropin-releasing hormone (TRH) from the

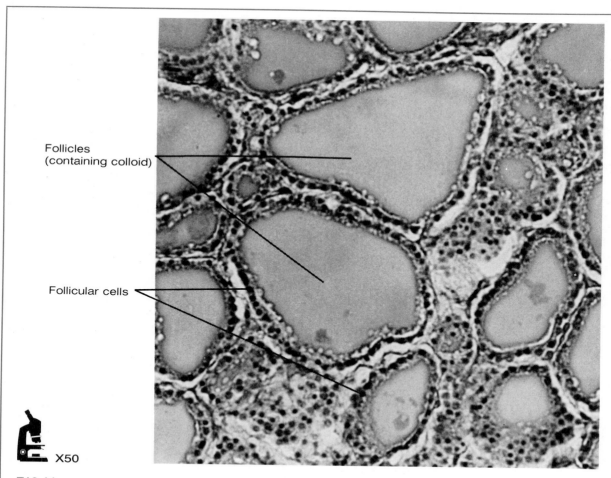

Follicles (containing colloid)

Follicular cells

X50

F10.11
Photomicrograph of the thyroid gland.

brain. This system is influenced by a negative feedback of the thyroid hormones themselves.

THYROID DISORDERS

In the young, hypothyroidism (lowered thyroid function) can have dire consequences. It can lead to an abnormal development of bones, connective tissue, and reproductive structures. In addition, the development of the nervous system may be faulty, leading to mental retardation. The condition caused by severe hypothyroidism beginning in infancy is called *cretinism*. If the hypothyroid state is recognized early and thyroid hormone replacement therapy is begun, the hypothyroid condition can be markedly improved. If treatment is delayed, the mental retardation becomes permanent.

Severe hypothyroidism in the adult is called *myxedema*. This condition, which is considerably more common in women than in men, is characterized by a puffiness of the face and eyelids and a swelling of the tongue and larynx. The skin becomes dry and rough, the hair scant, and the individual has both a low basal metabolic rate and a low body temperature. The sufferer also has poor muscle tone, lacks strength, and fatigues easily. His or her mental activity is generally sluggish and retarded. Myxedema can be alleviated by the administration of thyroid hormones.

An excess of thyroid hormones acting on the tissues is called *thyrotoxicosis*. One form of thyrotoxicosis is *hyperthyroidism* (thyroid overfunction). Hyperthyroidism is characterized by an elevated basal metabolism, elevated body temperature, and a rapid heartbeat. Despite an increased appetite, there may be a large weight loss. The sufferer perspires freely and may be nervous, emotionally unstable, and unable to sleep. Hyperthyroidism can be treated surgically, with drugs that impair thyroid function, or with radioactive iodine that is taken up by the gland and destroys some of the thyroid cells.

A *goiter* is simply an enlargement of the thyroid gland. It may or may not be associated with hypothyroidism or hyperthyroidism. In one form of thyrotoxicosis known as Graves' disease, the thyroid usually exhibits a diffuse enlargement (goiter). Simple goiter is glandular enlargement without either hypothyroidism or thyrotoxicosis.

CALCITONIN

In addition to the thyroid hormones, the thyroid releases a hormone called **calcitonin,** which lowers blood calcium and phosphate levels. Calcitonin is a polypeptide consisting of 32 amino acids. It is produced by parafollicular cells ("C" cells), which are located between or adjacent to the thyroid follicles. The plasma calcium level is

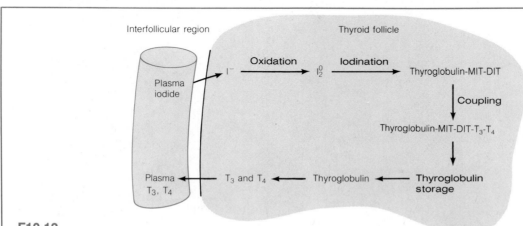

F10.12

Diagrammatic representation of the steps involved in the synthesis of the thyroid hormones. T_3 is triiodothyronine, and T_4 is tetraiodothyronine (thyroxine).

the major controller of calcitonin secretion: when the concentration of calcium ions in the plasma rises, calcitonin secretion increases.

PARATHYROID GLANDS

The **parathyroid glands** lie on the posterior surface of the thyroid (F10.14). They are composed of densely packed masses of epithelial cells interspersed with blood vessels.

PARATHYROID HORMONES AND THEIR EFFECTS

Parathyroid hormone (PTH, or **parathormone)** is a polypeptide, and it is currently believed that two and possibly three forms of the hormone may appear in the blood. Parathyroid hormone is a principal controller of calcium and phosphate metabolism. It increases the plasma calcium concentration, and decreases the plasma phosphate concentration. It increases phosphate excretion in the urine but decreases calcium excretion. Parathyroid hormone also increases the movement of calcium and phosphate from bone into the extracellular fluid, and it enhances a step in the metabolic transformation of vitamin D_3, which occurs in the kidneys. In addition, parathyroid hormone promotes the absorption of calcium from the gastrointestinal tract, but this effect is mediated by way of the effect of the hormone on the metabolic transformation of vitamin D_3. The plasma calcium level is the major controller of parathyroid hormone secretion: when the plasma calcium level falls, parathyroid hormone secretion increases.

PARATHYROID DISORDERS

Hyperparathyroidism, which results in an excess of parathyroid hormone, causes extensive bone decalcification and may lead to deformities and fractures. The plasma calcium level rises, and the calcification of soft tissues—especially the kidneys—may occur.

Hypoparathyroidism, which results in a deficiency of parathyroid hormone, leads to a lowered plasma calcium level, which greatly increases neuromuscular excitability. The end result may be muscular twitching and spasms (tetany).

F10.13
Structures of the thyroid hormones and structures important in their synthesis.

ADRENAL GLANDS

The **adrenal glands,** which are located superior to the kidneys, are dual structures that consist of medullary and cortical portions (F10.15).

ADRENAL MEDULLA

The central portion of each adrenal, the **adrenal medulla,** can be regarded as a modified portion of the sympathetic division of the autonomic nervous system. The cells of the adrenal medulla function in a manner similar to postganglionic sympathetic cells. The adrenal medulla is composed of cells arranged in groups or cords surrounding small blood vessels (capillaries and venules).

Adrenal Medullary Hormones and Their Effects

The adrenal medulla produces two catecholamine hormones: epinephrine (adrenaline) and norepinephrine (noradrenaline) (F10.16). (Recall that norepinephrine is also released from end terminals of postganglionic neurons of the sympathetic nervous system; thus, it may be present from sources other than the adrenal medulla.) Although epinephrine and norepinephrine are similar in structure and do exert a number of common effects, they are not identical in function. Norepinephrine works mainly on the vascular system, whereas epinephrine has important metabolic as well as cardiovascular effects.

Epinephrine. **Epinephrine** counteracts the hypoglycemic effects of insulin by elevating blood-glucose levels. It mobilizes liver carbohydrate stores and stimulates the production of lactic acid from glycogen in muscle. The lactic acid can be used by the liver to manufacture new carbohydrate (glucose or glycogen).

Epinephrine increases the rate, force, and amplitude of heartbeat. It constricts blood vessels in a number of body areas, including the skin, mucous membranes, and kidneys, but it may induce vessels to dilate in some areas such as skeletal muscles. The effects of epinephrine on blood vessels, however, are probably not very important compared to the effects of norepinephrine released from the endings of postganglionic sympathetic neurons.

Norepinephrine. **Norepinephrine** increases the heart rate and the force of contraction of cardiac muscle. It also constricts blood vessels in almost all areas of the body.

Release of the Adrenal Medullary Hormones

In general, the adrenal medulla releases a mixture of about 80% epinephrine and 20% norepinephrine, but these percentages vary considerably under different physiological conditions. The release of the catecholamine hormones of the adrenal medulla is controlled by preganglionic neurons to the medulla from the sympathetic division of the autonomic nervous system. A variety of conditions leads to the release of adrenal medullary hormones, including emotional excitement, injury, exercise, and low blood-glucose levels. Because catecholamines are liberated by both the adrenal medulla and the sympathetic nervous system, it is often convenient to consider the two as a single sympatheticoadrenal system. Together the divisions of this system maintain blood pressure and help regulate carbohydrate metabolism.

ADRENAL CORTEX

The outer portion of each adrenal gland, the **adrenal cortex,** is organized into three rather indistinct layers (F10.17). The outermost region of the cortex is called the *zona glomerulosa.* Beneath the zona glomerulosa is the *zona fasciculata.* Oc-

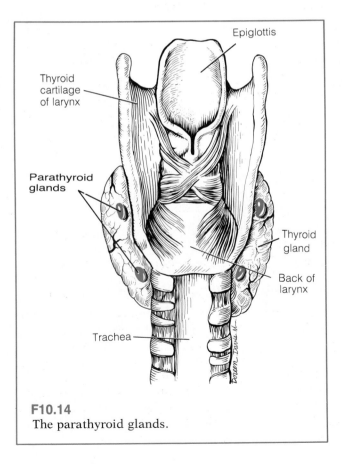

F10.14
The parathyroid glands.

Labels: Epiglottis; Thyroid cartilage of larynx; Parathyroid glands; Thyroid gland; Back of larynx; Trachea

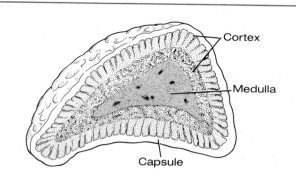

F10.15
An adrenal gland. The gland is a dual structure that consists of a central medulla and a surrounding cortex.

F10.16
Structures of the adrenal medullary catecholamine hormones epinephrine and norepinephrine.

cupying the deepest region of the cortex and lying adjacent to the adrenal medulla is the *zona reticularis.*

Adrenal Cortical Hormones and Their Effects

Mineralocorticoids. The zona glomerulosa produces adrenal cortical hormones that help regulate sodium and potassium metabolism. These are the **mineralocorticoids,** such as **aldosterone.** Aldosterone promotes the reabsorption of sodium and the excretion of potassium by the urine-forming structures of the kidneys.

Glucocorticoids. The zona fasciculata produces adrenal cortical hormones that affect carbohydrate metabolism. These are the **glucocorticoids,** such as **cortisol.** The glucocorticoids supplement and conserve the energy derived from circulating glucose. In response to the glucocorticoids, glucose utilization in peripheral tissues is inhibited, fatty acids from adipose tissue are mobilized, and muscle shifts from glucose to fatty acids as a source of metabolic energy. The glucocorticoids promote gluconeogenesis (carbohydrate production from noncarbohydrate precursors) and glycogen deposition in the liver, and the elevation of blood glucose. They also accelerate the breakdown of proteins and inhibit amino

acid uptake and protein synthesis by many tissues other than the liver. In the liver, however, amino acid uptake is enhanced, as is the utilization of amino acids for protein and glucose synthesis. At pharmacological concentrations (that is, concentrations higher than those normally present in the body), cortisol has an anti-inflammatory action.

The adrenal cortex also produces some androgenic substances that resemble male sex hormones, as well as a much smaller quantity of estrogenic materials that resemble female sex hormones. The adrenal cortical hormones (and the male and female sex hormones) are steroids that are synthesized from cholesterol.

Release of the Adrenal Cortical Hormones

The release of the mineralocorticoids (such as aldosterone) from the adrenal cortex is influenced by several factors. One of these is a substance called **renin,** which is released by specialized cells of the kidneys. Renin acts on the precursor substance **angiotensinogen,** which is manufactured by the liver and is present in the blood. Renin converts angiotensinogen to angiotensin I. In turn, angiotensin I is enzymatically converted into a series of other angiotensins (II, III, and so forth), but these molecules will be considered simply as **angiotensin.** Among the effects of angiotensin is the stimulation of aldosterone release from the adrenal cortex. Renin secretion is itself controlled by a number of factors, including the activity of sympathetic nerves to the kidneys and the renal arterial blood pressure (decreased pressure increases renin secretion).

Aldosterone release is also stimulated by an elevation in the potassium concentration of the fluids bathing the adrenal glands, as well as by a decrease in the sodium content of the body.

The release of the glucocorticoids (such as cortisol) is controlled primarily by ACTH from the pituitary, and ACTH, in turn, is influenced by corticotropin-releasing factor (CRF) from the brain. The glucocorticoids exert an inhibitory influence over ACTH and CRF release by way of negative feedback. Various stressful situations (such as trauma or emotional stress) cause increased release of ACTH and consequently increased glucocorticoid secretion. This effect is probably neurally mediated by way of CRF. Once

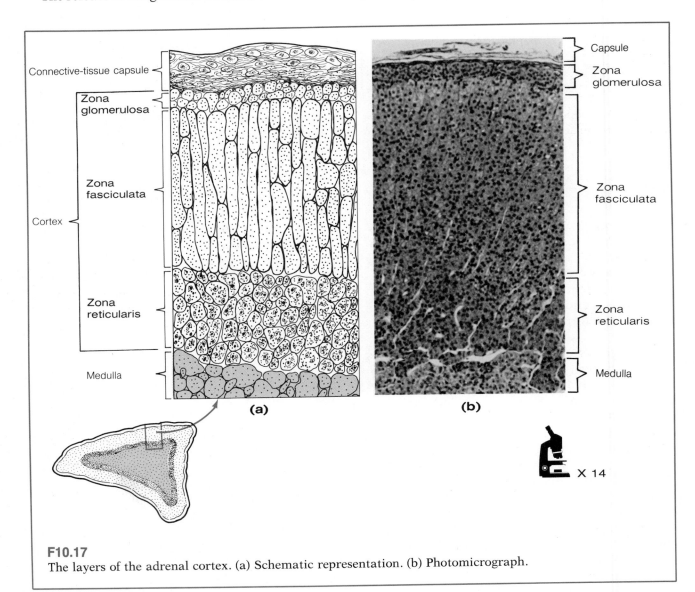

F10.17
The layers of the adrenal cortex. (a) Schematic representation. (b) Photomicrograph.

released, cortisol is bound to a plasma glycoprotein called *transcortin* for transport through the circulation.

ADRENAL DISORDERS

Primary adrenal cortical hypofunction in humans is called *Addison's disease.* Inadequate amounts of aldosterone impair the body's ability to conserve sodium and excrete potassium. This impairment may lead to a decreased extracellular fluid volume, weight loss, decreased plasma volume, low blood pressure, decreased cardiac size and output, general weakness, and shock. A deficiency of glucocorticoids in Addison's disease may result in loss of appetite (anorexia), fasting hypoglycemia, apathy, weakness, and a diminshed ability to withstand various types of physiological stress. Hormone administration can alleviate the symptoms of Addison's disease.

Adrenal cortical hyperfunction can result in a number of disorders. Hypercortisolism may produce *Cushing's syndrome,* which is characterized by increased blood-glucose levels, increased protein breakdown, osteoporosis (softening of bones), weakness, and hypertension. Hyperaldosteronism is characterized by potassium depletion and expansion of the extracellular fluid compartment, which may result in hypertension or edema (excessive accumulation of fluid in the tissue spaces). Certain adrenal tumors can secrete excessive quantities of androgenic substances, producing masculinizing effects that are particularly evident in females.

PANCREAS

The **pancreas,** which is located behind the stomach between the spleen and the duodenum of the small intestine, is both an endocrine gland that produces hormones and an exocrine gland that produces digestive enzymes. The endocrine portion of the pancreas consists of aggregations of cells clustered into groups called **islets of Langerhans,** which contain several functionally different cell types (F10.18).

PANCREATIC HORMONES AND THEIR EFFECTS

Alpha or A cells of the islets of Langerhans produce the hormone **glucagon,** beta or B cells produce the hormone **insulin,** and delta or D cells produce somatostatin—the same substance that is produced in the brain and inhibits the release of growth hormone. There is some evidence that somatostatin, which is also secreted by the mucosa of the upper gastrointestinal tract and has been proposed as a possible neurotransmitter substance, inhibits the release of both glucagon and insulin. Thus, somatostatin exemplifies the fact that a given chemical substance may be synthesized at different sites and may perform different functions.

Insulin

Insulin is composed of two linked polypeptide chains. One chain consists of 21 amino acids and the other consists of 30 amino acids. Circulating insulin molecules are, for the most part, bound to protein carriers in the plasma.

Insulin facilitates the uptake and utilization of glucose by many cells—a notable general exception is the brain—and it prevents the excessive breakdown of glycogen in liver and muscle. As a result, insulin is a powerful hypoglycemic agent. Insulin favors lipid formation by promoting the uptake and metabolic utilization of glucose (an important precursor of fatty acids and fat) by liver and adipose tissue cells. It inhibits the breakdown and mobilization of stored fat, and it favors the synthesis of proteins, in part by facilitating the movement of amino acids into cells.

Glucagon

The activities of glucagon, which is a polypeptide molecule composed of 29 amino acids, are generally opposite to those of insulin. Glucagon decreases glucose oxidation and promotes hyperglycemia. Its main action seems to be to stimulate the breakdown of liver glycogen. Glucagon also stimulates liver carbohydrate production from noncarbohydrate precursors. In addition, it stimulates the breakdown of lipids in liver and adipose tissue. This breakdown increases the substances available for carbohydrate production and also results in the formation of substances called ketone bodies. Glucagon may also have a mild stimulating effect on protein breakdown.

RELEASE OF PANCREATIC HORMONES

The release of the pancreatic hormones is under chemical, hormonal, and neural control. The blood-glucose level appears to be the major factor governing insulin release: the higher the blood-glucose level, the greater the insulin release. The blood-glucose level also influences the release of glucagon, but in opposite fashion: the lower the blood-glucose level, the greater the glucagon release. Amino acids stimulate the simultaneous secretion of both insulin and glucagon.

Several hormones from the gastrointestinal

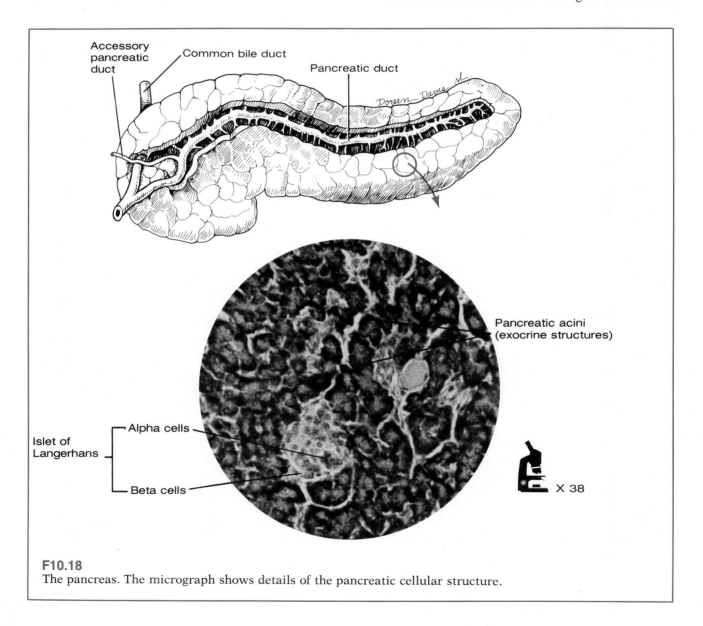

F10.18
The pancreas. The micrograph shows details of the pancreatic cellular structure.

tract (such as secretin, gastrin, cholecystokinin, gastric inhibitory peptide) have been reported to promote insulin release, and glucagon itself has also been found to promote insulin secretion.

Acetylcholine, the chemical released at the nerve endings of the parasympathetic division of the autonomic nervous system, stimulates insulin release. Epinephrine and norepinephrine from the adrenal medulla and terminal endings of the neurons of the sympathetic division of the autonomic nervous system inhibit insulin release. Conversely, glucagon output appears to be increased by the activation of the sympathetic division of the autonomic nervous system. These neural controls are thought to be continuously active in modulating the basal secretion of insulin and glucagon.

PANCREATIC DISORDERS

The most common pancreatic endocrine disorder is the condition known as *diabetes mellitus*. This condition is caused by a relative insulin deficiency that can be due to a deficient pancreatic secretion of insulin or to a decreased sensitivity to insulin. For example, in many obese diabetics, insulin levels are normal or even above normal, but insulin sensitivity is decreased. This insulin insensitivity appears to have multiple causes, but in some cases various target cells apparently have a deficiency of insulin receptors. There is also some evidence that a relative or absolute excess of glucagon, as well as a deficiency of insulin, occurs in certain forms of diabetes.

Diabetes mellitus is characterized by increased levels of blood glucose coupled with a reduced entry of glucose into cells and an impairment of cellular ability to use glucose. As blood glucose rises, it appears in the urine. Extra water is osmotically required to excrete the glucose, and a large volume of urine is produced (polyuria). This situation can lead to dehydration and ultimately to circulatory difficulties, including hypotension and decreased brain bloodflow that can cause brain damage and death.

Free fatty acids are mobilized and released from adipose tissue in diabetes mellitus, and they provide energy sources for the cells. Such energy sources are necessary because of the deficient cellular uptake and utilization of glucose. The oxidation of fatty acids in the liver produces substances known as *ketone bodies* (for example, acetoacetic acid and beta-hydroxybutyric acid). These may be produced faster than other body tissues can metabolize them, and thus they may appear in the urine. In severe cases of diabetes, the body's buffering and excretory mechanisms are not able to cope with these acidic substances, and the pH of the body fluids falls, resulting in acidosis. The acidosis, in turn, can cause altered respiration, nervous system depression, coma, and death.

Protein synthesis decreases and protein breakdown increases in diabetes mellitus, and this can impair the body's ability to combat infections and to repair injured tissues. In the case of a diabetic whose pancreas remains functional but does not produce sufficient insulin, drugs taken orally have been used to stimulate insulin production. However, the effectiveness and safety of these drugs remains controversial, and they are not used as commonly as they once were. If the individual's pancreas cannot secrete insulin, the insulin deficiency must be made up by injection. In many obese diabetics, weight loss can eliminate many of the manifestations of the disease.

GONADS

The male gonads (*testes*) and the female gonads (*ovaries*) produce hormones as well as gametes (reproductive cells, that is, sperm or ova). The testes produce the male sex hormones—that is, the androgens such as testosterone—and the ovaries produce the female sex hormones—the estrogens and progesterone. The endocrine as well as the reproductive roles of the gonads are discussed in Chapter 20.

OTHER ENDOCRINE TISSUES AND HORMONES

The digestive tract is the source of a number of hormonal substances (such as gastrin, secretin, cholecystokinin). These are discussed in Chapter

16. The placenta is also a source of hormones, as is discussed in Chapter 20.

The **pineal body** of the brain is widely believed to be an endocrine structure. In rats, the pineal has an antigonadal activity, and it is influenced by the lighting cycle to which an animal is exposed. In the light, light strikes the retinas of the eyes, giving rise to nerve impulses that ultimately reach the pineal. These impulses are believed to inhibit the secretion of pineal hormones, whose activity would inhibit gonadal function. The exact nature of the pineal hormone remains in dispute. Some investigators feel that it is a substance called melatonin. Others believe it to be a polypeptide. The relevance of various studies of pineal function in animals to the function of the human pineal is uncertain at the present time.

The **thymus gland,** which is discussed in relation to immune responses (Chapter 14), is also regarded as a source of hormonal material. A group of polypeptides collectively called thymosin, which appear to be involved in the development of immunologically competent white blood cells, have been isolated from the thymus.

PROSTAGLANDINS

The prostaglandins are a specialized group of fatty acids that serve as chemical messengers (F10.19). They occur in very small quantities in the body, and probably only a few tenths of a milligram of prostaglandins are synthesized per day. Nevertheless, prostaglandins have been isolated from a wide variety of tissues including intestine, liver, kidney, pancreas, heart, lung, brain, and male and female reproductive structures.

Some investigators believe that prostaglandins travel through the bloodstream and thus act as hormones. However, most researchers feel that

F10.19
Structures of some prostaglandins. The prostaglandins are lipids, specifically 20-carbon fatty acids.

the prostaglandins are primarily local messengers that exert their actions in the tissues that synthesize them. In any event, they have a wide range of effects. Prostaglandins PGE_2 and PGF_2, for example, stimulate uterine contractility and may be important in the process of giving birth and in the fertilization of the ovum. They also inhibit the secretion of progesterone by the corpus luteum of the ovary. Other prostaglandins inhibit gastric secretion, relax smooth muscle in the airways to the lungs, increase urine flow, and increase or lower blood pressure.

STUDY OUTLINE

BASIC ENDOCRINE FUNCTIONS pp. 245–250

1. Endocrine glands are generally ductless glands composed of epithelial cells that secrete their products into the extracellular space around the cells and from which the products enter the bloodstream.
2. Hormones are chemical messengers (such as proteins, polypeptides, steroids) produced by endocrine structures; they are carried by the blood to the body's cells where they exert their effects.

Transport of Hormones   Many hormones, particularly the steroid hormones, are bound to specific carrier proteins within the bloodstream. The reversible association between a hormone and a carrier protein provides both a storage and buffer system for the hormone.

Mechanisms of Hormone Action   A particular hormone does not necessarily affect all cells but only its target cells. The target cells of a hormone possess receptors to which molecules of the hormone can attach.

    UTILIZATION OF INTRACELLULAR MEDIATORS The binding of some hormone molecules to receptors on plasma membranes alters the activity of the enzyme adenylate cyclase, leading to changes in the level of cyclic AMP within the cells that can affect various cellular functions. Calcium ions and cyclic GMP may also mediate the effects of hormones at the cellular level.

    ACTIVATION OF GENES   Hormone molecules enter a cell and combine with receptors. The hormone-receptor complex interacts either directly or indirectly with the genetic material to activate certain genes. This gene activation leads to the synthesis of m-RNA and ultimately to the production of proteins (for example, enzymes) that influence cellular reactions or processes.

Hormonal Interrelationships   Almost every activity influenced by the endocrine system is affected by a balance among different hormones acting either together or in sequence. Moreover, some hormones influence the activities or the rates of production of other hormones.

PITUITARY GLAND   Composed of neurohypophysis (infundibulum and lobus nervosus) and ade-

nohypophysis (pars distalis, pars tuberalis, and pars intermedia). **pp. 250–256**

Neurohypophyseal Hormones and Their Effects   The lobus nervosus releases two peptide hormones.

    ANTIDIURETIC HORMONE   Promotes reabsorption of water from urine-forming structures of kidneys; constricts arterioles.

    OXYTOCIN   Stimulates smooth muscles of uterus; promotes contraction of myoepithelial cells that surround alveoli of mammary glands.

Release of Neurohypophyseal Hormones   Hormones synthesized in brain by neurosecretory cells; transported to lobus nervosus; neural inputs to brain influence release.

Adenohypophyseal Hormones and Their Effects   The adenohypophysis releases several hormones.

    GONADOTROPINS   Follicle-stimulating hormone (FSH) and luteinizing hormone (LH) are involved in:
    1. Ovarian follicle development and estrogen production.
    2. Ovulation and formation of corpus luteum.
    3. Spermatogenesis and androgen production.

    THYROTROPIN   Stimulates synthesis and release of thyroid hormones.

    ADRENOCORTICOTROPIN   Stimulates release of hormones from cortical region of adrenal gland, particularly cortisol and other glucocorticoids.

    GROWTH HORMONE   Affects growth of skeletal system; enhances entrance of amino acids into cells, and incorporation into proteins; increases release of fatty acids into blood.

    PROLACTIN   Involved in initiation and maintenance of milk secretion in females.

Other Adenohypophyseal Substances   Beta-lipotropin and several forms of melanocyte-stimulating hormone.

Pro-opiomelanocortin and the Production of Adenohypophyseal Hormones   Pro-opiomelanocortin is a polypeptide molecule that contains the amino acid sequences of ACTH and β-LPH. It is apparently a precursor molecule from which ACTH, β-LPH, and other substances can be split.

Release of Adenohypophyseal Hormones Brain exercises good deal of control.

HORMONE RELEASING AND INHIBITING SUBSTANCES Manufactured in brain by neurosecretory cells and travel through circulation by way of hypothalamic-hypophyseal portal veins to adenohypophysis, where they influence release of adenohypophyseal hormones.

NERVOUS SYSTEM EFFECTS ON ADENOHYPOPHYSEAL HORMONE RELEASE Neural activity influences many endocrine functions.

Pituitary Disorders Can be caused by disorders of gland itself, or difficulties involving releasing or inhibiting substances from the brain.

FEEDBACK CONTROL OF HORMONE RELEASE
Negative feedback inhibits hormone release; long and short feedback loops. **p. 257**

THYROID GLAND Well vascularized; composed of hollow balls of cells called follicles. **pp. 257–261**

The Thyroid Hormones: Triiodothyronine and Thyroxine Iodine attached to tyrosine molecules; involved in wide range of metabolic activities; increase oxygen consumption and heat production; increase carbohydrate oxidation; stimulate many aspects of lipid metabolism.

Release of the Thyroid Hormones TSH from pituitary controls synthesis and release of thyroid hormones; thyrotropin-releasing hormone and negative feedback involved in control.

Thyroid Disorders
1. Cretinism—Severe hypothyroidism beginning in infancy.
2. Myxedema—Severe hypothyroidism in the adult.
3. Thyrotoxicosis—Excess of thyroid hormones acting on tissues; hyperthyroidism, for example.
4. Goiter—Enlarged thyroid gland.

Calcitonin Hormone from the thyroid that lowers blood calcium and phosphate levels.

PARATHYROID GLANDS Located on posterior surface of lateral lobes of thyroid. **p. 261**

Parathyroid Hormones and Their Effects Two or three forms of parathyroid hormone; controller of calcium and phosphate metabolism.

Parathyroid Disorders
1. Hyperparathyroidism—Bone decalcification; calcification of soft tissue.
2. Hypoparathyroidism—Lowered plasma calcium level; increased neuromuscular excitability.

ADRENAL GLANDS Consist of medullary and cortical portions. **pp. 261–265**

Adrenal Medulla Central portion; function similar to postganglionic sympathetic cells.

ADRENAL MEDULLARY HORMONES AND THEIR EFFECTS Catecholamines released by adrenal medulla and sympathetic nervous system.

Epinephrine (adrenaline) Has metabolic and cardiovascular effects.

Norepinephrine (noradrenaline) Primarily affects vascular system.

RELEASE OF THE ADRENAL MEDULLARY HORMONES Controlled by preganglionic neurons from sympathetic division of autonomic nervous system; may be caused by emotional excitement, injury, exercise, low blood-glucose levels.

Adrenal Cortex Outer portion of adrenal gland.

ADRENAL CORTICAL HORMONES AND THEIR EFFECTS

Mineralocorticoids Such as aldosterone, regulate sodium and potassium metabolism.

Glucocorticoids Such as cortisol, inhibit glucose utilization in peripheral tissues; promote gluconeogenesis and glycogen deposition in liver and elevation of blood glucose; influence protein metabolism.

RELEASE OF THE ADRENAL CORTICAL HORMONES Renin-angiotensin system and concentration of potassium in the fluids bathing the adrenal glands influence release of mineralocorticoids, as does sodium content of body; ACTH controls release of glucocorticoids; CRF influences ACTH release.

Adrenal Disorders
1. Addison's disease—Hypofunction of adrenal cor-

tex; inadequate aldosterone; impaired ability to conserve sodium and excrete potassium, inadequate glucocorticoids, fasting hypoglycemia, weakness.

2. Cushing's syndrome—Increased blood-glucose levels; increased protein breakdown.
3. Hyperaldosteronism—Potassium depletion; edema.
4. Adrenal tumors—May produce virilizing or feminizing effects.

PANCREAS Both an endocrine and an exocrine gland; endocrine portion composed of islets of Langerhans.   **pp. 265–267**

Pancreatic Hormones and Their Effects Alpha cells produce glucagon, beta cells produce insulin, delta cells produce somatostatin.

INSULIN Facilitates uptake and utilization of glucose by many cells; prevents excessive glycogen breakdown in liver and muscle; influences lipid and protein metabolism.

GLUCAGON Activities are generally opposite to those of insulin; stimulates breakdown of liver glycogen.

Release of Pancreatic Hormones Chemical, hormonal, and neural control.
1. Increased plasma glucose levels increase insulin release and decrease glucagon release.
2. Amino acids stimulate secretion of insulin and glucagon.
3. Several gastrointestinal hormones promote insulin release.
4. Acetylcholine stimulates insulin release; epinephrine and norepinephrine inhibit insulin release.

Pancreatic Disorders Diabetes mellitus caused by relative insulin deficiency.

GONADS Testes and ovaries produce hormones as well as gametes.   **p. 267**

OTHER ENDOCRINE TISSUES AND HORMONES Digestive tract and placenta are sources of hormonal substances; *pineal body* believed to be a source of endocrine material; *thymus gland* produces thymosin.   **p. 267**

PROSTAGLANDINS Chemical messengers that stimulate uterine contractility; may function in parturition and ovum fertilization, as well as many other actions.   **p. 268**

SELF-QUIZ

1. Endocrine glands are generally ductless glands composed of epithelial cells that release their secretions directly into the target organ. True or False?
2. The effects of some hormones on cellular function appear to be at least partly mediated intracellularly by cyclic AMP. True or False?
3. ADH and oxytocin are manufactured in the: (a) lobus nervosus; (b) hypothalamus; (c) adenohypophysis.
4. Gonadotropins are: (a) produced by the ovaries and testes; (b) neurohypophyseal hormones; (c) adenohypophyseal hormones.
5. Surging levels of LH together with FSH are important in: (a) stimulation of the thyroid gland; (b) ovulation; (c) stimulation of the cortical region of the adrenal glands.
6. A hormone secreted by the adenohypophysis of the pituitary gland is: (a) thyrotropin; (b) antidiuretic hormone; (c) oxytocin.
7. Match the hormones with their appropriate effects.

ADH	Oxytocin
LH	TSH
ACTH	GH

   (a) Stimulates the cortical region of the adrenal gland
   (b) Stimulates the reabsorption of water from the urine-forming structures of the kidneys
   (c) Stimulates growth
   (d) Stimulates the smooth muscle of the uterus
   (e) Stimulates the synthesis of the thyroid hormones
   (f) Acts on the interstitial cells of Leydig of the testes to stimulate the production of the male sex hormones

8. Specialized neurosecretory cells within the brain manufacture a group of compounds known collectively as: (a) gonadotropins; (b) luteinizing hormone; (c) releasing and inhibiting substances.
9. Brain activities of the sort that are responsible for emotions are generally unable to influence hormonally controlled events. True or False?
10. Match the pituitary gland disorders with their possible symptoms.
    Lack of ACTH
    Gonadotropin deficiency
    TSH deficiency
    ACTH overproduction
    Growth hormone deficiency
    Growth hormone overproduction

    Antidiuretic hormone deficiency
    (a) Underactivity of thyroid
    (b) Adrenal cortical hyperfunction
    (c) Menstrual failure
    (d) Pituitary dwarfs
    (e) Acromegaly
    (f) Adrenal cortical insufficiency
    (g) Diabetes insipidus

11. Hormones that contain iodine are produced by which glands? (a) pituitary; (b) adrenal medulla; (c) thyroid.
12. The follicular storage form of the thyroid hormones is: (a) thyroglobulin; (b) thyroid-binding globulin; (c) free thyroxine.
13. The thyroid hormones generally travel freely in the circulation. True or False?
14. In the young, hypothyroidism may lead to: (a) elevated body temperature; (b) elevated basal metabolism; (c) abnormal bone development.
15. Enlargement of the thyroid gland is known as: (a) Addison's disease; (b) folliculitis; (c) goiter.
16. When the plasma calcium level falls, parathyroid hormone secretion: (a) decreases; (b) remains constant; (c) increases.
17. Hypoparathyroidism: (a) increases neuromuscular excitability; (b) raises plasma calcium levels; (c) leads to calcification of the kidneys.
18. The release of epinephrine is controlled by nerves to the adrenal medulla from the sympathetic division of the autonomic nervous system. True or False?
19. Epinephrine is able to counteract the hypoglycemic effects of insulin by: (a) elevating blood-glucose levels; (b) lowering renal potassium levels; (c) raising renal sodium levels.
20. Steroid hormones: (a) are released only by the cortex of the adrenal glands; (b) may attach to carrier proteins within the bloodstream; (c) exert their effects solely by attaching to receptors at cell membranes.
21. Insulin release is stimulated by: (a) low blood-sugar levels; (b) the activity of sympathetic nerves to the pancreas; (c) gastrointestinal hormones.
22. A major effect of insulin is to facilitate the uptake and utilization of glucose by many cells and to prevent the excessive breakdown of glycogen in liver and muscle. True or False?
23. Which hormone would most likely be released in increased amounts in response to elevated blood-glucose levels? (a) epinephrine; (b) glucagon; (c) insulin.

LEARNING OBJECTIVES

After completing this chapter, you should be able to:

- Describe four functions of the blood.
- Describe three functions of the plasma proteins.
- Explain the function of each of the formed elements of the blood.
- Describe the formation and fate of erythrocytes.
- Describe the structure and function of hemoglobin.
- Discuss iron metabolism.
- Explain the control of erythrocyte production.
- Compare and contrast anemia and polycythemia.
- Discuss the formation of a platelet plug as a hemostatic mechanism.
- Explain the process of blood clotting, and distinguish between the extrinsic and intrinsic clotting mechanisms.
- Explain why clots do not usually form within normal blood vessels, and describe the difficulties that can result from intravascular clotting.
- Compare and contrast hemophilia and thrombocytopenia.

CHAPTER CONTENTS

THE BLOOD 11

The cells and tissues of the body are linked by the blood vessels and the blood that flows within them. The blood transports materials throughout the body. It carries oxygen and food materials to and carbon dioxide away from the cells. It also transports cellular products such as hormones between cells. In addition, the blood and its components help maintain homeostasis by buffering the tissues against extremes of pH, by aiding in the regulation of body temperature, and by protecting the tissues from toxic foreign materials or organisms.

The volume of blood in both lean males and lean females varies almost directly with body weight and averages approximately 79 milliliters per kilogram of body weight ($\pm 10\%$). However, fat tissue has little vascular volume, and the volume of blood per unit of body weight declines as the proportion of adipose tissue in the body increases. Since the average female has a greater fat/lean-tissue ratio than the average male, the average female tends to have a lower blood volume per kilogram of body weight than the average male (about 67 ml/ kg body weight for average females and about 75 ml/kg body weight for average males). Because of this difference, as well as differences in body size, the general range of total blood volume is 4 to 5 liters in females and 5 to 6 liters in males.

The blood consists of both a liquid component, called plasma, and a nonliquid portion whose structures are collectively called formed elements.

PLASMA

Plasma, which is approximately 90% water, accounts for about 60% of the total blood volume in males and 64% in females. The portion of the plasma that is not water consists of various dissolved or colloidal materials (Table 11.1). Hormones and other cellular products are present in the plasma, as are metabolic end products such as urea. The plasma contains plasma proteins, including various albumins, fibrinogen (which is involved in blood clotting), and globulins (some of which act as antibodies in immune responses, while others serve as transport molecules). The plasma proteins act as buffers that help stabilize the pH of the internal environment, and they contribute importantly to the osmotic pressure and viscosity of the plasma. Various ions, including sodium (Na^+), chloride (Cl^-), and bicarbonate (HCO_3^-), are present within the plasma. These ions also contribute to plasma osmotic pressure, which averages about 300 milliosmoles per liter. The plasma contains food materials such as carbohydrates (for example, glucose), amino acids, and lipids, as well as gases such as oxygen, nitrogen, and carbon dioxide.

FORMED ELEMENTS

The portion of the blood that is not plasma consists of **formed elements** (Table 11.2). The formed elements include erythrocytes, or red blood cells; various types of leukocytes, or white blood cells; and platelets.

ERYTHROCYTES

Erythrocytes or **red blood cells** are small, circular, biconcave discs, approximately 7.5 microns in di-

Table 11.1 *Principal Constituents of Plasma**	
Constituent	Approximate Amount
Water	90 percent
Plasma Proteins	
Albumins	3.2 to 5.0 g/100 ml
Fibrinogen	0.20 to 0.45 mg/100 ml
Globulins	
Alpha	0.40 to 0.98 g/100 ml
Beta	0.56 to 1.06 g/100 ml
Gamma	0.44 to 1.04 g/100 ml
Glucose	61 to 130 mg/100 ml
Cholesterol	128 to 347 mg/100 ml
Bilirubin (total, indirect reacting)	0 to 1.1 mg/100 ml
Urea	13.8 to 39.8 mg/100 ml
Sodium	310 to 356 mg/100 ml
Potassium	12 to 21 mg/100 ml
Calcium (total)	8.2 to 11.6 mg/100 ml
Iron	0.04 to 0.21 mg/100 ml
Chloride	355 to 381 mg/100 ml

*Plasma also contains a wide variety of other substances too numerous to list here, including vitamins, hormones, enzymes, and metabolic end products.

ameter that have no nuclei (F11.1; Table 11.2). Their principal function is to transport oxygen and carbon dioxide.

Erythrocytes are the most numerous formed elements contained in the blood, and they contribute importantly to total blood viscosity (which is normally 3.5 to 5.5 times that of water). Although their numbers are quite variable, a cubic millimeter of peripheral venous blood usually contains about 5.1 to 5.8 million erythrocytes in males and about 4.3 to 5.2 million erythrocytes in females. The proportion of erythrocytes in a sample of blood is called the **hematocrit.** It is determined by centrifuging a blood sample in a hematocrit tube until the erythrocytes are packed at the bottom of the tube. The hematocrit is then measured as a ratio or percentage of the packed erythrocyte volume to the total sample volume. Although there is considerable variability, normal hematocrits of samples of peripheral venous blood average about 46.2% for males and 40.6% for females. The hematocrit of arterial blood is generally slightly lower than that of venous blood, and the hematocrit of blood in the very small vessels of the body is considerably lower than the hematocrit of blood in the large arteries and veins.

Hemoglobin

A substance called **hemoglobin** is essential to the ability of erythrocytes to transport oxygen and carbon dioxide, and a single erythrocyte contains up to 300 million hemoglobin molecules. Males generally have more hemoglobin (14 to 18 grams per 100 ml of peripheral venous blood) than females (12 to 16 grams per 100 ml of peripheral venous blood).

A molecule of hemoglobin is composed of the protein *globin*, combined with four nonprotein groups called *hemes* (F11.2). The globin protein consists of four polypeptide chains, each of which has a heme group bound to it. Each heme group contains an iron atom (Fe^{++}) that can combine reversibly with one molecule of oxygen. Therefore, a hemoglobin molecule can potentially associate with four oxygen molecules. When hemoglobin is combined with oxygen, it is called *oxyhemoglobin;* when it is not carrying oxygen, it is called *reduced hemoglobin.* Under normal circumstances, a liter of blood traveling to the tissues contains approximately 198 ml of oxygen. Of this amount, about 195 ml are associated with hemoglobin molecules within erythrocytes. The remaining 3 ml are in physical solution dissolved in the plasma. Thus, erythrocytes and hemoglobin account for almost all of the blood's ability to transport oxygen.

Hemoglobin can also combine with carbon dioxide and thus aid in the transport of this substance. However, in contrast to oxygen, which is carried in association with the iron of the heme groups of hemoglobin, carbon dioxide is carried in reversible association with the protein portion of the hemoglobin molecule The transport of oxygen and carbon dioxide by hemoglobin is discussed further in Chapter 15.

Iron

The production of hemoglobin requires iron, and the body normally contains about 4 grams of iron. Approximately 65% of the body's iron is found in

Table 11.2 *Formed Elements of the Blood*

Formed Element	Approximate Diameter (microns)	Approximate Abundance (per mm³)	Function
ERYTHROCYTES – *RBC*	7.5	Males: 5.1 to 5.8 million Females: 4.3 to 5.2 million	Transport oxygen and aid in the transport of carbon dioxide
PLATELETS – *Thrombocytes*	2.5	250,000 to 400,000	Involved in the processes of hemostasis and blood coagulation
LEUKOCYTES – *WBC* **Granulocytes** Neutrophils	12 to 14 ①	3000 to 7000 (60 to 70 percent of total leukocytes)	Phagocytic cells that are capable of ameboid movement
Eosinophils	12 ②	50 to 400 (1 to 4 percent of total leukocytes)	Phagocytic cells that are believed to destroy antigen–antibody complexes
Basophils	9 ③	0 to 50 (0 to 1 percent of total leukocytes)	Believed to release chemicals such as histamine and heparin
Agranulocytes Monocytes	20 to 25	100 to 600 (2 to 6 percent of total leukocytes)	Develop into large phagocytic cells called macrophages within the tissue spaces
Lymphocytes	9 (small) 12 to 14 (large)	1000 to 3000 (20 to 30 percent of total leukocytes)	Involved in specific immune responses, including antibody production

hemoglobin, and about 15% to 30% is stored in the liver, spleen, bone marrow, and elsewhere, mainly in the form of intracellular iron–protein complexes called *ferritin* and *hemosiderin*. In the blood, iron forms a loose combination with a beta globulin called *transferrin,* and it is in this form that iron is transported in the plasma. Iron can be released from transferrin for use or storage.

Small amounts of iron are lost each day through the feces, urine, sweat, and cells sloughed from the skin. In women, additional iron is lost as a result of menstrual bleeding. The average daily loss of iron is about 0.9 mg in males and about 1.7 mg in females.

Iron losses must be replaced if iron homeostasis is to be maintained. Ingested iron is absorbed into the body by active processes, primarily in the upper portion of the small intestine, and total body iron is largely regulated by alterations in its intestinal absorption rate. When the body stores of iron are high, the intestinal absorption rate of iron is low, but when iron stores are depleted, the rate of iron absorption increases. In general, the rate of iron absorption is slow, and a maximum of a few milligrams of iron are absorbed per day.

Production of Erythrocytes

Erythrocytes develop from cells located in the red marrow of bones that include the ribs, the sternum, the vertebrae, and the pelvis. The red marrow contains stem cells called *hemocytoblasts* that can differentiate into all the different formed elements of the blood. During erythrocyte production, some of the hemocytoblasts give rise to cells called *proerythroblasts* (F11.3). The proerythroblasts, in turn, form cells called *basophilic erythroblasts,* which synthesize hemoglobin. As hemoglobin synthesis continues, the basophilic erythroblasts differentiate into cells called *polychromatophilic erythroblasts* which, in turn, give

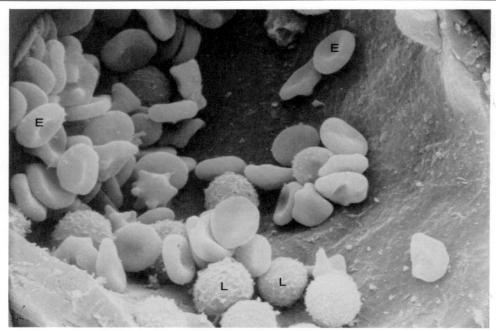

F11.1
Formed elements of the blood within a blood vessel. E: erythrocytes, or red blood cells. L: leukocytes, or white blood cells (×6130). (From *Tissues and Organs: A Text Atlas of Scanning Electron Microscopy* by Richard G. Kessel and Randy H. Kardon, W. H. Freeman & Co., © 1979.)

rise to cells called *normoblasts* (*acidophilic erythroblasts*). When the normoblast cytoplasm has attained a hemoglobin concentration of about 34%, the nucleus is pinched off, enclosed in a portion of the cell membrane and a thin layer of cytoplasm. The nonnucleated cells that result are called *reticulocytes* (*polychromatophilic erythrocytes*). Reticulocytes are essentially young erythrocytes that contain small numbers of residual ribosomes. It is these cells that are usually released from the bone marrow into the blood. Reticulocytes generally become mature erythrocytes within one or two days after their release from the bone marrow, and the normal proportion of reticulocytes in the blood is less than 0.5% to 1.0%.

The process of erythrocyte formation is called **erythropoiesis.** During erythropoiesis, the various cells continue to divide through the normoblast stage of development so that greater and greater numbers of cells are formed. Occasionally, when erythrocytes are being manufactured very rapidly, nucleated cells appear in the blood, but this is not a usual occurrence.

Among the substances required for the production of erythrocytes is vitamin B_{12}. Vitamin B_{12} is required for DNA formation and, as a result, is necessary for normal nuclear maturation and division. Folic acid, another vitamin, is also required for DNA formation and thus for erythrocyte production.

Control of Erythrocyte Production

A substance called **erythropoietin** stimulates erythrocyte production. Erythropoietin is a glycoprotein produced in the kidneys and possibly elsewhere. The kidneys are believed to release an enzyme called renal erythropoietic factor (erythrogenin) into the blood. Renal erythropoietic factor is then thought to act on a plasma protein precursor to form erythropoietin.

The production of erythropoietin, and thus erythrocyte production, is regulated by a negative feedback mechanism that is sensitive to the amount of oxygen delivered to the tissues. An inadequate supply of oxygen—particularly to the kidneys—leads to a greater production of erythropoietin and thereby to increased erythrocyte production. Conversely, an increased oxygen supply

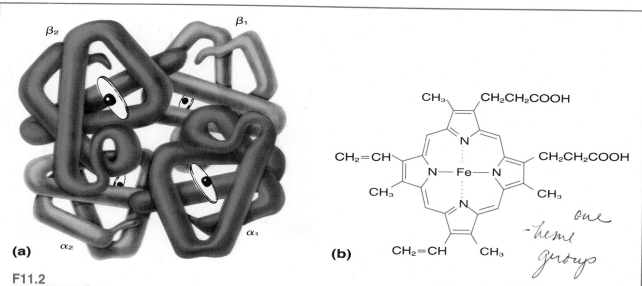

F11.2

Hemoglobin. (a) A hemoglobin molecule showing its four polypeptide chains (2 α chains and 2 β chains). The structure in the center of each chain represents an iron-containing heme group. (b) Structure of a single iron-containing heme group. *4 = 8 Oxygen atoms*

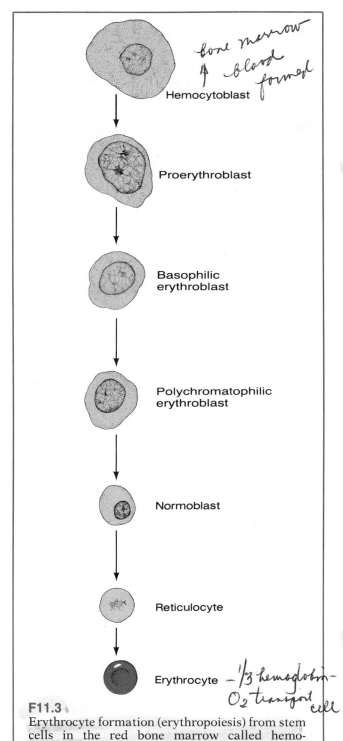

bone marrow & blood former

Hemocytoblast

Proerythroblast

Basophilic erythroblast

Polychromatophilic erythroblast

Normoblast

Reticulocyte

Erythrocyte — *1/3 hemoglobin — O₂ transport cell*

F11.3
Erythrocyte formation (erythropoiesis) from stem cells in the red bone marrow called hemocytoblasts.

to the tissues leads to a decrease in erythropoietin levels and, as a result, to a decrease in erythropoiesis. The male sex hormone testosterone enhances erythropoietin production, and the estrogenic female sex hormones tend to depress it. This could account at least in part for the greater numbers of erythrocytes and higher levels of hemoglobin in males than in females.

Fate of Erythrocytes

The average life of an erythrocyte is approximately 120 days in males and 109 days in females. Aged, abnormal, or damaged erythrocytes are disposed of by phagocytic cells of the reticuloendothelial system called macrophages (F11.4). These cells are found in the spleen, liver, bone marrow, and other tissues. The spleen is particularly important in the destruction of pathologic or defective erythrocytes, and the number of abnormal erythrocytes circulating in the blood increases considerably if the spleen is removed.

During erythrocyte destruction, the macrophages degrade hemoglobin (F11.4). The globin protein is broken down into its component amino acids, and the amino acids can be used in the synthesis of new proteins. The iron is liberated from the nonprotein heme, and the remainder of the heme is converted to *biliverdin*. Biliverdin is in turn converted to *bilirubin*, which enters the blood and binds to albumin. Bilirubin is eventually taken up by the liver, conjugated to glucuronic acid, and secreted in the bile. The liberated iron also enters the blood where it combines with transferrin. The iron can be used in the synthesis of new hemoglobin during erythropoiesis in the bone marrow or it can be stored.

Anemia

Anemia is a condition characterized either by a decreased number of erythrocytes in the blood or by a decreased concentration of hemoglobin. Whatever the cause, anemia decreases the blood's ability to transport oxygen to the tissues. Because tissues cannot function at optimum levels without adequate supplies of oxygen, anemia is frequently associated with a lack of energy, listlessness, and fatigue. Anemia can result from a number of conditions, several of which are discussed in the following sections.

Hemorrhagic Anemia.

Hemorrhagic anemia is the result of the loss of substantial quantities of blood. When blood is lost as the result of such occurrences as large wounds, stomach ulcers, or excessive menstrual bleeding, the lost volume is rather quickly replaced by fluid from the tissue spaces. The lost erythrocytes, however, are replaced by the slower processes of erythropoiesis. Until erythrocyte replacement is complete (which may take several weeks), there are fewer circulating erythrocytes than normal. Repeated or chronic hemorrhage can produce severe anemia, even if there is an increase in erythropoiesis.

Iron Deficiency Anemia.

Iron deficiency anemia is due to an excessive loss, deficient intake, or poor absorption of iron. In iron deficiency anemia, the erythrocytes produced are smaller than usual, and they contain less hemoglobin than normal.

Pernicious Anemia.

Pernicious anemia is the result of an inability to absorb adequate amounts of vitamin B_{12} from the digestive tract. Vitamin B_{12}, which is required for the normal maturation of erythrocytes, is found in such foods as liver, kidney, milk, eggs, cheese, and meat. The absorption of vitamin B_{12} requires the presence of a glycoprotein substance called *intrinsic factor*, which is produced by cells in the epithelial lining of the stomach. Individuals who suffer from pernicious anemia lack sufficient quantities of functional intrinsic factor and cannot adequately absorb vitamin B_{12}. As a result, they produce fewer erythrocytes than normal, and those that are produced are larger and more fragile than normal. Pernicious anemia can be treated by injections of vitamin B_{12}, thus bypassing the absorption problem.

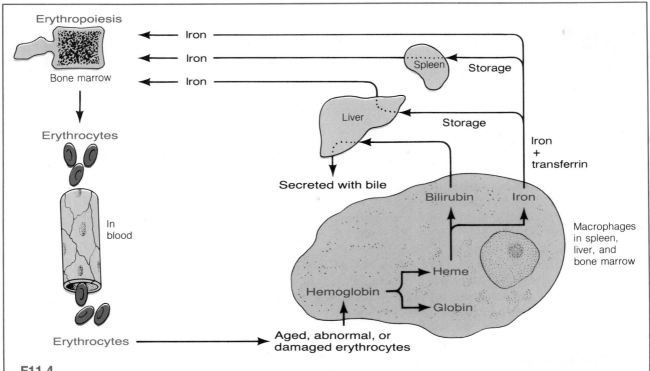

F11.4
Aged, abnormal, or damaged erythrocytes are disposed of by macrophages. The iron that is released may be either stored or reused in the synthesis of new hemoglobin.

Sickle-Cell Anemia.   *Sickle-cell anemia* is the result of the production of an abnormal hemoglobin (hemoglobin S; HbS) due to a genetic defect. In this condition, the globin portion of the molecule is abnormal. When the abnormal hemoglobin molecules are exposed to low concentrations of oxygen, they form fibrous precipitates within the erythrocytes, distorting them into the sickle shape characteristic of the disease. The misshapen erythrocytes are very fragile and often rupture as they pass through the capillaries, and especially through the spleen. Sickled cells can also become trapped in small blood vessels, impeding blood-flow. In its severest form, sickle-cell anemia can be fatal.

Aplastic Anemia.   *Aplastic anemia* is the result of an inadequate production of erythrocytes due to the inhibition or destruction of the red bone marrow. Destroyed marrow may be replaced by fatty tissue, fibrous tissue, or tumor cells. Aplastic anemia can be caused by radiation, various toxins, and certain medications.

Polycythemia

Polycythemia is a condition in which there is a net increase in the total circulating erythrocyte mass of the body. There are several types of polycythemia.

Secondary Polycythemia.   *Secondary polycythemia* occurs when decreased oxygen supplies to the tissues or increased levels of erythropoietin formation lead to an increased production of erythrocytes. In secondary polycythemia there may be 6 to 8 million and occasionally 9 million erythrocytes per cubic millimeter of blood. A common type of secondary polycythemia, called physiologic polycythemia, occurs in individuals living at high altitudes (4275 to 5200 meters) where oxygen availability is less than at sea level. Such people may have 6 to 8 million erythrocytes per cubic millimeter of blood.

Polycythemia Vera.   *Polycythemia vera* (*erythremia*) occurs when excess erythrocytes are produced as a result of tumorous abnormalities of the tissues that produce blood cells. Often, excess white blood cells and platelets are also produced. In polycythemia vera there may be 8 to 9 and occasionally 11 million erythrocytes per cubic millimeter of blood, and the hematocrit may be as high as 70% to 80%. In addition, the total blood volume sometimes increases to as much as twice normal. The entire vascular system can become markedly engorged with blood, and circulation times for blood throughout the body can increase to twice the normal value. The increased numbers of erythrocytes can increase the viscosity of the blood to as much as five times normal. Capillaries can become plugged by the very viscous blood, and the flow of blood through the vessels tends to be extremely sluggish.

LEUKOCYTES

Leukocytes, or **white blood cells,** are formed elements that defend the body against invasion by foreign organisms or chemicals and remove debris that results from dead or injured cells. Leukocytes act primarily in the tissues; those within the blood are mainly being transported by the circulation. Leukocytes are present within the blood in much smaller numbers than erythrocytes: a cubic millimeter of blood usually contains between 5,000 and 10,000 leukocytes. Many leukocytes are found in lymphoid tissues such as the thymus, lymph nodes, spleen, and lymphoid areas in the linings of the gastrointestinal tract. There are two major classes of leukocytes: granulocytes and agranulocytes (F11.5; Table 11.2).

Granulocytes

Granulocytes have clearly evident granules in their cytoplasm. They are also called polymorphonuclear leukocytes because their nuclei are generally divided into two or more lobes and have a variety of shapes. They are formed from stem cells in the red bone marrow. Three types of granulocytes are distinguished based on their reactions to certain stains. They are neutrophils, eosinophils, and basophils.

Neutrophils.   **Neutrophils** possess small cytoplasmic granules that appear light pink to blue-black when stained with Wright's stain. Neutrophils are phagocytic cells that are capable of ameboid movement. They are able to leave the blood vessels and enter the tissues where they protect the body by ingesting bacteria and other foreign substances.

Eosinophils. **Eosinophils** possess coarse cytoplasmic granules that appear reddish-orange when stained with Wright's stain. Eosinophils are capable of both phagocytosis and ameboid movement, and they are believed to ingest and destroy antigen–antibody complexes (see Chapter 14). The number of eosinophils in the blood and in the tissues increases during certain parasitic infections and in conditions involving allergic hypersensitivity (for example, asthma, hay fever).

Basophils. **Basophils** possess relatively large cytoplasmic granules that appear reddish-purple to blue-black when stained with Wright's stain. Their function is not known for certain, but they are believed to be functionally similar to cells found in connective tissue called *mast cells*. Mast cells contain the chemical substances *histamine* and *heparin*. Histamine causes vascular dilation and increased blood vessel permeability in inflammation and contributes to allergic responses. Heparin can prevent blood coagulation and can also enhance the removal of fat particles from the blood after a fatty meal.

Agranulocytes

In addition to the granulocytes, two agranular leukocyte types are present in the blood. They are monocytes and lymphocytes.

Monocytes. **Monocytes** are formed in the red bone marrow, and they are capable of ameboid movement. Monocytes can leave the blood vessels and enter the tissues where they develop into large, phagocytic cells called *macrophages* that can ingest bacteria and other foreign substances.

Lymphocytes. Only a few of the total number of **lymphocytes** are found in the blood; most are lodged in the lymphoid tissues. Lymphocytes are

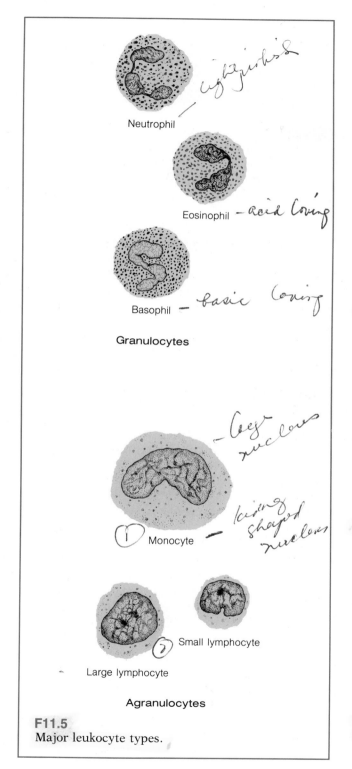

Neutrophil

Eosinophil

Basophil

Granulocytes

Monocyte

Small lymphocyte

Large lymphocyte

Agranulocytes

F11.5
Major leukocyte types.

important in the body's specific immune responses, including antibody production. Their development and specific functions are discussed in Chapter 14.

Leukemia

Leukemia is a cancerous condition in which an uncontrolled proliferation of leukocytes leads to a diffuse and almost total replacement of the red bone marrow with leukemic cells. These cells often replace the cells that form erythrocytes, and anemia results. In addition, there is frequently a decrease in blood platelets, which are also formed in the red bone marrow. Since platelets are involved in blood clotting and other hemostatic processes, leukemia may be accompanied by bleeding and hemorrhage. In fact, one of the causes of death in leukemia is internal hemorrhage, especially cerebral hemorrhage. Many of the leukocytes that are formed in the red bone marrow in leukemia are immature or abnormal. These abnormal leukocytes are not able to defend the body adequately against invasion by foreign organisms, and leukemia victims can also die from infection.

Infectious Mononucleosis

Infectious mononucleosis, which occurs mostly in children and young adults, is caused by a virus called the Epstein-Barr virus. The disease is characterized by an increase in the relative and absolute numbers of lymphocytes in the blood, and many of the lymphocytes are atypical. The symptoms of infectious mononucleosis include fatigue, sore throat, and slight fever. Affected individuals are usually watched carefully for complications during the course of the disease, which may last several weeks. Recovery is usually complete without any ill effects.

PLATELETS

Platelets are formed elements that are small cytoplasmic fragments about 2.5 microns in diameter having no nuclei and containing many granules. Platelets are formed in the red bone marrow as pinched-off portions of large cells called *megakaryocytes*. There are about 250,000 to 400,000 platelets per cubic millimeter of blood. Platelets

are involved in blood clotting, and they have also been implicated in other hemostatic processes.

HEMOSTASIS

Hemostasis is the arrest of bleeding or the stopping or slowing of the flow of blood through a vessel or a part. When a blood vessel is severed or damaged, hemostasis is achieved by mechanisms that include vascular constriction, the formation of a platelet plug, and blood clotting.

VASCULAR CONSTRICTION

When a blood vessel is cut or ruptured, one of the first noticeable responses is the constriction of the vessel. This decreases or stops bloodflow and allows time for other hemostatic activities to occur. In general, the greater the damage to a vessel, the greater the degree of constriction. Thus, a vessel ruptured by crushing usually bleeds less than a cleanly cut vessel.

FORMATION OF A PLATELET PLUG

Within an undamaged vessel, platelets generally circulate freely and do not stick to the normal endothelial cells that line the vessel. However, in an injured vessel where the deeper lying subendothelial tissues are exposed, platelets adhere to the underlying connective tissues, specifically collagen, and release several chemicals, among which is ADP (adenosine diphosphate). The ADP causes the surfaces of nearby platelets to become "sticky" and, as a result, still more platelets adhere and form a platelet plug or clump. The formation of a platelet plug may slow or stop bleeding completely from a damaged blood vessel that has a relatively small hole in it, but if the hole is large, blood clotting may be necessary to stop the bleeding. The ability of platelets to plug small vascular holes is important in closing minute tears that occur many times daily in the capillaries and other small vessels, and a person with an insufficient number of platelets often develops numerous, small hemorrhagic areas under the skin and throughout the internal tissues. In addition to

ADP, other chemicals released by platelets during the formation of a platelet plug facilitate vasoconstriction (for example serotonin) and contribute to blood clotting.

BLOOD CLOTTING

Blood clotting or **coagulation** is a complex process that involves a number of different factors, many of which are present in the plasma (Table 11.3). The formation of a blood clot requires the conversion of a soluble protein called *fibrinogen* (normally present in the plasma) into an insoluble, threadlike polymer called *fibrin* (F11.6). The insoluble fibrin threads form a network that entraps blood cells, platelets, and plasma to form the clot itself (F11.7). The fibrin threads can adhere to damaged blood vessels and thus anchor the clot in place.

The conversion of fibrinogen to fibrin during clot formation normally requires the enzymatic action of a factor called *thrombin*, which is produced from the inactive plasma protein precursor *prothrombin*. Minute amounts of thrombin may be produced continually from prothrombin, but this thrombin is generally inactivated or destroyed relatively rapidly so that its concentration in the

Table 11.3 *Factors Involved in Blood Coagulation*

Factor Number	Name	Type of Factor and Origin
I	Fibrinogen	A plasma protein produced in the liver
II	Prothrombin	A plasma protein produced in the liver
III	Tissue thromboplastin	A complex mixture of lipoproteins containing one or more phospholipid substances that is released from damaged tissues
IV	Calcium ions	An ion in the plasma that is acquired in the diet and from bones
V	Proaccelerin (labile factor, accelerator globulin)	A plasma protein produced in the liver
VI	Not utilized	
VII	Serum prothrombin conversion accelerator (stable factor, proconvertin)	A plasma protein produced in the liver
VIII	Antihemophilic factor (antihemophilic globulin)	A plasma protein produced in the liver
IX	Plasma thromboplastin component (Christmas factor)	A plasma protein produced in the liver
X	Stuart factor (Stuart–Prower factor)	A plasma protein produced in the liver
XI	Plasma thromboplastin	A plasma protein produced in the liver
XII	Hageman factor	A plasma protein
XIII	Fibrin stabilizing factor	A protein present in plasma and in platelets

blood does not rise high enough to promote clotting. During cell, tissue, or platelet disruption, however, the formation of thrombin increases considerably and its concentration may rise high enough to lead to clot formation.

The exact sequence of events that leads to the formation of thrombin during clotting is currently an area of much study and debate. However, researchers believe that the production of thrombin from prothrombin requires a *prothrombin activator* (*prothrombin-converting factor*). The production of prothrombin activator occurs either as a result of the action of an extrinsic mechanism that is activated when blood vessels are ruptured and tissues are damaged, or as the result of the action of an intrinsic mechanism that is present within the blood itself and that is activated when the blood is traumatized.

Extrinsic Mechanism

The extrinsic mechanism is activated when a substance called tissue thromboplastin (actually a complex mixture of lipoproteins that contains one or more phospholipid substances) is released from damaged tissues. Tissue thromboplastin, together with factor VII from the plasma and calcium ions, activates factor X (Table 11.3). Activated factor X, together with the effective form of factor V and a phospholipid substance, form prothrombin activator. Prothrombin activator acts enzymatically

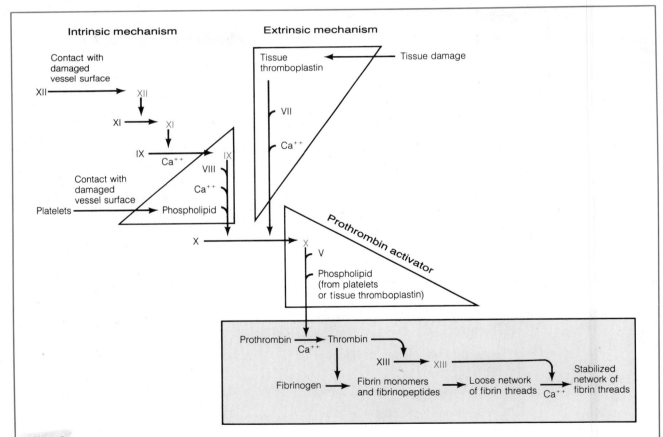

F11.6

The sequence of events that leads to the formation of a fibrin clot. The factors indicated in color are activated factors. (See the text for a detailed discussion.)

to catalyze the formation of thrombin from prothrombin. Thrombin, in turn, converts fibrinogen into fibrin monomers and fibrinopeptides. The fibrin monomers associate (polymerize) with one another, forming a loose network of fibrin threads. Thrombin also activates factor XIII, which augments the bonding between fibrin monomers, thereby strengthening and stabilizing the fibrin network.

Intrinsic Mechanism

The intrinsic mechanism is triggered when inactive factor XII in the plasma is activated by contact with a damaged vessel surface, perhaps by contact with underlying collagen fibers. Activated factor XII activates factor XI, and activated factor XI, in turn, activates factor IX. Activated factor IX, together with the effective form of factor VIII, calcium ions, and a phospholipid substance, then activates factor X. From this point on, the sequence of events occurs in the same fashion as in the extrinsic mechanism. The intrinsic clotting mechanism, however, is generally a slower acting mechanism than the extrinsic mechanism. In the intrinsic clotting sequence, the phospholipid substance required as a cofactor for some of the steps

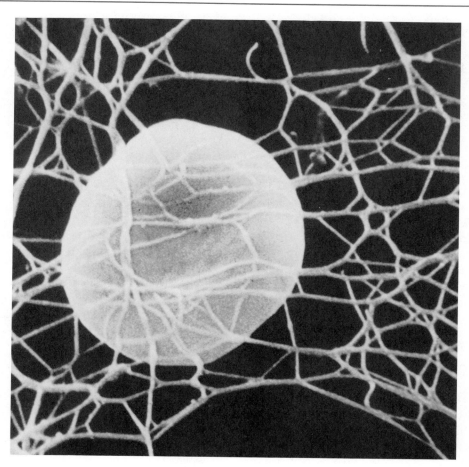

F11.7
Portion of a blood clot showing fibrin network and an entrapped erythrocyte.

becomes exposed on the surfaces of platelets (as platelet factor III) during platelet adhesion and clumping (agglutination), which is initiated when platelets contact collagen fibers.

Many of the reactions of the clotting process occur in what is called a cascade fashion. One factor becomes activated and this factor, in turn, activates another that activates still another and so on. In this way, inactive factors in the blood are changed into active clotting factors.

Positive Feedback Nature of Clot Formation

Once a blood clot starts to develop, thrombin acts in a positive feedback fashion to promote the development of the clot. For example, thrombin acts enzymatically on prothrombin to generate still more thrombin. In addition, thrombin accelerates the actions of some of the other clotting factors, and it enhances platelet adhesion and agglutination.

Limiting Clot Growth

Despite the positive feedback nature of the clotting process, clot formation normally occurs only locally at a site of damage. In general, once the various clotting factors are activated, they are rapidly inactivated or carried away by the blood so that their overall concentrations do not rise high enough to induce widespread clotting. For example, thrombin is removed from the blood in the region of a clot by being adsorbed onto the fibrin threads and by being inactivated by combination with an alpha globulin fraction of the plasma proteins called *antithrombin III*.

Clot Retraction

After the formation of a clot, a phenomenon known as clot retraction (syneresis) occurs. The fibrin meshwork shrinks and becomes denser and stronger. This activity helps pull the edges of a damaged vessel closer together. Clot retraction requires large numbers of platelets that, during clot formation, become trapped within the fibrin meshwork and attached to the fibrin strands. The platelets contain an actomyosinlike contractile protein, and they contract and pull the fibrin threads closer together.

During clot retraction, a fluid known as *serum* is extruded from the fibrin meshwork. Serum is essentially plasma that lacks fibrinogen and some of the other clotting factors that are removed during the clotting process.

Clot Dissolution

An enzyme called **plasmin** (**fibrinolysin**), which can decompose fibrin and dissolve clots, is present in the blood in an inactive form (plasminogen). Activated factor XII as well as thrombin and certain substances found in damaged tissues are able to convert plasminogen to plasmin. Thus, when clotting is initiated, so too is a mechanism for the eventual dissolution of the clot. In general, within a few days after blood has leaked into the tissues and clotted, the plasmin activators cause the formation of enough plasmin to dissolve the clot. Plasmin may also remove clots that form spontaneously in small peripheral blood vessels (for example, capillaries), although clots that block large vessels are less likely to be removed.

Prevention of Clotting within Normal Blood Vessels

Under most circumstances, blood does not clot as it flows through the blood vessels. This is due in part to the fact that the endothelial cells that line normal, undamaged blood vessels form a smooth surface that tends to prevent the contact activation of the intrinsic clotting mechanism. In addition, a monomolecular layer of negatively charged protein is adsorbed onto the endothelial cell lining, and this layer is believed to repel various clotting factors and platelets.

Anticlotting factors within the blood also contribute to clot prevention within normal blood vessels. For example, heparin, which is secreted by mast cells and basophils, inhibits the clotting process by accelerating the activity of antithrombin III, which inactivates thrombin. In addition, plasmin can destroy a number of clotting factors, as well as fibrin. Because of the presence of anticlotting factors, clotting normally does not occur until the rate of formation of thrombin rises above a critical level.

Intravascular Clotting

In spite of the various anticlotting mechanisms, clots may occur occasionally within the vascular system itself. If a clot blocks bloodflow in a vessel that supplies tissue critical to survival, the result can be very serious.

If the clot is fixed or adheres to a vessel wall, it is called a *thrombus*. A thrombus that forms in the coronary circulation may block the blood supply to heart tissue, leading to heart damage and ultimately to heart failure and death. The formation of a thrombus in a coronary artery is called a coronary thrombosis. Just what causes such clots to form is currently the subject of intense investigation, and several theories have been presented. One line of evidence suggests that continuous, unreversed damage to a blood vessel wall that removes the inner endothelium and exposes deeper lying tissues establishes conditions favorable to clot formation. Other investigations have revealed alterations in blood clotting characteristics in persons prone to thromboses that suggest a hyperreactivity of the clotting mechanism itself. Still others indicate that changes occur in a blood vessel wall as the result of the activity of viruses, and that the altered vessel wall may promote clot formation at the site. None of these occurrences is mutually exclusive, and all may play a role in clot formation.

If an intravascular clot is not fixed but floats within the blood, it is called an *embolus*. Emboli are potentially dangerous because they can become lodged in smaller vessels, thus blocking them and leading to tissue damage due to blood-flow restriction. Emboli that lodge in the vessels of a lung can damage the lung itself and elicit cardiovascular reflexes leading to hypotension (low blood pressure) and death.

Conditions Leading to Excessive Bleeding

One of the factors required in the blood-clotting sequence is factor VIII (antihemophilic factor). An inherited genetic deficiency of this factor results in a clotting defect called *hemophilia A*. A genetic deficiency of another clotting factor, factor IX (plasma thromboplastin component), results in the less common defect, *hemophilia B*. The blood of individuals who suffer from hemophilia does not clot properly, and even minor damage to blood vessels can result in substantial bleeding.

Nutritional deficiencies or liver disease can also lead to clotting problems and excessive bleeding. Vitamin K is required for the normal synthesis of prothrombin in the liver, and a deficiency of this vitamin interferes with prothrombin production. As a result, decreased levels of prothrombin may be present in the blood leading to clotting difficulties. The liver is also concerned with the production of a number of other clotting factors, and diseases that impair liver function can severely affect the clotting system.

Reduced numbers of platelets, a condition known as *thrombocytopenia*, causes bleeding states in which blood loss occurs through capillaries and other small vessels. Low platelet numbers can result from increased platelet destruction or from depressed platelet production due to pernicious anemia, certain drug therapies, or radiation.

STUDY OUTLINE

PLASMA   Approximately 90% water; transports hormones and metabolic end products, contains proteins (such as albumin, fibrinogen, globulins), ions (such as sodium, bicarbonate, chloride), food materials, and gases.   **pp. 275–276**

FORMED ELEMENTS   pp. 276–284

Erythrocytes (red blood cells)   Small, circular, biconcave discs with no nuclei that function in the transport of oxygen and carbon dioxide; the proportion of erythrocytes in a sample of blood is the hematocrit.

HEMOGLOBIN   Compound found in erythrocytes that can bind with oxygen and carbon dioxide. Hemoglobin is composed of a protein called globin and four nonprotein groups called hemes, which each contain an iron atom. Oxygen binds reversibly with the iron of the heme groups. Carbon dioxide binds reversibly with the globin portion of hemoglobin, which is composed of four polypeptide chains.

IRON   Component of hemoglobin. Iron is transported in the plasma in loose combination with a beta globulin called transferrin. It can be stored in the liver, spleen, and elsewhere in the form of intracellular iron–protein complexes called ferritin and hemosiderin. Iron is absorbed into the body in the upper part of the small intestine. Total body iron is regulated to a large extent by alterations in the intestinal absorption rate.

PRODUCTION OF ERYTHROCYTES   Erythrocytes develop from cells located in red bone marrow. Stem cells called hemocytoblasts give rise to proerythroblasts, which become basophilic erythroblasts that synthesize hemoglobin. Basophilic erythroblasts differentiate into polychromatophilic erythroblasts, which become normoblasts. When the hemoglobin content of the normoblast cytoplasm reaches about 34%, the nucleus is pinched off. The resulting reticulocytes mature into erythrocytes. The process of erythrocyte formation is called erythropoiesis. Vitamin B_{12} and folic acid are required for erythrocyte production.

CONTROL OF ERYTHROCYTE PRODUCTION   A substance called erythropoietin stimulates erythrocyte production. It is produced in kidneys by action of renal erythropoietic factor on plasma protein precursor. The production of erythropoietin and thus erythrocyte production is regulated by a negative feedback mechanism that is sensitive to the amount of oxygen delivered to the tissues. An inadequate supply of oxygen leads to increased erythropoiesis.

FATE OF ERYTHROCYTES   Aged, abnormal, or damaged erythrocytes are disposed of by phagocytic macrophages (especially in the spleen, liver, and bone marrow). Hemoglobin is degraded. The amino acids of globin can be used in the synthesis of new protein. The iron is liberated from the heme. The remainder of the heme is converted to biliverdin, which is in turn converted to bilirubin. Bilirubin is eventually taken up by the liver, conjugated to glucuronic acid, and secreted with the bile.

ANEMIA   Condition characterized by either a decreased number of erythrocytes in the blood or by a decreased percentage of hemoglobin.

Hemorrhagic Anemia   Result of substantial blood loss and relatively slow replacement of lost erythrocytes.

Iron Deficiency Anemia   Due to an excessive loss, deficient intake, or poor absorption of iron. Red blood cells are smaller than normal and they contain a lower than normal level of hemoglobin.

Pernicious Anemia   Result of inability to absorb adequate amounts of vitamin B_{12} due to insufficient quantities of functional intrinsic factor. Erythrocytes are larger and more fragile than normal.

Sickle-Cell Anemia   Result of production of abnormal hemoglobin due to a genetic defect. Abnormal hemoglobin causes distortion of erythrocytes into sickle shape at low concentrations of oxygen. Misshapen erythrocytes can rupture or block blood vessels.

Aplastic Anemia   Inadequate production of erythrocytes due to inhibition or destruction of red bone marrow.

POLYCYTHEMIA Condition in which there is a net increase in the total circulating erythrocyte mass of the body.

Secondary Polycythemia Increased production of erythrocytes due to decreased oxygen supplies to the tissues or increased levels of erythropoietin formation. Physiologic polycythemia is seen in people living at high altitudes.

Polycythemia Vera Production of excess erythrocytes due to tumorous abnormalities of the tissues that produce blood cells.

Leukocytes Formed elements of the blood that act primarily in the tissues and are mainly being transported by the circulation.

GRANULOCYTES Leukocytes that have clearly evident granules in their cytoplasm and a nucleus that is generally divided into two or more lobes. Formed in red bone marrow.

Neutrophils Phagocytic cells that are capable of ameboid movement. They can leave blood vessels and enter the tissues where they ingest bacteria and other foreign substances.

Eosinophils Ameboid, phagocytic cells that are believed to ingest and destroy antigen–antibody complexes.

Basophils May be functionally related to mast cells of connective tissue that contain histamine and heparin.

AGRANULOCYTES Two types.

Monocytes Ameboid cells formed in red bone marrow that can leave blood vessels and enter the tissues where they develop into large, phagocytic cells called macrophages.

Lymphocytes Important in the body's specific immune responses including antibody production. Many lymphocytes are found in lymphoid tissues.

LEUKEMIA A cancerous condition in which there is an excessive, uncontrolled proliferation of leukocytes that leads to a diffuse and almost total replacement of the red bone marrow with leukemic cells.

INFECTIOUS MONONUCLEOSIS Disease caused by the Epstein-Barr virus that is characterized by an increase in the relative and absolute numbers of lymphocytes in the blood.

Platelets Cytoplasmic fragments of megakaryocytes that are involved in blood clotting and other hemostatic processes.

HEMOSTASIS The arrest of bleeding or the stopping or slowing of the flow of blood through a vessel or a part. pp. 284–289

Vascular Constriction Occurs when a blood vessel is cut or ruptured. It decreases or stops bloodflow.

Formation of a Platelet Plug In a damaged vessel, platelets adhere to exposed collagen and release ADP, which leads to further platelet adhesion and the formation of a platelet plug. The plug may slow or stop bleeding if vessel damage is not too great. Platelets also release chemicals that facilitate vasoconstriction and contribute to blood clotting.

Blood Clotting Involves the conversion of a soluble plasma protein called fibrinogen into an insoluble, threadlike polymer called fibrin by the enzymatic action of thrombin. The fibrin threads entrap blood cells, platelets, and plasma to form the clot. Thrombin can be formed from an inactive plasma protein precursor by two mechanisms.

EXTRINSIC MECHANISM Tissue thromboplastin, which is released from damaged tissues, acts with other clotting factors to form prothrombin activator. Prothrombin activator catalyzes the formation of thrombin from prothrombin.

INTRINSIC MECHANISM Activation of factor XII and platelet adhesion and clumping may occur when inactive factor XII and platelets contact a damaged vessel surface. Phospholipid of platelets and active factor XII act with other clotting factors to form prothrombin activator.

POSITIVE FEEDBACK NATURE OF CLOT FORMATION Once clot formation begins, thrombin promotes clot development by acting on prothrombin to generate more thrombin, by accelerating the actions of other clotting factors, and by enhancing platelet adhesion and agglutination.

LIMITING CLOT GROWTH Once the various clotting factors are activated, they are rapidly inactivated or carried away by the blood so that their overall concentrations do not rise high enough to induce widespread clotting.

CLOT RETRACTION After clot formation, the fibrin meshwork shrinks and becomes denser and stronger. Clot retraction requires large numbers of platelets that apparently contain an actomyosin-like contractile protein.

CLOT DISSOLUTION An enzyme called plasmin can decompose fibrin and dissolve clots.

PREVENTION OF CLOTTING WITHIN NORMAL BLOOD VESSELS In normal blood vessels, clotting is prevented in part by the smooth endothelial lining of the vessels and by a layer of negatively charged protein adsorbed onto the surface of the endothelial cells. Anticlotting factors in the blood (such as heparin, antithrombin III, plasmin) may also inhibit clotting.

INTRAVASCULAR CLOTTING Clots may occur within the vascular system and block bloodflow to tissues critical for survival. A clot that is fixed or adheres to a vessel wall is called a thrombus and one that floats within the blood is called an embolus. Clots may form in blood vessels as a result of continuous, unreversed damage to vessel walls, hyperreactivity of the clotting mechanisms, or viral activity.

CONDITIONS LEADING TO EXCESSIVE BLEEDING An inherited genetic deficiency of factor VIII (antihemophilic factor) or factor IX (plasma thromboplastin component) may result in blood that does not clot properly and substantial bleeding. Nutritional deficiencies and liver diseases can also cause clotting problems and excessive bleeding. Thrombocytopenia (reduced numbers of platelets) may result in bleeding from capillaries.

SELF-QUIZ

1. As the amount of adipose tissue increases, the volume of blood per unit of body weight: (a) increases; (b) declines; (c) remains constant.
2. The hematocrit of a sample of peripheral venous blood is usually higher in females than in males. True or False?
3. The majority of the oxygen that is transported by the blood is in physical solution dissolved in the plasma. True or False?
4. Total body iron is regulated to a large extent by alterations in the intestinal absorption rate of iron. True or False?
5. Match the following terms with their appropriate descriptions:

Bilirubin   Globin
Heme   Ferritin
Erythropoietin

(a) A substance that stimulates erythrocyte production
(b) A component of hemoglobin that contains iron
(c) A component of hemoglobin that can bind reversibly with carbon dioxide
(d) A storage form of iron
(e) A breakdown product of hemoglobin that is taken up by the liver, conjugated to glucuronic acid, and excreted with the bile

6. Pernicious anemia is a condition that results from: (a) an inability to absorb adequate amounts of vitamin B_{12} from the digestive tract due to insufficient quantities of functional intrinsic factor; (b) a deficiency of iron in the blood; (c) the destruction of the red bone marrow.
7. Leukocytes act primarily in the tissues, and those observed within the blood are mainly being transported by the circulation. True or False?
8. Which of the following leukocyte types leave the blood vessels and enter the tissues where they protect the body by ingesting bacteria and other foreign substances? (a) neutrophil; (b) basophil; (c) lymphocyte.
9. Mast cells and basophils: (a) contain heparin; (b) produce antibodies; (c) secrete renal erythropoietic factor.
10. In the tissues, monocytes are converted to: (a) mast cells; (b) macrophages; (c) megakaryocytes.
11. The leukocyte type that is involved in specific immune responses, including antibody production is the: (a) lymphocyte; (b) neutrophil; (c) megakaryocyte.
12. The Epstein-Barr virus is believed to cause: (a) sickle cell anemia; (b) infectious mononucleosis; (c) leukemia.
13. Platelets are formed in the red bone marrow as pinched-off portions of cells called: (a) reticulocytes; (b) mast cells; (c) megakaryocytes.
14. ADP released by platelets facilitates: (a) vasoconstriction; (b) the conversion of prothrombin to thrombin; (c) the formation of a platelet plug.
15. During clot formation, fibrinogen is converted to fibrin by: (a) thrombin; (b) thromboplastin; (c) antithrombin III.
16. Blood clotting cannot occur unless tissue thromboplastin is released from damaged tissues. True or False?
17. Once it is formed, thrombin may contribute to clot development by acting enzymatically to convert serotonin to thromboplastin. True or False?
18. Clot retraction requires large amounts of: (a) antithrombin III; (b) plasmin; (c) platelets.
19. Fibrin can be decomposed and clots dissolved by: (a) thromboplastin; (b) plasmin; (c) fibrinogen.
20. A deficiency of vitamin K interferes with the normal production of prothrombin. True or False?

THE HEART

The circulatory system can be separated into two divisions: the **cardiovascular system** and the **lymphatic system.** The cardiovascular system is composed of the **heart,** which serves as a pump for the blood, and the **blood vessels,** which carry the blood throughout the body. The lymphatic system is a system of vessels that collect excess tissue fluid from between the cells of the body and carry it to the cardiovascular system. Along the pathways of the lymphatic vessels are structures called lymph nodes, which are involved in protecting the body against bacteria and other foreign substances. In addition, the spleen is generally considered to be a lymphatic structure. This chapter examines the function of the heart. The functions of the blood vessels and the lymphatic system are considered in Chapter 13.

The cardiovascular system is a continuous, closed system. Confined to the heart and the numerous vessels, the blood repeatedly travels through the heart, into arteries, then to capillaries, into veins, and back to the heart. Normally, blood does not leave this system, although some of the fluid part of the blood does pass through the walls of the capillaries to join the tissue fluid between the cells. However, this fluid is returned to the cardiovascular system either directly or by way of the lymphatic system (see Chapter 13). The heart is the pump that keeps the blood flowing through the system of vessels.

CIRCULATION THROUGH THE HEART

The adult heart is a cone-shaped, muscular organ about the size of a fist (F12.1). It is located within the thoracic cavity between the lungs, and is enclosed within a double-walled membranous sac called the *pericardium.* The heart is a four-chambered structure that is divided longitudinally into right and left halves (F12.2). Each half contains an upper, receiving chamber called an **atrium** and a lower, propulsion chamber called a **ventricle.** The right and left atria are separated by the *interatrial septum*, and the right and left ventricles are separated by the *interventricular septum.*

Venous blood that has passed through the vessels of the body is returned to the right atrium by the superior and inferior venae cavae. From the right atrium, the blood enters the right ventricle, which pumps it into the pulmonary trunk to the lungs. In the lungs the blood gives up carbon dioxide and receives oxygen (see Chapter 15). The newly oxygenated blood from the lungs is returned to the left atrium by pulmonary veins. From the left atrium the blood enters the left ventricle, which pumps it into the aorta to the body.

Four valves keep the blood flowing in the proper direction through the chambers of the heart.

Atrioventricular (AV) valves are located at the openings between the atria and the ventricles. The valve between the right atrium and the right ventricle has three cusps (flaps) and is therefore

known as the **tricuspid valve.** The valve between the left atrium and the left ventricle has only two cusps and is called the **bicuspid (mitral) valve.** Strong fibrous strands called chordae tendinae run from the cusps of the atrioventricular valves to cone-shaped papillary muscles that project from the walls of the ventricles. When the ventricles contract and the pressure within them increases, the atrioventricular valves are forced shut. This action prevents blood in the ventricles from flowing back into the atria. Tension exerted on the valve cusps by the papillary muscles prevents the cusps from being forced into the atria

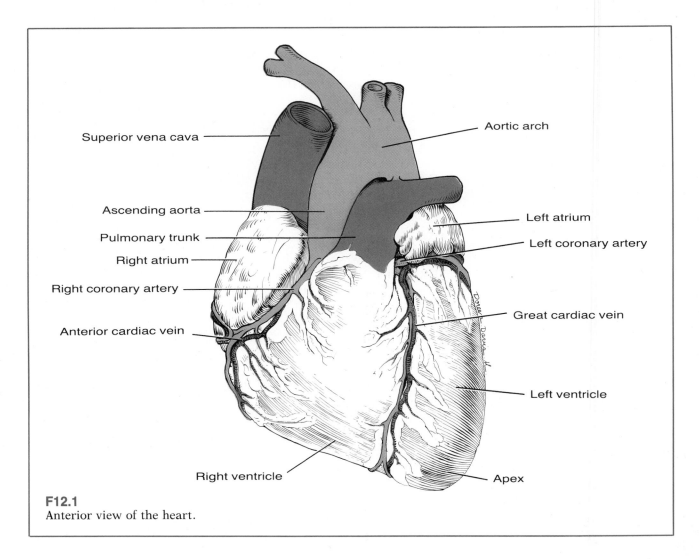

F12.1
Anterior view of the heart.

during ventricular contraction, thus helping keep the valves closed.

Semilunar valves prevent blood in the pulmonary trunk and aorta from returning to the ventricles when the ventricles complete their contraction and relax. The **pulmonary semilunar valve** is located in the pulmonary trunk where it leaves the right ventricle, and the **aortic semi-** **lunar valve** is located in the ascending aorta where it leaves the left ventricle. Both semilunar valves have three cusps that each resemble a shallow pocket (F12.3). When a ventricle contracts and exerts force on the blood within it, the blood pushes the cusps against the vessel wall and flows past them into the vessel. When the ventricle relaxes, the blood starts to flow back toward it, filling the

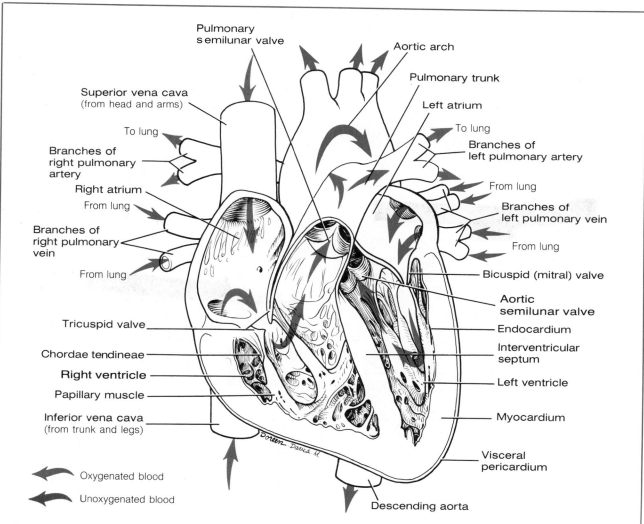

F12.2

Frontal section of the heart. The arrows indicate the path of the bloodflow through the chambers, the valves, and the major vessels. Note that the branches of the right pulmonary vein pass behind the heart and enter the left atrium.

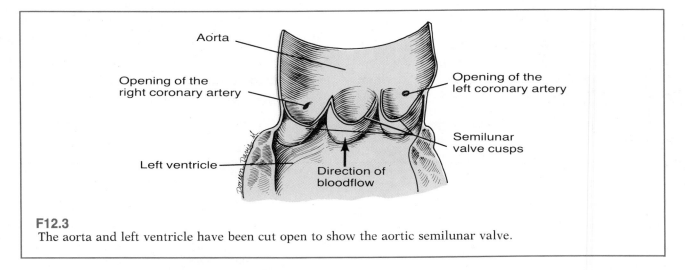

F12.3
The aorta and left ventricle have been cut open to show the aortic semilunar valve.

cusps and causing their unattached margins to meet in the middle of the vessel. This action creates a seal that prevents the backflow of blood into the ventricle.

PUMPING ACTION OF THE HEART

Because the chambers on the right side of the heart are separated from those on the left, the heart is really a double pump (F12.4). The right pump receives blood that has passed through the vessels of the body and sends it to the lungs—that is, through the *pulmonary circuit*. The left pump receives blood from the lungs and sends it to the body—that is, through the *systemic circuit*. The right and left pumps work in unison. When the heart beats, both atria contract simultaneously, and then both ventricles do the same.

The period from the end of one heartbeat to the end of the next is called the **cardiac cycle.** During this cycle, blood from the superior and inferior venae cavae moves through the right atrium, past the open tricuspid valve, and into the right ventricle (F12.5). At the same time, blood from the pulmonary veins moves through the left atrium, past the open bicuspid valve, and into the left ventricle. The simultaneous contraction of both atria then squeezes more blood into the ventricles. Subsequently, the simultaneous contraction of both

ventricles closes the atrioventricular valves and forces blood past the semilunar valves and into the pulmonary trunk (from the right ventricle) and the aorta (from the left ventricle).

Note from this discussion that the contraction of the atria is not essential for the movement of blood into the ventricles. In fact, even if the atria fail to function, the ventricles can still pump considerable quantities of blood.

CARDIAC MUSCLE

The wall of the heart is composed almost entirely of cardiac muscle, and it is this muscle that is responsible for the pumping action of the heart.

CELLULAR ORGANIZATION

The individual cells of cardiac muscle interconnect with one another to form branching networks (F12.6). Where adjoining cells meet end to end, their junctions form structures called *intercalated discs*. Within the intercalated discs are two types of cell-to-cell membrane junctions: *desmosomes*, which attach one cell to another, and *gap junctions*, which allow electrical impulses to spread from cell to cell. Cardiac muscle cells are cross-striated, and they possess thick, myosin-containing filaments and thin, actin-containing filaments that are arranged into regularly ordered sarcomeres and myofibrils. The basic contractile

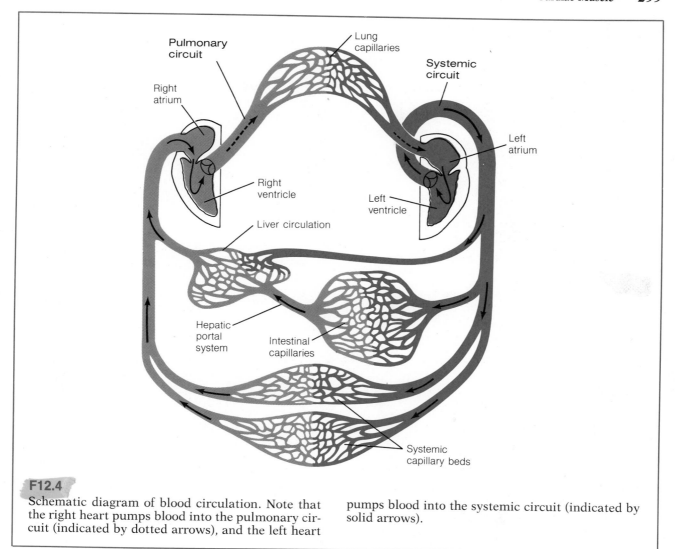

Pulmonary circuit

Lung capillaries

Systemic circuit

Right atrium

Left atrium

Right ventricle

Left ventricle

Liver circulation

Hepatic portal system

Intestinal capillaries

Systemic capillary beds

F12.4

Schematic diagram of blood circulation. Note that the right heart pumps blood into the pulmonary circuit (indicated by dotted arrows), and the left heart pumps blood into the systemic circuit (indicated by solid arrows).

events that occur in cardiac muscle cells are believed to be similar to those that occur in skeletal muscle cells (Chapter 6). However, cardiac muscle cells have larger *t* tubules and a less developed sarcoplasmic reticulum than skeletal muscle cells, and some of the calcium ions involved in contraction are obtained from the extracellular fluid.

AUTOMATIC CONTRACTION

Like some types of smooth muscle, certain cells of cardiac muscle undergo spontaneous, rhythmical, self-excitation. Moreover, a stimulatory impulse can spread from cell to cell through gap junctions. Thus, cardiac muscle is self-excitable, and it contracts automatically without external stimulation. However, the spontaneous activity of cardiac muscle is continually influenced by neurons of the autonomic nervous system that supply the heart and by certain chemicals and hormones that circulate within the blood.

DEGREE OF CONTRACTION

In general, when cardiac muscle contracts, it contracts as much as it can for the existing conditions.

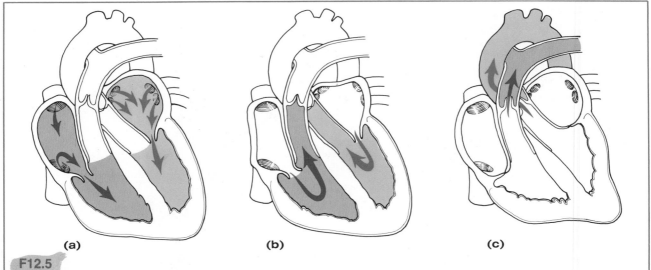

F12.5
Bloodflow through the heart during a single cardiac cycle. (a) Blood fills both atria and enters both ventricles. (b) The atria contract, squeezing more blood into the ventricles. (c) The ventricles contract, forcing blood into the aorta and the pulmonary trunk.

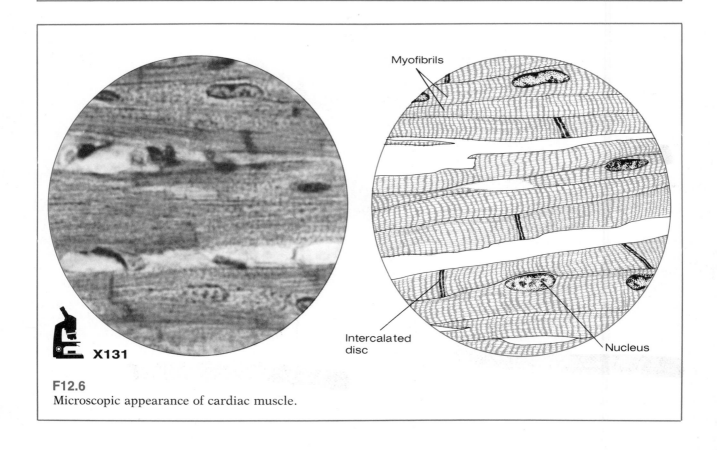

X131

F12.6
Microscopic appearance of cardiac muscle.

The individual cells of skeletal muscles respond in basically the same way, but a stimulatory impulse that excites one skeletal muscle cell does not spread to adjacent cells. Thus, a skeletal muscle, which is composed of thousands of cells, shows graded contractions depending on the number of cells stimulated. In contrast, a stimulus that excites a cardiac muscle cell can be transmitted to other cardiac muscle cells through gap junctions. Thus, an entire mass of interconnected cardiac muscle cells responds as a unit, contracting as much as it can for the existing conditions whenever a single cell is stimulated. This is not to imply that all contractions of the cardiac muscle of the heart are identical. Indeed, there is considerable variation depending on the conditions at the time of contraction.

REFRACTORY PERIOD

Following the excitation of a cardiac muscle cell, there is a relatively long period of time—called the *refractory period*—during which the cell cannot normally be reexcited. This period is particularly evident in ventricular cardiac muscle cells, where the refractory period extends well into the relaxation phase of the contractile cycle (F12.7). The relatively long refractory period normally prevents the heart from undergoing a prolonged tetanic contraction or a spasm, which would halt bloodflow and therefore cause death.

THE MYOCARDIUM

The cardiac muscle that is the principal constituent of the wall of the heart is referred to as the **myocardium** (F12.8). The myocardium is covered externally by a portion of the pericardium—the epicardium, or visceral pericardium—and it is lined internally by the endocardium.

The myocardium varies considerably in thickness from one heart chamber to another, in proportion to the force needed to overcome the resistance encountered in pumping blood from the different chambers. The muscle of the atria meets little resistance in pushing blood into the ventricles, and the walls of the atria are the thinnest part of the myocardium. In contrast, the ventricles must move the blood through either the

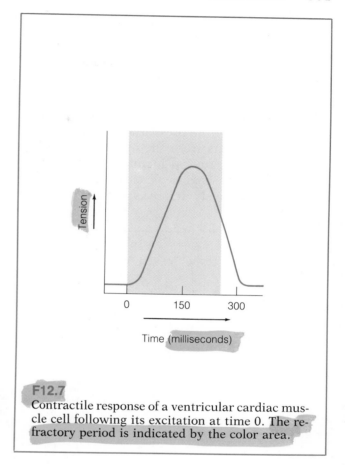

F12.7
Contractile response of a ventricular cardiac muscle cell following its excitation at time 0. The refractory period is indicated by the color area.

systemic or pulmonary circuits and back into a receiving chamber. These activities require considerable force, and the myocardium of the ventricles is thicker than that of the atria. The left ventricle, which propels blood through the entire systemic circuit to all parts of the body and back into the right atrium, has thicker walls and pumps with greater force than the right ventricle, which moves the blood only through the pulmonary circuit and back into the left atrium.

METABOLISM OF CARDIAC MUSCLE

The myocardium can obtain only insignificant amounts of energy—that is, ATP—from anaerobic metabolism. Therefore, cardiac muscle depends on aerobic metabolism for the continuous supply of energy required to support its contractile activity. For this reason, a substantial and continuous delivery of oxygen to the myocardium is essential, and the myocardium receives an abundant blood supply through coronary arteries, which arise from the aorta near its junction with the left ventricle.

In a resting individual the myocardium obtains most of its energy from the oxidation of fatty acids and small amounts from the oxidation of lactic acid and glucose. During increased activity—for example, during heavy physical exercise—the myocardium actively removes lactic acid from coronary blood and oxidizes it directly for the production of energy. In this regard, it should be noted that the anaerobic metabolic processes utilized by skeletal muscles during vigorous activity produce lactic acid that is released into the blood.

EXCITATION AND CONDUCTION IN THE HEART

The heart contains specialized cardiac muscle cells that initiate the impulses that cause it to beat (F12.9). It also contains specialized cardiac mus-

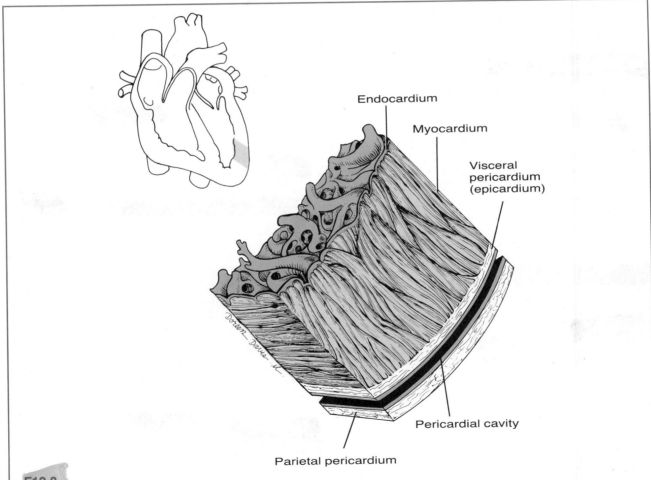

Endocardium

Myocardium

Visceral pericardium (epicardium)

Pericardial cavity

Parietal pericardium

F12.8 Section through the wall of the heart showing the parietal and visceral layers of the pericardium (heart sac), the myocardium (cardiac muscle), and the endocardium (inner lining of the heart chamber).

cle cells that conduct the impulses throughout the myocardium. This conducting system coordinates the heartbeat, producing an efficient pumping action.

EXCITATION

Like other cells, cardiac muscle cells maintain an unequal distribution of ions on either side of their cell membranes, and they are electrically polarized. If the cell membrane of a cardiac muscle cell becomes depolarized to threshold, an action potential results and the cell is stimulated to contract.

In the wall of the right atrium, near the entrance of the superior vena cava, is a small mass of specialized cardiac muscle cells called the **sinoatrial node (sinuatrial node, SA node)**. In a resting adult, the cells of this node spontaneously depolarize to threshold and generate an action potential approximately 70 to 80 times each minute (that is, about every 0.8 second). The reasons for this depolarization are not entirely clear. Some investigators suggest that the membranes of the cells of the sinoatrial node are relatively permeable to sodium ions, and the entrance of sodium ions into the cells gradually decreases the membrane potential (normally about −55 to −70 millivolts, inside negative) to threshold. Others stress the importance of a decreasing membrane permeability to potassium and a consequent diminished

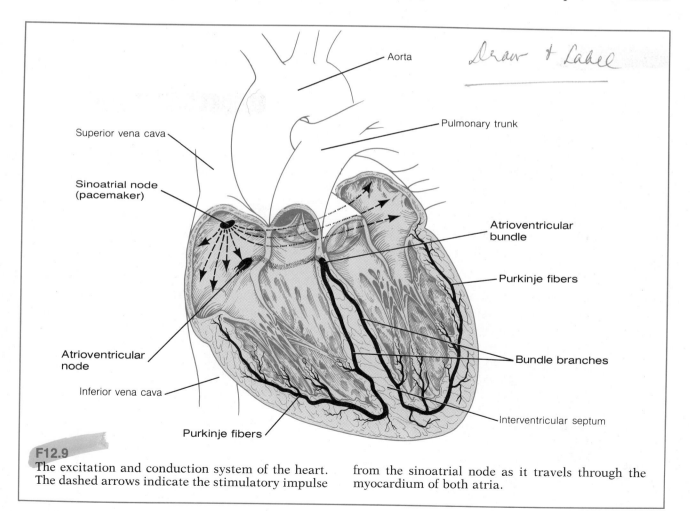

F12.9
The excitation and conduction system of the heart. The dashed arrows indicate the stimulatory impulse from the sinoatrial node as it travels through the myocardium of both atria.

outflow of potassium ions to the depolarization process. In any event, the action potentials developed by the cells of the sinoatrial node depend in large measure on changes in sodium and potassium permeabilities, and a coupled sodium-potassium pump is operative in these cells.

Other areas of the myocardium can also undergo spontaneous depolarization to threshold and generate action potentials. However, they do so at slower rates than the sinoatrial node. As a result, impulses from the sinoatrial node spread to these areas and stimulate them so frequently that they do not generate action potentials at their own inherent rates. Thus, the rate of discharge of the sinoatrial node sets the rhythm for the entire heart, and it is for this reason that the sinoatrial node is called the *pacemaker* of the heart.

CONDUCTION

An impulse generated in the sinoatrial node spreads from cell to cell throughout the myocardium of the atria, stimulating the atria to contract. However, the heart possesses a fibrous skeleton that surrounds the openings between the atria and the ventricles as well as the openings of the aorta and the pulmonary trunk (F12.10). This skeleton cannot depolarize, and, thus, a stimulatory impulse transmitted from the sinoatrial node throughout the atria cannot pass directly to the myocardium of the ventricles. Instead, the impulse reaches the ventricles by way of a specialized conducting system.

A group of specialized cardiac muscle cells called the **atrioventricular node (AV node)** is located within the interatrial septum just above the junction of the atria and the ventricles (see F12.9). From the atrioventricular node, a bundle of specialized muscle tissue called the **atrioventricular bundle (bundle of His)** passes to the ventricles. The atrioventricular bundle enters the interventricular septum, where it divides into **right** and **left bundle branches** that travel down the septum. Terminal branches called **Purkinje fibers** are given off to the cells of the ventricular myocardium by the bundle branches.

As a stimulatory impulse from the sinoatrial node spreads throughout the atrial myocardium,

it reaches the atrioventricular node. After a delay of approximately 0.1 second, the impulse is transmitted from the atrioventricular node through the atrioventricular bundle and the Purkinje fibers to the cells of the ventricles. One reason for the delay in the transmission of the impulse at the atrioventricular node is that the cells in the region are quite small, and they transmit impulses slowly. This delay is important to a coordinated heartbeat because it ensures that the atria will contract and squeeze blood into the ventricles before the ventricles contract.

In contrast with the delay in the conduction of the stimulatory impulse at the atrioventricular node, the rate of conduction of the impulse from the node to the ventricles along the specialized conducting fibers is quite rapid. Thus, the entire ventricular myocardium is stimulated almost simultaneously, producing a strong, effective pumping action.

THE CARDIAC CYCLE

The main function of the heart is to receive blood from the veins at low pressure and pump it into the arteries at a pressure that is high enough to propel it through the vessels and back again to the heart.

In order to function as a pump, the heart repeats two alternating phases: (1) the chambers contract, forcing the blood within them out of the chambers. This phase is called **systole.** (2) The chambers relax, allowing them to refill with blood. This phase is known as **diastole.** The cardiac cycle, therefore, consists of a period of contraction and emptying (systole) followed by a period of relaxation and filling (diastole). Although both the atria and the ventricles undergo systole and diastole, the terms usually refer to ventricular contraction or relaxation.

The pressure changes, volume changes, and valve actions that occur during the cardiac cycle are compared in Figure 12.11. Although this figure is concerned with the left chambers of the heart, the right chambers follow a similar pattern. However, the peak pressure within the right ventricle is considerably lower. The systolic pressure in the

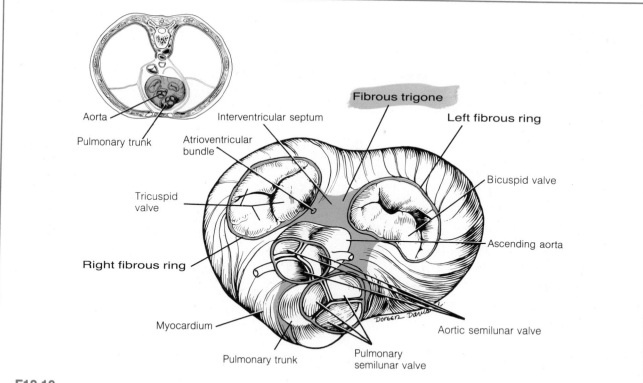

Aorta

Pulmonary trunk

Interventricular septum

Atrioventricular bundle

Fibrous trigone

Left fibrous ring

Tricuspid valve

Bicuspid valve

Right fibrous ring

Ascending aorta

Myocardium

Pulmonary trunk

Pulmonary semilunar valve

Aortic semilunar valve

Doreen Davis

F12.10
Superior view of the heart with the atria removed, showing the fibrous skeleton (in color) that surrounds the valve openings.

right ventricle—which has to pump the blood only through the pulmonary circuit—reaches about 30 millimeters of mercury (mm Hg), whereas as the figure shows, the systolic pressure in the left ventricle reaches 120 mm Hg.

Each cardiac cycle lasts approximately 0.8 second. Atrial systole requires about 0.1 second, and ventricular systole about 0.3 second. Thus, during each cardiac cycle the atria are in diastole for 0.7 second and the ventricles for 0.5 second. It is important to emphasize that during ventricular diastole, blood flows into the atria and through the open atrioventricular valves into the ventricles prior to the contraction of the atria. In fact, the ventricles are 70% filled before the atria contract. When the atria do contract, they force a small amount (approximately 30% of the ventricular capacity) of additional blood into the ventricles.

However, since the ventricles are almost full prior to atrial contraction, defects that impair atrial pumping may not seriously impair cardiac function, at least in resting individuals.

PRESSURE CURVE OF THE LEFT ATRIUM

During most of ventricular diastole, the pressure within the left atrium is slightly higher than the pressure within the left ventricle. Moreover, the pressure within the left atrium gradually increases as blood flows into it (and continues into the left ventricle) from the pulmonary veins (F12.11). When the left atrium contracts, there is a sudden further increase in the atrial pressure (curve *a*). Shortly thereafter, ventricular systole begins, and the pressure within the left ventricle rises above that in the left atrium, causing the left

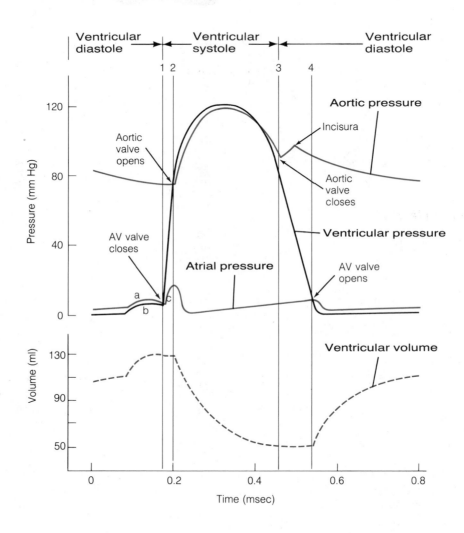

F12.11

Comparisons of pressure changes, volume changes, and valve actions within the left chambers of the heart during a single cardiac cycle. Note that the duration of a complete cycle is 0.8 second.

atrioventricular valve to close (line 1). The sudden increase in the intraventricular pressure is so great that it causes the left atrioventricular valve to bulge slightly into the left atrium. This activity contributes to a further increase in the atrial pressure (curve *c*) that lasts only until the aortic semilunar valve opens (line 2). The pressure within the left atrium, which is now in diastole, then quickly drops to almost 0 mm Hg. However, as blood from the pulmonary veins again fills the atrium, the atrial pressure shows a steady increase until the left atrioventricular valve reopens during the following ventricular diastole (line 4). The reopening of the left atrioventricular valve allows blood from the left atrium to enter the left ventricle. The flow of blood from the atrium into the ventricle causes the atrial pressure to drop again to almost 0 mm Hg. The atrium is now ready to begin another cycle.

PRESSURE CURVE OF THE LEFT VENTRICLE

During most of ventricular diastole, the pressure within the left ventricle is slightly lower than the pressure within the left atrium (F12.11). The contraction of the left atrium causes a slight increase in the pressure within the left ventricle (curve *b*) because atrial systole increases the volume of blood within the ventricle, as shown in the ventricular volume curve. When ventricular systole begins, the pressure within the left ventricle rises rapidly (lines 1 to 2). When the pressure in the left ventricle exceeds the pressure in the left atrium, the left atrioventricular valve closes. When the pressure within the left ventricle becomes greater than the pressure within the aorta, the aortic semilunar valve is forced open (line 2), allowing blood to flow into the aorta. The pressure within the ventricle continues to rise to about 120 mm Hg and then begins to fall. At the end of ventricular systole, the intraventricular pressure drops below the pressure in the aorta. As a result, blood from the aorta begins to flow back toward the left ventricle, causing the aortic semilunar valve to close (line 3). The intraventricular pressure continues to fall precipitously until it becomes less than the atrial pressure (line 4), thus allowing blood in the

left atrium to push open the left atrioventricular valve and flow into the left ventricle. Notice that at this point (line 4) the volume of blood within the left ventricle is at its lowest and that it immediately begins to increase as the blood that had collected within the left atrium flows into the left ventricle. This increase in blood volume occurs prior to atrial systole. Notice also that there is only a slight increase in ventricular volume following atrial systole.

PRESSURE CURVE OF THE AORTA

The aorta is an elastic artery. As blood is pumped into the aorta during ventricular systole, the vessel distends, or expands. Then, as the bloodflow from the left ventricle into the aorta decreases and stops during late systole and diastole, the aorta recoils, and squeezes down on the blood within it. This action helps maintain a certain amount of pressure within the vessel (F12.11). If the aorta and the other elastic arteries had rigid walls that were not distensible, the pressure within them would rise very abruptly during ventricular systole. If they were not capable of elastic recoil, the pressure within them would fall just as abruptly during ventricular diastole.

When the left ventricle is in diastole and no blood is being ejected into the aorta, the aortic pressure curve shows a steady, but gradual, decrease as the blood pumped into the aorta during the previous systole flows into other vessels of the body (F12.11). The elastic recoil of the aortic wall maintains some pressure on the blood (the aortic blood pressure never drops much below 80 mm Hg), but it is not sufficient to keep the pressure constant. When the left ventricle contracts and the aortic semilunar valve opens (line 2), the aortic pressure increases rapidly as a large volume of blood is pumped into the aorta. The pressure peaks at about 120 mm Hg, and then decreases as the flow of blood from the left ventricle into the aorta diminishes. However, the ventricular pressure falls more rapidly than the aortic pressure, and as it drops below the aortic pressure, the aortic semilunar valve closes (line 3). The closure of the aortic semilunar valve produces a brief rise in the aortic pressure, which forms the incisura on the diagram.

CARDIAC OUTPUT

Each cardiac cycle moves blood out of the ventricles into the pulmonary circuit and the systemic circuit. The term **cardiac output** is used to indicate the volume of blood pumped per minute by either ventricle; for this reason, it is sometimes referred to as **cardiac minute output.**

Generally, it is the left ventricle that is referred to when cardiac output is discussed. However, over any period of time, the normal heart ejects the same amount of blood from each ventricle, although some variation may occur on a beat-to-beat basis. If this were not the case and the left ventricle pumped a greater volume than the right, blood would accumulate in the systemic circuit. Similarly, if the right ventricle pumped more blood than the left, blood would accumulate in the pulmonary circuit. As is discussed later, unequal pumping by the two ventricles can be a problem in some diseases of the heart.

Cardiac output is equal to the heart rate multiplied by the stroke volume. The **stroke volume** is the amount of blood pumped by one side of the heart per beat. The average stroke volume of a human heart is about 75 ml, and the heart rate of a resting individual is generally between 70 and 80 beats per minute. Therefore, the cardiac output under resting conditions is between 5 and 6 liters per minute (for example, 75 ml per beat times 70 beats per minute = 5250 ml per minute = 5.25 l per minute). Under stressful conditions, the cardiac output can be greatly increased—up to 30 liters per minute in a trained athlete. The difference between the volume of blood moved per minute during rest and the volume that the heart is capable of moving per minute is called the **cardiac reserve.**

Changes in either the heart rate or the stroke volume can alter the cardiac output, and in general an increase in either factor increases the cardiac output. However, at very high heart rates—usually beginning between 170 and 250 beats per minute—cardiac output actually decreases. This decrease is due to the fact that at very high heart rates, the stroke volume declines, offsetting the effects of the increased rate. A major reason for the decline in stroke volume is that, as the heart rate increases, the length of diastole decreases, leaving a shorter time for the ventricles to fill with blood. Thus, less blood is present within the ventricles at the end of diastole to be pumped out during the following systole.

The cardiac output changes to meet the varying needs of the body, and since it is equal to the heart rate multiplied by the stroke volume, the mechanisms that control the heart rate and the stroke volume also control the cardiac output (F12.12).

CONTROL OF HEART RATE

The most important factor in the control of heart rate is the effect of the autonomic nervous system. The sinoatrial node and the atrioventricular node are particularly well supplied by sympathetic and parasympathetic neurons, and the atria also receive both sympathetic and parasympathetic innervation. The ventricles are supplied mainly by sympathetic neurons, with far fewer parasympathetic neurons present.

Stimulation of the sympathetic neurons increases the heart rate, and maximal sympathetic stimulation can almost triple the rate. The sympathetic neurons release norepinephrine, which increases the rate of discharge of the sinoatrial node, decreases the conduction time through the atrioventricular node, and increases the excitability of all portions of the heart. The exact mechanisms by which norepinephrine exerts its effects are not clearly understood, but it is believed to either decrease membrane permeability to potassium ions or increase membrane permeability to sodium ions (and calcium ions), thereby enhancing the depolarization of excitable cells.

Parasympathetic fibers innervate the heart by way of the vagus nerves. The right vagus is distributed predominantly to the sinoatrial node, and the left vagus innervates principally the atrioventricular node. Stimulation of the parasympathetic neurons to the heart decreases the heart rate. The parasympathetic neurons release acetylcholine, which decreases the rate of discharge of the sinoatrial node and increases the conduction time through the atrioventricular node. Acetylcholine increases membrane permeability to potassium ions, thereby causing hyperpolarization and making excitable tissue less excitable. Intense parasympathetic stimulation can stop impulse generation in the sinoatrial node and block the transmission of impulses through the atrioventricular node to the ventricles. However, soon after the ventricles stop contracting, some other portion of the conducting system—often in the atrioventricular bundle—begins to discharge spontaneously and stimulate ventricular contraction. This phenomenon is known as *ventricular escape.*

The heart rate at any given moment is largely determined by the balance between the stimulatory effects of norepinephrine released from sympathetic neurons and the inhibitory effects of acetylcholine released from parasympathetic neurons. If the sympathetic nerves to the heart are blocked, the heart rate decreases (due to the unopposed parasympathetic activity), and if the parasympathetic nerves are blocked, the heart rate increases (due to the unopposed sympathetic activity). In a resting individual, parasympathetic activity is dominant, and if all autonomic innervation is blocked, the heart rate increases from the normal 70 to 80 beats per minute to about 100 beats per minute, which is the inherent rate of discharge of the sinoatrial node. During exercise, the activity of the sympathetic neurons increases, the activity of the parasympathetic neurons declines, and the heart beats faster.

CONTROL OF STROKE VOLUME

The amount of blood that moves out of a ventricle with each beat—that is, the stroke volume—is equal to the difference between (1) the volume of

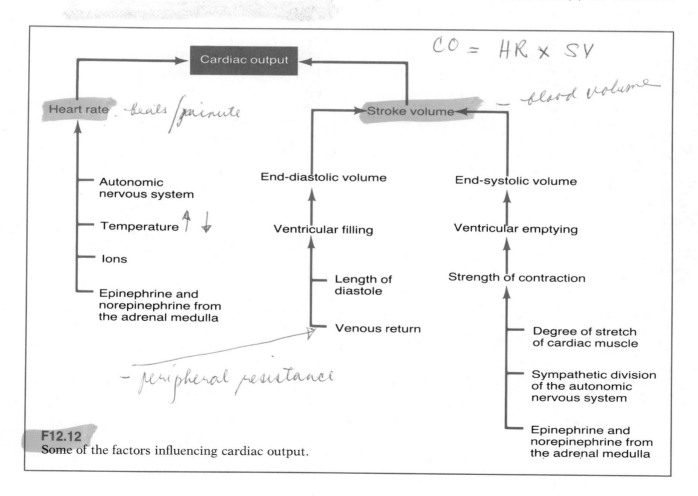

F12.12
Some of the factors influencing cardiac output.

blood within the ventricle at the end of diastole just as systole begins (the **end-diastolic volume**), and (2) the volume of blood remaining within the ventricle when the semilunar valves close at the end of systole (the **end-systolic volume**). A number of factors influence the end-diastolic and end-systolic volumes and, therefore, the stroke volume.

End-Diastolic Volume

The end-diastolic volume depends on the amount of blood that enters a ventricle during diastole, and this, in turn, is determined by two main factors: the length of diastole and the venous return.

Length of Diastole. The longer the period of diastole, the more time is available for the ventricles to fill with blood. However, as the heart rate increases, the length of diastole becomes shorter, and less time is available for ventricular filling. In this regard, it should be noted that norepinephrine from the sympathetic nervous system not only increases the heart rate, but it also accelerates the rates at which cardiac muscle fibers shorten and relax. Thus, as norepinephrine increases the heart rate, the decreased time required for cardiac muscle fibers to shorten and relax provides somewhat more time for ventricular filling during diastole than would otherwise be the case. Moreover, the rapid relaxation of the ventricles after contraction causes the intraventricular pressure to fall rapidly, which enhances the pressure gradient for the flow of blood from the atria into the ventricles. Nevertheless, in a very rapidly beating heart, the length of diastole may be so shortened that ventricular filling is significantly reduced.

Venous Return. The volume of blood that returns from the veins to one side of the heart in a given time has a major influence on the end-diastolic volume. (The term *venous return* usually refers to the volume of blood flowing from the systemic veins into the right side of the heart in a given time, but under normal circumstances the same volume of blood flows from the pulmonary veins into the left side of the heart.) In general, if the length of diastole is constant, an increased venous return leads to increased ventricular filling

and a relatively large end-diastolic volume, which stretches the ventricle wall. If the length of diastole decreases (for example, when the heart rate increases), a ventricle may still fill completely if, as is often the case, the venous return increases. As is discussed in Chapter 13, conditions in the peripheral blood vessels have a major influence on the magnitude of the venous return.

End-Systolic Volume

The end-systolic volume depends on the amount of blood ejected from a ventricle during its contraction. A ventricle does not eject all the blood it contains when it contracts, and the degree to which a ventricle is emptied during systole is determined primarily by the strength of the ventricular contraction.

The Frank–Starling Law of the Heart. The Frank–Starling law of the heart states that, within limits, as cardiac muscle is stretched, its force of contraction increases. Thus, when ventricular filling produces a comparatively large end-diastolic volume that stretches the ventricle wall, the ventricle contracts more forcefully and ejects a larger volume of blood than when ventricular filling produces a smaller end-diastolic volume that does not stretch the ventricle wall as greatly. This phenomenon provides an inherent, self-regulating mechanism by which the heart is able to adjust stroke volume to changing end-diastolic volumes. However, if a ventricle is overly stretched (that is, if the end-diastolic volume is too great), the effectiveness of the ventricular contraction diminishes.

The influence of stretch on the force of contraction of cardiac muscle helps ensure that over any period of time the normal heart ejects the same amount of blood from each ventricle, even though some variation may occur on a beat-to-beat basis. For example, suppose the stroke volume of the right ventricle increases while that of the left ventricle remains unchanged. The increased right ventricular stroke volume causes an increased volume of blood to enter the pulmonary circuit, which, in turn, leads to an increase in the volume of blood returned to the left ventricle. The increase in the volume of blood returned to the left ventricle produces a greater left ventricular end-diastolic volume, which stretches the ventricle wall. Consequently, the left ventricle contracts more forcefully, and its stroke volume increases, thereby restoring the balance in the amount of blood pumped by each ventricle.

Sympathetic Nervous System. Norepinephrine from the sympathetic nervous system is an important factor that influences the strength of ventricular contraction. Norepinephrine, which increases heart rate and accelerates the rates at which cardiac muscle fibers shorten and relax, also increases the force of contraction of cardiac muscle. At any particular end-diastolic volume (that is, degree of stretch), a more forceful contraction ejects a larger volume of blood from the ventricles, which in turn decreases the end-systolic volume and increases the stroke volume. Under normal conditions, sympathetic activity maintains the strength of ventricular contraction at a level about 20% greater than would be the case without stimulation, and maximal sympathetic stimulation can increase contractile strength to about 100% greater than normal. Norepinephrine is believed to increase contractile strength at least in part by increasing membrane permeability to calcium ions, which directly enhances the contractile activity of cardiac muscle cells.

In contrast to the effects of sympathetic activity, parasympathetic activity has comparatively little effect on the strength of ventricular contraction, and maximal parasympathetic stimulation decreases ventricular contractile strength only about 30%.

FACTORS THAT INFLUENCE CARDIAC FUNCTION

Cardiac function can be influenced by a number of factors.

CARDIAC CENTER

Both the sympathetic and the parasympathetic neurons that supply the heart are under the control of a center within the brain stem called the **cardiac center.** This center receives neural input from higher centers in the brain and from various receptors associated with the cardiovascular system. In general, when the sympathetic nerves to the heart are stimulated, the parasympathetic nerves are inhibited. Conversely, sympathetic inhibition and parasympathetic stimulation are usually elicited simultaneously. The role of various inputs to the cardiac center in the control of cardiovascular activity is considered in Chapter 13.

EXERCISE

Chronic, heavy exercise generally causes cardiac muscle to hypertrophy and the ventricular chambers to enlarge, leading to an increase in stroke volume. Compared to a heart with a smaller stroke volume, a heart with a larger stroke volume can pump the same amount of blood per minute with fewer beats, and it can pump more blood when the rates of the two hearts are the same. Thus, both at rest and during exercise, an individual in good physical condition can generally maintain the cardiac output required for a particular activity level with a lower heart rate than an individual in poor physical condition. In addition, the maximum cardiac output and cardiac reserve are greater in a person in good physical condition.

TEMPERATURE

When the heart is warmed, the discharge rate of the sinoatrial node increases, and a rise in body temperature of 1° C. increases the heart rate about 12 to 20 beats per minute. The influence of temperature on heart rate may account for the rapid heart rate that accompanies fevers. However, an increased activity of the sympathetic nerves that supply the heart may also be involved. When the heart is cooled, the heart rate declines and the heart ultimately stops. During some surgical procedures, the body temperature is artificially lowered, and the reduced heart rate that results makes it possible to perform operations that cannot be performed on a rapidly beating heart.

IONS

The levels of various inorganic ions in the blood and interstitial fluid can also influence cardiac function. If the level of potassium ions increases substantially, the heart rate drops and the heart becomes extremely dilated, flaccid, and weak. An elevated level of potassium ions is believed to decrease membrane potentials, which in turn decreases the intensities of action potentials, thereby weakening heart contractions. If the level of calcium ions increases excessively, the heart contracts spastically, probably because of the direct involvement of calcium ions in the contractile process. A substantial increase in the level of sodium ions slows the heart and depresses cardiac function, presumably by interfering with the normal role of calcium ions in the contractile process. The levels of calcium and sodium ions, however, rarely rise sufficiently to alter cardiac function to any great degree.

CATECHOLAMINES

Epinephrine and norepinephrine from the adrenal medulla increase both the heart rate and the force of contraction, and they can provide a relatively small, slow-acting, but effective adjunct to the sympathetic innervation of the heart.

CARDIAC ABNORMALITIES

VALVULAR MALFUNCTIONS

Valvular malfunctions can interfere with the normal movement of blood through the heart. They can decrease the amount of blood pumped out of a ventricle with each contraction, and thereby make it necessary for the heart to work harder to maintain a given cardiac output.

Valvular Regurgitation

Valvular regurgitation occurs when the cusps of a valve do not form a tight seal when the valve is closed. As a result, blood leaks back or regurgitates into the chamber from which it came. If an atrioventricular valve does not close completely, blood flows back into the atrium when the ventricle contracts, and less than the normal amount of blood may be moved into the aorta or pulmonary trunk. Similarly, if a semilunar valve does not close completely, blood that moves into the aorta or pulmonary trunk during ventricular con-

traction flows back into the ventricle when it relaxes. Growths or scar tissue that form on a valve as a result of diseases such as rheumatic fever can prevent the valve from closing securely and cause valvular regurgitation.

Valvular Stenosis

Valvular stenosis is a condition in which the opening of a valve becomes so narrowed that it interferes with the flow of blood through it. If the opening of an atrioventricular valve is too narrow, the ventricle may not fill completely with blood, and a lower than normal amount of blood may be pumped out when the ventricle contracts. If the opening of a semilunar valve is too narrow the ventricle may not eject a normal amount of blood into the aorta or pulmonary trunk when it contracts. Growths or scar tissue on a valve can cause valvular stenosis as well as valvular regurgitation. In many cases valvular regurgitation and valvular stenosis occur in the same valve.

CONGESTIVE HEART FAILURE

Congestive heart failure is a condition in which the heart fails to pump enough blood to meet the body's needs. In congestive heart failure, there is an abnormal increase in blood volume and interstitial fluid, and the heart is generally dilated with blood (as are the veins and capillaries). Congestive heart failure may be due to an impaired contractile ability of cardiac muscle (for example, as a result of a heart attack), or it may be due to an increased workload placed on the heart (for example, as a result of valvular malfunction). In any event, the heart is unable to perform normally the work demanded of it.

In congestive heart failure, compensatory mechanisms may initially allow the cardiac output to be maintained—at least in a resting individual. In response to impaired cardiovascular function, the activity of sympathetic nerves to the heart increases, and the kidneys retain fluid in the body. The increased sympathetic activity increases the contractile strength of cardiac muscle, thereby helping to maintain cardiac output. The retention of fluid by the kidneys increases the blood volume (and the interstitial fluid volume).

The increased blood volume leads to an increased venous return and end-diastolic volume that stretches the ventricular muscle. As described by the Frank–Starling law of the heart, the stretched muscle contracts more forcefully, increasing the stroke volume and thereby helping to maintain cardiac output. In addition, the myocardium may hypertrophy (particularly in cases of valvular malfunction), and although the contractile activity per unit weight of hypertrophied muscle may be below normal, the increased muscle mass permits an overall increase in work capacity.

If the heart continues to fail, the compensatory mechanisms may become detrimental. The continued retention of fluid can cause the end-diastolic volume to increase to the point where the cardiac muscle is stretched excessively, and its contractile strength declines. In addition, hypertrophied cardiac muscle may not receive a sufficient blood supply to meet its needs. Thus, the stroke volume decreases and the cardiac output declines.

In congestive heart failure, only one side of the heart may fail initially. For example, if the left heart is failing, the right ventricle continues to pump blood normally into the pulmonary circuit, but blood returning from the lungs is not pumped efficiently into the systemic circuit by the left ventricle. As a consequence, blood accumulates in the pulmonary circuit, and the pressure within the lung capillaries can rise to the point where fluid is forced out of the vessels. The accumulation of fluid in the lung tissues can result in potentially fatal pulmonary edema. Moreover, since the cardiovascular system is a closed circuit, the failure of the left side of the heart eventually produces an excessive strain on the right side of the heart that can result in total heart failure.

In addition to correcting the cause—for example, by surgically repairing a defective valve—congestive heart failure is treated with drugs that increase the contractile strength of the failing cardiac muscle (for example, digitalis) and eliminate excess fluid (for example, diuretics).

HEART SOUNDS

NORMAL HEART SOUNDS

There are two principal sounds that normally occur as blood moves through the heart during a cardiac cycle. These sounds are best described as

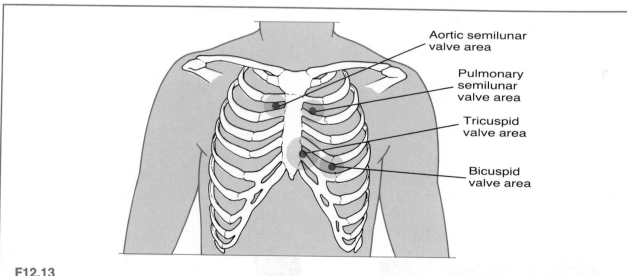

F12.13
Areas of the chest where sounds associated with the different heart valves can be best detected.

"lub-dup." The first heart sound (the "lub") is associated with the closure of the atrioventricular valves at the beginning of ventricular contraction. It is largely due to vibrations of the taut atrioventricular valves immediately after closure and to the vibration of the walls of the heart and major vessels around the heart. The second sound (the "dup") is associated with the closure of the semilunar valves as the ventricles begin to relax following their contraction. This sound is due largely to vibrations of the taut, closed semilunar valves and to the vibration of the walls of the pulmonary artery, the aorta, and to some extent the ventricles. The areas of the chest where a stethoscope can be placed to detect most effectively the sounds associated with the different valves are indicated in Figure 12.13.

ABNORMAL HEART SOUNDS

By listening to the heart, a trained person can obtain considerable information about its condition. Abnormal sounds known as *heart murmurs* can be indicative of particular problems. These sounds, which are described as blowing or vibrating sounds, are caused by a turbulent flow of blood as it passes through the heart.

Both valvular regurgitation and valvular stenosis can cause heart murmurs. In valvular regurgitation, the backward movement of the regurgitated blood interferes with the normal pattern of bloodflow through the heart, causing a detectable turbulence. In valvular stenosis, there is a rapid, turbulent flow of blood through the narrowed valvular opening. In addition, the walls around a narrowed valve are often roughened, further contributing to the turbulence.

Some heart murmurs, called *functional murmurs*, are not pathological, but are considered to be normal. For example, the rapid movement of blood through the heart during heavy exercise may result in turbulence that produces a functional murmur. Functional murmurs are particularly common in young people.

ELECTROCARDIOGRAPHY

The pattern of electrical activity associated with the contraction of cardiac muscle during a heart-

beat can be recorded by the procedure of **electrocardiography.** This procedure, which produces a recording called an **electrocardiogram (ECG)**, is useful in detecting conditions that interfere with the normal conduction of impulses through the heart.

As an impulse generated by the sinoatrial node travels through the heart it produces electrical currents that spread through the body fluids surrounding the heart and then continue onward to the body surface. By placing electrodes on the body surface, it is possible to detect and record the electrical potentials generated by the heart.

The electrodes used to obtain electrocardiograms are placed in various positions on the body

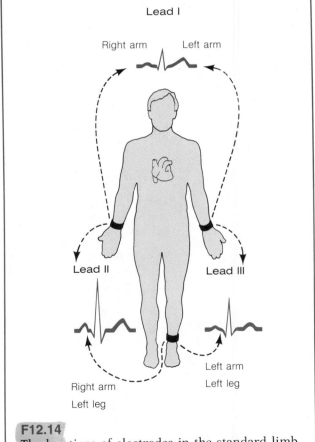

F12.14
The locations of electrodes in the standard limb leads for an electrocardiogram. The leads are designated I, II, and III.

surface. Commonly, electrocardiograms are obtained using three "standard" limb leads or electrode placements (F12.14). Although all three leads give similar patterns of recordings, there are differences in the amplitudes of the waves. These differences are often important in diagnosing various heart conditions.

NORMAL ELECTROCARDIOGRAM

A normal electrocardiogram of a single heartbeat consists of a regularly spaced series of waves designated **P, Q, R, S,** and **T** (F12.15). (The letters simply indicate the order of appearance of the waves.) The *P wave* is caused by electrical currents that are produced as the atria depolarize prior to contraction. The *QRS complex*—actually three separate waves: a Q wave, an R wave, and an S wave—is caused by the depolarization of the ventricles. The *T wave* is caused by currents that are generated as the ventricles repolarize—that is, as the ventricles recover from being depolarized. The repolarization of the atria occurs during ventricular depolarization, and the atrial recovery wave is generally obscured by the QRS complex.

The time interval between the beginning of the P wave and the beginning of the QRS complex indicates the length of time between the beginning of the contraction of the atria and the beginning of the contraction of the ventricles. This period of time is referred to as either the P–Q interval or the P–R interval (because the Q wave is often absent). In a similar manner, the time between the beginning of the Q wave and the end of the T wave (the Q–T interval) provides a general indication of the duration of ventricular contraction.

ABNORMAL ELECTROCARDIOGRAMS

The appearances of the different waves and the durations of the various intervals between the waves of the electrocardiogram are useful in diagnosing abnormalities that alter the conduction of impulses through the heart. The following are a few of the more common cardiac abnormalities detectable from electrocardiograms.

Abnormal Heart Rates

In a resting adult, the heart normally beats about 70 to 80 times a minute. When the rate drops below about 60 beats per minute, it is referred to as *bradycardia*. Bradycardia is generally not considered to be pathological. Much more serious and often associated with cardiovascular pathology is a resting heart rate of over 100 beats per minute, which is called *tachycardia*. When the heart rate is very fast, the ventricles do not have time to fill properly, and the movement of blood through the heart is impaired.

Somewhat coordinated atrial contractions that occur at a very rapid rate (often between 200 and 350 per minute) are called *atrial flutter* (F12.16a). During atrial flutter, the atria pump almost no blood into the ventricles. Atrial flutter may occur in rheumatic and coronary heart disease. Extremely rapid, uncoordinated contractions of the atrial myocardium are called *atrial fibrillation* (F12.16b). In atrial fibrillation, numerous impulses spread through the atria in all directions, and no P wave is evident on the electrocardiogram. The uncoordinated contractions of the atrial myocardium during atrial fibrillation are ineffective in pumping blood into the ventricles. Like atrial fibrillation, *ventricular fibrillation* is characterized by rapid, uncoordinated contractions of the ventricular myocardium. During

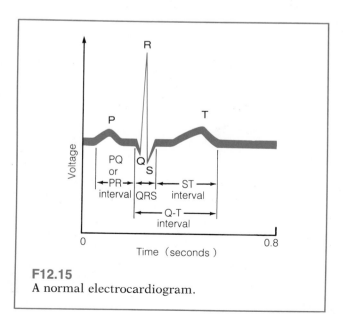

F12.15
A normal electrocardiogram.

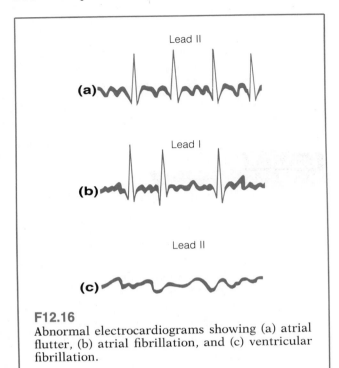

F12.16
Abnormal electrocardiograms showing (a) atrial
flutter, (b) atrial fibrillation, and (c) ventricular
fibrillation.

ventricular fibrillation, a total irregularity of the
QRS complex is evident on the electrocardiogram
(F12.16c). The uncoordinated contractions of the
ventricular myocardium are ineffective in pumping blood from the ventricles, and ventricular
fibrillation is generally fatal unless emergency
measures (such as electrical defibrillation) are undertaken immediately.

Heart Block

Occasionally, the conduction of a stimulatory impulse through the heart is blocked at some point.

Atrioventricular Block.   Damage to or depression
of the atrioventricular node or the atrioventricular bundle can impair the conduction of stimulatory impulses from the atria to the ventricles. In
first-degree (incomplete) heart block, there is a
longer than normal delay in the conduction of the
impulse from the atria to the ventricles. In
first-degree heart block, the electrocardiogram of
a resting person's heart shows a P–R interval of
greater than 0.20 second (possibly up to 0.35 to

0.45 second) compared to a normal P–R interval of
approximately 0.16 second (F12.17b). In second-
degree (incomplete) heart block, some of the stimulatory impulses from the sinoatrial node fail to
be transmitted to the ventricles. As a result, only
every second (or third, or fourth, and so on) atrial
contraction is followed by a ventricular contraction (F12.17c). In this condition, the heart retains a definite, though altered, rhythm. In
third-degree (complete) heart block, the im-

F12.17
(a) Normal electrocardiogram. (b,c,d) Abnormal
electrocardiograms. (b) First-degree (incomplete)
heart block showing prolonged P–R interval.
(c) Second-degree (incomplete) heart block showing two atrial contractions for every ventricular
contraction. (d) Third-degree (complete) heart
block showing no correlation between the occurrence of the P wave and the QRS–T complex.

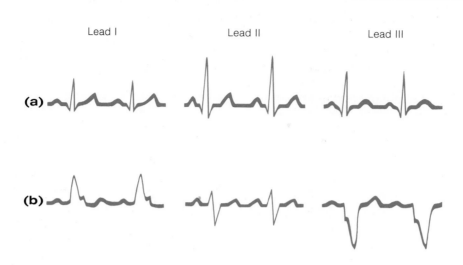

F12.18
(a) Normal electrocardiograms. (b) Abnormal electrocardiograms indicating left bundle branch block. Notice that the QRS complex is prolonged considerably.

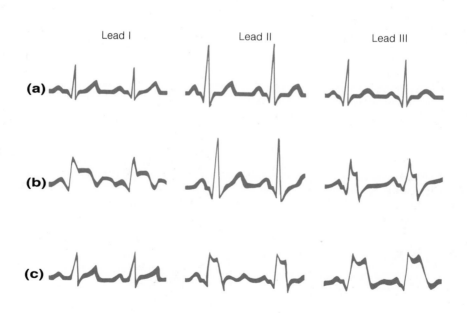

F12.19
(a) Normal electrocardiograms. (b, c) Abnormal electrocardiograms due to heart attacks. (b) Acute anterior wall infarction. (c) Acute posterior wall apical infarction.

pairment of conduction is so severe that no stimulatory impulses are conducted from the atria to the ventricles. In this condition, the ventricles contract at a rate that is slower than and completely independent of the rate of the atria. Often, impulses that stimulate the ventricles originate spontaneously in the atrioventricular bundle (whose inherent rhythm is slower than that of the sinoatrial node). Thus, the atrioventricular bundle functions as an ectopic (that is, "out of place") pacemaker for the ventricles in place of the normal pacemaker (the sinoatrial node). In the electrocardiogram of a complete heart block, there is no correlation between the occurrences of the P wave and the QRS–T complex (F12.17d).

Bundle Branch Block.   If there is an impairment of conduction of the stimulatory impulse along one of the major branches of the atrioventricular bundle, the ventricle supplied by the impaired bundle branch depolarizes more slowly than the other ventricle. Such a block shows up on the electrocardiogram as a prolonged QRS complex (F12.18). By varying the positions of the leads of the electrocardiograph, it is possible to determine whether the left or right bundle branch is blocked.

Heart Attacks

An insufficient flow of blood through the coronary arteries—due to an obstruction of a vessel, for example—can lead to myocardial damage and, if severe enough, to the deaths of cardiac muscle cells. An area of dead cardiac muscle cells is called a *myocardial infarct*, and a person who suffers such damage has suffered a heart attack. Because damaged or dead cardiac muscle cells do not conduct impulses normally, the presence of such areas produces altered ECG tracings (F12.19).

STUDY OUTLINE

CIRCULATION THROUGH THE HEART
pp. 295–298
1. The superior and inferior venae cavae return blood from the body to the right atrium. Blood then moves into the right ventricle, which pumps it into the pulmonary trunk to the lungs. Blood from the lungs returns to the left atrium by way of pulmonary veins. Blood then moves into the left ventricle, which pumps it into the aorta to the body.
2. Valves keep blood flowing in the proper direction through the heart. Atrioventricular valves are between the left atrium and left ventricle (bicuspid valve), and between the right atrium and right ventricle (tricuspid valve). Semilunar valves are in the pulmonary trunk and aorta.

PUMPING ACTION OF THE HEART
p. 298
1. The heart functions as a double pump. The right pump receives blood from the systemic circuit and pumps it into the pulmonary circuit. The left pump receives blood from the pulmonary circuit and pumps it into the systemic circuit.
2. Both atria contract simultaneously, followed by the simultaneous contraction of both ventricles.
3. The contraction of the atria is not essential for the movement of blood into the ventricles, and even if the atria fail to function, the ventricles still pump considerable quantities of blood.

CARDIAC MUSCLE   pp. 298–302

Cellular Organization Cardiac muscle cells form branching networks. The cells are connected end to end by intercalated discs. The contractile events in cardiac muscle cells are believed to be similar to those in skeletal muscle cells.

Automatic Contraction Cardiac muscle contracts automatically without external stimulation; the spontaneous activity of cardiac muscle is continually influenced by neurons of the autonomic nervous system and by certain chemicals and hormones.

Degree of Contraction When cardiac muscle contracts, it contracts as much as it can for the existing conditions.

Refractory Period Cardiac muscle cells have relatively long refractory periods that normally prevent the heart from undergoing a tetanic contraction or spasm.

The Myocardium The cardiac muscle of the heart wall is called the myocardium. The myocardium varies in thickness from one heart chamber to another in proportion to the force needed to overcome the resistance encountered in pumping blood from the different chambers.

Metabolism of Cardiac Muscle The myocardium can obtain only insignificant amounts of energy from anaerobic metabolism, and cardiac muscle depends primarily on aerobic metabolism for the continuous supply of energy required to support its contractile activity.

EXCITATION AND CONDUCTION IN THE HEART pp. 302–304

Excitation The sinoatrial node spontaneously generates stimulatory impulses approximately 70 to 80 times per minute in a resting individual. Other areas of the myocardium may also exhibit spontaneous activity, but the sinoatrial node dominates them and is the pacemaker of the heart.

Conduction A stimulatory impulse from the sinoatrial node spreads throughout the myocardium of the atria, stimulating atrial contraction and reaching the atrioventricular node. After a delay of about 0.1 second, the impulse is transmitted from the atrioventricular node through the atrioventricular bundle and the Purkinje fibers to the cells of the ventricles.

THE CARDIAC CYCLE The heart repeats two alternating phases: systole (contraction) and diastole (relaxation). pp. 304–308

Pressure Curve of the Left Atrium
1. The pressure increases gradually as blood from the pulmonary veins enters.
2. The pressure increases suddenly when the atrium contracts.
3. The pressure increases when ventricular systole begins and the atrioventricular valve bulges into the atrium.
4. The pressure then drops quickly.

Pressure Curve of the Left Ventricle
1. There is a slight pressure increase when the atrium contracts.
2. There is a rapid rise in pressure during ventricular systole.
3. The pressure falls at the end of ventricular systole to a value below that of the pressure in the aorta.

Pressure Curve of the Aorta
1. The pressure decreases gradually during ventricular diastole.
2. The pressure increases rapidly when the ventricle contracts and ejects blood into the aorta.

3. The pressure then decreases but rises briefly when the aortic semilunar valve closes.

CARDIAC OUTPUT Cardiac output is equal to the heart rate multiplied by the stroke volume. The stroke volume is the amount of blood pumped by one side of the heart per beat. At very high heart rates the stroke volume decreases. A major reason for this decrease is diminished time for ventricular filling due to a decrease in the length of diastole. pp. 308–311

Control of Heart Rate The most important factor is the effect of the autonomic nervous system. Sympathetic stimulation increases heart rate and parasympathetic stimulation decreases heart rate.

Control of Stroke Volume The stroke volume is equal to the difference between the end-diastolic volume and the end-systolic volume.

END-DIASTOLIC VOLUME Determined mainly by:

Length of Diastole

Venous Return

END-SYSTOLIC VOLUME Determined mainly by the strength of ventricular contraction, which is influenced by:

The Frank–Starling Law of the Heart Stretching the ventricle wall increases the strength of ventricular contraction.

Sympathetic Nervous System Sympathetic stimulation increases the strength of ventricular contraction.

FACTORS THAT INFLUENCE CARDIAC FUNCTION pp. 311–312

Cardiac Center Both sympathetic and parasympathetic neurons to the heart are under the control of a cardiac center in the brain stem.

Exercise Generally causes cardiac muscle to hypertrophy and the ventricular chambers to enlarge, leading to an increase in stroke volume.

Temperature Warming the heart increases the heart rate and cooling the heart decreases the heart rate.

Ions If the level of potassium ions increases substantially, the heart rate drops and the heart becomes extremely dilated, flaccid, and weak. If the level of calcium ions increases excessively, the heart exhibits spastic contraction. A substantial increase in the level of sodium ions slows the heart and depresses cardiac function.

Catecholamines Epinephrine and norepinephrine from the adrenal medullae can increase the heart rate and the force of contraction.

CARDIAC ABNORMALITIES pp. 312–313

Valvular Malfunctions Can make it necessary for the heart to work harder to maintain a given cardiac output.

VALVULAR REGURGITATION Occurs when the cusps of a valve do not form a tight seal when the valve is closed. As a result, blood leaks into the chamber from which it came.

VALVULAR STENOSIS A condition in which the opening of a valve becomes so narrowed that it interferes with the flow of blood through it.

Congestive Heart Failure A condition in which the heart fails to pump enough blood to meet the body's needs. In congestive heart failure there is an abnormal increase in blood volume and interstitial fluid, and the heart is generally dilated with blood. The kidneys retain fluid, the activity of sympathetic nerves to the heart increases, and the myocardium may hypertrophy.

HEART SOUNDS pp. 313–314

Normal Heart Sounds The first heart sound is associated with the closure of the atrioventricular valves at the beginning of systole. The second heart sound is associated with the closure of the semilunar valves as the ventricles begin to relax following their contraction.

Abnormal Heart Sounds Heart murmurs can indicate cardiac problems. Both valvular regurgitation and valvular stenosis can cause heart murmurs.

ELECTROCARDIOGRAPHY By placing electrodes on the body surface, it is possible to detect and record the electrical potentials generated by the heart. pp. 314–318

Normal Electrocardiograms A normal electrocardiogram consists of a regularly spaced series of waves. The P wave represents atrial depolarization. The QRS complex represents ventricular depolarization. The T wave represents ventricular repolarization. The P–R interval represents the length of time between the beginning of the contraction of the atria and the beginning of the contraction of the ventricles. The Q–T interval provides a general indication of the duration of ventricular contraction.

Abnormal Electrocardiograms

ABNORMAL HEART RATES Bradycardia: below 60 beats per minute; tachycardia: above 100 beats per minute. Somewhat coordinated atrial contractions that occur at a very rapid rate are called atrial flutter. Extremely rapid, uncoordinated atrial contractions are called atrial fibrillation (no evident P wave on the electrocardiogram). Extremely rapid, uncoordinated ventricular contractions are called ventricular fibrillation (total irregularity of the QRS complex on the electrocardiogram).

HEART BLOCK

Atrioventricular Block May delay or prevent impulse transmission from the atria to the ventricles. First-degree block: delayed impulse transmission (P–R interval exceeds 0.20 second on the electrocardiogram). Second-degree block: transmission of some impulses from the sinoatrial node to the ventricles is prevented (two, three, or more P waves for each QRS complex on the electrocardiogram). Third-degree block: the transmission of impulses from the atria to the ventricles is prevented (no correlation between P wave and QRS–T complex on the electrocardiogram).

Bundle Branch Block An impairment of conduction in one of the major branches of the atrioventricular bundle (a prolonged QRS complex on the electrocardiogram).

HEART ATTACKS An area of dead cardiac muscle cells is called a myocardial infarct; areas of damaged or dead cardiac muscle cells result in altered electrocardiograms.

SELF-QUIZ

1. Blood within the pulmonary veins returns to the: (a) right atrium; (b) right ventricle; (c) left atrium; (d) left ventricle.

2. The left atrium contracts at the same time as the: (a) right atrium; (b) right ventricle; (c) left ventricle.

3. The cardiac muscle of the heart: (a) commonly undergoes prolonged tetanic contractions; (b) does not contract unless stimulated by the nervous system; (c) obtains only an insignificant amount of energy from anaerobic metabolism.

4. The sinoatrial node is the only area of the myocardium that can undergo spontaneous depolarization to threshold and generate action potentials. True or False?

5. The stimulatory impulse from the sinoatrial node is normally delayed for a short time at the: (a) atrioventricular node; (b) atrioventricular bundle; (c) Purkinje fibers.

6. During left ventricular systole, which event occurs first? (a) the atrioventricular valve closes; (b) the semilunar valve opens; (c) the pressure within the ventricles peaks.

7. During ventricular diastole, the ventricles are about 70% filled before the atria contract. True or False?

8. During most of ventricular diastole, the pressure within the left ventricle is slightly lower than the pressure within the left atrium. True or False?

9. Cardiac output is equal to the heart rate multiplied by the: (a) aortic pressure; (b) stroke volume; (c) ventricular end-systolic volume.

10. Stimulation of the parasympathetic neurons to the heart: (a) decreases the heart rate; (b) decreases the membrane permeability to potassium ions; (c) decreases the conduction time through the atrioventricular node.

11. Within limits, as cardiac muscle is stretched, its force of contraction increases. True or False?

12. Norepinephrine: (a) decreases the contractile strength of the ventricles; (b) decreases the heart rate; (c) accelerates the rates at which cardiac muscle fibers shorten and relax.

13. Chronic, heavy exercise most likely will lead to: (a) an increased heart rate at rest; (b) an increased stroke volume; (c) a decreased cardiac reserve.

14. In general, when the level of potassium ions is increased excessively, the heart rate increases and the heart exhibits spastic contraction. True or False?

15. Growths or scar tissue that form on a valve as a result of rheumatic fever can cause valvular regurgitation. True or False?

16. In congestive heart failure: (a) the blood volume decreases; (b) the kidneys excrete large volumes of fluid; (c) the activity of the sympathetic nerves to the heart increases.

17. The first heart sound is due largely to the: (a) vibrations of the taut atrioventricular valves immediately after closure; (b) vibrations of the taut closed semilunar valves; (c) turbulent flow of blood past the open semilunar valves.

18. The P wave of a normal electrocardiogram indicates: (a) atrial depolarization; (b) ventricular depolarization; (c) atrial repolarization; (d) ventricular repolarization.

19. The Q–T interval of a normal electrocardiogram provides a general indication of: (a) the duration of atrial contraction; (b) the duration of ventricular contraction; (c) the length of time between the beginning of atrial contraction and the beginning of ventricular contraction; (d) the length of time that the stimulatory impulse is delayed at the atrioventricular node.

20. A P–R interval of 0.30 second on an electrocardiogram would most likely indicate: (a) ventricular fibrillation; (b) an atrioventricular block; (c) atrial flutter.

BLOOD VESSELS AND THE LYMPHATIC SYSTEM

13

Upon leaving the heart, the blood enters the vascular system, which is composed of numerous vessels. The vessels transport the blood to all parts of the body, permit the exchange of nutrients, metabolic end products, hormones, and other substances between the blood and the interstitial fluid, and ultimately return the blood to the heart.

CIRCULATION THROUGH THE VESSELS

PATTERN OF BLOODFLOW

Blood is carried away from the heart in vessels called **arteries.** The major arteries branch into progressively smaller arteries, which in turn give rise to very small arterial vessels called **arterioles.** From the arterioles, the blood flows into networks of tiny **capillaries,** then into larger **venules,** and finally into increasingly larger **veins,** which return it to the heart.

The walls of the large arteries—for example, the pulmonary trunk and the aorta and its major branches—contain many elastic fibers as well as smooth muscle. These arteries are called *elastic arteries*, and they can basically be considered to behave in a manner resembling that of elastic tubes. During ventricular systole, elastic arteries are stretched as blood is ejected from the heart. During diastole, the recoil of the elastic arteries helps maintain pressure within the vessels.

The walls of the arterioles are composed predominantly of smooth muscle, and these vessels play an important role in regulating the flow of blood into the capillaries. When the smooth muscle of the arterioles contracts, the internal cavities, or lumens, of the vessels are narrowed—that is, the vessels undergo **vasoconstriction,** which restricts the flow of blood into the capillaries. When the muscles relax, the lumens of the arterioles enlarge—that is, the vessels undergo **vasodilation,** which allows the blood to flow into the capillaries more freely.

In most tissues, a capillary network contains two types of vessels: thoroughfare channels, which directly connect arterioles and venules, and true capillaries, which branch from and join with the thoroughfare channels (F13.1). A ring of smooth muscle called a *precapillary sphincter* usually surrounds each true capillary at the point where it arises from a thoroughfare channel. The contraction and relaxation of the sphincters help regulate the flow of blood through the capillaries.

Capillaries have extremely thin walls, and they are sites at which the exchange of materials between the blood and the interstitial fluid takes place. Capillary structure varies from one part of the body to another, but in general, a capillary consists of a single layer of endothelial cells (which line all blood vessels) surrounded by a thin basement membrane. Water-filled clefts occur between adjacent endothelial cells, and in some capillaries the endothelial cells contain small, oval windows called fenestrations, which are usually covered by a very thin diaphragm. Although a single capillary is only about 0.5 to 1 mm long and

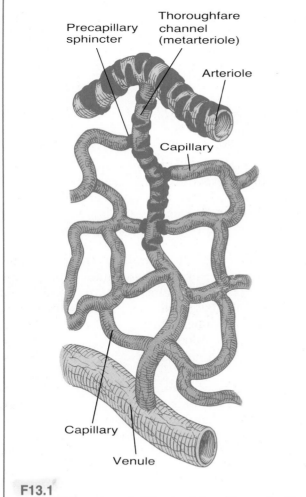

F13.1
A capillary network. Thoroughfare channels directly connect arterioles and venules, and true capillaries branch from and join with the thoroughfare channels.

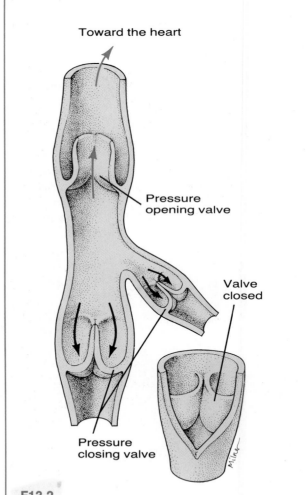

F13.2
Valves of a vein. The arrows indicate that the valves are forced open by pressure from below and shut by pressure from above. This allows blood to move in only one direction—toward the heart.

0.01 mm in diameter, capillaries are so numerous that their total surface area within the body has been estimated to be more than 600 square meters. This large area provides an extensive surface across which the exchange of materials can occur.

Veins generally have larger lumens and thinner walls than arteries of comparable sizes, and some veins—particularly those of the lower limbs—contain valves that ensure a one-way flow of blood toward the heart (F13.2). When blood within these veins attempts to flow backward, away from the heart, the cusps of the valves—which are similar to the pocketlike cusps of the aortic and pulmonary semilunar valves—fill with blood and block the vessel.

VELOCITY OF BLOODFLOW

In general, the velocity of bloodflow in any segment of the cardiovascular system is inversely related to the total cross-sectional area of the vessels of the segment. In other words, the larger the total cross-sectional area, the lower the velocity of flow.

Although the lumen of an arteriole is smaller than the lumen of an artery, there are so many arterioles that their total cross-sectional area exceeds that of the arteries (F13.3). Thus, the velocity of bloodflow in the arterioles is lower than that in the arteries. Similarly, the smallest individual vessels, the capillaries, are so numerous that they have the greatest total cross-sectional area of the

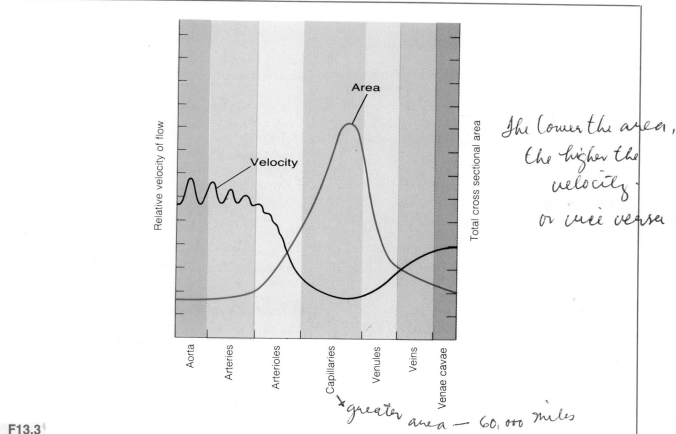

The lower the area, the higher the velocity or vice versa.

→ greater area — 60,000 miles

F13.3
Graph indicating the relationship between the velocity of bloodflow, and the total cross-sectional area of various segments of the cardiovascular system.

entire cardiovascular system. As a consequence, the velocity of bloodflow in the capillaries is lower than that in any other vessels. The slow bloodflow in the capillaries allows adequate time for the exchange of nutrients, metabolic end products, and other substances between the blood and the interstitial fluid.

PULSE

When blood is ejected from the heart during ventricular systole, the pressure within the arteries rises and the arteries expand. During diastole, the arterial pressure falls and the arteries recoil. The expansion and subsequent recoil of the arteries can be felt at various locations on the body surface as the **pulse.**

The pressure fluctuations within the arterioles are less extreme than those within the arteries. Within the capillaries and beyond, the pressures do not rise and fall greatly with the beating of the heart, but remain at relatively constant levels.

PRINCIPLES OF CIRCULATION

Efficient circulation requires an adequate volume of circulating blood, blood vessels that are in good condition, and a heart that functions smoothly to pump the blood through the vessels.

BLOODFLOW

The *bloodflow* is the actual volume of blood that passes through a vessel in a given time. Bloodflow is a function of the pressure forcing the blood through a vessel and the resistance of the vessel to the flow of blood through it.

PRESSURE

The pumping action of the heart imparts energy to the blood. This energy is evident as the *pressure* that drives the blood through the vessels.

RESISTANCE

As the blood flows through the vessels, it encounters varying degrees of *resistance*, which is essentially a measure of friction. As a consequence, the energy imparted to the blood by the pumping action of the heart is ultimately dissipated as heat.

This dissipation of energy is evident as a progressive drop in pressure as the blood moves through the vessels (F13.4). Thus, the pressure is highest in the arteries, and it gradually drops as the blood flows through the arterioles, capillaries, venules, and veins. In general, the greater the resistance the blood encounters, the harder the heart must pump and the more pressure it must generate in order to keep the blood circulating.

Factors That Affect Resistance

Several factors affect the magnitude of the resistance the blood encounters as it flows through the vessels.

Blood Viscosity.   The more viscous the blood, the greater the resistance to its flow through any given vessel. The more protein and cells within the blood, the greater its viscosity.

Vessel Length.   The longer a vessel, the greater the resistance encountered as blood flows through it.

Vessel Diameter.   The smaller the diameter of a vessel, the greater its resistance to the flow of blood through it. The resistance of a vessel is inversely related to the fourth power of the vessel's diameter:

$$\text{resistance} \propto \frac{1}{(\text{diameter})^4}$$

For example, if the diameter of a vessel is decreased by one-half, its resistance increases 16 times.

Alterations in Resistance

The lengths of the blood vessels are constant, and under normal circumstances the viscosity of the blood does not vary greatly. However, the diameters of blood vessels can change. For example, the diameters of the muscular arterioles vary according to the degree of constriction or relaxation of the arteriolar muscles, and the arterioles are of major importance in determining the amount of resistance the blood encounters as it flows through the vessels.

The resistance to the flow of blood offered by the entire systemic circulation is called the *total peripheral resistance*. The resistance to the flow of blood offered by the entire pulmonary circulation is called the *total pulmonary resistance*.

RELATIONSHIP AMONG FLOW, PRESSURE, AND RESISTANCE

The relationship among bloodflow, pressure, and resistance can be expressed as:

$$\text{flow } (F) = \frac{\text{pressure}}{\text{resistance}} \qquad 13.1$$

In this relationship, the pressure is the pressure that drives the blood through a particular vessel or vessels. This pressure is represented by the pressure drop (ΔP) that occurs as the blood flows through the vessel, and it is equal to the pressure at the beginning of the vessel (P_1) minus the pressure at the end of the vessel (P_2). Thus, $\Delta P = P_1 - P_2$, and

$$F = \frac{\Delta P}{\text{resistance}} \qquad 13.2$$

The resistance term of this relationship includes the components of resistance just discussed—that is, blood viscosity, vessel length, and vessel diameter. However, as previously pointed out, the component of resistance most likely to change, and thus affect the relationship, is vessel diameter, particularly arteriole diameter.

The relationship among bloodflow, pressure, and resistance expressed in equation 13.2 can be applied to the entire systemic or pulmonary circulation, as well as to individual vessels or groups of

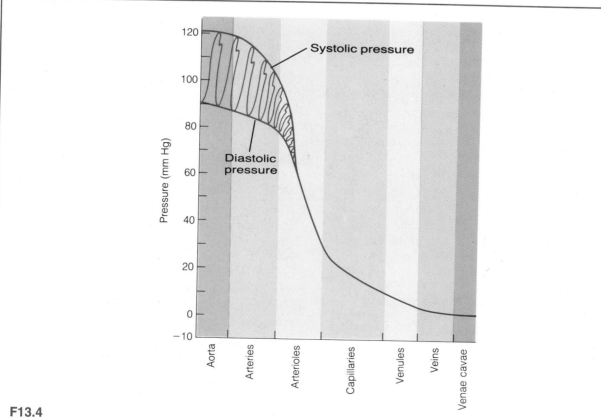

F13.4
Graph indicating pressures in various portions of the cardiovascular system.

vessels. For example, the volume of blood that flows through the systemic circulation in a given time is equal to the volume of blood ejected by the left ventricle during that time. If the time considered is one minute, the flow term of the relationship is equal to the cardiac output (recall that the cardiac output is the volume of blood pumped by a ventricle per minute). Thus,

$$\text{cardiac output} = \frac{\Delta P}{\text{resistance}} \qquad 13.3$$

The resistance to the flow of blood through the entire systemic circulation is the total peripheral resistance. Thus, the resistance term of the relationship is equal to the total peripheral resistance, and

$$\text{cardiac output} = \frac{\Delta P}{\text{total peripheral resistance}} \qquad 13.4$$

The pressure at the end of the systemic circuit (P_2) is the pressure within the veins at the right atrium. Since this pressure is almost zero, the pressure drop (ΔP) that occurs as blood flows through the systemic circuit is essentially equal to the pressure at the beginning of the circuit—that is, $\Delta P = P_1 - P_2 = P_1 - 0 = P_1$. This pressure ($P_1$) is the pressure in the aorta at the left ventricle due to the pumping action of the heart. However, as discussed in Chapter 12, the pressure within the aorta rises and falls with the beating of the heart. As such, the mean aortic pressure, rather than the aortic pressure during ventricular systole or diastole, provides the best indication of the pressure at the beginning of the systemic circuit. Thus,

$$\text{cardiac output} = \frac{\text{mean aortic pressure}}{\text{total peripheral resistance}} \qquad 13.5$$

The resistance to flow in the large systemic arteries is only slight, and the pressure changes very little as the blood flows through them. Therefore, the mean aortic pressure is essentially equal to the mean arterial pressure—which can be approximated as [systolic pressure + 2(diastolic pressure)]/3—and

$$\text{cardiac output} = \frac{\text{mean arterial pressure}}{\text{total peripheral resistance}} \qquad 13.6$$

This relationship can be rearranged into the form:

$$\text{mean arterial pressure} = \text{cardiac output} \times \text{total peripheral resistance} \qquad 13.7$$

Thus, the mean arterial pressure is influenced by changes in either the cardiac output or the total peripheral resistance. Moreover, as discussed later, the body possesses receptors involved in sensing and regulating arterial pressure, and the activity of these receptors leads to changes in heart rate and contractile strength—which alter cardiac output—and to changes in vessel diameter—which alter total peripheral resistance.

ARTERIAL PRESSURE

The pressure within the large systemic arteries (which is essentially equal to the pressure generated by the pumping action of the left ventricle of the heart) is the immediate driving force for bloodflow through the body's organs and tissues. Therefore, it must be carefully maintained in order to ensure adequate organ and tissue bloodflows. A number of factors affect the arterial pressure through mechanisms that influence vessel resistance and cardiac function.

NEURAL FACTORS

Vasomotor Nerve Fibers

Vasomotor nerve fibers are efferent fibers that regulate blood vessel diameters. With the exception of the capillaries, these fibers supply almost all types of vessels, particularly the arterioles. Changes in blood vessel diameters due to the activity of vasomotor nerve fibers alter the resistance of the vessels to bloodflow, which, in turn, can lead to a change in the mean arterial pressure (see equation 13.7).

Nerves of the sympathetic division of the autonomic nervous system contain both *vasoconstrictor* and *vasodilator* vasomotor fibers. The vasoconstrictor fibers are the most widely distributed and are the most important. At their junctions with blood vessels, they release norepinephrine, which stimulates the constriction of the vessels.

Vasodilator fibers—as well as vasoconstrictor fibers—are present in the sympathetic nerves that innervate skeletal muscles. At their junctions with blood vessels, they release acetylcholine, which leads to vessel dilation. The vasodilator fibers, however, are probably not very important in the overall control of vessel resistance. The major regulation of blood vessel resistance is accomplished by the vasoconstrictor fibers, and most vessel dilation occurs as a result of diminished vasoconstrictor activity.

Vasomotor Center

In the lower third of the pons and the medulla oblongata of the brain is an area known as the **vasomotor center.** The vasomotor center plays an important role in the regulation of blood vessel resistance and thus in the regulation of arterial pressure (F13.5). Nerve impulses from this center are ultimately transmitted by vasoconstrictor fibers to blood vessels, and the vasomotor center is continuously, or tonically, active in promoting some degree of vasoconstriction, which is called vasomotor tone.

The activity of the vasomotor center is influenced by nerve impulses that arrive at the center from receptors associated with the cardiovascular system, as well as by nerve impulses from sensory surfaces of the body and from higher brain centers. Hormones and other substances in the blood also affect the activity of the vasomotor center. For example, carbon dioxide strongly stimulates the vasoconstrictor activity of the vasomotor center. When carbon dioxide levels increase, greater vasoconstriction is noted, and this fact has been used to explain the rise in arterial pressure seen early in asphyxiation.

Cardiac Center

The cardiac center, described in Chapter 12, influences arterial pressure by virtue of its effects on cardiac function. Many of the inputs that affect the vasomotor center also influence the cardiac center.

Baroreceptors

In the arch of the aorta, in the slightly enlarged region where each common carotid artery divides into internal and external carotid arteries—that is, in the **carotid sinus**—and, to a lesser extent, in the walls of almost every large artery in the neck and thoracic regions, are receptors that respond to the distension or stretch of the vessel walls. Since the degree of stretch of the vessel walls is directly related to the arterial pressure, these receptors function as pressure receptors, or **baroreceptors** (*baro* = pressure) (F13.5). Nerve impulses from the baroreceptors ultimately inhibit the vasoconstrictor activity of the vasomotor center; they also influence the cardiac center in such a manner that the activity of the parasympathetic nerves to the heart increases and the activity of the sympathetic nerves to the heart decreases.

When the arterial pressure rises, the rate of nerve impulse transmission from the baroreceptors increases. As a consequence, blood vessels dilate, and the rate and contractile strength of the heart decrease, leading to a decline in cardiac output. These activities lower the arterial pressure.

When the arterial pressure falls, the rate of nerve impulse transmission from the baroreceptors diminishes, resulting in an increased vasoconstriction and an increased rate and contractile strength of the heart. These activities raise the arterial pressure.

The arterial baroreceptors protect the cardiovascular system against relatively short-term changes in arterial pressure. However, they do little to protect the system from sustained long-term pressure changes. This situation is due to the fact that the baroreceptors display adaptation, and after a few days their rates of discharge return to normal levels regardless of continued high or low arterial pressures. The baroreceptors still function, but in essence they become "reset" to operate at a different pressure level.

Aortic and Carotid Bodies

The **aortic** and **carotid bodies,** which are located at the aortic arch and at the branchings of the common carotid arteries, contain receptors that are sensitive to arterial oxygen, carbon dioxide, and hydrogen ion concentrations. Like the baroreceptors, these receptors send impulses to the vasomotor center. Impulses sent in response to decreased arterial oxygen cause an increase in the arterial pressure. Changes in carbon dioxide and hydrogen ion concentrations can also alter the arterial pressure, but the effects of these substances by way of the aortic and carotid body pathways are relatively small.

CHEMICALS AND HORMONES

A number of chemicals and hormones influence the arterial pressure.

Angiotensin

Angiotensin is a powerful vasoconstrictor that raises the arterial pressure. When the arterial pressure falls, the formation of angiotensin, which is produced from precursors in the plasma by enzymatic action, increases.

Epinephrine

Epinephrine produced by the medullae of the adrenal glands causes a transitory increase in the sys-

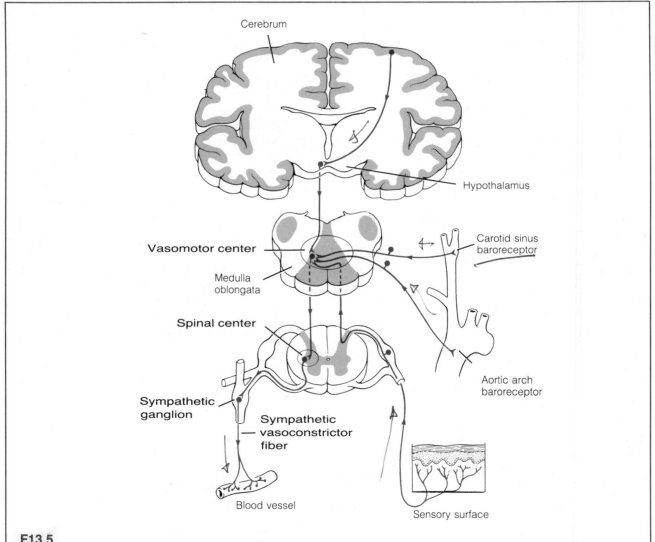

F13.5
Diagrammatic representation of the neural pathways involved in the regulation of blood vessel diameter.

tolic pressure within the arteries. Usually, when sympathetic nervous system activity causes widespread effects on blood vessels throughout the body, the adrenal medullae are stimulated and epinephrine, as well as norepinephrine, is released.

Vasopressin

Vasopressin, or antidiuretic hormone, raises the arterial pressure by stimulating arteriole constriction. When the arterial pressure falls, the release of vasopressin from the pars nervosa of the pituitary gland increases.

BLOOD VOLUME

The volume of blood within the circulatory system affects the arterial pressure. In general, an increase in blood volume tends to raise the arterial pressure, in large part by increasing the venous return and ventricular filling, which in turn leads to an increased cardiac output. Conversely, a decrease in blood volume tends to lower the arterial pressure. Thus, mechanisms that alter blood volume can help control arterial pressure. For example, a rise in arterial pressure that leads to an elevation in capillary pressure favors the movement of fluid out of the capillaries and into the interstitial spaces. This movement of fluid decreases the blood volume and lowers the arterial pressure. Conversely, a fall in arterial pressure that leads to a decrease in capillary pressure favors the movement of fluid out of the interstitial spaces and into the capillaries. This movement of fluid tends to increase the blood volume and raise the arterial pressure.

The kidneys exert a significant effect on blood volume by virtue of their ability to regulate salt and water excretion, and they are particularly important in the long-term regulation of arterial pressure. A slight increase in the arterial pressure can cause a large increase in the formation of urine, which tends to decrease the blood volume and lower the arterial pressure. Conversely, a slight fall in the arterial pressure can lead to a substantial decrease in the formation of urine. As a result, fluids taken into the body tend to remain within the body where they can increase the blood volume and thereby raise the arterial pressure.

INFLUENCE OF MULTIPLE FACTORS

From the preceding discussion, it is evident that a number of different factors affect the arterial pressure. Consequently, the arterial pressure at any moment is usually the result of the combined influence of several factors. Moreover, different mechanisms work together to maintain the arterial pressure and thus to maintain a circulation that can meet the body's needs. For example, if the arterial pressure falls, neural mechanisms that involve baroreceptors, the cardiac and vasomotor centers, and vasomotor nerves, as well as chemical and hormonal mechanisms, and mechanisms leading to changes in blood volume may all act together to return the arterial pressure to normal.

MEASUREMENT OF ARTERIAL PRESSURE

The arterial pressure, which is commonly called the **blood pressure,** is usually measured indirectly (F13.6). An inflatable cuff connected to a meter or mercury manometer is wrapped around the upper arm. The meter or manometer indicates the pressure within the cuff in millimeters of mercury (mm Hg). A pressure of 80 mm Hg, for example, is equal to the pressure exerted by a column of mercury 80 mm high. A stethoscope is placed below the cuff at the elbow in order to listen to the sounds in the brachial artery of the arm. The cuff is then inflated to a pressure sufficient to close the artery so no blood flows past the cuff, and no sound is heard in the stethoscope. The pressure is then reduced slowly until an intermittent thumping sound is heard. The pressure at which the sound is first heard is recorded as the systolic pressure. The sound is produced because the contraction of the ventricles raises the pressure within the artery enough to overcome the resistance of the cuff as it presses the artery closed. Consequently, blood flows past the cuff into the lower arm in a turbulent fashion. The sound of this turbulence can be heard in the stethoscope. When the ventricles relax and the pressure in the artery drops, the cuff again closes the artery and no blood flows into the lower arm. When this occurs no sound is heard in the stethoscpoe.

Upon further reducing the pressure in the cuff,

the intermittent thumping sound becomes dull and muffled. The pressure at which this occurs is recorded as one diastolic end point (diastolic pressure). This point corresponds to the point at which the pressure in the artery, even during the relaxation of the ventricles, is sufficient to overcome the resistance of the cuff, and thus the vessel never closes. However, enough pressure is exerted by the cuff on the artery to produce a turbulent bloodflow that can be heard with the stethoscope. If the pressure in the cuff is reduced further, the thumping sound eventually completely disappears. The pressure at which this occurs is recorded as a second diastolic end point. At this point, the flow of blood in the vessel is smooth, and no intermittent sound of turbulence can be heard in the stethoscope. Although the systolic pressure and both diastolic end points may be recorded—for example, 120/85/80—generally, only one diastolic end point is considered. Thus, arterial pressure is reported as a fraction such as 120/80, with the numerator being the systolic pressure and the denominator being the diastolic pressure. The difference between the systolic and diastolic pressures is called the *pulse pressure*. The pulse pressure tends to increase if the stroke volume of the ventricles increases or if the distensibility of the arteries decreases.

TISSUE BLOODFLOW

Even when the mean arterial pressure is relatively constant, the bloodflow through many body tissues varies from time to time. Changes in the

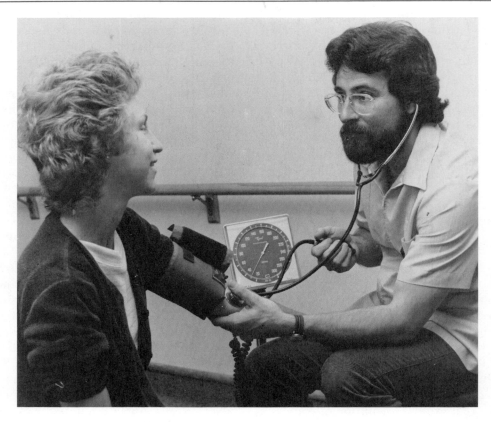

F13.6
Indirect measurement of blood pressure using a manometer.

bloodflow through a particular tissue are primarily due to alterations in the resistance of the tissue's blood vessels, and the resistance of the vessels is controlled by both local autoregulatory mechanisms and external factors.

LOCAL AUTOREGULATORY CONTROL

Local autoregulatory mechanisms alter bloodflow through a tissue according to the needs of the tissue. For example, in many tissues a decrease in the oxygen level leads to a local dilation of arterioles and to an increased relaxation of precapillary sphincters. These responses result in an increased bloodflow through the tissue, which increases the delivery of oxygen. In addition, researchers have proposed that the metabolic processes of a tissue receiving an inadequate blood supply, as well as the low availability of nutrients and oxygen, lead to the formation and accumulation of substances that dilate arterioles and precapillary sphincters in the tissue. This dilation decreases the resistance to bloodflow, leading to an increased flow through the tissue. Among the substances that have been proposed as local vasodilator substances are carbon dioxide, lactic acid, adenosine, adenosine phosphate substances, histamine, potassium ions, and hydrogen ions.

The autoregulatory abilities of some tissues are greater than others; this ability is particularly well developed in skeletal and cardiac muscle and in the gastrointestinal tract.

EXTERNAL FACTORS

The activity of external factors such as vasomotor nerve fibers and hormones can lead to the constriction or dilation of blood vessels in particular tissues, and, therefore, to changes in bloodflow through the tissues. However, external factors usually influence large segments of the vascular system, and they are generally more involved with mechanisms that function for the well-being of the entire body—such as those that maintain arterial blood pressure or body temperature—than they are with the control of tissue bloodflow according to the needs of particular tissues. For example, in a warm environment, the activity of sympathetic vasoconstrictor fibers to skin vessels decreases, and the vessels dilate. This dilation results in an increased bloodflow to the body surface, and thereby increases the loss of body heat from the blood to the environment.

CAPILLARY EXCHANGE

Capillaries are sites at which the exchange of materials between the blood and the interstitial fluid takes place. Most capillaries are permeable to water and small particles such as glucose, inorganic ions, urea, amino acids, and lactic acid; and these materials pass readily between the blood and the interstitial fluid. Protein permeability, however, is usually quite limited, although it varies from tissue to tissue. It is highest in the liver, less in muscle, and very limited in the central nervous system.

CAPILLARY BLOODFLOW

As previously described, blood flows slowly through the capillaries, allowing adequate time for the exchange of materials between the blood and the interstitial fluid. Moreover, the precapillary sphincter muscles at the entrances of the capillaries undergo cycles of contraction and relaxation at frequencies of 2 to 10 cycles per minute. Because of this cyclical activity, which is called **vasomotion,** capillary bloodflow is usually intermittent rather than continuous and steady.

Local autoregulatory factors, particularly oxygen levels, have a strong influence on vasomotion. For example, in many tissues, when the oxygen level in a particular region is low, the intermittent periods of bloodflow through the capillaries of the region occur more often, and each period of flow lasts longer. Thus, bloodflow through the region increases, leading to a greater delivery of oxygen and nutrients.

Although the bloodflow through any one capillary is intermittent, there are so many capillaries that their function on an organ level becomes averaged. Therefore, it is possible to consider capillary function in terms of average capillary pressures, rates of flow, and rates of transfer of substances between the blood and interstitial fluid. This approach is used in the following sections.

MOVEMENT OF MATERIALS BETWEEN THE BLOOD AND INTERSTITIAL FLUID

Several processes contribute to the movement of materials across capillary walls between the blood and the interstitial fluid.

Diffusion

Diffusion is the most important means by which substances such as nutrients and metabolic end products pass between the blood and the interstitial fluid. If a capillary is permeable to a particular substance, and if there is a higher concentration of the substance within the capillary than outside it, the net diffusion of the substance will be outward. If the substance is in higher concentration within the interstitial fluid than within the capillary, its net diffusion will be inward. Lipid-soluble substances, including oxygen and carbon dioxide, can penetrate plasma membranes and move across capillary endothelial cells. Water-soluble substances, such as glucose and amino acids, can pass through water-filled clefts between adjacent endothelial cells. However, the size of the clefts limits the size of the particles that can pass through them.

Endocytosis and Exocytosis

Substances are taken into one side of capillary endothelial cells by endocytosis and released from the other side of the cells by exocytosis. However, there is disagreement about the importance of this activity in the exchange of materials between the blood and the interstitial fluid.

Fluid Movement

A general movement of fluid—that is, water and dissolved particles to which a capillary is permeable—takes place across the capillary wall. As previously indicated, under certain circumstances, the movement of fluid between the blood and the interstitial fluid leads to alterations in blood volume (and arterial pressure). However, under normal circumstances, this movement causes little change in the volume of either the blood or the interstitial fluid. Several factors are involved in the movement of fluid across the capillary wall.

Fluid Pressures.   The pressure of the blood within the capillaries (that is, the *capillary pressure*) tends to force fluid out of the capillaries and into the tissue spaces by filtration through the capillary walls. Opposing this movement is the pressure of the interstitial fluid (that is, the *interstitial fluid pressure*), which tends to move fluid out of the tissue spaces and into the capillaries. The interstitial fluid pressure, however, is generally much less than the capillary pressure, and there is considerable evidence that it is even below atmospheric pressure.

Osmotic Pressures.   The presence of nondiffusible proteins in the interstitial fluid leads to the development of an osmotic force, called the *colloid osmotic pressure of the interstitial fluid*, which tends to draw water out of the capillaries by osmosis. This activity is opposed by the *colloid osmotic pressure of the plasma*, also called the *oncotic pressure*, which results from the presence of nondiffusible plasma proteins, particularly albumin. The oncotic pressure tends to draw water into the capillaries by osmosis.

When osmosis occurs across a barrier like the capillary wall, which contains pores or clefts that are relatively wide compared to the width of a water molecule, substantial quantities of water move rapidly across the barrier, dragging along solutes to which the barrier is freely permeable. Thus, the occurrence of osmosis across the capillary wall results in a movement of fluid similar to that due to the capillary and interstitial fluid pressures.

Direction of Fluid Movement.   If the forces tending to move fluid out of a capillary (the capillary pressure and the colloid osmotic pressure of the interstitial fluid) are greater than the forces tending to move fluid into the capillary (the interstitial fluid pressure and the oncotic pressure), fluid leaves the capillary and enters the tissue spaces. If the inward forces are dominant, fluid enters the capillary from the tissue spaces.

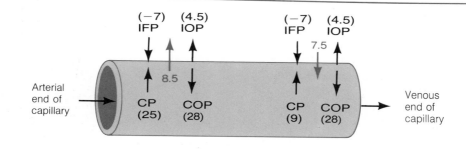

Out: CP   IOP
      25 + 4.5 = 29.5

In: IFP   COP
      −7 + 28 = 21.0
                 8.5 Out

In: IFP   COP
      −7 + 28 = 21.0

Out: CP   IOP
      9 + 4.5 = 13.5
                  7.5 In

F13.7

Forces involved in the movement of fluid into and out of capillaries. Values are general and may not apply to any particular capillary. They are expressed in mm Hg, with the atmospheric pressure considered to be 0 mm Hg. Thus, negative values represent pressures below atmospheric pressure. CP = capillary pressure; COP = colloid osmotic pressure of the plasma; IFP = interstitial fluid pressure; IOP = colloid osmotic pressure of the interstitial fluid.

The traditional view of fluid movement across the capillary wall is that under normal circumstances the capillary pressure at the arterial end of a systemic capillary is relatively high, and outward moving forces predominate (F13.7). Thus, fluid moves out of the capillary into the tissues. As the blood flows through the capillary, the capillary pressure drops. Gradually the balance of forces favoring movement outward is reversed, and fluid moves back into the capillary at the venous end.

An alternative view of fluid movement across the capillary wall is that when a precapillary sphincter muscle relaxes during the rhythmic cycling of vasomotion, the capillary pressure rises, and an outward movement of fluid occurs all along the capillary. Then, when the precapillary sphincter constricts, the capillary pressure falls, and an inward movement of fluid occurs all along the capillary.

In either case, not all of the fluid that moves out of the capillaries returns to the capillaries. Some of the fluid, including any protein that escapes from the capillaries, enters the lymphatic system to be ultimately returned to the bloodstream with the lymph.

VENOUS RETURN

Blood that leaves the capillaries flows into venules and then into veins on its way back to the heart. The greater the distance the blood travels, the greater the total resistance encountered, and thus the greater the pressure drop. Blood in the venules and veins has traveled through the arteries, arterioles, and capillaries, and consequently the pressures within the venous circulation are quite low. For example, the pressures within the veins of the arms or legs average about 6 to 8 mm Hg.

The volume of blood that flows from the systemic veins into the right atrium of the heart in a given time is called the **venous return.** (Under normal circumstances, the same volume of blood flows from the pulmonary veins into the left atrium.) The magnitude of the venous return depends on the pressure driving blood through the veins and on the resistance of the veins to the flow of blood through them (recall that flow = ΔP/resistance). The pressure driving blood through the veins is represented by the difference between the pressure in the peripheral veins (the *peripheral venous pressure*) and the pressure at the right atrium of the heart. Consequently, factors that increase the peripheral venous pressure tend to increase the venous return, and factors that decrease the peripheral venous pressure tend to decrease it. For example, a decrease in the blood volume—as may result from hemorrhage—can lower the peripheral venous pressure and decrease the venous return, and an increase in the blood volume—as may result from a transfusion—can raise the peripheral venous pressure and increase the venous return. In a similar manner, factors that increase the right atrial pressure (which is normally close to 0 mm Hg) can also decrease the venous return. For example, a leaky right atrioventricular valve can cause an increase in the right atrial pressure and a decrease in the venous return.

The activities of skeletal muscles have an important influence on the venous return. The veins are quite flexible, and the movements of the skeletal muscles that surround them act in a pumping fashion to put pressure on the veins and compress them. This compression forces blood out of the veins and into the heart. The valves of the large veins contribute to the effectiveness of this pumping action by permitting blood leaving a compressed portion of a vein to flow only toward the heart and not in the opposite direction.

Breathing movements also influence the venous return. When a person inhales, the pressure within the thoracic (chest) cavity decreases, and the pressure within the abdominal cavity increases. These pressure changes facilitate the return of blood from the veins to the heart.

The veins have a large capacity, and they contain a substantial amount of the total blood volume. In fact, about 64% of the total circulating blood volume is normally in the systemic venous circulation, about 7% is in the heart, about 15% is in the systemic arterial vessels, about 5% is in the

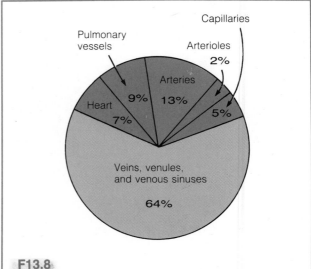

F13.8
Percentage of the total blood volume in different portions of the cardiovascular system.

systemic capillaries, and about 9% is in the pulmonary vessels (F13.8). Because of their large capacity, the veins serve as blood reservoirs. Although they are not heavily muscular, the veins can constrict and decrease their capacity, and even a slight venous constriction forces a good deal of blood into other portions of the cardiovascular system. Moreover, a constriction of the veins can decrease venous distensibility, raise the peripheral venous pressure, and increase the venous return. (In this regard it should be noted that the veins are relatively low-resistance vessels, and a moderate venous constriction does not greatly increase the overall resistance to bloodflow.)

EFFECTS OF GRAVITY ON THE CARDIOVASCULAR SYSTEM

Previous considerations of the pressures in various portions of the cardiovascular system generally assumed the body to be in a horizontal position in which the blood vessels are at approximately the same level as the heart and pressures due to the force of gravity acting on the blood are negligible. However, when the body is in an upright position, the weight of the blood affects these

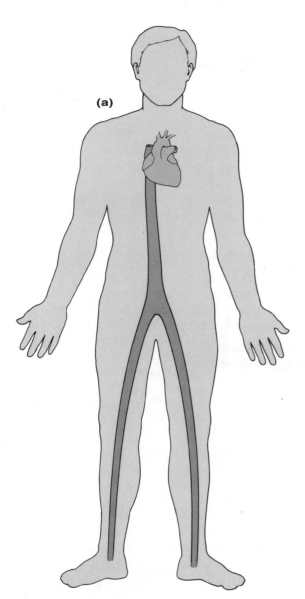

Venous pressure at foot = 90 mm Hg

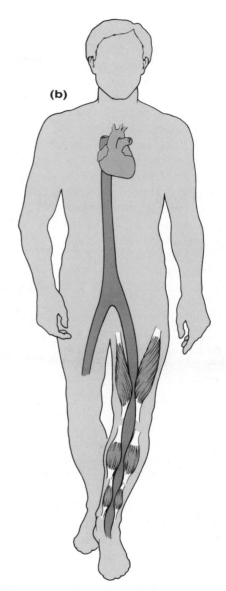

Venous pressure at foot = < 25mm Hg

F13.9

(a) When a person is standing upright and perfectly still, columns of blood extend within the blood vessels from the feet to the heart. The weights of these columns raise the pressure within the vessels of the lower regions of the body. (b) During movement, muscular contractions compress the veins and force blood toward the heart. This activity reduces the pressure within the veins of the lower regions of the body such as the veins of the feet.

pressures. For example, when a person moves from a supine position to a position in which he or she is standing upright and perfectly still, columns of blood extend within the blood vessels from the feet to the heart (F13.9*a*). The weights of these columns raise the pressure within the vessels of the lower regions of the body. The increased pressure distends the veins, increasing their capacity so they contain a greater amount of the total blood volume. Moreover, the increased pressure results in an increased filtration of fluid out of the capillaries and into the tissue spaces. The accumulation of blood in the veins and the increased filtration of fluid out of the capillaries reduces the effective circulating blood volume. If no adequate compensatory adjustments occur, the arterial pressure falls, the bloodflow to the brain declines, and the person becomes dizzy or faints. Under most circumstances, however, several mechanisms compensate for the effects of gravity when a person is upright.

One mechanism is the baroreceptor mechanism described earlier. For example, when a person sits up or stands after lying down, the force of gravity causes the blood to accumulate in the veins of the lower limbs and abdomen, and the arterial pressure in the head and upper body falls. However, the falling pressure is sensed by the baroreceptors, which initiate a generalized vasoconstriction and increase in heart rate and contractile strength, thus preventing a severe pressure drop that could result in a diminished bloodflow to the brain and a loss of consciousness.

A second mechanism that compensates for the effects of gravity is the pumping action exerted on the veins by the skeletal muscles of the legs. The contractions of the muscles compress the veins and force blood toward the heart (F13.9*b*). This activity enhances the venous return and reduces the pressure within the veins of the lower regions of the body. As a result, the upright position has less of an effect on venous distension and the accumulation of blood in the veins as well as on the capillary pressure and the filtration of fluid out of the cardiovascular system than would otherwise be the case. The importance of this mechanism is evident in the fact that when a person is standing perfectly still, the pressure within the veins of the feet can rise above 90 mm Hg, but the muscular activity of walking reduces the pressures within these veins to less than 25 mm Hg.

LYMPHATIC SYSTEM

As previously indicated, the volume of fluid that moves out of the capillaries and into the tissue spaces is slightly greater—by about 3 liters per day—than the volume of fluid that moves into the capillaries from the tissues. The lymphatic system returns this extra fluid (including any plasma protein that escapes from the capillaries) back to the bloodstream.

PATTERN OF LYMPH FLOW

The lymphatic system begins in the tissues as numerous dead-end vessels called lymphatic capillaries (F13.10). The edges of adjacent endothelial cells of lymphatic capillaries are only loosely attached and usually overlap. This arrangement forms a functional one-way valve, which operates as follows. When interstitial fluid outside a lymphatic capillary attempts to flow into it, the fluid pushes the edges of the endothelial cells inward and enters the capillary. Once within the capillary, the fluid cannot reenter the intercellular spaces because, as it attempts to do so, it forces the edges of the endothelial cells together, closing the valve.

Because of their structural arrangement, lymphatic capillaries are more permeable than most blood capillaries, and virtually all the components of the interstitial fluid, including proteins and other large particles, can enter these vessels. Thus, the fluid within the lymphatic capillaries has essentially the same composition as the interstitial fluid. From the lymphatic capillaries, the fluid, which is called **lymph** after it enters the lymphatic system, flows into progressively larger lymphatic vessels that ultimately return it to the bloodstream (F13.11). Although they are thinner, the walls of the larger lymphatic vessels are similar to the walls of veins, and the vessels contain valves that ensure a one-way flow of lymph.

MECHANISMS OF LYMPH FLOW

Lymph flows slowly; approximately 3 liters of it enter the bloodstream every 24 hours. Lymph flow is strongly influenced by the contractions of skeletal muscles, which act in a pumping fashion to put pressure on the lymph vessels and compress them. This action forces lymph along the vessels. In addition, it appears that when a section of a lymph vessel is distended by lymph it can contract and thereby propel the lymph along the vessel. The valves of the lymph vessels contribute to the effectiveness of these activities by permitting lymph within the vessels to flow only toward the bloodstream and not in the opposite direction.

REMOVAL OF FOREIGN PARTICLES FROM THE LYMPH

As it flows, the lymph generally passes through one or more accumulations of lymphatic tissue

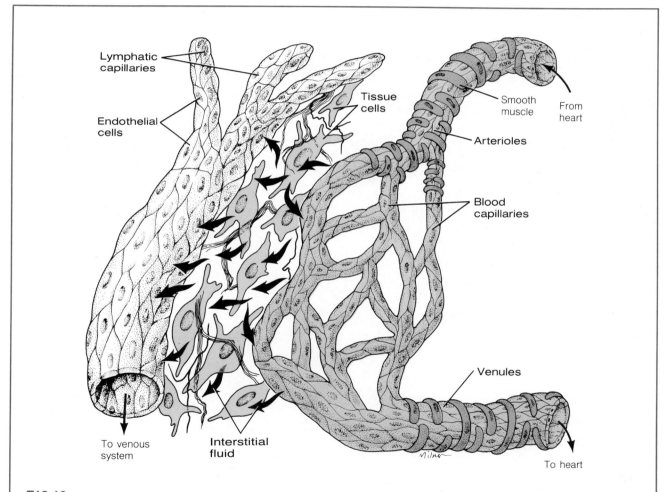

F13.10
Schematic representation of the lymphatic capillaries showing their relationship to interstitial fluid, the blood vascular system, and tissue cells. The arrows indicate the directions of fluid movement. Note that the lymphatic capillaries begin as dead-end vessels.

called **lymph nodes,** which are distributed along the course of the lymphatic vessels (F13.11). Bacteria and other foreign substances are removed from the lymph by phagocytes—primarily macrophages—that are present within the lymph nodes, and during infections the lymph nodes often enlarge and become tender. The lymph nodes are also important in specific immune responses to bacteria and other substances (see Chapter 14).

SPLEEN

The **spleen** carries on functions that are typical of lymphoid tissue, and it is generally considered to be a lymphoid organ. Macrophages of the spleen destroy over aged erythrocytes and remove foreign particles from the blood. The spleen is also involved in specific immune responses to bacteria and other substances within the blood.

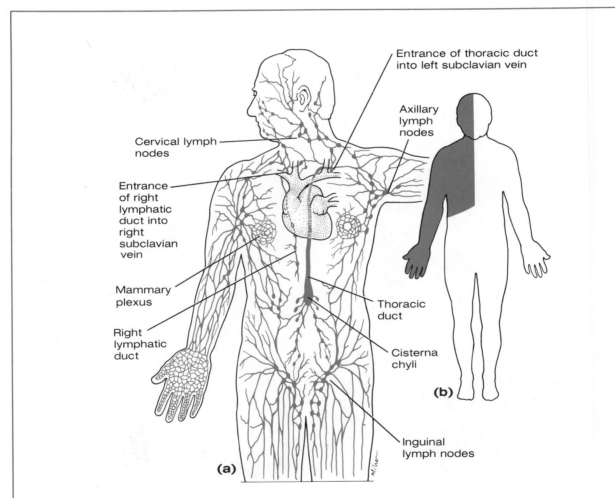

F13.11
The lymphatic system. (a) Major lymphatic vessels and groups of lymph nodes. (b) Lymph from the colored area returns to the blood vascular system through the right lymphatic duct. Lymph from the light area travels through the thoracic duct.

CIRCULATION IN SPECIAL REGIONS

PULMONARY CIRCULATION

The pulmonary arteries and arterioles generally have larger diameters and thinner walls than corresponding vessels of the systemic circulation, and the resistance to bloodflow through the pulmonary circulation is less than that through the systemic circulation. As a result, less pressure is required to move blood through the pulmonary circuit than through the systemic circuit, and the pressure within the pulmonary arteries (about 22 mm Hg systolic pressure; 8 mm Hg diastolic pressure) is considerably lower than the pressure within the systemic arteries (about 120 mm Hg systolic pressure; 80 mm Hg diastolic pressure).

The bloodflow through different local areas of the pulmonary circulation varies with the oxygen levels in the areas. A decreased oxygen level causes vasoconstriction, and an increased oxygen level causes vasodilation. Although the local effect of oxygen on pulmonary vessels is the opposite of its effect on many systemic arterioles, it is consistent with the role of the lungs in providing oxygen to the blood. For example, if a portion of the lungs is not functioning effectively, the level of oxygen in the region falls. The decline in the oxygen level causes the vessels supplying the area to constrict, which results in a greater bloodflow through the vessels in efficiently functioning areas of the lungs.

CORONARY CIRCULATION

The flow of blood through the coronary vessels is greatly influenced by the aortic pressure and by the pressures in the walls of the heart chambers (that is, the intramural pressures). The influence of these pressures is particularly evident in the left coronary artery (F13.12). When the left ventricle contracts (line 1), the intramural pressure increases greatly, squeezing the coronary vessels and causing a significant reduction in the coronary flow (curve *a*). With the opening of the aortic semilunar valve (line 2), the effect of the intramural pressure of the ventricles is overcome by the increased aortic blood pressure. The increased aortic pressure causes an increase in bloodflow through the left coronary artery (curve *b*). As ventricular systole progresses, the flow of blood into the aorta diminishes, and the pressure within the aorta begins to fall (line 3). However, the left ventricle is still contracting at this time, and the intramural pressure remains high. As a result, there is a secondary slowing of the coronary flow (curve *c*). As ventricular systole ends (line 4), the intramural pressure drops, and the flow through the coronary vessels increases once again (curve *d*). During ventricular diastole, the flow through the coronary vessels is determined primarily by the aortic pressure, and the flow gradually decreases as the aortic pressure drops (curve *e*). Also during this period, the atria contract, and the pressure within their walls increases. This increase in the atrial intramural pressure also tends to decrease the flow of blood through the coronary vessels.

Under resting conditions, the bloodflow through the coronary vessels is approximately 250 ml per minute, which is about 5% of the cardiac output. However, during strenuous exercise, the coronary vessels dilate, and the coronary bloodflow can increase to four or five times the resting level—that is, up to 1250 ml per minute. The dilation of the coronary vessels during exercise is believed to be primarily due to the influence of local autoregulatory factors (such as low oxygen levels resulting from increased cardiac activity).

An increased bloodflow to the heart during exercise is particularly important because, even in the resting condition, the heart removes about 65% of the oxygen in the arterial blood that flows to it. The removal of this much oxygen is very efficient compared to most other tissues, where as little as 25% of the oxygen is removed. However, it means that during activity, the heart cannot greatly increase the oxygen available to it by increasing the percentage of oxygen removed from the blood. Therefore, increasing the coronary bloodflow—and thus the delivery of oxygen—is the principal way in which additional oxygen is made available to the myocardium.

CEREBRAL CIRCULATION

Because the brain is unable to store glycogen, it must have a constant, dependable bloodflow in order to ensure that glucose is always available to it as an energy source. In addition, brain tissue must have a steady, adequate supply of oxygen. Irreversible damage occurs if the brain's bloodflow is stopped for over five minutes.

The bloodflow through the brain is about 750 ml per minute, which is 15% of the total cardiac output at rest. The overall metabolic rate of the brain varies little under widely different physiological conditions—for example, intense mental activity, muscular activity, or sleep—and although the bloodflow to different localized areas may increase in response to increased neuronal activity, the total cerebral bloodflow is remarkably constant over a very broad range of mean

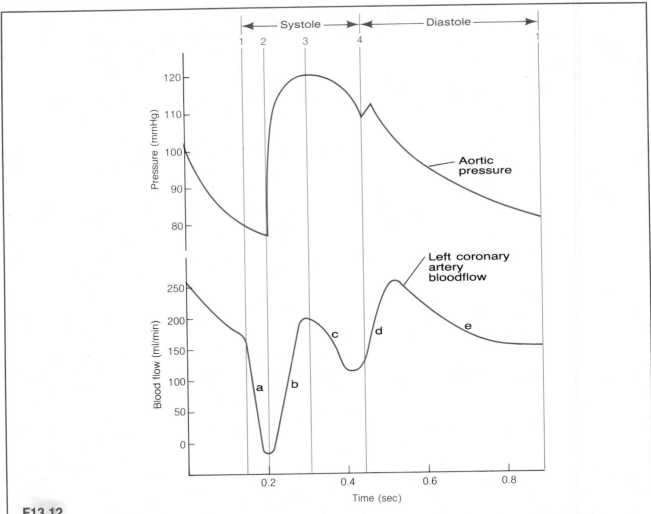

F13.12
Bloodflow through the left coronary artery correlated with aortic blood pressure and ventricular systole and diastole.

arterial pressures (F13.13). However, under certain circumstances, local autoregulatory mechanisms can cause changes in cerebral bloodflow. For example, an elevated level of carbon dioxide in the arterial blood or a decreased level of oxygen causes the dilation of cerebral blood vessels, which increases cerebral bloodflow. Of the two factors, the brain seems to be more sensitive to the carbon dioxide level than to the oxygen level.

When there is a temporary interference with the cerebral bloodflow, a person may faint, and factors that lower the arterial pressure below the level required to maintain an adequate cerebral bloodflow can cause fainting. For example, the fainting that often accompanies severe emotional upset is due to a slowing of the heart and to a strong stimulation of the vasodilator fibers to skeletal muscle vessels. As a result, the arterial pressure declines, bloodflow to the brain is impaired, and the person faints.

A more serious condition occurs when a cerebral vessel is obstructed or ruptured, and the flow of blood to the region of the brain supplied by the vessel is reduced. Such a condition is called a *cerebrovascular accident*, or *stroke*. The reduced flow of blood results in an inadequate oxygen supply to the cells of the affected area and may cause permanent brain damage. The neurological effects of a stroke vary, depending on the site of brain damage.

SKELETAL MUSCLE CIRCULATION

During exercise, the energy requirements of skeletal muscles increase so greatly that they can be met only by a tremendous increase in the bloodflow to the muscles. During strenuous exercise, muscle bloodflow can be as much as 20 times greater than the flow at rest. The increased bloodflow is made possible by the dilation of arterioles and the opening of previously closed capillaries within the muscles. Two separate mechanisms are involved in increasing the bloodflow through active skeletal muscles.

The first mechanism involves the central nervous system. When signals are sent to a muscle to initiate muscular activity, sympathetic vasodilator fibers to the arterioles of the muscle are also activated by way of a pathway that originates in the cerebral cortex and passes downward to the hypothalamus. As a result, the arterioles of the muscle dilate, thus increasing the bloodflow to the muscle. This vasodilation occurs even prior to exercise—that is, prior to any great decrease in the oxygen level within the muscle.

The second mechanism is the local autoregulatory response. The increased muscular activity during exercise decreases the oxygen level in the muscle. The decreased oxygen level leads to a vascular dilation that increases the bloodflow through the muscle.

Vasodilation and increased bloodflow occur in skeletal muscles in which all sympathetic nerve fibers have been inactivated, and it appears that the local autoregulatory response is the more important of the two control mechanisms. The function of the sympathetic vasodilator fibers seems to be to increase the flow of blood through a muscle at the onset of muscular activity until the lack of oxygen within the muscle leads to vasodilation by way of the local autoregulatory response.

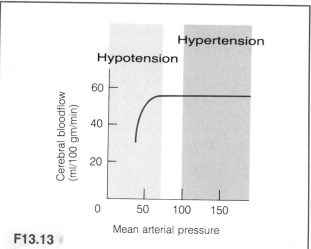

F13.13

Relationship of mean arterial pressure and cerebral bloodflow. The central area is the normal range of mean arterial pressure.

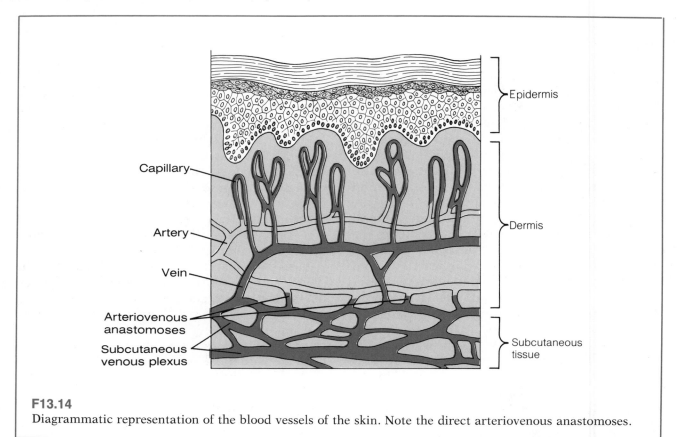

F13.14

Diagrammatic representation of the blood vessels of the skin. Note the direct arteriovenous anastomoses.

CUTANEOUS CIRCULATION

The bloodflow to the skin serves two functions: it provides nutrients to and carries metabolic end products away from the skin, and it brings heat from the internal structures of the body to the surface, where the heat can be lost from the body. The cutaneous bloodflow is normally 20 to 30 times greater than is needed to meet the nutrient needs of the skin, and most of the flow is for purposes of temperature regulation.

The bloodflow to the skin can vary tremendously. Under ordinary circumstances, cutaneous bloodflow is approximately 400 ml per minute. However, under extreme conditions, the bloodflow can increase to 2500 ml per minute. A unique arrangement of the blood vessels beneath the body surface makes such a large bloodflow possible (F13.14). In addition to the usual capillary beds, there are extensive subcutaneous venous plexuses (*plexus* = network) that can hold large volumes of blood. These plexuses are located close enough to the surface to allow heat to pass from the blood to the surface, and thus to be lost from the body. The bloodflow through the subcutaneous venous plexuses varies with the degree of constriction or dilation of the vessels that carry blood to the plexuses, and a temperature control center in the hypothalamus acts through sympathetic nerves to regulate the diameters of these vessels. In some locations—for example, the hands, feet, and ears—blood can bypass the capillary system and flow directly from arterial vessels into the venous plexuses by way of vascular shunts called arteriovenous anastomoses. The walls of these arteriovenous shunts are muscular, and their constriction decreases the flow of blood into the venous plexuses and reduces the loss of body heat.

CARDIOVASCULAR ADJUSTMENTS DURING EXERCISE

During exercise, the dilation of skeletal muscle vessels due to local autoregulatory responses and to some extent to sympathetic vasodilator activity greatly increases the bloodflow to skeletal muscles. At the same time, sympathetic vasoconstrictor activity causes a compensatory constriction of vessels elsewhere, particularly in the kidneys and gastrointestinal tract. The compensatory vasoconstriction does not fully offset the dilation of the skeletal muscle vessels, however, and the total peripheral resistance decreases.

Also during exercise, the activity of the sympathetic nerves to the heart increases, and the activity of the parasympathetic nerves decreases. In addition, venous return is enhanced by the increased pumping effects of the contracting skeletal muscles, and by the constriction of the veins due to sympathetic stimulation. As a consequence of these activities, both the heart rate and the stroke volume increase, resulting in an increase in the cardiac output. During most forms of exercise, the increase in cardiac output is somewhat greater than the decrease in total peripheral resistance, and the mean arterial pressure rises (recall that mean arterial pressure = cardiac output × total peripheral resistance).

It is believed that the cardiovascular changes characteristic of exercise—and also excitement, stress, or anger—are initiated by nerve impulses that originate within the cerebral cortex and pass to the hypothalamus. The hypothalamus, in turn, acts on the brain-stem centers that influence cardiovascular function and directly on the sympathetic vasodilator neurons to skeletal muscle vessels to produce the cardiovascular adjustments that occur during exercise.

DISORDERS OF THE CIRCULATORY SYSTEM

HIGH BLOOD PRESSURE

High blood pressure, or *hypertension*, can lead to heart attack, heart failure, brain stroke, or kidney damage. Approximately 20% of the population can expect to have high blood pressure at some time during their lives. In general, arterial blood pressures above 150/90 mm Hg are considered to be hypertensive. The degree of elevation of both the systolic pressure and the diastolic pressure is important in hypertension since both pressures affect the mean arterial pressure.

Hypertension due to an unknown cause is called *idiopathic* or *essential hypertension*, and approximately 95% of the cases of high blood pressure are classed as idiopathic. Many researchers believe that high blood pressure is frequently caused by an excessive dietary intake of sodium or an excessive retention of sodium by the body, which leads to an increase in blood volume and cardiac output. Researchers have also proposed that an increased peripheral resistance due to vascular constriction contributes to high blood pressure, and a number of studies suggest a sequence of initially high cardiac output followed by increased peripheral resistance.

High blood pressure is sometimes associated with kidney diseases or conditions that impair renal bloodflow. In some cases, the formation of renin in the kidneys leads to the formation of angiotensin and arteriolar constriction, although other factors are also likely to be involved. High blood pressure can also accompany the secretion of excessive amounts of norepinephrine and epinephrine by tumors of the adrenal medulla, and the secretion of excessive amounts of hormones by tumors of the adrenal cortex.

ARTERIOSCLEROSIS AND ATHEROSCLEROSIS

As they grow older, a great number of people experience degenerative changes in the walls of their arteries that are known as *arteriosclerosis*, or "hardening of the arteries." Arteriosclerosis can greatly reduce the elasticity of the vessel walls. As the walls become less elastic, the vessels cannot properly expand and recoil in response to the pressure changes produced by the beating heart. Consequently, in severe cases, the pressure within the vessels rises quite high during systole and falls unusually low during diastole.

One form of arteriosclerosis that causes serious disease is *atherosclerosis*. Atherosclerosis is characterized by deposits (plaques) of abnormal smooth muscle cells, lipid materials, and connective tissue in arterial walls. The atherosclerotic deposits can gradually narrow or occlude blood vessels. In addition, the deposits can promote the formation of blood clots within the vessels. In any case, the flow of blood through the vessels is often restricted or blocked. Frequently, the first indication of atherosclerosis is the dysfunction of an organ supplied by an affected vessel. For example, a restricted or inadequate bloodflow to the heart can cause pain during exertion or emotional stress that is called *angina pectoris*, and a blocked coronary vessel can lead to the death of a portion of the myocardium—that is, to a myocardial infarct. Similarly, the occlusion of a vessel that supplies the brain may kill brain tissue.

EDEMA

The accumulation of excess fluid in the tissues is called *edema*. Edema causes the tissues to swell, and it can impair tissue function because the excess fluid increases the distance that materials must diffuse between the capillaries and the cells. Many factors can contribute to edema. In general, any event that enhances the movement of fluid out of the blood vessels and into the tissues, or that retards the return of fluid from the tissues back into the blood vessels can lead to edema. Thus, edema may result from: (1) an increased capillary permeability, as may occur in certain allergic responses; (2) an elevated capillary pressure, due, for example, to blockage of a vein; (3) an increased interstitial fluid colloid osmotic pressure, caused, for example, by blocked lymphatic vessels; or (4) a decreased plasma colloid osmotic pressure, due, for example, to the loss of plasma proteins in the urine in certain kidney diseases.

CIRCULATORY SHOCK

Circulatory shock is a condition in which the cardiac output is so reduced that the body tissues fail to receive an adequate blood supply. Any condition that decreases the cardiac output can lead to circulatory shock. Thus, circulatory shock can result from: (1) diseases and weaknesses of the heart itself, caused, for example, by a heart attack; (2) reductions in blood volume, perhaps due to hemorrhage; or (3) vascular difficulties that reduce the venous return, due, for example, to extreme vasodilation and the accumulation of blood in the veins. In severe cases of shock, the inadequate blood supply to the tissues impairs tissue function so greatly that death results.

ANEURYSMS

An *aneurysm* is a localized dilation of an artery due to a weakness of the artery wall. The most frequent site of aneurysm formation is the abdominal aorta. The greatest danger from an aneurysm is that the affected vessel will rupture; however, aneurysms are also common sites of thrombus formation, and a portion of a thrombus may break off and block a smaller vessel. In addition, aneurysms can obstruct or erode neighboring organs or tissues.

PHLEBITIS

The inflammation of a vein is called *phlebitis*. Phlebitis can result from a number of conditions. One type of phlebitis is caused by bacteria that invade a vein, perhaps from an abscess or where the vein passes through an area of inflammation. In phlebitis, the inflammation of a vein can lead to the formation of a thrombus with all of its accompanying dangers. This condition is called thrombophlebitis.

VARICOSE VEINS

Varicose veins are veins that are dilated, lengthened, and tortuous. Varicose veins are often brought about by a congenital and inherited weakness of the walls and valves of the veins. Such veins are unable to carry blood efficiently toward the heart, thus allowing the blood to accumulate in the lower limbs when the body is in an upright position. Other varicosities may be caused by pressure on veins, such as might occur during pregnancy, in obesity, or from an abdominal tumor. In severe cases, varicose veins can interfere with the return of blood to the extent that muscle cramps and edema occur. Hemorrhage, phlebitis, and thrombosis are also possible complications.

EFFECTS OF AGING ON THE CARDIOVASCULAR SYSTEM

Aging apparently affects the heart and the blood vessels since cardiovascular problems cause more than half the deaths of the population over age 65. However, not all of these cardiovascular problems are necessarily due to aging alone. Evidence suggests that in many nonindustrialized societies there is little cardiovascular disease. By comparison, approximately 20% of the population of industrialized societies suffer from hypertension by the age of 65. Although age and genetic factors probably contribute to the changes in the walls of the blood vessels that lead to myocardial infarction or stroke, many investigators believe that in advanced societies environmental factors, and especially diet, play a more significant role than age or genetic inheritance.

Various studies have concluded that: (1) a substantial lowering of plasma cholesterol would greatly reduce the number of heart attacks; (2) maintaining diastolic blood pressures of under 75 mm Hg would result in a threefold reduction in the number of heart attacks; and (3) regular moderate exercise would reduce the incidence of fatal heart attacks by approximately half. In addition, nonsmokers have half the incidence of heart attacks as do smokers. These studies suggest that aging may not be the major cause of cardiovascular disorders.

STUDY OUTLINE

CIRCULATION THROUGH THE VESSELS
pp. 323–326

Pattern of Bloodflow Blood flows from the heart, to arteries, to arterioles, to capillaries, to venules, to veins, and back to the heart.

Velocity of Bloodflow In general, the velocity of flow in any segment of the cardiovascular system is inversely related to the total cross-sectional area of the vessels of the segment. The velocity of flow is lowest in the capillaries.

Pulse The elastic expansion and subsequent recoil of the arteries that occurs in association with the beating of the heart is the pulse. It can be felt at various locations on the body surface.

PRINCIPLES OF CIRCULATION Efficient circulation requires an adequate blood volume, blood vessels in good condition, and a smoothly functioning heart.
pp. 326–328

Bloodflow The volume of blood that flows through a vessel or vessels in a given time.

Pressure The pumping action of the heart imparts energy to the blood. This energy is evident as the pressure that drives the blood through the vessels.

Resistance Essentially a measure of friction.

FACTORS THAT AFFECT RESISTANCE

Blood Viscosity

Vessel Length

Vessel Diameter

ALTERATIONS IN RESISTANCE Vessel length is constant and blood viscosity normally does not vary greatly. Most alterations in resistance are due to changes in vessel, particularly arteriole, diameter.

Relationship among Flow, Pressure, and Resistance Generally,

$$\text{flow} = \frac{\text{pressure}}{\text{resistance}}$$

and:

$$\text{mean arterial pressure} = \text{cardiac output} \times \text{total peripheral resistance}$$

ARTERIAL PRESSURE A number of factors affect the arterial pressure through mechanisms that influence vessel resistance and cardiac function.
pp. 328–332

Neural Factors

VASOMOTOR NERVE FIBERS Vasomotor nerve fibers affect vessel resistance by altering vessel diameters. Sympathetic vasoconstrictor fibers release norepinephrine and cause vasoconstriction. Sympathetic vasodilator fibers to skeletal muscle vessels release acetylcholine and cause vasodilation.

VASOMOTOR CENTER A tonically active brainstem center that plays a role in the regulation of blood vessel resistance and thus in the regulation of arterial pressure. The vasomotor center is influenced by inputs from higher brain centers, cardiovascular system receptors, and receptors associated with sensory surfaces of the body.

CARDIAC CENTER Can influence arterial pressure by virtue of its effects on cardiac function.

BARORECEPTORS Baroreceptor activity influences vessel diameters, heart rate, and cardiac contractility, and thereby affects arterial pressure.

AORTIC AND CAROTID BODIES The aortic and carotid bodies (chemoreceptors) can also influence arterial pressure.

Chemicals and Hormones

ANGIOTENSIN An extremely powerful vasoconstrictor that raises arterial pressure.

EPINEPHRINE Causes a transitory increase in systolic pressure within the arteries.

VASOPRESSIN Raises arterial pressure by stimulating arteriolar constriction.

Blood Volume In general, an increase in blood volume tends to raise the arterial pressure and a decrease in blood volume tends to lower it. Kidney mechanisms that influence blood volume are particularly important in the long-term regulation of arterial pressure.

Influence of Multiple Factors A number of different factors affect arterial pressure, and the arterial pressure at any moment is usually the result of the combined influence of several factors. Different mechanisms work together to maintain the arterial pressure and thus to maintain a circulation that can meet the body's needs.

Measurement of Arterial Pressure Usually done indirectly using inflatable cuff and manometer. Blood pressure is reported as systolic pressure/diastolic pressure—for example, 120/80. The pulse pressure is the difference between the systolic and diastolic pressures.

TISSUE BLOODFLOW   pp. 332–333

Local Autoregulatory Control Local autoregulatory mechanisms alter bloodflow in accordance with tissue needs. Oxygen deficiency can lead to local dilation of arterioles and increased relaxation of precapillary sphincters. Carbon dioxide, lactic acid, adenosine, adenosine phosphate compounds, histamine, potassium ions, and hydrogen ions may act as local vasodilator substances.

External Factors External factors such as vasomotor nerve fibers and hormones usually influence large segments of the vascular system, and they are generally more involved with mehcanisms that function for the well-being of the entire body than they are with the control of tissue bloodflow according to the needs of particular tissues.

CAPILLARY EXCHANGE   Most capillaries are permeable to water and small particles such as glucose, inorganic ions, urea, amino acids, and lactic acid; and these materials pass readily between the blood and the interstitial fluid. Protein permeability is generally quite limited.   pp. 333–335

Capillary Bloodflow Blood flows slowly through the capillaries, allowing adequate time for the exchange of materials between the blood and the interstitial fluid. Precapillary sphincters contract and relax cyclically, and capillary bloodflow is usually intermittent rather than continuous and steady.

Movement of Materials Between the Blood and Interstitial Fluid Several processes are involved.

DIFFUSION Diffusion is the most important means by which substances such as nutrients and metabolic end products pass between the blood and the interstitial fluid.

ENDOCYTOSIS AND EXOCYTOSIS There is dis-agreement about the importance of this activity in the exchange of materials between the blood and interstitial fluid.

FLUID MOVEMENT Under normal circumstances this movement causes little change in the volume of either the blood or the interstitial fluid.

Fluid Pressures The capillary pressure tends to force fluid out of the capillaries and into the tissue spaces by filtration through the capillary walls. The interstitial fluid pressure opposes this movement.

Osmotic Pressures The colloid osmotic pressure of the interstitial fluid tends to draw water out of the capillaries by osmosis, and the colloid osmotic pressure of the plasma tends to draw water into the capillaries by osmosis.

Direction of Fluid Movement If the forces tending to move fluid out of a capillary are greater than forces tending to move fluid in, then fluid will leave, and vice versa.

VENOUS RETURN   Influenced by blood volume, contraction of skeletal muscles, breathing movements, valves of veins, constriction of veins.   pp. 335–336

EFFECTS OF GRAVITY ON THE CARDIO-VASCULAR SYSTEM   When a person moves from a supine position to a position in which he or she is standing upright and perfectly still, the weight of the blood causes high pressures in blood vessels of the lower regions of the body. Baroreceptors and muscular contractions during movement help compensate for effects of gravity when a person is upright.   pp. 336–338

LYMPHATIC SYSTEM   Returns extra fluid, including plasma protein, to bloodstream.   pp. 338–340

Pattern of Lymph Flow Fluid from the tissue spaces enters small dead-end lymphatic capillaries. The fluid—which is called lymph after it enters the lymphatic system—flows through progressively larger vessels that ultimately return it to the bloodstream.

Mechanisms of Lymph Flow Lymph flow is generally slow and depends on skeletal muscle contraction, contraction of lymphatic vessels, valves in lymph vessels.

Removal of Foreign Particles from the Lymph Lymph nodes contain phagocytes—primarily macrophages—that remove bacteria and other foreign substances from the lymph.

Spleen   Carries on functions that are typical of lymphoid tissue. Macrophages of the spleen destroy over-aged erythrocytes and remove foreign particles from the blood.

CIRCULATION IN SPECIAL REGIONS
pp. 341–344

Pulmonary Circulation   The pulmonary circuit is a low-pressure circuit. Low oxygen levels cause constriction of pulmonary vessels by local autoregulatory mechanisms.

Coronary Circulation   Greatly influenced by aortic pressure and by the pressures in the walls of the heart chambers.

Cerebral Circulation   Total cerebral bloodflow is remarkably constant over a very broad range of arterial pressures. Interference with cerebral bloodflow can lead to fainting or stroke.

Skeletal Muscle Circulation   During exercise bloodflow to skeletal muscles increases greatly, primarily due to local autoregulatory mechanisms, but sympathetic vasodilator neurons are also involved.

Cutaneous Circulation   Primarily for purposes of temperature regulation rather than skin nutrition. Bloodflow can vary tremendously.

CARDIOVASCULAR ADJUSTMENTS DURING EXERCISE   Skeletal muscle bloodflow increases greatly due to dilation of skeletal muscle vessels. Other vessels constrict to compensate, but total peripheral resistance generally decreases. Heart rate and stroke volume increase as does cardiac output. Mean arterial pressure usually rises.   **p. 345**

DISORDERS OF THE CIRCULATORY SYSTEM
pp. 345–347

High Blood Pressure (Hypertension)   Most cases of hypertension are due to unknown causes. Many cases of hypertension may be due to an excessive dietary intake of sodium or an excessive retention of sodium by the body, which leads to an increase in blood volume and cardiac output. An increased peripheral resistance due to vascular constriction may also be involved.

Arteriosclerosis and Atherosclerosis   Arteriosclerosis is a degenerative change in arterial walls that can lead to the loss of elasticity of the vessel walls. Atherosclerosis is a form of arteriosclerosis characterized by deposits of abnormal smooth muscle cells, lipid materials, and connective tissue. The deposits can narrow or occlude blood vessels or favor clot formation.

Edema   An accumulation of excess fluid in the tissues. In general, any event that enhances the movement of fluid out of the blood vessels and into the tissues, or that retards the return of fluid from the tissues to the bloodstream, can lead to edema.

Circulatory Shock   A condition in which the cardiac output is so reduced that the body tissues fail to receive an adequate blood supply. It can be due to diseases or weaknesses of the heart, a reduction in blood volume, or vascular difficulties that reduce venous return.

Aneurysms   Localized dilation of an artery due to a weakness of the artery wall.

Phlebitis   Inflammation of a vein that can lead to clotting.

Varicose Veins   Dilated, lengthened, and tortuous veins that can interfere with venous return and lead to clotting.

EFFECTS OF AGING ON THE CARDIOVASCULAR SYSTEM   Environment and life style as well as age and genetic inheritance may contribute to cardiovascular problems associated with aging.   **p. 347**

SELF-QUIZ

1. The exchange of substances between the blood and the interstitial fluid occurs primarily in the: (a) arteries; (b) arterioles; (c) capillaries; (d) veins.
2. The velocity of bloodflow is slowest in the: (a) arteries; (b) arterioles; (c) capillaries; (d) veins.
3. Blood pressure is generally lowest in the: (a) arteries; (b) capillaries; (c) veins.
4. In general, which of the following will cause an increased resistance to bloodflow? (a) a decrease in vessel length; (b) a decrease in vessel diameter; (c) a decrease in blood viscosity.
5. The mean arterial pressure is equal to the cardiac output multiplied by the: (a) total peripheral resistance; (b) total systemic bloodflow; (c) total venous return.
6. At their junctions with blood vessels, sympathetic vasoconstrictor neurons generally release: (a) acetylcholine; (b) norepinephrine; (c) vasopressin.
7. Which of the following is likely to lead to a fall in blood pressure? (a) decreased arterial oxygen concentration; (b) decreased activity of the carotid sinus baroreceptors; (c) stretching the wall of the aortic arch.
8. Angiotensin is an extremely powerful vasodilator that is produced from precursors in the plasma by enzymatic action. True or False?
9. The kidneys exert a significant effect on blood volume by virtue of their ability to regulate salt and water excretion, and they are particularly important in the long-term regulation of arterial pressure. True or False?
10. When a person's blood pressure is reported to be 120/80, the 120 is the: (a) pulse pressure; (b) systolic pressure; (c) diastolic pressure.
11. Local autoregulatory mechanisms alter bloodflow to a particular tissue in accordance with the needs of the tissue. True or False?
12. Which of the following forces favors the movement of fluid into the capillaries from the tissue spaces? (a) colloid osmotic pressure of the plasma; (b) colloid osmotic pressure of the interstitial fluid; (c) capillary pressure.
13. Venous return is likely to be enhanced by a decreased peripheral venous pressure and an increased right atrial pressure. True or False?
14. The lymphatic system returns extra fluid but not plasma proteins to the bloodstream. True or False?
15. The flow of blood through the coronary vessels is: (a) independent of the aortic pressure; (b) greatly influenced by the pressures in the walls of the heart chambers; (c) continuous and constant throughout the cardiac cycle.
16. The cerebral bloodflow: (a) can be interrupted for long periods without damage to brain tissue; (b) varies in accordance with variations in mean arterial pressure; (c) may be influenced under certain circumstances by local autoregulatory mechanisms.
17. The dilation of skeletal muscle vessels: (a) may occur in response to the activity of sympathetic vasodilator neurons; (b) occurs only in pathological conditions; (c) is not influenced by local autoregulatory mechanisms.
18. Most of the cutaneous bloodflow is for: (a) temperature regulation; (b) provision of nutrients to the skin; (c) removal of wastes from the skin.
19. During exercise: (a) the stroke volume of the heart generally decreases; (b) the contractility of the cardiac muscle generally decreases; (c) the cardiac output generally increases.
20. Many cases of hypertension are believed to be caused by a deficient dietary intake of sodium or a deficient retention of sodium by the body, which leads to a decrease in blood volume and cardiac output. True or False?

DEFENSE MECHANISMS OF THE BODY

14

The body is continually exposed to a wide variety of potentially harmful factors, including bacteria, viruses, and hazardous chemicals. Pathogenic bacteria that invade the body often release enzymes that break down cell membranes and organelles or give off toxins that disrupt the functions of organs and tissues. Viruses can enter cells and utilize cellular facilities to reproduce more virus particles. They can kill cells by depleting them of essential components or by causing them to produce toxic substances.

Fortunately, the body is able to resist many organisms and chemicals that can damage tissues. This ability is called **immunity.** Some of the body's immunity is provided by nonspecific defense mechanisms that do not require previous exposure to a particular foreign substance in order to react to it. These mechanisms include the barriers formed by body surfaces and inflammatory responses that occur when tissues are injured. Immunity is also provided by specific immune responses that depend on prior exposure to a specific foreign material, recognition of it upon subsequent exposure, and reaction to it. These responses are mediated by specialized lymphocytes and cells derived from them.

BODY SURFACES

The skin and mucous membranes that form the body surfaces are the first barriers to the invasion of the body by potentially damaging factors. The intact skin prevents many microorganisms and chemicals from entering the body. The skin's sweat glands and sebaceous glands secrete substances that are toxic to many types of bacteria, and the outer portion of the epidermis contains a tough, water-insoluble protein called keratin that is resistant to weak acids and bases and many protein-digesting enzymes.

The mucous membranes that line body cavities opening to the exterior—for example, the digestive, respiratory, urinary, and reproductive tracts—secrete mucus, which entraps small particles that can then be swept away or engulfed by phagocytic cells. The mucous membrane of the upper respiratory tract contains ciliated cells that move mucus and its entrapped particles toward the throat (pharynx). The tears that bathe the mucous membranes of the eyes contain the enzyme *lysozyme*, which has antimicrobial activity, and the acidic gastric juice secreted by the stomach can destroy many bacteria and bacterial toxins.

Although the surface barriers are quite effective, from time to time potentially damaging agents or organisms do gain entrance to the body. It is then necessary for other defense mechanisms to attempt to neutralize or destroy them.

INFLAMMATION

In its most basic form **inflammation** is an acute, nonspecific, physiological response of the body to tissue injury caused by factors such as chemicals, heat, mechanical trauma, or bacterial invasion. Although some differences may be evident depending on the causative agent, the site of injury, and the state of the body, the fundamental events of the *acute, nonspecific inflammatory response* are similar in virtually all cases (Table 14.1). These events include changes in bloodflow and vessel permeability, as well as the movement of leukocytes from the vessels into the tissues. The acute, nonspecific inflammatory response brings into the invaded or injured area plasma proteins and phagocytes that can inactivate or destroy the invaders, remove debris, and set the stage for tissue repair. The events of this response are mediated

Table 14.1   *Events of Inflammation*

Event	Significance	Contributing Mediators or Mechanisms
Dilation of small blood vessels	Increases bloodflow to injured area; increases delivery of plasma proteins and phagocytic leukocytes	Histamine, kinins, complement components, prostaglandins
Increased permeability of small vessels (particularly capillaries and venules)	Plasma fluid and solutes, including proteins and other large molecules, move out of the circulatory system and into inflamed tissues	Histamine, kinins, complement components, prostaglandins
Slowing of bloodflow	Associated with margination and pavementing of leukocytes	Increased blood viscosity and clumping of erythrocytes due to loss of fluid and solutes
Walling off of injured area	Limits spread of toxic products or bacteria	Formation of fibrin in injured tissues
Margination and pavementing of leukocytes	Associated with emigration of leukocytes	Clumping of erythrocytes and slowing of bloodflow
Emigration of leukocytes	Leukocytes leave the circulation and enter injured tissues	Amebalike activity of leukocytes
Aggregation of leukocytes and macrophages at site of injury	Provides large numbers of phagocytic cells to engulf invading agents or organisms	Amebalike activity of leukocytes and macrophages attracted by chemotactic factors (for example, kinins, complement components)
Phagocytosis by leukocytes and macrophages	Removes invading agents or organisms and debris	Certain surface characteristics of materials to be phagocytized; opsonins

by a number of chemical substances that are released or generated in the injured area.

CHANGES IN BLOODFLOW AND VESSEL PERMEABILITY

Although a transient vasoconstriction may occur immediately following an injury to the tissues, small vessels in the injured area soon dilate. This dilation leads to an increased bloodflow to the area, which increases the delivery of plasma proteins and phagocytic leukocytes. In addition, the permeability of capillaries and venules in the area increases, and plasma fluid and solutes, including proteins and other large molecules, move out of the circulatory system and into the inflamed tissues. This movement is aided by the vasodilation that occurs, because the dilation of arterioles in the area leads to an increased pressure within capillaries and venules that favors the filtration of fluid and solutes out of the vessels. Moreover, the movement of proteins from the blood into the interstitial fluid due to the increased vessel permeability diminishes the difference in protein concentration between the plasma and the interstitial fluid, and thus favors the accumulation of fluid within the tissues.

As fluid and solutes move out of the circulatory system and into the tissues during inflammation, the viscosity of the blood increases and erythrocytes clump together. This increases the resistance to bloodflow and, as inflammation progresses, bloodflow through small vessels in the injured area slows and sometimes even stops.

WALLING-OFF EFFECT OF INFLAMMATION

One of the substances that moves from the blood into the tissue spaces during inflammation is fibrinogen. In the tissues, fibrinogen is converted to fibrin, forming a clot that walls off the injured area. This walling-off effect can delay or limit the spread of toxic products or bacteria.

LEUKOCYTE EMIGRATION

As erythrocytes clump together and the bloodflow slows during inflammation, a process known as *margination* takes place, and leukocytes—particularly neutrophils and monocytes—are displaced to the periphery of the bloodstream where they contact the endothelial linings of capillaries and venules in the inflamed area. At first the leukocytes slowly tumble or roll along the endothelial surfaces, but soon a phenomenon called *pavementing* occurs, and the leukocytes adhere to the vessel surfaces. The adhered leukocytes begin to exhibit an amebalike activity, and they squeeze between the endothelial cells of the capillaries and venules into the tissue spaces. The first leukocytes to arrive in the tissues are generally neutrophils, followed some time later by monocytes. In the tissues, monocytes are transformed into macrophages, and some of the macrophages normally present in the tissues multiply and become motile.

CHEMOTAXIS AND LEUKOCYTE AGGREGATION

Leukocytes and macrophages migrate through the tissues at varying rates of speed and aggregate at the site of injury or invasion. Their direction of movement is determined by chemical mediators released at the site of injury, and they are attracted to the highest concentrations of certain chemical substances. Such a movement of cells in response to chemical factors is called *chemotaxis*. Positive chemotaxis is a movement toward a chemical substance, and negative chemotaxis is a movement away from a chemical substance.

In most acute, nonspecific inflammatory responses, leukocytes and macrophages aggregate at the site of injury or invasion in a fairly predictable sequence. Neutrophils usually predominate in the early stages of the response, and monocytes and macrophages predominate during the later stages. Neutrophils are present in greater numbers in the circulation and are more mobile than monocytes, and this probably contributes to their initial predominance. However, in the tissues, neutrophils die off faster than monocytes and macrophages. Moreover, neutrophils may produce or potentiate factors that facilitate monocyte emigration. These facts are believed to contribute to the predominance of monocytes and macrophages later in inflammation.

In contrast to the dominance of neutrophils, monocytes, and macrophages in most acute, nonspecific inflammatory responses, certain types of allergies and inflammatory responses to parasites are characterized by the presence of large numbers of eosinophils. In fact, a preponderance of eosinophils is often indicative of an allergic response.

PHAGOCYTOSIS

A major benefit of inflammation is the phagocytosis of microorganisms, foreign materials, and debris by leukocytes, particularly neutrophils, and macrophages. The phagocytic cells identify the materials to be engulfed by "recognizing" certain surface characteristics of the materials. Roughness and positive charges have been suggested as surface characteristics that enhance the likelihood of phagocytosis, but exactly which characteristics are important in the acute, nonspecific inflammatory response, and how they trigger phagocytosis is not certain. In any event, a phagocytic cell attaches to a particle and engulfs it by endocytosis, forming a membrane-bounded phagocytic vesicle that contains the engulfed particle within the phagocytic cell.

Lysosomes containing powerful digestive enzymes attach to the phagocytic vesicle, forming a digestive vacuole, and the lysosomal enzymes are released into the vacuole. Since the lysosomes appear microscopically as granules, this process is called *degranulation*. Neutrophils also produce a significant amount of hydrogen peroxide, which can enter the digestive vacuole where it is very destructive to many organic substances. The lysosomal enzymes and hydrogen peroxide break the engulfed material down into products of low molecular weight that can be utilized by the phagocytic cell. Macrophages, but apparently not neutrophils, can also extrude residual breakdown products from the cells.

Phagocytic cells continue to ingest and digest foreign particles until toxic substances from the foreign particles and the hydrolytic enzymes released by the lysosomes accumulate and kill the cell. Macrophages are larger cells that are less selective about what they engulf than are neu-trophils, and they generally phagocytize more particles before they die. As a result, macrophages are particularly important in clearing an inflamed area of foreign material and tissue debris.

The phagocytic activity of leukocytes and macrophages can be modified by many factors, including a number of proteins collectively called *opsonins*. Opsonins coat specific foreign particles and thereby render them more susceptible to phagocytosis.

In addition to their phagocytic activities, neutrophils can release lysosomal granules directly into the interstitial fluid. The lysosomal enzymes digest extracellular material, making it easier for macrophages to engulf it.

ABSCESS AND GRANULOMA FORMATION

The ultimate goal of the inflammatory response is to overcome the injury or invasion of the body and clear the area for tissue repair. Occasionally, however, the inflammatory response is unable to overcome an invasion, and an abscess or a granuloma forms. An *abscess* is a sac of pus that consists of microbes, leukocytes, macrophages, and liquified debris walled off by fibroblasts or collagen. It does not spontaneously diminish and must be drained. A *granuloma* can form when the invading agents are microbes that can survive within the phagocytes or are materials that cannot be digested by the phagocytes. In such cases, layers of phagocytic-type cells form and are surrounded by a fibrous capsule. The central cells contain the offending agent. In this fashion, a person may harbor live bacteria such as tuberculosis-causing bacteria for many years without displaying any overt symptoms.

CHEMICAL MEDIATORS OF INFLAMMATION

Although tissue injury precipitates the acute, nonspecific inflammatory response, a variety of chemical substances mediate it. Some of these substances are released from tissue cells in the injured area, some are released from leukocytes, and some are generated by enzyme-catalyzed reactions.

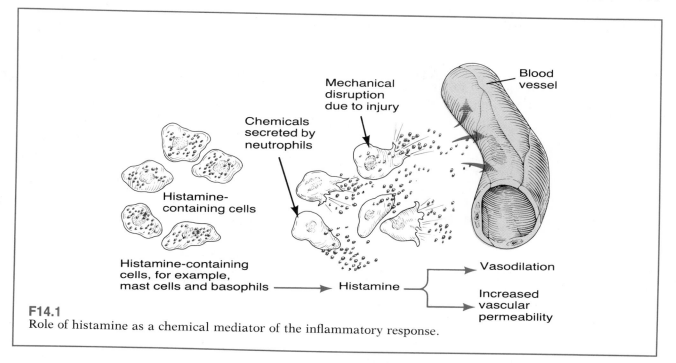

F14.1
Role of histamine as a chemical mediator of the inflammatory response.

Histamine

Histamine is present in many cells, particularly mast cells, basophils, and platelets (F14.1). It causes vasodilation (particularly of arterioles) and increased vascular permeability (particularly in venules). Several factors cause histamine release, including mechanical disruption of cells due to injury and chemicals secreted by neutrophils attracted to the site.

Kinins

The *kinins* are a group of polypeptides that dilate arterioles, increase vascular permeability, act as powerful chemotactic agents, and induce pain (F14.2). They are formed from inactive precursors called kininogens that normally circulate in the plasma. The enzyme *kallikrein*, which is also present in the plasma in inactive form (prekallikrein), acts on kininogens to form kinins. Multiple factors can convert prekallikrein to kallikrein. Among them is active factor XII, which is also involved in blood clotting. Moreover, kallikrein is present in many body tissues, as well as in neutrophils, and once inflammation has begun,

kallikrein from these sources contributes to kinin generation. In fact, the kinin-generating system displays a positive feedback in that kinins act chemotactically to attract neutrophils whose kallikrein can catalyze the production of still more kinins. In addition, tissue kallikrein cannot normally act on plasma kininogens because the kininogens are too large to pass out of the circulatory system. In inflammation, the kinins increase vascular permeability, allowing kininogens to enter the tissues and become activated by kallikrein.

Complement System

The *complement system* is particularly important in specific immune responses, but it can also contribute to acute, nonspecific inflammation. This system consists of a series of plasma proteins that normally circulate in the blood in inactive forms (F14.3). The activation of the system's initial components initiates a cascading series of reactions that ultimately produce active complement molecules.

Active complement components can mediate virtually every event of the acute, nonspecific inflammatory response. Complement components enhance vasodilation and vascular permeability by direct effects on blood vessels, by stimulating histamine release from mast cells and platelets, and by activating plasma kallikrein. They act as chemotactic agents in attracting neutrophils, and as opsonins that enhance phagocytosis by attaching to microbes, where they serve as structures to which phagocytic cells can bind. In addition, complement components directly attack invading microbes and kill them without prior phagocytosis. The complement components involved in this ac-

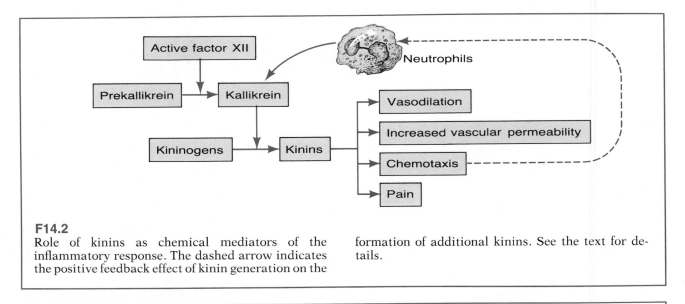

F14.2
Role of kinins as chemical mediators of the inflammatory response. The dashed arrow indicates the positive feedback effect of kinin generation on the formation of additional kinins. See the text for details.

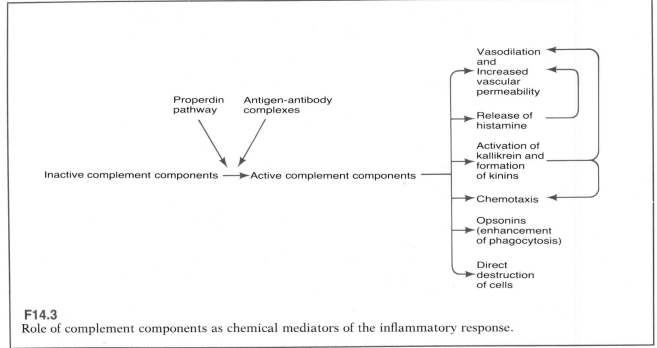

F14.3
Role of complement components as chemical mediators of the inflammatory response.

tivity become embedded in the surface of a microbe and kill it by forming tunnels that make the microbe leaky.

In acute, nonspecific inflammation, active complement components are generated by a pathway initiated by the protein *properdin*. Properdin, which is normally present in the plasma, can bind to particular carbohydrate chains that are present on certain bacterial or viral surfaces but not on the body's own cells. The binding alters the conformation of properdin, enabling it to initiate the enzymatic splitting and activation of complement component C_3. This, in turn, leads to the formation of other active complement components.

Prostaglandins

Prostaglandins may also play a role in inflammatory responses. For example, prostaglandin PGE_1 induces vasodilation and increased vascular permeability, and it may stimulate leukocyte migration through capillary walls. Prostaglandins are also believed to contribute to the pain associated with inflammation, and they may mediate the fever-producing effects of substances called endogenous pyrogens that are released from leukocytes during inflammatory responses.

Leukotrienes

Leukotrienes are a group of substances produced by leukocytes. They are chemically related to the prostaglandins. Certain leukotrienes apparently cause increased vascular permeability, and some seem to be chemotactic agents that attract neutrophils. Researchers are actively studying the leukotrienes, and, as more information about their actions is obtained, it seems likely that they will be found to play a significant role in inflammation and perhaps other activities.

SYMPTOMS OF INFLAMMATION

Inflammation can produce both local and systemic symptoms (Table 14.2).

Local Symptoms

The main local symptoms of inflammation are redness, heat, swelling, and pain. The redness and heat (increased temperature) are due to the dilation of blood vessels and increased bloodflow to

the inflamed area. The swelling (edema) is caused by the increased vessel permeability and the movement of fluid and solutes from the circulatory system into the tissue spaces. The pain has been attributed to the increased pressure exerted on sensory nerve endings in the swollen tissues and to the effects of kinins and perhaps other chemical mediators of the inflammatory response on afferent nerve terminals.

Systemic Symptoms

Fever is a common systemic symptom of inflammation, especially in inflammatory responses associated with the spread of organisms into the bloodstream. Neutrophils and perhaps other cells involved in the inflammatory response release endogenous pyrogens, which influence the body's

Table 14.2	*Symptoms of Inflammation*
Symptom	Contributing Causes
LOCAL Redness	Dilation of vessels, increased bloodflow to injured area
Heat (increased temperature)	Dilation of vessels, increased bloodflow to injured area
Swelling (edema)	Increased vessel permeability, and movement of fluid and solutes from the circulatory system into the tissue spaces
Pain	Increased pressure on sensory nerve endings in swollen tissues; effects of some chemical mediators of inflammation (for example, kinins) on afferent nerve terminals
SYSTEMIC Fever	Endogenous pyrogens released from neutrophils
Increased production and release of leukocytes	Granulopoietin released from monocytes and macrophages

temperature-regulating mechanism and induce fever. Although an extremely high fever can be harmful, phagocytosis is favored by higher body temperatures. Thus, fever (as well as the local increase in temperature at the site of inflammation) may enhance this activity. In addition to eliciting fever, the endogenous pyrogens also cause iron and other trace metals to move out of the plasma and into the liver. This action may retard the proliferation of bacteria that require a high concentration of iron to multiply.

Inflammation, particularly in response to many types of bacterial invasion, is often accompanied by an increased production and release of leukocytes. The leukocyte count in the peripheral blood may rise from about 7000 cells/mm³ to 25,000 or more, and in most nonspecific inflammations the increase is due to an absolute as well as a relative rise in the number of circulating neutrophils. Certain bacterial products induce monocytes and macrophages to release a substance called *granulopoietin*, which enhances the synthesis and release of granulocytes (particularly neutrophils) by the bone marrow. In contrast, infections caused by viruses and protozoa are often associated with decreased numbers of leukocytes in the blood.

INTERFERON

The protein **interferon** provides some protection to the body against invasion by viruses, particularly until the more slowly reacting specific immune responses can take over. In general, when a virus invades the body, it enters a cell and reproduces more virus particles by utilizing cellular facilities to synthesize viral components, including nucleic acids and proteins. The new viruses are released to infect other, healthy cells and repeat the process. However, viruses (as well as other substances) also induce cells to produce interferon. One of the most effective inducers of interferon production is double-stranded RNA, and it has been suggested that viruses and other interferon inducers cause the production of double-stranded RNA within cells. The double-stranded RNA is believed to trigger the DNA of a cell to produce a messenger RNA that directs the pro-

duction of interferon by the cell. The interferon leaves the cell and binds to receptors on the cell membranes of other cells. This binding triggers the synthesis of enzymatic proteins within the cells that can prevent viral reproduction by breaking down messenger RNAs and inhibiting protein synthesis. The newly synthesized enzymatic proteins are inactive until the cells are infected by a virus or exposed to double-stranded RNA. This activation requirement may help protect the normal nucleic acid and protein-synthesizing mechanisms of the cells from inhibition in the absence of viral infection.

It is thought that interferon inhibits the activity of some viruses by other mechanisms. Although interferon apparently cannot prevent the reproduction of certain viruses, the viruses are either not released from the cells or, if they are released, they are not capable of infecting other cells.

All cells can manufacture interferon when they are appropriately stimulated, and at least three different forms of interferon are produced. The principal form made by leukocytes is not the same as the major form produced by fibroblasts. A third form, T or immune interferon, is produced by T cells, which are involved in specific immune responses.

Interferon may play a role in protecting the body against some forms of cancer. Although viruses are suspected of causing some human cancers, interferon also seems to work against tumors thought to be caused by nonviral agents such as radiation and chemicals. In fact, interferon is increasingly viewed as a versatile agent whose effects on cells go beyond the inhibition of viral activity. For example, it slows cell division and inhibits the proliferation of both healthy and abnormal cells. It mobilizes and enhances the cell-killing properties of macrophages, and it stimulates the activity of a special group of effector cells called natural killer (NK) cells. Both macrophages and natural killer cells can attack tumor cells, and natural killer cells are believed to be important in *immune surveillance*—that is, in detecting and eliminating abnormal cells that may be cancerous before they multiply and cause clinical cancer. In fact, it has been suggested that clinical cancer is due to ineffective immune surveillance mechanisms. Immune interferon appears to be a more potent antitumor agent than the major types of interferon produced by leukocytes or fibroblasts.

SPECIFIC IMMUNE RESPONSES

The **specific immune responses** generally require previous exposure to a foreign agent or organism to be most effective. The specific immune responses are not always beneficial, however, and they are sometimes responsible for allergies and other harmful reactions.

Traditionally, two major aspects of the specific immune responses are recognized. One aspect, known as **humoral immunity,** is mediated by specialized proteins called antibodies. Antibodies are produced by plasma cells that are progeny of lymphoid cells called B cells. The second aspect, known as **cell-mediated immunity,** is mediated by certain populations of lymphoid cells called T cells.

ANTIGENS

An **antigen** is a substance that elicits a specific immune response when it is present in the body. Antigenic substances generally have molecular weights of 8,000 to 10,000 or more, and proteins, polysaccharides, complex lipids, and nucleic acids commonly act as antigenic materials. Lower molecular weight substances that are collectively called *haptens* can also elicit specific immune responses if they combine with larger molecules such as proteins.

B CELLS

B cells are lymphocytes that are committed to differentiate into the antibody-producing plasma cells involved in humoral immunity. Like all lymphocytes, B cells are derived from stem cell precursors in the red bone marrow (F14.4). Stem cells that are destined to give rise to B cells proliferate and develop within the bone marrow (or within the liver and perhaps the spleen during fetal life). B cells continually circulate through the blood,

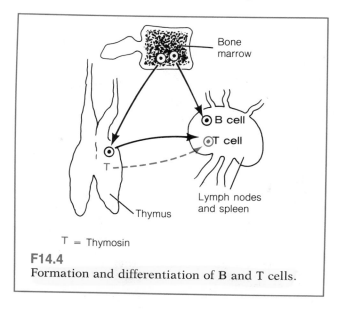

F14.4
Formation and differentiation of B and T cells.

tissues, and lymph, and at any one time large numbers of B cells are localized in the lymph nodes, spleen, and other lymphoid tissues.

T CELLS

T cells are a heterogeneous group of lymphocytes that include those cells committed to participating in cell-mediated immune responses. Lymphocyte stem cells from the red bone marrow that are destined to give rise to T cells proliferate and differentiate in the thymus gland (F14.4). In addition, hormonal substances from the thymus— for example, thymosin—are believed to influence T cells that have left the gland. Like B cells, T cells continually circulate through the blood, tissues, and lymph, and at any one time large numbers of T cells are localized in the lymph nodes, spleen, and other lymphoid tissues.

ANTIBODIES

An **antibody** is a specialized protein that is produced in response to the presence of an antigen. An antibody can combine with the specific antigen that stimulated its production to form an *antigen–antibody complex.* Antibodies belong to a family of proteins called globulins, and they are referred to as *immunoglobulins (Ig).* Five immunoglobulin classes of antibodies are distinguished.

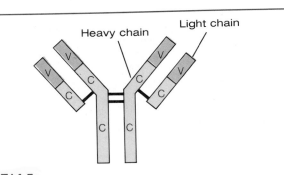

F14.5
Diagrammatic representation of the basic structure of an antibody. The structure is composed of four polypeptide chains, two heavy and two light. Many regions of the chains are of constant composition (C) among antibodies of a particular class. Only relatively small regions vary (V) in composition from antibody to antibody.

Immunoglobulin classes G and M (IgG and IgM) are involved in specific immunity against bacteria and viruses. Immunoglobulins of class E (IgE) are involved in certain allergic responses. Immunoglobulins of class A (IgA) are released from lymphoid tissues that line the gastrointestinal, urogenital, and respiratory tracts and are involved in protecting these areas. Immunoglobulins of class D (IgD) are present on the surfaces of many B cells and are believed to be involved in the initiation of humoral immune responses.

The basic structure of an antibody consists of two "heavy" and two "light" polypeptide chains (F14.5). One end of each chain contains a variable region whose amino acid sequence differs for each kind of antibody. The particular conformation of the variable regions enables an antibody to bind to a specific antigen in a manner that is analogous to the "lock and key" interaction between an enzyme and its substrate. The remainder of each polypeptide chain is called the constant region. The constant regions contain binding sites for molecules and cells that function as effectors of antibody-mediated activities. The constant regions of the heavy polypeptide chains differ from one class of antibody to another, and these differences provide the basis for distinguishing the various classes of antibodies.

Antibodies inactivate or destroy foreign substances in a number of ways. They can combine with bacterial toxins or destructive foreign enzymes and inactivate them by inhibiting their interactions with target cells or substrates. They can also bind to surface components of viruses, preventing the attachment of the viruses to target cells and thereby keeping the viruses from entering the cells.

A major function of antibodies is to enhance the basic inflammatory response. Antigen–antibody complexes, particularly those involving antibodies of the IgG or IgM classes, are powerful activators of the complement system (F14.3). As noted earlier, active complement components can mediate virtually every aspect of the acute, nonspecific inflammatory response, and they can also directly attack and kill invading microbes. Antibodies, particularly those of the IgG class, also act as opsonins that enhance phagocytosis by attaching to foreign substances where they serve as structures to which phagocytic cells can bind.

HUMORAL (ANTIBODY) IMMUNITY

Humoral immunity involves the production and release into the blood and lymph of antibodies to various antigens that the body recognizes as foreign. Humoral immune responses are important in providing specific immune resistance to most bacteria, bacterial toxins, and the extracellular phase of viral infections. The humoral system responds to an antigen as follows.

When an antigen reaches lymphoid tissues such as lymph nodes or the spleen, a tiny fraction of the B lymphocytes are stimulated to undergo rapid cell division (F14.6). Most of the stimulated cells develop into *plasma cells*, which produce antibodies that are released into the blood and lymph. The antibodies then combine with the specific antigen that stimulated their production, and responses such as those described in the previous section occur.

Some of the stimulated B cells do not differentiate into plasma cells, but instead form *memory cells*, which provide the humoral system with a "memory" of the exposure to the antigen. The sys-

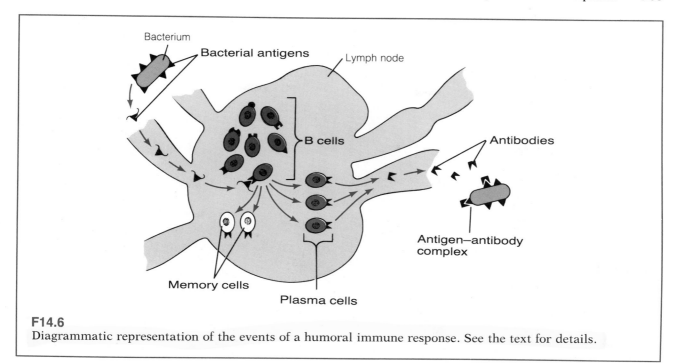

F14.6

Diagrammatic representation of the events of a humoral immune response. See the text for details.

tem responds rather slowly following an initial antigen exposure, and usually several days are required to build up substantial levels of antibodies (F14.7). Because of the memory component, however, a subsequent exposure to the same antigen produces a very rapid outpouring of antibodies.

According to one theory of antibody production, different antigens stimulate different populations, or clones, of B cells (see F14.6), and each clone can ultimately produce one specific antibody. The particular antibody a given clone of B cells can produce is displayed on the surface of the cells. When an antigen enters the system, it combines with an antibody specific for it that is displayed on the surfaces of particular B cells. This occurrence stimulates those B cells to undergo division and differentiation into plasma cells that manufacture the antibody. According to this theory, there are sufficient clones of B cells to allow for the production of antibodies to all of the millions of antigens to which a person might ever be exposed.

In actual fact, B cell stimulation and antibody production usually involve more than just an exposure to an antigen specific for the B cell's antibody. In most cases, macrophages and a group of T cells called *helper T cells* are also involved. According to some researchers, macrophages process and present the antigen to helper T cells, which in turn interact with B cells to promote their stimulation.

CELL-MEDIATED IMMUNITY

The cell-mediated immune responses depend on T cells, and particularly on a group of T cells called *effector T cells*. These responses are especially effective against cells that possess surface antigens that the body recognizes as foreign. For example, most virus-infected cells acquire viral antigens on their surfaces, and effector T cells produced by the cell-mediated immune system in response to the antigens can attack and destroy the virus-infected cells. Thus, cell-mediated immune responses are important in providing specific immune resistance to the intracellular phase of viral infections. They are also important in providing specific immune resistance to fungi, parasites, and intracellular bacteria, and they play a major role

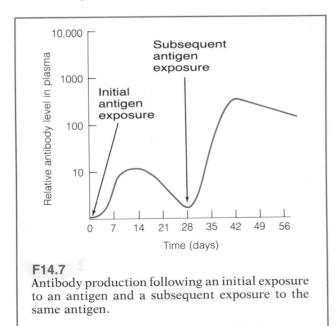

F14.7
Antibody production following an initial exposure to an antigen and a subsequent exposure to the same antigen.

in the rejection of solid-tissue transplants. The cell-mediated immune system responds to an antigen as follows.

Like the B cells involved in humoral immunity, T cells are clonal, and they possess specific surface receptors (F14.8). When an antigen combines with receptors displayed on the surfaces of particular T cells (macrophages and perhaps helper T cells are thought to be involved in the process), the T cells become sensitized and undergo division and differentiation. Some of the T cell progeny serve as a memory component of the system, and others—the effector T cells—participate directly in the cell-mediated immune response. In this response, effector T cells leave the lymphoid tissues and travel throughout the body. When they encounter cells bearing the specific antigen that initiated their production, they combine with them (F14.9). Some effector T cells can lyse and thereby kill cells that have the antigen on their surfaces. Others release a number of different chemicals that are collectively called *lymphokines*. These chemicals enhance the basic inflammatory response and subsequent phagocytosis. Some lymphokines are chemotactic, attracting to the area neutrophils and, especially, monocytes, which can be converted to macrophages. Others keep macrophages

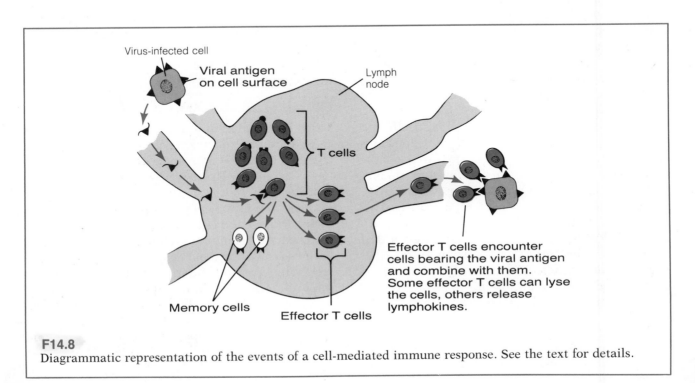

F14.8
Diagrammatic representation of the events of a cell-mediated immune response. See the text for details.

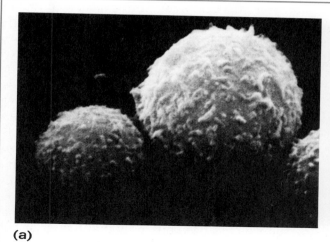
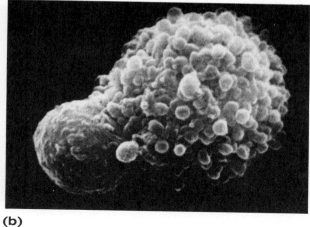

F14.9
(a) Effector T cell (smaller sphere at left) attaches to a cell recognized as foreign. (b) Destruction of the foreign cell is evidenced by the deep folds in its surface membrane (x 7250).

already in the area at an increased level of phagocytic activity (these active cells are called "angry" macrophages). Moreover, some lymphokines may act as cytotoxins and kill cells directly.

Transplant Rejection

The cell-mediated immune system plays an important role in the rejection of solid-tissue transplants (F14.10). Antigens called *histocompatibility antigens* are present on the surfaces of the nucleated cells of the body, and different individuals—except identical twins—possess different histocompatibility antigens. When a tissue from one individual is transplanted into another individual who is not an identical twin, the histocompatibility antigens on the cells of the donor's tissue are recognized as foreign, and T lymphocytes of the recipient become sensitized to them. This sensitization leads to the formation of effector T cells that attach to the antigens on the cells of the donor's tissue and destroy the cells by the mechanisms of cell-mediated immunity. In transplant surgery, the ability of T cells to respond in this fashion is weakened by drugs, antilymphocyte serum, or radiation. In many cases B cell functions are also suppressed, and the individual becomes very susceptible to infection.

ACTIVE IMMUNITY

A resistance to infection that results from an antigen-induced activation of an individual's specific immune responses is called *active immunity*. Most commonly, active immunity is acquired by infection. Alternatively, active immunity can be acquired through vaccination, in which a person is injected with a small amount of antigenic material. The material may be bacterial, a toxin, or other substance. It is usually pretreated by drying, ultraviolet light, or some other means so that it is not strong enough to cause disease, but still acts as an antigen that stimulates specific immune responses.

Following the activation of an individual's specific immune responses by infection or vaccination, antibodies or effector T cells may remain present to resist the disease-causing agent for some time. In addition, if an individual who has had an infection or has been vaccinated is exposed to the disease-causing agent at a later time, the memory components of the humoral and cell-mediated immune systems enable them to respond rapidly. It should be noted, however, that some microorganisms apparently do not activate the memory components of the systems, and the response to each exposure to one of these organisms is the same as the slowly developing response to the initial exposure.

PASSIVE IMMUNITY

Passive immunity is essentially "borrowed" immunity. For example, antibodies to a disease-causing agent can be transferred into an individual from sources (often nonhuman) that have been exposed to the agent. Passive immunity is of relatively short duration—generally only several weeks.

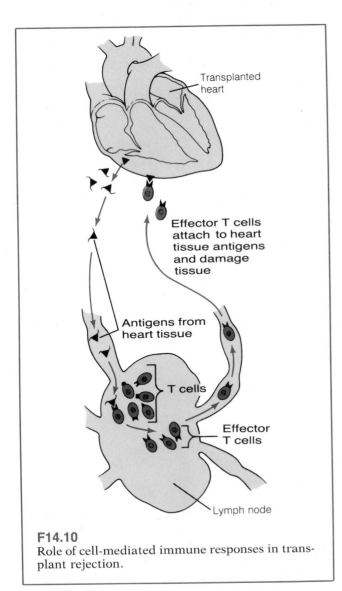

F14.10
Role of cell-mediated immune responses in transplant rejection.

Labels in figure:
Transplanted heart

Effector T cells attach to heart tissue antigens and damage tissue

Antigens from heart tissue

T cells

Effector T cells

Lymph node

Moreover, the recipient's own specific immune mechanisms are not stimulated to produce antibodies or effector T cells against the disease-causing agent. A common instance of passive immunity occurs during pregnancy. Antibodies of the IgG class formed by the mother move across the placenta to the fetus. These antibodies protect the infant during the period just after birth when the infant's own ability to produce antibodies is relatively poor. Passive immunity can also be provided by injecting antibodies into an individual. For example, antibodies to the venom of a particular type of snake can be built up in a horse, recovered from the horse plasma, and injected into a person who has been bitten by that type of snake in order to provide immediate protection against the venom. A possible danger of such a procedure, however, is that injected antibodies may themselves be antigenic in the recipient's system—especially if they were obtained from nonhuman sources. Thus, the injected antibodies may themselves stimulate the recipient to produce antibodies against them. This occurrence can lead to severe allergic responses.

HYPERSENSITIVITY

Hypersensitivities are immune responses that can be harmful to the body. Both the humoral and cell-mediated immune systems participate in these responses.

Role of Complement Components in Hypersensitivity

In some types of hypersensitivity—for example, immune complex hypersensitivity and cytotoxic hypersensitivity—complement components generated during a humoral immune response to an antigen damage body tissues and cells. For example, immune complexes (antigen–antibody complexes) can become deposited in blood vessel walls where they activate the complement system and initiate an inflammatory response that damages the vessels. The deposition of immune complexes in the kidneys can lead to an inflammatory condition known as glomerulonephritis (see Chapter 18).

Role of Effector T Cells in Hypersensitivity

In the type of hypersensitivity called cell-mediated (delayed) hypersensitivity, the activities of effector T cells during a cell-mediated immune response cause damage to body tissues. For example, the toxin of poison ivy is a hapten that can bind to proteins or cell membranes in the skin. Consequently, exposure to poison ivy toxin can sensitize an individual and cause the production of effector T cells against the toxin. If such a sensitized individual is again exposed to poison ivy toxin, effector T cell responses (including the release of lymphokines) induce a tissue-damaging inflammatory reaction at the site of exposure that is characterized by itching, swelling, and vesication. Hypersensitivity reactions involving effector T cells usually take from one to three days to reach their peak intensities.

Role of Class IgE Antibodies in Hypersensitivity

Class IgE antibodies are involved in a type of hypersensitivity called anaphylactic hypersensitivity. This type of hypersensitivity is the cause of many common allergic reactions. In these reactions, the exposure of a susceptible individual to a particular antigen—for example, dust, pollen, or a particular food—sensitizes the individual and causes the production of class IgE antibodies from plasma cells. These antibodies bind to mast cells and basophils. When such a sensitized person is again exposed to the antigen, the antigen attaches to the antibodies bound to the mast cells and basophils. This attachment stimulates the release of histamine and other chemicals that induce inflammatory responses such as vasodilation and increased vascular permeability as well as increased mucous secretion and the contraction of smooth muscles of the airways of the lungs.

Often, an anaphylactic-type allergic reaction is localized to a particular site in the body. For example, if the exposure to a particular antigen occurs in the nasal area, the result may be sneezing, runny nose, congestion, and other symptoms of hay fever due to the irritation, increased mucous secretion, increased bloodflow, and increased protein leakage in the area. Occasionally (particularly in response to injected antigens), a systemic rather than a localized anaphylactic allergic reaction occurs, and severe hypotension and airway constriction quickly develop. This reaction is called *anaphylactic shock*, and it can be fatal if countermeasures such as the injection of epinephrine are not taken promptly. Some sensitized people develop anaphylactic shock in response to the antigen in a single insect sting or in response to drugs such as penicillin.

The administration of antihistamines often provides some relief from anaphylactic-type allergic reactions, but the relief is frequently incomplete because chemicals other than histamine are also released from mast cells and basophils. When the particular antigen to which a person is sensitive has been identified, desensitization therapy may be utilized. Desensitization consists of injecting an individual with small but increasing quantities of the offending antigen. Some researchers think this treatment results in the production of class IgG rather than class IgE antibodies to the antigen. According to this view, when the person again encounters the particular antigen, the class IgG antibodies attach to it and prevent the antigen from attaching to class IgE antibodies bound to mast cells and basophils. Alternatively, it has been proposed that desensitization causes a group of T cells called *suppressor T cells* to suppress the synthesis of class IgE antibodies directed against the antigen. In either case, the antigen does not attach to class IgE antibodies that are bound to mast cells and basophils, and the anaphylactic-type allergic reaction does not occur.

TOLERANCE

Tolerance can be generally defined as the failure of the body to mount a specific immune response against a particular antigen. For example, the body is composed of proteins and other substances that are antigenic. Normally, however, the body exhibits self-tolerance and does not produce antibodies or effector T cells against its own antigens.

Tolerance is due to the elimination, inactivation, or suppression of specific B cells and T cells as a result of appropriate antigenic exposure. Thus, under proper conditions, exposure to an antigen can prevent rather than initiate a specific immune response. Tolerance is more easily established in the fetus than in the adult. In this regard it should be noted that during fetal development a person is exposed almost exclusively to his or her own antigens. During this time, these antigens are believed to react with and paralyze or eliminate the particular clones of B cells and T cells that could produce specific immune responses against them. However, B cells and T cells continue to differentiate in the bone marrow and thymus throughout a person's lifetime, and the maintenance of tolerance to antigens such as self-antigens requires an active and continuous inhibition of those B cells and T cells that could produce specific immune responses against them.

Two general mechanisms—clonal deletion and clonal suppression—are believed to be responsible for tolerance. In *clonal deletion*, exposure to an antigen eliminates or inactivates those clones of B cells and T cells specific for the antigen. Exactly how this occurs is not completely understood. However, under appropriate conditions, the exposure of B cells to an antigen apparently results in the removal or modulation of the receptors for the antigen on the surfaces of particular B cells. The removal or modulation of the receptors renders the cells unresponsive, or tolerant, to the antigen. Immature B cells appear to be much more sensitive to such an occurrence than mature B cells, and this may help explain why tolerance is more easily established in the fetus than in an adult.

In *clonal suppression*, an appropriate exposure to an antigen activates suppressor T cells, which in turn block the activity of helper T cells involved in initiating specific immune responses against the antigen. The suppressor T cells also block the development of effector T cells responsible for cell-mediated immune responses against the antigen, and they may directly suppress the activities of B cells that could produce a humoral immune response against the antigen. Thus, in clonal suppression, B cells and T cells that can produce specific immune responses against particular antigens remain present, but their activities are blocked by suppressor T cells.

AUTOIMMUNE RESPONSES

Occasionally, the body's tolerance to its own antigens breaks down, and *autoimmune responses* occur—that is, a person produces antibodies or effector T cells that attack his or her own tissues. There are several causes of such occurrences. Drugs, environmental chemicals, viruses, or genetic mutations may cause the formation of new or altered antigens on cell surfaces. The body treats these antigens as foreign, and produces antibodies or effector T cells against them. Alternatively, foreign antigens that are structurally very similar to some of the body's own antigens may stimulate the production of antibodies or effector T cells that cross-react with body antigens. For example, certain streptococcal bacteria possess antigens that induce the formation of antibodies that cross-react with heart tissue, and severe, recurrent infections caused by streptococci sometimes lead to the development of rheumatic fever several weeks after the infection has subsided, suggesting an autoimmune response. Still another possibility is that certain body antigens are not normally exposed to the humoral or cell-mediated immune systems, and the body does not become tolerant to them. If, at some time, tissue disruption following, for example, injury or infection exposes these antigens, the body treats them as foreign and produces antibodies or effector T cells against them. For example, antigens of the cornea of the eye do not seem to circulate in the body fluids. Should damage to the cornea expose these antigens, specific immune responses against them may cause corneal opacity.

Suppressor T cells are thought to be particularly important in autoimmunity. If suppressor T cells are involved in tolerance (and particularly in preventing humoral and cell-mediated immune responses against a person's own antigens) as was previously suggested, then a deficiency of these cells could contribute to the development of autoimmune responses.

Table 14.3	Summary of the A-B-O system	
Blood Type	Antigens (Agglutinogens) on Erythrocytes	Antibodies (Agglutinins) in Plasma
A	A	Anti-B
B	B	Anti-A
AB	Both A and B	Neither anti-A nor anti-B
O	Neither A nor B	Both anti-A and anti-B

BLOOD GROUPS AND TRANSFUSION

A-B-O SYSTEM

The surfaces of erythrocytes contain particular antigens (agglutinogens) that react with the appropriate antibodies (agglutinins). The reactions form the basis of various blood-typing classifications. The most often considered of these surface antigens are those of the A-B-O system (Table 14.3). The antigens of the A-B-O system, which are inherited, are designated A and B, and the lack of either A or B antigens on the erythrocytes is designated O. A given individual may have either antigen A, antigen B, both antigens A and B (AB), or neither antigen A nor B (O).

Antibodies against antigens A or B begin to build up in the plasma shortly after birth. The antibody levels peak at about 8 to 10 years of age, and the antibodies remain present in declining amounts throughout the rest of life. The mechanism that stimulates the production of anti-A and anti-B antibodies is unclear, but it has been suggested that antibody development is initiated by small amounts of A- and B-type antigens that enter the body in the food, in bacteria, or by other means. A person normally produces antibodies against those antigens that are not on his or her erythrocytes, but does not produce antibodies against those antigens that are present on his or her erythrocytes. Thus, a person with antigen A has anti-B antibodies; a person with antigen B has anti-A antibodies; a person with neither antigen A nor B (O) has both anti-A and anti-B antibodies; and a person who possesses both antigens A and B has neither anti-A nor anti-B antibodies. The individual's blood type indicates the antigens he or she possesses and not the antibodies.

TRANSFUSION

The mixing of incompatible blood types can cause erythrocyte destruction and other problems. For example, if a person with type A blood (antigen A on erythrocytes, anti-B antibodies in plasma) is transfused with type B blood (antigen B on erythrocytes, anti-A antibodies in plasma), the recipient's anti-B antibodies will attack the incoming type B erythrocytes. The type B erythrocytes will be agglutinated (clumped), and hemoglobin will be released into the plasma—that is, the cells will undergo hemolysis. Incoming anti-A antibodies of the type B blood may also attack the type A erythrocytes of the recipient, with similar results. However, unless large amounts of blood are transfused, this problem is usually not as serious because the incoming antibodies are diluted in the recipient's plasma.

During a transfusion reaction resulting from the mixing of incompatible blood types, agglutinated erythrocytes can plug small blood vessels. When hemolysis is rapid, the released hemoglobin can precipitate in the urine-forming structures of the kidneys and contribute to kidney failure.

Since type AB individuals possess neither anti-A nor anti-B antibodies to attack incoming erythrocytes, they are often called universal recipients. On the other hand, since the erythrocytes of type O individuals will not be attacked by either anti-A or anti-B antibodies, these people are called universal donors. However, these terms are misleading because other erythrocyte antigens and plasma antibodies can cause transfusion problems. If possible, therefore, blood for transfusion should be closely matched to the blood of the potential recipient.

Rh SYSTEM

Another example of an erythrocyte antigen–antibody system is the Rh system (so named because of its initial study in rhesus monkeys). The Rh system consists of a group of surface antigens on erythrocytes, and certain Rh antigens are very likely to cause transfusion reactions. An individual who possesses these antigens is designated Rh positive and an individual who lacks them is designated Rh negative. The antibody components of the system—the anti-Rh antibodies—are not normally present in the plasma, but anti-Rh antibodies can be produced upon exposure and sensitization to Rh antigens. In general, an individual does not produce anti-Rh antibodies against Rh antigens that are present on his or her erythrocytes, but upon sensitization a person can produce anti-Rh antibodies against Rh antigens that are not on his or her erythrocytes.

Sensitization can occur if Rh positive blood is transfused into an Rh negative recipient. It can also occur when an Rh negative mother carries a fetus who is Rh positive (due to the presence of Rh antigens inherited from the father). In this case, some of the fetal Rh antigens may enter the maternal circulation and sensitize the mother so that she begins to produce anti-Rh antibodies against the fetal antigens. The most likely time for sensitization to occur is near the time of birth, but because it takes some time for the mother to build up anti-Rh antibodies, the first Rh positive child carried by a previously unsensitized Rh negative mother is usually unaffected. However, if an Rh negative mother who has been sensitized by transfusion or by a previous Rh positive pregnancy subsequently carries an Rh positive fetus, maternal anti-Rh antibodies may enter the fetal circulation and cause the agglutination and hemolysis of fetal erythrocytes. This can result in an anemic condition known as hemolytic disease of the newborn (erythroblastosis fetalis). In this condition, the breakdown of the hemoglobin released during hemolysis results in the formation of sufficient bilirubin to give the infant's skin a yellow color—that is, jaundice. In some cases, bilirubin is deposited in nerve cells, causing brain damage. The usual treatment for severe hemolytic disease of the newborn resulting from Rh incompatibility is to remove the infant's Rh positive blood and replace it with Rh negative blood from an unsensitized donor. The Rh negative erythrocytes will not be attacked by the maternal anti-Rh antibodies in the infant's system, and the replacement of the infant's blood reduces the levels of these antibodies.

Since the sensitization of an Rh negative mother to the Rh antigens of an Rh positive fetus usually occurs near the time of birth, it is common to inject Rh negative mothers soon after the delivery of an Rh positive child with agents that prevent or limit sensitization. These agents contain anti-Rh antibodies that bind to any fetal Rh antigens that have entered the mother's system. This inhibits the antigens from sensitizing the mother and causing her to produce antibodies against them.

METABOLISM OF FOREIGN CHEMICALS

Many toxic foreign chemicals—for example, drugs and chemical pollutants in air, water, or food—can enter the body by way of the gastrointestinal tract, lungs, or skin. A number of these chemicals, particularly organic chemicals, are metabolized in the liver and also to some extent in the kidneys, skin, and other organs.

Often, the metabolic transformation of a hazardous chemical decreases its toxicity and enhances its excretion. For example, a nonpolar, lipid-soluble chemical may be converted into a more polar, less lipid-soluble substance that can be eliminated from the body in the urine.

On the other hand, the metabolic processing of some foreign chemicals enhances their toxicities. For example, it is believed that many chemicals linked to the occurrence of cancer become carcinogenic only after metabolic transformation. Moreover, because the pathways involved in the metabolism of foreign chemicals generally process materials normally required by the body, the presence of foreign chemicals may upset the metabolism of necessary substances.

RESISTANCE TO STRESS

A stressful event can be thought of as any event that leads to an increased release of glucocorticoids such as cortisol from the cortices of the adrenal glands resulting in glucocorticoid levels in the bood plasma that are higher than those present at the same hour of the day in undisturbed, normal individuals on a similar activity (sleep-wake) cycle. Among the events that can have this effect are noxious or potentially noxious occurrences such as physical trauma, prolonged heavy exercise, and various infections. Many stressful events are believed to enhance glucocorticoid release by either directly or indirectly promoting an increased release of corticotropin-releasing factor (CRF) from neurons of the hypothalamus of the brain. The CRF, in turn, enhances the release of adrenocorticotropin (ACTH) from the pituitary gland, and the ACTH increases the release of glucocorticoids from the adrenal cortices. Moreover, there is some evidence that certain infections may cause the production of ACTH or an ACTHlike substance by extrapituitary sources (possibly lymphocytes), and this substance may be at least partly responsible for an increased release of glucocorticoids in such cases.

Many investigators have suggested that the glucocorticoids are particularly important to a person's ability to combat physical stress. Since a major response to overwhelming stress is vasodilation and circulatory failure, it has been proposed that the glucocorticoids combat stress by permitting norepinephrine to induce vasoconstriction while at the same time preventing an excessive vasoconstriction that can lead to tissue ischemia (deficiency of blood) by acting directly on vascular smooth muscle and stimulating heart muscle. It has also been suggested that the metabolic effects of the glucocorticoids—for example, raising blood-glucose levels—are particularly important in mobilizing the body's resources to resist physical stress.

A common response to stress is the activation of the sympathetic nervous system and the adrenal medullae. This response elevates the levels of epinephrine and norepinephrine, which leads to responses such as increased blood-glucose levels, elevated blood pressure, increased heart rate, and increased bloodflow to skeletal muscles. These responses can also be useful in meeting physical stress.

The levels of other hormones are also frequently affected by stress. For example, the release of antidiuretic hormone (ADH) and aldosterone can increase during stress, resulting in a decreased urine output and in increased retention of water and sodium that leads to an increase in blood volume.

A number of psychosocial situations that elicit emotional responses such as fear, anger, or anxiety—for example, final exams for college students or awaiting a surgical operation—are also associated with an increased release of glucocorticoids and stress responses. In this regard it has been suggested that, although stress responses can be of benefit to an individual in meeting physical stress, they may not be as useful in resisting psychosocial stress. In fact, it has been proposed that stress responses elicited by chronic psychosocial stress—for example, high pressure employment or anxiety provoking economic situations—actually contribute to pathological conditions such as high blood pressure or heart disease.

STUDY OUTLINE

BODY SURFACES Skin and mucous membranes are first barriers to the invasion of body by potentially damaging factors.
1. Skin contains keratin, and sweat and sebaceous glands secrete chemicals toxic to many bacteria.
2. Mucus secreted by mucous membranes entraps small particles that may then be swept away or engulfed by phagocytic cells. **p. 353**

INFLAMMATION In its most basic form, inflammation is an acute, nonspecific physiological response of the body to tissue injury caused by factors such as chemicals, heat, mechanical trauma, or bacterial invasion. **pp. 354–360**

Changes in Bloodflow and Vessel Permeability
1. Small vessels in injured area dilate, leading to increased bloodflow that increases the delivery of plasma proteins and phagocytic leukocytes.
2. Small vessel permeability increases, and plasma fluid and solutes (including proteins) move out of the circulatory system and into the inflamed tissues.
3. As inflammation progresses, bloodflow through the small vessels slows and sometimes even stops.

Walling-off Effect of Inflammation
In the tissues, fibrinogen from the plasma is converted to fibrin, forming a clot that walls off the injured area. This walling-off effect can delay or limit the spread of toxic products or bacteria.

Leukocyte Emigration
As bloodflow slows during inflammation, leukocytes marginate and pavement. The leukocytes exhibit amebalike activity and squeeze between endothelial cells of capillaries and venules into the tissue spaces. The first leukocytes in the tissues are generally neutrophils, followed later by monocytes, which become macrophages.

Chemotaxis and Leukocyte Aggregation
Chemical mediators determine the direction of movement of leukocytes. In acute, nonspecific inflammation neutrophils usually predominate initially and monocytes and macrophages predominate during later stages.

Phagocytosis
Leukocytes (particularly neutrophils) and macrophages phagocytize microorganisms, foreign materials, and debris. Opsonins can coat specific foreign particles and render them more susceptible to phagocytosis.

Abscess and Granuloma Formation
Occasionally the inflammatory response is unable to overcome an invasion and an abscess or granuloma forms.
1. Abscess   A sac of pus consisting of microbes, leukocytes, macrophages, and liquefied debris walled off by fibroblasts or collagen.
2. Granuloma   Layers of phagocytic-type cells surrounded by a fibrous capsule. The central cells contain the offending agent—for example, microbes that can survive within the phagocytes or materials that cannot be digested by the phagocytes.

Chemical Mediators of Inflammation

HISTAMINE Present in mast cells, basophils, and platelets. Released in response to mechanical disruption of cells by injury and chemicals from neutrophils. Causes vasodilation and increased vascular permeability.

KININS A group of polypeptides formed from kininogens by kallikrein. Kinins dilate arterioles, increase vascular permeability, act as chemotactic agents, and induce pain.

COMPLEMENT SYSTEM A series of plasma proteins that normally circulate in an inactive form. Active complement components can mediate virtually every event of the acute, nonspecific inflammatory response, and they can attack and kill invading microbes without prior phagocytosis. The complement system can be activated by properdin.

PROSTAGLANDINS Prostaglandins may induce vasodilation and increased vascular permeability, stimulate leukocyte migration through capillary walls, and contribute to pain and fever of inflammation.

LEUKOTRIENES Substances produced by leukocytes. Certain leukotrienes apparently cause increased vascular permeability, and some seem to be chemotactic agents that attract neutrophils.

Symptoms of Inflammation

LOCAL SYMPTOMS Redness and heat due to vessel dilation and increased bloodflow; swelling due to increased vessel permeability and the movement of fluid and solutes out of the circulatory system; pain due to increased pressure on nerve endings and effects of some chemical mediators.

SYSTEMIC SYMPTOMS   Fever (neutrophils release endogenous pyrogens) and often an increased production and release of leukocytes.

INTERFERON   Provides some protection to the body against viral invasion. Viruses induce a cell to produce interferon. Interferon leaves the cells in which it is produced and binds to receptors on cell membranes of other cells. This binding triggers the synthesis of enzymatic proteins within the cells that can act to prevent viral reproduction. Interferon may also prevent the release of viruses from cells or make viruses incapable of infecting other cells. Interferon may also play a role in protecting the body against some forms of cancer, and it is increasingly seen as a versatile agent that has effects that go beyond the inhibition of viral activity.   **p. 360**

SPECIFIC IMMUNE RESPONSES   Two major aspects: humoral immunity and cell-mediated immunity.   **pp. 361–368**

Antigens   An antigen is a substance that elicits a specific immune response when it is present in the body. Antigenic substances generally have molecular weights of 8,000 to 10,000 or more. Lower molecular weight substances (haptens) may elicit specific immune responses if they combine with larger molecules.

B Cells   Lymphocytes that are committed to differentiate into antibody-producing plasma cells involved in humoral immunity.

T Cells   A heterogeneous group of lymphocytes that include those cells committed to participating in cell-mediated immune responses.

Antibodies   An antibody is a specialized protein that is produced in response to the presence of an antigen. An antibody can combine with the specific antigen that stimulated its production to form an antigen–antibody complex. Antibodies inactivate or destroy foreign substances in a number of ways:
1. Inhibit interactions of bacterial toxins or destructive foreign enzymes with target cells or substrates.
2. Prevent attachment of viruses to target cells.
3. Enhance the basic inflammatory response.

Humoral (Antibody) Immunity   An antigen stimulates some B lymphocytes to undergo division. Most develop into antibody-producing plasma cells that form antibodies. Antibodies are released and can combine with the antigen that stimulated their production to inactivate or destroy the antigen. Some stimulated B cells form memory cells, and upon sub-

sequent exposure to the antigen, there is a rapid outpouring of antibodies. In most cases, macrophages and helper T cells are involved in the stimulation of B cells by an antigen.

Cell-Mediated Immunity   When an antigen combines with receptors on the surfaces of particular T cells, the T cells are sensitized and undergo division and differentiation. Some T cell progeny serve memory functions and others—which are called effector T cells—participate in the cell-mediated immune response. Effector T cells leave the lymphoid tissues and travel throughout the body. When they encounter cells bearing the specific antigen that initiated their production, they combine with them. Some effector T cells can lyse cells that have the antigen on their surfaces. Others release lymphokines that enhance the basic inflammatory response and subsequent phagocytosis.

TRANSPLANT REJECTION   The cell-mediated immune system contributes importantly to the rejection of solid-tissue transplants.

Active Immunity   Resistance to infection that results from an antigen-induced activation of an individual's specific immune responses is called active immunity. Active immunity may be acquired by infection or by vaccination.

Passive Immunity   Essentially "borrowed" immunity. For example, antibodies to a disease-causing agent can be transferred into an individual from sources (often nonhuman) that have been exposed to the agent. Maternal antibodies of the IgG class may pass into the fetus during pregnancy to provide passive immunity.

Hypersensitivity   Immune responses that can be harmful to the body.

ROLE OF COMPLEMENT COMPONENTS IN HYPERSENSITIVITY   In some types of hypersensitivity, the generation of complement components during a humoral immune response to an antigen causes damage to body tissues and cells.

ROLE OF EFFECTOR T CELLS IN HYPERSENSITIVITY   The activities of effector T cells during a cell-mediated immune response can, in some cases, cause damage to body tissues.

ROLE OF CLASS IgE ANTIBODIES IN HYPERSENSITIVITY   The exposure of a susceptible individual to a particular antigen can sensitize the individual and cause the production of class IgE antibodies that bind to mast cells and basophils. Upon subsequent exposure to the antigen, the antigen attaches to the antibodies that are bound to mast cells and basophils, stimulating the release of chemicals that induce inflammatory responses, increased mucus secretion, and the contraction of smooth muscles of the airways of the lungs.

Tolerance   Can be generally defined as the failure of the body to mount a specific immune response against a particular antigen. The body normally exhibits self-tolerance and does not produce antibodies or effector T cells against its own antigens.

Autoimmune Responses   Occasionally, the body's tolerance to its own antigens breaks down, and a person produces antibodies or effector T cells that act against his or her own tissues.

BLOOD GROUPS AND TRANSFUSION

A-B-O System   Based on inherited antigens on erythrocyte surfaces and antibodies that build up against antigens not on a person's erythrocytes.

Transfusion   Mixing of incompatible blood types during transfusion can cause agglutination and hemolysis of erythrocytes.   **pp. 369–370**

Rh System   Based on inherited antigens on erythrocyte surfaces. Antibodies can be produced against Rh antigens not on a person's erythrocytes upon exposure and sensitization to the antigens.

METABOLISM OF FOREIGN CHEMICALS   Many potentially toxic or poisonous foreign chemicals are metabolically transformed into less toxic, more easily excreted, substances. In some cases, metabolic transformation may enhance the toxicity of foreign chemicals.

RESISTANCE TO STRESS   A stressful event can be thought of as any event that leads to an increased release of glucocorticoids such as cortisol from the cortices of the adrenal glands resulting in glucocorticoid levels in the blood plasma that are higher than those present at the same hour of the day in undisturbed, normal individuals on a similar activity (sleep-wake) cycle. The glucocorticoids may help combat physical stress. Stress can also activate the sympathetic nervous system and adrenal medullae, and their activities may also be useful in combating physical stress. Some psychosocial situations elicit stress responses, but the responses may not be useful in resisting the stress.
p. 371

SELF-QUIZ

1. The skin and mucous membranes serve as barriers to the invasion of the body by potentially damaging factors. True or False?

2. During an acute, nonspecific inflammatory response: (a) the permeability of small vessels decreases; (b) plasma proteins are unable to leave the circulatory system; (c) vessels of the microcirculation dilate.

3. Which cells usually predominate in the early stages of most acute, nonspecific inflammatory responses? (a) neutrophils; (b) monocytes; (c) eosinophils.

4. Histamine: (a) kills invading cells directly; (b) causes dilation of arterioles and venules; (c) decreases vessel permeability, particularly in venules.

5. Kininogens are converted to kinins by: (a) kallikrein; (b) opsonin; (c) chemotaxin.

6. The complement system can be activated by: (a) kallikrein; (b) properdin; (c) histamine.

7. Active complement components can mediate virtually every event of the acute, nonspecific inflammatory response. True or False?

8. The fever that often accompanies inflammation may be due to: (a) endogenous pyrogens released by neutrophils; (b) the effects of kinins on afferent nerve terminals; (c) the release of properdin into tissue fluids.

9. One of the most effective inducers of interferon production is double-stranded DNA. True or False?

10. Haptens: (a) commonly elicit specific immune responses in their free, uncombined forms; (b) are relatively low-weight molecular substances that can elicit specific immune responses if they combine with larger molecules such as proteins; (c) generally have molecular weights greater than 10,000 in their free, uncombined forms.

11. B cells: (a) are committed to participating in cell-mediated immune responses; (b) release lymphokines; (c) can differentiate into plasma cells.

12. The constant regions of the polypeptide chains of an antibody molecule contain binding sites that allow the antibody to attach to a specific antigen. True or False?

13. Antibodies are produced by: (a) effector T cells; (b) mast cells; (c) plasma cells.

14. Lymphokines are released by certain: (a) effector T cells; (b) B cells; (c) monocytes.

15. Active immunity can be acquired through: (a) the transfer of antibodies from mother to fetus; (b) the injection of a pretreated bacterial toxin; (c) the injection of antibodies obtained from another person.

16. Anaphylactic hypersensitivity involves: (a) activated complement components; (b) effector T cells; (c) class IgE antibodies.

17. Some relief from an anaphylactic-type allergic reaction may be provided by the administration of: (a) opsonins; (b) antihistamines; (c) mast cells.

18. Normally, antibodies and effector T cells are not produced against a person's own tissues. True or False?

19. An adult with type A blood would normally have: (a) anti-A antibodies in the plasma; (b) anti-B antibodies in the plasma; (c) neither anti-A nor anti-B antibodies in the plasma.

20. During stress, which substance(s) is(are) released by the cortices of the adrenal glands? (a) glucocorticoids; (b) ACTH; (c) CRF.

21. Glucocorticoids may help a person combat physical stress. True or False?

LEARNING OBJECTIVES

After completing this chapter, you should be able to:

- Describe the forces that maintain the expansion of the lungs in normal individuals, and explain why the lungs collapse when air enters the pleural cavity.
- Describe the factors responsible for the increased lung volume and the flow of air into the lungs during inspiration.
- Describe the factors responsible for the reduced lung volume and the flow of air out of the lungs during expiration.
- Discuss the factors that influence pulmonary airflow.
- Describe the factors that influence alveolar ventilation.
- Explain how alveolar airflow and bloodflow are matched.
- Describe the process by which gas exchange occurs between the lungs and the blood, and between the blood and the body tissues.
- Explain how oxygen and carbon dioxide are transported in the blood.
- Discuss the neural mechanisms involved in the control of respiration.
- Describe the roles of oxygen, carbon dioxide, and hydrogen ions in the control of respiration.

CHAPTER CONTENTS

THE RESPIRATORY SYSTEM

The metabolic activities of the body's cells consume oxygen and produce carbon dioxide. Consequently, oxygen must be continually supplied to the cells and carbon dioxide removed if the cells are to survive and if homeostasis is to be maintained. Oxygen is supplied and carbon dioxide is removed by the respiratory system with the help of the circulatory system.

STAGES OF RESPIRATION

Respiration is an integrated process, but for purposes of analysis it can be divided into five stages: (1) the movement of air into and out of the lungs; (2) the exchange of oxygen and carbon dioxide between the air in the lungs and the blood within the pulmonary capillaries; (3) the transport of oxygen and carbon dioxide throughout the body by the blood; (4) the exchange of oxygen and carbon dioxide between the blood and the interstitial fluid and cells; and (5) the utilization of oxygen and the production of carbon dioxide by metabolic processes within cells. This chapter deals with the first four of these respiratory stages. The fifth is considered in Chapter 17.

ORGANIZATION OF THE RESPIRATORY SYSTEM

The exchange of oxygen and carbon dioxide between the air and the blood occurs in the lungs. In order to reach the exchange sites in the lungs, the air must flow through a series of conducting passageways that branch from one another much like the branches of a tree. The blood supply of the passageways warms the inspired air, and the secretions of the mucous membranes that line the passageways humidify the air and entrap small particles. The mucous membranes of the respiratory tract contain ciliated cells that move mucus and the entrapped particles toward the throat (*pharynx*), where they can be removed by coughing or swallowing.

The two **lungs** are located within a sealed cavity called the **thoracic cavity** (F15.1). The floor of the thoracic cavity is formed by a dome-shaped muscle called the **diaphragm,** and the walls are formed by the rib cage and its associated muscles. Within the thoracic cavity, each lung is enclosed within a double-walled sac called the **pleura.** The portion of the pleura that adheres firmly to the lungs is the *visceral pleura;* the portion attached to the structures that form the thoracic cavity is the *parietal pleura.* Between the two layers of the pleura is an extremely narrow **pleural cavity,** which is filled with **pleural fluid.** The pleural fluid is secreted by the pleura, and it acts as a lubricant to reduce the friction between the two layers of the pleura during respiratory movements. The pleural fluid also couples the visceral pleura and parietal pleura (and therefore the lungs and the thoracic cavity structures) to one another. This coupling is analogous to the coupling of two sheets of glass by a thin film of water, in which it is extremely difficult to separate the sheets (the visceral pleura and the parietal pleura) by a force applied at right angles to their surfaces even though the sheets can slide readily past one another. Because of this coupling, the movements of the lungs closely follow the movements of the structures that form the thoracic cavity.

AIRFLOW IN THE RESPIRATORY SYSTEM

Air enters the respiratory system through the nose, or perhaps the mouth, and flows into the pharynx, which is used by both the respiratory system and the digestive system (F15.2). From the pharynx, the air enters the *larynx.* The larynx contains folds of tissue called vocal cords, which can vibrate and produce sounds as air passes over them. During swallowing, a flaplike structure, the *epiglottis,* deflects solids and fluids away from the laryngeal opening (the *glottis*) and into the esophagus to the stomach. Any solids or fluids that enter the larynx are generally expelled by violent coughing. From the larynx, the air flows toward the lungs along a tube called the *trachea,* which divides into left and right *primary bronchi* (F15.3). From the primary bronchi, the air passes through

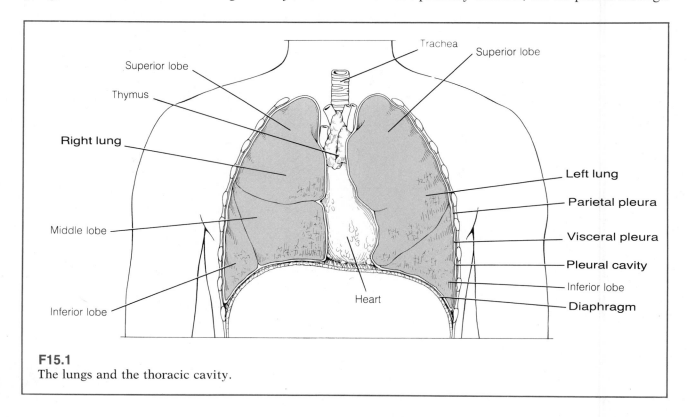

F15.1
The lungs and the thoracic cavity.

progressively smaller branching passageways (secondary bronchi, tertiary bronchi, bronchioles) to ultimately reach clusters of small, thin-walled air sacs called **alveoli** (singular: *alveolus*). The alveoli are sites of gas exchange between the air in the lungs and the blood. The air within the alveoli is separated from the blood by a very thin respiratory membrane formed by the alveolar epithelium and its basement membrane, and the endothelium of the pulmonary capillaries and its basement membrane (F15.4). In some places there is a small amount of connective tissue between the two

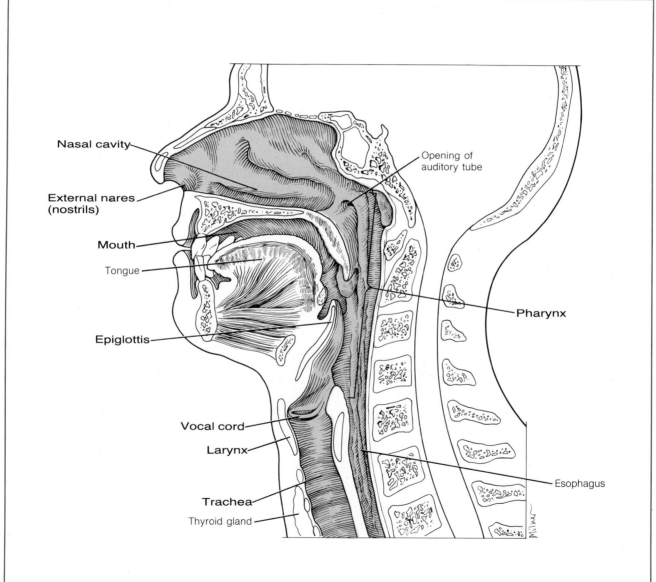

F15.2
Sagittal view of the head and neck, showing the nasal cavity, the mouth, and the pharynx.

basement membranes. Since, for the efficient diffusion of gases, the respiratory membrane must be moist, the alveolar surfaces exposed to the air are coated with a thin layer of fluid.

MECHANICS OF BREATHING

In order to maintain a concentration of oxygen and carbon dioxide within the alveolar air that is favorable to the diffusion of these gases across the respiratory membrane, it is necessary to bring fresh air in and remove the air already within the lungs. The movement of air into and out of the lungs depends on pressure differences between the air in the atmosphere and the air in the lungs.

ATMOSPHERIC PRESSURE

At sea level, the gases of the atmosphere under the influence of gravity exert a pressure on the surface of the earth equivalent to that which would be exerted by a 760-millimeter-thick layer of mercury (Hg). Consequently, at sea level at any point on the surface of the earth, the pressure exerted by the column of atmospheric gases above that point is the same as the pressure that would be exerted by a column of mercury 760 millimeters high. Thus, *atmospheric pressure* at sea level can be expressed as equal to 760 mm Hg.

PRESSURE RELATIONSHIPS IN THE THORACIC CAVITY

The lungs are stretched within the thoracic cavity, and the visceral pleura (which adheres to the lungs) is separated from the parietal pleura (which is attached to the structures forming the thoracic cavity) by only a thin layer of pleural fluid, which couples the visceral pleura and parietal pleura (and therefore the lungs and thoracic

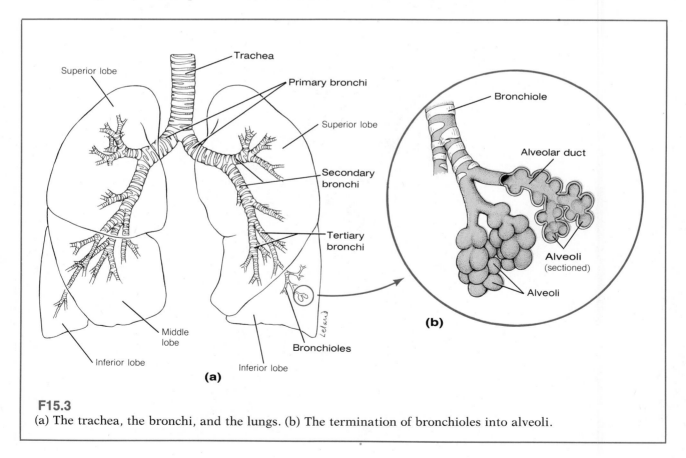

F15.3
(a) The trachea, the bronchi, and the lungs. (b) The termination of bronchioles into alveoli.

cavity structures) to one another. The walls and partitions of the lungs contain elastic connective tissue whose recoil tends to reduce the size of the lungs. In addition, as is discussed later, the surface tension of the fluid that coats the exposed surfaces of the alveoli also favors a reduction in the size of the lungs. Thus, the lungs tend to pull away from the structures that form the thoracic cavity and collapse. When the lungs are in the resting position following a normal exhalation, the collapsing force of the lungs is equal to a pressure of about 4 mm Hg (F15.5).

The pressure within the pleural cavity (the *intrapleural* or *intrathoracic pressure*) pushes against the outer walls of the lungs and also favors their collapse. However, the tendency of the lungs to pull away from the structures that form the thoracic cavity leads to a pressure within the pleural cavity that is somewhat less than the atmospheric pressure. When the lungs are in the resting position, the intrapleural pressure is approximately 4 mm Hg less than the atmospheric pressure—that is, approximately 756 mm Hg at sea level.

The alveoli and passageways within the lungs

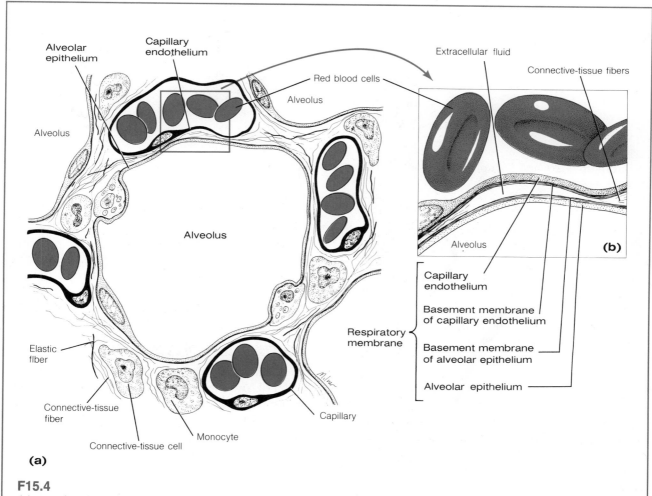

F15.4

(a) An alveolus surrounded by capillaries, showing the respiratory membrane that separates the air in the alveolus from the blood in the capillaries. (b) The respiratory membrane of the lungs.

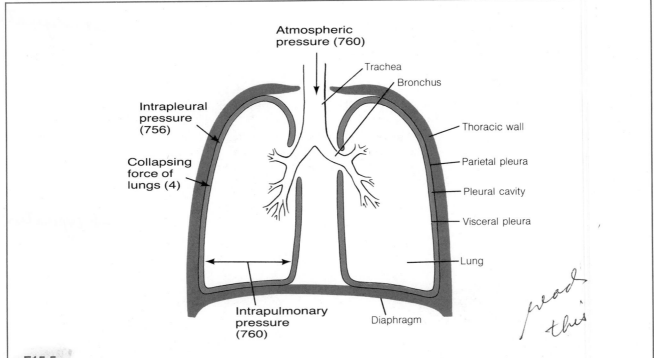

Atmospheric pressure (760)

Trachea

Bronchus

Intrapleural pressure (756)

Thoracic wall

Parietal pleura

Collapsing force of lungs (4)

Pleural cavity

Visceral pleura

Lung

Intrapulmonary pressure (760)

Diaphragm

read this

F15.5
Intrapulmonary and intrapleural pressures in the resting position. All pressures (in parentheses) are given in millimeters of mercury.

communicate with the atmosphere through the trachea, and the pressure within the lungs when they are in the resting position is equivalent to the atmospheric pressure. Thus, at sea level, the resting *intrapulmonary pressure* is 760 mm Hg. This pressure pushes against the inner walls of the lungs and opposes their collapse. Under normal circumstances, then, the forces favoring the collapse of the lungs (the 4 mm Hg collapsing force of the lungs and the 756 mm Hg intrapleural pressure) are balanced by a force that opposes their collapse (the 760 mm Hg intrapulmonary pressure). As a consequence, the lungs normally remain expanded within the thoracic cavity and do not collapse. However, if a lung or the thoracic wall is punctured or torn so that the pleural cavity is exposed to the atmosphere, air enters the pleural cavity and the intrapleural pressure becomes equal to the atmospheric pressure. As a consequence, the forces tending to reduce the size of

the lung (the 4 mm Hg collapsing force of the lung and the now 760 mm Hg intrapleural pressure) become greater than the force opposing a reduction in lung size (the 760 mm Hg intrapulmonary pressure), and the lung collapses. The presence of air in the pleural cavity is known as a *pneumothorax*.

VENTILATION OF THE LUNGS

Air flows from a region of higher pressure to a region of lower pressure. Thus, for air to flow into the lungs from the atmosphere, the pressure within the lungs—that is, the intrapulmonary pressure—must be less than the atmospheric pressure. Similarly, for air to flow out of the lungs, the intrapulmonary pressure must be greater than the atmospheric pressure. The pressure within the gas-filled spaces of the lungs can be altered by altering the volume of the spaces, and this can be accomplished by changing the volume of the thoracic cavity (recall that under normal circumstances, the pleural fluid couples the visceral pleura to the parietal pleura, and the movements of the lungs closely follow the movements of the structures that form the thoracic cavity). Thus, when the volume of the thoracic cavity increases, so does the volume of the lungs. The increase in lung volume lowers the intrapulmonary pressure below the atmospheric pressure, and air flows into the lungs until the pressure within the lungs again equals the atmospheric pressure. Conversely, when the volume of the thoracic cavity decreases, so does the volume of the lungs. The decrease in lung volume compresses the air within the lungs, and the intrapulmonary pressure rises above the atmospheric pressure. As a consequence, air flows out of the lungs until the pressure within the lungs becomes equal to the atmospheric pressure.

Inspiration

Inspiration, or **inhalation,** refers to the movement of air into the lungs. As previously indicated, air moves into the lungs when the volume of the thoracic cavity—and thus the lungs—increases, and the intrapulmonary pressure falls. During quiet breathing, the intrapulmonary pressure during inspiration decreases about 1 mm Hg below the resting intrapulmonary pressure, which is the same as the atmospheric pressure.

There are two ways of increasing the volume of the thoracic cavity during inspiration. One way is by contracting the diaphragm. When it contracts, the diaphragm flattens, lowering its dome. This movement increases the longitudinal dimension of the thoracic cavity. The second way is by elevating the ribs. In the resting position, the ribs slant downward and forward from the vertebral column. The contraction of muscles such as the external intercostal muscles, which are located between the ribs, pulls the ribs upward (F15.6a). This movement increases the anterior-posterior dimension of the thoracic cavity. The contraction of the diaphragm is the dominant means of increasing the volume of the thoracic cavity and lowering the intrapulmonary pressure during normal, quiet inspiration. The elevation of the ribs is most evident during forced inspiration.

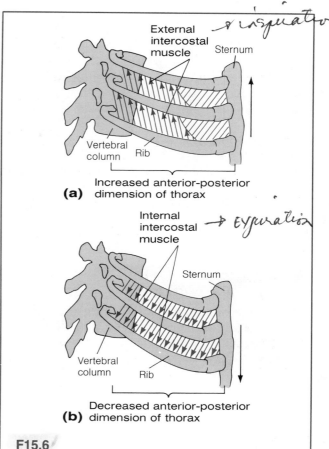

(a) Increased anterior-posterior dimension of thorax

(b) Decreased anterior-posterior dimension of thorax

F15.6
Mechanisms by which the volume of the thoracic cavity is altered during respiration. (a) Inspiration: the volume of the thoracic cavity is increased by the contraction of the diaphragm and the external intercostal muscles. (b) Forced expiration: the volume of the thoracic cavity is decreased by the contraction of the internal intercostal muscles. In both figures, the arrows indicate the direction in which the muscles pull.

Expiration

Expiration, or **exhalation,** refers to the movement of air out of the lungs. As previously indicated, air moves out of the lungs when the volume of the thoracic cavity—and thus the lungs—decreases, and the intrapulmonary pressure rises. During quiet breathing, the intrapulmonary pressure during expiration rises about 1 mm Hg above the resting intrapulmonary pressure.

During quiet breathing, the volume of the thoracic cavity is decreased by passive processes that do not involve muscular contractions. When the muscles involved in inspiration relax, the elastic recoil of the lungs, chest wall, and abdominal structures returns the ribs and diaphragm to their resting positions. This activity reduces the volume of the thoracic cavity and raises the intrapulmonary pressure.

During forced expiration, the volume of the thoracic cavity is further reduced by the contraction of internal intercostal muscles, which pull the ribs downward (F15.6b). Moreover, the muscles of the anterior abdominal wall aid in forced expiration by exerting pressure on the abdominal viscera, thus forcing the diaphragm upward.

FACTORS THAT INFLUENCE PULMONARY AIRFLOW

The volume of air that flows in a given time between the atmosphere and the alveoli is influenced by the pressure that moves the air through the respiratory passageways and by the resistance the air encounters as it flows through the passageways. The relationship between these factors can be expressed as:

$$\text{flow} = \frac{\text{pressure}}{\text{resistance}}$$

The pressure that moves the air through the respiratory passageways is represented by the difference between the atmospheric pressure and the pressure within the alveoli. A major factor influencing resistance is passageway diameter: the smaller the diameter of a passageway, the greater its resistance. Smooth muscles associated with the respiratory passageways, particularly the bronchioles, can alter passageway diameters.

Under normal circumstances, the resistance of the respiratory passageways is slight, and a substantial airflow occurs in response to only a slight difference in atmospheric and alveolar pressures. During quiet inspiration, for example, the pressure difference between the atmosphere and the alveoli is only 1 mm Hg; yet approximately 500 ml of air enter the lungs with each breath.

Under certain circumstances, the constriction of the passageways (particularly the bronchioles), or the accumulation of fluid or mucus within the passageways, increases the resistance sufficiently to impede the flow of air between the atmosphere and the alveoli. Since the passageways tend to widen as the lung volume increases during inspiration and narrow as the lung volume decreases during expiration, an increased resistance can particularly affect the flow of air out of the lungs. If the resistance increases, the pressure difference between the atmosphere and the alveoli must increase in order to maintain the airflow. Within limits, the pressure difference can be increased by an increased contraction of the respiratory muscles.

Parasympathetic nervous stimulation, histamine, and leukotrienes contract smooth muscles of the respiratory passageways and increase passageway resistance. Histamine also increases mucous secretion. Sympathetic nervous stimulation and epinephrine relax smooth muscles of the passageways and decrease passageway resistance.

SURFACE TENSION AND PULMONARY SURFACTANT

At an air-liquid interface, attractive forces between the molecules of a liquid such as water cause the molecules to be attracted more strongly into the liquid than into the air. This differential attraction gives rise to a tension at the surface of the liquid—the *surface tension*—that tends to reduce the liquid's surface area to the smallest possible value and to oppose any increase in the surface area. Thus, the surface tension of the fluid that coats the exposed surfaces of the alveoli tends to reduce the size of the alveoli and oppose the expansion of the lungs. If this fluid were pure water, which has a relatively high surface tension, the tendency of the alveoli to collapse would be so great that the expansion of the lungs during inspiration would require exhausting muscular effort. However, specialized alveolar cells produce a phospholipoprotein complex called **pulmonary surfactant,** which lowers the surface tension of the fluid that coats the alveoli, significantly decreasing the muscular effort required to expand the lungs. In fact, even during heavy exercise, the ventilation of the lungs requires only about 3% to 4% of the total energy expenditure. Some children who are born prematurely do not produce sufficient quantities of pulmonary surfactant. The lack of adequate pulmonary surfactant production is believed to contribute to (or perhaps cause) an often fatal condition known as *respiratory distress syndrome of the newborn*, or *hyaline membrane disease*, in which breathing is difficult and labored.

LUNG VOLUMES

In young males, the lungs have a total capacity of approximately 5900 ml. However, a person cannot exhale all the air from the lungs, and about 1200 ml of air always remains no matter how forced the expiration. This remaining volume is called the **residual volume** (F15.7).

The volume of air that moves into and out of the lungs with each breath is called the **tidal volume.** During normal, quiet breathing, it is about 500 ml. The **inspiratory reserve** is the volume of air (approximately 3000 ml) that can be inspired in addition to the normal, quiet tidal volume. Following a normal passive expiration of the quiet tidal volume, additional air can be forced out. This **expiratory reserve** is about 1200 ml. The **vital capacity** is the maximum amount of air that can be moved into and out of the lungs, from the deep-est possible inspiration to the most forced expiration. Therefore, the vital capacity represents the sum of the inspiratory reserve volume; the tidal volume during normal, quiet breathing; and the expiratory reserve volume. Note that it does not include the residual volume. In healthy young men the vital capacity is about 4700 ml. It is somewhat less in women because they tend to have smaller thoracic cages and smaller lung capacities (Table 15.1).

ANATOMICAL DEAD SPACE

The exchange of oxygen and carbon dioxide between air in the lungs and the blood occurs within the alveoli and, to some extent, within the small bronchioles and ducts that lead to the alveoli. However, all these gas exchange areas will be considered together as the alveoli. No gas exchange takes place within the remaining respiratory passageways (nose, trachea, bronchi, and so on). These air-filled passageways are referred to, therefore, as *anatomical dead space*. The volume of the anatomical dead space is approximately 150 ml.

MINUTE RESPIRATORY VOLUME

The volume of air moved into the respiratory passageways in one minute is called the **minute respiratory volume.** It is equal to the respiratory rate

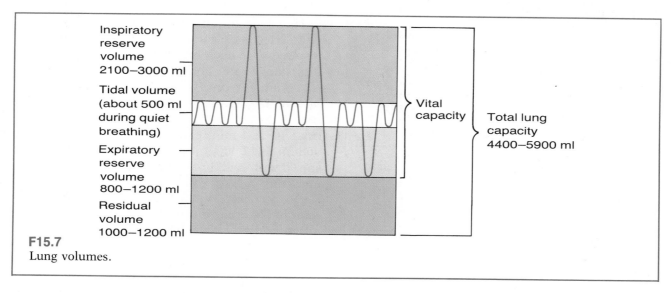

F15.7
Lung volumes.

Inspiratory reserve volume 2100–3000 ml

Tidal volume (about 500 ml during quiet breathing)

Expiratory reserve volume 800–1200 ml

Residual volume 1000–1200 ml

Vital capacity

Total lung capacity 4400–5900 ml

(expressed as number of breaths per minute) multiplied by the volume of air that enters the respiratory passageways with each breath (the tidal volume). For example, during quiet breathing the respiratory rate is about 12 breaths per minute, and approximately 500 ml of air enters the respiratory passageways with each breath. Thus, the minute respiratory volume is 6000 ml/min (12 breaths/min × 500 ml/breath). The minute respiratory volume can be altered by altering either the respiratory rate or the amount of air entering the respiratory passageways with each breath.

ALVEOLAR VENTILATION

Because the exchange of gases between the lungs and the blood occurs within the alveoli and not within the respiratory passageways, the volume of air that moves into the alveoli from the atmosphere in a given time is more important than the volume of air that enters the respiratory passageways. The volume of *atmospheric air* that enters the alveoli in one minute is called the **alveolar ventilation** (per minute).

With each inspiration, the first air to reach the alveoli is air that was left in the respiratory passageways—that is, in the anatomical dead space—from the previous expiration. Also, the last air to be inspired never reaches the alveoli. It remains within the respiratory passageways and is the first air to be exhaled. The volume of air within

the respiratory passageways is approximately 150 ml. Therefore, of the approximately 500 ml of atmospheric air that enters the respiratory passageways with each inspiration during quiet breathing, only 350 ml reaches the alveoli.

Note that although only 350 ml of atmospheric air enters the alveoli with each inspiration during quiet breathing, a total of 500 ml of air actually moves into the alveoli (the first 150 ml is air that remained in the respiratory passageways from the previous expiration). Note, too, that the air within the alveoli is not completely replaced with each breath. For example, during quiet breathing, an expiratory reserve volume of 800 to 1200 ml and a residual volume of 1000 to 1200 ml remain within the lungs at the end of expiration, and only about 500 ml of air is taken into the lungs with each breath.

The alveolar ventilation is equal to the respiratory rate (expressed as breaths per minute) multiplied by the volume of atmospheric air that enters the alveoli with each breath (that is, the tidal volume minus the dead space volume). Thus, during quiet breathing at a rate of 12 breaths/min, the alveolar ventilation is 4200 ml/min (12 breaths/min × 350 ml of atmospheric air entering the alveoli/breath). Under conditions of increased demand, the alveolar ventilation can be increased by increasing either the respiratory rate or the volume of air inspired with each breath.

The alveolar ventilation is generally increased more effectively by increasing the volume of air inspired with each breath than by increasing the respiratory rate. For example, if the respiratory rate doubles from its resting value (that is, doubles to 24 breaths/min) but the amount of air moved during each inspiration remains the same (that is, 500 ml/breath), the minute respiratory volume is 12,000 ml/min. Of the 500 ml of air inspired with each breath, 350 ml enters the alveoli and 150 ml remains in the respiratory passageways. Thus, the alveolar ventilation is 8400 ml/min (24 breaths/min × 350 ml of atmospheric air entering the alveoli/breath). If the respiratory rate remains the same (that is, 12 breaths/min) but the volume of air moved during each inspiration doubles (that is, to 1000 ml/breath), the minute

Table 15.1	*Average Lung Volumes in Young Males and Females (Age 20–30 Years)*	
	Volume (in ml)	
	Male	Female
Total Lung Capacity	5900	4400
Vital capacity	4700	3400
Inspiratory reserve	3000	2100
Tidal volume (at rest)	500	500
Expiratory reserve	1200	800
Residual volume	1200	1000

respiratory volume is still 12,000 ml/min. Of the 1000 ml of air inspired with each breath, 850 ml reaches the alveoli and 150 ml remains in the respiratory passageways. Thus, the alveolar ventilation is 10,200 ml/min (12 breaths/min × 850 ml/breath). During exercise and in most other situations where an increased supply of oxygen and an increased elimination of carbon dioxide are required, the increase in breathing depth—that is, the volume of air moved with each breath—is proportionally greater than the increase in respiratory rate.

MATCHING OF ALVEOLAR AIRFLOW AND BLOODFLOW

For efficient gas exchange, the airflow and bloodflow to particular alveoli must be well matched. Poor gas exchange occurs when the alveolar airflow is either inadequate or excessive in relation to the bloodflow. Local autoregulatory mechanisms contribute to matching alveolar airflow and bloodflow. One of these mechanisms was considered in Chapter 13, where it was pointed out that pulmonary vessels constrict and bloodflow decreases in areas where the oxygen level is low (and pulmonary vessels dilate and bloodflow increases in areas where the oxygen level is high). In addition, the respiratory passageways dilate and the airflow to the alveoli increases in areas where the carbon dioxide level is high (and the respiratory passageways constrict and the airflow to the alveoli decreases in areas where the carbon dioxide level is low).

These autoregulatory mechanisms act in the following manner to match alveolar airflow and bloodflow. In an area of alveoli whose airflow is inadequate in relation to their bloodflow, the oxygen level will be low and the carbon dioxide level will be high. As a consequence, the pulmonary blood vessels in the area constrict, and the bloodflow decreases. In addition, the respiratory passageways dilate, and the airflow increases. Thus, the alveolar airflow and bloodflow become more closely matched. In an area of alveoli whose airflow is excessive in relation to their bloodflow, the oxygen level will be high and the carbon dioxide level will be low. As a consequence, the pul-

monary blood vessels in the area dilate, and the bloodflow increases. In addition, the respiratory passageways constrict, and the airflow decreases. Again, the alveolar airflow and bloodflow become more closely matched.

TRANSPORT OF GASES BY THE BLOOD

PARTIAL PRESSURE

The atmosphere is a mixture of several gases, including nitrogen, oxygen, and carbon dioxide. The total atmospheric pressure at sea level is 760 mm Hg. According to Dalton's law of gases, the total pressure exerted by a mixture of gases is equal to the sum of the separate pressures exerted by each individual gas. Each individual gas in a mixture exerts a *partial pressure* according to its percentage concentration in the mixture that contributes to the total pressure exerted by the mixture. Thus, each atmospheric gas acts independently of the others in generating the total atmospheric pressure, and each contributes partially to the total. For example, oxygen, which makes up approximately 21% of the atmospheric mixture of gases, contributes approximately 21% of the total pressure. Thus, the partial pressure of oxygen in the atmosphere at sea level is approximately 21%, or 160 mm Hg, of the total 760 mm Hg sea-level atmospheric pressure.

GASES IN LIQUIDS

When a liquid is exposed to a gaseous atmosphere, gases enter the liquid and dissolve in it in proportion to their individual gas pressures. The number of molecules of a particular gas that enters a liquid, however, depends not only on the pressure of that gas in the atmosphere, but also on the solubility of the gas in the liquid. Nevertheless, by increasing the pressure of the gas, more molecules of it will dissolve in the liquid. A gas that is dissolved in a liquid can be treated mathematically under specific conditions as though it were in a free state and able to exert an independent pressure or partial pressure. Thus, each of the gases of the atmosphere can, while in solution, be considered to exert a pressure or partial pressure. Once a gas dissolves in a liquid, the gas molecules can diffuse, and the net diffusion is from regions of high pressure to regions of lower pressure. Moreover, if a liquid contains a particular gas at a higher pressure than the pressure of that gas in the surrounding atmosphere, molecules of the gas will leave the liquid and enter the atmosphere.

OXYGEN AND CARBON DIOXIDE EXCHANGE AMONG LUNGS, BLOOD, AND TISSUES

The partial pressure of oxygen (PO_2) in the atmosphere at sea level, as previously noted, is approximately 160 mm Hg. The atmospheric partial pressure of carbon dioxide (PCO_2) is only about 0.3 mm Hg. However, due to the humidification of the inhaled air and to the mixing of atmospheric air with air remaining in the lungs during each respiratory cycle, the lungs do not contain gases at the same partial pressures that exist in the atmosphere. The actual PO_2 and PCO_2 in the alveoli of the lungs, as well as the partial pressures of these gases in the blood entering and leaving the alveolar capillaries and in the body tissues of a resting person at sea level, are given in Figure 15.8. As can be seen from the figure, within the alveoli the PO_2 is about 104 mm Hg, and the PCO_2 is approximately 40 mm Hg. These pressures remain fairly constant throughout the average respiratory cycle.

As the figure also shows, blood entering the alveolar capillaries from the tissues contains oxygen at a partial pressure of only 40 mm Hg and carbon dioxide at a partial pressure of 45 mm Hg. As a result of the partial pressure differences between the oxygen and carbon dioxide in the blood and the oxygen and carbon dioxide in the alveoli of the lungs, there is a net diffusion of oxygen into the blood from the alveoli and a net diffusion of carbon dioxide into the alveoli from the blood. As a consequence of these activities, the partial pressures of oxygen and carbon dioxide in the blood leaving the alveolar capillaries are essentially equal to the partial pressure of these gases in the alveoli (PO_2: 104 mm Hg; PCO_2: 40 mm Hg).

Within the tissues, metabolic reactions utilize oxygen and produce carbon dioxide. Thus, the PO_2

in the tissues is approximately 40 mm Hg, whereas the PCO_2 is about 45 mm Hg. When the blood from the lungs, which contains oxygen at a relatively high partial pressure, reaches the tissues, there is a net diffusion of oxygen from the blood to the tissues. In a similar fashion, there is a net diffusion of carbon dioxide from the tissues to the blood. The end result is that the venous blood leaving the tissues contains oxygen at a partial pressure of about 40 mm Hg and carbon dioxide at a partial pressure of about 45 mm Hg. It is these partial pressures that were previously described for blood entering the alveolar capillaries of the lungs.

Thus, the exchanges of oxygen and carbon dioxide that occur between the blood and alveoli on the one hand, and the blood and tissues on the other, take place by simple diffusion and depend on the pressure differences of the respective gases that exist between the blood and the alveoli, and the blood and the tissues.

OXYGEN TRANSPORT IN THE BLOOD

Since oxygen is relatively insoluble in water, relatively little actually dissolves in the fluid of the blood. In fact, under normal circumstances, each liter of systemic arterial blood contains only about 3 ml of dissolved oxygen. However, the total oxygen content of a liter of systemic arterial blood is about 197 ml. The additional 194 ml of oxygen is bound to the iron of the heme groups of hemoglobin molecules in erythrocytes.

The combination of hemoglobin with oxygen is reversible, and whether hemoglobin binds with or releases oxygen depends in large part on the oxygen partial pressure. The degree of saturation of hemoglobin with oxygen at any given PO_2 is indicated by the oxygen-hemoglobin dissociation curve (F15.9). When the PO_2 is relatively high, hemoglobin binds with much oxygen and is essentially completely saturated. At lower oxygen partial pressures, hemoglobin binds with less oxygen and is only partially saturated.

The PO_2 is relatively high in the alveolar capillaries of the lungs. Here, oxygen from the alveoli diffuses into the plasma, and then into the erythrocytes, where it binds with hemoglobin. Within the various body tissues, however, metabolic activities utilize oxygen, and there is a net diffusion of oxygen out of the blood. Thus, the P_{O_2} in the tissue capillaries is lower than that in the alveolar capillaries, and hemoglobin cannot bind with as much oxygen in the tissue capillaries as it can in the alveolar capillaries. As a result, the oxygen-rich hemoglobin that leaves the alveolar capillaries releases some of its oxygen when it reaches the tissue capillaries. This oxygen can then diffuse into the tissues for their use.

In addition to the P_{O_2}, several other factors affect the binding of oxygen to hemoglobin. Under acidic conditions, the amount of oxygen that binds to hemoglobin at any given oxygen partial pressure is diminished. Thus, the higher the hydrogen ion concentration—that is, the lower the pH—the less oxygen is bound to hemoglobin at any given P_{O_2} (F15.10). This effect is due to the fact that hydrogen ions bind with the hemoglobin molecules, altering their molecular structure and thereby decreasing the amounts of oxygen they can carry.

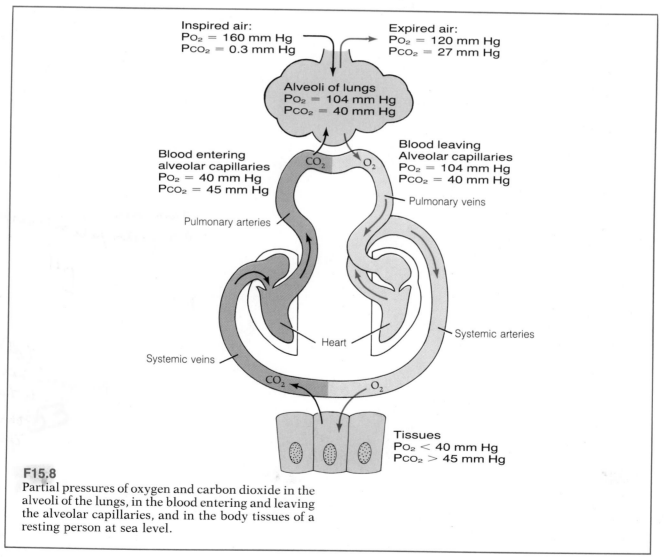

F15.8
Partial pressures of oxygen and carbon dioxide in the alveoli of the lungs, in the blood entering and leaving the alveolar capillaries, and in the body tissues of a resting person at sea level.

The partial pressure of carbon dioxide has basically the same effect on the binding of oxygen and hemoglobin as the hydrogen ion concentration. That is, the higher the P_{CO_2}, the less oxygen is bound to hemoglobin at any given P_{O_2} (F15.11). This effect is due in large measure to the fact that the P_{CO_2} can influence the pH in the following manner:

$$CO_2 + H_2O \rightleftharpoons H_2CO_3 \rightleftharpoons H^+ + HCO_3^-$$

| carbon dioxide | water | carbonic acid | hydrogen ion | bicarbonate ion |

As these reactions show, carbon dioxide can combine with water to form carbonic acid. (Although not absolutely required for this reaction to occur, an enzyme within erythrocytes, *carbonic anhydrase*, greatly increases the rate of the reaction.) The carbonic acid can dissociate into hydrogen ions and bicarbonate ions, and the hydrogen ions contribute to the hydrogen ion concentration. Thus, an increase in the P_{CO_2} tends to increase the hydrogen ion concentration and therefore lower the pH.

Temperature is another factor that influences the binding of oxygen and hemoglobin, and the higher the temperature, the less oxygen is bound to hemoglobin at any given P_{O_2} (F15.12).

The influences of pH, P_{CO_2}, and temperature on the binding of oxygen by hemoglobin operate to ensure adequate deliveries of oxygen to active tissues that need it most. Active tissues tend to have higher hydrogen ion concentrations in their vicinities than do less active tissues (active skeletal muscles, for example, can produce lactic acid). They also tend to produce more carbon dioxide and to have higher temperatures as a result of their metabolic activity than do less active tissues. Consequently, hemoglobin binds less oxygen in active tissues (and thus provides more for delivery) than it does in less active tissues.

A substance produced by erythrocytes called *DPG* (2,3-diphosphoglyceric acid) binds reversibly with hemoglobin and alters the hemoglobin molecule so it releases oxygen. Thus, DPG promotes the release of oxygen from hemoglobin. DPG production increases in conditions associated with decreased oxygen supplies to tissues, and some investigators believe this response helps maintain oxygen delivery to the tissues. However, other investigators point out that an increased level of

F15.9
Oxygen-hemoglobin dissociation curve. The curve indicates the percent saturation of hemoglobin at different partial pressures of oxygen (P_{O_2}) when the pH is 7.4 and the temperature is 38° C.

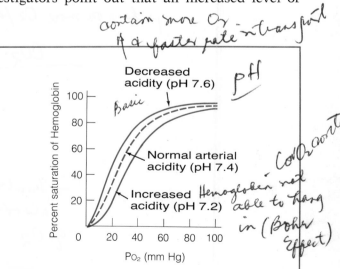

F15.10
Graph indicating the percent saturation of hemoglobin at different partial pressures of oxygen, showing the effects of acidity.

DPG not only favors the release of oxygen from hemoglobin in the tissues, but it also makes it more difficult for oxygen to combine with hemoglobin in the lung capillaries. Therefore, they question whether increased DPG production in conditions associated with decreased oxygen supplies to the tissues is always beneficial.

CARBON DIOXIDE TRANSPORT IN THE BLOOD

Carbon dioxide is transported by the blood in several different ways. A small amount of carbon dioxide—perhaps 8%—is transported in physical solution dissolved in the plasma. About 20% is transported in reversible association with various blood proteins. These carbon dioxide–protein complexes are called *carbamino compounds*. Carbon dioxide can combine with many proteins in the blood. However, the most abundant protein is the globin of hemoglobin, and many carbamino unions form between it and carbon dioxide. Thus, oxygen is carried by hemoglobin in association with the iron of the heme groups, and carbon dioxide is carried in reversible association with the globin protein portion of the hemoglobin molecule. Hemoglobin that is not carrying oxygen (*reduced hemoglobin*) is able to form carbamino compounds more readily than hemoglobin that is carrying oxygen (*oxyhemoglobin*). Thus, in the tissues, the presence of hemoglobin molecules that have given up their oxygen favors the combination of carbon dioxide with hemoglobin. In the lungs, the binding of oxygen to reduced hemoglobin molecules decreases their affinity for carbon dioxide, and favors the displacement of carbon dioxide from the molecules. The displaced carbon dioxide can diffuse into the alveoli for elimination from the body.

Most of the carbon dioxide in the blood—about 72%—is transported as bicarbonate ions (F15.13). Carbon dioxide produced in the tissues diffuses into the plasma and from there into erythrocytes. Within the erythrocytes, the enzyme carbonic anhydrase facilitates the combination of carbon dioxide and water to form carbonic acid. The carbonic acid can dissociate into hydrogen ions and bicarbonate ions, and most of the hydrogen ions associate with hemoglobin molecules. This association eliminates free hydrogen ions and thus helps prevent a substantial increase in the hydrogen ion

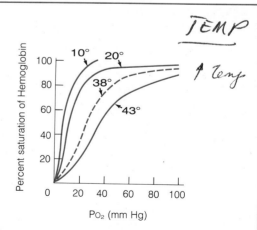

→ hang on more O_2 [heary extra?]

CO_2

Decreased carbon dioxide (PCO_2 20 mm Hg)

Normal arterial carbon dioxide (PCO_2 40 mm Hg)

Increased carbon dioxide (PCO_2 80 mm Hg) → *Greater unloading of O_2*

F15.11
Graph indicating the percent saturation of hemoglobin at different partial pressures of oxygen, showing the effects of carbon dioxide.

TEMP

↑ Temp

10° 20° 38° 43°

F15.12
Graph indicating the percent saturation of hemoglobin at different partial pressures of oxygen, showing the effects of temperature (°C).

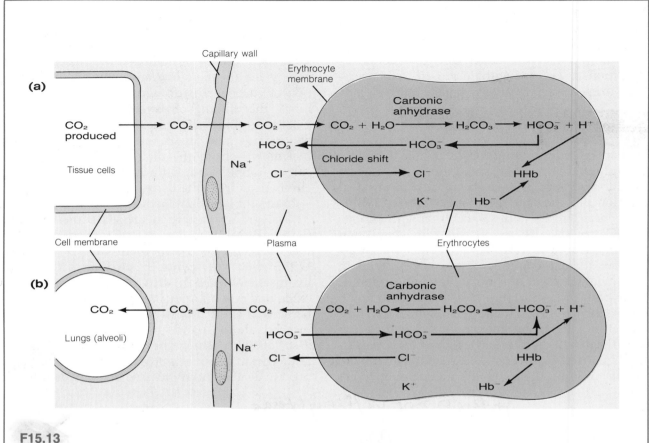

F15.13
Transport of carbon dioxide in the blood as bicarbonate ions. (a) Sequence of events that occurs in the

tissues. (b) Sequence of events that occurs in the lungs. (Hb is hemoglobin.)

concentration of the blood. In fact, because of the ability of hemoglobin molecules (and other substances in the blood) to combine with hydrogen ions, the pH of venous blood leaving the tissues (pH = 7.34) is only slightly lower than that of arterial blood flowing to the tissues (pH = 7.4). The bicarbonate ions resulting from the dissociation of carbonic acid molecules diffuse out of the erythrocytes into the plasma, and in response to this movement of negatively charged bicarbonate ions, chloride ions from the plasma enter the erythrocytes. This exchange is known as the chloride shift. As a result of these events, carbon dioxide from the tissues is transported as bicarbonate ions in the blood.

At the lungs, the reverse of the preceding events occurs. Carbon dioxide diffuses out of the blood into the alveoli, and the concentration of carbon dioxide within the plasma and the erythrocytes declines. Within the erythrocytes, hydrogen ions combine with bicarbonate ions to form carbonic acid, which is split by carbonic anhydrase into carbon dioxide and water. As the concentration of bicarbonate ions within the erythrocytes diminishes, bicarbonate ions from the plasma diffuse in and chloride ions leave. The carbon dioxide that is produced leaves the erythrocytes and diffuses into the alveoli for elimination from the body.

The combination of hydrogen ions and bicarbonate ions to form carbonic acid and ultimately carbon dioxide and water is enhanced by the binding of oxygen to hemoglobin in the lung capillaries. When hemoglobin combines with oxygen, its ability to bind hydrogen ions decreases, and more hydrogen ions leave the hemoglobin molecules than would otherwise be the case. These hydrogen ions can combine with bicarbonate ions to form carbonic acid and, ultimately, carbon dioxide and water.

As the preceding discussion indicates, carbonic anhydrase can catalyze the reaction

$$CO_2 + H_2O \overset{\text{carbonic}}{\underset{\text{anhydrase}}{\rightleftharpoons}} H_2CO_3 \rightleftharpoons H^+ + HCO_3^-$$

in either direction. The overall direction of the reaction depends on the partial pressure of carbon dioxide as well as on the pH and the bicarbonate ion concentration. In the tissues, where the P_{CO_2} is relatively high (and hydrogen ions are tied up by hemoglobin molecules), the formation of bicarbonate ions is favored. In the lungs, where the P_{CO_2} is lower (and hemoglobin molecules release hydrogen ions), the reaction moves from bicarbonate ions and carbonic acid toward carbon dioxide and water.

PROBLEMS OF GAS TRANSPORT

The transport of oxygen and carbon dioxide by the blood can be affected adversely by a variety of conditions. These include malfunctions of the cardiovascular system that interfere with the transport of gases by the blood, respiratory system problems that upset the gas exchange process, and deficiencies in the composition of the air breathed. The following are only a few of the possible conditions that can cause problems.

Cyanosis

Cyanosis is a bluish discoloration of the skin and mucous membranes. Reduced hemoglobin—that is, hemoglobin that is not combined with oxygen—appears darker than oxyhemoglobin, and cyanosis occurs when there is a significant concentration of reduced hemoglobin in the arterial blood. Cyanosis can result from inadequacies of either the respiratory or circulatory systems.

Hypoxia

Any state in which an inadequate amount of oxygen is delivered to the tissues is called *hypoxia*. Hypoxia can be caused by a number of factors.

In some cases, not enough oxygen reaches the blood, and the arterial P_{O_2} is below normal. Hypoxia due to such an occurrence is referred to as *hypoxemic hypoxia* (*arterial hypoxia*). Hypoxemic hypoxia can result from insufficient oxygen in the air breathed or from respiratory problems—such as pneumonia, emphysema, or paralysis of the respiratory muscles—that prevent sufficient oxygen from reaching the blood within the lungs.

In other cases, enough oxygen reaches the blood, but there is not enough hemoglobin available to carry it to the tissues. Since this problem is often associated with anemia, hypoxia due to a deficiency of hemoglobin is referred to as *anemic hypoxia*.

In still other cases, the bloodflow to the tissues is less than normal. The blood may be able to carry a normal amount of oxygen, but oxygen is not delivered to the tissues fast enough for the cells to maintain their normal metabolism. Hypoxia due to such an occurrence is called *stagnant hypoxia* (*hypokinetic hypoxia*). Stagnant hypoxia can result from heart failure or the presence of emboli in the blood.

Carbon Monoxide Poisoning

Carbon monoxide (CO) is a colorless, odorless gas. It is present in the exhaust fumes of automobiles and is produced when carbon products such as wood and coal are burned. Carbon monoxide is hazardous because it and oxygen compete for the same hemoglobin binding sites. Hemoglobin has a much greater affinity for carbon monoxide than it does for oxygen, and when even small amounts of carbon monoxide are present in the air breathed, the carbon monoxide preferentially occupies the oxygen-binding sites of the hemoglobin molecules. Moreover, the hemoglobin–carbon monoxide bond is so strong that very little carbon monoxide is removed from the blood. Consequently, carbon monoxide causes drowsiness, coma, and ultimately death due to hypoxia.

CONTROL OF RESPIRATION

The muscles responsible for inspiration—for example, the diaphragm and external intercostals—are skeletal muscles that require neural stimulation to initiate their contractions. When nerve impulses activate the inspiratory muscles, the thorax expands and the lungs fill with air.

Following inspiration, nerve impulses to the inspiratory muscles diminish greatly, and the diaphragm and other muscles involved in inspiration relax. When the inspiratory muscles relax, expiration occurs as a result of the elastic recoil of structures such as the lungs and chest wall, which returns the thorax to its resting position. During forced expiration, nerve impulses activate expiratory muscles—such as the internal intercostals and muscles of the anterior abdominal wall—that help depress the ribs and diminish the size of the thoracic cavity.

Even when the thorax is at rest at the end of expiration, some nerve impulses are still transmitted to the respiratory muscles, and the muscles remain in a state of slight contraction (tonus), which helps maintain normal body posture.

GENERATION OF RHYTHMICAL BREATHING MOVEMENTS

The rhythmic pattern of inspiration and expiration characteristic of normal breathing depends on the cyclical stimulation of the respiratory muscles. This cyclical stimulation is due primarily to the activity of neurons called inspiratory neurons, whose cell bodies are found in the medulla oblongata of the brain (F15.14). The medullary inspiratory neurons connect either directly or indirectly with neurons that supply the inspiratory muscles.

Medullary Inspiratory Neurons

The discharges of the **medullary inspiratory neurons** are synchronized with inspiration, and these neurons cease discharging during expiration. It is currently believed that the medullary inspiratory neurons possess an inherent automaticity and rhythmicity. As a result, they undergo spontaneous, cyclical self-excitation and thus provide the underlying rhythm for respiration. Since expiration is basically a passive process that occurs when the muscles controlling inspiration relax, the cyclical activity of the inspiratory neurons can account for alternating cycles of inspiration and expiration.

Inputs from other neurons apparently modify the basic rhythm of the medullary inspiratory neurons. Among these inputs are connections with neurons of the pons, and the influence of afferent impulses from stretch receptors in the lungs.

Neurons of the Pons

Within the pons are two areas called the **pneumotaxic area** and the **apneustic area.** The pneumotaxic area continually sends inhibitory signals to the medullary inspiratory neurons that tend to limit the period of inspiration. When the signals from the pneumotaxic area are strong, the period of inspiration is short; when the signals are weak, the inspiratory period lasts longer. The apneustic area sends stimulatory signals to the medullary inspiratory neurons that tend to prolong inspiration. However, the pneumotaxic area normally overrides the apneustic area.

Lung Stretch Receptors

The lungs contain stretch receptors that increase their activity as the lungs expand during inspiration. These receptors send afferent impulses to the medullary inspiratory neurons to inhibit the activity of the inspiratory neurons and aid in terminating inspiration. This pathway, however, does not appear to be of great importance in the control of respiration during normal breathing, and the response is weak unless the lungs are distended to a considerable extent. Thus, the lung stretch receptors are believed to be mainly a protective mechanism that prevents the lungs from overfilling.

Medullary Expiratory Neurons

The medulla also contains neurons called expiratory neurons. During most normal, quiet breathing, the **medullary expiratory neurons** are not active, and expiration is a passive process that occurs when the inspiratory muscles relax. However, when the respiratory drive for increased pulmonary ventilation becomes greater than normal, the expiratory neurons are activated, apparently by the medullary inspiratory neurons. The expiratory neurons send impulses to the expiratory muscles to facilitate expiration.

In summary, a major factor initiating inspiration is the spontaneous discharge of the inspiratory neurons in the medulla. This activity ultimately stimulates the diaphragm and other muscles involved in inspiration. Expiration occurs when the medullary inspiratory neurons cease firing both as a result of their own self-limitation and also as a result of inhibitory im-

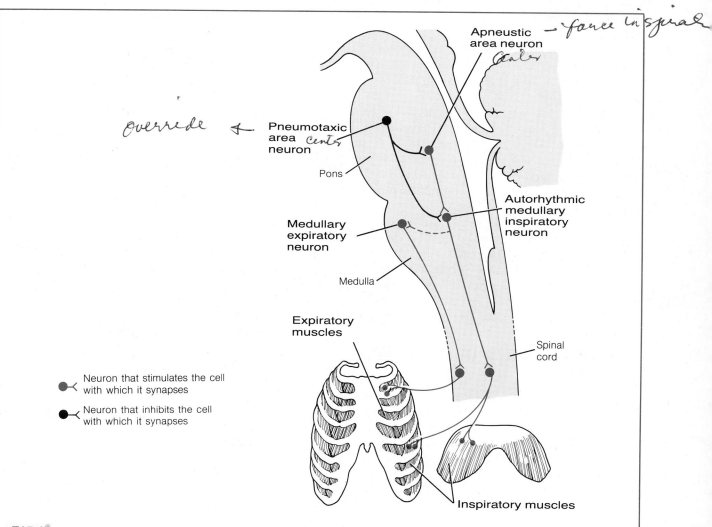

F15.14

Diagrammatic representation of the pontine and medullary neural pathways involved in the control of respiration. The dashed line indicates that the medullary inspiratory neurons apparently activate the expiratory neurons when the respiratory drive for increased pulmonary ventilation becomes greater than normal.

pulses from such sources as the pneumotaxic area in the pons and lung stretch receptors. Following expiration, the inspiratory neurons spontaneously discharge again, and the cycle repeats itself. In some cases, expiration is facilitated by the activation of neurons that cause the contraction of expiratory muscles.

CONTROL OF RATE AND DEPTH OF BREATHING

The rate and depth of respiratory movements are important because they determine the alveolar ventilation. The rate of respiration is determined by the length of time that elapses between bursts of nerve impulses to the respiratory muscles. The more often the respiratory muscles are activated, the more rapid will be the rate of respiration. The depth of respiration is determined both by the number of motor units of the respiratory muscles that are active and by the frequency of nerve impulses to the muscle cells of each motor unit. The more motor units that are active, the greater will be the expansion of the thoracic cavity, and the more the lungs will expand and fill with air.

Breathing supplies the body with oxygen and eliminates carbon dioxide. Moreover, changes in ventilation can influence the plasma concentration of hydrogen ions because carbon dioxide can, as described earlier, undergo the following reaction:

$$CO_2 + H_2O \rightleftharpoons H_2CO_3 \rightleftharpoons H^+ + HCO_3^-$$

Thus, a change in ventilation that increases or decreases the partial pressure of carbon dioxide in the plasma is likely to affect the plasma hydrogen ion concentration. As a result, it seems reasonable to consider oxygen, carbon dioxide, and hydrogen ions as possible regulators of respiratory activity.

Arterial P_{O_2}

If the partial pressure of oxygen in the arterial blood is reduced below normal while the P_{CO_2} and pH are held constant, an increase in alveolar ventilation occurs (F15.15). However, within the normal range of arterial partial pressures of oxygen (the normal arterial P_{O_2} is about 95 mm Hg), the effect of oxygen on alveolar ventilation is relatively slight. It is not until low partial pressures of oxygen are present that a major stimulatory effect of oxygen on alveolar ventilation is evident. This response is reasonable because hemoglobin remains almost completely saturated with oxygen until there is a considerable drop in the P_{O_2} (see F15.9). Thus, the transport of oxygen to the tissues is not substantially diminished until the arterial P_{O_2} falls quite low.

The receptors sensitive to the partial pressure of oxygen are located at the bifurcations of the common carotid arteries, as well as at the arch of the aorta, in structures called the **carotid** and **aortic bodies.** Nerve impulses from these receptors are ultimately transmitted to the respiratory neurons to stimulate respiration. The receptors of the carotid and aortic bodies are sensitive to the P_{O_2} within their local tissue environment. However, the carotid and aortic bodies receive such an abundant blood supply that their oxygen needs are normally met without creating a significant difference between the P_{O_2} in their local environment and the P_{O_2} in the arterial blood. That is, under normal circumstances, the P_{O_2} in the local environment of the carotid and aortic bodies is almost identical to that in the arterial blood. Thus, in essence, the receptors of the carotid and aortic bodies respond to the arterial P_{O_2}. A decrease in the P_{O_2} increases the discharge rates of the receptors, and the nerve impulses that are generated stimulate respiration.

Despite the presence of these receptors, conditions of moderate anemia generally do not alter ventilation even though the total oxygen content of the blood may be below normal. In anemia, the reduced oxygen content of the blood is due to a reduced number of functional hemoglobin molecules, and the arterial P_{O_2}, which is determined solely by the amount of dissolved oxygen, is not altered. Thus, as long as the carotid and aortic bodies receive a sufficient supply of oxygen, the P_{O_2} in their local tissue environment remains essentially normal, and the discharge rates of the receptors are not significantly altered. However, in certain conditions, such as severe low blood pressure, the oxygen supply to the carotid and aortic bodies is so deficient that the utilization of ox-

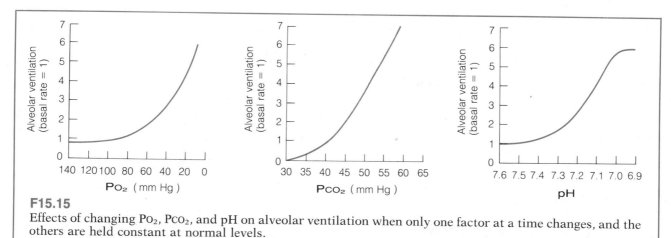

F15.15

Effects of changing P_{O_2}, P_{CO_2}, and pH on alveolar ventilation when only one factor at a time changes, and the others are held constant at normal levels.

ygen by the cells causes the P_{O_2} in the local tissue environment to fall below that in the arterial blood. In such cases, the receptors increase their rates of discharge even though the arterial P_{O_2} may still be normal.

Arterial P_{CO_2}

The partial pressure of carbon dioxide in the plasma is a major regulator of alveolar ventilation. An elevation of only 5 mm Hg in the arterial P_{CO_2}, with the P_{O_2} and pH held constant, increases ventilation 100% (F15.15). Conversely, a lower than normal arterial P_{CO_2} inhibits respiration.

The respiratory effects of the arterial P_{CO_2} are due to associated changes in hydrogen ion concentrations, particularly in the cerebrospinal fluid and the interstitial fluid of the brain. Carbon dioxide can readily diffuse from the plasma to the cerebrospinal fluid and the interstitial fluid of the brain, where it combines with water to form carbonic acid. The carbonic acid can dissociate into hydrogen ions and bicarbonate ions, thus increasing the hydrogen ion concentration of the cerebrospinal fluid and the interstitial fluid of the brain. The increase in hydrogen ion concentration stimulates respiration by acting on chemosensitive cells in the medulla (central chemoreceptors) that have synaptic input to the respiratory neurons.

The carotid and aortic bodies are also sensitive to the arterial P_{CO_2}, apparently because of the influence of carbon dioxide on the hydrogen ion concentration within their local tissue environment. However, the carotid and aortic bodies are much less important than the central chemoreceptors in mediating the respiratory effects of the arterial P_{CO_2}.

Arterial pH

The hydrogen ion concentration of the plasma can also affect respiration. Increasing the arterial concentration of hydrogen ions—that is, decreasing the arterial pH—with the P_{O_2} and P_{CO_2} held constant substantially increases alveolar ventilation (F15.15). Hydrogen ions can stimulate both the central chemoreceptors and the carotid and aortic bodies, but there is disagreement about which of these pathways is most important in mediating the respiratory effects of the arterial pH. In any case, compared to carbon dioxide, hydrogen ions diffuse rather poorly from the blood to the cerebrospinal fluid, and the arterial P_{CO_2} is generally more important than the arterial pH in normal respiratory regulation.

Interaction of Different Factors in the Control of Respiration

The various factors that influence respiration interact with one another to regulate breathing. For example, if the arterial P_{O_2} is decreased without keeping the P_{CO_2} or the pH constant, any increase in respiration due to the stimulatory effect of the low P_{O_2} can cause additional carbon dioxide to be expelled. The expulsion of additional carbon dioxide can lower the arterial P_{CO_2} and decrease the arterial concentration of hydrogen ions. Thus, the stimulatory effect of the decreased arterial P_{O_2} may be quickly counterbalanced by a lowered P_{CO_2} and a decreased concentration of hydrogen ions, both of which act to inhibit ventilation. At any instant, therefore, respiration is controlled by multiple factors, and the sensitivity to some factors, such as oxygen, may not be as great as the sensitivity to others, such as carbon dioxide.

EFFECTS OF EXERCISE ON THE RESPIRATORY SYSTEM

INCREASED PULMONARY VENTILATION

The muscular activity associated with exercise rapidly consumes oxygen and produces carbon dioxide. Thus, during exercise, the body requires more oxygen and must eliminate more carbon dioxide than when at rest, and it is not surprising that during exercise both the rate and depth of breathing increase. During heavy exercise, the volume of air that reaches the alveoli may increase to as much as 20 times the resting volume.

The precise changes responsible for increased respiration during exercise are still not clearly understood, but several factors are thought to be involved. It is believed that when brain areas controlling movement transmit signals to exercising muscles, signals are also transmitted to the respiratory neurons to stimulate respiration. Consequently, increases in respiration occur even before muscular activity causes substantial changes in blood P_{O_2}, P_{CO_2}, or pH levels. Neural signals from joints and muscles involved in physical movements may also contribute to the stimulation of respiration during exercise.

If the neural mechanisms do not correctly adjust respiration to meet body requirements during exercise, the levels of chemical factors such as carbon dioxide may change, and these factors can then contribute to the overall control of respiration. The mechanisms that control respiration during exercise are quite effective, and in all but very heavy exercise the arterial P_{O_2}, P_{CO_2}, and pH remain almost exactly normal.

INCREASED DELIVERY OF OXYGEN TO THE TISSUES

Both at rest and during exercise, the blood that leaves the lungs through the pulmonary veins is approximately 97% saturated with oxygen, and the factor that seems to be most important in determining how much oxygen can be delivered to the tissues during exercise is not the pulmonary ventilation but the amount of blood the heart is capable of pumping. If a person can increase his or her cardiac output through training, the pulmonary ventilation appears to be capable of increasing enough so that the extra volume of blood is maximally saturated with oxygen.

COMMON DISEASES OF THE RESPIRATORY SYSTEM

COMMON COLD

The *common cold*, or *acute coryza*, is an inflammation of the mucous membrane of the upper respiratory tract that is familiar to most people. The initial inflammation is caused by various viruses and is often followed by bacterial infections of the sinuses, ears, or bronchi. When inflamed, the mucous membrane becomes engorged and swollen, causing discomfort and difficulty in breathing. Later the mucous membrane discharges a watery fluid that makes its presence known in the form of a "runny nose." Such discharges from the mucous membranes that line the paranasal sinuses can irritate the larynx and trachea, producing a cough.

BRONCHIAL ASTHMA

Bronchial asthma is an allergic (antigen-antibody) response to foreign substances that generally enter the body by way of either the air breathed or the food eaten. It is characterized by episodes of wheezing and difficult breathing. In response to the foreign substances, the mucous membranes of the respiratory system secrete excessive amounts of mucus, and the smooth muscles that surround the smaller bronchi and the bronchioles go into spasms. These responses narrow the passageways, making it difficult to move air into or out of the alveoli.

BRONCHITIS

Bronchitis is an acute or chronic inflammation of the bronchial tree. It is caused by bacterial infection or by irritants such as smoke or chemicals in the inhaled air. The mucous membranes of the respiratory system produce a sticky secretion that inhibits the normal protective function of the macrophages of the respiratory tract and hinders the self-cleaning actions of the cilia of the cells that line the bronchi. As the secretions accumulate within the bronchi, they are removed by coughing, which is annoying but serves a useful purpose by helping to keep the lungs clear.

TUBERCULOSIS

Tuberculosis, an infection caused by the tubercle bacillus (*Mycobacterium tuberculosis*), can affect many parts of the body. However, because the bacterium most commonly enters the body by inhalation, tuberculosis of the lungs is the most prevalent form. Even if not inhaled, the bacilli may enter the lymphatic system or the bloodstream and thus reach the lungs. When the bacilli reach the lungs, the lung tissue reacts by forming small clumps (tubercles) around the bacilli. Many of the bacilli are engulfed by phagocytes, and fibrous walls form around them. If the bacilli are not successfully walled off, the lung tissue is destroyed and the site of infection spreads. The process may continue until both lungs are extensively destroyed. Even if such a massive involvement of the lungs does not occur, the fibrosis in the affected portions interferes with the diffusion of gases and causes the lungs to lose their elasticity, thereby reducing the vital capacity.

EMPHYSEMA

Emphysema is a condition that develops slowly as a secondary response to other respiratory problems, such as chronic bronchitis or tuberculosis, or to environmental irritants such as cigarette smoke or industrial pollutants. The alveoli become overdistended, and the walls of the alveoli break down and are often replaced by fibrous tissue. Consequently, the surface area across which gaseous exchange can occur is greatly reduced. As a result, the partial pressure of oxygen in the blood is lowered, and even mild exercise increases the breathing rate. In addition, the elastic tissues of the overexpanded lungs are reduced, making expiration difficult. A reduced expiratory volume is an early symptom of emphysema.

There are, therefore, two basic problems facing the victim of emphysema: (1) the lungs are "fixed" in inspiration; and (2) the respiratory surfaces of the lungs have deteriorated so much that they are no longer adequate to accomplish normal gas exchange. Unfortunately, the disease is progressive and irreversible.

PNEUMONIA

The inflammation of *pneumonia* causes a fibrinous exudate to be produced within the alveoli. The lung, or a part of it, becomes solid and airless, which makes it very difficult for gaseous exchange to occur within the alveoli. Most cases of pneumonia are probably caused by one of several viruses. Another common cause is the pneumococcus bacterium. However, pneumonia can also be caused by the inhalation of foods or other foreign bodies that cause obstruction of a bronchus. This obstruction can lead to collapse of the lung, accumulation of fluids within the lung, and subsequent infection.

PLEURISY

Pleurisy is an inflammation of the pleural membranes that surround the lungs. The pleura are most commonly infected with the pneumococcus, the streptococcus, or the tubercle bacillus. In the early stages, the inflamed pleural membranes are "dry" and are covered with fibrous material. This condition causes pain during breathing. Adhesions between the layers of the pleura may develop as a result of pleurisy, and in severe cases, surgery may be necessary to remove them. In the later stages of pleurisy there is often an excessive secretion of pleural fluid into the pleural cavity.

EFFECTS OF AGING ON THE RESPIRATORY SYSTEM

Tissue changes that occur in the respiratory system with increased age cause the chest wall to become more rigid and the lungs to become less elastic. Thus, although the total lung volume does not change with age, the rigidity of the chest wall and the loss of elasticity in the lungs results in a diminished ventilating capacity. Because of these changes, the vital capacity decreases in males from about 4700 ml at age 20 to about 4000 ml at age 70. Accompanying these changes is a decrease in the arterial Po_2, which is quite pronounced when a person is lying in a supine position where breathing is more difficult. For this reason elderly people tend to become hypoxic during sleep and they are often more comfortable if supported by several pillows. Carbon dioxide diffuses through tissues much more rapidly than oxygen; therefore, in contrast to the Po_2, the Pco_2 of the arterial blood is not affected much by age.

With increasing age, there is a decrease in the phagocytic activity of macrophages and in the activity of the cilia of the epithelial linings of the respiratory tract. This decreased activity makes the cleaning of the respiratory tract lining less efficient. The rigidity of the chest wall, the loss of elasticity in the lungs, and the diminished phagocytic activity and ciliary action all cause elderly people to be more susceptible to pneumonia and other respiratory infections.

STUDY OUTLINE

STAGES OF RESPIRATION   For purposes of analysis, respiration can be divided into five different stages:
1. the movement of air into and out of the lungs
2. the exchange of oxygen and carbon dioxide between the air in the lungs and the blood within the pulmonary capillaries
3. the transport of oxygen and carbon dioxide throughout the body by the blood
4. the exchange of oxygen and carbon dioxide between the blood and the interstitial fluid and cells
5. the utilization of oxygen and the production of carbon dioxide by metabolic processes within cells
   p. 377

ORGANIZATION OF THE RESPIRATORY SYSTEM
1. To reach exchange sites in the lungs, air must flow through conducting passageways. As it flows, the air is warmed and humidified and small particles are removed.
2. The lungs are located within the thoracic cavity and are enclosed in a double-walled sac called the pleura.
3. The pleural fluid couples the visceral pleura and the parietal pleura (and therefore the lungs and the thoracic cavity structures) to one another. Because of this coupling, the movements of the lungs closely follow the movements of the structures that form the thoracic cavity.   **pp. 377–378**

AIRFLOW IN THE RESPIRATORY SYSTEM   Air enters through the nose or mouth and flows into the pharynx and then the larynx. From the larynx the air flows along the trachea and through progressively smaller branching passageways (bronchi, bronchioles) to reach alveoli, which are sites of gas exchange between the air in the lungs and the blood.   **pp. 378–380**

MECHANICS OF BREATHING

Atmospheric Pressure   The atmospheric pressure at sea level is 760 mm Hg.   **pp. 380–387**

Pressure Relationships in the Thoracic Cavity
1. The elastic connective tissue of the lungs and the surface tension of the fluid that coats the alveoli favor a reduction in size of lungs, as does the intrapleural pressure. These forces are balanced by the intrapulmonary pressure, and the lungs normally remain expanded and do not collapse.
2. If air enters the pleural cavity, the balance of forces maintaining the expansion of the lungs is upset and the lungs collapse.

Ventilation of the Lungs   Air flows from an area of higher pressure to an area of lower pressure, and the pressure within the lungs is altered by altering the volume of the thoracic cavity.

   INSPIRATION   The volume of the thoracic cavity is increased by the contraction of the diaphragm and the elevation of the ribs. This increase, in turn, increases the lung volume and the intrapulmonary pressure drops below the atmospheric pressure. As a consequence, air flows into the lungs.

EXPIRATION When muscles involved in inspiration relax, the volume of the thoracic cavity—and thus the lungs—decreases due to the elastic recoil of respiratory structures such as the lungs and chest wall. The decrease in lung volume raises the intrapulmonary pressure above the atmospheric pressure and air flows out of the lungs. During forced expiration, the contraction of expiratory muscles further reduces the volume of the thoracic cavity.

Factors That Influence Pulmonary Airflow The volume of air that flows in a given time between the atmosphere and the alveoli is influenced by the pressure that moves the air through the respiratory passageways and by the resistance that the air encounters as it flows through the passageways. The diameters of the passageways affect resistance. Parasympathetic stimulation, histamine, and leukotrienes contract smooth muscles of the respiratory passageways and increase resistance. Sympathetic stimulation and epinephrine relax smooth muscle of passageways and decrease resistance.

Surface Tension and Pulmonary Surfactant Pulmonary surfactant produced by alveolar cells lowers the surface tension of the fluid that coats the exposed surfaces of the alveoli, allowing the lungs to be expanded with a reasonable muscular effort.

Lung Volumes
RESIDUAL VOLUME About 1000–1200 ml of air that cannot be exhaled from the lungs.
TIDAL VOLUME The volume of air that moves in and out of lungs with each breath (about 500 ml during normal, quiet breathing).
INSPIRATORY RESERVE Extra volume of air that can be inspired in addition to the normal, quiet tidal volume (about 2100–3000 ml).
EXPIRATORY RESERVE Extra volume of air that can be expired following a normal, passive expiration of the quiet tidal volume (about 800–1200 ml).
VITAL CAPACITY Sum of inspiratory reserve volume, normal, quiet tidal volume, and expiratory reserve volume (3400 to 4700 ml).

Anatomical Dead Space Air-filled respiratory passageways in which no gas exchange takes place. The volume of the anatomical dead space is about 150 ml.

Minute Respiratory Volume The volume of air moved into the respiratory passageways in one minute. It is equal to the respiratory rate multiplied by the volume of air that enters the respiratory passageways with each breath (the tidal volume).

Alveolar Ventilation The volume of atmospheric air that enters the alveoli (per minute). It is equal to the respiratory rate multiplied by the volume of atmospheric air that enters the alveoli per breath. In general, the alveolar ventilation is increased more effectively by increasing the volume of air inspired with each breath than by increasing the respiratory rate.

Matching of Alveolar Airflow and Bloodflow Local autoregulatory mechanisms are involved.
1. Low oxygen levels cause pulmonary vessels to constrict and high oxygen levels cause them to dilate.
2. Low carbon dioxide levels cause respiratory passageways to constrict and high carbon dioxide levels cause them to dilate.

TRANSPORT OF GASES BY THE BLOOD
pp. 387–394

Partial Pressure Oxygen, 21% of atmospheric gas mixture, contributes 21% of total atmospheric pressure.

Gases in Liquids Gases enter a liquid and dissolve in it in proportion to their individual gas pressures.

Oxygen and Carbon Dioxide Exchange among Lungs, Blood, and Tissues
LUNG Partial pressure differences between O_2 and CO_2 in blood and in alveoli lead to a net diffusion of O_2 into blood from alveoli and a net diffusion of CO_2 into alveoli from blood.
TISSUES There is a net diffusion of O_2 from blood to tissues and a net diffusion of CO_2 from tissues to blood.

Oxygen Transport in the Blood Most oxygen is transported bound to the iron of hemoglobin in erythrocytes. The binding of oxygen to hemoglobin is affected by the oxygen partial pressure, hydrogen ion concentration, carbon dioxide partial pressure, and temperature.

Carbon Dioxide Transport in the Blood
1. Dissolved in plasma.
2. Carbamino compounds (CO_2–protein complexes).
3. Bicarbonate ions.

Problems of Gas Transport
CYANOSIS A bluish discoloration of the skin and

mucous membranes. It occurs when there is a significant concentration of reduced hemoglobin in the arterial blood.

HYPOXIA   Inadequate amounts of oxygen delivered to tissues.

CARBON MONOXIDE POISONING   Carbon monoxide occupies oxygen-binding sites of hemoglobin.

CONTROL OF RESPIRATION   Inspiration occurs when nerve impulses activate diaphragm and external intercostals; thorax expands; lungs fill with air. Expiration occurs when nerve impulses to inspiratory muscles diminish and respiratory structures such as the lungs and chest wall recoil. During forced expiration, nerve impulses activate expiratory muscles. **pp. 394–398**

Generation of Rhythmical Breathing Movements

MEDULLARY INSPIRATORY NEURONS   Believed to be inherently rhythmic.

NEURONS OF THE PONS
PNEUMOTAXIC AREA   Tends to limit period of inspiration.
APNEUSTIC AREA   Tends to prolong inspiration; normally overriden by pneumotaxic area.

LUNG STRETCH RECEPTORS   May aid in terminating inspiration when lungs are considerably distended.

MEDULLARY EXPIRATORY NEURONS   When respiratory drive for increased pulmonary ventilation becomes greater than normal, these neurons send impulses to expiratory muscles to facilitate expiration.

Control of Rate and Depth of Breathing

ARTERIAL P_{O_2}   Low P_{O_2} in arterial blood (with P_{CO_2} and pH constant) increases alveolar ventilation.

ARTERIAL P_{CO_2}   Elevated arterial P_{CO_2} (with P_{O_2} and pH constant) increases ventilation.

ARTERIAL pH   Increased arterial hydrogen ion concentration (with P_{O_2} and P_{CO_2} constant) increases ventilation.

INTERACTION OF DIFFERENT FACTORS IN THE CONTROL OF RESPIRATION   At any instant, respiration is controlled by multiple factors and the sensitivity to some factors such as oxygen may not be as great as the sensitivity to other factors, such as carbon dioxide.

EFFECTS OF EXERCISE ON THE RESPIRATORY SYSTEM   p. 398

Increased Pulmonary Ventilation   During exercise, both the rate and depth of breathing increase.

Increased Delivery of Oxygen to the Tissues   Both at rest and during exercise, the blood that leaves the lungs is approximately 97% saturated with oxygen, and the factor that seems to be most important in determining how much oxygen is delivered to the tissues during exercise is not the pulmonary ventilation, but the amount of blood that the heart is capable of pumping.

COMMON DISEASES OF THE RESPIRATORY SYSTEM   pp. 398–399

Common Cold   Viral inflammation and sometimes subsequent bacterial infection of mucous membranes of upper respiratory tract.

Bronchial Asthma   Allergic response characterized by wheezing caused by excessive mucus and spasms of smooth muscles of bronchioles.

Bronchitis   Acute or chronic inflammation of bronchial tree caused by bacterial infection or irritants.

Tuberculosis   Infection caused by *Mycobacterium tuberculosis*, which most commonly affects the lungs.

Emphysema   Progressive condition that is a secondary response to other respiratory problems; overdistended alveoli and reduced elastic tissue of lungs produce reduced expiratory volume.

Pneumonia   Fibrinous exudate produced within alveoli, which causes part of lung to become solid and airless.

Pleurisy   Inflammation of pleural membranes.

EFFECTS OF AGING ON THE RESPIRATORY SYSTEM   p. 400
1. Chest wall rigidity.
2. Loss of lung elasticity.
3. Decreased ciliary activity of cells located on respiratory tract linings.
4. Decreased phagocytic activity, which increases susceptibility to infection.

SELF-QUIZ

1. The mucous membranes of the respiratory tract contain ciliated cells that move mucus and entrapped particles toward the throat. True or False?
2. The pleural cavity contains: (a) fluid; (b) air; (c) the lungs.
3. The resting intrapulmonary pressure of an individual at sea level is: (a) 140 mm Hg; (b) 756 mm Hg; (c) 760 mm Hg.
4. When air is moving into the lungs during inspiration: (a) the intrapulmonary pressure is higher than the atmospheric pressure; (b) the atmospheric pressure is higher than the intrapulmonary pressure; (c) the intrapleural pressure is higher than the atmospheric pressure.
5. Contracting the diaphragm is the dominant means of lowering the intrapulmonary pressure during normal quiet inspiration. True or False?
6. During expiration: (a) the volume of the thoracic cavity increases; (b) the diaphragm contracts; (c) the intrapulmonary pressure is higher than the atmospheric pressure.
7. The volume of gas that remains in the lungs even after the most forceful expiration is the: (a) residual volume; (b) tidal volume; (c) anatomical dead space volume.
8. Which has the greatest volume? (a) expiratory reserve volume; (b) vital capacity; (c) inspiratory reserve volume.
9. Smooth muscles of respiratory system passageways are relaxed and the resistance to the flow of air through the passageways is decreased by: (a) parasympathetic nervous stimulation; (b) histamine; (c) epinephrine.
10. Pulmonary surfactant increases the surface tension of the pleural fluid. True or False?
11. The minute respiratory volume is equal to the volume of air that enters the respiratory passages with each breath, multiplied by the: (a) respiratory rate; (b) alveolar ventilation; (c) anatomical dead space volume.

12. The lungs do not contain gases at the same partial pressures that exist in the atmosphere. True or False?
13. Venous blood that leaves the tissues contains oxygen at a partial pressure of about: (a) 40 mm Hg and carbon dioxide at a partial pressure of about 45 mm Hg; (b) 45 mm Hg and carbon dioxide at a partial pressure of about 40 mm Hg; (c) 40 mm Hg and carbon dioxide at a partial pressure of about 40 mm Hg.
14. In general, the binding of oxygen to hemoglobin will be increased by: (a) an increase in the oxygen partial pressure; (b) a decrease in pH; (c) an increase in temperature.
15. Which of the following is a true statement about the transport of respiratory gases? (a) Most of the CO_2 is carried as bicarbonate ions within the blood. (b) CO_2 is never carried as bicarbonate ions. (c) On the average, 27% of the CO_2 is carried as bicarbonate ions within the blood.
16. The filling of the lungs and the rhythmic movements of respiration depend on nerve impulses that travel to the diaphragm and chest muscles involved in respiration. True or False?
17. The activity of the apneustic area tends to shorten the period of inspiration. True or False?
18. In moderate anemia: (a) the carotid and aortic bodies generally provide a greatly increased stimulus to the respiratory center; (b) the amount of dissolved oxygen in the blood is abnormally low; (c) the reduced oxygen content of the blood is due to a reduced number of functional hemoglobin molecules.
19. An elevation of 5 mm Hg of arterial P_{CO_2}, with P_{O_2} and pH held constant, will: (a) inhibit ventilation; (b) increase ventilation; (c) have no effect on ventilation.
20. The alveolor ventilation increases in response to: (a) increased arterial P_{O_2}; (b) lowered arterial pH; (c) decreased arterial P_{CO_2}.

LEARNING OBJECTIVES

After completing this chapter, you should be able to:

- Describe the factors involved in the control of chewing and swallowing.
- Describe the factors that influence gastric motility.
- Explain the movements that occur in the large intestine.
- Explain the functions and control of the salivary secretions.
- Discuss the phases of gastric secretion.
- Discuss the inhibition of gastric secretion.
- List the principal digestive fluids that are present in the small intestine, and describe the factors that stimulate the secretion of each.
- Describe the processes involved in the digestion and absorption of carbohydrates, proteins, and lipids.
- Discuss the absorption of vitamins.
- Discuss the absorptive function of the large intestine.

CHAPTER CONTENTS

THE DIGESTIVE SYSTEM 16

The cells of the body require a continual supply of energy in order to perform their particular functions: contraction, secretion, synthesis, and so on. The ingested food provides the materials from which this energy is derived and from which components for the synthesis of new molecules are obtained. Most food, however, must be broken down into simpler molecules before it can be absorbed into the body and utilized by the cells. The digestive system mechanically and chemically processes the food and provides for its absorption into the body.

ORGANIZATION OF THE DIGESTIVE SYSTEM

The digestive system consists of a tube called the **gastrointestinal tract,** or **alimentary canal,** which extends from the mouth to the anus (F16.1). As long as food remains within the gastrointestinal tract, it is technically outside the body. To enter the body it must cross the wall of the tract. Secretions of the salivary glands, gastric glands, intestinal glands, liver, and pancreas—all of which assist in digestion—empty into the gastrointestinal tract. Although the tract is a continuous tube, it is divided into specialized regions, each of which performs specific functions. These regions include the mouth, esophagus, stomach, small intestine, and large intestine.

BASIC ORGANIZATION OF THE GASTROINTESTINAL TRACT

Starting with the esophagus and continuing to the anus, the wall of the gastrointestinal tract has the same basic arrangement of four layers, or tunics (F16.2). The four layers, from the lumen (cavity) of the tract outward, are the *tunica mucosa, tunica submucosa, tunica muscularis*, and *tunica serosa* or *adventitia*.

Tunica Mucosa

The **tunica mucosa** is the mucous membrane that lines the gastrointestinal tract. It consists of three layers: (1) an inner epithelial layer that borders the lumen; (2) a middle lamina propria, which is composed of loose connective tissue and generally contains blood vessels, lymph nodules, and small glands; and (3) an outer mus-

405

cularis mucosae, which is basically a thin layer of smooth-muscle fibers.

Tunica Submucosa

The **tunica submucosa** is a thick layer of either dense or loose connective tissue located deep to the mucosa. It contains blood vessels, lymph vessels, nerve fibers, and in some regions, glands.

Tunica Muscularis

In most regions of the gastrointestinal tract, the **tunica muscularis** is a double layer of muscle tissue. The muscle fibers of the inner layer are arranged circularly around the tract, whereas the outer fibers are oriented longitudinally along its long axis. At several points along the tract, the fibers of the circular layer are thickened, forming

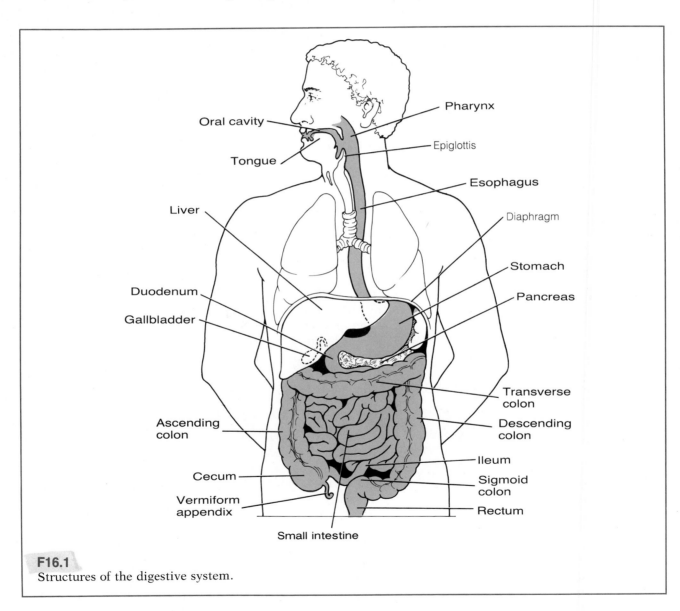

F16.1
Structures of the digestive system.

sphincters that control the movement of food from one region of the tract to another.

Tunica Serosa or Adventitia

The outermost tunic of the gastrointestinal tract is composed primarily of a layer of connective tissue. In the esophagus, this connective-tissue layer merges into the connective tissue of surrounding structures and is called the **adventitia.** Along the rest of the tract, the connective tissue is covered with a single layer of epithelial cells. In this case, the outer tunic is called the **serosa.**

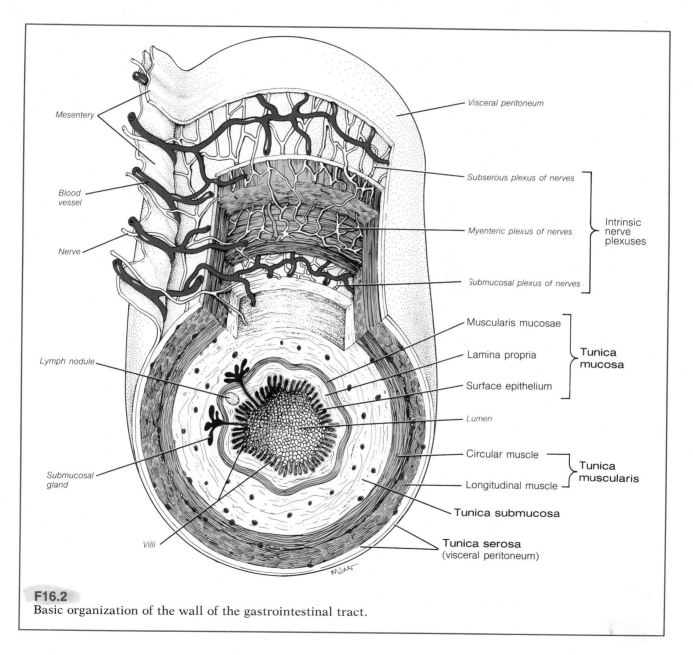

F16.2
Basic organization of the wall of the gastrointestinal tract.

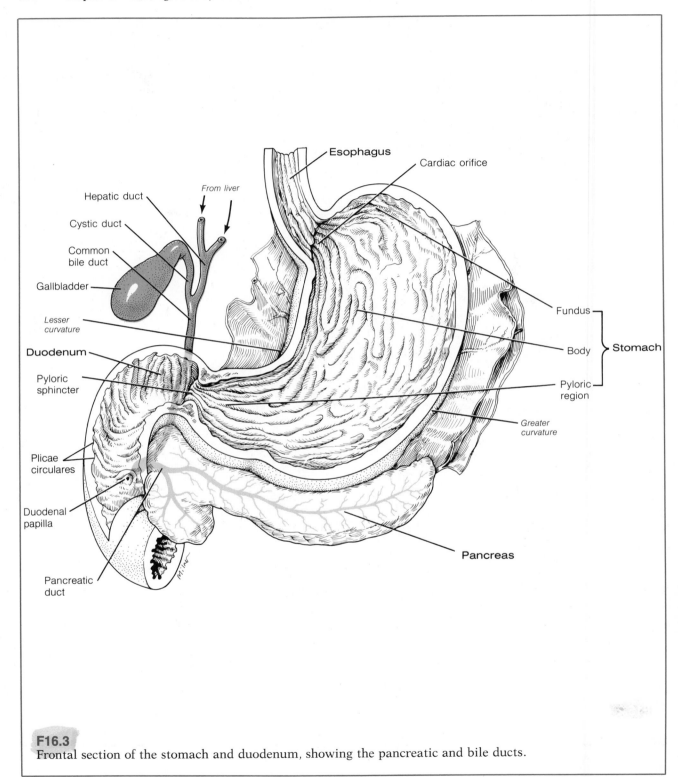

Frontal section of the stomach and duodenum, showing the pancreatic and bile ducts.

MODIFICATIONS OF THE BASIC ORGANIZATION OF THE GASTROINTESTINAL TRACT

Various regions of the gastrointestinal tract contain modifications of the basic structure just described.

Stomach

The esophagus opens into the stomach at the cardiac orifice, and the stomach is divided into several regions (F16.3). The main portion of the stomach is the *body;* the portion that bulges above the opening of the esophagus is the *fundus;* and the portion that tapers inferiorly to join the small intestine is the *pyloric region.* The lamina propria of the tunica mucosa of the stomach contains many gastric glands (*gastric* = stomach) (F16.4). These glands secrete gastric juice onto the surface of the mucosa through small invaginations called gastric pits. The gastric glands within the fundus and body of the stomach (fundic glands or gastric glands proper) contain several types of cells, including: (1) mucous neck cells, which are mucous-secreting cells that are especially prevalent around the gastric pits; (2) parietal (oxyntic) cells, which produce hydrochloric acid; and (3) chief (zymogenic) cells, which secrete pepsinogen, a precursor of the digestive enzyme pepsin. The glands of the cardiac region (cardiac glands) and the pyloric region (pyloric glands) secrete mainly mucus.

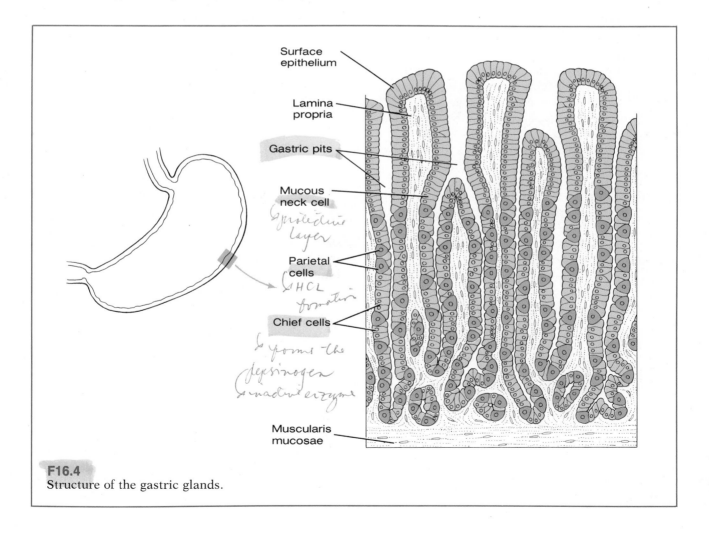

Surface epithelium
Lamina propria
Gastric pits
Mucous neck cell
Parietal cells
Chief cells
Muscularis mucosae

F16.4
Structure of the gastric glands.

Small Intestine

The small intestine is divided into three regions. The first 25 cm or so is the duodenum; the next 2.5 m the jejunum; and the final 3.6 m the ileum. Throughout much of the small intestine, the tunica mucosa and tunica submucosa form permanent, circular, shelflike folds (the plicae circulares) that extend into the intestinal lumen. The intestinal surface, including the plicae circulares, is studded with many tiny leaflike, tonguelike, or fingerlike villi, which are projections of the tunica mucosa (F16.5). Each villus is covered with a single layer of epithelial cells and contains a lymphatic capillary called a lacteal, which is surrounded by a network of blood capillaries. Two main types of epithelial cells cover the villi: goblet cells, which secrete mucus, and more numerous absorptive cells, which participate in the absorption (and also the digestion) of food materials. The free surfaces of the absorptive cells are folded into microscopic projections called microvilli. The microvilli (together with the plicae circulares and the villi) greatly increase the surface area of the small intestine across which absorption can occur.

The tunica mucosa of the small intestine contains many tubular intestinal glands (crypts of Lieberkuhn) that empty their secretions between the villi. The tunica submucosa of the duodenum contains mucous glands called Brunner's glands.

Large Intestine

The large intestine, which is about 1.5 m long, is divided into regions that include the cecum, ascending colon, transverse colon, descending colon, sigmoid colon, and rectum (F16.1). A narrow, blind tube called the veriform appendix extends downward from the cecum, and the external opening of the large intestine is called the anus. The tunica mucosa of the large intestine contains a large number of mucous cells and lacks villi.

INTRINSIC NERVE PLEXUSES

Beginning in the esophagus and continuing to the anus, the wall of the gastrointestinal tract contains complex interconnections of neurons organized into **intrinsic nerve plexuses** (F16.2). The intrinsic nerve plexuses coordinate much of the activity of the tract, and they make possible the occurrence of intratract reflexes. In these reflexes, the stimulation of receptors within the wall of the gastrointestinal tract neurally influences the activities of effector cells of the tract (for example, muscle or secretory cells) independently of the central nervous system. In addition, hormones and reflexes that operate by way of the central nervous system also influence gastrointestinal tract activity.

MECHANICAL PROCESSES OF THE DIGESTIVE SYSTEM

Ingested food must be continually moved along the gastrointestinal tract so it can be acted on by digestive enzymes secreted into various regions of the tract. Moreover, the contents of the tract must be constantly churned and mixed so all the food particles come into contact with the enzymes. This churning also brings the food into contact with the wall of the tract, which allows for the absorption of digested food.

The moving and mixing of food within the gastrointestinal tract is accomplished by the contraction and relaxation of the muscles associated with the tract. The muscles of the mouth, pharynx, and tunica muscularis in the upper portion of the esophagus (as well as the muscle of the external anal sphincter, which surrounds the anus) are skeletal muscles that require nerve impulses to initiate their contractions. The muscles of the remainder of the gastrointestinal tract are smooth muscles whose contractions are strongly influenced by hormones and neural signals.

Two basic forms of movement—mixing and propulsion—occur within the gastrointestinal tract. A major mixing movement is **segmentation,** in which stationary muscular contractions that occur at intervals along a portion of the gastrointestinal tract divide the tract into constricted and unconstricted regions (F16.6). As segmentation proceeds, constricted regions relax and unconstricted regions constrict, thoroughly mixing the contents of the tract.

An important propulsive movement is **peristalsis,** in which the muscles surrounding a portion of the gastrointestinal tract undergo a wavelike contraction (F16.7). This contraction produces a ring of constriction that moves along the tract, forcing materials within the tract to move ahead of it. Peristaltic waves normally travel toward the anus, and it has been suggested that this is due to the organization and influence of the intrinsic nerve plexuses.

CHEWING

Chewing (*mastication*) is accomplished by opening, closing, and lateral movements of the jaws,

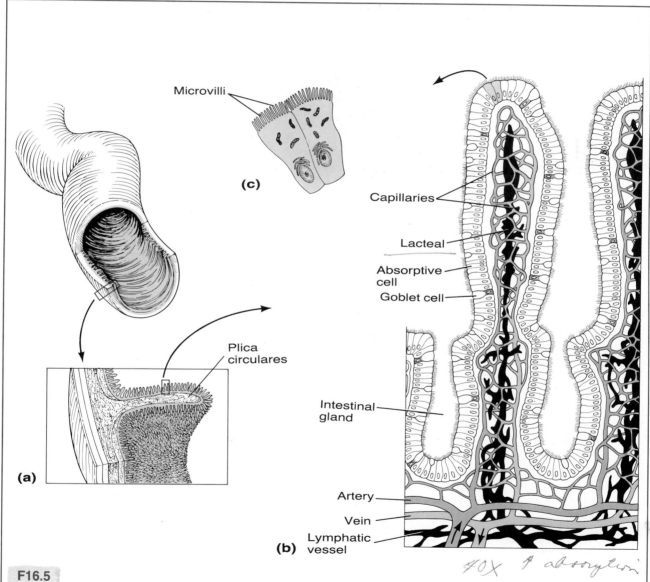

F16.5
Microscopic structure of the small intestine. (a) Enlargement showing villi extending from a plica circulares. (b) Higher magnification of a section through two villi showing blood vessels and lacteals. (c) Enlargement of two absorptive cells showing microvilli on their free surfaces.

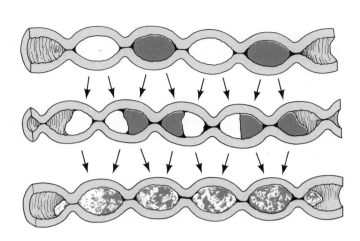

F16.6
The mixing action of segmentation in the small intestine.

accompanied by the continual movement of food between the teeth by the tongue and the muscles of the cheeks. Chewing is part voluntary and part reflex. The reflex aspect of chewing is initiated by tactile stimulation due to a mass of food (a bolus) pressing against the teeth, gums, and anterior portion of the roof of the mouth. This stimulation inhibits the muscles that raise the lower jaw. When the lower jaw drops, stretch reflexes are initiated, and the muscles contract and raise the jaw. This movement again presses the bolus against the teeth, gums, and anterior portion of the roof of the mouth, and the cycle repeats itself. Chewing reduces ingested materials into smaller, more easily digestible pieces, and thoroughly mixes them with the saliva secreted into the mouth by the salivary glands. The mixing of food with saliva moistens the food and enhances its formation into a manageable mass.

SWALLOWING

Swallowing moves the bolus along the gastrointestinal tract from the mouth to the stomach. Swallowing is initiated by placing the tip of the tongue against the roof of the mouth and forcing the bolus into the pharynx (F16.8). The bolus stim-

ulates pressure receptors around the opening of the pharynx, and this stimulation elicits a complex swallowing reflex that is coordinated by a swallowing center in the brain stem.

The swallowing reflex generates a peristaltic contraction of pharyngeal constrictor muscles that moves the bolus along the pharynx toward the esophagus. Muscles surrounding the upper portion of the esophagus are normally in a state of tonic contraction, and they act as a sphincter (the *upper esophageal sphincter*) that keeps the entrance to the esophagus closed. However, during swallowing, the upper esophageal sphincter relaxes, allowing the bolus to enter the esophagus.

Instead of entering the esophagus, the bolus could go into the larynx or even upward into the nasal cavity. However, food is prevented from entering the larynx during swallowing by the approximation of the vocal cords, which blocks the laryngeal opening. In addition, the larynx is moved upward, and in this position the epiglottis tips backward and covers the laryngeal opening as the bolus passes by. The bolus is prevented from entering the nasal cavity by the elevation of the soft palate against the wall of the pharynx. Sometimes the coordination of these events by the swallowing center fails, and food does enter the larynx or nasal cavity. The usual response to food that enters the larynx is a violent cough, which expels the food back into the pharynx.

The bolus is moved along the esophagus by a primary peristaltic wave; that is, by a peristaltic contraction of the muscle of the tunica muscularis of the esophagus that is basically a continuation of the peristaltic wave that began in the pharynx. About 2 to 5 cm above the cardiac orifice, the smooth muscle of the esophagus is normally in a state of tonic contraction, and it acts as a physiological sphincter (the *lower esophageal sphincter*) that keeps the lower portion of the esophagus closed. As the peristaltic wave moves along the esophagus during swallowing, the lower esophageal sphincter relaxes, allowing the bolus to enter the stomach. Food generally requires between 5 and 10 seconds to move from the mouth to the stomach.

The distension of the esophagus by materials within it stimulates receptors in the esophageal wall and elicits secondary peristaltic waves that are not directly associated with the act of swallowing. These waves are of value in continuing the movement of any food material that may not have been moved completely through the esophagus by the primary peristaltic wave.

GASTRIC MOTILITY

The mechanical activities of the stomach include: (1) storing ingested food until it can be utilized by the remainder of the gastrointestinal tract; (2) mixing the food with gastric secretions; and (3) moving the food into the duodenum of the small intestine at a rate that is consistent with efficient intestinal digestion and absorption.

Within limits, the stomach can expand to accommodate increasing amounts of food without a great increase in the pressure within it. This ability is due in part to neurally mediated reflex activity, and in part to the ability of gastric smooth muscle to lengthen, when stretched, without a marked increase in tension. As the food within the stomach is mixed with gastric juice, the mixture assumes a semifluid consistency and is referred to as *chyme*.

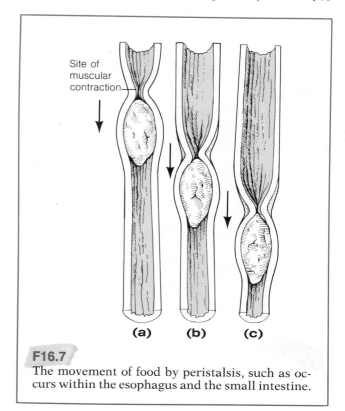

Site of muscular contraction

(a)   (b)   (c)

F16.7

The movement of food by peristalsis, such as occurs within the esophagus and the small intestine.

The opening between the stomach and duodenum is surrounded by the *pyloric sphincter*, which is generally partially open and thus offers only limited resistance to the movement of the stomach contents into the duodenum. However, this resistance is usually sufficient to prevent chyme from entering the duodenum in the absence of gastric muscular activity.

The stomach undergoes peristaltic contractions that begin in the body of the stomach and proceed toward the duodenum. The contractions are relatively weak in the upper region of the stomach, but they increase in strength and speed of transmission in the lower region where the muscle of the stomach wall is quite thick. A strong peristaltic contraction in the pyloric region of the stomach forces some of the chyme out of the stomach and into the duodenum. In addition, the arrival of a peristaltic wave at the terminal pyloric region constricts this area and forces much of the chyme back toward the body of the stomach. This

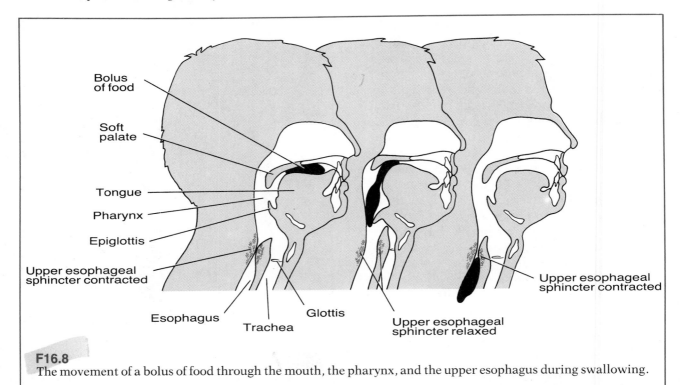

Bolus
of food

Soft
palate

Tongue

Pharynx

Epiglottis

Upper esophageal
sphincter contracted

Esophagus

Trachea

Glottis

Upper esophageal
sphincter relaxed

Upper esophageal
sphincter contracted

F16.8
The movement of a bolus of food through the mouth, the pharynx, and the upper esophagus during swallowing.

activity provides an effective mixing action.

The gastric smooth muscle contains pacemaker cells, which exhibit a basic electrical rhythm—that is, a spontaneous, cyclical depolarization and repolarization—that has a frequency of about 3 oscillations per minute. This rhythm is important in determining the frequency of the gastric muscular contractions.

Gastric motility (particularly the intensity of gastric muscular contractions) and the rate of gastric emptying are under the combined influence of a variety of excitatory and inhibitory mechanisms. For example, the volume of material contained within the stomach exerts a moderate influence on gastric motility and emptying. As the stomach fills and the volume of its contents increases, the stomach wall is distended, and mechanoreceptors in the wall are stimulated. The stimulation of these receptors initiates reflexes that enhance gastric motility both by way of the intrinsic nerve plexuses—that is, the intratract reflex pathway—and by way of the vagus nerves and

central nervous system. In addition, the hormone **gastrin** is released by cells in the pyloric region in response to distension (and also in response to the presence of partially digested protein and caffeine). Gastrin enters the bloodstream and ultimately returns to the stomach where it stimulates gastric motility.

The volume and composition of the chyme that enters the duodenum exert a major influence on gastric motility and the rate of gastric emptying (F16.9). As the duodenum fills with chyme, its wall is distended, and a reflex called the **enterogastric reflex** (*entero* = intestine) is initiated. In this reflex, nerve impulses that inhibit gastric motility and slow gastric emptying are transmitted from the duodenum to the stomach both by way of the intrinsic nerve plexuses and by way of the vagus nerves and central nervous system. Irritants in the duodenum, an excessive acidity of the duodenal chyme—that is, a pH below 3.5—a relative hypertonicity or hypotonicity of the chyme, or the presence of breakdown products of protein digestion also activate the enterogastric reflex and inhibit gastric motility and emptying. Moreover, the acidity of the duodenal chyme and the presence of certain amino acids and fatty acids trigger the release of hormones such as **secretin, chole-cystokinin,** and **gastric inhibitory peptide** from cells of the duodenum and jejunum. These hormones travel through the bloodstream to the stomach where they exert an inhibitory effect on gastric motility.

The neural and hormonal duodenal mechanisms that inhibit stomach emptying tend to prevent additional chyme from entering the small in-

testine until the chyme already present has been effectively processed. The amount of gastric contents that enters the small intestine is therefore largely regulated by the small intestine itself.

MOTILITY OF THE SMALL INTESTINE

Both segmentation and peristalsis take place in the small intestine. Segmentation may cause some movement of materials along the intestine, but its primary contributions are to mix the chyme thoroughly with the digestive juices and to expose different portions of the chyme to the intes-

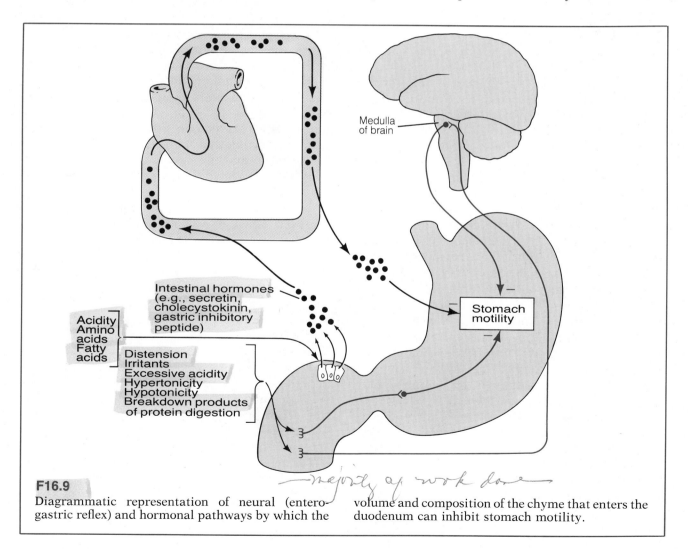

F16.9

Diagrammatic representation of neural (entero-gastric reflex) and hormonal pathways by which the volume and composition of the chyme that enters the duodenum can inhibit stomach motility.

tinal mucosa. Peristaltic contractions are primarily responsible for moving the chyme along the small intestine. In general, the peristaltic waves of the small intestine are weak, and they die out after traveling only a short distance. Therefore, the movement of the chyme through the small intestine tends to be relatively slow, allowing ample time for digestion and absorption.

Local mechanical and chemical stimulation by the intestinal contents are believed to be particularly important in the initiation and continuation of the movements of the small intestine. The motility of the small intestine is increased by distension; by hypertonic, hypotonic, or excessively acidic contents; and by the presence of certain products of digestion.

Like its gastric counterpart, the smooth muscle of the small intestine exhibits a basic electrical rhythm that is important in determining the frequency of intestinal muscular contractions. However, this rhythm varies in different portions of the intestine. It is most rapid in the duodenum—about 11 or 12 per minute—and it decreases progressively along the length of the intestine to about 6 or 7 per minute in the terminal ileum.

For 2 or 3 centimeters preceding the entrance of the ileum into the cecum of the large intestine, the muscle of the ileal wall is thickened, forming the *ileocecal sphincter*. The distension of the cecum initiates a reflex by way of the intrinsic nerve plexuses that stimulates the muscle of the ileocecal sphincter, and most of the time the sphincter is at least mildly constricted. The constriction of the sphincter facilitates absorption in the small intestine by slowing the movement of the chyme from the ileum into the large intestine. When food enters the stomach, a reflex called the *gastroileal reflex* intensifies peristalsis in the ileum. In addition, the hormone gastrin, which is released by cells in the pyloric region of the stomach, relaxes the ileocecal sphincter. These events enhance the movement of chyme from the ileum into the large intestine.

The *ileocecal valve*, which is composed of two folds of tissue, is located at the opening of the ileum into the cecum of the large intestine. When the contents of the ileum attempt to move into the cecum, the folds of tissue are pushed apart and the valve opens, allowing the ileal contents to enter the cecum. However, when the contents of the cecum attempt to move backward into the ileum, the folds are forced together and the valve closes. Thus, the ileocecal valve helps prevent regurgitation of the contents of the cecum back into the ileum.

MOTILITY OF THE LARGE INTESTINE

The movements of the large intestine are generally quite sluggish, and it frequently takes 18 to 24 hours for material to pass through the colon. Mixing movements of a segmental type occur within the colon at a much lower frequency than do those within the small intestine. At infrequent intervals, perhaps three to four times per day, substantial segments of the colon undergo several strong contractions called *mass movements*, which propel material along the colon for considerable distances. Mass movements often occur following a meal, indicating that the presence of food within the stomach and the duodenum activates *gastrocolic* and *duodenocolic reflexes*, which increase colon motility.

The distension of the wall of the rectum by a mass movement of fecal material initiates the *defecation reflex*, which tends to move material out of the lower colon and rectum. Nerve impulses associated with the defecation reflex are transmitted by way of the intrinsic nerve plexuses, and these impulses are strongly reinforced by impulses that travel to and from the sacral region of the spinal cord. The defecation reflex stimulates peristaltic contractions of the descending colon, sigmoid colon, and rectum. It also causes the relaxation of the internal anal sphincter, which surrounds the anal opening. This combination of contraction and relaxation tends to propel the feces through the anus. However, a second sphincter, the external anal sphincter, also surrounds the anal opening. The external anal sphincter is composed of skeletal muscle, and it can be voluntarily controlled. If the external anal sphincter is voluntarily relaxed, defecation will occur. If the external anal sphincter is voluntarily constricted, defecation can be prevented. If this is done, the defecation reflex usually dies out after a few minutes, but it generally returns in several hours when another mass movement propels additional fecal material into the rectum.

The defecation reflex is usually assisted through voluntary efforts that include taking a deep inspiration and then closing the glottis and contracting the abdominal muscles. These actions raise the pressure within the abdomen, which aids defecation by increasing the pressure on the contents of the large intestine.

ALTERATIONS IN GASTROINTESTINAL TRACT MOTILITY

Alterations in gastrointestinal tract motility can produce unpleasant changes in the digestive process such as vomiting, diarrhea, and constipation.

Vomiting

Vomiting, which ejects the contents of the stomach through the mouth, is a complex reflex coordinated by a vomiting center within the medulla of the brain. During vomiting, the diaphragm and abdominal muscles contract, raising the intra-abdominal and intragastric pressures. Then, the lower esophageal sphincter relaxes, and the contents of the stomach are expelled through the esophagus and out of the mouth. The vomiting center receives input from many areas, both within and at the surface of the body. Thus, vomiting can result from dizziness, unpleasant odors or sights, and so forth, as well as from disturbances within the gastrointestinal tract itself. Excessive vomiting can lead to severe disturbances in the fluid, salt and acid–base balances of the body.

Diarrhea

Diarrhea is characterized by watery stools, generally accompanied by frequent defecation. Diarrhea can be due to either an increased secretion of fluid into or a decreased absorption of fluid from the gastrointestinal tract. In either case, an increased volume of material within the large intestine can lead to more frequent defecation. Like vomiting, prolonged diarrhea can cause disturbances in the fluid, salt, and acid–base balances of the body.

Constipation

If the motility of the large intestine is decreased, digested materials remain within the colon and rectum for prolonged periods of time. The longer fecal materials remain within the colon, the more water is absorbed. As the feces lose water, they become drier and harder, making defecation more difficult and sometimes painful. This condition is referred to as *constipation*. Constipation can be caused by insufficient bulk in the diet, lack of exercise, or even by emotions.

CHEMICAL DIGESTION AND ABSORPTION

The chemical breakdown of large food molecules into smaller molecules that can be absorbed from the gastrointestinal tract is accomplished by digestive enzymes. These enzymes function mainly by hydrolysis; that is, they split large molecules into smaller ones by introducing water into the molecular structures (F16.10). In this fashion, large carbohydrate molecules such as starches are digested to monosaccharides such as glucose; proteins are split into their constituent amino acids; and fats are broken primarily to monoacylglycerols and free fatty acids. The smaller molecules that result from digestion are absorbed from the lumen of the gastrointestinal tract, and ultimately enter the bloodstream to be carried to the cells of the body.

Many of the digestive enzymes and other secretory products of the gastrointestinal tract are produced and released by specialized epithelial cells that are commonly organized into exocrine secretory glands (F16.11). These exocrine glands release their products by way of ducts onto the surface of the tract.

MOUTH AND SALIVA

Three pairs of large salivary glands—the **parotid, sublingual,** and **submandibular (submaxillary)** glands—produce most of the saliva (F16.12). The output of these glands is supplemented slightly by

F16.10

Hydrolytic cleavage of an amino acid chain to liberate a single amino acid. The peptide bond is broken by the introduction of water.

smaller buccal glands. Approximately two-thirds of the 1 to 1.5 liters of saliva produced per day by an average adult comes from the submandibular glands. Approximately one-quarter comes from the parotid glands, and the remainder is produced by the other glands.

Secretions of the Salivary Glands

The salivary glands produce two different secretions: (1) a mucous secretion that contains mucins; and (2) a serous secretion that contains the enzyme salivary amylase (ptyalin). The parotid glands produce serous secretions, the submandibular glands produce both mucous and serous secretions, and the sublingual and buccal glands primarily produce mucous secretions.

Mucins. The **mucins** are the major proteins of the saliva, and they have large polysaccharides attached to them. When mixed with water, the mucins form a highly viscous solution known as mucus, which lubricates the mouth and food.

Salivary Amylase. **Salivary amylase** splits starch molecules, which are composed of thousands of glucose units, into smaller fragments. Like other enzymes, salivary amylase has an optimal pH at which it functions best (6.9), but it is stable at pH values from 4 to 11. After food has been swallowed, the digestive action of salivary amylase continues in the stomach until the enzyme is inactivated by the acidic gastric juices.

Composition of Saliva

The exact composition of saliva depends on the glands from which it comes, on the secretion rate, and to some degree on the stimulus that evokes secretion. Generally, saliva is 97% to 99.5% water, and the pH of mixed saliva is usually from 6.0 to 7.0. Sodium, potassium, chloride, and bicarbonate ions are some of the main electrolytes of saliva. Saliva also contains kallikrein as well as specific soluble blood group substances of the sort responsible for blood type. Kallikrein acts enzymatically on plasma protein precursors to produce bradykinin, which is a vasodilator substance. The release of kallikrein from active salivary glands and the subsequent production of bradykinin is believed to increase bloodflow to the glands by promoting local vasodilation.

Functions of Saliva

In addition to its lubricating and digestive functions, saliva assists in bolus formation and in swallowing by moistening food particles and holding them together. It also helps dissolve food so it can be tasted. Saliva aids in speech by moistening the mouth and throat, and it has bacteriostatic properties.

Control of Salivary Secretion

Salivary secretion occurs in response to nerve impulses. Salivatory nuclei in the medulla–pons region of the brain receive impulses from the mouth and pharynx as well as from higher brain centers.

Nerve impulses travel to the salivary glands by way of the autonomic nervous system. Both parasympathetic and sympathetic stimulation cause salivary secretion, with parasympathetic stimulation having the greatest effect. The flow of saliva is enhanced by the presence of food in the mouth as well as by the odor of food, the sight of food, the thought of food, or the presence of irritating food in the stomach and upper intestine. Chewing stimulates salivary secretion, whereas intense mental effort, dehydration, fear, and anxiety tend to reduce it.

STOMACH

Gastric Secretions

The gastric secretions include mucus, hydrochloric acid, and pepsinogen, which is a precursor of the protein-digesting enzyme pepsin.

Mucus. Cells that secrete a viscous, alkaline mucus line the surface of the stomach. The mucus, which adheres to the stomach walls in a layer 1 to 1.5 mm thick, lubricates the walls and protects the gastric mucosa. Even the slightest irritation of the mucosa directly stimulates these cells to secrete large amounts of mucus.

Glands that secrete a thin mucus and little else are found in the region of the cardiac orifice of the stomach (cardiac glands) and in the pyloric region (pyloric glands). Gastric glands in the body and fundus of the stomach produce some mucus as well as additional secretions.

Hydrochloric Acid. Cells in the gastric glands called *parietal (oxyntic) cells* produce hydrochloric acid, which is a strong acid that dissociates into hydrogen ions and chloride ions. The hydrochloric acid facilitates protein digestion in the stomach and also kills many of the bacteria that enter the digestive tract with the food. The parietal cells produce a hydrochloric acid solution that is about 0.16 molar (pH 0.8). Actual stomach acidity, however, depends on the rate of hydrochloric acid secretion, as well as on the rate of neutralization and dilution of the acid by other secretions.

The precise details by which the parietal cells

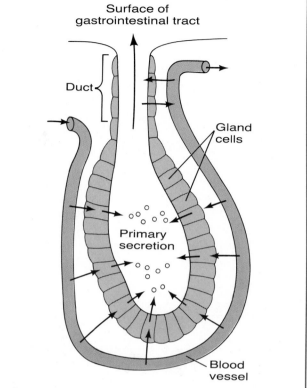

F16.11
Diagrammatic representation of an exocrine gland. Raw materials obtained from the blood are transformed by the gland cells into a primary secretion that is released into the glandular lumen. As the primary secretion moves along the glandular duct, active transport or diffusion across the duct walls may alter its composition.

produce hydrochloric acid are unknown. The prevalent theory holds that chloride ions are actively transported from the blood to the stomach lumen across the parietal cells, and hydrogen ions from dissociated water molecules are actively transported from the interior of the parietal cells to the stomach lumen, leaving behind hydroxide ions (OH^-) (F16.13). Carbon dioxide from the plasma diffuses into the cells where it (and also carbon dioxide from cellular metabolism) combines with water to form carbonic acid. This reaction is facilitated by the enzyme carbonic anhydrase. Carbonic acid can dissociate into hydrogen and bicar-

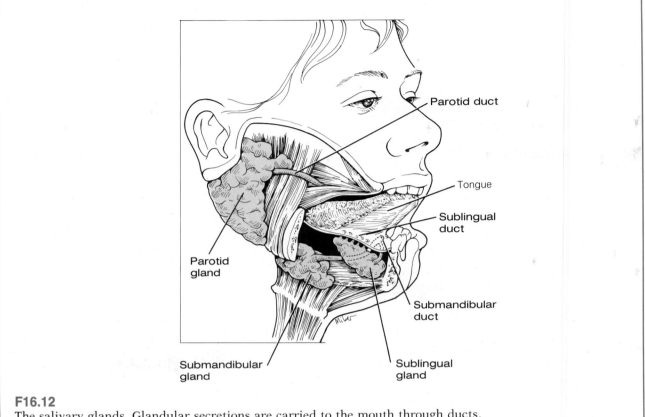

F16.12
The salivary glands. Glandular secretions are carried to the mouth through ducts.

bonate ions. Thus, hydrogen, bicarbonate, and hydroxide ions are present in the cells. The hydrogen ions and hydroxide ions combine to form water, and the bicarbonate ions diffuse out of the cells into the blood. As a result of the basic bicarbonate ions entering the blood, the pH of the venous blood that leaves the actively secreting stomach is higher than that of the arterial blood that flows to the stomach.

Pepsinogen.   Cells in the gastric glands of the stomach called *chief cells* produce pepsinogen, which is a precursor of the active enzyme pepsin. In the stomach, pepsinogen is converted to pepsin by contact with hydrochloric acid and previously formed pepsin. Pepsin, which is optimally active at acid pHs, digests proteins by breaking peptide bonds that involve certain amino acids such as

tryptophan, phenylalanine, and tyrosine. This activity results in smaller *peptide chains* of amino acids. The production of peptide chains is about as far as protein digestion in the stomach proceeds.

Control of Gastric Secretion

The stomach glands produce as much as 3 liters of secretions (gastric juice) per day. Diet affects the amount of juice secreted, and a typical meal results in the secretion of up to 700 ml of gastric juice.

Gastric secretion is stimulated by both neural and hormonal mechanisms. Neural mechanisms involve the activity of both local, intratract reflexes and the vagus nerves from the medulla of the brain. Parasympathetic impulses transmitted by way of the vagus nerves to the stomach increase both pepsin and acid secretion and also

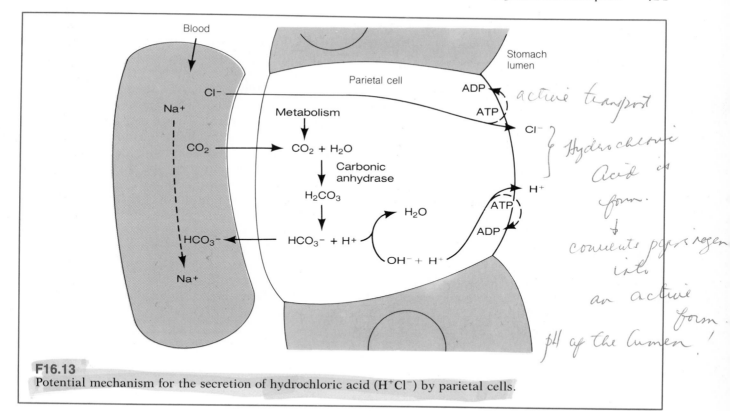

Handwritten notes:
active transport
Hydrochloric Acid is form.
converts pepsinogen into an active form.
pH of the lumen!

F16.13
Potential mechanism for the secretion of hydrochloric acid (H⁺Cl⁻) by parietal cells.

cause some increase in mucus secretion. Hormonal mechanisms involve the hormone gastrin, which stimulates gastric secretion, particularly the release of hydrochloric acid. The neural and hormonal mechanisms of control are interrelated because vagal activity causes gastrin release.

The control of gastric secretion is divided into several regulatory phases.

Cephalic or Reflex Phase. Gastric secretion can be initiated by the sight, smell, or taste of food. This response is due to nerve impulses that travel from sensory receptors to the central nervous system and then by way of the vagus nerves to the stomach, where gastric secretion is stimulated. The secretory response elicited by the sight or smell of food is basically a conditioned reflex that is not elicited when a person is afraid, depressed, or has no desire for food (appetite). If the vagus nerves to the stomach are severed, secretion due to the cephalic or reflex phase ceases.

Gastric Phase. The distension of the stomach by the presence of food stimulates the flow of gastric juice and also potentiates its acid and pepsin content (F16.14). One way this response occurs is as a result of local reflexes that stimulate gastric secretion. The distension of the stomach also results in a vagovagal reflex by which impulses are sent to the brain stem and then back to the stomach by way of the vagus nerves to stimulate the flow of gastric juice (and the release of gastrin). In addition, local responses to the distension of the pyloric region of the stomach or to the presence of substances called *secretagogues*—for example, partially digested protein and caffeine—lead to the release of gastrin.

Gastrin release is inhibited by the presence of high concentrations of hydrogen ions (acid) in the stomach. This activity provides a negative feedback mechanism that helps prevent excessive stomach acidity and helps maintain an optimal pH for the function of the peptic enzymes. When the pH of the stomach contents reaches 2.0, the gastrin mechanism for the stimulation of gastric secretion becomes totally blocked.

Intestinal Phase.   A slight stimulation of gastric secretion occurs in response to the presence of chyme within the upper portion of the small intestine (particularly the duodenum) after the stomach has nearly emptied and gastric secretory activity is relatively low following a meal. It is believed that this stimulation of gastric secretion is mainly due to the production of small amounts of gastrin by the duodenal mucosa in response to distension or chemical stimuli. However, other hormones or reflexes are probably also involved.

Intestinal Factors That Inhibit Gastric Secretion.   Although the presence of chyme in the intestine stimulates gastric secretion during the intestinal phase when the stomach is almost empty and gastric secretory activity is relatively low, it frequently has an inhibitory effect on gastric secretion during the gastric phase when the stomach contains food and gastric secretion is strongly stimulated (F16.15). The distension of the duodenum, irritants in the duodenum, an excessive acidity of the duodenal chyme, a relative hypertonicity or hypotonicity of the chyme, or the presence of breakdown products of protein digestion leads to a neural inhibition of gastric secretion by way of an enterogastric reflex. In this reflex, nerve impulses that inhibit gastric secretion are transmitted from the duodenum to the stomach both by way of the intrinsic nerve plexuses and by way of the vagus nerves and central nervous system.

In addition, the acidity of the duodenal chyme

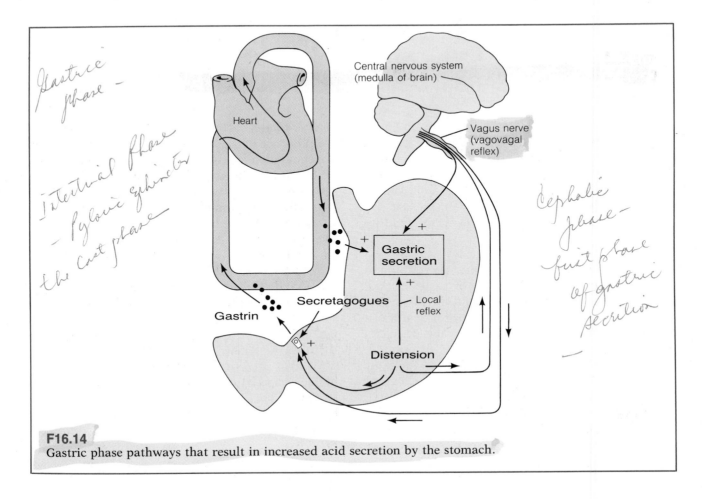

F16.14
Gastric phase pathways that result in increased acid secretion by the stomach.

and the presence of certain amino acids and fatty acids leads to the release of intestinal hormones such as secretin, cholecystokinin, and gastric inhibitory peptide that enter the blood and travel to the stomach. When the stomach is actively secreting and the mechanisms that cause gastrin release are strongly stimulated, these hormones have an inhibitory effect on gastric secretion.

Protection of the Stomach

The mucosal surface of the stomach is composed of epithelial cells that are connected by tight junctions. Together with the alkaline mucous secretions, these cells form a protective barrier that prevents the hydrochloric acid and enzymes released into the stomach from digesting the stomach itself. Damaged cells are rapidly replaced, and the lining of the stomach is renewed about every three days.

SMALL INTESTINE AND ASSOCIATED STRUCTURES

The most important portion of the gastrointestinal tract as far as digestion and absorption are concerned is the small intestine and its associated structures. Food that arrives at the small intestine has not been completely digested and is

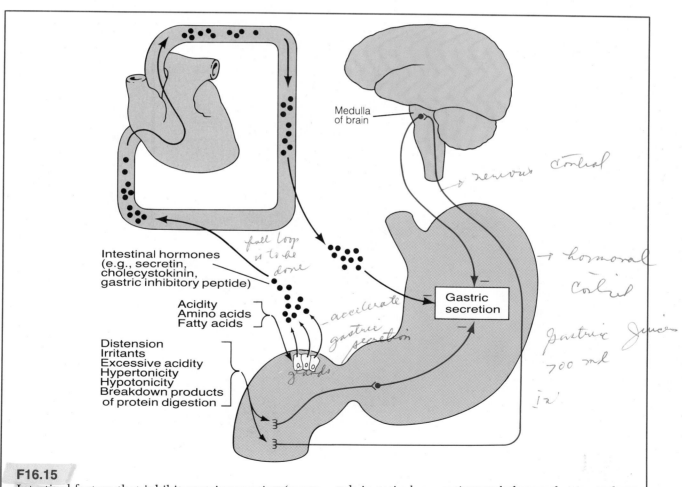

F16.15
Intestinal factors that inhibit gastric secretion (particularly hydrochloric acid release) when the stomach is actively secreting and the mechanisms that cause gastrin release are strongly stimulated.

not yet prepared for absorption. Some carbo-hydrate digestion is begun in the mouth by pty-alin, and the protein-digesting enzymes of the stomach produce polypeptide chains of amino acids, but carbohydrate and protein digestion re-main incomplete. Lipids that arrive at the intes-tine are essentially undigested. Thus, most di-gestive activity occurs within the small intestine, as does practically all of the absorption.

Three fluids are present in the small intestine: pancreatic juice, bile, and intestinal juice. Sub-stances within these fluids continue the processes of digestion. Many of the digestive processes are ultimately completed by enzymes associated with the membranes of the microvilli of the intestinal epithelium.

Pancreatic Juice

The pancreas is a dual structure that is both an exocrine gland that produces digestive enzymes and an endocrine gland that produces hormones (F16.16). The exocrine portion of the pancreas is composed of acinar cells and collecting ducts. The ducts from the pancreas join and open into the duodenal portion of the intestine, a short distance below the pyloric sphincter.

The exocrine portion of the pancreas produces an aqueous, isotonic fluid with a high bicarbonate ion concentration and a basic pH (about 8.0). In the intestine, this fluid, particularly the bicar-bonate ions, helps neutralize the acidic chyme from the stomach. The exocrine pancreas also pro-

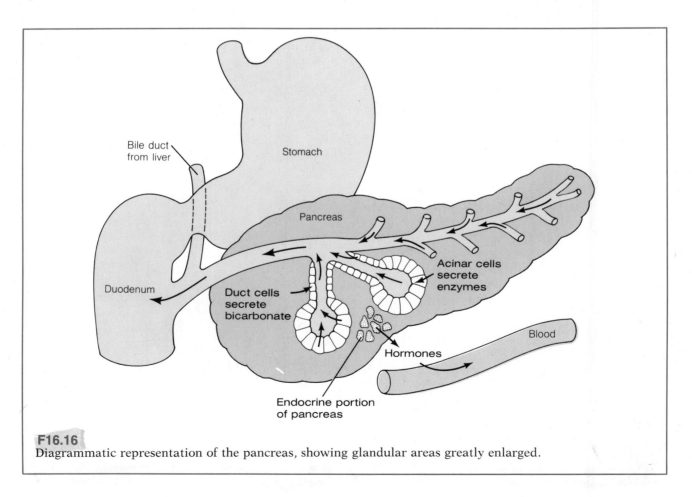

F16.16
Diagrammatic representation of the pancreas, showing glandular areas greatly enlarged.

duces carbohydrate, protein, and lipid-digesting enzymes as well as ribonuclease and deoxyribonuclease, which are enzymes that digest RNA and DNA, respectively. The combined secretions of the exocrine portion of the pancreas make up the pancreatic juice.

The exocrine secretory activities of the pancreas are controlled both hormonally and neurally (F16.17). The intestinal hormone secretin, which is released primarily in response to the presence of hydrochloric acid, stimulates the release from the pancreas of a watery fluid that contains large

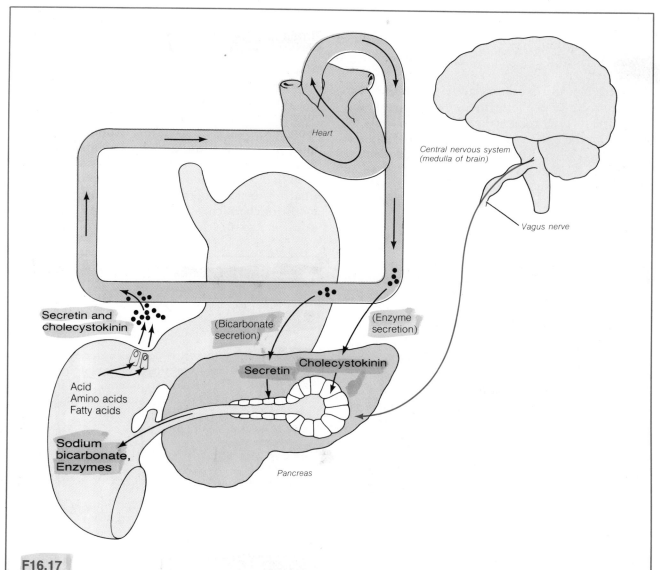

F16.17

Schematic representation showing the regulation of pancreatic secretion. Hormonal regulation is shown in black, and neural regulation is shown in color.

amounts of bicarbonate ions. A second intestinal hormone, cholecystokinin, which is released principally in response to the presence of certain amino acids and fatty acids, mainly promotes the release of digestive enzymes from the pancreas.

Neurally, pancreatic secretion is stimulated by way of the vagus nerves, and the effect is mostly on enzymatic secretion. This response is most evident during the cephalic and gastric phases of stomach secretion.

Bile — emulsify the fats.

The secretory and excretory activities of the liver continually produce bile in amounts averaging about 600 to 1000 ml per day. The bile travels by ducts from the liver, where it is produced, and the gallbladder, where it is stored, to the duodenum (see F16.3). Bile is an aqueous solution of water and electrolytes such as sodium and bicarbonate that contains bile salts (of cholic and chenodeoxycholic acids), bile pigments (such as bilirubin from the breakdown of hemoglobin), cholesterol, neutral fats, and lecithin. Approximately 94% of the bile salts released into the duodenum are reabsorbed in the ileum. They are then returned to the liver by the blood and resecreted. This cycle is known as *enterohepatic circulation.*

The rate of bile secretion is chemically, neurally, and hormonally controlled (F16.18). Bile salts present in the plasma as the result of enterohepatic circulation are powerful choleretics, that is, chemical substances that stimulate bile flow. Stimulation of the vagus nerves can also increase the rate of hepatic bile secretion. The hormone secretin increases the secretion of bile but not the rate of bile acid production. The hepatic bloodflow also influences bile secretion and, within limits, the flow of bile increases as the hepatic bloodflow increases.

During periods when large quantities of bile are not required for digestion, bile is stored in the gallbladder, which has an approximate capacity of 40 to 70 ml. Within the gallbladder, water and electrolytes are rapidly absorbed, and the concentration of bile salts and pigments may increase five to ten times. Shortly after the ingestion of a meal, the gallbladder contracts and bile is re-

leased into the duodenum. The contraction of the gallbladder is stimulated primarily by the hormone cholecystokinin. Vagal stimulation can also cause weak contractions of the gallbladder. A summary of the activities of the gastrointestinal hormones (gastrin, secretin, cholecystokinin, and gastric inhibitory peptide) is presented in Table 16.1.

Intestinal Secretions

Mucus. The first part of the duodenum contains **Brunner's glands,** which secrete mucus. The mucus provides a protective coat for the intestinal mucosa. Brunner's glands produce mucus in response to tactile or irritating stimulation of the mucosa, vagal stimulation, and intestinal hormones, particularly secretin. Cells called **goblet cells** that are located on the surface of the intestinal mucosa also produce mucus in response to tactile or chemical stimulation of the mucosa. The intestinal glands (crypts of Lieberkuhn) contain

Table 16.1 *Representative Activities of the Gastrointestinal Hormones*
Gastrin Stimulates gastric secretion, particularly the secretion of hydrochloric acid
Secretin Inhibits gastric acid secretion when the stomach is actively secreting and the mechanisms that cause gastrin release are strongly stimulated Stimulates the release of a watery fluid that contains bicarbonate from the pancreas Stimulates the secretion of bile
Cholecystokinin Inhibits gastric acid secretion when the stomach is actively secreting and the mechanisms that cause gastrin release are strongly stimulated Stimulates the release of enzymes from the pancreas Stimulates the contraction of the gallbladder
Gastric inhibitory peptide Inhibits gastric acid secretion when the stomach is actively secreting and the mechanisms that cause gastrin release are strongly stimulated

mucus-producing goblet cells whose secretory activity is probably controlled by local nervous reflexes.

Intestinal Juice. Intestinal glands, which are located over the entire surface of the small intestine, produce an intestinal juice that has a pH of 6.5 to 7.5 and is isotonic with the plasma. The actual fluid secretions of the intestinal glands contain almost no digestive enzymes. Almost all of the enzymes in the intestinal juice are derived from the disintegration of enzyme-rich epithelial cells that

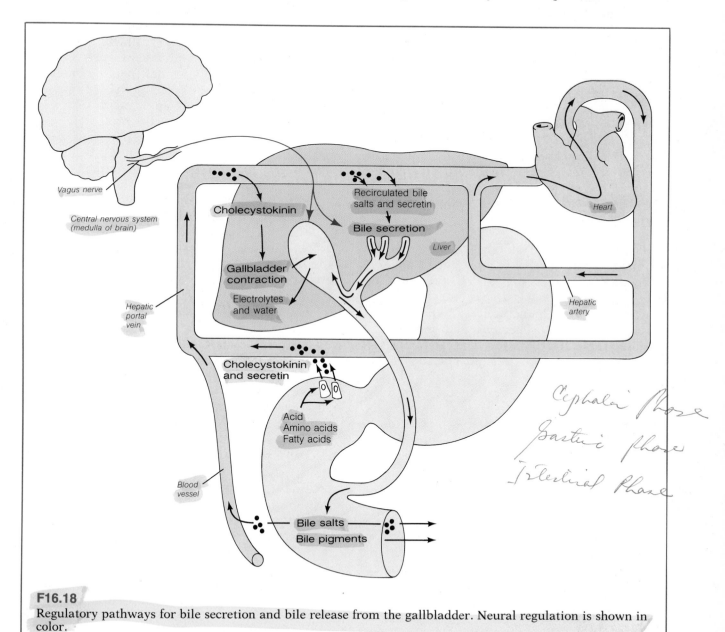

F16.18

Regulatory pathways for bile secretion and bile release from the gallbladder. Neural regulation is shown in color.

are constantly being discharged from the tips of the intestinal villi.

The main stimuli for the production of intestinal juice are local reflexes initiated by such things as tactile or irritating stimulation of the intestinal mucosa. Secretin and cholecystokinin also stimulate the secretion of intestinal juice, but these hormones are not considered to be major controlling factors.

The membranes of the microvilli of the intestinal epithelium contain hydrolytic enzymes that are important in digestion. Among these are several different disaccharidases, which participate in carbohydrate digestion, and a number of peptidases, which are involved in protein digestion. The microvilli also play an important role in the absorption of materials from the intestinal lumen.

Digestion and Absorption in the Small Intestine

Carbohydrates.   Plant starches and sucrose are the major dietary carbohydrates. Small amounts of lactose (milk sugar) may also be present. Amylases in both the saliva (salivary amylase) and the pancreatic juice (pancreatic amylase) break starch into the disaccharides maltose and isomaltose (Table 16.2). Maltose and isomaltose are each composed of two monosaccharide glucose units. Sucrose is a disaccharide composed of the monosaccharides glucose and fructose. The disaccharide lactose is made up of glucose and galactose. The final digestion of these disaccharides to absorbable monosaccharide units is completed by the disaccharidase enzymes (maltase, isomaltase, sucrase, and lactase) of the microvilli of the small intestine.

Sugars are absorbed primarily in the duodenum and upper jejunum of the small intestine, and their absorption is practically complete by the time the chyme reaches the ileum. Glucose is absorbed by active transport, and galactose is transported by the same carrier as glucose. Fructose is apparently absorbed by passive processes.

Proteins.   The pancreatic juice contains the inactive protein-digesting enzymes trypsinogen, chy-

motrypsinogen, and procarboxypeptidase. In the intestine, trypsinogen is converted to active **trypsin** by a substance called *enterokinase* from the intestinal mucosa and by previously formed trypsin. In turn, trypsin converts chymotrypsinogen to active **chymotrypsin,** and procarboxypeptidase to active **carboxypeptidase.** These active enzymes further the protein digestion begun in the stomach by pepsin (Table 16.2). Trypsin splits peptide bonds involving basic amino acids such as lysine and arginine, producing small peptide chains of amino acids. Chymotrypsin splits amino acid bonds involving aromatic amino acids such as tyrosine and phenylalanine, also producing small peptide chains. Carboxypeptidase frees the terminal amino acid from the carboxyl (acid) end of an amino acid chain. Together, the actions of these enzymes (as well as pepsin in the stomach) degrade proteins into an assortment of individual amino acids and small peptide chains.

A number of peptidase enzymes of the microvilli of the mucosal epithelium continue the process of protein digestion. Aminopeptidases liberate the terminal amino acid from the amino end of a small peptide chain. Tetrapeptidases split the final amino acid from tetrapeptides (chains of four amino acids), producing tripeptides and free amino acids, and tripeptidases split tripeptides into dipeptides and free amino acids.

Much protein digestion occurs in the upper portion of the small intestine, and protein is 60% to 80% digested and absorbed by the time the chyme reaches the ileum. During absorption, amino acids, dipeptides, and even tripeptides are taken into absorptive intestinal epithelial cells by active transport. Within the cells, dipeptidase and tripeptidase enzymes split the dipeptides and tripeptides into their component amino acids, and individual amino acids leave the cells and enter the bloodstream.

Lipids.   Dietary fat, which consists mostly of triacylglycerols, is essentially undigested when it reaches the small intestine. The first step in fat digestion is emulsification; that is, the dispersal of large water-insoluble fat droplets into a suspension of fine droplets that provide an increased

Table 16.2 *Principal Enzymes Involved in the Digestion of Carbohydrates, Proteins, and Lipids** *

Enzyme	Source	Action
Carbohydrates		
Salivary amylase	Salivary glands	Splits starch into the disaccharides maltose and isomaltose
Pancreatic amylase	Pancreas	Splits starch into the disaccharides maltose and isomaltose
Disaccharidases	Intestinal epithelium	Split disaccharides into monosaccharides
Maltase		Splits maltose into glucose
Isomaltase		Splits isomaltose into glucose
Sucrase		Splits sucrose into glucose and fructose
Lactase		Splits lactose into glucose and galactose
Proteins		
Pepsin (Pepsinogen)	Chief cells of stomach	Splits proteins into smaller peptide chains of amino acids
Trypsin (Trypsinogen)	Pancreas	Splits proteins and peptides into smaller peptides
Chymotrypsin (Chymotrypsinogen)	Pancreas	Splits proteins and peptides into smaller peptides
Carboxypeptidase (Procarboxypeptidase)	Pancreas	Frees terminal amino acid from the carboxyl (acid) end of an amino acid chain
Aminopeptidases	Intestinal epithelium	Free terminal amino acid from the amino end of an amino acid chain
Tetrapeptidases	Intestinal epithelium	Split final amino acid from tetrapeptides, producing tripeptides and free amino acids
Tripeptidases	Intestinal epithelium	Split tripeptides into dipeptides and free amino acids
Dipeptidases	Intestinal epithelium	Split dipeptides into free amino acids
Lipids		
Pancreatic lipase	Pancreas	Splits triacylglycerols mainly into monoacylglycerols and free fatty acids
Epithelial cell lipase	Intestinal epithelium	Splits monoacylglycerols into glycerol and fatty acids
Cholesterol esterase	Pancreas	Splits cholesterol esters into free cholesterol and free fatty acids

*Substances in parentheses are precursors of active enzymes.

surface area upon which the water-soluble, fat-digesting enzymes can act (F16.19). (Bile salts produced by the liver and released into the intestine with the bile are important in this process.) The small, emulsified fat particles are digested by lipase enzymes, primarily from the pancreas (Table 16.2). This activity results mainly in the formation of monoacylglycerols and free fatty acids.

Bile salt molecules are amphipathic, and they can aggregate with one another to form small, water-soluble structures called micelles (see Chapter 2). The monoacylglycerols and free fatty acids formed during fat digestion associate with bile salt micelles, and it is in this form that they reach the intestinal epithelium. During absorption, the monoacylglycerols and free fatty acids are believed to leave the micelles and enter absorptive epithelial cells by simple diffusion. The bile salts, which can be reused, are ultimately absorbed in the ileum as part of the enterohepatic circulation process.

During their entry into absorptive intestinal epithelial cells, many of the monoacylglycerols are digested to glycerol and fatty acids by an epithelial cell lipase. Within the endoplasmic reticulum of the epithelial cells, the free fatty acids combine with newly synthesized glycerol and small amounts of glycerol from the digested monoacylglycerols to form triacylglycerols. In addition, the cells synthesize phospholipids, cholesterol, and proteins. Small protein-coated globules of triacylglycerols, synthesized or absorbed phospholipids and cholesterol, and some free fatty acids leave the epithelial cells and enter the lacteals of the lymphatic system as minute droplets known as **chylomicrons.** Chylomicrons are, by weight, about 90% triacylglycerols, 5% phospholipids, 4% free fatty acids, 1% cholesterol, and a small amount of protein. Fat absorption usually takes place in the duodenum and jejunum and is completed in the ileum.

Ingested cholesterol occurs both as free cholesterol and as cholesterol esters (cholesterol combined with a fatty acid). The cholesterol esters are digested to free cholesterol and free fatty acids by a pancreatic cholesterol esterase (Table 16.2). Free

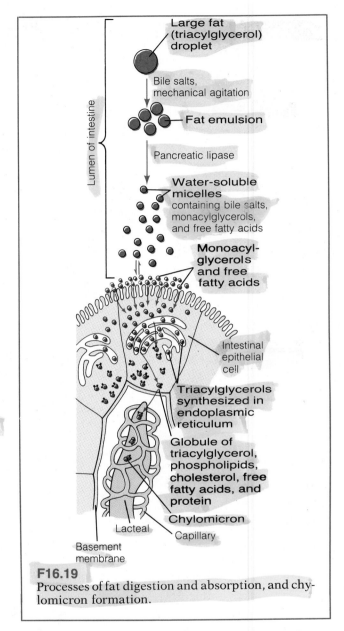

F16.19
Processes of fat digestion and absorption, and chylomicron formation.

cholesterol can associate with micelles, and it is ultimately absorbed at the intestinal epithelium.

Vitamins.   Fat-soluble vitamins such as A, D, E, and K associate with micelles, and they are absorbed in conjunction with fat digestion. Water-soluble vitamins, such as vitamin C and the B vita-

Defecation Reflex

mins with the exception of vitamin B_{12}, are absorbed by passive transport in the proximal portion of the intestine. Vitamin B_{12} combines with a glycoprotein substance, intrinsic factor, from the stomach and is actively absorbed across the intestinal wall at specific sites in the ileum.

Water and Electrolytes. The small intestine can absorb about 200 to 400 ml of water per hour. Five to ten liters of water derived from food, drink, and digestive secretions enter the small intestine each day, but only about 0.5 liter enters the large intestine. The rest is absorbed, primarily through the upper small intestine. Water can move across the intestinal wall in both directions, and the absorption of water from the intestine occurs osmotically in association with the absorption of particulate materials. In fact, the chyme within the duodenum becomes isosmotic with the blood plasma as a result of the osmotic removal or addition of water.

An active transport mechanism for sodium is present within the epithelial cells of the mucosa, particularly in the jejunum, and sodium is actively absorbed. Potassium, magnesium, and phosphate can also be actively absorbed through the mucosa. Chloride movement passively follows that of sodium in the upper part of the small intestine. However, chloride is actively transported in the ileum. Calcium absorption occurs actively along the entire small intestine (particularly the duodenum) and requires vitamin D. The amount of calcium absorbed varies with body requirements and is controlled in part by hormones from the parathyroid glands.

LARGE INTESTINE — *E. Coli*

The free surface of the large intestine is covered with epithelial cells, but no villi are present. The mucosa possesses abundant mucus-secreting goblet cells and intestinal glands. The epithelial cells of the intestinal glands produce almost no enzymes, and the intestinal glands contain a large number of mucus-secreting goblet cells. The rate of mucous production within the large intestine increases in response to direct tactile stimulation

of the surface goblet cells and as a result of local reflexes that involve the goblet cells of the intestinal glands. Extrinsic innervation also influences mucous secretion.

Sodium is actively absorbed in the large intestine, and chloride absorption also occurs. Bicarbonate ions are actively secreted by the large intestine. In addition, the large intestine absorbs 300 to 400 ml of water per day as a consequence of sodium (and chloride) transport. Many microorganisms inhabit the large intestine, and some of the bacteria produce vitamins, such as vitamin K.

The fecal material that ultimately leaves the digestive tract consists of water and solids such as undigested food residue, microorganisms, and sloughed-off epithelial cells.

blood
Clot
75 % of feces thr
Bili Pigments
indole
phenole
enole
hydrogen sulfi

COMMON PATHOLOGIES OF THE DIGESTIVE SYSTEM

In addition to conditions such as vomiting, diarrhea, and constipation, which were discussed previously, other abnormal conditions of the gastrointestinal tract or its accessory structures commonly occur.

PEPTIC ULCER

A *peptic ulcer* is an erosion of the wall of the gastrointestinal tract in an area of the tract exposed to gastric juice containing acid and pepsin (F16.20). Peptic ulcers are most commonly found in the stomach (gastric ulcers) and the duodenum (duodenal ulcers). They may be caused by an excessive acid-pepsin secretion, or they may result from an insufficient secretion of mucus, which normally protects the gastrointestinal mucosa from being digested by the acid-pepsin gastric juice.

The most common symptom of a peptic ulcer is pain. However, in some cases, the pain can be temporarily relieved by the ingestion of food, apparently because the food provides some protection by coating the ulcer or by acting as a buffer.

If the erosion due to a gastric ulcer is sufficiently severe, blood vessels in the stomach wall are damaged, and bleeding occurs into the stomach itself (a bleeding ulcer). In extreme cases,

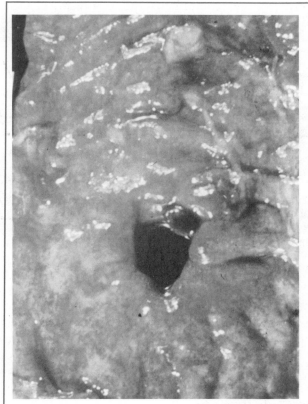

F16.20

A peptic ulcer. A punched-out sharply defined ulcer is seen in the mucosa of the duodenum.

an ulcer can lead to perforation; that is, to a hole entirely through the wall of the gastrointestinal tract. A perforation allows the contents of the gastrointestinal tract to pass into the abdominal cavity. Such an occurrence is very serious and requires immediate surgery.

GASTROENTERITIS

Gastroenteritis is an acute or chronic inflammation of the mucosa of the stomach and intestine. It is often caused by irritants such as excessive alcohol or cathartics (medicines that cause bowel movements), but can have any of a wide variety of other causes including viral infections, food allergies, overeating, and so forth.

GALLSTONES

Gallstones are particles containing cholesterol and bile salts and pigments that sometimes form in bile. The stones can block the cystic duct from the gallbladder or the common bile duct to the duodenum. Gallstones are often painful and those that block the cystic duct are especially so because the contractions of the gallbladder exert force on them.

In the case of cystic gallstones bile is still able to reach the duodenum directly from the liver through the common bile duct. Blockage of the common bile duct, however, prevents bile from reaching the intestine and thus interferes with the proper absorption of fat. In addition, the bile pigments cannot reach the intestine to be excreted. These accumulate in the blood and are eventually deposited in the skin. This occurrence produces a yellow color in the skin, a condition which is called *jaundice*.

PANCREATITIS

Pancreatitis is an inflammation of the pancreas that is often caused by the digestion of parts of the organ by pancreatic enzymes that are normally carried within the pancreatic ducts to the small intestine. In pancreatitis, the enzymes become activated within the ducts, and they may destroy the ducts and the pancreatic cells. Since pancreatic enzymes are very important in the digestion of carbohydrates, proteins, and fats, pancreatitis can produce severe nutritional problems.

HEPATITIS

Hepatitis is an infection of the liver (F16.21). Most commonly, it is a viral infection transmitted either by virus-infected blood or by contaminated food or water. The infected liver becomes enlarged, and its functioning is impaired, which can lead to jaundice. In some cases, liver function is depressed for a year or more.

CIRRHOSIS

Cirrhosis is a chronic inflammation of the liver that is progressive and diffuse. In the affected areas of the liver, some cells are replaced by fibrous

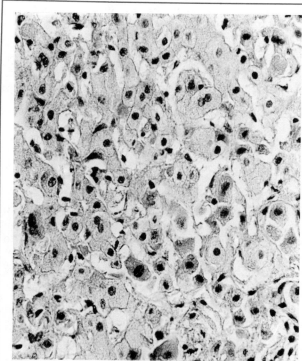

F16.21
Photomicrograph of acute hepatitis. The infected liver cells have beome balloon-shaped, and necrosis is occurring in cells with dark nuclei (×400).

connective tissue, thereby interfering with liver function. Cirrhosis can cause a reduction in the production of bile, in the excretion of bile pigments, and in the production of blood-clotting factors and plasma albumin. Cirrhosis also reduces the liver's ability to detoxify the blood, and toxins accumulate.

APPENDICITIS

The appendix is a blind-ending pouch extending from the cecum. If it becomes obstructed—for example, with hardened fecal material—its venous circulation may be interfered with. Such interference reduces the oxygen supply to the area and permits bacteria to flourish. The appendix then becomes inflamed and filled with pus, a condition called *appendicitis*.

If an inflamed appendix is not removed soon enough, it can rupture and release its contents into the abdominal cavity. This occurrence can cause peritonitis, which is an inflammation of the lining of the abdominopelvic cavity.

STUDY OUTLINE

ORGANIZATION OF THE DIGESTIVE SYSTEM
Specialized regions include mouth, esophagus, stomach, small intestine, large intestine. **pp. 405–410**

Basic Organization of the Gastrointestinal Tract

TUNICA MUCOSA Epithelial layer borders lumen, then lamina propria of loose connective tissue, and muscularis mucosae of smooth muscle.

TUNICA SUBMUCOSA Connective tissue.

TUNICA MUSCULARIS Inner circular and outer longitudinal layers of muscle.

TUNICA SEROSA OR ADVENTITIA Primarily connective tissue.

Modifications of the Basic Organization of the Gastrointestinal Tract

STOMACH Divided into fundus, body, and pyloric region; gastric glands secrete mucus, hydrochloric acid, pepsinogen.

SMALL INTESTINE Divided into duodenum, jejunum, ileum; contains fingerlike projections called villi; epithelial cell surfaces folded into microvilli; contains intestinal glands, Brunner's glands.

LARGE INTESTINE Divided into cecum, ascending colon, transverse colon, descending colon, sigmoid colon, rectum. Contains many mucous cells; no villi.

Intrinsic Nerve Plexuses Interconnections of neurons that coordinate much of the activity of the gastrointestinal tract and make possible the occurrence of intratract reflexes.

MECHANICAL PROCESSES OF DIGESTION
pp. 410–417
1. Segmentation—major mixing movement.
2. Peristalsis—major propulsive movement.

Chewing Part voluntary, part reflex; tactile stimulation of food against teeth, gums, and anterior portion of roof of mouth leads to reflex relaxation of muscles that raise jaw; when jaw drops, stretch reflexes lead to muscle contractions that raise jaw.

Swallowing Movement of bolus into pharynx elicits swallowing reflex.
1. Peristaltic contraction of pharyngeal constrictor muscles
2. Relaxation of upper esophageal sphincter
3. Approximation of vocal cords, upward movement of larynx, elevation of soft palate
4. Primary peristaltic wave in esophagus
5. Relaxation of lower esophageal sphincter

Gastric Motility
1. Stomach can expand to accommodate food without great increase in intragastric pressure.
2. Stomach undergoes peristaltic contractions that move chyme into the duodenum and provide effective mixing.
3. Basic electrical rhythm is important in determining frequency of gastric contraction.
4. Gastric motility and emptying are influenced by distension of stomach (via neural reflexes and gastrin) and by volume and composition of chyme in duodenum (via enterogastric reflex and intestinal hormones).

Motility of the Small Intestine Both segmentation and peristalsis occur in response to local mechanical and chemical stimulation; gastroileal reflex intensifies peristalsis in ileum; gastrin relaxes ileocecal sphincter.

Motility of the Large Intestine Movements are generally sluggish and include mixing movements of a segmental type and mass movements; defecation reflex leads to emptying of lower portion of large intestine.

Alterations in Gastrointestinal Tract Motility

VOMITING Ejects contents of stomach through mouth; during vomiting, the diaphragm and abdominal muscles contract and the lower esophageal sphincter relaxes.

DIARRHEA Watery stools generally accompanied by frequent defecation; can be due to either increased secretion of fluid into or decreased absorption of fluid from gastrointestinal tract.

CONSTIPATION Decreased motility of large intestine leads to increased water absorption and dry, hard feces.

CHEMICAL DIGESTION AND ABSORPTION
Accomplished by digestive enzymes that function mainly by hydrolysis; digested materials absorbed from lumen of gastrointestinal tract ultimately enter bloodstream; enzymes and other secretory products are in large measure produced by specialized epithelial cells that are commonly organized into exocrine secretory glands. **pp. 417–431**

Mouth and Saliva Saliva is produced by parotid, sublingual, submandibular, and buccal glands.

SECRETIONS OF THE SALIVARY GLANDS

Mucins Major proteins of saliva. Form mucus with water. Mucus lubricates mouth and food.

Salivary Amylase An enzyme that breaks down large starch molecules.

COMPOSITION OF SALIVA 97% to 99.5% water; pH 6.0 to 7.0; contains electrolytes (sodium, potassium, chloride, bicarbonate ions); kallikrein (leads to production of bradykinin, a vasodilator).

FUNCTIONS OF SALIVA Lubrication; digestion; bolus formation; dissolves foods, aids taste; moistens mouth and throat; aids speech.

CONTROL OF SALIVARY SECRETION Both parasympathetic and sympathetic stimulations cause secretion, with parasympathetic stimulation having the greatest effect.

Stomach

GASTRIC SECRETIONS

Mucus Lubricates stomach walls; protects gastric mucosa; produced by cardiac, pyloric, and gastric glands.

Hydrochloric Acid Produced by parietal cells; facilitates protein digestion; kills bacteria.

Pepsinogen Produced by chief cells; digests proteins when activated to pepsin.

CONTROL OF GASTRIC SECRETION

Cephalic or Reflex Phase Elicited by sight, smell, or taste of food.
1. Sensory receptors to CNS to vagus nerves to stomach for stimulation of gastric secretion.
2. Response elicited by sight and smell of food is basically a conditioned reflex.

Gastric Phase Elicited by distension of stomach; local reflexes; vagovagal reflex, gastrin.

Intestinal Phase Elicited by presence of chyme in upper portion of small intestine; gastrin production.

Intestinal Factors That Inhibit Gastric Secretion An enterogastric reflex and, under proper conditions, intestinal hormones.

PROTECTION OF THE STOMACH
Tall, columnar epithelial cells; alkaline mucous secretions.

Small Intestine and Associated Structures Most digestion and almost all absorption.

PANCREATIC JUICE
1. Basic—helps neutralize acid chyme from stomach
2. Carbohydrate, protein, and lipid-digesting enzymes
3. Ribonuclease and deoxyribonuclease
4. Hormonal control by secretin, cholecystokinin
5. Neural control by vagus nerves

BILE Produced in liver; stored in gallbladder; travels to duodenum.
1. Contains bile salts, bile pigments, cholesterol, neutral fats, and lecithin
2. Chemical control by choleretics
3. Neural control by vagus nerves
4. Hormonal control by secretin
5. Gallbladder contraction stimulated by cholecystokinin

INTESTINAL SECRETIONS

Mucus Produced by Brunner's glands, goblet cells, and intestinal glands; protects intestinal mucosa.

Intestinal Juice Produced by intestinal glands; isotonic with plasma; contains almost no digestive enzymes; local reflex control. Microvilli of intestinal epithelial cells contain digestive enzymes.

DIGESTION AND ABSORPTION IN THL SMALL INTESTINE

Carbohydrates
1. Amylases break starch into the disaccharides maltose and isomaltose.
2. Disaccharidase enzymes of microvilli break disaccharides into monosaccharides.
3. Glucose and galactose are absorbed by active transport.

Proteins
1. Enterokinase and previously formed trypsin convert trypsinogen to trypsin.
2. Trypsin activates chymotrypsinogen and procarboxypeptidase.
3. Trypsin, chymotrypsin, and carboxypeptidase produce an assortment of individual amino acids and small peptide chains.
4. Peptidase enzymes of intestinal epithelium continue the process of protein digestion.
5. Amino acids are absorbed by active transport.

Lipids (mostly undigested when reach small intestine)
1. Emulsification; bile salts are important.
2. Lipase enzymes digest small, emulsified fat particles mainly to monoacylglycerols and free fatty acids, which associate with bile salt micelles.
3. Absorption is probably by simple diffusion.
4. Triacylglycerols resynthesized in intestinal epithelial cells.
5. Chylomicrons leave cells and enter lymph.

Vitamins
1. Fat-soluble vitamins associate with micelles; absorbed in conjunction with fat digestion.
2. Water-soluble vitamins except B_{12} absorbed by passive transport in proximal intestine.
3. B_{12} combines with intrinsic factor; actively absorbed at specific ileum sites.

Water and Electrolytes Water absorbed mainly in small intestine; sodium can be actively absorbed; potassium, magnesium, and phosphate can also be actively absorbed; calcium absorption requires vitamin D.

Large Intestine Epithelial cells, but no villi; essentially no enzymes; mucus-secreting goblet cells. Absorption: sodium, chloride, and small amount of water. Fecal matter: composed of water, mucus, undigested food residue, microorganisms, and sloughed-off epithelium. Bacteria produce certain vitamins and digest cellulose.

COMMON PATHOLOGIES OF THE DIGESTIVE SYSTEM   pp. 431–433

Peptic Ulcer   Erosion of the wall of the gastrointestinal tract in an area of the tract exposed to gastric juice containing acid and pepsin; may be caused by excessive gastric juice secretion or insufficient mucus secretion.

Gastroenteritis   Acute or chronic inflammation of mucosa of stomach and intestine; caused by irritants, viral infection, or food allergy.

Gallstones   Particles containing cholesterol and bile salts. The particles can block the cystic or common bile ducts.

Pancreatitis   Inflammation of pancreas; often caused by digestion of parts of the organ by pancreatic enzymes.

Hepatitis   Infection of the liver, usually viral; causes enlargement of liver and impaired liver function.

Cirrhosis   Chronic liver inflammation; may result in reduced bile production, reduced bile pigment excretion, reduced blood-clotting factor production, and blood toxin accumulation.

Appendicitis   Inflammation of appendix; can be due to obstruction that interferes with venous circulation.

SELF-QUIZ

1. Gastric glands secrete: (a) mucus; (b) cholecystokinin; (c) trypsin.
2. Secondary peristaltic contractions in the esophagus: (a) move food from the stomach to the mouth during vomiting; (b) occur only during swallowing; (c) are generated in response to the distension of the esophagus.
3. Peristaltic contractions that occur in the stomach are generally strongest in the upper portion of the stomach and weakest in the lower portion near the duodenum. True or False?
4. The enterogastric reflex: (a) inhibits gastric motility; (b) stimulates the secretion of saliva; (c) causes defecation.
5. Gastric motility can be increased by: (a) gastrin; (b) mucins; (c) salivary amylase.
6. Digestive enzymes degrade fats into: (a) amino acids; (b) monoacylglycerols; (c) monosaccharides.
7. Salivary amylase is produced primarily by which glands? (a) buccal; (b) Brunner's; (c) parotid.
8. The optimal pH at which salivary amylase splits large starch molecules into smaller fragments is: (a) 6.9; (b) 4.2; (c) 8.4; (d) 11.7.
9. Saliva contains: (a) pepsin; (b) vitamin K; (c) kallikrein.
10. Chewing stimulates salivary secretion, whereas severe mental effort, dehydration, fear, and anxiety tend to reduce the flow of saliva. True or False?
11. The salivary substance that converts protein precursors to bradykinin is: (a) kallikrein; (b) salivary amylase; (c) mucin.
12. Parietal cells produce: (a) pepsinogen; (b) mucus; (c) hydrochloric acid.
13. Hydrochloric acid functions primarily in the facilitation of lipid digestion in the stomach and also kills many of the bacteria that enter the digestive tract with the food. True or False?
14. The chloride ions of the hydrochloric acid of the stomach are believed to come from: (a) chloride ions derived from carbonic acid; (b) chloride ions from plasma that are transported across parietal cells to the stomach lumen; (c) water molecules that combine with bicarbonate ions.
15. Protein digestion in the stomach proceeds to the stage of the formation of: (a) monoacylglycerols; (b) chylomicrons; (c) peptide chains.
16. A slight stimulation of gastric secretion may occur in response to the presence of chyme within the upper portion of the small intestine (particularly the duodenum) after the stomach has nearly emp-

tied and gastric secretory activity is relatively low following a meal. True or False?

17. Match the following digestive agents with their appropriate lettered descriptions.

Pancreatic juice *e*   Bile salts *f*
Tetrapeptidases *d*   Cholecystokinin *a*
Amylases *f*   Trypsin *e*

(a) Inhibits gastric acid secretion when the stomach is actively secreting
(b) Aids in the emulsification of fats
(c) Breaks certain peptide bonds that involve basic amino acids such as lysine and arginine
(d) Enzymes of the microvilli of the mucosal epithelium
(e) Contains the inactive protein-digesting enzymes trypsinogen, chymotrypsinogen, and procarboxypeptidase
(f) Contained in both the saliva and the pancreatic juice

18. The secretion of pancreatic enzymes occurs in response to: (a) cholecystokinin; (b) salivary amylase; (c) mucin.

19. The absorption of which element occurs actively along the entire small intestine (particularly the duodenum) and requires vitamin D? (a) potassium; (b) calcium; (c) sodium.

20. Sodium is actively absorbed in the large intestine, and chloride absorption also occurs there. True or False?

LEARNING OBJECTIVES

After completing this chapter, you should be able to:

- Outline the events of glycolysis and the Krebs cycle.
- Describe the movement of electrons through the electron transport system.
- Discuss the interconversion of carbohydrates, proteins, and triacylglycerol.
- Explain the utilization of carbohydrate, protein, and triacylglycerol during the absorptive and postabsorptive metabolic states.
- Distinguish between the two major classes of vitamins.
- Describe the roles of minerals in the body.
- Describe the concept of and the determination of the basal metabolic rate.
- Explain three theories of nutritional regulation.
- Explain three ways in which heat is transferred between the body and the environment.
- Distinguish between heat stroke and heat exhaustion.

CHAPTER CONTENTS

METABOLISM, NUTRITION, AND TEMPERATURE REGULATION 17

The chemical reactions that constantly occur within the body are grouped together under the classification of metabolism. The metabolic reactions produce heat, and this heat is important in body temperature regulation. Many metabolic reactions process absorbed nutrients such as glucose, amino acids, and tri-acylglycerols for use by the body. Portions of these substances are utilized for the synthesis of new structural materials, but the majority are used for the production of adenosine triphosphate (ATP), which provides energy to support the body's activities.

UTILIZATION OF GLUCOSE AS AN ENERGY SOURCE

One of the principal food materials utilized in the formation of ATP is the six-carbon sugar glucose. The breaking of the glucose molecule's chemical bonds releases energy that can be trapped in the form of ATP. In the body, glucose is broken down in a precisely controlled series of reactions that releases energy in a stepwise fashion. Each reaction or step is catalyzed by specific enzymes, often with the help of additional agents such as coenzymes or inorganic ions. The individual reactions are linked together into metabolic pathways in which the product of one reaction is a substrate for the next.

A metabolic pathway known as **glycolysis** breaks a glucose molecule into two molecules of pyruvic acid, and the energy released in this process is used to produce a small amount of ATP (F17.1). In the absence of oxygen, glycolysis converts the pyruvic acid molecules to lactic acid. However, when oxygen is available, as is usually the case, pyruvic acid is broken down, and carbon dioxide and water are produced. This process, which is accomplished by a pathway called the Krebs cycle and a system known as the electron transport system, releases a considerable amount of energy, which is trapped in the form of ATP.

GLYCOLYSIS

The enzymes for the glycolytic reactions are located within the cytoplasm of the cells. (F17.1 depicts the chemical transformations of glycolysis.)

The first step of glycolysis adds a phosphate group to the number six carbon of a glucose molecule, producing a molecule of glucose-6-phosphate. The second step rearranges the glucose-6-phosphate into a molecule of fructose-6-phosphate. The third step adds a phosphate group to the number one carbon of the fructose-6-phosphate to produce a molecule of fructose-1,6-diphosphate. The first and third steps each require a molecule of ATP, which provides a phosphate group and leaves behind a molecule of adenosine diphosphate (ADP). Thus, the initial stages of glycolysis do not produce energy, but actually use energy in the form of ATP.

The fourth step of glycolysis converts the six-carbon molecule of fructose-1,6-diphosphate to two three-carbon molecules. These are 3-phosphoglyceraldehyde and dihydroxyacetone phosphate. The fifth step transforms the dihydroxyacetone phosphate to 3-phosphoglyceraldehyde. Thus, each original glucose molecule forms two molecules of 3-phosphoglyceraldehyde, and the remaining steps of glycolysis occur twice for every glucose molecule used. For simplicity's sake, however, each of the steps beyond 3-phosphoglyceraldehyde is shown only once in Figure 17.2.

The sixth glycolytic step converts 3-phosphoglyceraldehyde to a molecule of 1,3-diphosphoglyceric acid. In this step, an inorganic phosphate group is added to 3-phosphoglyceraldehyde. Also, two electrons and a proton (a hydrogen ion, H^+, is a proton) are transferred to the coenzyme nicotinamide adenine dinucleotide (NAD^+), forming NADH, and another proton is released as H^+. The loss of electrons by a molecule is called *oxidation*, and the gain of electrons is called *reduction*. Thus, in this step, the 3-phosphoglyceraldehyde is oxidized, and the NAD^+ is reduced.

In terms of energy considerations, the structure of 1,3-diphosphoglyceric acid is especially important. Many high-energy compounds contain phosphate groups and have similar basic configurations. These can be written:

$$\begin{array}{ccc} Y & & O \\ \| & & \| \\ X-O & \sim & P-O^- \\ & & | \\ & & O^- \end{array}$$

with the $\sim$ indicating a high-energy bond. The Y is usually a carbon or an oxygen, and the X is a carbon or phosphorus. As can be seen from Figure 17.2, 1,3-diphosphoglyceric acid possesses this high-energy configuration. Thus, up to this point, the glycolytic reactions have produced from a single glucose molecule two three-carbon molecules of 1,3-diphosphoglyceric acid with high-energy configurations.

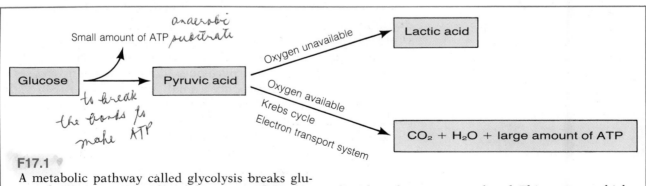

F17.1

A metabolic pathway called glycolysis breaks glucose down to pyruvic acid, and the energy released in this process is used to produce a small amount of ATP. When oxygen is unavailable, glycolysis converts pyruvic acid to lactic acid. When oxygen is available, pyruvic acid is broken down, and carbon dioxide and water are produced. This process, which is accomplished by the Krebs cycle and the electron transport system, releases a considerable amount of energy, which is used to produce a large amount of ATP.

In the seventh glycolytic step, 1,3-diphospho-glyceric acid transfers its high-energy phosphate to a molecule of ADP, converting the ADP to ATP. A molecule of 3-phosphoglyceric acid results from the reaction.

Steps eight and nine of glycolysis rearrange the 3-phosphoglyceric acid into a molecule of phosphoenolpyruvate, which also possesses the high-energy configuration. In step 10, phosphoenolpyruvate transfers its high-energy phosphate to ADP, generating another molecule of ATP. A three-carbon molecule of pyruvic acid results from the reaction.

Up to this point, the glycolytic breakdown of a single glucose molecule to two molecules of pyruvic acid utilizes two ATP molecules and generates four for a net profit of two. Along the way two NADH + H$^+$ are produced. (Recall that the steps

beyond 3-phosphoglyceraldehyde occur twice for each glucose molecule used.)

NAD$^+$ and other coenzymes are present in cells in only small amounts. Consequently, the NAD$^+$ that forms NADH + H$^+$ in the sixth glycolytic step must be regenerated if glycolysis is to continue processing glucose molecules. In order to regenerate NAD$^+$, NADH + H$^+$ must get rid of two electrons (and the accompanying protons) by transferring them to some acceptor molecule. When oxygen is unavailable, pyruvic acid acts as an electron acceptor, and the glycolytic pathway is completed by step 11, which converts pyruvic acid to lactic acid and regenerates NAD$^+$. Thus, glycolysis, which breaks a glucose molecule into two molecules of lactic acid, is an anaerobic pathway that requires no molecular oxygen. However, much of the energy contained within the glucose

F17.2
The steps of glycolysis by which a molecule of glucose is transformed into two molecules of lactic acid.

See the text for a detailed discussion. Pi is inorganic phosphate.

molecule remains within the lactic acid molecules, and the anaerobic breakdown of glucose to lactic acid by glycolysis yields a net profit of only two ATP.

Glycolysis alone cannot produce enough ATP to support the body's activities. However, it does allow cells to produce some ATP under anaerobic conditions. This ability is particularly important in skeletal muscle cells during heavy exercise.

To produce enough ATP to support the body's activities, oxygen-requiring, aerobic processes are necessary. When oxygen is available, as is usually the case, it acts as an electron acceptor by way of the electron transport system, and the pyruvic acid produced in step 10 of glycolysis is not converted to lactic acid. Instead it enters the Krebs cycle and is broken down. These aerobic events release a substantial amount of energy, which is used to produce a large amount of ATP.

KREBS CYCLE

The enzymes of the **Krebs cycle,** or **citric acid cycle,** as it is also called, are contained within the mitochondria of the cells. In order to enter the Krebs cycle, a molecule of pyruvic acid enters a mitochondrion and combines with a molecule of coenzyme A to form acetyl-coenzyme A (F17.3). During the reaction, one of the three carbons of pyruvic acid is lost as carbon dioxide, and an $NADH + H^+$ is formed. The two-carbon, acetyl-coenzyme A molecule enters the Krebs cycle by combining with a four-carbon molecule of oxaloacetic acid to form a six-carbon molecule, citric acid. The Krebs cycle itself involves the stepwise rearrangement of citric acid to reform oxaloacetic acid. Along the way, two carbons are lost as carbon dioxide. At the conclusion of the Krebs cycle, three additional $NADH + H^+$ have been produced, and electrons and protons have been transferred to the coenzyme flavin adenine dinucleotide (FAD), forming $FADH_2$. In addition, a molecule of ATP has been formed from a molecule of guanosine triphosphate (GTP), which was itself formed from guanosine diphosphate (GDP).

Since two pyruvic acid molecules are produced from one glucose molecule, the results of two pyruvic acid molecules that each form acetyl-coenzyme A and enter the Krebs cycle are: six carbon dioxide molecules, two ATP, eight $NADH + H^+$, and two $FADH_2$. As was the case for glycolysis, the coenzymes NAD^+ and FAD must be regenerated from $NADH + H^+$ and $FADH_2$ if the Krebs cycle is to continue. This regeneration is accomplished by the oxygen-requiring processes of the electron transport system.

ELECTRON TRANSPORT SYSTEM

The final steps in the aerobic generation of ATP involve the transfer of electrons from the coenzymes $NADH + H^+$ and $FADH_2$ to oxygen, and the combination of protons (H^+) with the oxygen to form water and regenerate NAD^+ and FAD. These activities are accomplished by the **electron transport system,** which is also located within the mitochondria. The electron transport system consists of a series of electron carriers, each of which can accept and donate electrons. Although not all the carriers are known with certainty, a likely sequence is shown in Figure 17.4. Some of the carriers are coenzymes that incorporate vitamins such as thiamine (vitamin B_1) or riboflavin (vitamin B_2). Others are proteins called cytochromes, which contain a heme group whose iron atom acts as an electron carrier. After being reduced by accepting electrons, each carrier becomes reoxidized by passing the electrons to the next carrier. Ultimately, the electrons are passed to oxygen, which combines with protons to form water.

Each $NADH + H^+$ produced within the mitochondria during the breakdown of pyruvic acid donates a pair of electrons to the electron transport system. As the electrons are passed from one carrier to another and finally to oxygen, energy is released. At three steps in this process, sufficient energy is released to synthesize ATP from ADP and inorganic phosphate. Thus, three molecules of ATP are produced from each of these $NADH + H^+$.

The $NADH + H^+$ generated in the cytoplasm during glycolysis cannot enter the mitochondria to donate electrons directly to the electron transport system. Instead, they donate their electrons to shuttle molecules that can penetrate the mitochondrial membrane and in turn donate the electrons to the electron transport system. One shuttle molecule, malate, donates the electrons before the first site of ATP synthesis, but another, glycerol phosphate, passes electrons to the carriers at a point beyond the first ATP generation point. Thus, sometimes three and sometimes only two ATP are produced from each of the NADH + H$^+$ generated during glycolysis, and it is not yet certain which shuttle system predominates in normal living tissue. The electrons from FADH$_2$ also enter the electron transport system after the first ATP-

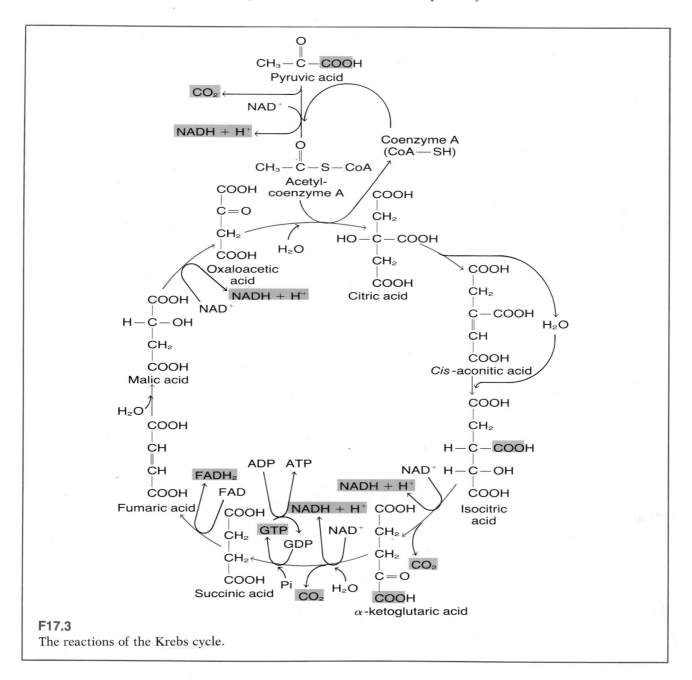

F17.3
The reactions of the Krebs cycle.

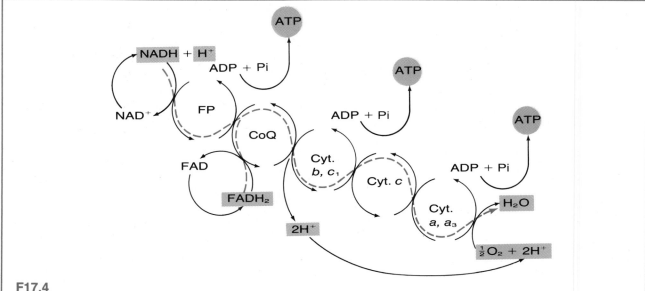

F17.4

The electron transport system. Electrons are passed to various electron acceptors (flavoprotein [FP], coenzyme Q [CoQ], cytochromes b and c_1, cytochrome c, and cytochromes a and a_3). Each electron acceptor is reduced when it receives electrons and reoxidized when it passes the electrons on to the next acceptor. The dashed line traces the path of electrons. At the end of the chain, oxygen is reduced, and water is formed.

generating step, and only two ATP molecules are produced from each $FADH_2$.

Overall, the aerobic breakdown of a single molecule of glucose produces a net profit of 36 to 38 ATP (F17.5). Two ATP are generated from glycolysis and two from the Krebs cycle. The eight $NADH + H^+$ produced within the mitochondria yield 24 ATP in the electron transport system, and the two $FADH_2$ yield four ATP. The two $NADH + H^+$ from glycolysis yield four to six ATP, depending on the shuttle molecules involved. The process of forming ATP during glycolysis and the Krebs cycle is called *substrate-level phosphorylation*, and the formation of ATP during electron transport is known as *oxidative phosphorylation*.

The uncontrolled combustion of glucose in the presence of oxygen produces carbon dioxide and water and releases about 686,000 calories of energy per mole (1 calorie is the amount of heat energy required to raise the temperature of 1 gram of water from 15 to 16°C). Although dependent on pH, the production of a mole of ATP from ADP traps about 7300 calories of energy. Therefore, the net profit to the cell from the aerobic breakdown of a mole of glucose is 277,400 calories of energy ($38 \times 7300 = 277,400$). Thus, by breaking down glucose in a controlled series of reactions, the aerobic processes can trap in the form of ATP about 40% of the energy released. The other 60% appears as heat. Remaining at the end of the reactions are carbon dioxide and water. In actual fact, however, the ATP yield in living cells is often lower than the maximum (38) used in these calculations.

UTILIZATION OF AMINO ACIDS AND TRIACYLGLYCEROLS AS ENERGY SOURCES

In addition to glucose, which is a carbohydrate, molecules of other substances—most importantly amino acids and triacylglycerols—can undergo metabolic conversions to intermediates that can enter the glycolytic and Krebs cycle pathways, and thereby provide energy for body activities.

AMINO ACIDS

Amino acids can lose their nitrogen-containing amino groups and be converted to α-keto acids that can ultimately enter the Krebs cycle—for example, by way of pyruvic acid or the Krebs cycle intermediate oxaloacetic acid, both of which are α-keto acids. (An α-keto acid is similar to an amino acid except that it has oxygen rather than an amino group bonded to its alpha carbon.)

The process of converting an amino acid to an α-keto acid generally involves a reaction known as *transamination* in which the amino group of an amino acid is exchanged with the oxygen of an α-keto acid, most commonly α-ketoglutaric acid (F17.6a). The acquisition of an amino group by α-ketoglutaric acid in exchange for its oxygen produces a molecule of glutamic acid (an amino acid), and the amino acid that provides the amino group is converted to an α-keto acid. Glutamic acid can then undergo a reaction called *oxidative deamination* in which its amino group is removed, with the resultant production of ammonia, and α-ketoglutaric acid is reformed (F17.6b). The overall result of the transamination reaction involving α-ketoglutaric acid and the subsequent deamination of glutamic acid is to transform an amino acid into an α-keto acid, with the resultant production of ammonia. As explained previously, the α-keto acids produced in this process can ultimately enter the Krebs cycle. The ammonia is converted to urea by the liver, and the urea enters the bloodstream and is excreted by the kidneys.

TRIACYLGLYCEROLS

Triacylglycerols (fat) can be broken into their components, glycerol and fatty acids. The glycerol can be converted to 3-phosphoglyceraldehyde, which is one of the substances of the glycolytic pathway. The fatty acids can undergo a stepwise breakdown into molecules of acetyl-coenzyme A by the process of *beta oxidation* (F17.7). In this process, a fatty acid molecule is activated by com-

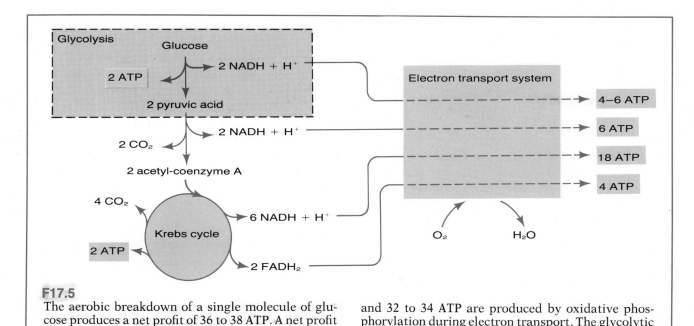

F17.5

The aerobic breakdown of a single molecule of glucose produces a net profit of 36 to 38 ATP. A net profit of four ATP is produced by substrate-level phosphorylation during glycolysis and the Krebs cycle, and 32 to 34 ATP are produced by oxidative phosphorylation during electron transport. The glycolytic reactions occur in the cytoplasm of the cell. The other reactions take place within the mitochondria.

Transamination

$$R_1-CH-COOH + HOOC-CH_2-CH_2-C-COOH \xrightarrow{\text{Enzymes}} R_1-C-COOH + HOOC-CH_2-CH_2-CH-COOH$$

(with NH_2 on the amino acid, O on α-ketoglutaric acid, O on α-keto acid, NH_2 on glutamic acid)

Amino acid   α-ketoglutaric acid   α-keto acid   Glutamic acid

(a)

$$HOOC-CH_2-CH_2-CH-COOH + H_2O \xrightarrow{\text{Enzyme}} HOOC-CH_2-CH_2-C-COOH + NH_3$$

(with NH_2 on glutamic acid, O on α-ketoglutaric acid; NAD → NADH + H⁺)

Glutamic acid   NAD   NADH + H⁺   α-ketoglutaric acid   Ammonia

(b)

Deamination

F17.6

(a) Transamination reaction involving α-ketoglutaric acid and an amino acid. In this reaction, the amino group of the amino acid is exchanged with the oxygen of α-ketoglutaric acid. This exchange converts α-ketoglutaric acid to glutamic acid, and the amino acid to an α-keto acid. (Note that an α-keto acid is similar to an amino acid, except that it has an oxygen rather than an amino group bonded to its alpha carbon.) (b) Oxidative deamination of glutamic acid. This reaction forms α-ketoglutaric acid and ammonia. The overall result of the transamination reaction involving α-ketoglutaric acid and the subsequent oxidative deamination of the glutamic acid that is produced is to transform an amino acid into an α-keto acid, with the resultant production of ammonia.

bination with a molecule of coenzyme A. After a series of reactions, two carbons of the fatty acid carbon chain are broken away as a molecule of acetyl-coenzyme A, which can enter the Krebs cycle. The remainder of the fatty acid carbon chain can be recycled through the reactions and another two carbons can be split off as acetyl-coenzyme A. This process can be repeated until the entire carbon chain of the fatty acid is utilized.

OTHER USES OF FOOD MOLECULES

In addition to being broken down and utilized as sources of energy, food molecules are also built into larger structures. For example, amino acids can be assembled into proteins that function as enzymes or structural cellular components, and glucose molecules can be linked together to form large glycogen molecules, which are a storage form of carbohydrate. Moreover, when considering the uses of the various food molecules, it is important to realize that the metabolic pathways followed by carbohydrates (such as glucose), proteins (amino acids), and triacylglycerols (fat) are interrelated (F17.8). Thus, pathways exist by which amino acids or glucose, for example, can be converted to triacylglycerols, or by which amino acids can form glucose. The liver is especially active in such interconversions; so is adipose tissue.

The ability to perform interconversions enables the body to synthesize many of the substances it needs. However, some essential substances cannot be manufactured by the body, and these must be obtained from external sources. For example, the amino acids of proteins are essential sources of nitrogen. Although the body can form some amino acids, it cannot synthesize all of the amino acids it needs. Therefore, certain amino acids, called essential amino acids, must be obtained in the diet. In addition, some fatty acids cannot be synthesized, and these must also come from dietary sources. The same is largely true of vitamins, many of which serve as coenzymes in chemical reactions.

NUTRIENT POOLS

In any consideration of metabolism, it must be remembered that a person is not a static entity but rather exists in a dynamic steady state. The processes involved in the synthesis of protoplasm from simpler food molecules (anabolism) occur simultaneously with the breakdown of molecules to

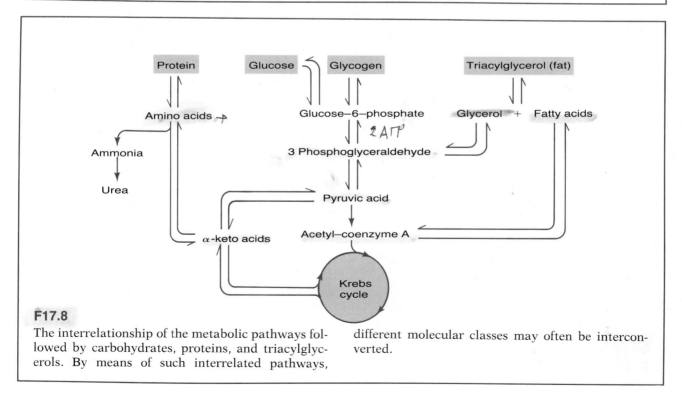

F17.7

Breakdown of a fatty acid molecule by beta oxidation.

F17.8

The interrelationship of the metabolic pathways followed by carbohydrates, proteins, and triacylglycerols. By means of such interrelated pathways, different molecular classes may often be interconverted.

provide energy (catabolism). Thus, almost all the body's atoms and molecules are in a constant state of flux; even the atoms and molecules of bone are continually being exchanged.

In view of the constant turnover of the body's atoms and molecules, it is useful to consider the existence of nutrient pools within the body that can be drawn upon to meet its needs. For example, ingested protein (as well as the body's own protein) can be broken down to amino acids that contribute to an amino acid pool (F17.9). Amino acids from this pool can be used for the synthesis of proteins and other nitrogen-containing structures required by the body. They can also have their nitrogen removed and be used as energy sources or be converted to carbohydrate or triacylglycerols.

Since carbohydrates are readily converted to triacylglycerols, carbohydrate and triacylglycerol pools are often considered together (F17.10). Carbohydrates and triacylglycerols are major energy sources for the body. The body is able to store excess triacylglycerol as well as some excess carbohydrate. Triacylglycerol stores, however, provide the major energy reserves for the body, and comparatively small amounts of energy are stored as carbohydrate (glycogen).

METABOLIC STATES

A person actually exists in two different metabolic states. After eating, the body is in an *absorptive state* in which food substances are being absorbed from the gastrointestinal tract into the bloodstream. When absorption is complete, the body is in a *postabsorptive*, or fasting, *state*. In the postabsorptive state, a person's needs must be met by the use of materials already present within the body.

ABSORPTIVE STATE

Carbohydrate

During the absorption of a typical meal (65% carbohydrate, 25% protein, and 10% triacylglycerol),

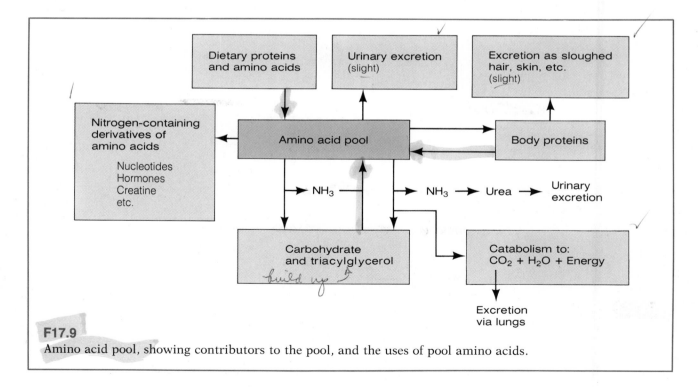

F17.9
Amino acid pool, showing contributors to the pool, and the uses of pool amino acids.

the body usually utilizes carbohydrate (glucose) as its major energy source, although some energy is derived from absorbed amino acids and triacylglycerols (F17.11). During the absorptive state, the body stores as glycogen some carbohydrate not used for energy, and it converts much to adipose tissue triacylglycerol.

Carbohydrate is absorbed from the gastrointestinal tract largely in the form of the monosaccharides glucose, galactose, and fructose. However, glucose is the principal monosaccharide, and most of the galactose and fructose are either converted to glucose or enter essentially the same metabolic pathways as glucose. Therefore, the absorbed monosaccharides can be considered simply as glucose.

Glucose absorbed from the gastrointestinal tract enters the blood and is carried to the liver, where a large portion of it enters liver cells. The liver itself uses little glucose for energy during the absorptive state, and it converts much into triacylglycerol or glycogen. Some of the triacylglycerol is stored in the liver, but most is packaged into protein-containing particles called very low density lipoproteins (VLDL), which are released into the blood. As is discussed later, many of the fatty acids of the VLDL triacylglycerols are utilized in the formation of adipose tissue triacylglycerol. Moreover, adipose tissue cells can take up and convert to triacylglycerol absorbed glucose in the blood that is not taken up by the liver. Skeletal muscle and certain other tissues can also take up some glucose and store it as glycogen.

Protein

During the absorptive state, the body uses amino acids for protein synthesis, and amino acids are also converted into adipose tissue triacylglycerol (F17.11). Some of the amino acids from digested protein are taken up by the liver, where their nitrogen is removed and they are converted to α-keto acids. The α-keto acids can ultimately be metabolized by the Krebs cycle to provide energy for the liver cells. They can also be converted to fatty acids, which can be synthesized into triacylglycerol, much of which is packaged into very low density lipoproteins and released into the blood. The liver also uses amino acids to synthesize proteins such as plasma proteins. Skeletal

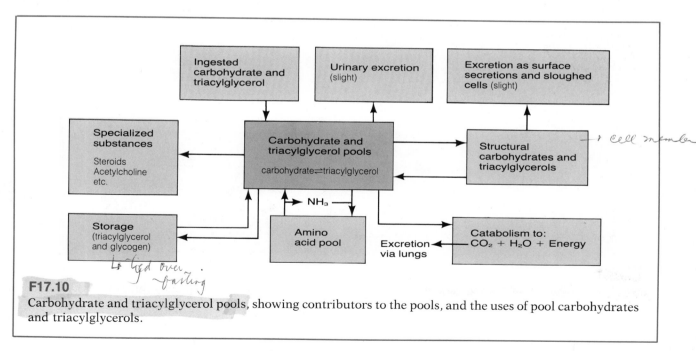

F17.10

Carbohydrate and triacylglycerol pools, showing contributors to the pools, and the uses of pool carbohydrates and triacylglycerols.

muscle cells and other cells take up many amino acids and synthesize proteins from them.

Triacylglycerol

During the absorptive state, the body uses some triacylglycerol for synthesis, and some is used as an energy source. However, much triacylglycerol is stored in adipose tissue (F17.11). Triacylglycerols from the digestive system enter the lacteals of the lymphatic system as a major component of chylomicrons. The lymphatic system delivers the chylomicrons to the bloodstream. The enzyme lipoprotein lipase, which is associated with the endothelium of capillaries, particularly in skeletal muscle and adipose tissue, breaks down the triacylglycerols of the chylomicrons (and also the triacylglycerols of the very low density lipoproteins from the liver), releasing free fatty acids. Some of the free fatty acids are taken up and utilized as an energy source by the muscle

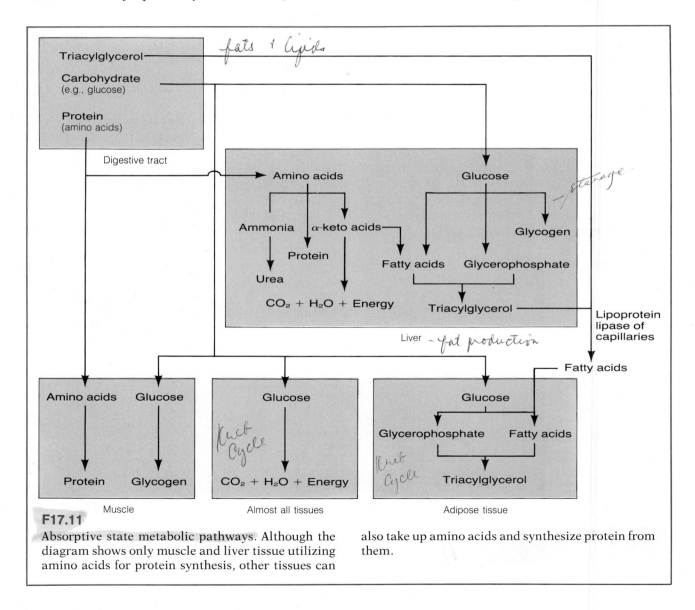

F17.11

Absorptive state metabolic pathways. Although the diagram shows only muscle and liver tissue utilizing amino acids for protein synthesis, other tissues can also take up amino acids and synthesize protein from them.

cells (not shown in F17.11), and many are taken up and resynthesized into triacylglycerols by the adipose tissue cells. The glycerol portion of the resynthesized triacylglycerols is derived from glucose.

When the absorptive state uses of carbohydrates and amino acids as well as triacylglycerols are considered, it is evident that the components of adipose tissue triacylglycerols are derived primarily from: (1) chylomicron triacylglycerols, (2) triacylglycerols produced as a result of the synthetic activities of the liver, and (3) glucose.

POSTABSORPTIVE STATE

During the postabsorptive state, the blood glucose level must be maintained, especially for the support of the nervous system, even though no glucose is being absorbed. The level is maintained by using body glucose sources and glucose-sparing, fat-utilizing metabolic pathways (F17.12).

Glucose Sources

The liver contains stores of glycogen that can be broken down to glucose. These glycogen stores, however, can support the body's activities for only about four hours. Other tissues, including skeletal muscle, also contain glycogen reserves, but these too are limited. Moreover, skeletal muscle lacks the enzyme required to form free glucose from glycogen, and it cannot directly release glucose to maintain the blood glucose level. Instead, skeletal muscle breaks glycogen down to glucose-6-phosphate, which it can catabolize for energy by way of glycolysis and the Krebs cycle. In addition, the pyruvic acid and, under anaerobic conditions, lactic acid that are produced by muscle glycolysis can enter the blood and travel to the liver where they can be converted to glucose. As a result, muscle glycogen can serve as an indirect source of blood glucose.

The catabolism of triacylglycerols to glycerol and fatty acids, particularly in adipose tissue, can also provide glucose because glycerol can be converted to glucose by the liver.

Protein, however, is the major blood glucose source during prolonged fasting. Large quantities of muscle protein that are not essential to cellular function (and to a lesser extent similar protein in other tissues) can be catabolized to amino acids, which in turn can be converted to glucose by the liver. During prolonged fasting, the kidneys as well as the liver produce substantial amounts of glucose for the body.

Glucose Sparing and Triacylglycerol Utilization

During the postabsorptive state, almost all the organs and tissues of the body—the nervous system is the major exception—diminish their use of glucose for energy and depend primarily on triacylglycerol. This adjustment spares glucose from the liver for use by the nervous system, which normally requires glucose as its principal energy source. During the postabsorptive state, adipose tissue triacylglycerol is broken down and fatty acids are released into the blood. Almost all tissues except the nervous system can take up the fatty acids, which can ultimately enter the Krebs cycle to provide energy for body activities.

During the postabsorptive state, the liver not only utilizes fatty acids as an energy source but it also processes them into a group of substances called ketone bodies (acetoacetic acid, β-hydroxybutyric acid, and acetone), which are released into the blood. Many tissues can take up these ketone bodies (specifically acetoacetic acid and β-hydroxybutyric acid) and, by way of the Krebs cycle, utilize them as an important source of energy. Thus, the breakdown of adipose tissue triacylglycerol during fasting ultimately provides energy for the body and spares glucose for the nervous system.

As a result of these activities, an average person can fast for many weeks as long as water is provided. With prolonged fasting, however, a noteworthy change in brain metabolism occurs. The requirement that the brain must utilize glucose for its energy supply no longer seems to hold true after four or five days of fasting and, in addition to glucose, the brain begins to use large quantities of ketone bodies as energy sources. Since the major source of blood glucose during prolonged fasting is protein, this change conserves body protein that would otherwise have to be converted to glucose for the support of the nervous system.

CONTROL OF ABSORPTIVE AND POSTABSORPTIVE STATE EVENTS

Several hormonal and neural mechanisms control and regulate metabolism in both the absorptive and postabsorptive states. For example, when glucose is being absorbed following a meal—that is, during the absorptive state—the blood glucose level may rise from about 85 mg per 100 ml to about 120 mg per 100 ml. The rise in the blood glucose level causes an increased release of insulin, which influences both membrane transport and enzymatic activity. With the notable exception of most areas of the brain, insulin facilitates the entry of glucose into many cells, particularly skeletal muscle and adipose tissue cells. It favors the oxidation of glucose and the formation of glycogen within cells, and it inhibits glycogen breakdown. Insulin stimulates the active transport of amino acids into most cells, and it promotes pro-

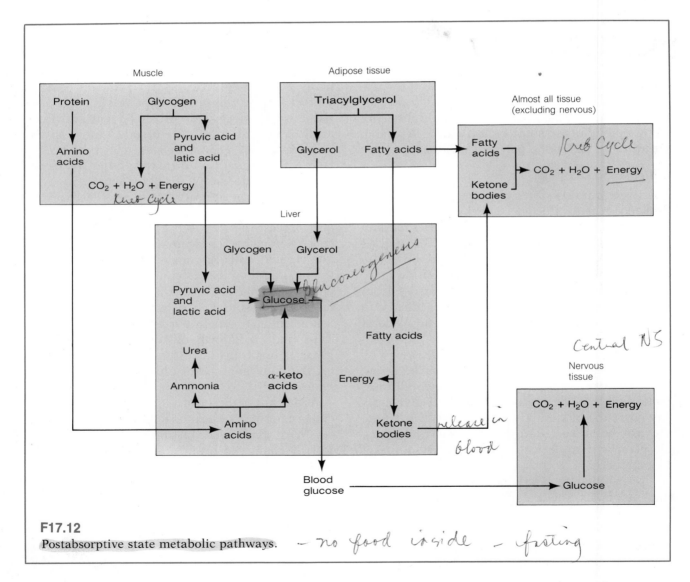

F17.12
Postabsorptive state metabolic pathways. — no food inside — fasting

25% decrease of blood glucose after one month fasting.

tein synthesis and inhibits protein breakdown. Insulin also promotes the synthesis and inhibits the breakdown of triacylglycerols in adipose tissue cells. Thus, in general, insulin promotes metabolic activities of the sort that occur during the absorptive state.

When absorption is completed and blood glucose levels begin to fall—that is, during the postabsorptive state—insulin release diminishes. This favors the breakdown of glycogen, proteins, and triacylglycerols and the release of glucose, amino acids, glycerol, and fatty acids into the blood. A declining blood glucose level also leads to an increased release of the hormone glucagon. Glucagon stimulates the breakdown of glycogen and the production of carbohydrate from non-carbohydrate precursors (such as amino acids) in the liver. Glucagon also stimulates the breakdown of triacylglycerols in adipose tissue. When the blood glucose level is falling rapidly (but apparently not during prolonged fasting), glucose receptors in the brain are affected and sympathetic neural activity to the adrenal medullae and to adipose tissue cells increases. The adrenal medullary hormones—principally epinephrine—stimulate the breakdown of glycogen in liver and skeletal muscle as well as the breakdown of triacylglycerols in adipose tissue. The sympathetic neural stimulation to adipose tissue cells also promotes the breakdown of triacylglycerols. The overall effect of these activities is to increase the concentrations of glucose, glycerol, and fatty acids in the blood. Thus, during the postabsorptive state, hormonal and neural mechanisms mobilize carbohydrates, proteins, and triacylglycerols for the maintenance of blood glucose levels and the support of body activities.

LIVER FUNCTION

In addition to its important metabolic role, the liver performs several other important functions. The liver:

1. Produces albumin and other plasma proteins
2. Produces prothrombin and other blood clotting factors

3. Excretes bilirubin, a breakdown product of hemoglobin
4. Produces bile, which aids in fat digestion
5. Carries out many reactions of carbohydrate, protein, and lipid metabolism
6. Stores glycogen
7. Synthesizes urea and ketone bodies
8. Stores iron (as ferritin), vitamins A and D, and other substances
9. Detoxifies many harmful drugs and toxins
10. Phagocytizes microorganisms and other foreign material

Many of these functions are discussed in other chapters.

VITAMINS

Normal metabolism and growth require not only adequate supplies of carbohydrates, proteins, and fats, but also minute amounts of other organic nutrients called **vitamins.** Vitamins do not provide energy or serve as structural components. Rather, most vitamins function either directly or after chemical alteration as coenzymes or components of coenzymes involved in various metabolic reactions.

In general, vitamins cannot be synthesized by the body; they must be obtained from external sources. The principal source of vitamins is ingested food, although some vitamins, such as vitamin K, can be produced by bacteria in the gastrointestinal tract. If raw materials, called provitamins, are provided, the body can assemble some vitamins. For example, the provitamin carotene found in foods such as spinach, carrots, liver, and milk can be used by the body in the production of vitamin A.

Vitamins are divided into two major groups: water-soluble vitamins and fat-soluble vitamins. Water-soluble vitamins, which include the B vitamins and vitamin C, are generally absorbed with water from the gastrointestinal tract. (Vitamin B_{12} is absorbed in association with intrinsic factor.) Fat-soluble vitamins, which include vitamins A, D, E, and K, are absorbed with digested dietary triacylglycerols. As a general rule, the storage of the water-soluble vitamins by the body is slight.

The fat-soluble vitamins, with the exception of vitamin K, are usually stored to a greater extent. Vitamins A and D, in particular, are stored by the liver. Specific vitamins, examples of their functions in or importance to the body, and some of the results of their deficiencies are listed in Table 17.1.

MINERALS

The body needs not only organic compounds but also certain inorganic elements, or **minerals.** Minerals make up 4% to 5% of the body by weight, and many of them are essential for life. Minerals are involved in numerous body functions, and the activity of many enzymes depends on their presence. They are frequently found in chemical unions with organic compounds—for example, the iron of the hemoglobin molecule. The minerals used by the body are generally present in ionized form. Specific minerals and examples of their functions in or importance to the body are listed in Table 17.2.

METABOLIC RATE

The **metabolic rate** is the total energy expended by the body per unit time. This energy is derived from the breakdown of carbohydrates, proteins, triacylglycerols, and other organic molecules that are originally obtained in the diet. The energy is used for biological work such as muscle contraction, active transport, or molecular synthesis, and it also appears as heat. Actually, only a relatively small amount of the energy is used for work; most appears as heat. However, the heat can be valuable in body temperature regulation.

A reasonable estimate of the metabolic rate can be obtained by measuring the body's rate of oxygen consumption (F17.13). For this purpose it is generally assumed that for every liter of oxygen consumed, the body produces 4.825 kilocalories of energy (1 kilocalorie = 1000 calories).

BASAL METABOLIC RATE

Many factors including eating, exercise, and exposure to extreme environmental temperatures significantly influence a person's metabolic rate. Therefore, for purposes of comparison between individuals, the metabolic rate is often determined under specific, standard conditions, which eliminate many of these variables. The metabolic rate determined under these conditions is called the **basal metabolic rate,** and it is usually expressed in terms of kilocalories of energy expended per square meter of body surface area per hour.

The conditions under which the basal metabolic rate is determined include having the person awake but in a supine position at mental and physical rest in a room with a temperature between 20°C and 25°C (68°F to 77°F). The person should also be in a postabsorptive state and have had nothing to eat for 12 to 18 hours. Under these conditions, no energy is expended in digesting, absorbing, and processing ingested nutrients, and none is expended by muscular contractions that do external work such as moving objects in the environment. Moreover, the person is in moderate thermal surroundings.

The basal metabolic rate is not the body's lowest metabolic rate. The metabolic rate during sleep is 10% to 15% lower than basal, and this lower rate is presumably due, at least in part, to a more complete muscular relaxation during sleep.

FACTORS THAT AFFECT METABOLIC RATE

Both sex and age influence metabolic rates. Basal metabolic rates tend to be lower in females than in males, and they tend to increase with age to approximately five years and decrease with age thereafter.

The ingestion of a typical meal containing carbohydrate, protein, and fat increases the metabolic rate 10% to 20% over basal. The effect of food consumption on metabolism is called the *specific dynamic action*. Protein ingestion has the greatest effect, and the ingestion of protein alone increases metabolism as much as 30%. The ingestion of carbohydrate alone may increase metabolism only 5%, and the ingestion of lipid alone increases metabolism approximately 8%. The energy expended in the digestion and absorption of food accounts for only a small portion of the increase. Most of the increase appears to be due to the processing of exogenous nutrients by the liver.

The greatest changes in metabolic rates are pro-

Table 17.1 *Vitamins*

Vitamin	Examples of Functions or Importance	Some Results of Deficiency
WATER SOLUBLE VITAMINS		
Vitamin B$_1$ (Thiamine)	Involved in the metabolism of carbohydrates and many amino acids.	Deficiency leads to beriberi, which is characterized in some cases by cardiac muscle weakness and heart failure, and in other cases by peripheral neuritis, with muscle atrophy and occasionally paralysis.
Vitamin B$_2$ (Riboflavin)	Component of FAD, which is involved in Krebs cycle reactions.	Deficiency leads to cracking at the corners of the mouth.
Niacin (Nicotinic acid)	Component of NAD$^+$, which is involved in glycolytic, Krebs cycle, and other reactions.	Deficiency gives rise to pellagra, which is characterized by diarrhea, dermatitis, and mental disturbances.
Vitamin B$_6$ (Pyridoxine)	Involved in amino acid and protein metabolism.	Deficiency leads to dermatitis and perhaps convulsions and gastrointestinal disturbances in children.
Vitamin B$_{12}$ (Cyanocobalamin)	Involved in erythrocyte maturation and nucleic acid metabolism.	Deficiency results in macrocytic anemia, neurological disturbances, loss of peripheral sensation, and, in severe cases, paralysis.
Pantothenic acid	Component of coenzyme A, which functions in the entry of pyruvic acid into the Krebs cycle and in the degradation of fatty acids.	
Folic acid	Involved in the synthesis of purines and thymine, which are required for DNA formation; important in erythrocyte maturation.	Deficiency leads to macrocytic anemia, which is characterized by the production of abnormally large erythrocytes.
Biotin	Involved in amino acid and protein metabolism.	Deficiency leads to muscle pain, fatigue, and poor appetite.
Vitamin C (Ascorbic acid)	Important in the formation of collagen and in the maintenance of normal intercellular substances throughout the body.	Deficiency leads to scurvy, which is characterized by poor wound healing, defective bone formation and maintenance, and fragile blood vessel walls.
FAT-SOLUBLE VITAMINS		
Vitamin A	Important in the formation of photopigments; apparently needed for the maintenance of certain epithelial surfaces such as the mucous membranes of the eyes.	Deficiency results in night blindness and corneal opacity.

Table 17.1   *Vitamins (Continued)*

Vitamin	Examples of Functions or Importance	Some Results of Deficiency
Vitamin D	Needed for the proper absorption of calcium from the gastrointestinal tract.	Deficiency leads to rickets in children and osteomalacia in adults.
Vitamin E	Inhibits the breakdown of certain fatty acids.	
Vitamin K	Required for the synthesis of pro-thrombin and certain other clotting factors in the liver.	Deficiency leads to retarded blood clotting and excessive bleeding.

Table 17.2   *Minerals*

Mineral	Examples of Functions or Importance
Calcium	Needed for the formation of bones and teeth; required for blood clotting; necessary for normal skeletal and cardiac muscle activity; important in normal nerve function.
Sodium	Exerts a major influence on the osmotic pressure of the extracellular fluid; important in muscle and nerve function; major cation (positively charged ion) of the extracellular fluid.
Potassium	Important in normal muscle and nerve function; major cation of the intracellular fluid.
Phosphorus	Needed for the formation of bones and teeth; a component of energy substances (such as ATP), which are involved in energy storage and transfer; a component of nucleic acids (DNA, RNA).
Magnesium	Important in normal muscle and nerve function; activates a number of enzymes.
Iron	A component of hemoglobin, which is important in oxygen transport; a component of myoglobin; involved in the formation of ATP as a component of cytochromes of the electron transport system.
Iodine	Essential for the formation of the thyroid hormones thyroxine and triiodothyronine, which help govern metabolism.
Copper	Required in the manufacture of hemoglobin and the pigment melanin.
Zinc	A component of several enzymes such as carbonic anhydrase and carboxypeptidase, and thus important for reactions catalyzed by these enzymes.
Fluorine	Normally present in teeth and bones; appears to provide protection against dental caries (cavities).
Manganese	Needed for the formation of urea; activates some enzymes.
Cobalt	A component of vitamin B_{12}, which is necessary for the normal maturation of erythrocytes.

duced by muscular activity. Strenuous exercise can increase metabolism as much as 15 times.

REGULATION OF TOTAL BODY ENERGY BALANCE

The concept of energy balance implies a relationship between energy intake and energy expenditure or outflow. If an energy balance is to be maintained, the energy acquired through food must equal the energy expended by the body. If the energy taken in exceeds the body's needs, some is stored—for example, as adipose tissue triacylglycerol—and the person gains weight. If the energy taken in is less than the body needs, stored forms of energy are utilized, and the person loses weight. If the body's energy requirements

and energy intake are balanced, body weight remains stable.

Since the body weights of most adults do remain relatively stable, there must be some sort of regulation to maintain a reasonable balance between energy intake and energy expenditure. In this regard, it appears that one of the major means of maintaining an energy balance is to control food (energy) intake in accordance with the energy expenditure requirements of the body. (See Table 17.3, Recommended Daily Dietary Allowances.)

CONTROL OF FOOD INTAKE

A number of brain areas, including certain areas of the hypothalamus, have been implicated in the control of food intake, and there are several theories about the nature of the signals that influence

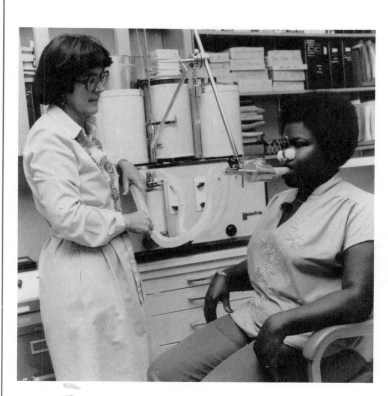

1 liter O₂ + Glucose, fat or protein

Liberates approximately 4.825 kcal energy

O₂ consumed =	15.4 liters/hr.
	× 4.825 kcal/liter
	74.3 kcal/hr.

F17.13

Measurement of the metabolic rate by measuring oxygen consumption. The utilization of 1 liter of oxygen is approximately equal to the liberation of 4.825 kilocalories (kcal) of energy.

Table 17.3 *Recommended Daily Dietary Allowances**

	Age (years)	Weight (kg)	Weight (lb)	Height (cm)	Height (in)	Energy (kcal)[†]	Protein (g)	Fat-soluble vitamins Vitamin A activity (RE)[‡]	(IU)	Vitamin D (IU)	Vitamin E activity[§] (IU)
Infants	0.0–0.5	6	14	60	24	kg × 117	kg × 2.2	420[††]	1,400	400	4
	0.5–1.0	9	20	71	28	kg × 108	kg × 2.0	400	2,000	400	5
Children	1–3	13	28	86	34	1,300	23	400	2,000	400	7
	4–6	20	44	110	44	1,800	30	500	2,500	400	9
	7–10	30	66	135	54	2,400	36	700	3,300	400	10
Men	11–14	44	97	158	63	2,800	44	1,000	5,000	400	12
	15–18	61	134	172	69	3,000	54	1,000	5,000	400	15
	19–22	67	147	172	69	3,000	54	1,000	5,000	400	15
	23–50	70	154	172	69	2,700	56	1,000	5,000		15
	51+	70	154	172	69	2,400	56	1,000	5,000		15
Women	11–14	44	97	155	62	2,400	44	800	4,000	400	12
	15–18	54	119	162	65	2,100	48	800	4,000	400	12
	19–22	58	128	162	65	2,100	46	800	4,000	400	12
	23–50	58	128	162	65	2,000	46	800	4,000		12
	51+	58	128	162	65	1,800	46	800	4,000		12
Pregnant						+300	+30	1,000	5,000	400	15
Lactating						+500	+20	1,200	6,000	400	15

*Reproduced with permission of the Food and Nutrition Board, National Academy of Sciences—National Research Council Recommended Daily Dietary Allowances. Revised 1974. The allowances are intended to provide for individual variations among most normal persons as they live in the United States under usual environmental stresses. Diets should be based on a variety of common foods in order to provide other nutrients for which human requirements have been less well defined.
[†]Kilojoules (kJ) = 4.2 × kcal.
[‡]Retinol equivalents.
[§]Total vitamin E activity estimated to be 80% as α-tocopherol and 20% other tocopherols.
[††]Assumed to be all as retinol in milk during the first six months of life. All subsequent intakes are assumed to be half as retinol and half as β-carotene when calculated from international units. As retinol equivalents, three fourths are as retinol and one fourth as β-carotene.

these areas and thus alter feeding activity. Some of the signals are believed to be concerned with maintaining normal levels of nutrient stores within the body. These signals are principally involved with **nutritional regulation,** and they could be particularly important in maintaining a balance between energy intake and energy expenditure. Other signals are believed to reflect the immediate effects of feeding on the gastrointestinal tract. These signals are principally involved with **alimentary regulation,** and they could be important in providing enough time for the gastrointestinal tract to process ingested food.

Table 17.3 *(Continued)*

	Water-soluble vitamins							Minerals					
Ascorbic acid (mg)	Folic acid (µg)	Niacin (mg)	Riboflavin (mg)	Thiamin (mg)	Vitamin B$_6$ (mg)	Vitamin B$_{12}$ (µg)	Calcium (mg)	Phosphorus (mg)	Iodine (µg)	Iron (mg)	Magnesium (mg)	Zinc (mg)
35	50	5	0.4	0.3	0.3	0.3	360	240	35	10	60	3
35	50	8	0.6	0.5	0.4	0.3	540	400	45	15	70	5
40	100	9	0.8	0.7	0.6	1.0	800	800	60	15	150	10
40	200	12	1.1	0.9	0.9	1.5	800	800	80	10	200	10
40	300	16	1.2	1.2	1.2	2.0	800	800	110	10	250	10
45	400	18	1.5	1.4	1.6	3.0	1,200	1,200	130	18	350	15
45	400	20	1.8	1.5	2.0	3.0	1,200	1,200	150	18	400	15
45	400	20	1.8	1.5	2.0	3.0	800	800	140	10	350	15
45	400	18	1.6	1.4	2.0	3.0	800	800	130	10	350	15
45	400	16	1.5	1.2	2.0	3.0	800	800	110	10	350	15
45	400	16	1.3	1.2	1.6	3.0	1,200	1,200	115	18	300	15
45	400	14	1.4	1.1	2.0	3.0	1,200	1,200	115	18	300	15
45	400	14	1.4	1.1	2.0	3.0	800	800	100	18	300	15
45	400	13	1.2	1.0	2.0	3.0	800	800	100	18	300	15
45	400	12	1.1	1.0	2.0	3.0	800	800	80	10	300	15
60	800	+2	+0.3	+0.3	2.5	4.0	1,200	1,200	125	18+[§§]	450	20
80	600	+4	+0.5	+0.3	2.5	4.0	1,200	1,200	150	18	450	25

[**]The folic acid allowances refer to dietary sources as determined by *Lactobacillus casei* assay. Pure forms of folic acid may be effective in doses less than one fourth of the recommended dietary allowance.

[††]Although allowances are expressed as niacin, it is recognized that on the average 1 mg of niacin is derived from each 60 mg of dietary tryptophan.

[§§]This increased requirement cannot be met by ordinary diets; therefore, the use of supplemental iron is recommended.

Nutritional Regulation

Glucostatic Theory. According to the *glucostatic theory* of hunger and feeding regulation, the brain contains glucose receptor cells (neurons) that are sensitive to their own rate of glucose utilization. It is believed that when their rate of glucose utilization is high (as may occur when blood glucose levels increase, perhaps as a result of eating), the glucose receptor cells increase their activity and feeding is depressed. Conversely, when their rate of glucose utilization is low (as may occur when blood glucose levels decrease, perhaps as a result of fasting), the glucose receptor cells decrease their activity and feeding is enhanced. In support of this theory is the fact that a decrease in the concentration of blood glucose can often be associated with the development of hunger. In addition, an increase in the blood glucose level increases the electrical activity of certain regions of the hypothalamus.

Increased levels of amino acids in the blood also tend to depress feeding, whereas decreased amino acid levels tend to enhance it. Although this effect is not nearly as great as that involving blood glucose, it may contribute to food intake regulation in a manner similar to that proposed for blood glucose.

Lipostatic Theory. According to the *lipostatic theory* of hunger and feeding regulation, some substance or substances (perhaps fatty acids or glycerol) are released from fat stores in adipose tissue in direct proportion to the total adipose tissue mass. These substances act in a manner similar to that of glucose to depress feeding. Thus, the greater the total adipose tissue mass, the less food is consumed. It is believed that such a mechanism, if it exists, would provide a reasonable form of long-term feeding regulation. In support of the lipostatic theory is the observation that increases in the quantity of adipose tissue in the body decrease the overall degree of feeding. In addition, the basal rate of glycerol release from adipose tissue appears to be directly related to the size of the adipose tissue cells, and the long-term average concentration of free fatty acids in the blood is directly proportional to the amount of adipose tissue in the body.

Thermostatic Theory. According to the *thermostatic theory* of hunger and feeding regulation, brain areas involved in the control of body temperature interact with areas that control food intake. As a result of this interaction, exposure to a cold environment tends to enhance feeding, and exposure to a warm environment tends to depress it. The thermostatic theory is supported by the observation that animals exposed to cold tend to overeat, whereas animals exposed to heat tend to undereat.

Hormonal Influences. The plasma concentrations of various hormones that affect the metabolism of glucose, amino acids, and triacylglycerols may also influence brain areas that control food intake. In support of this concept are a number of experiments that indicate that hormones such as insulin (which is secreted in increased amounts during food absorption) and glucagon (which is secreted in increased amounts during fasting) are able to influence hunger.

Alimentary Regulation

It has been proposed that the distension of the gastrointestinal tract, and especially the stomach, by food depresses feeding. Moreover, researchers have suggested that the gastrointestinal hormone cholecystokinin, which is secreted in association with the digestion of food, depresses feeding by either directly or indirectly activating vagal afferent nerve fibers. In addition, the body may be able to meter food intake in association with the acts of salivation, tasting, chewing, and swallowing. Such metering could limit food consumption. The regulation of short-term food intake—for example, the amount of food eaten at one meal—by such mechanisms as the distension of the gastrointestinal tract, the influence of the hormone cholecystokinin, and the metering of the amount of food eaten would help to coordinate feeding with the ability of the gastrointestinal tract to process the food.

OBESITY

Obesity (the condition of being overweight) is a serious problem that often accompanies and contributes to a number of degenerative diseases, including high blood pressure, coronary artery disease, arteriosclerosis, and diabetes. One view of obesity stresses that it is the result of consuming too much energy (food) in relation to energy expenditure (physical activity). Indeed, there is evidence that many sedentary people eat more in relation to their degree of physical activity than active people, and thus tend to become obese. Most cases of obesity, however, can be only partly explained by lack of exercise. It is only during the time that obesity is developing that energy intake exceeds energy expenditure, and in many obese people, body weight eventually reaches some higher than normal value at which it remains relatively stable. Thus, the mechanisms that control energy intake in relation to energy expenditure appear to be operative, but they maintain body weight at a higher than normal level. Perhaps the brain areas that control food intake malfunction or the incoming signals are somehow upset. Some researchers think habits and social customs as well as genetic factors contribute to the problem.

REGULATION OF BODY TEMPERATURE

Humans maintain a relatively constant body temperature, which is an important aspect of homeostasis because it allows the heat-sensitive chemical reactions of the body to proceed in a stable fashion. Under most conditions, the body temperature is higher than the environmental temperature. Thus, the heat produced by metabolism is of value in maintaining body temperature.

When determined in the morning under carefully controlled conditions, the oral temperature is approximately 36.7°C (98.1°F). However, not all parts of the body are the same temperature. The core or internal body temperature is usually considerably higher than the temperatures of the skin and external body parts (F17.14). Moreover, the body temperature varies somewhat with activity, and there is even a 24-hour diurnal or circadian body temperature cycle, with the lowest temperature generally occurring in the early morning after a night's rest and the highest temperature occurring in the evening.

Despite these variations, however, body temperature does remain relatively constant. For this constancy to occur, heat gain must equal heat loss, and both the production of heat by the body and the loss of heat from the body are subject to some degree of control.

MUSCLE ACTIVITY AND HEAT PRODUCTION

When a person is exposed to cold, muscle tone increases. This increase in muscle tone ultimately leads to oscillating, rhythmic muscle tremors (10 to 20 per second) known as *shivering*, which can increase body heat production several times.

When a person is exposed to a warm environment, muscle tone decreases. This decrease in tone may not cause much of a decline in body heat

production, however, because muscle tone is normally quite low. Moreover, high environmental temperatures tend to raise body temperature, which increases the rates of the body's chemical reactions and consequently results in the production of additional heat.

Chronic cold exposure causes some organisms to increase their heat production by means other than increased muscular activity. This phenomenon, called *nonshivering (chemical) thermogenesis*, is believed to be due mainly to an increased secretion of epinephrine, which substantially increases metabolism in certain tissues. However, it is not certain if nonshivering thermogenesis is a significant phenomenon in humans.

HEAT TRANSFER MECHANISMS

There are three principal means by which heat is transferred between the body and the environment: radiation, conduction and convection, and evaporation.

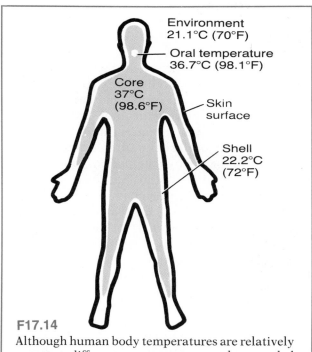

F17.14
Although human body temperatures are relatively constant, different temperatures may be recorded in different regions of the body.

Radiation

All dense objects, including the body, continually emit heat in the form of infrared rays, which are a type of electromagnetic radiation. Thus, the body is constantly exchanging heat by **radiation** with objects in the environment. The net direction of the heat exchange between objects by radiation depends on the surface temperatures of the objects, with the net direction being from objects with warmer surfaces to objects with cooler ones. Since the surface temperature of the body is usually higher than the surface temperatures of most objects in the environment, the net movement of heat by radiation is usually away from the body. However, the reverse may be true in a very warm environment or in the case of an exchange that involves a warm individual object in the environment.

Conduction and Convection

Conduction is the transfer of thermal energy from atom to atom or molecule to molecule as the result of direct contact between two objects. In this transfer, heat moves from the object with the higher surface temperature to the object with the lower one. Most commonly, body surfaces are in direct contact with air, and conductive heat exchanges occur between air molecules and the body. Usually, the air temperature is lower than the temperature of the body surface, and the body loses heat to the air by conduction. When heat is conducted from the body surface to air molecules, the air molecules are warmed. Since warm air is less dense than cool air, the warm air molecules rise away from the body surface and are replaced by cooler air molecules. This process, which is called **convection,** greatly enhances the conductive exchange of heat by constantly bringing cool air molecules into contact with the body surface. In fact, without convection, the conductive exchange of heat between the body surface and the air would be negligible. Convection, and thus the loss of heat from the body to the air by conduction, is enhanced by wind.

Evaporation

The **evaporation** of water requires the input of heat (heat of vaporization). Thus, when water evaporates from body surfaces, it carries heat away with it, and the evaporation of 1 gram of water removes about 580 calories of heat from the body. Water is lost from body surfaces in two ways: by *insensible water loss*, which is not controlled for purposes of temperature regulation, and by *sweating*, which is an important thermoregulatory activity.

Insensible Water Loss.   The skin is not perfectly impermeable to water, and some evaporative water loss continually occurs across external body surfaces. Moreover, respiration continually passes air in and out of the moist respiratory passages and the lungs. The inhaled air becomes humidified, and as a result, water is lost from the body with each expiration. Insensible water loss, which is so named because a person is not usually aware of its occurrence, can be as great as 600 ml per day.

Sweating.   Sweating requires active fluid secretion by sweat glands in the skin. Sweat is a dilute solution of sodium chloride that also contains urea, lactic acid, and potassium. In hot environments, sweat can be produced at rates up to 1.5 liters per hour. The sweat covers the body surface and evaporates, carrying heat with it. In fact, sweat must evaporate to produce its cooling effects. If it simply drips from the body, little benefit occurs. If the humidity (water vapor concentration of the air) is high and the air is already saturated with water, no additional water can evaporate into the air. Thus, sweating is a more effective means of cooling in low-humidity environments than in high-humidity environments.

BLOODFLOW AND HEAT EXCHANGE

The rate of heat exchange between the body and the environment by radiation and conduction depends in large measure on the temperature gra-

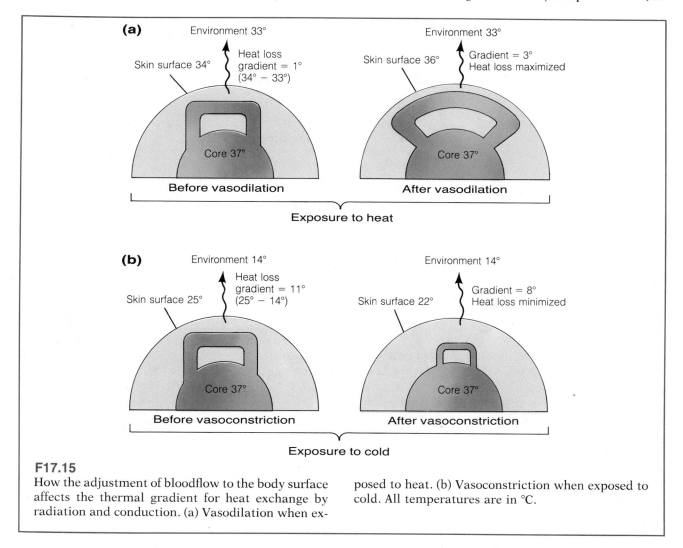

F17.15

How the adjustment of bloodflow to the body surface affects the thermal gradient for heat exchange by radiation and conduction. (a) Vasodilation when ex-posed to heat. (b) Vasoconstriction when exposed to cold. All temperatures are in °C.

dient (difference) between the body surface and the environment, and this gradient is influenced by the bloodflow to the body surface (F17.15). In warm environments, where the body must lose much heat in order to maintain a constant temperature, vessels to the skin dilate and a large amount of warm blood flows from the core of the body to the surface for heat exchange with the environment. This large bloodflow increases the temperature of the body surface and creates a more favorable gradient for heat loss. Conversely, in cool environments, where heat conservation is important, blood vessels to the skin constrict, reducing bloodflow to the body surface. This reduction in bloodflow retains needed heat within the core of the body and reduces the temperature of the body surface, which decreases the gradient for heat loss to the environment.

Adjusting bloodflow to the body surface is a generally effective means of balancing heat production and heat loss within the moderate temperature range of 20 to 28°C (68 to 82°F). At temperatures lower than about 20°C, large heat losses require increased heat production—for example,

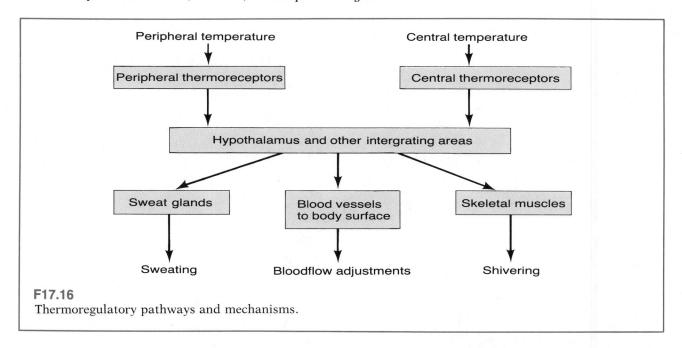

F17.16
Thermoregulatory pathways and mechanisms.

by shivering—for compensation. At temperatures above about 28°C, additional heat loss by the body as the result of the evaporation of water—that is, sweating—becomes necessary. In fact, when the environmental temperature is higher than the body temperature, the body gains heat by radiation and conduction, and evaporation is the only significant avenue of heat loss.

Humans do, of course, utilize other means of adjusting to high and low environmental temperatures. A person can, for example, reduce the body surface area over which heat loss occurs by curling into a ball. In addition, clothing helps insulate the body from both heat and cold, as does shelter.

BRAIN AREAS INVOLVED IN THERMOREGULATION

A number of brain areas, including certain areas of the hypothalamus, are involved in the integration of thermoregulatory activities (F17.16). These areas receive input from temperature receptors called *peripheral thermoreceptors* in the skin and certain mucous membranes, as well as from *central thermoreceptors* within the hypothalamus itself, the spinal cord, abdominal organs, and other internal structures. The output of the integrating

areas controls thermoregulatory activities such as adjustments in bloodflow to the body surface, shivering, and sweating.

The operation of the body's temperature control system appears to be analogous to that of a thermostat in which a particular temperature is set or called for (see Chapter 1). If inputs to the integrating areas from various thermoreceptors indicate that the actual body temperature differs from the called-for temperature, heat gain or heat loss adjustments are instituted as necessary to bring the actual temperature to the called-for temperature. Thus, if the body temperature is lower than called for, heat-conserving and heat-producing activities such as vasoconstriction and shivering are stimulated. If the temperature is higher than called for, heat loss activities such as vasodilation and sweating are initiated.

Researchers have suggested that inputs from the peripheral thermoreceptors enable the body temperature control system to make thermoregulatory adjustments before there is a change in the central or core body temperature. For example, if a person is exposed to cold, inputs from the peripheral thermoreceptors are believed to cause vasoconstriction and shivering before there is a decrease in the core body temperature. This type of response would be particularly useful in maintaining a stable body temperature in environments where the temperature fluctuates.

FEVER

The preceding view of temperature regulation envisions *fever* as a resetting of the body temperature thermostat to a new, higher level. This resetting can be accomplished by substances called pyrogens, which are released during conditions such as infection or inflammation. The effects of the pyrogens are believed to be mediated by prostaglandins. (The synthesis of prostaglandins is inhibited by aspirin, which may explain why aspirin reduces fever.) When the body temperature thermostat is set to a higher level in fever, the actual body temperature is initially below that level, and shivering, vasoconstriction, and chills can occur as the body attempts to raise its temperature to the new higher setting. By the same token, when a fever breaks, the thermostat is reset to its original level and the body temperature is now above that level. Consequently, sweating and vasodilation can occur as the body brings its temperature back to the lower setting.

HEAT STROKE AND HEAT EXHAUSTION

Occasionally, under heat stress, the body's temperature control system breaks down. In such a case, the body temperature rises rapidly, accompanied by a dry skin and the absence of sweating. This condition is called *heat stroke*.

A more common condition experienced by individuals exposed to heat is *heat exhaustion*. In this condition, the body temperature control system remains functional, but as a result of extreme sweating (fluid loss) and vasodilation to lose heat, the individual collapses and has low blood pressure (hypotension) and a cool, clammy skin with little rise in body temperature. Heat exhaustion is usually less dangerous than heat stroke.

STUDY OUTLINE

UTILIZATION OF GLUCOSE AS AN ENERGY SOURCE ATP is immediate source of energy for body's activity. One of principal food materials utilized in ATP formation is glucose. **pp. 439–444**

Glycolysis
1. Important pathway for glucose breakdown; enzymes in cytoplasm; no molecular O_2 required.
2. Glucose molecule broken into two molecules of pyruvic acid; 2 ATP spent, 4 ATP generated; 2 $NADH + H^+$ produced.
3. Pyruvic acid can be further degraded to lactic acid. Uses 2 $NADH + H^+$, regenerating NAD^+.

Krebs Cycle
1. Enzymes in mitochondria.
2. Two molecules of pyruvic acid each form acetyl-coenzyme A and enter Krebs cycle; produce: $6CO_2$, 2ATP, 8 $NADH + H^+$, and 2 $FADH_2$.

Electron Transport System
1. Transfer of electrons from $NADH + H^+$ and $FADH_2$ to oxygen, and the combination of protons with the oxygen to form water and regenerate NAD^+ and FAD. ATP is generated in this process.

2. Oxygen is required.
3. Glycolysis, Krebs cycle, electron transport together yield net profit of 36 to 38 ATP per glucose molecule.

UTILIZATION OF AMINO ACIDS AND TRI-ACYLGLYCEROLS AS ENERGY SOURCES Amino acids and triacylglycerols can undergo metabolic conversions to intermediates that can enter the glycolytic and Krebs cycle pathways and thereby provide energy for body activities. **pp. 444–446**

Amino Acids Lose amino groups and form α-keto acids that can ultimately enter glycolytic or Krebs cycle pathways.

Triacylglycerols Split into glycerol and fatty acids; glycerol enters glycolytic pathway as 3-phosphoglyceraldehyde; fatty acids broken by beta oxidation to acetyl-coenzyme A, which can enter Krebs cycle.

OTHER USES OF FOOD MOLECULES p. 446
1. Built into larger structures.
2. Interconverted—for example, amino acids or glucose form triacylglycerols.
3. Essential amino acids, some fatty acids and vitamins must be obtained in diet.

NUTRIENT POOLS   p. 447

Amino Acid Pool   Protein synthesis; energy sources; conversion to carbohydrate or triacylglycerol.

Carbohydrate and Fat Pools   Energy sources; storage.

METABOLIC STATES   pp. 448–453

Absorptive State

CARBOHYDRATE   Usually major energy source (glucose); converted to adipose tissue triacylglycerol; some stored as glycogen.

PROTEIN   Some protein synthesis; converted to adipose tissue triacylglycerol.

TRIACYLGLYCEROL   Some used for synthesis; energy; most stored as adipose tissue triacylglycerol.

Postaborptive State

GLUCOSE SOURCES   Liver glycogen supply; skeletal muscle glycogen supply; catabolism of triacylglycerols; protein provides major blood glucose source.

GLUCOSE SPARING AND TRIACYLGLYCEROL UTILIZATION   Glucose reserved for nervous system; body energy largely from triacylglycerols.

Control of Aborptive and Postabsorptive State Events   Hormonal and neural mechanisms involved; insulin tends to promote absorptive state metabolic activities; glucagon and epinephrine favor mobilization of stored carbohydrates and triacylglycerols.

LIVER FUNCTION   Storage and synthesis.   p. 453

VITAMINS   Most vitamins function either directly or after chemical alteration as coenzymes or components of coenzymes.   p. 453

Water-Soluble Vitamins   Examples are B vitamins and vitamin C. Water-soluble vitamins are generally absorbed with water from gastrointestinal tract.

Fat-Soluble Vitamins   Examples are vitamins A, D, E, and K. Fat-soluble vitamins absorbed with digested dietary triglycerides.

MINERALS   Involved in numerous body functions; enzyme activity.   p. 454

METABOLIC RATE   Total energy liberated by the body per unit time; measured by determining rate of oxygen consumption.   p. 454

Basal Metabolic Rate   Metabolic rate determined on awake, supine person; 20 to 25°C environment; postabsorptive state (12–18 hour fast period).

Factors That Affect Metabolic Rate   Age, sex, specific dynamic action, muscle activity.

REGULATION OF TOTAL BODY ENERGY BALANCE   If energy requirements and energy intake are balanced, body weight remains stable. pp. 457–461

Control of Food Intake   A number of brain areas have been implicated, and there are several theories about the nature of the signals that influence these areas.

NUTRITIONAL REGULATION

Glucostatic Theory   Glucose receptor cells located in brain are sensitive to their own rate of glucose utilization; high utilization leads to increased activity, and feeding is depressed.

Lipostatic Theory   Adipose tissue releases substance that ultimately inhibits feeding; greater total adipose tissue mass results in less feeding.

Thermostatic Theory   Brain areas involved in the control of body temperature interact with areas that control food intake. As a result of this interaction, exposure to a cold environment tends to enhance feeding, and exposure to a warm environment tends to depress it.

Hormonal Influences   The plasma concentrations of hormones that affect the metabolism of glucose, amino acids, and triacylglycerols may influence brain areas that control food intake.

ALIMENTARY REGULATION   Distension of gastrointestinal tract, influence of cholecystokinin, or ability to meter food intake may depress feeding.

Obesity   Serious problem that often accompanies and contributes to a number of degenerative diseases.

REGULATION OF BODY TEMPERATURE   Humans generally maintain a relatively constant body temperature. Thus heat gain equals heat loss. pp. 461–465

Muscle Activity and Heat Production   In cold, shivering can increase body heat production several times.

Heat Transfer Mechanisms

RADIATION   Emission of heat in the form of infrared rays; warmer objects radiate more heat than cooler objects.

CONDUCTION AND CONVECTION Conduction is transfer of thermal energy as the result of direct contact between two objects; it is aided by convection in which air molecules that exchange heat with body surface move away to be replaced by other molecules.

EVAPORATION OF WATER Requires input of heat.

   Insensible Water Loss Achieved through exhaled air and through skin.

   Sweating Active fluid secretion; cools body by evaporation.

Bloodflow and Heat Exchange Adjusting bloodflow to body surface is a generally effective means of balancing heat production and heat loss within the moderate temperature range of 20 to 28°C.

Brain Areas Involved in Thermoregulation

1. Brain areas involved in integration of thermoregulatory activities receive input from peripheral and central thermoreceptors.
2. If inputs to integrating centers indicate that actual body temperature differs from called-for temperature, heat gain or loss adjustments are instituted as necessary.

Fever Body temperature thermostat set higher; aspirin may inhibit synthesis of prostaglandins.

Heat Stroke and Heat Exhaustion

   HEAT STROKE Failure of body temperature control system results in rapid rise of body temperature, dry skin, and no sweating.

   HEAT EXHAUSTION Extreme sweating and vasodilation lead to collapse, with low blood pressure, cool skin, and little rise in body temperature.

SELF-QUIZ

1. The breakdown of glucose by glycolysis: (a) occurs within mitochondria; (b) does not liberate energy in the form of ATP; (c) requires no molecular oxygen.
2. When oxygen is not available, which substance serves as an electron acceptor to regenerate NAD^+ from $NADH + H^+$? (a) glucose; (b) pyruvic acid; (c) lactic acid.
3. Lactic acid molecules are degraded to carbon dioxide and water during glycolysis. True or False?
4. Unlike glucose, α-keto acids are unable to enter the Krebs cycle. True or False?
5. The fatty acids of triacylglycerols can be split by beta oxidation into molecules of: (a) glycerol; (b) acetyl-coenzyme A; (c) 3-phosphoglyceraldehyde.
6. During the absorptive state, the major energy source of the body is usually: (a) carbohydrate; (b) amino acids; (c) triacylglycerols.
7. Lipoprotein lipase breaks down chylomicron triacylglycerols, releasing free fatty acids. True or False?
8. During the postabsorptive state, glycogen in skeletal muscle cells is broken down to glucose, which is released from the cells into the blood. True or False?
9. Pyruvic acid and lactic acid produced by muscle glycolysis can be transported by the blood to the liver, where they can be converted to glucose. True or False?
10. During periods of fasting, which tissue has the greatest need for glucose to supply itself with energy? (a) nervous system; (b) muscular system; (c) liver.
11. Which hormone tends to promote metabolic activities of the sort that occur during the absorptive state? (a) epinephrine; (b) glucagon; (c) insulin.
12. Vitamins function primarily as: (a) energy sources; (b) structural components of the body; (c) coenzymes.
13. For a determination of the basal metabolic rate, a person must be: (a) in a postabsorptive state; (b) asleep; (c) in a room with a temperature of 5°C.
14. The greatest change in metabolic rate occurs as the result of: (a) protein ingestion; (b) muscular activity; (c) carbohydrate intake.
15. The basal metabolic rate is determined when a person is asleep and in a postabsorptive state. True or False?
16. In general, if the body's energy requirements and energy intake are balanced, body weight remains stable. True or False?
17. The transfer of thermal energy from atom to atom or molecule to molecule that occurs as the result of direct contact between two objects is called: (a) conduction; (b) radiation; (c) convection.
18. In warm environments, vessels to the body surface: (a) constrict; (b) dilate; (c) remain essentially the same as when in a cold environment.
19. A resetting of the body's thermostat to a new, higher level may be accomplished by: (a) aspirin; (b) pyrogens; (c) thermostatin.
20. A person suffering from heat exhaustion is likely to exhibit: (a) an absence of sweating; (b) low blood pressure; (c) an extremely high body temperature.

THE URINARY SYSTEM 18

To maintain homeostasis, the body must regulate the concentrations of water, acids, bases, and inorganic electrolytes (mineral ions) such as sodium, potassium, and calcium ions in its internal environment. Moreover, it must eliminate a variety of metabolic products, including urea. The kidneys and the other components of the urinary system are critically important in these activities.

This chapter examines the basic mechanisms of kidney function. The role of the kidneys in maintaining the fluid, electrolyte, and acid-base balances of the body is considered in Chapter 19.

BASIC RENAL PROCESSES

As blood flows through the kidneys, some of the plasma is filtered out of the vascular system and into structures called renal tubules (*renal* = kidney) (F18.1). This filtration process, which is primarily due to the pressure within the renal blood vessels, is not very selective, and almost all plasma components—with the exception of large protein molecules—can enter the renal tubules. As the filtrate flows along the tubules, water, electrolytes, glucose, and other substances required by the body are reabsorbed from them and returned to the blood by both active and passive transport processes. Moreover, materials within the blood that were not originally filtered into the renal tubules can be secreted into them. The reabsorption and secretion processes are selective, and hormonal mechanisms enable the kidneys to exert a high degree of control over the amounts of many materials—particularly water and electrolytes—that are reabsorbed and returned to the blood or secreted into the tubular fluid. As a result, the kidneys play a major role in maintaining the composition of the body's internal environment within a range that is optimal for cell survival and function.

When the reabsorption and secretion processes are completed, the fluid remaining within the renal tubules is transported to other components of the urinary system to be excreted as urine. Thus, urine consists of water and materials that were filtered or secreted into the renal tubules but not reabsorbed.

ORGANIZATION OF THE URINARY SYSTEM

An understanding of kidney function requires an understanding of the organization of the urinary system and particularly of the structure and arrangement of the renal tubules within the kidneys and the pattern of bloodflow through the kidneys.

The urinary system consists of the *kidneys*, which produce urine; the *ureters*, which carry the urine to the urinary bladder; the *urinary bladder*, which temporarily stores the urine; and the *urethra*, which transports the urine to the outside of the body (F18.2).

THE KIDNEYS

The **kidneys** are paired organs situated on the posterior abdominal wall, one on each side of the vertebral column. Each kidney is covered by a fibrous renal capsule, and the medial border of each kid-

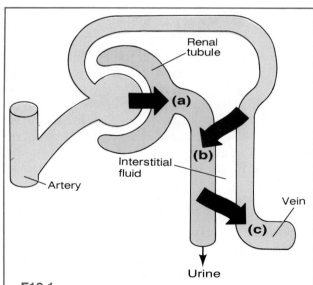

F18.1

(a) As blood flows through the kidneys, some of the plasma is filtered out of the vascular system and into renal tubules. (b) Some materials within the blood that are not filtered into the renal tubules can be secreted into them. (c) As fluid flows along the renal tubules, water, electrolytes, glucose, and other substances required by the body are reabsorbed from the tubules and returned to the blood.

ney contains an indentation where the renal arteries enter and the renal vein and the ureter leave.

Three regions can be distinguished within each kidney: the cortex, the medulla, and the pelvis (F18.3). The **cortex** is the outer layer of the kidney, just under the renal capsule. The **medulla** is located beneath the cortex. It consists of from 8 to 18 triangular *renal pyramids*. The pyramids are oriented so that their bases are covered by the cortex and their tips (papillae) point away from the cortex. The papilla of each pyramid projects into a funnel-shaped *minor calyx*. Several minor calyces join together to form a *major calyx*. The major calyces join with one another to form the **renal pelvis**, which is the expanded upper end of the ureter. Urine formed in the renal tubules passes as droplets from tiny pores in the papillae into the minor calyces, and from there into the major calyces, the renal pelvis, and finally into the ureter, which carries it to the urinary bladder.

THE RENAL TUBULES

The **renal tubules,** which are the functional units of the kidneys where urine is formed, consist of structures called **nephrons** and **collecting tubules** (F18.4). There are approximately 1 million nephrons in each kidney, and each nephron consists of (1) a network of parallel capillaries called a *glomerulus* and (2) a *tubule*. The tubule begins as a double-walled cuplike structure known as *Bowman's capsule*, which surrounds the glomerulus. The capsule and glomerulus together make up a *renal corpuscle* (F18.5). It is at the renal corpuscles, which are located within the cortical regions of the kidneys, that materials are filtered out of the blood and into the renal tubules. In this process, some of the blood plasma (except for most proteins) passes out of the glomerular capillaries and into the capsular space between the inner and outer layers of a Bowman's capsule. Because the capillaries involved are glomerular capillaries, the process is referred to as *glomerular filtration*.

From Bowman's capsule, the glomerular filtrate flows along the tubular portion of the nephron. Beyond the capsule, the nephron forms a tightly looping *proximal convoluted tubule* whose lumen is continuous with the capsular space. Beyond the proximal convoluted tubule, the nephron forms a hairpinlike *loop of Henle*. The descending limb of the loop of Henle passes into a pyramid of the medulla of the kidney, and the ascending limb of the loop of Henle passes out of the medulla and back into the cortex. (The loops of Henle of some nephrons—called *juxtamedullary nephrons*—ex- tend deep into the medulla, but the loops of Henle of other nephrons—called *cortical nephrons*—do not penetrate as deeply into the medulla.) Beyond the ascending limb of the loop of Henle, the neph- ron again forms a tightly looping tubule, the *distal convoluted tubule*. The distal convoluted tubules of

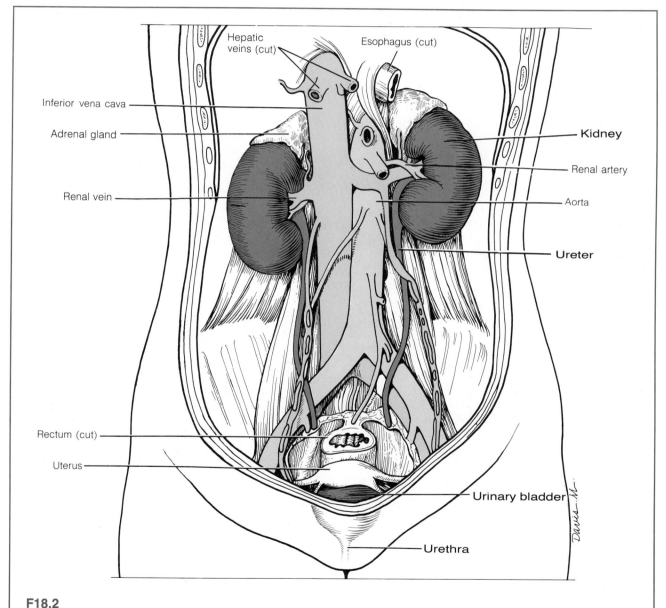

F18.2
Organs of the urinary system. The anterior abdominal wall and most of the abdominal organs have been removed.

several nephrons join with a collecting tubule, which passes back into a pyramid of the medulla. The collecting tubules ultimately open on the papillae of the pyramids and drain into the minor calyces. The processes of reabsorption and secretion occur while the glomerular filtrate is flowing along the tubular portion of the nephron and the collecting tubule. Consequently, these processes are referred to as *tubular reabsorption* and *tubular secretion*.

BLOODFLOW THROUGH THE KIDNEYS

In young adults, approximately 1100 ml of blood (approximately 20% of the cardiac output at rest) pass through the two kidneys each minute. This is a much greater bloodflow than is required to supply the nutritive needs of the kidneys. The large bloodflow is necessary, however, for the kidneys to excrete large amounts of waste products and regulate the concentration of many substances in the internal environment.

From the renal artery, blood flows into progressively smaller arteries (interlobar arteries, arcuate arteries, interlobular arteries) and then into vessels called *afferent arterioles* (F18.4). Each afferent arteriole supplies a renal corpuscle and gives rise to glomerular capillaries. Blood leaves the glomerulus through an *efferent arteriole*, which passes out of the renal corpuscle and divides into *peritubular capillaries* that surround the proximal and distal convoluted tubules of the nephrons.

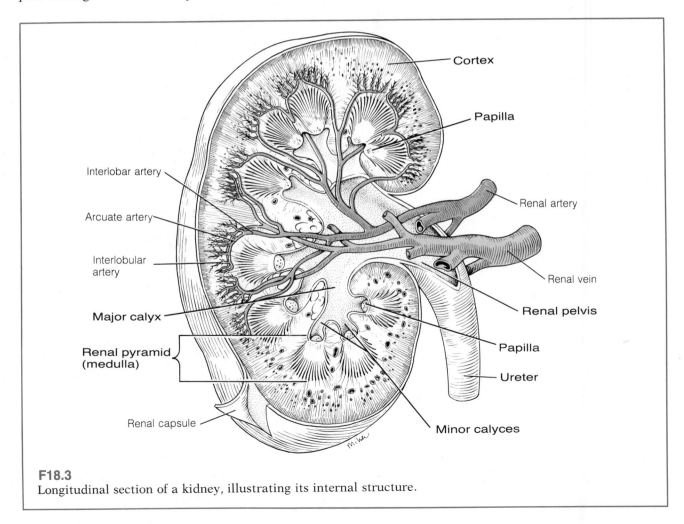

F18.3
Longitudinal section of a kidney, illustrating its internal structure.

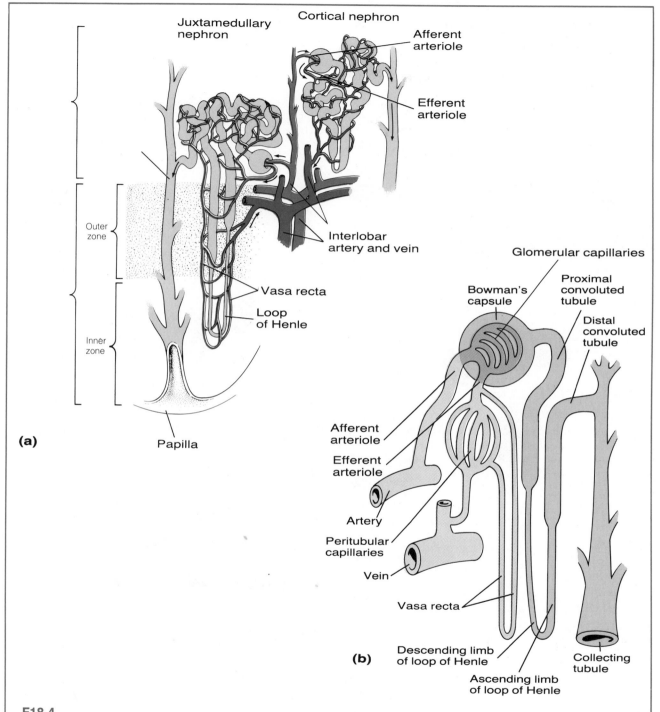

F18.4
(a) A cortical and a juxtamedullary nephron. (b) Diagrammatic representation of a juxtamedullary nephron and a collecting tubule.

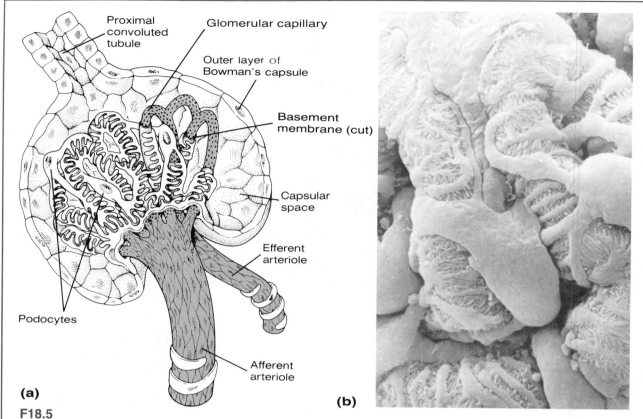

F18.5

(a) Longitudinal section of a renal corpuscle. (b) Scanning electron micrograph of podocytes surrounding glomerular capillaries (×2710). (From *Tissues and Organs: A Text-Atlas of Scanning Electron Microscopy* by Richard G. Kessel and Randy H. Kardon. W. H. Freeman and Company. Copyright © 1979.)

Peritubular capillaries also surround the loops of Henle of the cortical nephrons, and thin-walled vessels called *vasa recta* extend from the efferent arterioles of the juxtamedullary nephrons to supply their loops of Henle and the collecting tubules. Materials that are reabsorbed from the renal tubules return to the peritubular capillaries, and materials can be secreted into the renal tubules from the peritubular capillaries. Beyond the peritubular capillaries, blood flows into progressively larger veins (interlobular veins, arcuate veins, interlobar veins) and finally out of the kidney and back to the body through the renal vein.

KIDNEY FUNCTION

GLOMERULAR FILTRATION

Approximately 16% to 20% of the blood plasma that enters the kidneys is filtered from the glomerular capillaries into the Bowman's capsules of the nephrons as the glomerular filtrate. The remaining plasma continues into the efferent arterioles and on into the peritubular capillaries and vasa recta. Under normal circumstances, the kidneys produce approximately 180 liters (about 45 gallons) of glomerular filtrate per day. However, approximately 99% of the fluid volume of the filtrate is reabsorbed from the renal tubules and only about 1% (1 to 2 liters per day) is excreted as urine.

Between the plasma and the internal portion of a Bowman's capsule is a barrier composed of (1) the endothelial cells of the glomerular capillaries; (2) a mucopolysaccharide basement membrane; and (3) the specialized epithelial cells (podocytes) that form the inner portion of Bowman's capsule (F18.5). Despite the multilayered nature of this barrier, it is from 100 to 1000 times more permeable than a typical capillary, and water and solutes of small molecular dimension pass freely from the plasma into Bowman's capsules. However, this barrier is relatively impermeable to large molecules such as plasma proteins. As a result, the glomerular filtrate contains most plasma components at essentially the same concentrations as they are found within the plasma, but it is basically protein free. (Actually, a small amount of protein does appear in the glomerular filtrate, but this is normally reabsorbed by pino-cytosis in the proximal convoluted tubules so that essentially no protein is present in the urine.)

Net Filtration Pressure

The pressure of the blood within the glomerular capillaries (that is, the glomerular capillary pressure) is a major factor in glomerular filtration. This pressure—about 60 mm of mercury (Hg) or more—is higher than the pressure in most capillaries, and it favors filtration out of the glomerular capillaries and into Bowman's capsules. The glomerular capillary pressure is opposed by the pressure of the fluid within Bowman's capsules (about 18 mm Hg), and by the osmotic force exerted by unfiltered plasma proteins that remain within the glomerular capillaries (about 32 mm Hg). Thus, a *net filtration pressure* of about 10 mm Hg (60 − 18 − 32 = 10) favors the movement of materials out of the glomeruli and into Bowman's capsules (F18.6). This net filtration pressure is responsible for glomerular filtration, which occurs as a bulk flow of water and small dissolved par-

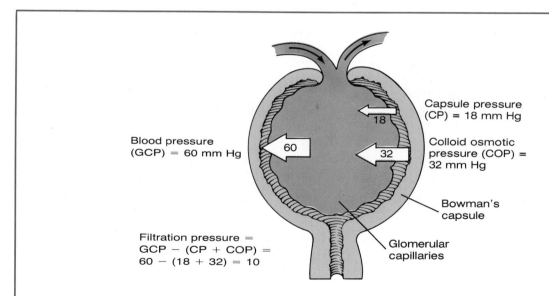

Blood pressure (GCP) = 60 mm Hg

Capsule pressure (CP) = 18 mm Hg

Colloid osmotic pressure (COP) = 32 mm Hg

Bowman's capsule

Glomerular capillaries

Filtration pressure = GCP − (CP + COP) = 60 − (18 + 32) = 10

F18.6
The pressure of the blood within the glomerular capillaries (GCP) favors the filtration of materials into Bowman's capsules. The pressure of the fluid within Bowman's capsules (CP) and the osmotic pressure exerted by unfiltered plasma proteins that remain within the glomerular capillaries (COP) oppose filtration.

ticles from the glomerular capillaries into Bowman's capsules.

Factors That Influence Glomerular Filtration

The rate of glomerular filtration at any moment depends on the relationship between the blood pressure within the glomerular capillaries that favors filtration and the forces that oppose filtration, which include the pressure of the fluid within Bowman's capsules and the osmotic force exerted by the plasma proteins within the glomerular capillaries. Variations in any of these factors can alter the glomerular filtration rate.

Influence of Arterial Blood Pressure and Autoregulatory Mechanisms.
It might be expected that changes in the arterial blood pressure would lead to comparable changes in the pressure within the glomerular capillaries and, as a result, to alterations in the glomerular filtration rate. However, the influence of arterial blood pressure on the glomerular filtration rate is not as great as expected. The kidneys possess intrinsic autoregulatory mechanisms that maintain a fairly stable glomerular filtration rate over a rather wide range of systemic blood pressures (F18.7). For example, when the arterial pressure rises, the afferent arterioles constrict, preventing a major rise in the

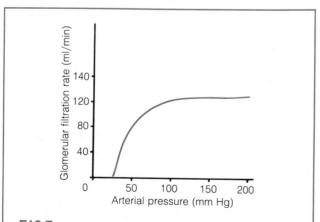

F18.7

Graph indicating the effects of changes in arterial blood pressure on the glomerular filtration rate.

blood pressure within the glomerular capillaries. Thus, even when the arterial pressure increases from 100 to 160 mm Hg, the glomerular filtration rate increases only a few percent.

Even though autoregulatory mechanisms can minimize the effects of changes in arterial blood pressure on the glomerular filtration rate, arterial pressure changes nevertheless do have a substantial influence on the output of urine (F18.8). This influence is due in part to the fact that autoregulation is not perfect, and an increase in the arterial pressure, for example, does cause some increase in the glomerular filtration rate. Since the kidneys normally form such a large volume of glomerular filtrate (about 180 liters per day), even a relatively small percentage increase in the glomerular filtration rate can cause a considerable amount of additional material to enter the renal tubules. Moreover, an elevation of the arterial pressure also causes a slight increase in the pressure within the peritubular capillaries, and this increase in pressure tends to decrease the rate at which fluid is reabsorbed from the renal tubules. Consequently, an increase in the arterial blood pressure from 100 to 120 mm Hg approximately doubles the output of urine, and an increase in the arterial blood pressure from 100 to 200 mm Hg increases the urine output six to eight times.

Influence of Sympathetic Neural Stimulation.
In addition to the influence of arterial blood pressure and intrinsic autoregulatory mechanisms, the activity of the sympathetic neurons that supply the kidneys can also influence the glomerular filtration rate. For example, an increase in sympathetic neural stimulation preferentially constricts the afferent arterioles. This constriction results in a lowered glomerular capillary pressure (and in a decreased rate of bloodflow into the glomeruli) and, as a consequence, in a lowered glomerular filtration rate.

TUBULAR REABSORPTION

Many plasma components that are filtered from the glomerular capillaries into Bowman's capsules are reabsorbed from the renal tubules to greater or lesser degrees. The reabsorption of a

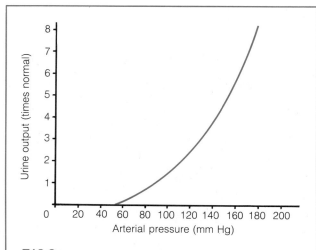

F18.8
Graph indicating the effect of changes in arterial blood pressure on the output of urine.

number of substances (especially water and many inorganic ions) is under hormonal control, and the amounts of such substances reabsorbed from the renal tubules vary with body needs.

Although filtration, as previously indicated, occurs by bulk flow, tubular reabsorption is the result of both diffusion and discrete transport mechanisms. Tubular reabsorptive transport can be either active or passive. Carrier molecules are often involved, and a single transport mechanism is sometimes responsible for the movement of several different substances with similar structures. Since carrier molecules can become saturated, tubular transport systems that utilize carriers (for example, the transport system for glucose) can transport only limited amounts of material per unit time. If the amount of glucose or some other substance transported by such a carrier system increases substantially in the plasma and subsequently in the glomerular filtrate, not all of the substance is reabsorbed and some appears in the final urine.

Active Tubular Reabsorption

A number of substances, including glucose, amino acids, and vitamins, as well as sodium, calcium, and chloride ions, are actively reabsorbed from the renal tubules. Active tubular reabsorption requires the expenditure of energy for the transport of a substance across the tubule wall. For example, in the proximal portion of a tubule, sodium diffuses passively from the tubular lumen into the tubular cells; then, energy-requiring active transport processes move sodium out of the cells into the interstitial fluid. From the interstitial fluid, the sodium enters the peritubular capillaries.

Note that in this example, the movement of sodium across the wall of the renal tubule requires the sodium to cross more than one membrane. Sodium must first enter the tubular cells through one membrane and then exit from the cells through another. Although only the exit of sodium from the cells is an active energy-requiring process, if any one step is active, the whole process is considered to be active. Thus, sodium is actively reabsorbed from the renal tubules. In subsequent sections of this chapter, the processes of tubular reabsorption (and tubular secretion) are considered as if only a single membrane is involved, and if any transport step is active, the whole process is considered to be active.

Passive Tubular Reabsorption

Some substances are passively reabsorbed from the renal tubules. For example, urea is freely filtered from the glomerular capillaries into Bowman's capsules and appears in the glomerular filtrate at a concentration that is essentially equal to that of urea in the plasma. As the filtrate moves through the renal tubules, however, water is reabsorbed. The reabsorption of water increases the concentration of urea in the tubular fluid to a level above that of the urea concentration in the interstitial fluid that surrounds the renal tubules. As a consequence, urea diffuses from the renal tubules to the interstitial fluid. Thus, urea reabsorption is passive and depends on water reabsorption and also on the permeability of the renal tubules to urea. Since the renal tubules are generally not as permeable to urea as they are to water, only about 40% to 60% of the filtered urea is normally reabsorbed, even though about 99% of the filtered water is reabsorbed.

Many foreign chemicals are also passively reabsorbed from the renal tubules. The epithelial cells of the tubules are fairly permeable to lipid-soluble substances, and many drugs and environmental pollutants are lipid-soluble, nonpolar materials. Therefore, they are relatively easily reabsorbed from the tubules as a consequence of water reabsorption in a manner similar to that described for urea. As a result, such substances can be difficult to excrete. However, many of these substances are converted by the liver to progressively more polar, less lipid-soluble materials that are not as readily reabsorbed from the renal tubules. Such materials are excreted more easily.

TUBULAR SECRETION

Substances enter the renal tubules not only by glomerular filtration but also by tubular secretion. Tubular secretion can move substances that leave the peritubular capillaries into the renal tubules. As with tubular reabsorption, the processes of tubular secretion can be either active or passive, and they are often hormonally controlled. Hydrogen and potassium ions are secreted into the renal tubules as are penicillin and many other chemicals not normally present in the body.

EXCRETION OF WATER AND ITS EFFECT ON URINE CONCENTRATION

A person who has just consumed a large quantity of water must eliminate a considerable volume of fluid without losing excessive amounts of electrolytes and other vital substances if homeostasis is to be maintained. Under such conditions, it is beneficial to produce a large volume of dilute urine. Conversely, a dehydrated individual who has had nothing to drink for some time must still produce urine for the elimination of wastes. In this case, it is advantageous to produce a small volume of concentrated urine that requires the excretion of only small amounts of water. Thus, if the kidneys are to be effective as regulatory organs, they must be able to produce a range of urine volumes and concentrations according to the body's needs. In fact, the kidneys can produce urine as dilute as 65 to 70 milliosmoles per liter or as concentrated

as 1200 and sometimes 1400 milliosmoles per liter compared to the concentration of the blood plasma, which is about 300 milliosmoles per liter.

The ability to produce a low-volume, concentrated urine depends on the presence of a highly concentrated interstitial fluid within the medullary region of each kidney. Therefore, the mechanisms responsible for maintaining this high interstitial fluid concentration will be considered first, and then the processes involved in producing different volumes and concentrations of urine will be examined.

Maintenance of a Concentrated Medullary Interstitial Fluid

Within the medullary region of a kidney, the interstitial fluid concentration increases from about 300 milliosmoles per liter at the cortex to as much as 1200 and sometimes 1400 milliosmoles per liter at the tips of the pyramids. The high medullary interstitial fluid concentration is due to the activities and anatomical arrangement of the loops of Henle of the nephrons (particularly those of the juxtamedullary nephrons), the collecting tubules, and the vasa recta.

Activities of the Loops of Henle and the Collecting Tubules. The loops of Henle of the juxtamedullary nephrons extend deep into the medulla, and the collecting tubules pass through the medulla to the tips of the pyramids. As fluid flows along the loops of Henle and the collecting tubules, the reabsorption of solutes raises the concentration of the medullary interstitial fluid. For example, negatively charged chloride ions are actively transported out of the fluid in the upper portions (the *thick segments*) of the ascending limbs of the loops of Henle into the interstitial fluid of the outer medulla, and because of electrical considerations, positively charged sodium ions passively follow the chloride ions (F18.9). This movement of sodium and chloride ions not only raises the concentration of the outer medullary interstitial fluid but, as is discussed later, sodium and chloride ions are carried downward into the inner medulla by the blood flowing through the vasa recta. In addition, sodium ions are actively transported out of the fluid in the collecting tubules, with chloride ions following, and sodium and chloride ions can move out of the lower portions (the *thin segments*) of the ascending limbs of the loops of Henle into the interstitial fluid of the inner medulla. The reabsorption of sodium and chloride ions—particularly in the thick segments of the ascending limbs of the loops of Henle—contributes importantly to the high concentration of the medullary interstitial fluid.

Water can be reabsorbed from the fluid within the collecting tubules and, as is discussed later, the amount of water reabsorbed is controlled hormonally by antidiuretic hormone. When a large

amount of water is reabsorbed from the collecting tubules, a substantial amount of urea can diffuse out of the final portions of the tubules into the interstitial fluid of the inner medulla and thus contribute to the high concentration of the medullary interstitial fluid. This movement of urea occurs as follows. The first portions of the collecting tubules are quite impermeable to urea. Consequently, when a large amount of water is reabsorbed, the concentration of urea in the tubular fluid increases considerably, and much of it diffuses out of the final portions of the collecting tubules into the interstitial fluid of the inner medulla (F18.9). This activity raises the urea

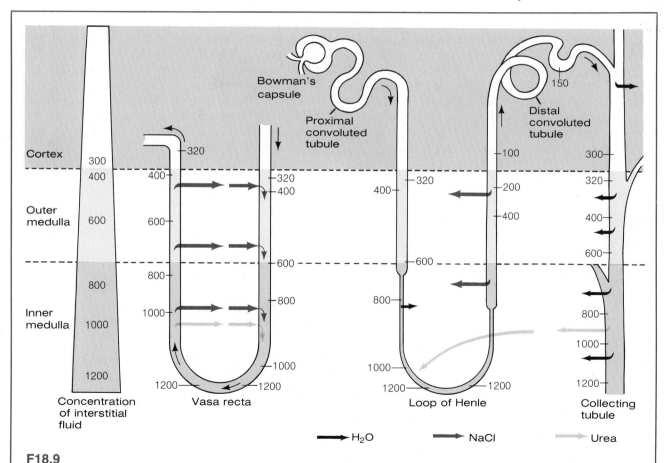

F18.9
Mechanisms involved in maintaining a concentrated medullary interstitial fluid and producing a low-volume, concentrated urine. The concentrations indicated are in milliosmoles per liter. They are the total concentrations of all solutes present, not the concentrations of individual substances.

concentration of the interstitial fluid above that of the tubular fluid in the bottoms of the loops of Henle. As a result, some of the urea diffuses into the tubular fluid at the bottoms of the loops, raising its concentration. The tubular fluid is then carried through the ascending limbs of the loops of Henle and the distal convoluted tubules (both of which are essentially impermeable to urea) to the collecting tubules, where the reabsorption of water can further increase the fluid's urea concentration. Even more urea then diffuses out of the final portions of the collecting tubules, and the cycle is repeated. Thus, when much water is reabsorbed from the collecting tubules, this cycling process can multiply the urea concentration of the interstitial fluid of the inner medulla (and the urea concentration of the fluid within the final portions of the collecting tubules) to quite high levels.

Together, the mechanisms that raise the concentration of the medullary interstitial fluid are frequently referred to as the *countercurrent mechanism* of the loop of Henle because some aspects of the different mechanisms—for example, the cycling process involved in urea transport—depend on the fact that fluid flows in opposite directions in the essentially parallel ascending and descending limbs of the loops of Henle and the collecting tubules.

Effect of Vasa Recta Bloodflow.

If a high medullary interstitial fluid concentration is to be maintained, the reabsorbed solutes (for example, sodium, chloride and urea) that raise the concentration of the fluid must not be removed by the blood. The removal of the solutes is minimized by the pattern of bloodflow in the vasa recta, which extend into the medulla.

The vasa recta form a countercurrent system. In this system, blood flows in opposite directions within the descending and ascending limbs of the vasa recta, and most solutes can pass readily between the blood and the interstitial fluid. The system is believed to operate essentially as follows.

Blood with a concentration of about 300 milliosmoles per liter flows from the cortex into the medulla along the descending limbs of the vasa recta, which extend into regions of increasingly

concentrated interstitial fluid (F18.9). As the blood flows along the descending limbs, sodium, chloride, urea, and other solutes that are in higher concentration in the interstitial fluid than in the blood diffuse into the blood, raising its concentration. The concentrated blood then flows into the ascending limbs of the vasa recta, which extend into regions of decreasing concentrations of interstitial fluid. As the blood flows along the ascending limbs, sodium, chloride, urea, and other solutes that are now in higher concentration in the blood than in the interstitial fluid diffuse out of the vasa recta. As a result, by the time the blood leaves the vasa recta, its concentration is only slightly higher than it was when the blood initially entered the vessels. Thus, the blood flowing through the vasa recta removes only a small amount of solutes from the medullary interstitial fluid. In addition, the blood within the vasa recta contains unfiltered proteins that exert an osmotic force, the colloid osmotic pressure of the plasma. This force causes water that is reabsorbed from the loops of Henle and collecting tubules to be picked up and carried away by the blood, thus preventing it from severely diluting the medullary interstitial fluid.

Together, the activities and arrangement of the loops of Henle of the juxtamedullary nephrons, the collecting tubules, and the vasa recta maintain a high medullary interstitial fluid concentration that increases progressively from about 300 milliosmoles per liter at the cortex to as much as 1200 and sometimes 1400 milliosmoles per liter at the tips of the pyramids. The kidneys utilize this highly concentrated interstitial fluid to produce a low-volume, concentrated urine.

Production of Different Urine Volumes and Concentrations

As previously indicated, the vast majority of the approximately 180 liters of glomerular filtrate formed per day is reabsorbed from the renal tubules and returned to the body. Moreover, the hormonal mechanism involving antidiuretic hormone that controls water reabsorption in the collecting tubules strongly influences the actual volume of filtrate reabsorbed. Consequently, this mechanism also influences the volume and concentration of the urine that is produced.

The concentration of the glomerular filtrate within Bowman's capsules is essentially the same as that of the plasma; that is, about 300 milliosmoles per liter. As the filtrate flows along the proximal portion of a renal tubule, sodium ions are actively transported out of the tubule into the interstitial fluid, with chloride ions following passively (F18.10). In addition, a number of other substances such as glucose and amino acids are reabsorbed by various transport mechanisms. The proximal tubule is permeable to water, and water leaves osmotically since the loss of sodium, chloride, and other substances transiently lowers the concentration of the tubular fluid and increases the concentration of the interstitial fluid. The net result of the removal of solutes and water from the proximal tubule is to reduce the volume of the original glomerular filtrate about 65% to 70%. Its concentration, however, remains at approximately that of the plasma. The materials that en-

ter the interstitial fluid from the proximal portion of the renal tubule ultimately return to the peritubular capillaries and are carried away from the kidney with the blood.

The fluid remaining within the proximal tubule proceeds into the descending limb of the loop of Henle, which dips into regions of increasingly concentrated interstitial fluid (F18.9). The descending limb is relatively permeable to water, and since the interstitial fluid outside the limb is more concentrated than the fluid within it, water moves out of the tubule. By the time the tubular fluid reaches the bottom of the loop of Henle, the loss of water has further reduced its volume and also raised its concentration above that of the plasma. In fact, at the bottom of the loop of Henle of a juxtamedullary nephron, the loss of water—together with the previously discussed entry of urea into the tubule—can raise the concentration of the tubular fluid to approximately 1200 milliosmoles per liter.

As already noted, within the ascending limb of

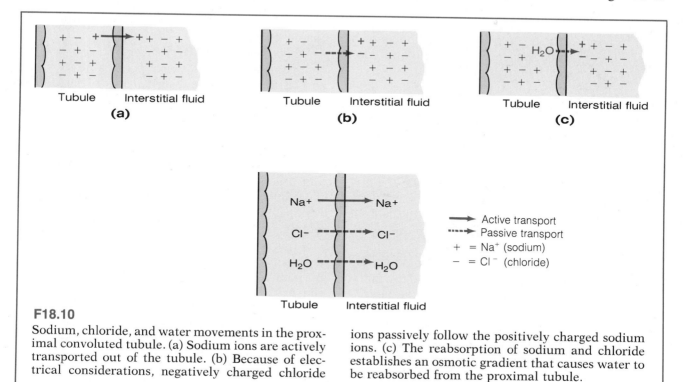

F18.10

Sodium, chloride, and water movements in the proximal convoluted tubule. (a) Sodium ions are actively transported out of the tubule. (b) Because of electrical considerations, negatively charged chloride ions passively follow the positively charged sodium ions. (c) The reabsorption of sodium and chloride establishes an osmotic gradient that causes water to be reabsorbed from the proximal tubule.

the loop of Henle, sodium and chloride ions move out of the tubule into the interstitial fluid (F18.9). In the thin segment of the ascending limb, the outward movement of sodium and chloride ions apparently occurs passively by diffusion as the concentrated tubular fluid from the bottom of the loop of Henle flows up the ascending limb, which extends into regions of decreasing concentrations of interstitial fluid. However, some researchers believe the outward movement of sodium and chloride ions in this portion of the limb occurs by active transport. In the thick segment of the ascending limb, chloride ions are actively transported out of the tubule and, because of electrical considerations, sodium ions follow passively.

The ascending limb of the loop of Henle— particularly the thick segment—is essentially impermeable to water. Consequently, as sodium and chloride ions move out of the tubular fluid within the limb, the concentration of the fluid drops, but its volume is not greatly altered. By the time the fluid completes its passage through the ascending limb and enters the distal convoluted tubule, the reabsorption of sodium and chloride ions has lowered its concentration below that of the plasma.

The processes occurring within the distal convoluted tubule are complex. Certain ions are reabsorbed from the fluid within the tubule while others are secreted into the fluid, and much of this activity is influenced by physiological regulatory mechanisms. In monkeys and presumably in other primates such as humans, the distal convoluted tubule is relatively impermeable to water. In general, the concentration of the fluid within the distal convoluted tubule remains below that of the plasma (and it can even fall below that of the fluid leaving the ascending limb of the loop of Henle), but its volume does not change greatly.

Up to this point, much of the fluid volume of the glomerular filtrate has been reabsorbed, primarily in the proximal tubule. Moreover, many nutrients required by the body such as glucose and amino acids are almost completely reabsorbed in the proximal tubule. In the case of a juxtamedullary nephron, the sodium and chloride ions that move out of the ascending limb of the loop of Henle contribute to the high concentration of the

medullary interstitial fluid, which is required for the production of a low-volume, concentrated urine.

From the distal convoluted tubule, the fluid enters a collecting tubule, which extends from the cortex into the medullary region of the kidney where there are increasing concentrations of interstitial fluid. The permeability of the collecting tubule to water is controlled by antidiuretic hormone (ADH), which is manufactured in the hypothalamus of the brain and released from the pituitary gland. When the ADH concentration of the plasma is high, the collecting tubule is very permeable to water. As a result, water leaves the tubular fluid and enters the concentrated interstitial fluid. (Note that even though the permeability of the collecting tubule is high, the movement of water out of the tubule depends on the presence of the concentrated medullary interstitial fluid to establish the proper osmotic gradient.) The outward movement of water reduces the volume of the fluid within the collecting tubule and increases its concentration. This low-volume, concentrated fluid then leaves the tubule at the tip of a pyramid to be excreted as urine.

When the plasma ADH concentration is low, the collecting tubule is not very permeable to water, and little water is reabsorbed from the tubular fluid. Consequently, the dilute fluid that flows along the distal convoluted tubule into the collecting tubule remains dilute, and its volume is not significantly reduced. (Moreover, ions can be actively reabsorbed from the collecting tubule, and this activity can lower the concentration of the tubular fluid even further.) In this case a high-volume, low-concentration fluid leaves the collecting tubule as urine.

As is evident from this discussion, the reabsorption of water from the renal tubules occurs by passive transport (osmosis), and depends on the reabsorption of solutes such as sodium and chloride ions to establish the proper osmotic gradient. However, once a gradient is established, the volume of water reabsorbed from the collecting tubules varies with their water permeability, which is controlled by ADH. Consequently, ADH is important in regulating fluid volume excretion and in providing the kidneys with the ability to produce a range of urine volumes and concentrations according to the body's needs. The factors influencing ADH release are discussed in Chapter 19.

Minimum Urine Volumes

The body must excrete about 600 milliosmoles of waste products such as urea, sulfates, phosphates, and the like each day. The elimination of these materials, even at a maximum urine concentration of 1400 milliosmoles per liter, requires a certain degree of water loss:

$$\frac{600 \text{ milliosmoles/day}}{1400 \text{ milliosmoles/liter}} = 0.444 \text{ liters/day}$$

Regardless of water intake or availability, this volume of fluid must be excreted simply to eliminate wastes. Beyond this requirement, however, physiological mechanisms come into play to vary the actual volume, concentration, and composition of the urine that is produced in order to compensate for altered intakes of fluids, electrolytes, and other substances, as well as to compensate for varying losses of these materials by other pathways, including the lungs and sweat glands. The composition of a representative sample of urine is given in Table 18.1.

PLASMA CLEARANCE

The *plasma clearance* of a substance is the rate at which the substance is eliminated or cleared from the plasma by the kidneys. By determining the plasma clearance of certain specific substances, the glomerular filtration rate and the rate of plasma flow through the kidneys can be determined. These types of information are useful in assessing kidney function.

The measurement of the plasma clearance of a substance requires the determination of (1) the rate of urine formation, (2) the concentration of the substance in the arterial plasma flowing to the kidneys, and (3) the concentration of the substance in the urine. With these values, the plasma clearance of the substance can be determined according to the formula:

Table 18.1 *Representative Components and Characteristics of a 24-hour Sample of Urine*

Component	Value
Calcium	0.01 to 0.30 g
Chloride	6.0 to 9.0 g
Creatinine	1.0 to 1.5 g
Potassium	2.5 to 3.5 g
Sodium	4.0 to 6.0 g
Urea	20.0 to 35.0 g

Characteristic	Value
pH	4.8 to 7.4
Specific gravity	1.016 to 1.022
Volume	1000 to 1600 ml

$$\text{plasma clearance of X (ml/min)} = \frac{\text{urine formation rate (ml/min)} \times \text{urine concentration of X}}{\text{plasma concentration of X}}$$

For example, suppose the concentration of urea in the plasma is 0.30 mg/ml. Suppose further that in one hour 66 ml of urine are formed (a formation rate that equates to 1.1 ml/min), and that this urine has a urea concentration of 17 mg/ml. The plasma clearance of urea would then be:

$$\text{plasma clearance of urea (ml/min)} = \frac{(1.1 \text{ ml/min}) \times (17 \text{ mg/ml})}{(0.30 \text{ mg/ml})}$$
$$= 62 \text{ ml/min}$$

This plasma clearance value means that in one minute an amount of urea is removed from the plasma that is equivalent to the amount of urea in 62 ml of plasma. (It does not mean that any single milliliter of plasma has all of the urea removed from it during one transit through the renal circulation.)

The clearance concept can be used to determine the glomerular filtration rate by measuring the plasma clearance of the carbohydrate inulin. Inulin is injected into the plasma, and its plasma concentration determined. Inulin is freely filtered from the glomerular capillaries into Bowman's capsules with the glomerular filtrate, but it is neither reabsorbed nor secreted by the renal tubules. As a result, the amount of inulin in a timed sample of urine is essentially equal to the amount of inulin filtered from the plasma into Bowman's capsules in the same amount of time. For example, suppose that the one-hour, 66-ml urine sample of the previous example contained 1500 mg of inulin (22.7 mg/ml) and that the plasma inulin concentration was 0.2 mg/ml. Since inulin is neither reabsorbed nor secreted by the renal tubules, all of the inulin in this one-hour sample of urine must have been filtered into Bowman's capsules from the plasma with the glomerular filtrate in the same hour. Moreover, because the plasma inulin concentration is only 0.2 mg/ml, in order to get 1500 mg of inulin into the urine in one hour, 7500 ml of plasma must have been filtered at the glomeruli in that hour. This activity equates to a glomerular filtration rate of 125 ml/min (7500 ml per hr/60 min per hr). The glomerular filtration rate could also have been obtained using the plasma clearance formula as follows:

$$\text{plasma clearance of inulin (ml/min)} =$$
$$\frac{(1.1 \text{ ml/min}) \times (22.7 \text{ mg/ml})}{(0.2 \text{ mg/ml})} = 125 \text{ ml/min}$$

In the case of inulin, the plasma clearance (of inulin) and the glomerular filtration rate are essentially equal.

The concept of plasma clearance can also be employed to determine the rate of plasma flow through the kidneys. The substance utilized for this purpose is para-aminohippuric acid (PAH). PAH is both freely filtered at the glomerulus and secreted into the renal tubules, but it is not reabsorbed from the tubules. As a result, the plasma that passes through the kidneys is almost completely (actually about 91%) cleared of PAH. Suppose, to continue the previous examples, that sufficient PAH is injected into the plasma to achieve a plasma concentration of 0.01 mg/ml. Suppose also that the 1.1 ml of urine produced in one minute contains 6.25 mg of PAH (5.68 mg/ml).

This means that a minimum of 625 ml (6.25 mg per min/0.01 mg per ml) of plasma must have passed into the kidneys in one minute to provide the amount of PAH that appears in the urine in one minute, even if all the PAH is removed from the plasma. The same conclusion can also be reached by using the plasma clearance formula:

$$\text{plasma clearance of para-aminohippuric acid}$$
$$\text{(ml/min)} =$$
$$\frac{(1.1 \text{ ml/min}) \times (5.68 \text{ mg/ml})}{(0.01 \text{ mg/ml})} = 625 \text{ ml/min}$$

This value is called the effective renal plasma flow (ERPF). In fact, since only about 91% of the para-aminohippuric acid is in reality cleared from the plasma, the actual renal plasma flow is about 687 ml/min (625 ml per min/0.91) or 989 liters per day. If the hematocrit is known (for example, 40%), then the total renal bloodflow (1648 liters per day) can also be determined.

FUNCTION OF THE URETERS, URINARY BLADDER, AND URETHRA

The urine that drips from the collecting tubules at the tips of the papillae enters the minor calyces. From there it flows into the major calyces and then into the renal pelvis. From the renal pelvis it is transported to the urinary bladder by ureters, one from each kidney. The urine is propelled along the ureters by peristaltic contractions of smooth muscles of the ureteral walls. Before opening into the bladder, the ureters travel obliquely through the posterior bladder wall. As a result, pressure in the bladder can compress the ureters and help prevent urine from flowing back into the ureters from the bladder, especially during bladder emptying (micturition).

The urinary bladder, which temporarily stores the urine, is a hollow organ whose walls contain smooth muscle (F18.11). As it fills with urine, the pressure within the bladder initially increases somewhat and then remains fairly constant up to a volume of about 300 to 400 ml (F18.12). Beyond this point the pressure rises rapidly. The bladder can hold 600 to 800 ml of urine, but it is generally emptied before it reaches this capacity.

A tube called the urethra carries urine from the bladder to the exterior of the body. At the junction of the urethra and bladder, the smooth muscle of the bladder surrounds the urethra and acts as a sphincter (the internal urethral sphincter) that tends to keep the urethra closed. During bladder emptying, the contraction of the bladder and the resulting changes in its shape pull the sphincter open. Thus, no special mechanism is required to relax this sphincter. As the urethra passes through the pelvic floor, it is surrounded by skeletal muscles that form the external urethral sphincter. When constricted, this sphincter is able to hold the urethra closed against strong bladder contractions.

MICTURITION

As the bladder fills with urine, its walls are stretched and mechanoreceptors within the walls transmit increasing numbers of sensory impulses to the sacral region of the spinal cord. These impulses ultimately stimulate parasympathetic neurons that supply the smooth muscles of the bladder wall and inhibit somatic motor neurons that supply the external urethral sphincter. As a consequence, when approximately 300 ml of urine has accumulated within the bladder, the muscles of the bladder wall contract, the external urethral sphincter relaxes, and *micturition* (urination or voiding) occurs.

Although micturition, as just described, is essentially the result of a local spinal reflex, it is also influenced by higher brain centers, particularly centers in the brain stem and cerebral cortex. In addition to being transmitted to the spinal cord, sensory impulses from mechanoreceptors within the bladder walls are sent to higher brain centers. These impulses can lead to a sensation of a full bladder and to a feeling of a need to urinate. Moreover, impulses sent from the brain can either facilitate or inhibit the reflex emptying of the bladder, and with training it is possible to gain a high degree of voluntary control over micturition. As a result, urination can be either voluntarily induced or postponed until an opportune time. However, until control is developed and training is complete, the reflex response is the dominant factor governing bladder emptying. A baby, therefore, urinates whenever its bladder is sufficiently full to activate the spinal reflex.

COMMON PATHOLOGIES OF THE URINARY SYSTEM

PRESSURE-RELATED PATHOLOGIES

Factors that upset the pressure relationships determining glomerular filtration rates can interfere with normal kidney function. (Recall that the glomerular filtration rate depends on the net filtration pressure, which equals glomerular blood pressure minus capsular pressure minus oncotic pressure.)

Prostate Hypertrophy

In males, it is not unusual for the prostate gland, which surrounds the urethra just below the bladder, to hypertrophy (enlarge) and compress the urethra so that it becomes difficult for urine to leave the bladder. As urine accumulates within the bladder, the pressure within the bladder rises. If the pressure increases enough, it causes a backing up of urine in the ureters and a dilation of the renal pelvis and calyces which, in turn, leads to an increased pressure within Bowman's capsules. The increased capsular pressure reduces the glomerular filtration rate and thus interferes with the kidney's regulation of body fluid composition.

Low Arterial Blood Pressure

Blood pressure changes associated with heart failure, hemorrhage, or shock can cause renal failure. In all of these conditions, there is a drop in arterial blood pressure, which reduces the ability of the kidneys to form glomerular filtrate. In addition, a substantial drop in blood pressure can activate reflexes involved in maintaining normal pressure within the major arteries of the body. These reflexes cause the afferent arterioles of the kidneys to constrict. The constriction of the afferent arte-

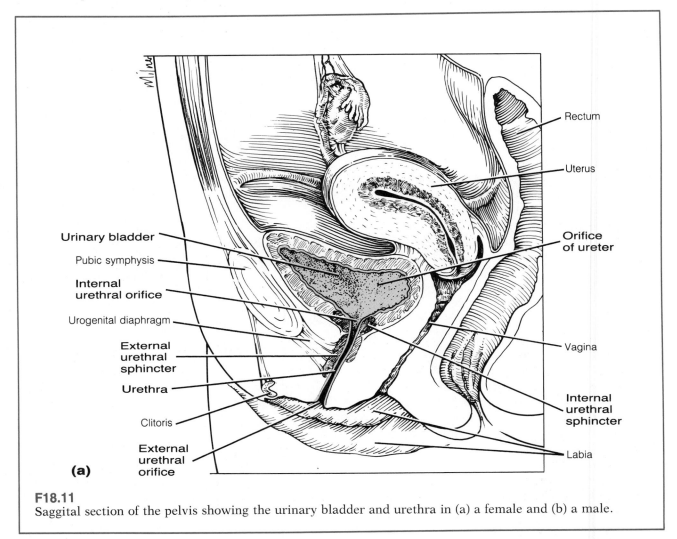

F18.11
Saggital section of the pelvis showing the urinary bladder and urethra in (a) a female and (b) a male.

rioles helps elevate the general body blood pressure, but it diminishes the blood pressure within the glomeruli and further reduces the formation of glomerular filtrate. The reduction in the formation of glomerular filtrate diminishes the kidneys' ability to excrete wastes and regulate body fluid composition.

ACUTE GLOMERULONEPHRITIS

Acute glomerulonephritis is an inflammatory condition of the glomeruli that is the result of a hyper-

sensitivity response in which immune complexes (antigen–antibody complexes) become deposited in the kidneys. It is frequently associated with streptococcal infections such as throat infections. In acute glomerulonephritis, many of the glomeruli become blocked by the inflammatory response, and others become so permeable that they allow erythrocytes and large amounts of plasma proteins to pass into Bowman's capsules with the glomerular filtrate. In severe cases, total or almost total renal shutdown can occur. However, the vast

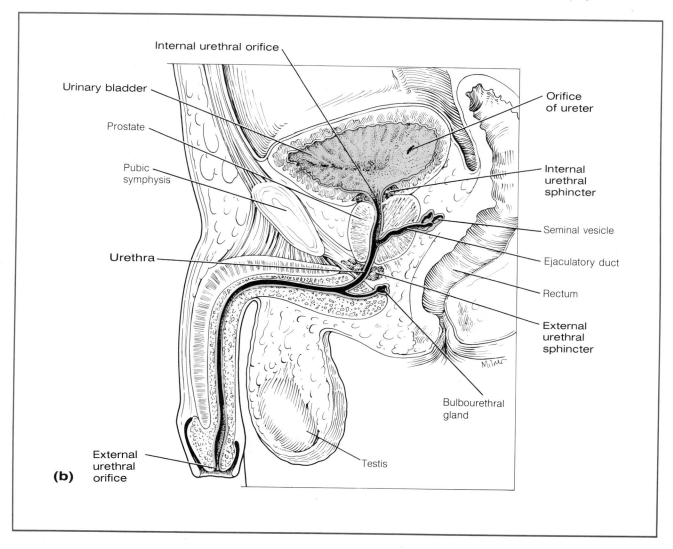

Internal urethral orifice

Urinary bladder

Prostate

Pubic symphysis

Urethra

External urethral orifice

(b)

Orifice of ureter

Internal urethral sphincter

Seminal vesicle

Ejaculatory duct

Rectum

External urethral sphincter

Bulbourethral gland

Testis

majority of individuals afflicted with acute glomerulonephritis return to normal renal function within a few months.

KIDNEY STONES

Kidney stones (*renal calculi*) sometimes form within the renal pelvis. The stones generally consist of uric acid, calcium oxalate, calcium phosphate, and certain other substances. What causes the stones to form is not known. However, there seems to be some correlation between their formation and kidney infections, a high concentration of salts in the urine, vitamin A deficiency, and hyperparathyroidism caused by a tumor.

A stone formed in the renal pelvis may remain there, or it may enter the ureter and pass to the bladder. The stone often causes severe, painful contractions of the ureter as it travels through it. A more serious condition results if a stone becomes lodged within the ureter, obstructing the flow of urine to the bladder. In addition to the retention of urine, kidney stones can cause ulcer-

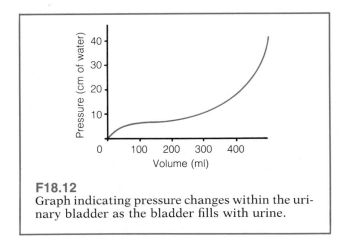

F18.12
Graph indicating pressure changes within the urinary bladder as the bladder fills with urine.

ations in the lining of the urinary tract, which makes the tract more prone to infections.

CYSTITIS

Cystitis is an inflammation of the urinary bladder accompanied by frequent and burning urination and blood in the urine. In the acute form of cystitis, the mucous-membrane lining of the bladder becomes swollen and some bleeding occurs. In the chronic condition, the wall of the bladder can become thickened and its capacity reduced.

The bladder is generally quite resistant to bacterial infections, but under certain conditions bacteria become established in the bladder lining, thus producing cystitis. Cystitis can also be caused by chemicals or by mechanical irritation, such as catheterization. Women have a higher incidence of cystitis than men, probably due to their short urethra, which makes it easier for bacteria to reach the bladder from outside the body. It is not uncommon for *E. coli* bacteria from a woman's anal region to infect the urethra as a result of improper cleansing of the area.

THE ARTIFICIAL KIDNEY

Effective kidney function is necessary for survival, and severe kidney disease can be fatal. The development of the artificial kidney machine, however, has greatly alleviated a number of kidney disease problems. The artificial kidney employs the principle of dialysis (see Chapter 2) to remove waste materials from the blood. In the artificial kidney, the patient's blood is passed through a dialysis tubing that allows urea, electrolytes, and other small molecules to move freely across its wall, but it does not allow the movement of large protein molecules. The dialysis tubing is immersed in a bath that contains various substances. If urea or some other small molecule is present in the blood but not in the bath, it diffuses out of the dialysis tubing and into the bath and thus is removed from the blood. If a specific low molecular weight substance is present in the bath at the same concentration as within the blood, then it diffuses into the tubing as fast as it diffuses out, and there is no net loss of the material from the blood. Thus, by regulating the composition and concentration of materials within the bath, the types and amounts of materials that leave the blood by net diffusion out of the dialysis tubing can be regulated. In this way, waste products are removed from the blood while needed constituents are retained.

EFFECTS OF AGING ON THE KIDNEYS

Kidney function declines progressively with aging. At age 70 the glomerular filtration rate is only about 50% of the rate at age 40. Renal bloodflow decreases from approximately 1100 ml per minute at ages 20 to 45 to only about 475 ml per minute at 80 to 89 years of age. There is a corresponding decrease in the function of the renal tubules and in their ability to concentrate the tubular fluid. However, the kidneys do retain their ability to regulate the acid–base balance of the body, although they respond less quickly to a sudden, large acid load.

STUDY OUTLINE

BASIC RENAL PROCESSES Filtration, reabsorption, secretion. **p. 469**

ORGANIZATION OF THE URINARY SYSTEM
The urinary system consists of the kidneys, ureters, urinary bladder, and urethra. **pp. 470–474**

The Kidneys Paired organs situated on posterior abdominal wall; three internal regions: cortex, medulla, and pelvis.

The Renal Tubules Structures where urine is formed.
Glomerulus: network of parallel capillaries.
Tubule: begins as Bowman's capsule.
Renal corpuscle: capsule and glomerulus.
Proximal convoluted tubule.
Loop of Henle: descending and ascending limbs.
Distal convoluted tubule.
Collecting tubule.

Bloodflow Through the Kidneys Substantial bloodflow allows kidneys to excrete large amounts of waste products and regulate concentration of many substances in the internal environment. Blood flows from renal artery, to smaller arteries, to afferent arteriole, to glomerulus, to efferent arteriole, to peritubular capillaries and vasa recta, to small veins, to, renal vein.

KIDNEY FUNCTION pp. 474–483

Glomerular Filtration
1. 16% to 20% of the blood plasma that enters kidneys is filtered from glomerular capillaries into Bowman's capsules.
2. Glomerular filtrate contains most plasma components at essentially same concentrations as plasma, but is basically protein-free.

NET FILTRATION PRESSURE Filtration is favored by the pressure of the blood within glomerular capillaries and opposed by the pressure of the fluid within Bowman's capsules and by the osmotic force exerted by unfiltered plasma proteins in glomerular capillaries.

FACTORS THAT INFLUENCE GLOMERULAR FILTRATION

Influence of Arterial Blood Pressure and Autoregulatory Mechanisms Arterial blood pressure influences glomerular filtration rate, but effect is not as great as expected due to intrinsic autoregulatory mechanisms, which maintain a fairly stable glomerular filtration rate over a rather wide range of systemic blood pressures.

Influence of Sympathetic Neural Stimulation Increased sympathetic stimulation preferentially constricts afferent arterioles, lowering glomerular capillary pressure and glomerular filtration rate.

Tubular Reabsorption Removes materials from renal tubules.

ACTIVE TUBULAR REABSORPTION Energy-requiring; glucose, amino acids, and vitamins, as well as sodium, calcium, and chloride ions are actively reabsorbed.

PASSIVE TUBULAR REABSORPTION Does not require energy; urea and many foreign chemicals are passively reabsorbed.

Tubular Secretion Moves substances that leave peritubular capillaries into renal tubules; may be active or passive; hydrogen ions, potassium ions, penicillin are secreted into renal tubules.

Excretion of Water and Its Effects on Urine Concentration Kidneys can produce urine as dilute as 65 to 70 milliosmoles per liter or as concentrated as 1200 or sometimes 1400 milliosmoles per liter.

MAINTENANCE OF A CONCENTRATED MEDULLARY INTERSTITIAL FLUID Activities and anatomical arrangement of the loops of Henle of the nephrons (particularly those of the juxtamedullary nephrons), the collecting tubules, and the vasa recta are important.

Activities of the Loops of Henle and the Collecting Tubules As fluid flows along the loops of Henle and the collecting tubules, the reabsorption of solutes raises the concentration of the medullary interstitial fluid.

Effect of Vasa Recta Bloodflow The pattern of bloodflow in the vasa recta minimizes solute removal from the medullary interstitial fluid.

PRODUCTION OF DIFFERENT URINE VOLUMES AND CONCENTRATIONS
1. Sodium ions are actively transported out of proximal tubule, with chloride ions following

passively. Substances such as glucose and amino acids are also reabsorbed. Water leaves osmotically. Volume of glomerular filtrate reduced 65% to 70%; concentration remains essentially that of plasma.

2. Water moves out of descending limb of loop of Henle. Tubular fluid is reduced in volume, and its concentration is increased.
3. Sodium and chloride ions move out of ascending limb of loop of Henle. This portion of tubule is relatively impermeable to water. Concentration of tubular fluid is diminished; volume is not greatly altered.
4. Certain ions are reabsorbed from the fluid within the distal convoluted tubule and others are secreted into the fluid. In monkeys and presumably in other primates such as humans, the distal convoluted tubule is relatively impermeable to water. In general, the concentration of the fluid within the distal convoluted tubule remains below that of plasma, and its volume does not change greatly.
5. Permeability of collecting tubule to water controlled by ADH. If much ADH is present, water permeability is high; water leaves and tubular fluid becomes concentrated and reduced in volume.

MINIMUM URINE VOLUMES   Need to eliminate waste products such as urea, sulfates, phosphates requires that about 600 milliosmoles of these materials be excreted daily.

PLASMA CLEARANCE   Rate at which a particular substance is eliminated or cleared from the plasma by the kidneys. Plasma clearance of inulin indicates glomerular filtration rate.   **pp. 483–484**

FUNCTION OF THE URETERS, URINARY BLADDER, AND URETHRA   **pp. 484–485**

1. Ureters carry urine from kidneys to urinary bladder; the urine is propelled by peristaltic contractions of ureteral smooth muscle.
2. Urinary bladder—hollow, muscular organ where urine is temporarily stored; bladder contracts to expel urine.
3. Urethra—tube that carries urine from bladder to exterior of body, surrounded by internal and external urethral sphincters.

MICTURITION   **p. 485**

1. As bladder fills, local spinal reflex causes bladder to contract and external urethral sphincter to relax, resulting in micturition.
2. Impulses from brain can facilitate or inhibit reflex emptying of bladder, and with training a high degree of voluntary control can be exercised over micturition.

COMMON PATHOLOGIES OF THE URINARY SYSTEM   **pp. 485–488**

Pressure-Related Pathologies

PROSTATE HYPERTROPHY   Can cause fluid backup that interferes with regulation of body fluid composition.

LOW ARTERIAL BLOOD PRESSURE   Reduces ability of kidneys to form glomerular filtrate.

Acute Glomerulonephritis   Inflammatory condition that affects glomeruli; result of hypersensitivity response; frequently associated with streptococcal bacterial infections.

Kidney Stones (renal calculi)   Formed in renal pelvis or urinary bladder; consist primarily of uric acid, calcium oxalate, and calcium phosphate; may cause urine retention, pain, and infection due to blockage of ureters and ulceration of urinary tract lining.

Cystitis   Urinary bladder inflammation that produces frequent, burning urination and blood in the urine; more common in women, probably due to short urethra.

ARTIFICIAL KIDNEY   **p. 488**

1. Dialysis.
2. Removal of wastes from blood without loss of needed constituents.

EFFECTS OF AGING ON THE KIDNEYS   **p. 488**

1. Progressive decline in function: decreased glomerular filtration rate; decreased renal bloodflow; decreased renal tubule function.
2. Acid–base regulation: ability retained by aging kidneys.

SELF-QUIZ

1. The kidneys are supplementary excretory organs that are important but not essential in maintaining the balance of substances required for internal constancy. True or False?

2. Urine passes as droplets from tiny pores in the papillae of the renal pyramids into: (a) minor calyces; (b) renal corpuscles; (c) collecting tubules.

3. The very large renal bloodflow is necessary to supply the nutritive needs of the kidneys. True or False?

4. Essentially 100% of the plasma that enters the kidneys is filtered from the glomerular capillaries into the Bowman's capsules of the nephrons as the glomerular filtrate. True or False?

5. Under normal circumstances, which of the following is present in the glomerular filtrate in the smallest amount? (a) sodium ions; (b) protein; (c) glucose.

6. Which force favors the movement of fluid out of the glomerular capillaries and into Bowman's capsules? (a) the osmotic force exerted by unfiltered plasma proteins within the glomerular capillaries; (b) the pressure of the fluid within Bowman's capsules; (c) the glomerular capillary pressure.

7. In general, an increase in the activity of the sympathetic neurons that supply the kidneys leads to a dilation of the afferent arterioles. True or False?

8. A rise in the arterial blood pressure: (a) reduces the glomerular filtration rate; (b) increases the output of urine; (c) has no effect on kidney function because of the influence of the kidney's intrinsic autoregulatory mechanisms.

9. Under normal circumstances, about 70% of the glomerular filtrate is reabsorbed from the renal tubules, and about 30% is excreted as urine. True or False?

10. The processes of tubular reabsorption: (a) are always passive; (b) never involve carrier molecules; (c) sometimes require the expenditure of cellular energy.

11. Urea reabsorption from the renal tubules is passive and depends on water reabsorption. True or False?

12. Most urea reabsorption occurs in the distal convoluted tubule. True or False?

13. At the proximal tubule of a nephron: (a) sodium ions are actively transported out of the tubule; (b) there is a net movement of water into the tubule; (c) there is a net movement of chloride ions into the tubule.

14. As fluid flows along the descending limb of the loop of Henle of a juxtamedullary nephron, there is a net movement of water out of the tubule. True or False?

15. Fluid at the bottom of the loop of Henle of a juxtamedullary nephron is: (a) less concentrated than the glomerular filtrate; (b) less concentrated than fluid within the distal convoluted tubule; (c) more concentrated than fluid within the proximal convoluted tubule.

16. Which portion of the nephron is relatively impermeable to water? (a) the proximal convoluted tubule; (b) the descending limb of the loop of Henle; (c) the ascending limb of the loop of Henle.

17. With no antidiuretic hormone: (a) large volumes of urine are produced; (b) large amounts of protein appear in the urine; (c) glucose is not reabsorbed from the kidney tubules.

18. In adults, micturition occurs solely as the result of a local spinal reflex, and it is essentially uninfluenced by higher brain centers. True or False?

19. In the artificial kidney, the patient's blood is passed through a dialysis tubing that allows the free movement of: (a) urea; (b) protein; (c) erythrocytes.

20. During aging, which of the following is least affected? (a) glomerular filtration rate; (b) renal bloodflow rate; (c) renal regulation of acid–base balance.

LEARNING OBJECTIVES

After completing this chapter, you should be able to:

- List similarities and differences in the composition of the plasma, the interstitial fluid, and the intracellular fluid.
- Describe the factors that influence the renal excretion of sodium
- Cite possible causes for a sensation of thirst.
- Describe the factors that influence the renal excretion of water.
- Explain how body potassium is regulated, and cite the regulatory factors involved.
- Explain the actions of the hormones involved in calcium regulation.
- State what distinguishes a strong acid from a weak acid.
- Explain the function of buffer systems in the body.
- Explain how the respiratory system can have an important influence on the pH of the internal environment.
- Explain how the kidneys act as essential regulators of acid–base balance.

CHAPTER CONTENTS

FLUID AND ELECTROLYTE BALANCE AND ACID–BASE REGULATION 19

As pointed out in the previous chapter, the body must regulate the concentrations of water, acids, bases, and electrolytes such as sodium, potassium, and calcium ions in its internal environment in order to maintain homeostasis. The kidneys are particularly important in the regulation of these substances, as are the lungs and the gastrointestinal tract.

BODY FLUIDS

There are three principal body fluids: (1) the **plasma,** or fluid within the blood vessels; (2) the **interstitial fluid,** or fluid that surrounds the cells; and (3) the **intracellular fluid,** or fluid within the cells (F19.1). Together, the plasma and the interstitial fluid make up the **extracellular fluid,** or fluid outside the cells. It is this fluid that forms the body's internal environment.

COMPOSITION OF PLASMA AND INTERSTITIAL FLUID

The different body fluids are not isolated from one another, and a continuous exchange of materials occurs between them. As can be seen from F19.2, the compositions of the plasma and the interstitial fluid are quite similar. These fluids are separated from one another by the walls of the capillaries, and the relatively free exchange of low molecular weight solutes that occurs across capillary walls accounts for the similarity between the two fluids. The major difference between the plasma and the interstitial fluid is that the plasma contains more proteins. This is due to the fact that plasma proteins tend to remain within the blood vessels since the capillaries are generally not very permeable to them. Under physiological conditions, these proteins normally exist as anions—that is, as negatively charged ions. For this reason, there is somewhat more sodium (a positively charged ion, or cation) and somewhat less chloride (an anion) within the plasma than within the interstitial fluid.

Since a relatively free exchange of materials does occur between the plasma and the interstitial fluid, the regulation of the fluid, electrolyte, and acid–base composition of the plasma by the kidneys (and to some degree by the lungs and gastrointestinal tract) also serves to regulate the fluid, electrolyte, and acid–base composition of the interstitial fluid.

COMPOSITION OF INTRACELLULAR FLUID

The composition of the intracellular fluid is considerably different from that of the plasma and the interstitial fluid (Figure 19.2). The intracellular fluid is separated from the interstitial fluid by the membranes of the cells, which are relatively impermeable to proteins. Consequently, cellular proteins tend to remain within the cells. Moreover, the action of the cell membrane's sodium-potassium pump moves sodium ions out of the cells and accumulates potassium ions within them.

REGULATION OF THE BODY'S INTERNAL ENVIRONMENT

In order to maintain homeostasis, the addition of substances such as water and electrolytes to the body's internal environment must be balanced by the removal of equal amounts of these materials. Ingestion, of course, adds materials to the internal environment, and excretion by organs such as the kidneys removes materials. In fact, the kidneys' role in controlling the excretion of water and elec-

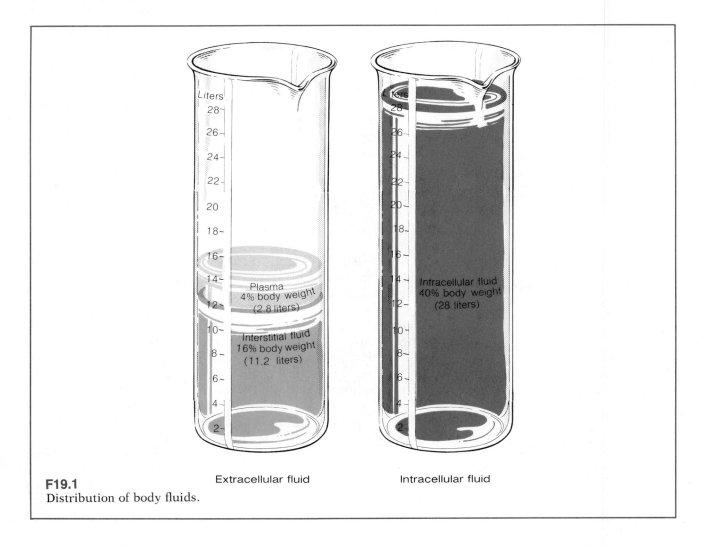

F19.1
Distribution of body fluids.

trolytes is one of their most important regulatory functions.

SODIUM REGULATION

Sodium is the major extracellular cation and, because of its osmotic effects, variations in the body sodium content can cause changes in the extracellular fluid volume, including the plasma volume. Changes in the plasma volume, in turn, can lead to changes in blood pressure, and mechanisms involved in blood pressure regulation are important in controlling body sodium content.

Sodium Ingestion

Animal studies indicate there are two components involved in determining the amount of sodium (in

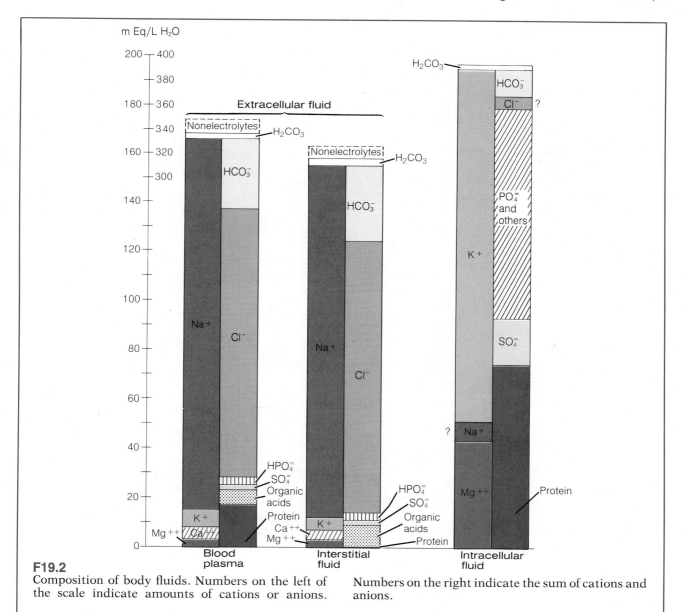

F19.2

Composition of body fluids. Numbers on the left of the scale indicate amounts of cations or anions. Numbers on the right indicate the sum of cations and anions.

the form of salt) that is ingested. One is the regulatory component, which governs the intake of sodium in such a way that a balance between sodium intake and outflow is maintained. The other is the hedonistic component, by which an animal demonstrates a strong preference for salt regardless of regulatory requirements. The degree to which regulatory and hedonistic components operate in humans is not entirely clear. Although persons severely depleted of sodium chloride often develop a desire for salt, people also seem to have a strong hedonistic salt appetite, and they consume large amounts of it whenever it is inexpensive and easily available. For example, in the United States, an average person consumes 10 to 15 g of salt per day even though less than 0.5 g per day is normally needed to meet regulatory requirements. Consequently, it appears that, for humans, the regulation of the body's sodium content is achieved mainly by controlling sodium excretion by the kidneys rather than by controlling sodium ingestion.

Renal Excretion of Sodium

The relationship between the glomerular filtration rate of sodium and the tubular reabsorption rate of sodium is very important in determining the amount of sodium excreted (F19.3). For example, if much sodium is filtered into Bowman's capsules but only little is reabsorbed, then a good deal of sodium will be excreted.

Since sodium is freely filtered from the plasma into Bowman's capsules, any factors that alter the general rate of glomerular filtration will also alter the glomerular filtration rate of sodium. Among the factors that influence the glomerular filtration rate are (1) the arterial blood pressure, and (2) the activity of sympathetic nerves to the kidneys that alter the diameters of the afferent arterioles.

The reabsorption of sodium from the renal tubules is subject to physiological control, and the control of tubular sodium reabsorption is believed to be particularly important in the long-term regulation of sodium excretion. A major factor in this control is the adrenal cortical hormone *aldosterone*. Aldosterone stimulates sodium reabsorption, particularly from the last portions of the distal convoluted tubules and the collecting tubules. When much aldosterone is present, almost all the sodium that enters the renal tubules is reabsorbed and as little as 0.1 g of sodium is excreted per day. When little aldosterone is

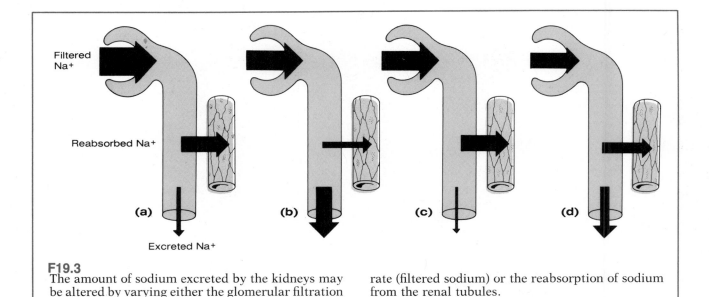

Filtered Na+

Reabsorbed Na+

(a)   **(b)**   **(c)**   **(d)**

Excreted Na+

F19.3
The amount of sodium excreted by the kidneys may be altered by varying either the glomerular filtration rate (filtered sodium) or the reabsorption of sodium from the renal tubules.

present as much as 30 to 40 g of sodium can be excreted per day.

The release of aldosterone is enhanced indirectly by the substance *renin*, which is secreted into the blood by specialized cells of kidney arterioles in response to a number of factors. Although it is currently impossible to assign a precise quantitative role to each of these factors, a decrease in the renal arterial pressure and/or an increase in the activity of sympathetic nerves to the kidneys apparently stimulate renin secretion. Renin is necessary for the conversion of the precursor substance *angiotensinogen*, which is manufactured by the liver and is normally present in the blood, into *angiotensin*, and one of the effects of angiotensin is the stimulation of aldosterone release from the adrenal glands. Thus, an increased release of renin leads to an increased formation of angiotensin, which in turn enhances the release of aldosterone. The aldosterone then stimulates sodium reabsorption from the last portions of the distal convoluted tubules and the collecting tubules.

Factors That Influence Sodium Excretion

As previously indicated, a change in the body sodium content can lead to a change in both the plasma volume and the blood pressure, and mechanisms involved in blood pressure regulation are important in controlling body sodium content. For example, an increase in the systemic blood pressure, due perhaps to an increase in the plasma volume, acts directly on the kidneys to cause some increase in the glomerular filtration rate (F19.4). Moreover, an increase in the systemic blood pressure (including the renal arterial pressure) acts on the kidneys to cause a decreased release of renin, which ultimately leads to a decrease in the level of aldosterone. Together, these responses increase the amount of sodium filtered into Bowman's capsules and decrease the amount of sodium reabsorbed from the renal tubules. As a result, the urinary excretion of sodium increases, and the total body mass of sodium declines. Because sodium, by virtue of its osmotic effects, has a direct influence on extracellular volume, the loss of sodium leads to a decrease in the plasma volume,

which tends to return the blood pressure to normal.

Some researchers believe that neural mechanisms involved in blood pressure regulation help mediate the effects of changes in the systemic blood pressure on the renal excretion of sodium. According to this view (and to continue the preceding example), an increase in the systemic blood pressure causes an increase in the rate at which nerve impulses are transmitted to the central nervous system from circulatory pressure receptors such as the aortic arch and carotid sinus baroreceptors and receptors in the atria of the heart (F19.4). The increased rate of nerve impulse transmission from these receptors leads to a decrease in the activity of the sympathetic nerves to the kidneys. The decrease in sympathetic activity, in turn, leads to a dilation of the afferent arterioles and, consequently, to an increase in the blood pressure within the glomerular capillaries, which contributes to the increase in the glomerular filtration rate. Moreover, the decrease in the activity of the sympathetic nerves to the kidneys also contributes to the decreased release of renin.

Sodium Content and Sodium Concentration

The loss or gain of sodium by the body changes the total sodium *content* of the body, but it does not necessarily have a great effect on the actual *concentration* of sodium in the internal environment. As is discussed in the next section, the body possesses receptors called **osmoreceptors,** which are sensitive to the osmotic pressure of the extracellular fluid. Regulatory mechanisms brought into play by these receptors compensate for changes in the osmotic concentration of the extracellular fluid (which is largely due to sodium) by stimulating or retarding water ingestion and water excretion in order to maintain the proper osmotic concentration. As a result, these receptors are of considerable importance in the regulation of the actual sodium concentration of the extracellular fluid.

WATER REGULATION

The water content of the body must be regulated

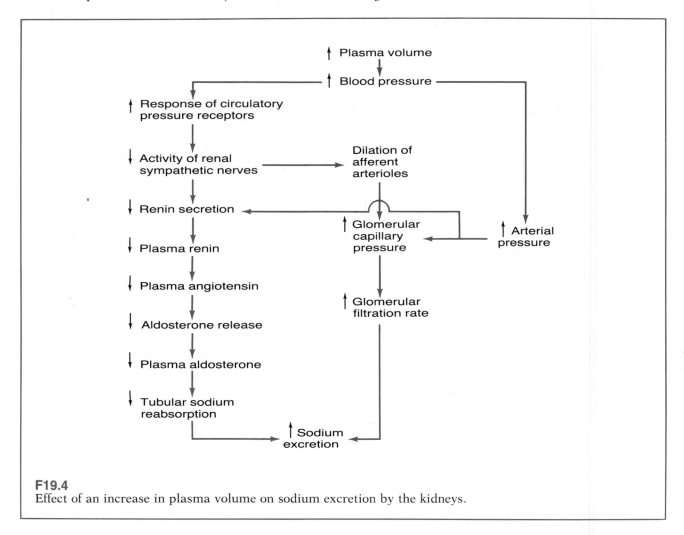

F19.4
Effect of an increase in plasma volume on sodium excretion by the kidneys.

because of its effects on both the osmotic pressure and the fluid volume of the internal environment.

Water Ingestion and Thirst

Although fluid intake is often influenced more by habit and social factors than by the need to regulate body water, water ingestion does seem to depend at least in part on the regulatory needs of the body. A sensation of thirst results from an increased osmotic concentration of the extracellular fluid and also from a substantial reduction in the plasma volume. It is thought that osmoreceptors in a hypothalamic thirst center in the brain, and perhaps also circulatory pressure receptors, stimulate thirst. Moreover, angiotensin stimulates thirst by a direct effect on the brain.

Often when a dehydrated person drinks water, the amount ingested approximates the amount required to bring the extracellular fluid back to the proper osmotic concentration. This fact is interesting because the person usually stops drinking before the ingested water has had time to be absorbed from the gastrointestinal tract and actually return the extracellular osmolarity to normal. Thus, the gastrointestinal tract may possess some sort of water metering system that is able to regulate water intake in accordance with the needs of the body. Moreover, the control of thirst may have a learned or anticipatory component that enables a person to anticipate his or her fluid needs and therefore drink sufficient water to prevent dehydration. For example, the amount of fluid consumed in association with eating appears to be a learned activity that may prevent dehydration.

Renal Excretion of Water

Body water is also controlled by the excretory pathways of the kidneys, and the relationship between the glomerular filtration rate of water and the tubular reabsorption rate of water is of major importance in determining the amount of water excreted. For example, if much water is filtered into Bowman's capsules but only little is reabsorbed, then a good deal of water will be excreted.

Since water is freely filtered from the plasma into Bowman's capsules, any factors that alter the general rate of glomerular filtration will also alter the glomerular filtration rate of water. Some of these factors—particularly some associated with blood pressure regulation—were discussed in the previous section on sodium regulation. For example, an increase in the systemic blood pressure, due perhaps to an increase in the plasma volume, leads to an increase in the glomerular filtration rate. (Conversely, a decrease in the systemic blood pressure leads to a decrease in the glomerular filtration rate.)

As explained in Chapter 18, the amount of water reabsorbed from the renal tubules depends in large part on the water permeability of the collecting tubules, which is controlled by *antidiuretic hormone (ADH)*. When the plasma ADH level is low, the water permeability of the collecting tubules is low, and relatively little water is reabsorbed. Consequently, a large volume of urine is produced. Conversely, when the plasma ADH level is high, collecting tubule permeability is high, much water is reabsorbed, and only a relatively small volume of urine is produced.

Inputs to neural centers that influence ADH release come both from osmoreceptors located in the hypothalamus and from circulatory pressure receptors, particularly those in the left atrium of the heart. The osmoreceptors, which are stimulated by increased extracellular osmolarity, stimulate the cells that secrete ADH. The pressure receptors, which are stimulated by increased atrial blood pressure (possibly resulting from increased plasma volume), inhibit the ADH-producing cells. Thus, maximum ADH release and substantial water reabsorption results from an increased extracellular osmolarity and a decreased blood pressure (plasma volume). Conversely, maximum ADH inhibition and substantial water excretion results from a decreased extracellular osmolarity and an increased blood pressure. These responses, especially when coupled with the previously discussed changes in glomerular filtration rate that occur in response to changes in blood pressure, tend to return the extracellular osmolarity and the blood pressure (plasma volume) to normal when they either increase or decrease.

Factors That Influence Water Excretion

From the preceding discussion it is evident that the osmotic pressure of the extracellular fluid and the blood pressure (plasma volume) are major factors influencing water excretion. Moreover, the factors that influence water excretion are very closely related to the factors that influence sodium excretion, and both are involved in the regulation of blood pressure as well as in the maintenance of the proper volume and concentration of the body's internal environment. Indeed, the activities of the mechanisms that influence water excretion and the mechanisms that influence sodium excretion can rarely be separated completely. For example, when an increase in the systemic blood pressure causes an increase in the glomerular filtration rate, both sodium and water are affected. Similarly, an increase in the plasma volume that increases the atrial and systemic blood pressure can decrease the reabsorption of sodium by way of the renin–angiotensin mechanism and reduce the reabsorption of water by way of the ADH mechanism. Moreover, a change in the body's sodium content due to a change in the excretion of sodium causes a change in the osmotic pressure of the extracellular fluid. The change in osmotic pressure is sensed by osmoreceptors that, in turn, cause the ADH mechanism to increase or decrease the reabsorption of water as necessary to reestablish the proper osmotic concentration of the extracellular fluid.

POTASSIUM REGULATION

Potassium is important in the excitability of nerve and muscle, and the potassium concentration of the extracellular fluid is closely regulated. Humans normally excrete potassium in amounts equal to those ingested, and thus remain in potassium balance.

Potassium is freely filtered from the plasma into Bowman's capsules, and it can be both reabsorbed and secreted by the renal tubules. Normally, most of the filtered potassium is actively reabsorbed, and adjustments in the amount of potassium excreted depend mainly on how much is secreted into the renal tubules.

The amount of potassium secreted into the renal tubules is related to the potassium concentration of the tubular cells. When the potassium concentration of the extracellular fluid is high, more potassium will be present in the tubular cells and, as a consequence, more potassium will be secreted than when the potassium concentration of the extracellular fluid is low. A second factor that influences the elimination of potassium is aldosterone. Besides promoting the tubular reabsorption of sodium, this hormone enhances the tubular secretion of potassium. In fact, the potassium concentration of the extracellular fluid that bathes the adrenal glands is a major stimulus for the release of aldosterone from the adrenal cortex. When the potassium concentration of the extracellular fluid rises, so does the release of aldosterone. When the potassium concentration falls, the release of aldosterone decreases. Once aldosterone is released, however, it exerts both its effects and enhances sodium reabsorption as well as potassium secretion.

CALCIUM REGULATION

The calcium level of the extracellular fluid is closely regulated, and low levels of calcium increase nerve and muscle excitability. Calcium regulation involves bone, which contains 99% of the body's calcium, as well as the kidneys and the gastrointestinal tract. The calcium-regulating activities of all three sites are influenced either directly or indirectly by *parathyroid hormone (parathormone, or PTH)* from the parathyroid glands.

Parathyroid hormone increases the plasma calcium concentration (and decreases the plasma phosphate concentration). It increases the movement of calcium and phosphate from bone into the extracellular fluid, and this activity is believed to be due at least in part to the ability of parathyroid hormone to stimulate the activity of cells called osteoclasts, which break down bone. In the kidneys, parathyroid hormone decreases the urinary excretion of calcium and increases the excretion of phosphate. Parathyroid hormone also enhances a step in the metabolic transformation of vitamin D_3 to a substance called 1,25-dihydroxycholecalciferol. This substance stimulates active calcium absorption from the intestine.

The release of parathyroid hormone is controlled by the calcium concentration of the extracellular fluid that bathes the parathyroid glands. When the calcium concentration is low, parathyroid hormone release increases, and this response tends to raise the extracellular fluid calcium concentration toward normal. Conversely, when the calcium concentration is high, parathyroid hormone release decreases, and this response tends to lower the extracellular fluid calcium concentration toward normal.

Calcitonin, a hormone from the thyroid gland, lowers the plasma calcium level, primarily by inhibiting the removal of calcium from bone. Thus, the influence of calcitonin on the plasma calcium level is opposite to that of parathyroid hormone. The secretion of calcitonin is controlled by the calcium concentration of the fluid that bathes the thyroid gland; high calcium concentrations enhance secretion. Calcitonin, however, appears to play a less important role in calcium regulation than parathyroid hormone does.

HYDROGEN ION AND ACID–BASE REGULATION

Because hydrogen ions affect enzyme action, most metabolic reactions are sensitive to the hydrogen ion concentration of the fluid in which they occur. Consequently, the proper regulation of hydrogen ion concentration is essential for effective cellular function.

Acids and Bases

As explained in Chapter 2, *acids* are substances that liberate hydrogen ions, and *bases* are substances that accept them. The concentration of free hydrogen ions (H$^+$) determines the acidity of a solution, and a solution's acidity is measured on the pH scale, with lower pH values indicating higher hydrogen ion concentrations (acidity).

Acids can be grouped as either strong acids or weak acids. Strong acids are those that dissociate virtually completely in solution, providing large numbers of free hydrogen ions. Hydrochloric acid (HCl), which in solution dissociates into hydrogen ions and chloride ions, is an example of a strong acid. Weak acids are those that do not dissociate completely and do not provide as many free hydrogen ions when they are placed in solution. For example, when dissolved in water, some of the molecules of the weak acid carbonic acid dissociate into hydrogen ions and bicarbonate ions. Substantial numbers of carbonic acid molecules, however, do not dissociate into hydrogen ions and bicarbonate ions. Weak acids such as carbonic acid generally dissociate in a predictable fashion, with a certain proportion of the molecules dissociating to provide free hydrogen ions and a certain proportion not undergoing this dissociation.

Acidosis and Alkalosis

The pH of the arterial blood is normally 7.4, whereas the pH of the venous blood and interstitial fluid is 7.35. A higher hydrogen ion concentration than normal that decreases the pH of the arterial blood is referred to as **acidosis,** and a lower hydrogen ion concentration than normal that increases the pH of the arterial blood is referred to as **alkalosis.**

The body's metabolic activities generate acidic products that tend to raise the hydrogen ion concentration of the internal environment. The phosphorus and sulfur of proteins, for example, are potential sources of phosphoric and sulfuric acids,

and metabolic reactions produce organic acids such as fatty acids and lactic acid. Thus, under most circumstances, the body must eliminate excess hydrogen ions in order to maintain the proper pH of the internal environment and prevent acidosis.

Buffer Systems

Although excess hydrogen ions must ultimately be eliminated from the body, principally by the kidneys, the body contains buffer systems that help stabilize the pH of the body fluids. A buffer system consists of a solution containing two or more chemical substances that can prevent extreme changes in the hydrogen ion concentration (pH) of the solution when either an acid or a base is added to it. A buffer system works by chemically combining with hydrogen ions as their concentration starts to rise and releasing them as their concentration starts to fall.

A solution of carbonic acid and sodium bicarbonate, for example, acts as a buffer system. When carbonic acid and sodium bicarbonate are placed together in solution, the sodium bicarbonate dissociates into sodium ions and bicarbonate ions. The carbonic acid, however, is a weak acid, and it does not dissociate fully into hydrogen ions and bicarbonate ions (F19.5). Thus, the carbonic acid–sodium bicarbonate buffer system contains undissociated carbonic acid, sodium ions, hydrogen ions, and bicarbonate ions. Moreover, a certain proportion of the carbonic acid is always dissociated into hydrogen ions and bicarbonate ions, and a certain proportion does not undergo this dissociation.

If hydrogen ions (for example, in the form of hydrochloric acid) are added to a buffer system such as the carbonic acid–sodium bicarbonate buffer system, many of the hydrogen ions do not remain in the free state to affect pH as they would if they were added to pure water (F19.6). In the carbonic acid–sodium bicarbonate buffer system, the addition of hydrogen ions (from hydrochloric acid) results in the presence of many hydrogen ions and many bicarbonate ions (from sodium bicarbonate) in the solution. These are the equivalent of carbonic acid molecules that have dis-

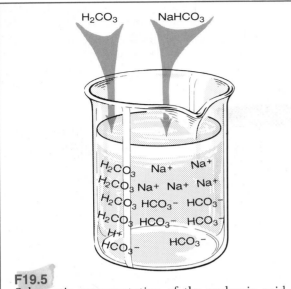

F19.5
Schematic representation of the carbonic acid–sodium bicarbonate buffer system. In solution, the sodium bicarbonate dissociates into sodium ions (Na^+) and bicarbonate ions (HCO_3^-). Because it is a weak acid, only a relatively small amount of the carbonic acid dissociates into hydrogen ions (H^+) and bicarbonate ions.

sociated into hydrogen ions and bicarbonate ions. Consequently, the presence of these ions upsets the normal proportion that must be maintained between carbonic acid molecules that dissociate into hydrogen ions and bicarbonate ions, and carbonic acid molecules that do not undergo this dissociation. In order to reestablish the normal proportion, some of the hydrogen ions and bicarbonate ions combine into undissociated carbonic acid molecules. This combination removes from the solution some of the hydrogen ions that result from the addition of the hydrochloric acid. These hydrogen ions would otherwise have been free to increase the hydrogen ion concentration of the solution and lower the pH. Instead, many of the hydrogen ions do not remain free, and the effect of the hydrogen ions from the hydrochloric acid on the hydrogen ion concentration (pH) of the solution is minimized.

Conversely, the removal of hydrogen ions from the carbonic acid–sodium bicarbonate buffer system—due, for example, to the addition of a base—also upsets the normal proportion that must be maintained between carbonic acid molecules that dissociate into hydrogen ions and bicarbonate ions, and carbonic acid molecules that do not undergo this dissociation. In this case, some of the carbonic acid molecules dissociate into hydrogen ions and bicarbonate ions in order to restore the normal proportion. This dissociation minimizes the effect of removing hydrogen ions on the pH of the solution.

In the body, the carbonic acid–sodium bicarbonate buffer system acts in the manner just described to stabilize the pH of the body fluids. Moreover, other body substances also act as buffers to resist changes in pH. Among these are large anions such as plasma proteins, intracellular phosphate complexes, and hemoglobin molecules (reduced hemoglobin has a much greater affinity for hydrogen ions than oxyhemoglobin does).

Respiratory Regulation of Acid–Base Balance

The respiratory regulation of acid–base balance makes use of the fact that in solution carbonic acid exists in reversible equilibrium with carbon dioxide and water (dissolved carbon dioxide) as follows:

$$CO_2 + H_2O \rightleftharpoons H_2CO_3 \rightleftharpoons H^+ + HCO_3^-$$

Thus, the higher the concentration of carbon dioxide in the internal environment of the body, the more carbonic acid is formed. As more carbonic acid is formed, the normal proportion that must be maintained between carbonic acid that dissociates into hydrogen ions and bicarbonate ions, and carbonic acid that does not undergo this dissociation is upset, and some of the carbonic acid dissociates into hydrogen ions and bicarbonate ions. This dissociation tends to increase the acidity (lower the pH) of the internal environment. Conversely, the lower the concentration of carbon dioxide in the internal environment, the more carbonic acid forms carbon dioxide and water. This activity also upsets the normal proportion that must be maintained between carbonic acid molecules that dissociate into hydrogen ions and bicarbonate ions, and carbonic acid molecules that do not undergo this dissociation. As a result, some of the dissociated hydrogen ions and bicarbonate ions of carbonic acid combine into undissociated carbonic acid molecules. This combination removes hydrogen ions and tends to decrease the acidity (raise the pH) of the internal environment.

Since the carbon dioxide concentration of the body's internal environment influences the pH of the internal environment, and since carbon dioxide is eliminated from the body at the lungs, the respiratory system plays an important role in maintaining the pH of the internal environment. If much carbon dioxide is eliminated at the lungs, the CO_2 concentration in the internal environment decreases. This decrease results in a less acidic, more basic internal environment with a higher pH. If only a small amount of CO_2 is eliminated at

the lungs, its concentration in the internal environment increases. This increase results in a more acidic internal environment with a lowered pH. Thus, the control of the CO_2 concentration in the internal environment of the body by the respiratory system serves as a kind of buffering mechanism. For example, if the pH of the internal environment falls below normal values (acidosis), an increased elimination of carbon dioxide at the lungs tends to raise it. Conversely, if the pH of the internal environment rises above normal levels (alkalosis), a decreased elimination of carbon dioxide at the lungs tends to decrease it. In view of the ability of the respiratory system to influence acid–base balance, it is not surprising that respiration is itself sensitive to such factors as the carbon dioxide content and the pH of the system. The influence of these factors on respiration was discussed in Chapter 15.

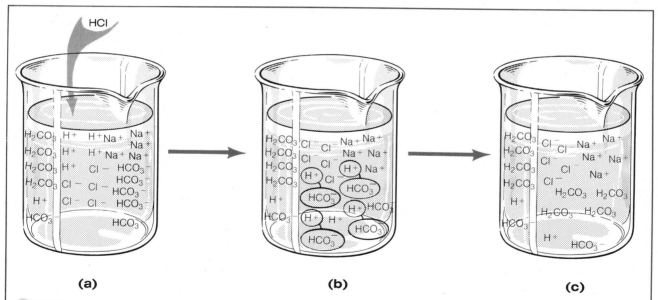

(a)   **(b)**   **(c)**

F19.6

Effect of adding strong acid (HCl) to the carbonic acid–sodium bicarbonate buffer system. (a) The hydrochloric acid added to the solution dissociates into hydrogen ions (H^+) and chloride ions (Cl^-). This dissociation upsets the proportion that must be maintained between carbonic acid molecules that are dissociated into hydrogen ions and bicarbonate ions, and carbonic acid molecules that have not undergone this dissociation. (b) As a result, some of the hydrogen ions and bicarbonate ions combine into undissociated carbonic acid molecules. (c) This combination ties up hydrogen ions from hydrochloric acid that would otherwise have been free to lower the pH of the solution.

Renal Regulation of Acid–Base Balance

Although the body's buffer systems can tie up some hydrogen ions and resist changes in pH to some degree, the continued production of acidic products by the body's metabolic reactions would eventually overcome this buffering capacity. Consequently, excess hydrogen ions must ultimately be eliminated from the body, and this is accomplished by the kidneys. The kidneys also regulate the extracellular concentration of bicarbonate ions, which is normally quite high.

Secretion of Hydrogen Ions.

Hydrogen ions are secreted into the renal tubules by tubular cells. Although the secretion mechanism is not well understood, it appears that the hydrogen ions are derived from carbonic acid (F19.7). The enzyme carbonic anhydrase is present within the tubular cells, and it is thought to catalyze the formation of carbonic acid from carbon dioxide and water.

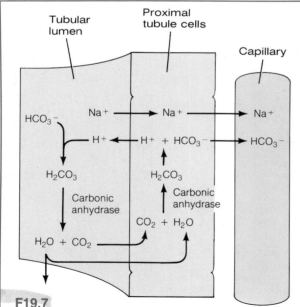

F19.7
Secretion of hydrogen ions by cells of the proximal tubule. Also shown is the pathway by which hydrogen ions in the tubular lumen participate in the reabsorption of bicarbonate ions from the tubule.

Some of the carbonic acid dissociates into hydrogen ions and bicarbonate ions. The hydrogen ions are actively transported into the lumen of the tubule in exchange for sodium ions. The bicarbonate ions that remain in the tubular cells, as well as the sodium ions, pass out of the cells and into the blood.

Reabsorption of Bicarbonate Ions.

The secretion of hydrogen ions as just described provides a way of recovering bicarbonate ions that are filtered from the plasma into Bowman's capsules (F19.7). Bicarbonate ions within the tubular lumen combine with secreted hydrogen ions to form carbonic acid molecules. The carbonic acid molecules are then split into carbon dioxide and water by carbonic anhydrase associated with the membranes of the tubular cells in the proximal convoluted tubule. The carbon dioxide enters the tubular cells, where carbonic anhydrase combines it with water to form carbonic acid. Some of the carbonic acid dissociates into hydrogen ions and bicarbonate ions. The hydrogen ions are actively transported into the lumen of the tubule in exchange for sodium ions. The bicarbonate ions that remain in the tubular cells, as well as the sodium ions, pass out of the tubular cells and into the blood.

Control of Hydrogen Ion Secretion and Bicarbonate Ion Reabsorption.

As is evident from the preceding discussion, the tubular reabsorption of bicarbonate ions that are filtered from the plasma into Bowman's capsules depends on the presence of hydrogen ions within the renal tubules. Moreover, the rate of secretion of hydrogen ions into the renal tubules varies directly with the acidity of the internal environment.

Normally, the balance between the rate of filtration of bicarbonate ions into Bowman's capsules and the rate of secretion of hydrogen ions into the renal tubules is such that there are slightly more hydrogen ions secreted than there are bicarbonate ions filtered. As a result, essentially all the bicarbonate ions are reabsorbed, and the slight excess of hydrogen ions that remain after bicarbonate ion reabsorption is excreted in the urine.

If the concentration of hydrogen ions in the internal environment falls, the rate of secretion of hydrogen ions into the renal tubules diminishes relative to the filtration rate of bicarbonate ions into Bowman's capsules. As a consequence, not all of the bicarbonate ions are reabsorbed, and bicarbonate ions are excreted in the urine. The urinary excretion of bicarbonate ions diminishes the concentration of bicarbonate ions in the internal environment. The diminished concentration of bicarbonate ions leads to a decrease in the number of hydrogen ions tied up by the carbonic acid–sodium bicarbonate buffer system (as well as by other buffer systems), and the hydrogen ion concentration of the internal environment tends to rise toward normal.

If the concentration of hydrogen ions in the internal environment rises, the rate of secretion of hydrogen ions into the renal tubules increases considerably in comparison to the rate of filtration of bicarbonate ions into Bowman's capsules. Many of the secreted hydrogen ions are not needed for the reabsorption of bicarbonate ions, and they are excreted in the urine. Moreover, as previously described, the process of secreting hydrogen ions into the renal tubules provides sodium ions and bicarbonate ions that are returned to the body. The return of these ions increases the concentration of bicarbonate ions in the internal environment, and thereby leads to an increase in the number of hydrogen ions tied up by the carbonic acid–sodium bicarbonate buffer system (as well as by other buffer systems). These responses tend to decrease the hydrogen ion concentration of the internal environment toward normal.

Buffering of Secreted Hydrogen Ions. The mechanisms that secrete hydrogen ions into the renal tubules can achieve a maximum concentration of free hydrogen ions in the tubular fluid that is equivalent to a pH of 4.5. At normal rates of urine flow this is not sufficient to excrete all of the hydrogen ions that must be eliminated from the

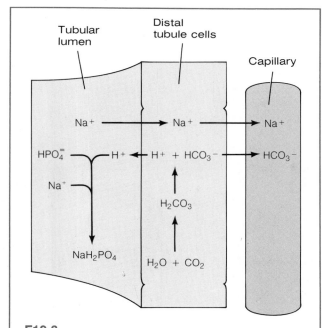

F19.8
Secretion of hydrogen ions into the tubular lumen by cells of the distal tubule. The hydrogen ions are excreted in combination with phosphate compounds.

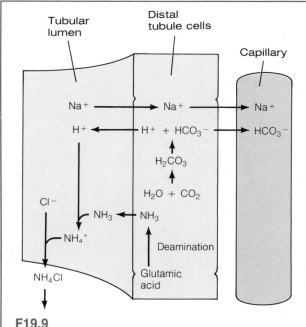

F19.9
Secretion of hydrogen ions and ammonia into the tubular lumen by cells of the distal tubule. The hydrogen ions are excreted in combination with ammonia in the form of ammonium salts.

body. Consequently, buffers must be present within the tubular fluid to tie up secreted hydrogen ions so that the pH of the tubular fluid remains above 4.5, thereby allowing additional hydrogen ions to be secreted. Phosphate compounds ($HPO_4^=$) and ammonia (NH_3) act as buffers to tie up hydrogen ions in the tubular fluid. The phosphate compounds are filtered at the glomeruli and are only poorly reabsorbed (F19.8). They combine with hydrogen ions ($H^+ + HPO_4^= \rightarrow H_2PO_4^-$) and are excreted in combination with a cation such as sodium (Na^+) in the form of a weakly acidic substance ($Na^+ + H_2PO_4^- \rightarrow NaH_2PO_4$). This same type of excretion can occur with other anions such as lactate.

Ammonia is formed in the tubular cells by the deamination of certain amino acids, particularly glutamic acid (F19.9). The ammonia diffuses from the tubular cells into the tubules where it combines with hydrogen ions to form ammonium ions ($H^+ + NH_3 \rightarrow NH_4^+$). This removes ammonia (as well as hydrogen ions) from the tubular fluid and allows the diffusion of still more ammonia from the tubular cells into the tubular fluid. The ammonium ions are excreted in combination with anions such as chloride (Cl^-) in the form of very weakly acidic or neutral substances ($NH_4^+ + Cl^- \rightarrow NH_4Cl$) that do not significantly lower the pH of the urine. It is interesting to note that the production rate of ammonia increases when the hydrogen ion concentration has been elevated for two or three days. This increase in ammonia production provides additional buffering capacity for hydrogen ion excretion and allows extra hydrogen ions to be excreted without exceeding the maximum allowable acidity of the tubular fluid.

Disorders of Acid–Base Balance

Despite the abilities of the kidneys and the lungs to regulate the pH of the extracellular fluids, a

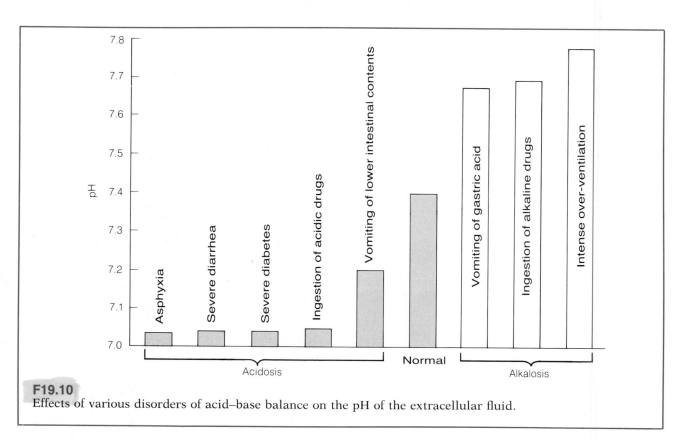

F19.10
Effects of various disorders of acid–base balance on the pH of the extracellular fluid.

wide variety of diseases and dysfunctions can upset the hydrogen ion concentration of the internal environment, producing either acidosis or alkalosis (F19.10). However, all of these can be grouped into two broad categories: respiratory acidosis or alkalosis and metabolic acidosis or alkalosis.

Respiratory acidosis or *alkalosis* is due to a change in the carbon dioxide and, consequently, the carbonic acid content of the body fluids. Respiratory acidosis frequently results from asphyxia or diseases that interfere with the elimination of carbon dioxide by the lungs. Respiratory alkalosis, which is much less common, can result from intense overventilation, which eliminates an excessive amount of carbon dioxide from the body.

Metabolic acidosis or *alkalosis* is acidosis or alkalosis due to all other factors besides an excess or deficiency of carbon dioxide in the body fluids. Metabolic acidosis can be caused by severe diarrhea or the vomiting of the lower intestinal contents. Both events cause the loss of bicarbonate-rich fluids from the gastrointestinal tract. This loss has the same effect as the excretion of bicarbonate ions by the kidneys—that is, a tendency to decrease the pH of the internal environment. Metabolic acidosis can also be caused by the production of an excessive amount of acidic ketone bodies—for example, acetoacetic acid—during diabetes mellitus or by the ingestion of acidic drugs. Metabolic alkalosis can be caused by vomiting the acidic gastric contents alone, which rarely occurs, or by the ingestion of alkaline drugs.

Both acidosis and alkalosis are serious problems, and a person generally cannot live more than a few hours if the pH of the arterial blood falls to about 7.0 or rises to about 8.0. Acidosis depresses the central nervous system, and a person suffering from severe acidosis usually enters a coma. Conversely, alkalosis causes an overexcitability of the nervous system that often leads to muscular tetany and convulsions.

STUDY OUTLINE

BODY FLUIDS pp. 493–494

1. Plasma—fluid within blood vessels.
2. Interstitial fluid—fluid that surrounds the cells.
3. Intracellular fluid—fluid within the cells.
4. Extracellular fluid—fluid outside of cells; made up of plasma and interstitial fluid; forms the body's internal environment.

Composition of Plasma and Interstitial Fluid Plasma and interstitial fluid are separated by capillary walls and are of similar composition.

Composition of Intracellular Fluid Different in composition from plasma and interstitial fluid.

REGULATION OF THE BODY'S INTERNAL ENVIRONMENT In order to maintain homeostasis, the addition of substances such as water and electrolytes to the body's internal environment must be balanced by the removal of equal amounts of these materials. pp. 494–507

Sodium Regulation Mechanisms involved in blood pressure regulation are important in controlling body sodium content.

SODIUM INGESTION Two components—regulatory and hedonistic; control of sodium ingestion does not appear to be major means of regulating body sodium content.

RENAL EXCRETION OF SODIUM Relation between glomerular filtration rate and tubular reabsorption rate of sodium is important.

1. Glomerular filtration rate of sodium is influenced by arterial blood pressure and sympathetic nerves to kidneys.
2. Reabsorption of sodium is stimulated by aldosterone; aldosterone release influenced by renin–angiotensin mechanism; renin secretion is apparently stimulated by a decrease in the renal arterial pressure and/or an increase in the activity of sympathetic nerves to kidneys.

FACTORS THAT INFLUENCE SODIUM EXCRETION Excretion of sodium is influenced by mechanisms involved in the regulation of blood pressure.

SODIUM CONTENT AND SODIUM CONCENTRATION Loss or gain of sodium by body changes the overall sodium content of the body but does not necessarily have a great effect on the actual concentration of sodium in the internal environment.

Water Regulation   Water content of body must be regulated because it affects osmotic pressure and fluid volume of internal environment.

WATER INGESTION AND THIRST   Sensation of thirst results from increased osmotic concentration of extracellular fluid or substantial reduction in plasma volume; osmoreceptors and perhaps circulatory pressure receptors stimulate thirst.

RENAL EXCRETION OF WATER   Relation between glomerular filtration rate and tubular reabsorption rate of water is important.
1. Glomerular filtration rate of water is influenced by changes in systemic blood pressure.
2. Amount of water reabsorbed from renal tubules depends in large part on the water permeability of the collecting tubules, which is controlled by antidiuretic hormone (ADH) from the pituitary gland.
3. ADH release is influenced by osmoreceptors in hypothalamus and circulatory pressure receptors. Maximum ADH release and substantial water reabsorption results from increased extracellular osmolarity and decreased blood pressure (plasma volume).

FACTORS THAT INFLUENCE WATER EXCRETION   Osmotic pressure of extracellular fluid and blood pressure are major factors; factors that influence sodium excretion are very closely related to factors that influence water excretion.

Potassium Regulation   Ingestion and excretion are normally in balance; most filtered potassium is actively reabsorbed and amount excreted depends mainly on secretion into renal tubules. Amount of potassium secreted is related to potassium concentration of renal tubular cells and influenced by aldosterone.

Calcium Regulation   Involves bone, kidneys, and gastrointestinal tract. The calcium regulating activities of all three sites are influenced either directly or indirectly by parathyroid hormone.
1. Parathyroid hormone increases movement of calcium and phosphate from bone into the extracellular fluid.
2. Parathyroid hormone decreases urinary excretion of calcium and increases excretion of phosphate.

3. Parathyroid hormone enhances a step in the metabolic transformation of vitamin D_3 to 1,25-dihydroxycholecalciferol, which stimulates active calcium absorption from intestine.
Calcitonin inhibits the removal of calcium from bone.

Hydrogen Ion and Acid–Base Regulation

ACIDS AND BASES   Acids liberate hydrogen ions, bases accept them; concentration of free hydrogen ions determines the acidity of a solution.

ACIDOSIS AND ALKALOSIS
1. pH of arterial blood is 7.4; pH of venous blood and interstitial fluid is 7.35.
2. Acidosis is a higher arterial blood hydrogen ion concentration than normal (decreased pH).
3. Alkalosis is a lower arterial blood hydrogen ion concentration than normal (increased pH).
4. Body's metabolic activities produce acidic products and excess hydrogen ions must be eliminated.

BUFFER SYSTEMS
1. Solution containing two or more chemical substances that can prevent extreme changes in pH of the solution when an acid or a base is added.
2. Body substances that act as buffers include carbonic acid–sodium bicarbonate buffer system, plasma proteins, intracellular phosphate complexes, and hemoglobin.

RESPIRATORY REGULATION OF ACID–BASE BALANCE
1. Makes use of the fact that in solution carbonic acid exists in reversible equilibrium with carbon dioxide and water.
2. Large carbon dioxide elimination at lungs tends to reduce acidity (raise pH) of internal environment and vice versa.
3. Respiration is sensitive to carbon dioxide content and pH of the system.

RENAL REGULATION OF ACID–BASE BALANCE

Secretion of Hydrogen Ions   Hydrogen ions are secreted into the renal tubules by tubular cells.

Reabsorption of Bicarbonate Ions   Bicarbonate ions filtered into Bowman's capsules can be reabsorbed by tubular cells.

Control of Hydrogen Ion Secretion and Bicarbonate Ion Reabsorption

1. Bicarbonate ion reabsorption depends upon the presence of hydrogen ions in renal tubules.
2. Rate of secretion of hydrogen ions into renal tubules varies directly with the acidity of the internal environment.

Buffering of Secreted Hydrogen Ions Phosphate compounds and ammonia buffer hydrogen ions within tubular fluid.

DISORDERS OF ACID–BASE BALANCE

1. Respiratory acidosis or alkalosis is due to a change in the carbon dioxide and consequently the carbonic acid content of the body fluids.
2. Metabolic acidosis or alkalosis is due to all other factors besides an excess or deficiency of carbon dioxide in the body fluids.

SELF-QUIZ

1. There are only slight differences in composition between the: (a) plasma and interstitial fluid; (b) intracellular fluid and plasma; (c) intracellular fluid and interstitial fluid.
2. It appears that humans normally regulate their body sodium content by carefully controlling the amount of sodium they ingest. True or False?
3. The arterial blood pressure can influence the glomerular filtration rate. True or False?
4. Angiotensin is a factor that stimulates the release of: (a) renin; (b) sodium; (c) aldosterone.
5. The loss or gain of sodium by the body does not necessarily have a great effect on the actual concentration of sodium in the internal environment. True or False?
6. A sensation of thirst results from: (a) an increased osmotic concentration of the extracellular fluid; (b) a substantial increase in plasma volume; (c) an increased volume of extracellular fluid.
7. The permeability of the kidney's collecting tubules to water is controlled hormonally by: (a) aldosterone; (b) ADH; (c) angiotensin.
8. ADH release is likely to increase in response to an increase in: (a) blood pressure; (b) plasma volume; (c) extracellular osmolarity.
9. In most cases, an increase in the glomerular filtration rate of water is accompanied by a simultaneous decrease in the glomerular filtration rate of sodium. True or False?
10. Adjustments in the amount of potassium excreted depend mainly on how much is: (a) filtered into Bowman's capsules; (b) reabsorbed from the renal tubules; (c) secreted into the renal tubules.
11. Which hormone enhances the tubular secretion of potassium? (a) aldosterone; (b) ADH; (c) parathyroid hormone.
12. Parathyroid hormone decreases the: (a) movement of calcium from bone into the extracellular fluid; (b) movement of phosphate from bone into the extracellular fluid; (c) urinary excretion of calcium.
13. The normal pH of arterial blood is about: (a) 7.0; (b) 7.7; (c) 7.4.
14. The principal organ for the elimination of excess hydrogen ions from the body is the: (a) liver; (b) lungs; (c) kidneys.
15. When the CO_2 concentration of the internal environment rises, the pH of the internal environment also tends to rise. True or False?
16. If a large amount of carbon dioxide is eliminated at the lungs, the carbon dioxide concentration in the internal environment decreases, leading to a more acidic, less basic internal environment with a higher pH. True or False?
17. In the case of alkalosis, the rate of secretion of hydrogen ions into the renal tubules: (a) diminishes in comparison to the rate of filtration of bicarbonate ions into Bowman's capsules; (b) increases in comparison to the rate of filtration of bicarbonate ions into Bowman's capsules.
18. Within the renal tubules, hydrogen ions are buffered by: (a) chloride; (b) ammonia; (c) sodium.
19. Metabolic acidosis can be caused by: (a) asphyxia; (b) severe diarrhea; (c) vomiting the acidic gastric contents alone.
20. Severe acidosis often results in: (a) muscular tetany; (b) convulsions; (c) coma.

REPRODUCTION AND INHERITANCE 20

The activities of the reproductive system include developmental changes that lead to puberty, gametogenesis (spermatogenesis and oogenesis), sexual intercourse, pregnancy, the development of the embryo, parturition, and lactation. These reproductive activities depend on complex interactions between the reproductive structures, the central nervous system, and endocrine glands such as the pituitary, testes, and ovaries.

ORGANIZATION OF THE MALE REPRODUCTIVE SYSTEM

TESTES

The male gonads—the **testes**—are located outside the abdominopelvic cavity in a skin-covered sac called the scrotum (F20.1). The male gametes, which are called **spermatozoa** or **sperm,** develop within the two testes in highly convoluted seminiferous tubules (F20.2). Between the seminiferous tubules are clusters of interstitial cells of Leydig, which produce the male sex hormone **testosterone.**

EPIDIDYMIS

A system of ducts carries sperm from each testis to the exterior of the body. The first portion of this duct system is contained within the **epididymis,** which fits against the posterior surface of a testis (F20.1; F20.2). Within the epididymis, a network of tubules from the testes join to form the epididymal duct (ductus epididymis). During their passage through the epididymis, the sperm undergo a maturational process without which they would be nonmotile and infertile when they entered the female reproductive tract. During sexual stimulation, smooth muscles in the wall of the epididymal duct contract. These contractions move sperm into the next portion of the duct system, the ductus deferens.

DUCTUS DEFERENS

The **ductus deferens** (**vas deferens**) is a continuation of the epididymal duct (F20.1). A ductus deferens ascends from each epididymis and enters the abdominopelvic cavity. Beneath the urinary bladder; each ductus deferens is joined by the duct of a

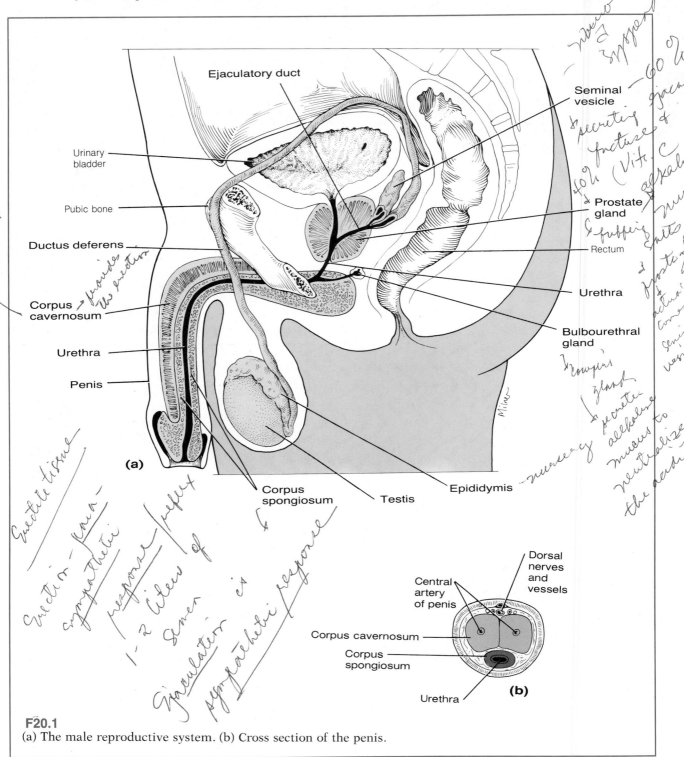

F20.1
(a) The male reproductive system. (b) Cross section of the penis.

seminal vesicle to form a short *ejaculatory duct,* which passes through the prostate gland to join the urethra. The wall of the ductus deferens contains a heavy layer of smooth muscle that undergoes peristaltic contractions during sexual stimulation, propelling sperm into the ejaculatory duct. Contractions of the ejaculatory duct propel spermatozoa and secretions from the seminal vesicles into the urethra.

PENIS

The **penis,** through which the urethra passes, is the copulatory organ by which a sperm-containing fluid called **semen** is placed in the female reproductive tract (F20.1). The penis contains three cylindrical bodies (two corpora cavernosa and one corpus spongiosum), which are each composed of tissue that contains numerous, sponge-like spaces. These spaces fill with blood during sexual stimulation, causing the penis to enlarge and become firm. This occurrence is referred to as an erection.

STRUCTURES THAT CONTRIBUTE TO THE FORMATION OF SEMEN

Semen is a mixture of spermatozoa from the testes and fluids from structures that include the seminal vesicles, the prostate gland, and the bulbourethral glands.

Seminal Vesicles

The **seminal vesicles** are two membranous pouches located behind and beneath the urinary bladder (F20.1). They secrete a viscous fluid (containing fructose and prostaglandins), which con-

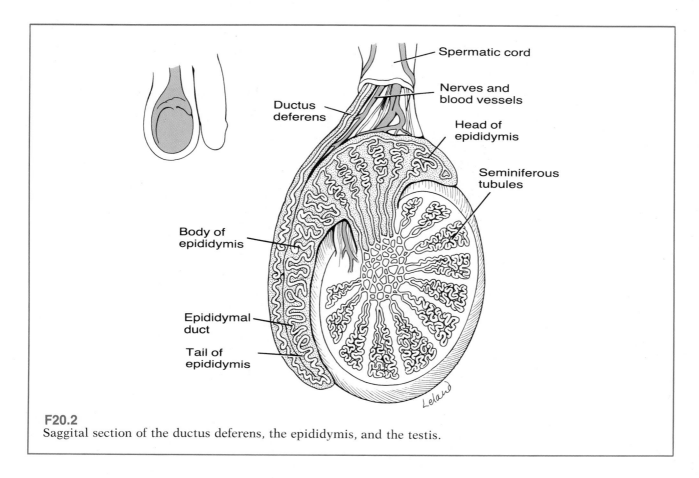

F20.2
Saggital section of the ductus deferens, the epididymis, and the testis.

tributes about 60% of the bulk of the semen. Fructose is a major energy source for ejaculated spermatozoa, which contain very little cytoplasm, and therefore have only limited intracellular glycogen available from which to derive energy. The prostaglandins are thought to facilitate the process of fertilization by reacting with mucus at the cervix of the female uterus to make it more receptive to sperm and by stimulating reverse peristaltic contractions that enhance the movement of sperm along the uterus and uterine tubes.

Prostate Gland

The **prostate gland** surrounds the urethra just below the bladder (F20.1). It secretes a thin, milky, alkaline fluid into the urethra through a number of small ducts. This fluid counteracts the mild acidity of other portions of the semen so that the semen itself is slightly alkaline, with a pH of about 7.5. The alkaline nature of the semen helps counteract the acid secretions of the female vagina (pH 3.5 to 4.0). This activity is believed to be important because sperm do not become optimally active until the pH of their environment rises to about 6.0 to 6.5.

Bulbourethral Glands

The **bulbourethral glands** are a pair of small mucous glands located below the prostate, on either side of the urethra (F20.1). Their secretion is carried to the urethra by a duct from each gland.

PUBERTY IN THE MALE

The time of life at which reproduction becomes possible is called **puberty.** Its age of onset varies considerably among individuals, but it most commonly occurs between 10 and 15 years of age. In males, puberty is marked by the first ejaculate of semen that contains mature spermatozoa, as well as by more gradual and subtle changes in personality and body form. The penis, scrotum, and testes enlarge, and hair grows on the face, axilla (armpit region), and pubic area. The voice deepens, and there is a tendency toward increased development of skeletal muscles accompanied by a decrease in the amount of body fat.

The events that occur at puberty are the result of an increased production of testosterone by the testes, and they require the continued production of adequate levels of testosterone to sustain them. Prior to puberty, the hypothalamus of the brain secretes only small amounts of gonadotropin-releasing hormone (GnRH). Consequently, there is little release of the gonadotropins—follicle-stimulating hormone (FSH) and luteinizing hormone (LH)—from the pituitary, and the testes are not stimulated to produce large amounts of testosterone. At puberty, an alteration in brain function occurs that leads to an increased release of GnRH, which ultimately results in an increased production of testosterone. Researchers believe that, prior to puberty, the hypothalamus is extremely sensitive to the negative-feedback effects of testosterone. Thus, even though only low levels of testosterone are present, GnRH release is suppressed. But at puberty, the hypothalamus decreases in sensitivity to the negative-feedback effects of testosterone, and GnRH is released in greater quantities.

SPERMATOGENESIS

The production of spermatozoa within the testes is called **spermatogenesis.** During spermatogenesis a unique type of cell division called meiosis takes place. (Meiosis is described in Chapter 3, and the specific events of this process should be reviewed at this time.) The purpose of meiosis, which is aptly called a reduction division, is to reduce the number of chromosomes in each gamete by half. The cells of the human body are *diploid;* that is, they contain 23 chromosomes from each parent, for a total of 46. Cells undergoing meiosis lose half of their total number of chromosomes, retaining 23 chromosomes. Such cells are said to be *haploid.* The formation of haploid spermatozoa (as well as haploid ova in the female) ensures that at fertilization when a male and a female gamete unite, the resulting cell will be a diploid cell that has 46 chromosomes.

Spermatogenesis occurs within the seminiferous tubules of the testes (F20.3). Located within the outer portion of each tubule are cells

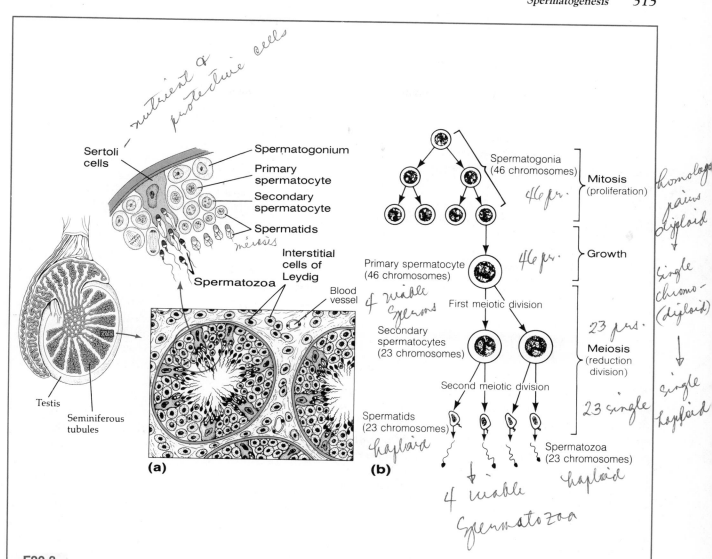

The handwritten annotations read:
- nutrient & protective cells
- meiosis
- 4 viable sperms
- haploid
- 4 viable spermatozoa
- 46 pr.
- 46 pr.
- 23 prs.
- 23 single
- homologous pairs diploid
- single chromo- (diploid) → single haploid

F20.3

(a) Section of testis showing spermatogenesis within the seminiferous tubules. (b) Diagrammatic representation of the process of spermatogenesis.

called *spermatogonia*, which have the diploid number of chromosomes (44 autosomes plus one X and one Y sex chromosome). The spermatogonia divide mitotically (see Chapter 3), thereby providing a continuous source of new cells that are used for the production of spermatozoa. Some spermatogonia move toward the lumen of the tubule and undergo a period of growth. These enlarged

cells are called *primary spermatocytes*. Each primary spermatocyte undergoes a first meiotic division, which results in the formation of two *secondary spermatocytes*. One of the secondary spermatocytes contains 22 autosomes plus the X sex chromosome; the other has the remaining 22 autosomes and the Y sex chromosome.

The second meiotic division divides each secondary spermatocyte into two *spermatids*. Thus, meiosis results in the formation of four haploid spermatids from one diploid spermatogonium. Each spermatid undergoes a complex series of changes by which it develops into a spermatozoon. During these changes, much of the cytoplasm of the cell is discarded and a tail containing a group of contractile proteins forms.

During spermatogenesis, the developing gametes are closely associated with *Sertoli cells*, which extend from the outer portion of a seminiferous tubule to the lumen. The Sertoli cells play an integral role in spermatogenesis, and they are believed to serve as a route by which nutrients and chemical (for example, hormonal) signals reach the developing gametes. In humans, the entire process of spermatogenesis, from spermatogonium to spermatozoa, requires about 72 days and occurs continuously beginning at puberty.

HORMONAL CONTROL OF TESTICULAR FUNCTION

The gonadotropins FSH and LH are required for normal spermatogenesis and testosterone secretion. Moreover, testosterone is itself necessary for sperm production. FSH stimulates spermatogenesis by acting on the Sertoli cells, which are in close contact with the spermatogenic cells at all stages of development. LH stimulates testosterone secretion by acting on the interstitial cells of Leydig. In addition, since testosterone is required for sperm production, LH can be regarded as important in this process as well.

The release of FSH and LH from the pituitary is stimulated by GnRH from the hypothalamus. Testosterone from the testes inhibits LH release by way of negative feedback, but it has little effect on FSH release. The inhibitory influence of testosterone on LH release is believed to be exerted mainly at the level of the pituitary rather than at the level of the hypothalamus. The release of FSH appears to be inhibited by a protein hormone called *inhibin*, which is thought to be released by the Sertoli cells of the seminiferous tubules. Although not all interactions are fully established, the general relationship among GnRH, FSH, LH, testosterone, and inhibin is depicted in F20.4.

MALE SEXUAL RESPONSES

ERECTION

Sexual intercourse (coitus) is the usual method by which male sperm are deposited within the vagina of the female. In order to accomplish sexual intercourse, the male penis must become enlarged and firm, a process referred to as *erection*.

Erection is a vascular phenomenon in which the spaces within the corpora cavernosa and corpus spongiosum of the penis become engorged with blood. In the absence of sexual stimulation, the arterioles that supply the corpora of the penis are partially constricted, which allows only limited bloodflow into the penis. Under these conditions the penis is flaccid. During sexual excitation, parasympathetic nerve impulses travel from the sacral region of the spinal cord to the arterioles of the penis, causing them to dilate. The dilation of the arterioles allows more blood to enter the vascular spaces of the erectile tissues. As the erectile tissues of the penis expand due to the increased blood volume, they compress the veins that drain the penis against the coverings of the penis. The compression of the veins reduces the flow of blood out of the penis, which further increases the engorgement of the erectile tissues. As a result, the penis becomes enlarged and firm.

An erection can result from visual or psychic stimuli, including thoughts or emotions that originate in the brain, or from the physical stimulation of touch receptors on the penis and various other regions of the body. Moreover, psychic factors can greatly influence the effectiveness of physical stimuli. Indeed, under adverse psychological conditions, the physical stimulation of the penis frequently will not cause an erection.

EMISSION AND EJACULATION

In addition to causing the penis to become erect, continued tactile stimulation of the penis ultimately results in a forceful expulsion of semen. When sexual stimulation becomes sufficiently intense, sympathetic nerve impulses from the lumbar region of the spinal cord cause rhythmic contractions of the epididymis, the ductus deferens, the seminal vesicles, and the prostate gland. These contractions propel the contents of the ducts and glands into the urethra, where they form semen. Up to this point, the process is referred to as **emission.** Following emission, skeletal muscles at the base of the penis undergo rhythmic contractions

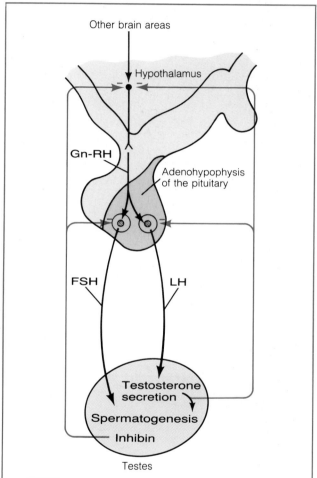

F20.4

Relationship among GnRH, FSH, LH, testosterone, and inhibin in the male. The negative-feedback influences of testosterone (and presumably inhibin) are believed to be exerted mainly at the level of the pituitary rather than at the level of the hypothalamus. (With regard to the relationships depicted in this figure, it should be noted that some researchers believe there are separate releasing substances for FSH and LH. However, present evidence favors the existence of a single gonadotropin-releasing hormone [GnRH] that stimulates the release of both FSH and LH.)

that expel, or ejaculate, the semen from the urethra. The rhythmic contractions that produce an **ejaculation** are accompanied by intensely pleasurable sensations and widespread muscular contractions. This reaction is called an *orgasm.*

Following ejaculation, the arterioles that supply the penis return to their usual partially constricted condition, which decreases the bloodflow into the penis. The decreased bloodflow relieves the vascular congestion within the corpora of the penis and allows the blood to return freely through the veins to the general circulation. As the accumulated blood leaves the corpora, the penis returns to a flaccid state.

ORGANIZATION OF THE FEMALE REPRODUCTIVE SYSTEM

OVARIES

The female gonads—the **ovaries**—are located within the abdominopelvic cavity (F20.5; F20.6). The female gametes, or **ova,** develop within the ovaries. The ovaries also produce the female sex hormones, the **estrogens** and **progesterone.** At birth, the two ovaries contain approximately 400,000 immature structures called *primary follicles* (see F20.8). Each primary follicle consists of an undeveloped ovum located within a small sphere that is composed of a single layer of cells. Beginning at the time of puberty and continuing throughout the 30 to 40 years during which a female is capable of reproduction, one follicle usually matures fully and ruptures each month (approximately every 28 days), releasing an ovum that is then available for fertilization. This being the case, only about 400 of the total number of follicles reach full maturity during a woman's lifetime. The rest reach various stages of development and then regress and degenerate.

UTERINE (FALLOPIAN) TUBES

An ovum that is released from a mature follicle at ovulation is carried to the uterus by a **uterine** or **fallopian tube,** which extends from the vicinity of the ovary to the upper portion of the uterus (F20.5; F20.6). The distal end of each uterine tube opens into the abdominopelvic cavity, very close to an ovary. This opening is surrounded by small, ciliated, fingerlike projections called fimbriae, one of which is generally attached to the ovary. The movements of the cilia generate currents that cause fluid to flow from the abdominopelvic cavity into the uterine tube and thus carry the ovum that has been released from its follicle into the tube. The uterine tube contains circular and longitudinal layers of smooth muscles and an inner mucosa that contains ciliated cells whose cilia beat rhythmically toward the uterus. Once within the uterine tube, the ovum is carried to the uterus by a weak fluid current caused by the beating of the cilia, possibly aided by peristaltic contractions of the smooth muscles of the uterine tube.

UTERUS

The **uterus** is a single, hollow, pear-shaped organ that lies within the pelvis behind the urinary bladder, and in front of the sigmoid colon and rectum (F20.5; F20.6). The wall of the uterus contains a thick layer of smooth muscle, the myometrium, and the mucosal lining of the cavity within the uterus is called the endometrium. The superficial, or functional, layer of the endometrium undergoes marked developmental changes during the menstrual cycle. Each month it thickens and becomes engorged with blood in preparation for receiving a fertilized ovum. If the ovum is not fertilized, the functional layer of the endometrium is sloughed off and discarded as the menstrual flow.

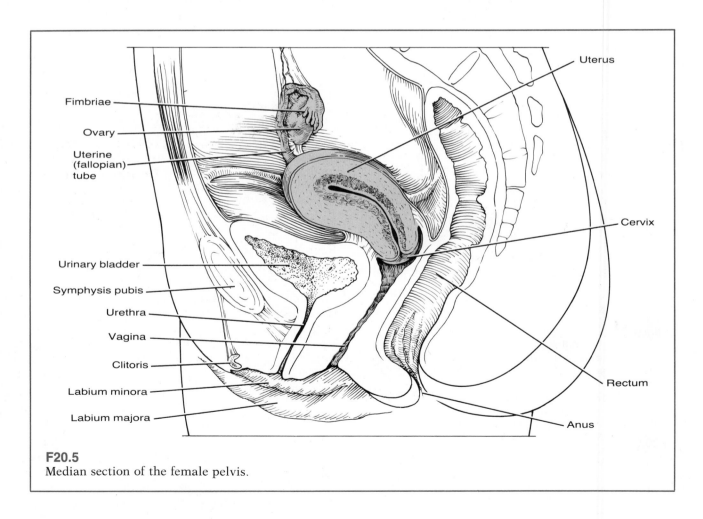

F20.5
Median section of the female pelvis.

VAGINA

The **vagina** is a canal that leads from the lower region, or cervix, of the uterus to the exterior of the body (F20.5; F20.6). The mucosa that lines the vagina proliferates during the menstrual cycle in a manner similar to the endometrial changes of the uterus, but there is considerably less development within the mucosa of the vagina than in the mucosa of the uterus.

PUBERTY IN THE FEMALE

On the average, females reach puberty a year earlier than males. In females, puberty is marked by the first episode of menstrual bleeding, which is called the *menarche*. At puberty, the vagina, uterus, and uterine tubes enlarge, and the deposition of fat in the breasts and hips causes them to become more prominent. In addition, the glandular portion of the breasts begins to develop. Pubic hair grows, and although the voice matures in the female, the change is not as dramatic as the voice change in males.

The events associated with puberty are largely the result of an increased production of female sex hormones (especially estrogens) by the ovaries, and they require the continued production of adequate levels of these hormones to sustain them. Prior to puberty, estrogens are secreted at very low levels, probably due mainly to the secretion of only small amounts of GnRH by the hypothalamus. At puberty, an alteration in brain function leads to increased GnRH release. The increased GnRH release stimulates the secretion of FSH and LH, which ultimately leads to an increased production of estrogens.

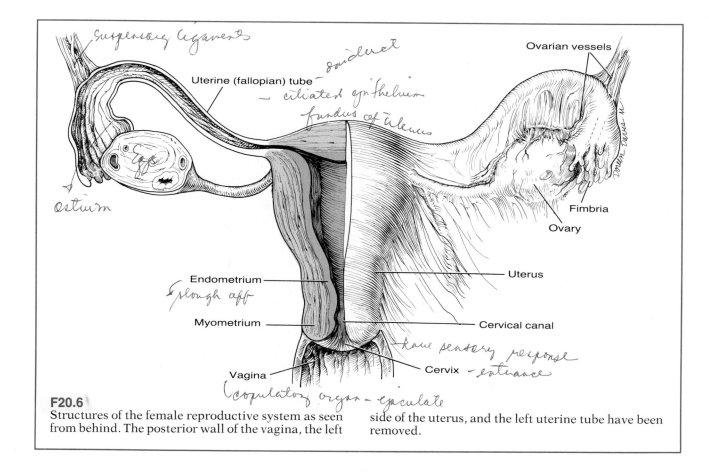

F20.6
Structures of the female reproductive system as seen from behind. The posterior wall of the vagina, the left side of the uterus, and the left uterine tube have been removed.

OOGENESIS

The production of ova within the ovaries is called **oogenesis,** and, as in the case of spermatogenesis, meiosis occurs during this process (F20.7). Initially, precursor cells called *oogonia* divide mitotically and thereby provide the cells that will develop into ova. Each oogonium is a diploid cell that contains 44 autosomes and two X sex chromosomes. By the time embryonic development is completed, several hundred thousand oogonia have undergone a growth phase and have become *primary oocytes*. The primary oocytes enter prophase of the first meiotic division, but do not complete it and are in first prophase at the time of birth. Each primary oocyte is located within a small sphere composed of a single layer of cells, thus forming a *primary follicle.*

It is not until puberty when the process of follicle growth and maturation begins to occur that a primary oocyte develops further. As a follicle matures, the primary oocyte within it completes the first meiotic division, producing two haploid cells of unequal size. The larger cell, called a *secondary oocyte*, receives half of the chromosomes and almost all of the cytoplasm. The smaller cell, called the first *polar body*, receives the other half of the chromosomes but very little of the cytoplasm. Thus, the secondary oocyte and the first polar body each contain 23 autosomes and one X sex chromosome. The secondary oocyte immediately begins the second meiotic division; however, it halts at metaphase and is released from the ovary in this stage when the follicle ruptures at ovulation.

The second meiotic division is not completed by the secondary oocyte unless it is fertilized. However, if fertilization does occur, meiosis is quickly completed. As was the case in the first meiotic division, the completion of the second meiotic division produces one large cell, the fully developed ovum, and one small cell, a second polar body. The first polar body also often undergoes the second meiotic division, producing two additional polar bodies. Therefore, during oogenesis, one ovum and as many as three polar bodies are produced from an oogonium. Beginning at puberty, one primary follicle generally matures fully, and consequently one secondary oocyte (commonly referred to as an ovum) is released at ovulation each month (approximately every 28 days).

THE OVARIAN CYCLE

The **ovarian cycle** consists of a series of events that occur within an ovary, including the development of follicles, the release of an ovum from a mature follicle at ovulation, and the formation of a structure called the corpus luteum. The length of the ovarian cycle varies from about 20 to 40 days,

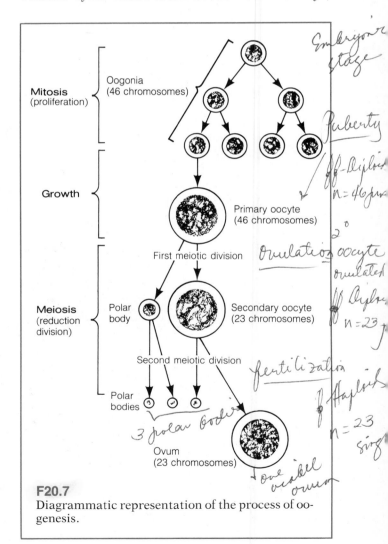

F20.7
Diagrammatic representation of the process of oogenesis.

averaging about 28 days. Therefore, the ovarian cycle is commonly considered to be a 28-day cycle, although in only about 30% of women is the cycle 27 or 28 days in duration. The ovarian cycle is closely correlated with another cycle called the menstrual cycle, which involves a series of changes that take place within the lining of the uterus (and to a lesser degree within the vagina).

DEVELOPMENT OF FOLLICLES

Under appropriate hormonal stimulation, some primary follicles undergo further development (F20.8). The single layer of cells that forms the wall of a primary follicle proliferates, and layers of ovarian cells called theca cells form around them. Such a follicle is referred to as a growing or maturing follicle. With continued growth, a clear, noncellular membrane called the *zona pellucida* forms within the growing follicle. The zona pellucida separates the developing ovum within the follicle from the surrounding follicle cells. As the solid, growing follicle becomes larger, a fluid-filled cavity called the *antrum* forms within

it. The antrum displaces the developing ovum and a few layers of follicle cells that surround it to one side of the follicle. As more fluid accumulates within the antrum, the follicle continues to enlarge and moves toward the surface of the ovary, where it produces a bulge. The follicle is now said to be mature and is called a *Graafian follicle.*

OVULATION

Under the proper hormonal conditions, the Graafian follicle ruptures and releases its ovum into the abdominopelvic cavity. This event is called **ovulation.** During ovulation, those follicle cells that directly surround the ovum remain with it. Therefore, the ovulated ovum is surrounded by a zona pellucida and a sphere of follicle cells that is now called the *corona radiata* (F20.9). The ovum, with its surrounding zona pellucida and corona radiata, is swept into the uterine tube by the flow of liquid from the abdominopelvic cavity into the tube. This flow is produced by the beating of the cilia attached to the fimbriae that surround the opening of the uterine tube.

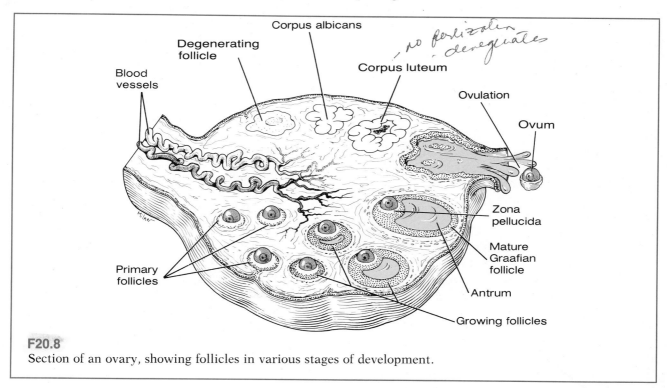

F20.8
Section of an ovary, showing follicles in various stages of development.

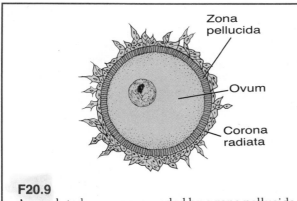

F20.9
An ovulated ovum surrounded by a zona pellucida and corona radiata.

It generally takes 10 to 14 days for a primary follicle to develop into a mature Graafian follicle. During this time, the developing ovum completes the first meiotic division and reaches metaphase of the second meiotic division, as described previously. The ovum is in this stage when it is released from the ovary at ovulation. Although a number of primary follicles begin to undergo further development during an ovarian cycle, only one usually matures fully and releases its ovum. The rest reach various stages of development and then regress and degenerate.

FORMATION OF THE CORPUS LUTEUM

Following ovulation, the ruptured Graafian follicle collapses. Shortly thereafter, the cells of the follicle increase in size and take on a yellowish color, due in part to the accumulation of lipid granules. The resulting structure is called a **corpus luteum** (that is, yellow body). The future of the corpus luteum depends on the fate of the ovum. If the ovum is not fertilized and pregnancy does not occur, the corpus luteum reaches its maximum development about eight to ten days after ovulation and then begins to degenerate. Eventually all that remains is a white connective-tissue scar called the *corpus albicans* (that is, white body). However, if the ovum is fertilized and pregnancy does occur, the corpus luteum continues to devel-

op, and it remains until near the end of the pregnancy.

HORMONAL CONTROL OF OVARIAN FUNCTION

Gonadotropin-releasing hormone (GnRH) from the hypothalamus and follicle-stimulating hormone (FSH) and luteinizing hormone (LH) from the pituitary are involved in the control of ovarian function. The release of GnRH is influenced by the female sex hormones (and probably by inputs from other brain areas). GnRH, in turn, stimulates the release of both FSH and LH.

The female sex hormones, the estrogens and progesterone, are produced by the ovaries. Although estrogens are secreted to some extent by various types of ovarian cells, the primary sources of estrogens are follicle cells and the corpus luteum. The corpus luteum is also the primary source of progesterone, and minute amounts of progesterone are secreted by follicle cells.

The plasma levels of FSH, LH, estrogens, and progesterone during a typical ovarian cycle are illustrated in F20.10. Follicle growth and maturation require FSH, LH, and estrogens, and appropriate levels of these hormones initiate the ovarian cycle and follicle development. During the first portion of the cycle, the level of estrogens is relatively low. However, as follicle development proceeds, the secretion of estrogens progressively increases (recall that follicle cells are a primary source of estrogens). At relatively low levels, such as are present during the first portion of the ovarian cycle, estrogens act in a negative-feedback fashion on the hypothalamus to inhibit the release of GnRH and on the pituitary to inhibit the release of FSH. This effect is thought to be responsible for the decline in the level of FSH that occurs during the second week of the cycle.

As follicle development continues and the level of estrogens increases rapidly, the effect of estrogens on the hypothalamus and pituitary changes. Relatively high levels—or perhaps rapidly rising levels—of estrogens, such as are present near the end of the second week of the

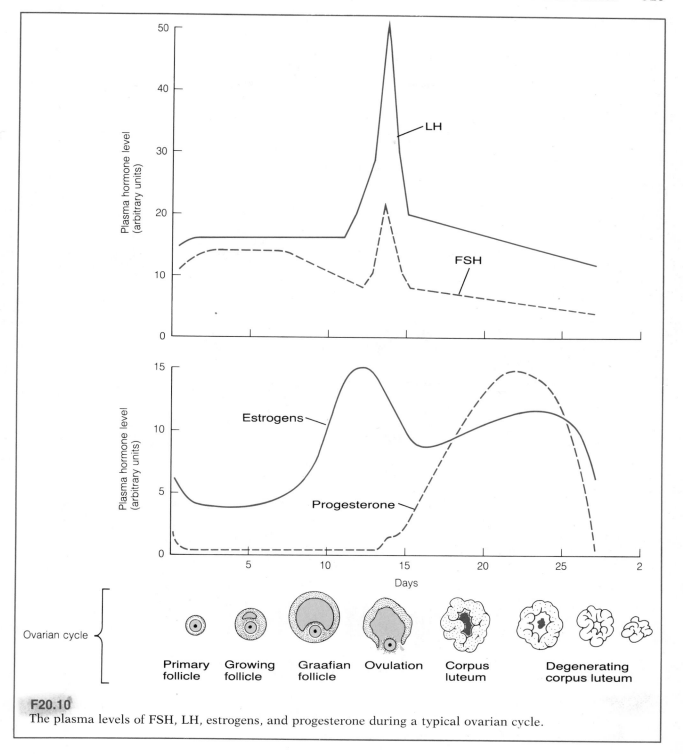

The plasma levels of FSH, LH, estrogens, and progesterone during a typical ovarian cycle.

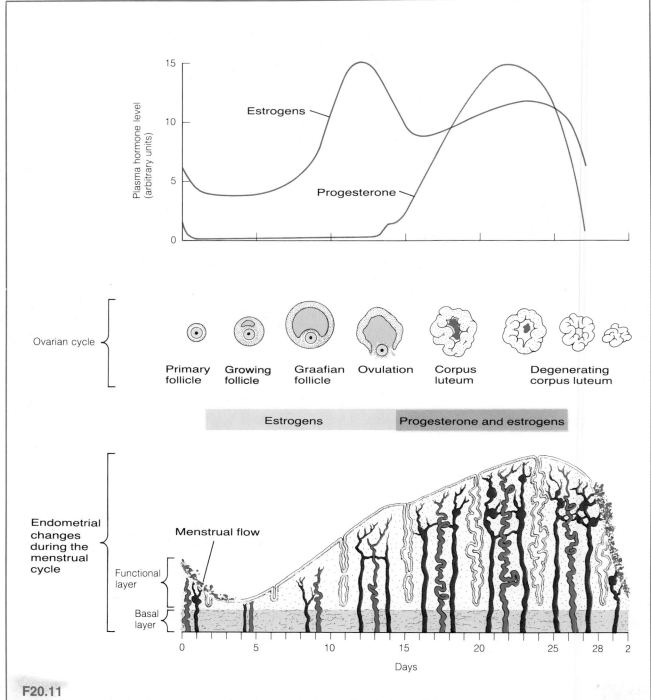

F20.11

The menstrual cycle, illustrating cyclic changes in the endometrium of the uterus.

cycle, stimulate the release of GnRH from the hypothalamus and enhance the sensitivity of the LH-releasing mechanisms of the pituitary to GnRH. As a consequence, there is a sharp increase in LH secretion (and a lesser increase in FSH secretion). The high level of LH leads to the final maturation of the follicle, the rupture of the follicle, and ovulation.

Near the time of ovulation, the level of estrogens declines, and some follicle cells secrete small but progressively increasing amounts of progesterone. These events appear to be due to changes that occur in the follicle before ovulation, which are caused at least in part by the sharp increase in LH secretion. In fact, some researchers propose that the rise in LH stimulates progesterone secretion, and that the progesterone in turn initiates a series of events that leads to the rupture of the follicle and ovulation. For example, progesterone is believed to stimulate the production of protein-digesting enzymes that weaken the follicle wall.

Following ovulation, the corpus luteum forms and begins to secrete substantial quantities of estrogens and progesterone. When present together, these hormones act in a negative-feedback fashion on the hypothalamus to inhibit the release of GnRH and perhaps also on the pituitary to inhibit the release of FSH and LH. As a result, the levels of FSH and LH decline during the latter portion of the ovarian cycle. If the ovum is not fertilized and pregnancy does not occur, the corpus luteum reaches its maximum development about eight to ten days after ovulation and then begins to degenerate. As it degenerates, the levels of estrogens and progesterone fall, removing the inhibition of FSH and LH secretion. As a consequence, the levels of FSH and LH increase, other follicles are stimulated to develop, and another ovarian cycle begins.

THE MENSTRUAL CYCLE

The **menstrual cycle** consists of a series of changes that occur in the endometrium of the uterus (and to a lesser degree within the vagina). It is closely associated with the ovarian cycle, and the estrogens and progesterone produced during the ovarian cycle control the events of the menstrual cycle (F20.11).

During the first portion of the ovarian cycle, the estrogens that are produced stimulate the functional layer of the endometrium of the uterus, causing it to proliferate. Straight tubular glands form, and blood vessels invade the new endometrial epithelium. Thus, the thickness and vascularity of the functional layer of the endometrium increase.

Following ovulation and the formation of the corpus luteum, both estrogens and progesterone are produced. Under the influence of these hormones, the endometrial glands continue to grow and begin to secrete small amounts of a fluid rich in glycogen. In addition, the arteries within the endometrium become enlarged and spiraled. In this condition, the endometrium is prepared to receive the embryo if fertilization should occur.

If fertilization and pregnancy do not occur, the corpus luteum degenerates and the levels of estrogens and progesterone decline. As the levels of these hormones decrease and their stimulatory effects are withdrawn, blood vessels of the endometrium undergo prolonged spasms (contractions) that reduce the bloodflow to the area of the endometrium supplied by the vessels. The resulting lack of blood causes the tissues of the affected region to degenerate. After some time, the vessels relax, which allows blood to flow through them again. However, capillaries in the area have become so weakened that blood leaks through them. This blood and the deteriorating endometrial tissue are discharged from the uterus as the menstrual flow. As a new ovarian cycle begins and the level of estrogens rises, the functional layer of the endometrium undergoes repair and once again begins to proliferate.

FEMALE SEXUAL RESPONSES

During sexual stimulation, the responses of the female, like those of the male, are governed both by psychological stimuli and by tactile stimulation of the genital organs and other areas of the body. The neural pathways over which sexual stimuli are carried in the female are basically the same as in the male, and the motor impulses that cause the various responses to sexual stimulation are carried over autonomic (that is, sympathetic and parasympathetic) nerves.

During sexual stimulation, the *clitoris* (which contains erectile tissue and which is located anterior to the opening of the vagina) becomes engorged with blood in a manner similar to that of the penis. Erectile tissue in the area of the vaginal opening also becomes engorged with blood. This occurrence narrows the vaginal orifice and facilitates the stimulation of the penis during intercourse. In addition, glands located around the cervix of the uterus and the area of the vaginal opening secrete mucus, which lubricates the epithelial lining of the vagina. The lubrication permits easy entrance of the penis into the vagina and facilitates rhythmic massaging stimulation of the female external genital organs as well as of the penis.

When sexual stimulation reaches sufficient intensity, the female undergoes an orgasm. The physiological changes that occur during an orgasm in the female are similar to those that occur in the male, except that there is no ejaculation in the female. The question as to whether a female orgasm in some way facilitates fertilization is not yet adequately answered. Certainly it is not necessary for a female to experience an orgasm for fertilization to occur.

PREGNANCY

For pregnancy to occur following sexual intercourse, spermatozoa must reach the upper region of the uterine tube, where fertilization usually takes place. The fertilized ovum must then be transported to the uterus and implant within the endometrium, and a placenta must form.

FERTILIZATION

The ovum remains capable of being fertilized for about 12 to 24 hours following ovulation, and sperm remain viable within the female reproductive tract for about 24 to 72 hours after ejaculation. Therefore, for fertilization to occur, sperm should be placed within the female tract no earlier than 72 hours before ovulation and no later than 24 hours after ovulation.

Once within the vagina, spermatozoa travel through the uterus and reach the upper portion of the uterine tubes rather quickly, often requiring only minutes to do so. The rate at which the sperm are transported is too rapid to be due solely to the motility of the sperm, and it has been suggested that the prostaglandins found in semen assist sperm transport by stimulating reverse peristaltic contractions of the smooth muscles of the uterus and uterine tubes. The semen deposited in the vagina during sexual intercourse usually contains several hundred million sperm. However, sperm mortality is high, and only a few thousand reach the uterine tubes. This high mortality rate is one reason why large numbers of sperm are required for pregnancy to occur.

During their passage through the uterus and the uterine tubes, the sperm are believed to undergo a process called *capacitation*, which makes them capable of fertilizing an ovum. The process of capacitation is not clearly understood, but it apparently involves an alteration of the sperm membrane that allows the sperm to release enzymes upon contact with the corona radiata surrounding the ovum. The enzymes—for example, hyaluronidase and proteases—are believed to contribute to the breakdown of portions of the zona pellucida and the material (hyaluronic acid) that holds the cells of the corona together, thus facilitating access to the ovum.

One of the sperm contacts the cell membrane of the ovum, fuses with it, and passes through into the cytoplasm (F20.12). Following this, the ovum completes the second meiotic division, and the nuclei (now called pronuclei) of the sperm and ovum unite to form a single nucleus. Thus, the ovum is fertilized, and the fertilized ovum, or *zygote*, contains the full complement of 46 chromosomes.

The interaction between the sperm that enters the ovum and the cell membrane of the ovum brings about changes in the zona pellucida (and perhaps in the cell membrane of the ovum) that prevent the entry of additional sperm. As a consequence of this interaction, enzymes are released from secretory vesicles of the ovum. Some of these enzymes prevent the movement of sperm through the zona pellucida by breaking down binding sites for sperm within the zona. Other enzymes may cause the zona pellucida (or perhaps the cell membrane of the ovum) to become impenetrable.

SEX DETERMINATION

The genetic sex of an individual is determined at the time of fertilization, even though the individual's sex does not become apparent until about the eighth week of development. If an ovum is fertilized by a sperm that contains an X sex chromosome, a female (XX) normally develops. If an ovum is fertilized by a sperm that contains a Y sex chromosome, a male (XY) develops.

DEVELOPMENT OF THE BLASTOCYST AND IMPLANTATION

The zygote, which begins to divide immediately, passes along the uterine tube to the uterus, a journey that requires about three to four days. By the time the uterus is reached, or shortly thereafter, enough cell divisions have occurred that the zygote has developed into a fluid-filled sphere of

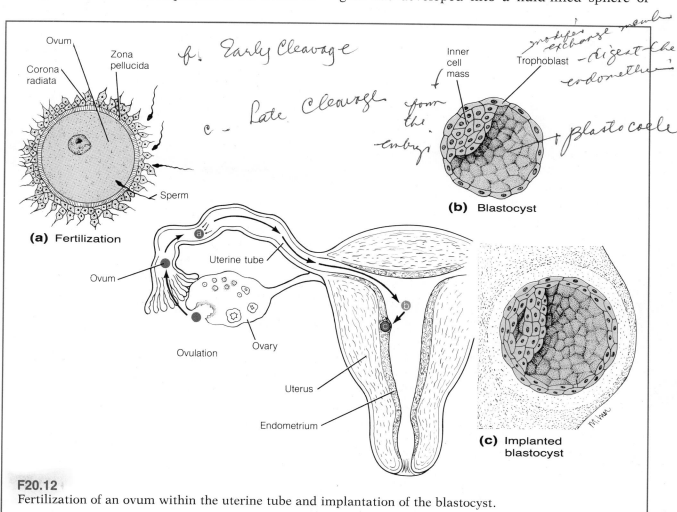

F20.12
Fertilization of an ovum within the uterine tube and implantation of the blastocyst.

cells called a **blastocyst** (F20.12). The outer sphere of cells of the blastocyst is called the *trophoblast*, and the blastocyst contains an accumulation of cells called the *inner cell mass*. The embryo develops from the inner cell mass.

The blastocyst remains free within the lumen of the uterus for several days before it implants in the endometrium. During this period, it obtains nourishment from the uterine fluids and continues to undergo cell division. At the same time, the zona pellucida disintegrates, allowing the trophoblast to expand. About seven days after ovulation, the blastocyst undergoes implantation into the endometrium (F20.12). By this time the endometrium of the uterus has been prepared for implantation by estrogens and progesterone. Upon contact with the endometrium, the cells of the trophoblast release enzymes that digest cells of the endometrium, thus eroding the endometrial surface. This erosion makes additional fluids and nutrients available to the blastocyst and allows the blastocyst to burrow into the endometrium. As the blastocyst erodes deeper, the endometrium grows over it, and within a few days the blastocyst is completely implanted within the endometrium. For about seven days following implantation, the embryo obtains nutrients from the destruction of endometrial cells by the trophoblast. Eventually, nutrients are obtained through the placenta, but erosion by the trophoblast continues to supply a significant fraction of the embryo's nutrients for the first two months of development.

THE PLACENTA

The **placenta** is an organ that, from the third month to the time of birth some six months later, supplies nutrients to and removes wastes from the developing individual (who is called a *fetus* after the second month). The placenta is also an endocrine gland that produces a number of hormones.

Following the implantation of the blastocyst in the uterine endometrium, the erosion of the endometrium (and vessels within the endometrium) by the trophoblast forms sinuses that become filled with maternal blood. With continued growth, trophoblastic projections become organized into fingerlike villi that extend into the blood-filled si-

nuses. Blood vessels develop within the villi, and blood from the developing individual begins to flow through the vessels (F20.13).

Under normal circumstances, there is no mixing of maternal and fetal blood within the placenta. The maternal blood remains within the blood-filled sinuses, and the fetal blood remains within the fetal blood vessels. However, oxygen, carbon dioxide, glucose, amino acids, fatty acids, various ions, and other substances can move in one direction or another between the maternal blood and fetal blood either by diffusion or by active transport. The fetal blood enters the placenta through vessels called umbilical arteries and passes through capillaries within the villi where the exchange of substances occurs (F20.13). It then leaves the placenta and returns to the fetus through umbilical veins. The umbilical vessels travel between the fetus and the placenta in the umbilical cord. The maternal blood enters the placenta through the uterine arteries, flows through the blood-filled sinuses, and leaves the placenta through the uterine veins.

MAINTENANCE OF THE ENDOMETRIUM

Because of its role in providing nutrients to and removing wastes from the developing individual, the endometrium of the uterus must be maintained in a highly developed state if pregnancy is to continue. Thus, the menstrual and ovarian cycles must be interrupted: menses must not occur, and the development of additional follicles within the ovaries must be suppressed. In order to maintain the endometrium, there must be high levels of estrogens and progesterone within the blood (F20.14). In addition to maintaining the endometrium, the estrogens and progesterone inhibit the secretion of FSH and LH, thus preventing the further development of any follicles.

During the early period of pregnancy, the required estrogens and progesterone are produced by the corpus luteum. In the absence of pregnancy, the corpus luteum begins to degenerate about eight to ten days after ovulation. However, if pregnancy occurs, the corpus luteum remains and continues to secrete estrogens and pro-

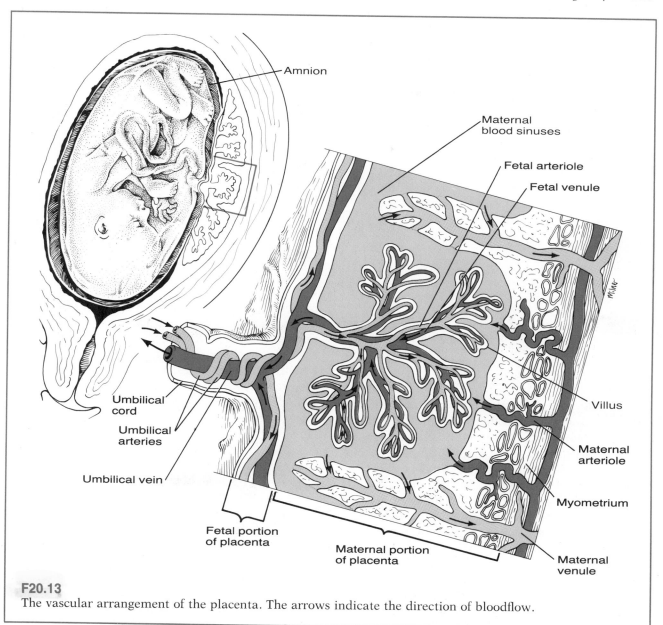

F20.13
The vascular arrangement of the placenta. The arrows indicate the direction of bloodflow.

gesterone. Although it is not completely understood why the corpus luteum persists during pregnancy, one factor that causes it to continue to secrete its hormones is *human chorionic gonadotropin* (*HCG*). This hormone, which is produced by the blastocyst and the placenta, has properties that are very similar to those of LH. HCG is se-

creted as early as the second week of pregnancy, reaches its peak during the third month, and then sharply declines and remains at a low level throughout the remainder of the pregnancy.

Another source of estrogens and progesterone during pregnancy is the placenta. By about the end of the second month, the placenta has devel-

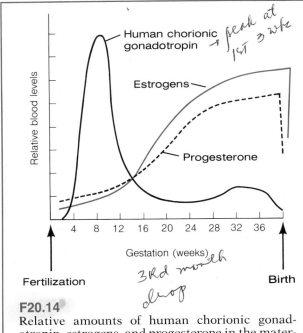

F20.14
Relative amounts of human chorionic gonadotropin, estrogens, and progesterone in the maternal blood during pregnancy.

oped to the point where it is secreting sufficient estrogens and progesterone to maintain the endometrium. Moreover, the substantial increases in the levels of estrogens and progesterone that occur as pregnancy progresses are due almost entirely to the secretion of these hormones by the placenta. Thus, beyond the second month, the continuance of the pregnancy no longer depends on the secretion of estrogens and progesterone by the corpus luteum.

GESTATION

The **gestation** (pregnancy) period usually lasts about 280 days from the beginning of the last menstrual period. Within the uterus, the fetus develops within a fluid-filled sac called the *amnion* (see F20.13). The *amniotic fluid* maintains a constant fetal temperature, protects the fetus against injury, and allows the fetus to move freely. During gestation, the total mass of the uterus increases to accommodate the growing fetus. In the later stages of pregnancy, the uterus occupies such a large portion of the abdominopelvic cavity that it may exert pressure on the rectum and the urinary bladder, causing constipation and frequent urination (F20.15).

The presence of a growing fetus places extra demands on a woman's systems. For example, a pregnant woman generally forms somewhat more urine than she did before she was pregnant because her kidneys must process the excretory products of the fetus as well as her own excretory products. Diet is especially important during pregnancy, and an expectant mother's diet must include sufficient amounts of vitamins, minerals, proteins, and other substances to supply both her own needs and those of the fetus.

TWINS

Twins occur in about 1 out of every 85 births. There are two types of twins, identical and fraternal, each the result of a different series of events. *Identical*, or *monozygotic*, *twins* occur when an inner cell mass that develops from a single fertilized ovum separates into two masses, each of which gives rise to a complete embryo. Because identical twins are the result of the fertilization of a single ovum by a single spermatozoon, they are always the same sex and have identical genes. However, they may have separate placentae, depending on the stage of development at which the inner cell mass separates.

Fraternal, or *dizygotic*, *twins* occur when two ova are released during a single ovarian cycle and both are subsequently fertilized by separate spermatozoa. Fraternal twins, therefore, can be of the opposite sex and they can differ just as much as any siblings. Fraternal twins always have separate placentae.

PARTURITION

Parturition is the process by which the fetus is expelled from the uterus. In the last three months of gestation, the uterus undergoes weak, irregular contractions. During the final weeks of pregnancy, the contractions gradually become stronger and more frequent. Then, within a few hours, the contractions become quite strong and occur about every 30 minutes. This marks the beginning of labor.

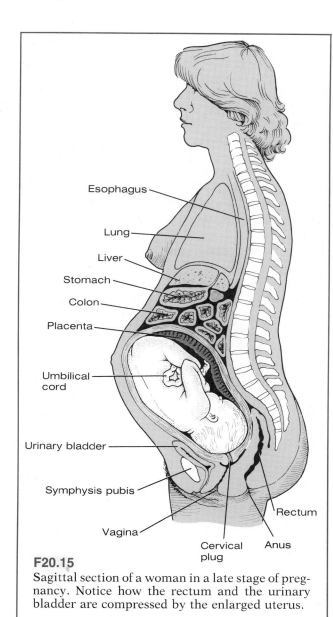

F20.15

Sagittal section of a woman in a late stage of pregnancy. Notice how the rectum and the urinary bladder are compressed by the enlarged uterus.

Labels in figure: Esophagus, Lung, Liver, Stomach, Colon, Placenta, Umbilical cord, Urinary bladder, Symphysis pubis, Vagina, Cervical plug, Anus, Rectum

LABOR

During *labor,* uterine contractions travel in peristaltic waves from the upper portion of the uterus downward toward the cervix, and each wave becomes progressively weaker as it approaches the cervix. As labor progresses, the contractions become stronger and more frequent, eventually oc-

curring only one to three minutes apart. The uterine contractions propel the fetus toward the cervical canal and cause the cervix of the uterus to dilate. In addition, the uterine contractions exert pressure on the amniotic sac. When the pressure becomes great enough, the amnion bursts, releasing the amniotic fluid, which escapes through the vagina.

DELIVERY

In over 90% of births, the head of the fetus is downward so that the head is delivered first. This manner of delivery allows the head to act as a wedge, forcing open the cervical and vaginal canals as contractions of the uterus (assisted by contractions of the abdominal muscles) move the fetus out of the uterus. The head-downward position also has the advantage of allowing the fetus to breathe safely even before it is completely free of the birth canal.

Within a few minutes after delivery, the uterus contracts to a very small size. This contraction separates the placenta from the wall of the uterus and expels it as the afterbirth. Bleeding is generally minimal during this process because the placental blood vessels completely constrict, and the contraction of the uterus squeezes closed the uterine vessels that supply blood to the placenta.

Within about five weeks following birth, the uterus returns to its normal size, and the endometrial surface heals. Menstrual cycles generally resume by about the end of the sixth week, provided the mother is not nursing her child.

FACTORS ASSOCIATED WITH PARTURITION

It is still not certain what initiates parturition. However, a number of different factors are believed to be involved, among which are the following:

1. The hormone oxytocin, which is released by the lobus nervosus of the pituitary gland, is particularly effective at stimulating uterine contractions late in pregnancy, although it is less effective early in pregnancy. Before the onset of labor, oxytocin levels are relatively low in both maternal and fetal (umbilical artery) blood. In very early labor, the oxytocin level

in maternal blood remains low, but the level in fetal blood rises. In advanced labor, the oxytocin levels in both maternal and fetal blood are elevated.

2. Prostaglandins that can stimulate uterine smooth muscle are synthesized by the placenta, and their release increases during labor. Moreover, there is evidence that oxytocin stimulates prostaglandin production by the placenta.

3. The concentration of oxytocin receptors in the uterus and the placenta increases during pregnancy and is high at parturition. In experimental animals, it has been found that estrogens and uterine distension can increase the concentration of oxytocin receptors. During pregnancy, the level of estrogens gradually increases (see F20.14), and near the time of the birth the fetus grows rapidly, accelerating uterine distension.

The preceding observations have resulted in the following theory to explain the initiation of parturition. During pregnancy, the concentration of oxytocin receptors increases greatly, probably under the influence of rising estrogen levels and uterine distension. Although there is no substantial elevation in circulating oxytocin at the onset of labor, the rising concentration of receptors increases the oxytocin sensitivity of the uterus to the point where relatively strong uterine contractions are stimulated. In addition, oxytocin (much of which may come from the fetus) binds to placental receptors, stimulating prostaglandin production. The released prostaglandins diffuse to the uterine muscle and enhance the oxytocin-induced contractions.

As uterine contractions move the fetus downward toward the birth canal, the fetus stimulates the cervix of the uterus. This stimulation triggers a neuroendocrine reflex that further strengthens the uterine contractions. Nerve impulses transmitted from the cervix to the hypothalamus cause the lobus nervosus of the maternal pituitary to

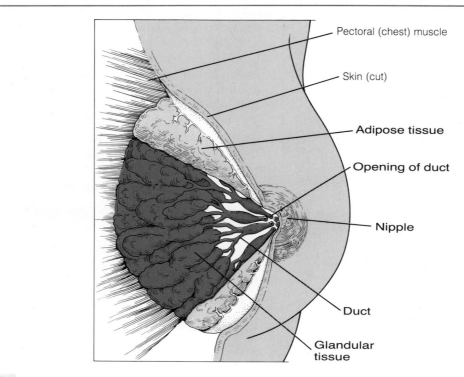

F20.16
Anterior view of a partially dissected left breast.

Pectoral (chest) muscle

Skin (cut)

Adipose tissue

Opening of duct

Nipple

Duct

Glandular tissue

release additional oxytocin, which enhances the contractions of the uterine muscle. Moreover, the stimulation of the cervix also seems to enhance uterine contractions by way of signals transmitted either neurally or along the uterine muscle to the body of the uterus. The enhanced uterine contractions push the fetus even more forcefully against the cervix, leading to the release of more oxytocin and still stronger uterine contractions. Thus, as described in Chapter 1, these activities act in a positive-feedback fashion to produce progressively stronger uterine contractions that ultimately expel the fetus from the uterus.

DISORDERS OF PREGNANCY

ECTOPIC PREGNANCY

The blastocyst normally implants in the upper portion of the uterine endometrium. However, in some cases, implantation occurs at a site other than the uterus. Such an occurrence is called an *ectopic pregnancy*. The most common ectopic pregnancy is a tubal pregnancy, which occurs within a uterine tube.

Ectopic pregnancies can endanger the life of the mother. They frequently cause internal hemorrhages because the trophoblast of the blastocyst destroys tissue at the site of implantation. The endometrium, which is the normal site of implantation, can withstand the destructive effects of the trophoblast, but most tissues simply break down. The lack of room for expansion is another danger in ectopic pregnancies. In the case of a tubal pregnancy, for example, the uterine tube may rupture as the embryo grows.

PLACENTA PREVIA

Occasionally, when implantation occurs in the lower regions of the uterus, the placenta covers the inner opening of the cervical canal. This condition is called *placenta previa*—that is, placenta leading the way. In this position, the placenta may irritate the cervix, which causes the uterus to contract and can result in a spontaneous abortion. If the pregnancy goes to term, the placenta is expelled before the baby is born, which causes the

mother to hemorrhage severely. The expulsion of the placenta prior to the expulsion of the fetus can also cause the fetus to begin breathing while still within the birth canal.

LACTATION

Lactation, or the production of milk by the breasts (mammary glands), requires the interaction of several hormones.

A fully developed breast is composed of glandular tissue surrounded by adipose tissue (F20.16). The glandular tissue consists of little sacs of secretory cells called *alveoli* that secrete milk into epithelium-lined ducts. The ducts carry the milk to the surface of the breast at the *nipple*. The alveoli and the initial portions of the ducts are surrounded by specialized contractile cells called *myoepithelial cells*.

PREPARATION OF THE BREASTS FOR LACTATION

Prior to puberty, the breasts are largely undeveloped. At puberty, the increased level of estrogens stimulates the growth and branching of the duct system within the breasts, and there is considerable breast enlargement due to fat deposition. The increased level of progesterone at puberty also contributes to breast growth. However, the alveoli do not develop completely, and the changes that occur at puberty do not result in milk production.

During pregnancy, the breasts enlarge still more, and the alveoli develop fully. These changes are due to the stimulatory effects of the high concentrations of estrogens and progesterone during pregnancy, as well as to the influence of the pituitary hormone *prolactin* and a hormone produced by the placenta called *human placental lactogen* (*chorionic somatomammotropin*). Prior to puberty, the prolactin level is low. However, estrogens stimulate prolactin secretion, and the prolactin level increases at puberty and rises still higher during pregnancy. (It is not known if estrogens influence prolactin secretion by way of the hypothalamus or by a direct action on the pituitary.) Human placental lactogen begins to be produced by about the fifth week of pregnancy, and the level of this hormone increases throughout pregnancy. Besides facilitating breast development, human placental lactogen exhibits properties similar to those of growth hormone. It promotes the incorporation of amino acids into protein, mobilizes fats for energy, and favors a relatively high blood-glucose level. These effects are believed to be useful in promoting the development of the fetus.

PRODUCTION OF MILK

Prolactin is the major hormone responsible for milk production. However, despite its relatively high levels (and the full development of the breasts) during pregnancy, significant milk production does not occur. This lack of milk production is due to the fact that the high levels of estrogens and progesterone present during pregnancy inhibit the milk-producing action of prolactin on the breasts (even though estrogens enhance prolactin secretion and act with it in promoting breast growth and development). At parturition, the loss of the placenta leads to a decline in the levels of estrogens and progesterone, and prolactin is able to stimulate milk production.

Following parturition, a reflex response is important in stimulating prolactin secretion. Mechanical stimulation of the nipples by the suckling infant initiates nerve impulses that are transmitted to the hypothalamus. The nerve impulses ultimately lead to prolactin release either by inhibiting the release of prolactin-inhibiting factor or by stimulating the release of a prolactin-releasing factor. Thus, even though the basal level of prolactin secretion returns to the prepregnancy level several months after parturition, an approximately tenfold surge in prolactin secretion lasting about one hour occurs each time the mother nurses her infant. The prolactin acts on the breasts to provide milk for the next nursing period. If the mother continues to nurse, milk production can continue for several years. However, lactation ceases if nursing is stopped.

MILK LET-DOWN REFLEX

The milk produced by the mammary glands tends to remain within the alveoli unless the specialized myoepithelial cells that surround the alveoli contract and force it into the ducts. The milk let-down reflex, which is initiated by suckling, leads to the contraction of the myoepithelial cells and the ejection of milk into the ducts of both breasts—not just the one being nursed—within a minute after suckling begins. In this reflex, the stimulation of the nipple by suckling sends nerve impulses to the hypothalamus, which leads to the release of oxytocin from the lobus nervosus of the pituitary gland (F20.17). Oxytocin causes the myoepithelial cells to contract, and this ejects the milk into the ducts, where it is easily obtainable by the nursing infant.

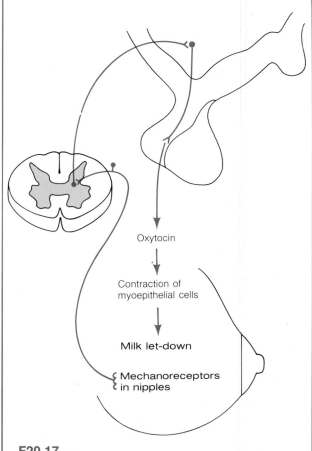

Oxytocin

Contraction of myoepithelial cells

Milk let-down

Mechanoreceptors in nipples

F20.17
The milk let-down reflex. Stimulation of the nipples elicits the release of oxytocin from the neurohypophysis of the pituitary.

Higher brain centers also influence milk letdown. For example, it is not unusual for a nursing mother to leak milk in response to the sound of a crying infant, even prior to physical contact between her and the child.

MENOPAUSE AND THE FEMALE CLIMACTERIC

At about age 50, the ovarian and menstrual cycles gradually become irregular. Ovulation fails to occur during many of the irregular cycles, and in most women the cycles cease altogether over the next several months or at most, a few years. The cessation of the menstrual cycle is referred to as **menopause,** and the entire period is called the **female climacteric.**

The female climacteric is thought to be caused by an inability of the ovaries to respond to hormonal signals, most probably due to a shortage of follicles resulting from their ovulation or degeneration during the reproductive years. As a result, the production of estrogens and progesterone is quite low, and estrogen-dependent tissues such as the genital organs and breasts gradually atrophy. During this period, the production of FSH and LH by the pituitary is quite high, apparently because of a partial release of these hormones from the negative-feedback influences of estrogens. Common symptoms during the female climacteric include sensations of warmth accompanied by sweating ("hot flashes"), irritability, fatigue, and anxiety.

COMMON PATHOLOGIES OF THE REPRODUCTIVE SYSTEMS

VENEREAL DISEASES

The most prevalent diseases of the reproductive systems of both males and females are the *venereal diseases.* These are infectious diseases that are spread through sexual contact.

Gonorrhea

Gonorrhea is an inflammation of the mucous membranes of the urogenital tract or the rectum caused by the bacterium *Neisseria gonorrhoeae.*

The inflammation usually results in a discharge of pus from the urethra and painful urination. In females, the cervix of the uterus and the uterine tubes can also become infected and inflamed, and *N. gonorrhoeae* is a frequent cause of pelvic inflammatory disease (see page 536). However, it is not unusual for a woman to have gonorrhea and transmit it through sexual contacts, yet have no apparent symptoms herself. In males, the spread of the gonococcus within the reproductive tract can lead to inflammation of the prostate, the seminal vesicles, and the epididymis.

Syphilis

Syphilis is a venereal disease caused by the bacterium *Treponema pallidum.* The bacterium usually produces a lesion called a chancre where it enters the body. Most commonly the chancre is on the penis or in the vagina. The chancre soon heals, and there may be no other symptoms for several weeks. However, during that time the infection spreads throughout the body by way of the bloodstream. After about six weeks, a skin rash accompanied by fever and aching joints develops. These secondary symptoms then disappear, and the disease enters a latent (inactive) period that can last for many years. During this latent period, the body may develop an immunity to the bacterium and destroy it, or the bacterium may spread to many different sites including the nervous and vascular systems, as well as various organs, causing damage to these structures. This latter occurrence produces severe and varied symptoms, depending on the structures affected. Syphilis can be detected by several different blood tests, one of which is called the Wassermann test.

Genital Herpes

Genital herpes is an increasingly common venereal disease. It is caused by the herpes simplex virus Type 2 (and sometimes Type 1), and is characterized by painful blisters on the reproductive organs. After some time the lesions heal, but they can recur periodically for years. In addition, many researchers believe the herpes simplex virus is implicated in the occurrence of cervical cancer.

MALE DISORDERS

Prostate Conditions

Although disorders of the male reproductive system are not restricted to the prostate gland, it is a structure that is frequently affected. Prostatitis is an inflammation of the prostate that is often due to bacterial infection. In this condition, the gland is swollen and tender, and in severe cases, abscesses can form. The prostate gland is also a common site of cancerous tumors, and an enlargement of the gland can compress the urethra, making urination difficult.

Impotence

Impotence is a fairly common male disorder, in which a man is unable to attain an erection of the penis, or to retain an erection long enough to complete sexual intercourse. Impotence can result from physical disorders of the vascular or nervous systems, and it is often a consequence of psychological or emotional problems.

Infertility

Male infertility is the inability of the male to fertilize the ovum. It is caused by inadequate production of spermatozoa, the production of abnormal or nonmotile spermatozoa, or an obstruction that prevents the delivery of spermatozoa from the testes to the female vagina. The production of normal sperm can be interfered with by exposure to x-rays, malnutrition, and certain diseases including mumps.

FEMALE DISORDERS

Abnormal Menstruation

Abnormalities of the menstrual cycle, which are among the most common disorders of the female reproductive system, can result from infections of the reproductive organs or malfunctions of the ovaries or the pituitary gland. Moreover, emotional or psychological factors are often involved.

Amenorrhea (the complete absence of menstrual periods) can be caused by disorders of the ovaries, the pituitary, or the hypothalamus. Dysmenorrhea (painful menstruation) is usually associated with strong contractions of the uterus. Its cause is not known with certainty, but prostaglandins appear to be involved, and antiprostaglandin drugs are often effective in treating this condition. Moreover, dysmenorrhea often disappears following the first pregnancy.

Pelvic Inflammatory Disease

Pelvic inflammatory disease (PID) is an infection of the pelvic organs of the female reproductive tract, including the uterus and uterine tubes. It is caused by bacteria—usually gonococci, streptococci, or staphylococci—that generally reach the pelvic structures by passing through the vagina to the uterus and the uterine tubes. In some cases, the bacteria reach the pelvic structures through the bloodstream from distant sites of infections.

Tumors

Tumors that develop within the female reproductive system can be either malignant (cancerous) or benign (noncancerous), and malignant tumors can metastasize—that is, spread to other locations within the body.

Ovarian tumors can be solid or they can occur as cysts (hollow sacs), which frequently contain fluid. Ovarian cysts are usually noncancerous, but a high percentage of solid ovarian tumors are malignant.

Tumors of the uterus and particularly the cervix are quite common. Fortunately, these tumors can be diagnosed early in their development by means of a simple test called a Pap smear. This test involves the removal (by a swab) of cells from the cervix and the surrounding area. The cells are then examined microscopically for signs of malignancy. With this procedure, cancerous cells can often be detected before any symptoms have appeared. If identified early, uterine tumors can be removed surgically or destroyed with radiation treatments. Since many women now have regular gynecological examinations that include Pap smears, the number of deaths due to uterine cancer has declined steadily.

Tumors are also common in the breasts, particularly after age 30, and breast cancer kills more women than any other type of cancer. Breast tumors, whether malignant or benign, can be best detected by regular manual self-examination. Breast tumors are often removed surgically, and they are also treated by radiotherapy and chemotherapy.

BIRTH CONTROL

Over the years much effort has been expended in attempting to develop a convenient, safe, and effective method of birth control, and a variety of birth control methods are now in widespread use.

BIRTH CONTROL METHODS AVAILABLE TO THE MALE

Coitus Interruptus

Coitus interruptus is the withdrawal of the penis just before the male orgasm so that ejaculation does not occur within the female tract. It is not a reliable method because it requires an unnatural response and perfect timing by the male.

Condom

A *condom* is a very thin sheath of rubber, plastic, or animal membranes that is pulled onto the erect penis just before intercourse. Condoms are an effective method of preventing fertilization as long as they do not tear or slip off after orgasm and allow semen to enter the vagina. They have the added advantage of providing protection against venereal disease.

Vasectomy

Vasectomy is a form of surgical sterilization in which each ductus deferens is tied and cut. This procedure does not interfere with normal ejaculation, but spermatozoa are prevented from entering the semen.

BIRTH CONTROL METHODS AVAILABLE TO THE FEMALE

Rhythm Method

The *rhythm method* of birth control relies upon a woman's ability to determine when she will ovulate and to abstain from sexual intercourse for several days preceding and following ovulation.

This can be an effective method if the menstrual cycle is regular and if the period of abstinence is long enough. However, menstrual cycles can vary in length, and ovulation is difficult to detect. One means that is used to detect ovulation is the measurement of body temperature. Generally, there is a slight rise in body temperature following ovulation.

Diaphragm

A *diaphragm* consists of a thin hemispherical dome of rubber or plastic with a spring margin. It covers the cervix, thereby preventing the entrance of sperm into the uterus. A diaphragm, which is initially fitted by a physician, is generally used in combination with a spermicidal cream or jelly. Although a properly fitted diaphragm is quite effective, its use requires the woman to predict sexual intercourse so the diaphragm will be in place when needed, or else she must interrupt the prelude to the act while she inserts it.

Intrauterine Device

An *intrauterine device* (IUD) consists of a small spiral, ring, or loop made of plastic, stainless steel, or copper that is inserted into the uterine cavity by a physician. Once in place, the IUD can remain within the uterus for a long period of time.

An IUD does not affect the menstrual cycle, but it does prevent the implantation of the blastocyst into the endometrium. Some women cannot tolerate an IUD and expel it from the uterus. There have been so many health problems with some forms of IUDs that they have been removed from public use.

Oral Contraceptives

Oral contraceptives, better known as "the pill," are extremely effective methods of preventing pregnancy. There are several types of pills available, most of which contain a combination of synthetic estrogenlike and progesteronelike substances. When taken daily for the first 20 or 21 days of the menstrual cycle, these substances apparently prevent the rise in luteinizing hormone that leads to ovulation. Even if ovulation should occur (as it apparently does in some cases), oral contraceptives can still prevent pregnancy, presumably by thickening and chemically altering the cervical mucus so that it is more hostile to sperm, and by making the uterine endometrium less receptive to implantation. In some women, oral contraceptives have undesirable side effects, the most serious of which is a tendency to form blood clots.

Chemical Contraception

Various types of douches, foams, suppositories, creams, and jellies that are introduced into the vagina either just before or after intercourse are available. Most of these prevent fertilization by acting as physical barriers to sperm and by serving as spermicides. They are most effective when used in combination with a diaphragm.

Tubal Ligation

Tubal ligation is a form of surgical sterilization in which the uterine tubes are tied and severed. This procedure prevents spermatozoa from reaching the ovum to fertilize it, and it prevents the ovum from reaching the uterus.

Induced Abortion

In recent years, induced abortion of the embryo or fetus has become a more common method of birth control, although the moral issues involved continue to be debated. Induced abortion involves using various means to tear loose the implanted embryo or fetus from the wall of the uterus. The detachment of the embryo or fetus is generally accomplished by scraping, by saline solution rinses, or by vacuum aspiration (suction).

HUMAN GENETICS

The study of heredity in humans is called **human genetics,** and a person's hereditary material is the DNA of the cells. The DNA provides a program, or set of instructions, that determines personal characters such as eye color or blood type. The DNA program is subdivided into informational units called **genes,** which control specific characters by presiding over the synthesis of chemical components of a person's cells. Each gene is a segment of a DNA chain, and most genes operate by specifying the sequences in which amino acids are joined to form polypeptides or proteins (see Chapter 3). Once they are formed, the polypeptides or proteins act as enzymes, as hormones, and in other ways to influence particular characters (the actual form of a character that is expressed in a person—for example, brown eyes or blue eyes—is referred to as a *trait*).

HUMAN CHROMOSOMES

Each human somatic cell (all cells except the reproductive cells) is a diploid cell that contains 46 chromosomes. Two of the chromosomes are **sex chromosomes** (two X chromosomes in females and one X and one Y chromosome in males). The remaining 44 chromosomes are called **autosomes.** The 44 autosomes can be grouped into 22 pairs of similar-appearing chromosomes (F20.18). One member of each pair contains genes derived from the person's father, and the other member of each pair contains genes derived from the person's mother. Each pair comprises a set of **homologous chromosomes.** The two sex chromosomes (XX) of the female are also homologous, but the two sex chromosomes of the male (XY) are not.

Genes are arranged linearly along the chromosomes, and each gene occupies a specific position, or **locus,** on a particular chromosome. Homologous chromosomes each possess genes that control the same characters.

Genes that control the same character and that occupy the same locus on homologous chromosomes are called **alleles.** If a gene (an allele) for a particular character on one chromosome of an homologous pair is identical to the gene (the allele) for that character on the other chromosome of the pair, the person is **homozygous** for the character. If the two genes differ from one another, the person is **heterozygous** for the character.

DOMINANCE

In some cases, one allele for a particular character can mask, or suppress, the effect of another allele for the character. In such a case, the first allele is said to be *dominant*, and the masked allele is referred to as *recessive*. In a homozygous person who possesses two dominant alleles for a particular character (a homozygous dominant), the effect of the dominant alleles is expressed, and the person is said to display a dominant trait with regard to the character. In a homozygous person who possesses two recessive alleles for the character (a homozygous recessive), the effect of these alleles is expressed, and the person displays a recessive trait. In a heterozygous person who possesses both dominant and recessive alleles for the character, the effect of the dominant allele is expressed, and the effect of the recessive allele is masked. Consequently, the person displays the dominant trait.

It is not always the case that one allele for a particular character masks the effects of a second allele for the character; that is, sometimes neither allele for a character on homologous chromosomes is completely dominant or completely recessive. In such a situation, the effects of both alleles are often expressed to some degree, and the person displays a trait that is intermediate between the traits that would be evident if one allele or the other were the only form of the gene present.

GENOTYPE AND PHENOTYPE

A person's actual genetic makeup is called his or her **genotype.** The outward expression of the genotype—that is, the expressed traits—is the person's **phenotype.** In the case of dominant and recessive alleles, for example, homozygous dominant, homozygous recessive, and heterozygous individuals have different genotypes, but homozygous dominant and heterozygous individuals have the same phenotype.

INHERITANCE PATTERNS

Genetic traits can be passed from one generation to the next. For example, the ability to roll one's tongue is inherited in the manner expected for a dominant allele located on an autosomal chromo-

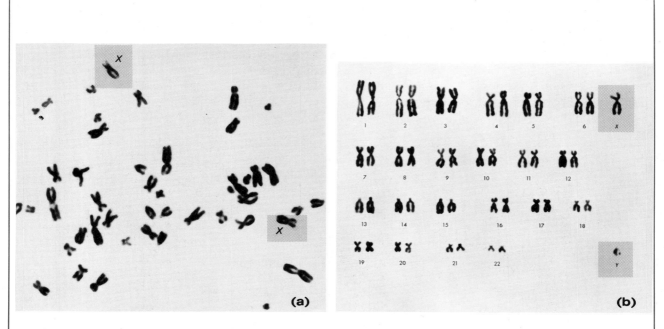

F20.18
Human chromosomes. (a) Chromosomes of a female with X sex chromosomes indicated. (b) Chromosomes of a male with homologous chromosomes arranged in pairs. X and Y are sex chromosomes. Note X and Y chromosomes in color boxes.

some. (The dominant allele for tongue rolling will be indicated by *T* and the recessive allele by *t*.) Thus, a person who is either homozygous dominant (*TT*) or heterozygous (*Tt*) will be able to roll his or her tongue, but a person who is homozygous recessive (*tt*) will not.

When a heterozygous individual who possesses both dominant and recessive alleles for tongue rolling forms haploid reproductive cells, the events of meiosis separate the two alleles in such a manner that half the gametes produced contain the dominant allele (*T*), and half contain the recessive allele (*t*). If a heterozygous male mates with a heterozygous female, there is a 50% probability that the sperm that fertilizes the ovum will contain the dominant allele (*T*) and a 50% probability it will contain the recessive allele (*t*). Similarly, there is a 50% probability that the ovum will contain the dominant allele and a 50% probability it will contain the recessive allele. This relationship and an indication of the possible genotypes of the fertilized ovum can be indicated by use of a *Punnett square*, as shown in F20.19. From this Punnett square, it is evident that there is a 25% probability that the person developing from the fertilized ovum will be homozygous dominant (*TT*) for tongue rolling, a 50% probability that the person will be heterozygous (*Tt*), and a 25% probability that the person will be homozygous recessive (*tt*). Phenotypically, there is a 75% proba-

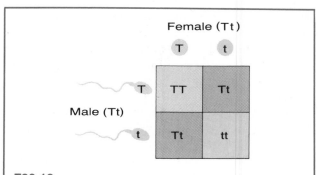

F20.19
Use of a Punnett square to indicate possible genotypes of a fertilized ovum when a male and female who are both heterozygous for tongue-rolling ability mate with one another.

bility that the person will display tongue-rolling ability (recall that a person who is either homozygous dominant or heterozygous for tongue rolling will be able to roll his or her tongue), and a 25% probability that the person will not.

The pattern of inheritance of a genetic trait in the members of different generations of a family can be recorded by constructing a *pedigree pattern*. Such patterns are used in human genetics to examine the manner in which particular traits are inherited. For example, the pedigree pattern for tongue-rolling ability illustrated in F20.20 indicates that this ability is inherited in the manner expected for a dominant allele.

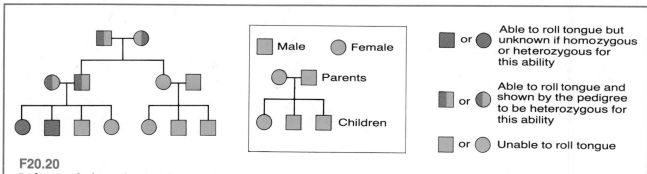

F20.20
Pedigree of a hypothetical family, illustrating the inheritance of tongue-rolling ability. The facts that parents with tongue-rolling ability can have children without this ability, and parents without this ability have only non-tongue-rolling children indicates that tongue-rolling ability is inherited in the manner expected for a dominant allele. Moreover, the fact that parents who both have tongue-rolling ability have a child without this ability shows that the parents are heterozygous for the ability.

X-LINKED TRAITS

Only a single X chromosome occurs in males. Consequently, the effect of a recessive allele located on the X chromosome of a male is generally expressed because there is no dominant allele to mask it (the Y chromosome apparently has few or none of the same genes as the X chromosome, and, therefore, does not mask recessive alleles on the X chromosome). As a result, the effects of recessive alleles located on an X chromosome are more frequently expressed in males than in females. This situation is evident in the case of the blood-clotting abnormality hemophilia A (see Chapter 11), which is inherited in the manner expected for a recessive allele located on the X chromosome. To display this abnormality, females must have the recessive allele on both X chromosomes, but males must have the allele only on their single X chromosome. If a heterozygous female who has the recessive allele for hemophilia A on one of her X chromosomes mates with a normal male, none of their female children will have hemophilia A. However, there is a 50% probability that their male children will have this disorder (F20.21). In this case, neither parent has hemophilia A, but the mother is termed a *carrier* because she possesses a recessive allele for hemophilia A that can be passed to her children.

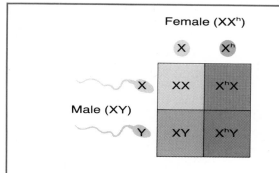

F20.21
Use of a Punnett square to indicate possible genotypes of a fertilized ovum when a normal male who does not have hemophilia A mates with a female who does not have hemophilia A, but who does have the recessive allele for hemophilia A on one of her X chromosomes (indicated by X^h).

EXPRESSION OF GENETIC POTENTIAL

In previous sections it was assumed that the effect of a dominant allele for a particular character is always expressed, and, consequently, that a person who possesses such an allele will display the dominant trait with regard to the character. Although this assumption is generally true, cases are known in which a person possessing a dominant allele for a character does not display the dominant trait. These cases serve to emphasize the fact that individual genes operate as part of a person's overall genetic makeup, and, at times, the activities of genes for other characters may modify, or otherwise influence, the expression of the effect of a particular gene. Moreover, environmental factors can also influence the expression of the effects of particular genes. For example, a person may have the genetic capacity to attain great height, but the actual attainment of this height depends on the availability of the proper amounts and types of food. Thus, particular genes represent po-

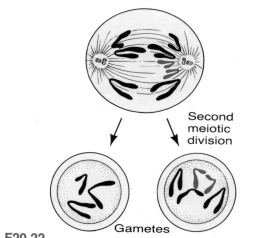

F20.22
Nondisjunction occurring during meiosis. During the second division sequence of meiosis, both chromatids of a double-stranded chromosome go to one pole of the dividing cell. This movement results in the production of some gametes that have an extra chromosome and others without this chromosome. The nondisjunct chromosomes are shown in color.

tentialities whose full effects and complete expression may be influenced by the activities of other genes and by environmental factors.

GENETIC DISORDERS

Many diseases and other abnormalities have genetic origins. Some of these are the result of abnormal numbers of chromosomes in cells, and others are due to malfunctioning or altered genes.

Nondisjunction

A major cause of abnormal numbers of chromosomes in cells is a process called *nondisjunction*, which can occur during meiosis (F20.22). Because of an abnormal segregation of chromosomes during the second division sequence of meiosis, a gamete may receive either both copies of a particular chromosome or neither copy. If one of these abnormal gametes participates in fertilization, the fertilized ovum can have either three chromosomes of one kind (a condition known as *trisomy*) or only one of that kind (*monosomy*).

Down's Syndrome

Down's syndrome (sometimes called "mongolism") is a genetic disorder that is characterized by a fairly consistent pattern of physical features and almost always by mental retardation. The most common and extreme form of Down's syndrome is due to the presence of three copies of autosome number 21, a condition known as *trisomy 21*. Trisomy 21 occurs most often in children born to mothers over 35 years old.

Abnormal Numbers of Sex Chromosomes

When gametes with abnormal numbers of sex chromosomes are involved in fertilization, the result is a fertilized ovum with an incorrect number of X and/or Y chromosomes. A person developing from such a fertilized ovum is often abnormal. Table 20.1 lists some of the conditions associated with abnormal numbers of sex chromosomes.

Galactosemia

Galactosemia is a genetic disease that is inherited in the manner expected for a recessive allele lo-

Table 20.1 *Normal and Abnormal Numbers of Sex Chromosomes in Humans*

Chromosomes	Sex	Characteristics
XX	Female	Fertile
XY	Male	Fertile
XO (no Y)	Female	Turner's syndrome*
XXX	Female	Fertile or sterile
XXXX	Female	Fertile or sterile
XXY	Male	Klinefelter's syndrome*
XXXY	Male	Klinefelter's syndrome*
XYY	Male	Fertile

*Victims of Turner's and Klinefelter's syndromes fail to mature sexually (are sterile) and often show mental impairment.

cated on an autosomal chromosome. Babies afflicted with this condition cannot tolerate the milk sugar lactose. If they drink milk, they develop cataracts of the eyes and show symptoms of damage to the brain and liver. However, if newborn galactosemic babies are taken off milk soon after birth and fed special lactose-free formulas, they can develop normally.

AMNIOCENTESIS

When it is suspected that a child may be born with a genetic (or other) abnormality, it is extremely useful to be able to determine the condition of the fetus before birth. This can often be accomplished by a procedure called *amniocentesis* (F20.23). In this procedure the locations of the fetus and placenta are first determined by recording the echoes of *ultrasound* waves directed at the uterus through the pregnant woman's abdominal wall. A hollow needle is then inserted through the abdominal wall and uterus into the amniotic fluid surrounding the fetus. Some of the amniotic fluid and fetal cells present in the fluid are collected through the needle. The fluid and cells can then be subjected to microscopic examination and biochemical testing that can detect abnormalities such as incorrect numbers of chromosomes.

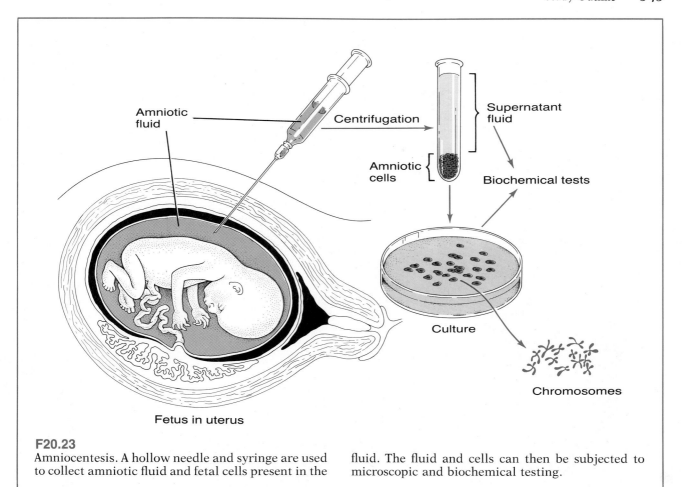

F20.23
Amniocentesis. A hollow needle and syringe are used to collect amniotic fluid and fetal cells present in the fluid. The fluid and cells can then be subjected to microscopic and biochemical testing.

STUDY OUTLINE

ORGANIZATION OF THE MALE REPRODUCTIVE SYSTEM pp. 511–514

Testes Male gonads in which sperm and testosterone are produced.

Epididymis Contains initial portion of duct system that transports sperm from a testis to the exterior of the body.

Ductus Deferens Continuation of epididymal duct; enters abdominopelvic cavity.

Penis Copulatory organ by which sperm-containing semen is placed within female reproductive tract.

Structures That Contribute to the Formation of Semen

SEMINAL VESICLES Membranous pouches that secrete a viscous fluid that contributes to the formation of semen.

PROSTATE GLAND Gland that secrets a thin, milky, alkaline fluid that contributes to the formation of semen.

BULBOURETHRAL GLANDS Small mucous glands whose secretion contributes to semen.

PUBERTY IN THE MALE p. 514
1. Puberty is the time of life at which reproduction becomes possible.
2. The events of puberty (for example, enlargement of the penis, scrotum, and testes; characteristic hair growth; voice changes) are a result of increased testosterone production by testes.
3. At puberty, an alteration in brain function occurs that leads to an increased release of GnRH, which ultimately results in an increased production of testosterone.

SPERMATOGENESIS p. 514
1. Spermatogenesis occurs within the seminiferous tubules of the testes.
2. Spermatogonia (diploid) develop into primary spermatocytes (diploid).
3. Primary spermatocytes undergo first meiotic division, forming secondary spermatocytes (haploid).
4. Secondary spermatocytes undergo second meiotic division, producing spermatids.
5. Spermatids develop into mature spermatozoa.

HORMONAL CONTROL OF TESTICULAR FUNCTION p. 516
1. FSH and LH release from the pituitary is stimulated by GnRH from the hypothalamus.
2. Testosterone inhibits LH release, mainly at level of pituitary.
3. FSH release is believed to be inhibited by inhibin from Sertoli cells.

MALE SEXUAL RESPONSES pp. 516–517

Erection The penis becomes enlarged and firm.
1. Erection results from vascular engorgement of spaces within the corpora cavernosa and corpus spongiosum of the penis.
2. Erection can be caused by physical stimulation (a reflex phenomenon), visual or psychic stimuli.

Emission and Ejaculation Results in the forceful expulsion of semen from the penis.

ORGANIZATION OF THE FEMALE REPRODUCTIVE SYSTEM pp. 517–519

Ovaries Female gonads in which ova, estrogens, and progesterone are produced.

Uterine (Fallopian) Tubes Tube that carries ovum from an ovary to the uterus.

Uterus An organ that receives a fertilized ovum and in which development normally takes place.

Vagina A canal that leads from the uterus to the exterior of the body.

PUBERTY IN THE FEMALE p. 519
1. The events of puberty (such as enlargement of the vagina, uterus, and uterine tubes; deposition of fat in the breasts and hips; characteristic hair growth) are largely a result of an increased production of female sex hormones (especially estrogens) by the ovaries.
2. At puberty, an alteration in brain function leads to an increased release of GnRH, which ultimately results in an increased production of estrogens.

OOGENESIS p. 520
1. Oogonia (diploid) develop into primary oocytes (diploid).
2. Primary oocytes undergo first meiotic division, forming secondary oocytes (haploid) and polar bodies.
3. Secondary oocyte begins second meiotic division, and completes it if fertilized to become a fully developed ovum.
4. Beginning at puberty, one secondary oocyte (commonly called an ovum) is usually released at ovulation each month.

THE OVARIAN CYCLE pp. 520–522

Development of Follicles
1. Under appropriate hormonal stimulation, some primary follicles undergo further development.
2. A zona pellucida forms within the growing follicle as does a fluid-filled cavity called the antrum.
3. A mature follicle is called a Graafian follicle.

Ovulation
1. Under proper hormonal conditions, the Graafian follicle ruptures and releases its ovum.
2. The ovulated ovum is surrounded by a zona pellucida and a sphere of follicle cells called the corona radiata.

Formation of the Corpus Luteum
1. Cells of ruptured Graafian follicle increase in size and take on a yellowish color, forming a corpus luteum.
2. If the ovum is not fertilized and pregnancy does not occur, the corpus luteum begins to degenerate about eight to ten days after ovulation.
3. If the ovum is fertilized and pregnancy does occur, the corpus luteum does not degenerate but remains until near the end of the pregnancy.

HORMONAL CONTROL OF OVARIAN FUNCTION pp. 522–525

1. Appropriate levels of FSH, LH, and estrogens initiate the ovarian cycle and follicle development.
2. As follicle development proceeds, the secretion of estrogens progressively increases.
3. At relatively low levels, estrogens inhibit the release of GnRH and FSH, which may explain the FSH decline during the second week of the cycle.
4. Relatively high levels—or perhaps rapidly rising levels—of estrogens stimulate GnRH release and enhance the sensitivity of the LH-releasing mechanism to GnRH. These occurrences result in a sharp increase in LH secretion, which leads to the final maturation of the follicle, the rupture of the follicle, and ovulation.
5. Estrogens and progesterone secreted by the corpus luteum inhibit GnRH and perhaps FSH and LH release.
6. As the corpus luteum degenerates, the levels of estrogens and progesterone fall, and the levels of FSH and LH rise. These occurrences stimulate other follicles to develop, and another cycle begins.

THE MENSTRUAL CYCLE p. 525

1. Estrogens produced during the first portion of the ovarian cycle stimulate the functional layer of the endometrium of the uterus to proliferate, and the thickness and vascularity of the layer increase.
2. Following ovulation, estrogens and progesterone from the corpus luteum cause further development of the endometrium.
3. As the corpus luteum degenerates and the levels of estrogens and progesterone fall, the functional layer of the endometrium degenerates and menstruation occurs.

FEMALE SEXUAL RESPONSES pp. 525–526

1. The clitoris becomes engorged with blood.
2. Erectile tissue in the area of the vaginal opening becomes engorged with blood.
3. The epithelial lining of the vagina becomes lubricated.

PREGNANCY pp. 526–530

Fertilization

1. Fertilization normally occurs in the upper portion of the uterine tube.
2. As they pass through the uterus and uterine tubes, the sperm are believed to undergo changes that make them capable of fertilizing an ovum.
3. When sperm contact the corona radiata, they release enzymes that help break down portions of the zona pellucida and the material that holds the cells of the corona together, thus permitting access to the ovum.
4. One sperm passes through the cell membrane of the ovum into the cytoplasm.
5. The ovum completes the second meiotic division and the nuclei of the sperm and ovum unite.

Sex Determination

1. If an ovum is fertilized by a sperm that contains an X sex chromosome, a female normally develops.
2. If an ovum is fertilized by a sperm that contains a Y sex chromosome, a male develops.

Development of the Blastocyst and Implantation

1. The blastocyst is a fluid-filled sphere of cells. The embryo develops from the inner cell mass of the blastocyst; the outer sphere of cells is the trophoblast.
2. After several days within the lumen of the uterus, the trophoblast cells release enzymes that digest the cells of the endometrium and the blastocyst implants in the uterine wall.

The Placenta

1. The placenta supplies nutrients to and removes wastes from the developing individual.
2. The placenta produces hormones.
3. There is normally no mixing of maternal and fetal blood in the placenta.

Maintenance of the Endometrium

1. High blood levels of estrogens and progesterone are required.
2. During the early period of pregnancy, the required estrogens and progesterone are produced by the corpus luteum.
3. By about the end of the second month, the placenta has developed to the point where it is secreting sufficient estrogens and progesterone to maintain the endometrium.

Gestation Normally lasts about 280 days from the beginning of the last menstrual period.

Twins

IDENTICAL TWINS Develop when an inner cell mass that develops from a single fertilized ovum separates.

FRATERNAL TWINS Develop when two ova are fertilized by separate spermatozoa.

PARTURITION   The process by which the fetus is expelled from the uterus.   **pp. 530–533**

Labor   Uterine contractions propel the fetus toward the cervical canal, and the cervix of the uterus dilates.

Delivery   Contractions of the uterus (assisted by contractions of the abdominal muscles) move the fetus out of the uterus.

Factors Associated with Parturition   It is still not certain what initiates parturition. Estrogens and prostaglandins are thought to be involved, and an increasing uterine sensitivity to oxytocin as pregnancy progresses is believed to be important.

DISORDERS OF PREGNANCY   p. 533

Ectopic Pregnancy   Implantation occurs at a site other than the uterus.

Placenta Previa   The placenta covers inner opening of cervical canal. The placenta is expelled before the baby is born, which causes the mother to hemorrhage.

LACTATION   pp. 533–535

Preparation of the Breasts for Lactation
1. At puberty, estrogens stimulate the growth and branching of the duct system within the breasts, and fat deposition leads to considerable breast enlargement. The increased level of progesterone at puberty also contributes to breast growth.
2. During pregnancy, the breasts become fully developed due to influences of high levels of estrogens and progesterone as well as prolactin and human placental lactogen.

Production of Milk
1. Prolactin is the major hormone responsible for milk production.
2. Mechanical stimulation of nipples by suckling initiates a reflex that stimulates prolactin secretion for some time after childbirth.

Milk Let-Down Reflex   Suckling reflexively causes oxytocin release, and oxytocin causes myoepithelial cells to contract, ejecting milk into ducts of breasts.

MENOPAUSE AND THE FEMALE CLIMACTERIC
1. Sexual cycles gradually cease; believed to be due to an inability of the ovaries to respond to hormonal signals.
2. Production of estrogens and progesterone is quite low and estrogen-dependent tissues gradually atrophy.

COMMON PATHOLOGIES OF THE REPRODUCTIVE SYSTEMS   pp. 535–537

Venereal Diseases

GONORRHEA   Inflammation of mucous membranes of genital tract or rectum caused by the bacterium *Neisseria gonorrhoeae*.

SYPHILIS   An infectious venereal disease caused by the bacterium *Treponema pallidum*.

GENITAL HERPES   Infection caused by herpes simplex virus Type 2 (and sometimes Type 1).

Male Disorders

PROSTATE CONDITIONS   Such as bacterial inflammation (prostatitis) and tumors.

IMPOTENCE   Inability to attain an erection or maintain it long enough for sexual intercourse to be accomplished.

INFERTILITY   Inability to fertilize an ovum.

Female Disorders

ABNORMAL MENSTRUATION   May be due to infections or glandular malfunctions, as well as to emotional and psychological factors.

PELVIC INFLAMMATORY DISEASE   Inflammation that can involve uterine tubes, ovaries, or peritoneum of abdominopelvic cavity.

TUMORS   Can occur in ovaries, uterus, or breasts.

BIRTH CONTROL   pp. 537–538

Birth Control Methods Available to the Male

COITUS INTERRUPTUS   Withdrawal of penis before ejaculation.

CONDOM   Cover pulled onto penis.

VASECTOMY   Ductus deferens is tied and cut.

Birth Control Methods Available to the Female

RHYTHM METHOD   Abstinence from sexual intercourse during fertile period.

DIAPHRAGM   Cover for cervix to prevent entrance of sperm into uterus.

INTRAUTERINE DEVICE   Device inserted into

uterus that prevents implantation of blastocyst.

ORAL CONTRACEPTIVES Pills containing synthetic estrogenlike and progesteronelike substances that inhibit ovulation, fertilization, or implantation.

CHEMICAL CONTRACEPTION Chemicals introduced into the vagina that act as physical barriers to sperm and as spermicides.

TUBAL LIGATION Tying and severing uterine tubes.

INDUCED ABORTION Removal of implanted embryo or fetus from uterus.

HUMAN GENETICS The study of heredity in humans. **pp. 538–542**

Human Chromosomes Each human somatic cell is a diploid cell that contains 46 chromosomes: 44 autosomes and 2 sex chromosomes. Genes that control the same character and that occupy the same locus on homologous chromosomes are called alleles.

Dominance In some cases, one allele for a particular character can mask, or suppress, the effect of another allele for the character. In such a case, the first allele is said to be dominant, and the masked allele is referred to as recessive.

Genotype and Phenotype A person's actual genetic makeup is called his or her genotype. The outward expression of the genotype—that is the expressed traits—is the person's phenotype.

Inheritance Patterns Genetic traits can be passed from one generation to the next, as exemplified by the inheritance of tongue-rolling ability.

X-Linked Traits Only a single X chromosome occurs in males, and the effects of recessive alleles located on the X chromosome are more frequently expressed in males than in females.

Expression of Genetic Potential Particular genes represent potentialities whose full effects and complete expression may be influenced by the activities of other genes and by environmental factors.

Genetic Disorders Many diseases and other abnormalities have genetic origins.

NONDISJUNCTION A major cause of abnormal numbers of chromosomes in cells. Occurs during meiosis.

DOWN'S SYNDROME The most common and extreme form is due to three copies of autosome 21. The syndrome is characterized by a fairly consistent pattern of physical features and almost always by mental retardation.

ABNORMAL NUMBERS OF SEX CHROMOSOMES Cause a number of different abnormalities.

GALACTOSEMIA Genetically based inability to tolerate the milk sugar lactose.

Amniocentesis A procedure in which amniotic fluid surrounding the fetus and fetal cells present in the fluid are collected. The fluid and cells can then be subjected to microscopic examination and biochemical testing that can detect abnormalities.

SELF-QUIZ

1. The interstitial cells of Leydig produce: (a) sperm; (b) testosterone; (c) luteinizing hormone.
2. The acidity of the semen counteracts the alkaline pH of the vagina. True or False?
3. In the male, prior to puberty: (a) the testes continually produce large amounts of testosterone; (b) there is little release of the follicle-stimulating hormone and luteinizing hormone from the pituitary; (c) the hypothalamus secretes large amounts of gonadotropin-releasing hormone.
4. Spermatids: (a) are haploid cells; (b) develop within the bulbourethral glands; (c) divide mitotically, giving rise to spermatogonia.
5. In the male, testosterone: (a) strongly stimulates FSH release; (b) inhibits LH release; (c) is not involved in sperm production.
6. The corpus luteum secretes: (a) luteinizing hormone; (b) follicle-stimulating hormone; (c) progesterone.
7. During the ovarian cycle, a high level of which hormone leads to the final maturation of the follicle, the rupture of the follicle, and ovulation? (a) prolactin;

(b) human chorionic gonadotropin; (c) luteinizing hormone.

8. During the first portion of the ovarian cycle, the functional layer of the endometrium of the uterus is stimulated to proliferate by: (a) progesterone; (b) estrogens; (c) endometriotropin.

9. Fertilization normally occurs within the: (a) ovary; (b) uterine tube; (c) uterus.

10. The interaction between the sperm that enters the ovum and the cell membrane of the ovum brings about changes in the zona pellucida that prevent the entry of additional sperm. True or False?

11. An ovum that is fertilized by a spermatozoon that contains a Y chromosome normally: (a) develops into a male; (b) develops into a female; (c) has an equal probability of developing into either a male or a female.

12. During implantation, the cells of the inner cell mass release enzymes that digest the cells of the trophoblast. True or False?

13. Maternal blood normally mixes with fetal blood within the placenta in order to exchange food materials and oxygen. True or False?

14. The placenta produces: (a) estrogens; (b) follicle-stimulating hormone; (c) luteinizing hormone.

15. During the early portion of pregnancy, the corpus luteum secretes: (a) progesterone; (b) human placental lactogen; (c) human chorionic gonadotropin.

16. During pregnancy, the corpus luteum degenerates about eight to ten days after implantation. True or False?

17. Oxytocin is particularly effective at stimulating uterine contractions early in pregnancy, but it is unable to do so late in pregnancy. True or False?

18. At puberty, the alveoli of the breasts develop fully and milk secretion begins. True or False?

19. Prolactin production: (a) occurs at a relatively high level prior to puberty; (b) ceases when the placenta is lost at parturition; (c) can be stimulated for some time following parturition by suckling.

20. During the female climacteric, the level of: (a) estrogens is relatively high; (b) FSH is relatively high; (c) progesterone is relatively high.

21. A person's actual genetic makeup is called his or her: (a) phenotype; (b) genotype; (c) heterotype.

22. In a male, the effects of recessive alleles located on the X chromosome are always masked by dominant alleles located on the Y chromosome. True or False?

APPENDIX 1
SELF QUIZ ANSWERS AND PAGE REFERENCES

CHAPTER 1

1. T *(p 3)*
2. F *(p 3)*
3. a *(p 4)*
4. c *(p 4)*
5. T *(p 4)*
6. c *(p 5)*
7. a *(p 7)*
8. F *(p 7)*
9. c *(p 7)*

CHAPTER 2

1. T *(p 13)*
2. c *(p 13)*
3. a *(p 13)*
4. b *(p 15)*
5. F *(p 19)*
6. c *(p 19)*
7. a *(p 21)*
8. T *(p 22)*
9. b *(p 25)*
10. c *(p 25)*
    g *(p 24)*
    b *(p 25)*
    e *(p 25)*
    i *(p 25)*
    a *(p 25)*
    f *(p 25, 26)*
    h *(p 26)*
    d *(p 26)*
11. F *(p 27)*
12. T *(p 30)*
13. F *(p 31)*
14. b *(p 32)*
15. c *(p 33)*
    d *(p 33)*
    a *(p 33)*
    b *(p 38)*
    e *(p 38, 39)*

16. c *(p 39)*
17. F *(p 41)*
18. b *(p 42)*
19. a *(p 43)*
20. c *(p 45)*

CHAPTER 3

1. c *(p 53)*
2. b *(p 54)*
   d *(p 54)*
   e *(p 54)*
   a *(p 56)*
   c *(p 56)*
3. b *(p 56)*
4. b *(p 57)*
5. a *(p 59)*
6. T *(p 59)*
7. b *(p 61)*
8. T *(p 61, 62)*
9. a *(p 63)*
10. c *(p 65)*
11. b *(p 66)*
12. a *(p 67)*
13. b *(p 67)*
14. d *(p 62)*
    e *(p 65)*
    a *(p 66)*
    b *(p 67)*
    c *(p 67)*
15. F *(p 69)*
16. c *(p 69)*
17. T *(p 60, 69)*
18. h *(p 72)*
    g *(p 72)*
    a *(p 72)*
    f *(p 72)*
    c *(p 73)*
    i *(p 72, 73)*
    e *(p 73)*
    b *(p 73)*

d *(p 73)*
19. F *(p 75)*
20. c *(p 76)*

CHAPTER 4

1. a *(p 83)*
2. b *(p 84)*
3. a *(p 83)*
4. b *(p 84)*
5. c *(p 84)*
6. F *(p 85)*
7. T *(p 85)*
8. b *(p 87)*
9. c *(p 88)*
10. F *(p 88)*
11. T *(p 88)*
12. a *(p 88, 89, 90)*
13. a *(p 90)*
14. c *(p 90)*
15. F *(p 90)*
16. c *(p 90)*
17. b *(p 91)*
18. T *(p 92)*
19. b *(p 92)*
20. F *(p 93)*

CHAPTER 5

1. a *(p 101)*
2. b *(p 101)*
3. c *(p 103)*
4. b *(p 102)*
5. b *(p 102)*
6. b *(p 104)*
7. c *(p 105)*
8. F *(p 105)*
9. a *(p 106, 107)*
10. F *(p 107)*
11. a *(p 107)*
12. b *(p 108)*

13. a *(p 109)*
14. F *(p 111)*
15. b *(p 111)*
16. T *(p 111)*
17. a *(p 114)*
18. T *(p 114)*
19. c *(p 115)*
20. F *(p 115)*

CHAPTER 6

1. T *(p 121)*
2. a *(p 121)*
3. c *(p 122)*
4. b *(p 124)*
5. a *(p 125)*
6. T *(p 127)*
7. b *(p 128)*
8. b *(p 129)*
9. F *(p 129)*
10. c *(p 129)*
11. c *(p 129)*
12. b *(p 131)*
13. c *(p 132)*
14. b *(p 132)*
15. c *(p 133)*
16. b *(p 133)*
17. T *(p 134)*
18. c *(p 141)*
19. b *(p 142)*
20. b *(p 121, 122, 143)*
21. a *(p 144)*
22. T *(p 142)*

CHAPTER 7

1. b *(p 152)*
2. c *(p 153)*
3. b *(p 153)*
4. T *(p 155)*
5. a *(p 157)*

6. b   *(p 157)*
7. b   *(p 158)*
8. b   *(p 158)*
9. c   *(p 160)*
10. h   *(p 155)*
    a   *(p 151, 152)*
    i   *(p 153)*
    b   *(p 155)*
    g   *(p 155)*
    c   *(p 160)*
    d   *(p 157)*
    e   *(p 168)*
    f   *(p 160, 161)*
11. T   *(p 161)*
12. b   *(p 166, 167)*
13. b   *(p 168)*
14. a   *(p 169)*
15. a   *(p 172)*
16. F   *(p 172)*
17. F   *(p 172)*
18. b   *(p 173)*
19. a   *(p 175)*
20. F   *(p 175)*
21. T   *(p 179)*

CHAPTER 8

1. a   *(p 186)*
2. F   *(p 186)*
3. b   *(p 187)*
4. c   *(p 188)*
5. a   *(p 190)*
6. F   *(p 190)*
7. b   *(p 192)*
8. a   *(p 195)*
9. F   *(p 196)*
10. T   *(p 198, 199)*
11. F   *(p 199)*
12. a   *(p 200)*
13. c   *(p 202)*
14. a   *(p 203)*
15. T   *(p 207)*
16. b   *(p 207)*
17. a   *(p 208)*
18. T   *(p 209)*
19. F   *(p 209)*
20. F   *(p 211)*
21. a   *(p 213)*
22. b   *(p 214)*
23. T   *(p 215)*
24. c   *(p 215, 216)*
25. b   *(p 216)*
26. T   *(p 218)*
27. F   *(p 219)*

CHAPTER 9

1. a   *(p 225)*
2. b   *(p 228)*
3. a   *(p 227)*
4. b   *(p 228)*
5. T   *(p 228)*
6. c   *(p 228)*
7. a   *(p 228)*
8. T   *(p 231)*
9. b   *(p 231)*
10. T   *(p 231)*
11. F   *(p 232)*
12. c   *(p 233)*
13. b   *(p 235)*
14. b   *(p 236)*
15. T   *(p 236, 237)*
16. a   *(p 237)*
17. a   *(p 237)*
18. T   *(p 238)*
19. b   *(p 239)*
20. c   *(p 239)*

CHAPTER 10

1. F   *(p 245)*
2. T   *(p 246)*
3. b   *(p 251)*
4. c   *(p 253)*
5. b   *(p 253)*
6. a   *(p 253)*
7. b   *(p 251)*
    f   *(p 253)*
    a   *(p 253)*
    d   *(p 251)*
    e   *(p 253)*
    c   *(p 253)*
8. c   *(p 254)*
9. F   *(p 256)*
10. f   *(p 256)*
    c   *(p 256)*
    a   *(p 256)*
    b   *(p 256)*
    d   *(p 256)*
    e   *(p 256)*
    g   *(p 256)*
11. c   *(p 257)*
12. a   *(p 258)*
13. F   *(p 258)*
14. c   *(p 260)*
15. c   *(p 260)*
16. c   *(p 261)*
17. a   *(p 261)*
18. T   *(p 262)*
19. a   *(p 262)*

20. b   *(p 264, 265)*
21. c   *(p 266, 267)*
22. T   *(p 265)*
23. c   *(p 266)*

CHAPTER 11

1. b   *(p 275)*
2. F   *(p 276)*
3. F   *(p 277)*
4. T   *(p 278)*
5. e   *(p 280)*
    b   *(p 276)*
    a   *(p 279)*
    c   *(p 276, 277)*
    d   *(p 278)*
6. a   *(p 281)*
7. T   *(p 282)*
8. a   *(p 283)*
9. a   *(p 283)*
10. b   *(p 283)*
11. a   *(p 283, 284)*
12. b   *(p 284)*
13. c   *(p 284)*
14. c   *(p 284)*
15. a   *(p 285)*
16. F   *(p 286–288)*
17. F   *(p 288)*
18. c   *(p 288)*
19. b   *(p 288)*
20. T   *(p 289)*

CHAPTER 12

1. c   *(p 296)*
2. a   *(p 298)*
3. c   *(p 301)*
4. F   *(p 304)*
5. a   *(p 304)*
6. a   *(p 307)*
7. T   *(p 305)*
8. T   *(p 307)*
9. b   *(p 308)*
10. a   *(p 308)*
11. T   *(p 310)*
12. c   *(p 310)*
13. b   *(p 311)*
14. F   *(p 311, 312)*
15. T   *(p 312)*
16. c   *(p 312)*
17. a   *(p 314)*
18. a   *(p 315)*
19. b   *(p 315)*
20. b   *(p 316)*

CHAPTER 13

1. c   *(p 324)*
2. c   *(p 325, 326)*
3. c   *(p 326)*
4. b   *(p 326)*
5. a   *(p 328)*
6. b   *(p 328)*
7. c   *(p 329)*
8. F   *(p 330)*
9. T   *(p 331)*
10. b   *(p 332)*
11. T   *(p 333)*
12. a   *(p 334)*
13. F   *(p 336)*
14. F   *(p 338)*
15. b   *(p 341)*
16. c   *(p 343)*
17. a   *(p 343)*
18. a   *(p 344)*
19. c   *(p 345)*
20. F   *(p 345)*

CHAPTER 14

1. T   *(p 353)*
2. c   *(p 355)*
3. a   *(p 355)*
4. b   *(p 357)*
5. a   *(p 357)*
6. b   *(p 359)*
7. T   *(p 357, 358)*
8. a   *(p 359, 360)*
9. F   *(p 360)*
10. b   *(p 361)*
11. c   *(p 361)*
12. F   *(p 362)*
13. c   *(p 362)*
14. a   *(p 364)*
15. b   *(p 365)*
16. c   *(p 367)*
17. b   *(p 367)*
18. T   *(p 367)*
19. b   *(p 369)*
20. a   *(p 371)*
21. T   *(p 371)*

CHAPTER 15

1. T   *(p 377)*
2. a   *(p 378)*
3. c   *(p 382)*
4. b   *(p 383)*
5. T   *(p 383)*

6. c *(p 383)*
7. a *(p 385)*
8. b *(p 385)*
9. c *(p 384)*
10. F *(p 384)*
11. a *(p 385, 386)*
12. T *(p 388)*
13. a *(p 388)*
14. a *(p 388)*
15. a *(p 391)*
16. T *(p 394)*
17. F *(p 394)*
18. c *(p 396)*
19. b *(p 397)*
20. b *(p 397)*

CHAPTER 16

1. a *(p 409)*
2. c *(p 413)*
3. F *(p 413)*
4. a *(p 414)*
5. a *(p 414)*
6. b *(p 417)*
7. c *(p 418)*
8. a *(p 418)*
9. c *(p 418)*
10. T *(p 419)*
11. a *(p 418)*
12. c *(p 419)*
13. F *(p 419)*
14. b *(p 419)*
15. c *(p 420)*
16. T *(p 422)*
17. e *(p 424, 425)*
    d *(p 428)*

f *(p 428)*
b *(p 428, 430)*
a *(p 422, 423)*
c *(p 428)*
18. a *(p 426)*
19. b *(p 431)*
20. T *(p 431)*

CHAPTER 17

1. c *(p 441)*
2. b *(p 441)*
3. F *(p 439)*
4. F *(p 445)*
5. b *(p 445)*
6. a *(p 448, 449)*
7. T *(p 450)*
8. F *(p 451)*
9. T *(p 451)*
10. a *(p 451)*
11. c *(p 453)*
12. c *(p 453)*
13. a *(p 454)*
14. b *(p 454, 457)*
15. F *(p 454)*
16. T *(p 457)*
17. a *(p 462)*
18. b *(p 463)*
19. b *(p 465)*
20. b *(p 465)*

CHAPTER 18

1. F *(p 469)*
2. a *(p 470)*

3. F *(p 472)*
4. F *(p 474)*
5. b *(p 475)*
6. c *(p 475)*
7. F *(p 476)*
8. b *(p 476)*
9. F *(p 474, 475)*
10. c *(p 477)*
11. T *(p 477)*
12. F *(p 480)*
13. a *(p 481)*
14. T *(p 481)*
15. c *(p 481, 482)*
16. c *(p 482)*
17. a *(p 482)*
18. F *(p 485)*
19. a *(p 488)*
20. c *(p 488)*

CHAPTER 19

1. a *(p 493)*
2. F *(p 496)*
3. T *(p 496)*
4. c *(p 497)*
5. T *(p 497)*
6. a *(p 498)*
7. b *(p 499)*
8. c *(p 499)*
9. F *(p 499)*
10. c *(p 500)*
11. a *(p 500)*
12. c *(p 500)*
13. c *(p 501)*
14. c *(p 504)*
15. F *(p 502)*

16. F *(p 503)*
17. a *(p 505)*
18. b *(p 506)*
19. b *(p 507)*
20. c *(p 507)*

CHAPTER 20

1. b *(p 511)*
2. F *(p 514)*
3. b *(p 514)*
4. a *(p 516)*
5. b *(p 516)*
6. c *(p 525)*
7. c *(p 525)*
8. b *(p 525)*
9. b *(p 526)*
10. T *(p 526, 527)*
11. a *(p 527)*
12. F *(p 528)*
13. F *(p 528)*
14. a *(p 529)*
15. a *(p 528)*
16. F *(p 528, 529)*
17. F *(p 531)*
18. F *(p 533)*
19. c *(p 534)*
20. b *(p 535)*
21. b *(p 539)*
22. F *(p 541)*

APPENDIX 2
WORD ROOTS, PREFIXES, SUFFIXES, AND COMBINING FORMS

PREFIXES AND COMBINING FORMS

a-, an- *absence or lack* acardia, lack of a heart; anaerobic, in the absence of oxygen

ab- *departing from; away from* abnormal, departing from normal

acou- *hearing* acoustics, the science of sound

acr-, acro- *extreme or extremity; peek* acrodermatitis, inflammation of the skin of the extremities

ad- *to or toward* adorbital, toward the orbit

aden-, adeno- *gland* adeniform, resembling a gland in shape

amphi- *on both sides; of both kinds* amphibian, an organism capable of living in water and on land

angi- *vessel* angiitis, inflammation of a lymph vessel or blood vessel

ant-, anti- *opposed to, preventing or inhibiting* anticoagulant, a substance that prevents blood coagulation

ante- *preceding; before* antecubital, in front of the elbow

arthr-, arthro- *joint* antropathy, any joint disease

aut-, auto- *self* autogenous, self-generated

bi- *two* bicuspid, having two cusps

bio- *life* biology, the study of life and living organisms

blast- *bud or germ* blastocyte, undifferentiated embryonic cell

broncho- *bronchus* bronchospasms, spasmodic contraction of bronchial muscle

bucco- *cheek* buccolabial, pertaining to the cheek and lip

caput- *head* decapitate, remove the head

carcin- *cancer* carcinogen, a cancer-causing agent

cardi-, cardio- *heart* cardiotoxic, harmful to the heart

cephal- *head* cephalometer, an instrument for measuring the head

cerebro- *brain, especially the cerebrum* cerebrospinal, pertaining to the brain and spinal cord

chondr- *cartilage* chondrogenic, giving rise to cartilage

circum- *around* circumnuclear, surrounding the nucleus

co-, con- *together* concentric, common center, together in the center

contra- *against* contraceptive, agent preventing conception

cost- *rib* intercostal, between the ribs

crani- *skull* craniotomy, a skull operation

crypt- *hidden* cryptomenorrhea, a condition in which menstrual symptoms are experienced but no external loss of blood occurs

cyt- *cell* cytology, the study of cells

de- *undoing, reversal, loss, removal* deactivation, becoming inactive

di- *twice, double* dimorphism, having two forms

dia- *through, between* diaphragm, the wall through or between two areas

dys- *difficult, faulty, painful* dyspepsia, disturbed digestion

ec, ex, ecto- *out, outside, away from* excrete, to remove materials from the body

en-, em- *in, inside* encysted, enclosed in a cyst or capsule

entero- *intestine* enterologist, one who specializes in the study of intestinal disorders

epi- *over, above* epidermis, outer layer of skin

eu- *well* euesthesia, a normal state of the senses

exo- *outside, outer layer* exophthalmos, an abnormal protrusion of the eye from the orbit

extra- *outside, beyond* extracellular, outside the body cells of an organism

gastr- *stomach* gastrin, a hormone that influences gastric acid secretion

glosso- *tongue* glossopathy; any disease of the tongue

hema-, hemato-, hemo- *blood* hematocyst, a cyst containing blood

hemi- *half* hemiglossal, pertaining to one half of the tongue

hepat- *liver* hepatitis, inflammation of the liver

hetero- *different or other* heterosexuality, sexual desire for a person of the opposite sex

hist- *tissue* histology, the study of tissues

hom-, homo- *same* homeoplasia, formation of tissue similar to normal tissue; homocentric, having the same center

hydr-, hydro- *water* dehydration, loss of body water

hyper- *excess* hypertension, excessive tension

hypno- *sleep* hypnosis, a sleeplike state

hypo- *below, deficient* hypodermic, beneath the skin; hypokalemia, deficiency of potassium

hyster-, hystero- *uterus or womb* hysterectomy, removal of the uterus; hysterodynia, pain in the womb

im- *not* impermeable, not permitting passage, not permeable

inter- *between* intercellular, between the cells

intra- *within, inside* intracellular, inside the cell

iso- *equal, same* isothermal, equal, or same, temperature

leuko- *white* leukocyte, white blood cell

lip-, lipo- *fat, lipid* lipophage, a cell that has taken up fat in its cytoplasm

macro- *large* macromolecule, large molecule

mal- *bad, abnormal* malfunction, abnormal functioning of an organ

mamm- *breast* mammary gland, breast

mast- *breast* mastectomy, removal of a mammary gland

meningo- *membrane* meningitis, inflammation of the membranes of the brain

meso- *middle* mesoderm, middle germ layer

meta- *beyond, between, transition* metatarsus, the part of the foot between the tarsus and the phalanges

metro- *uterus* metroscope, instrument for examining the uterus

micro- *small* microscope, an instrument used to make small objects appear larger

mito- *thread, filament* mitochondria, small, filamentlike structures located in cells

mono- *single* monospasm, spasm of a single limb

morpho- *form* morphology, the study of form and structure of organisms

multi- *many* multinuclear, having several nuclei

myelo- *spinal cord, marrow* myeloblasts, cells of the bone marrow

myo- *muscle* myocardium, heart muscle

narco- *numbness* narcotic, a drug producing stupor or numbed sensations

nephro- *kidney* nephritis, inflammation of the kidney

neuro- *nerve* neurophysiology, the physiology of the nervous system

ob- *before, against* obstruction, impeding or blocking up

oculo- *eye* monocular, pertaining to one eye

odonto- *teeth* orthodontist, one who specializes in proper positioning of the teeth in relation to each other

ophthalmo- *eye* ophthalmology, the study of the eyes and related disease

ortho- *straight, direct* orthopedic, correction of deformities of the musculoskeletal system

osteo- *bone* osteodermia, bony formations in the skin

oto- *ear* otoscope, a device for examining the ear

oxy- *oxygen* oxygenation, the saturation of a substance with oxygen

pan- *all, universal* panacea, a cure-all

para- *beside, near* paraphrenitis, inflammation of tissues adjacent to the diaphragm

peri- *around* perianal, situated around the anus

phago- *eat* phagocyte, a cell that engulfs and digests particles or cells

phleb- *vein* phlebitis, inflammation of the veins

pod- *foot* podiatry, the treatment of foot disorders

poly- *multiple* polymorphism, multiple forms

post- *after, behind* posterior, places behind a (specific) part

pre-, pro- *before, ahead of* prenatal, before birth

procto- *rectum, anus* proctoscope, an instrument for examining the rectum

psycho- *mind, psyche* psychogram, a chart of personality traits

pseudo- *false* pseudotumor, a false tumor

pyo- *pus* pyrocyst, a cyst that contains pus

retro- *backward, behind* retrogression, to move backward in development

sclero- *hard* sclerodermatitis, inflammatory thickening and hardening of the skin

semi- *half* semicircular, having the form of half a circle

steno- *narrow* stenocoriasis, narrowing the pupil

sub- *beneath, under* sublingual, beneath the tongue

super- *above, upon* superior, quality or state of being above others or a part

supra- *above, upon* supracondylar, above a condyle

sym-, syn- *together, with* synapse, the region of communication between two neurons

tachy- *rapid* tachycardia, abnormally rapid heartbeat

therm- *heat* thermometer, an instrument used to measure heat

tox- *poison* antitoxic, effective agaist poison

trans- *across, through* transpleural, through the pleura

tri- *three* trifurcation, division into three branches

viscero- *organ, viscera* visceroinhibitory, inhibiting the movements of the viscera

SUFFIXES

-able *able to, capable of* viable, ability to live or exist

-ac *referring to* cardiac, referring to the heart

-algia *pain in a certain part* neuralgia, pain along the course of a nerve

-ary *associated with, relating to* coronary, associated with the heart

-atresia *imperforate* proctatresia, an imperforate condition of the rectum or anus

-cide *destroy or kill* germicide, an agent that kills germs

-ectomy *cutting out, surgical removal* appendectomy, cuting out of the appendix

-emia *condition of the blood* anemia, deficiency of red blood cells

-ferent *carry* efferent nerves, nerves carrying impulses away from the CNS

-fuge *driving out* vermifuge, a substance that expels worms of the intestine

-gen *an agent that initiates* pathogen, any agent that produces disease

-gram *data that are systematically recorded, a record* electrocardiogram, a recording showing action of the heart

-graph *an instrument used for recording data or writing* electrocardiograph, an instrument used to make an electrocardiogram

-ia *condition* insomnia, condition of not being able to sleep

-iatrics *medical speciality* geriatrics, the branch of medicine dealing with disease associated with old age

-itis *inflammation* gastritis, inflammation of the stomach

-logy *the study of* pathology, the study of changes in structure and function brought on by disease

-lysis *loosening or breaking down* hydrolysis, chemical decomposition of a compound into other compounds as a result of taking up water

-malacia *soft* osteomalacia, a process leading to bone softening

-mania *obsession, compulsion* erotomania, exaggeration of the sexual passions

-odyn *pain* coccygodynia, pain in the region of the coccyx

-oid *like, resembling* cuboid, shaped as a cube

-oma *tumor* lymphoma, a tumor of the lymphatic tissues

-opia *defect of the eye* myopia, nearsightedness

-ory *referring to, of* auditory, referring to hearing

-pathy *disease* psychopathy, any disease of the mind

-phobia *fear* acrophobia, fear of heights

-plasty *reconstruction of a part, plastic surgery* rhinoplasty, reconstruction of the nose through surgery

-plegia *paralysis* paraplegia, paralysis of the lower half of the body or limbs

-rrhagia *abnormal or excessive discharge* metrorrhagia, uterine hemorrhage

-rrhea *flow or discharge* diarrhea, abnormal emptying of the bowels

-scope *instrument used for examination* stethoscope, instrument used to listen to sounds of various parts of the body

-statis *arrest, fixation* hemostasis, arrest of bleeding

-stomy *establishment of an artificial opening* enterostomy, the formation of an artificial opening into the intestine through the abdominal wall

-tomy *to cut* appendectomy, surgical removal of the appendix

-ty *condition of, state* immunity, condition of being resistant to infection or disease

-uria *urine* polyuria, passage of an excessive amount of urine

UNITS OF THE METRIC SYSTEM

Unit	Metric equivalent	Symbol	English equivalent
LINEAR MEASURE			
1 kilometer	= 1000 meters	km	0.62137 mile
1 meter	= 10 decimeters	m	39.37 inches
1 decimeter	= 10 centimeters	dm	3.937 inches
1 centimeter	= 10 millimeters	cm	0.3937 inch
1 millimeter	= 1000 microns	mm	
1 micron	= 1/1000 millimeter or 1000 millimicrons	μ	
1 millimicron	= 1 nanometer = 10 angstrom units	nm, mμ	no English equivalents
1 angstrom unit	= 1/100,000,000 centimeter	Å	
MEASURES OF CAPACITY			
1 kiloliter	=1000 liters	kl	35.15 cubic feet or 264.16 gallons
1 liter	=10 deciliters	l	1.0567 U.S. liquid quarts
1 deciliter	=100 milliliters	dl	0.03 fluid ounce
1 milliliter	volume of 1 g of water at standard temperature and pressure (STP)	ml	
MEASURES OF MASS			
1 kilogram	= 1000 grams	kg	2.2046 pounds
1 gram	= 100 centigrams	g	15.432 grains
1 centigram	= 10 milligrams	cg	0.1543 grain
1 milligram	= 1/1000 gram	mg	0.01 grain (about)
MEASURES OF VOLUME			
1 cubic meter	= 1000 cubic decimeters	m^3	
1 cubic decimeter	= 1000 cubic centimeters	dm^3	
1 cubic centimeter	= 1000 cubic millimeters	cm^3	
1000 cubic millimeters	= 1 milliliter (ml)	mm^3	

Abdomen *ab'-do-men* The portion of the body between the diaphragm and the pelvis.

Abortion *a-bor'-shun* Termination of a pregnancy before the embryo or fetus is viable outside the uterus.

Abscess *ab'-ses* A localized accumulation of pus and disintegrating tissue.

Absolute refractory period *ree-frac'-to-ree* Period following stimulation during which no additional action potential can be evoked.

Absorption *ab-sorp'-shun* Passage of a substance into or across a membrane or blood vessel.

Accommodation (1) Adaptation or adjustment by an organ or organism in response to differences or needs; (2) adjustment of the eye for seeing objects at various distances.

Acetylcholine *a-see-til-ko'-lene* A chemical transmitter substance released by certain nerve endings.

Acetylcholinesterase *es'-ter-ase* The enzyme that inactivates acetylcholine; present in muscle and nervous tissue.

Achondroplasia *a-kon-dro-play'-zee-ah* A condition, sometimes influenced by hereditary factors as well as hormonal levels, that produces a dwarf with short arms and legs but normal trunk and head.

Acid A substance that dissociates into hydrogen ions and anions when in an aqueous solution.

Acidosis *as-i-do'-sis* A condition in which the blood has a higher hydrogen ion concentration than normal with a decreased pH.

Acne Inflammatory disease of the skin.

Acromegaly *ak-ro-meg'-a-lee* An abnormal pattern of bone and connective-tissue growth characterized by enlarged hands, face, and feet and associated with excessive pituitary growth hormone that is secreted after growth in height has ceased.

Acrosome *ak'ro-some* Crescent-shaped structure molded to the nucleus and forming the anterior sperm head.

Actin *ak'-tin* A contractile protein of muscle fiber; myosin is another protein found in muscle.

Action potential An event occurring when a stimulus of sufficient intensity is applied to a neuron, allowing sodium ions to move into the cell and reverse the polarity.

Active site The catalytically active part of an enzyme.

Active state The development of tension by the contractile proteins of muscle.

Active transport Net movement of a substance across a membrane against a concentration gradient; requires release and use of energy.

Adaptation (1) Any change in structure, form, or habits to suit a new environment; (2) decline in the frequency of sensory nerve excitation when a receptor is stimulated continuously.

Addison's disease Condition resulting from abnormal, deficient secretion of adrenal cortical hormones.

Adenine *ad'-e-neen* A nucleic acid base; pairs with thymine in DNA and uracil in RNA.

Adenohypophysis *ad-e-no-high-pof'-i-sis* The part of the pituitary gland that includes the pars distalis, pars intermedia, and pars tuberalis.

Adenoids *ad'-e-noidz* Enlargement of the pharyngeal tonsils in children.

Adenosine diphosphate (ADP) *a-den'-o-sene di-fos'-fate* The substance formed when ATP is split apart and releases energy.

Adenosine triphosphate (ATP) *a-den'-o-sene tri-fos'-fate* The compound that is an important intracellular energy source.

Adipose *ad'-i-pos* Fatty.

Adrenal glands *ad-reen'-al* Hormone-producing glands located superior to the kidneys; each consists of medulla and cortex areas.

Adrenalin *ad-ren'-a-lin* Trademark name for epinephrine.

Adrenergic fibers *ad-ren-ur'-jick* Nerve fibers whose

terminals release norepinephrine or epinephrine upon stimulation.

Adrenocorticotropic hormone (ACTH) *a-dree'-no-kort-i-ko-tro'-pik* A hormone that influences the activity of the adrenal cortex and is released by the adenohypophysis of the pituitary.

Adventitia *ad-ven-tish'-yah* The outermost layer or covering of an organ.

Aerobic *ay-er-o'-bick* Requiring oxygen to live or grow.

Afferent *af'-er-ent* Carrying to or toward a center.

Afferent neuron *noo'-ron* Nerve cell that carries impulses toward the central nervous system.

Afterbirth *af'-ter-berth* The expelled placenta after a baby has been born.

Agglutinin *ag-gloo'-tin-in* An antibody in blood plasma that causes clumping of corpuscles or bacteria.

Agglutinogen *ag-gloo-tin'-o-jen* (1) An antigen that stimulates the formation of a specific agglutinin; (2) an antigen found on red blood cells that is responsible for determining the ABO blood group classification.

Albumin *al-bu'-min* A protein found in virtually all animal tissue and fluid.

Albuminuria *al-bu-min-oor'-ee-ah* The presence of albumin in the urine.

Alcohol A substance with one or more hydroxyl (OH) groups.

Aldehyde *al'-de-hyde* A substance that contains a —CHO group.

Aldosterone *al-dos'-ter-own* A hormone produced by the adrenal cortex that is important in sodium retention and reabsorption.

Alimentary *al-im-en'-ta-ree* Pertaining to nourishment or the digestive organs.

Alkalosis *al-kal-o'-sis* A condition in which the blood has a lower hydrogen ion concentration than normal with an increased pH.

Allele *al'-eel* One of two or more alternate forms of a gene.

Allergy *al'-ler-jee* A condition in which there is hypersensitive response to a particular substance or environmental element.

Allosteric effect *al-o-ster'-ik* Modification of an enzyme activity by an action at a site other than the active site.

Alpha helix *al'-fah heel'-iks* A regular helical arrangement of amino acids in proteins.

Alveolus *al-ve'-o-lus* (1) A general anatomic term referring to a small cavity or depression; (2) an air sac in the lungs.

Amenorrhea *a-men-or-ree'-ah* Absence of menstruation.

Amino acid *a-mee'-no* An organic compound containing nitrogen, carbon, hydrogen, and oxygen; the building block of protein.

Amino group The —NH_2 group.

Amnion *am'-nee-on* The innermost fetal membrane; it forms a fluid-filled sac for the embryo.

Amniocentesis *am'-nee-o-sen-tee'-sis* A procedure for obtaining amniotic fluid. Amniotic fluid is removed (by inserting a needle through a pregnant woman's abdomen into the sac containing the fetus). Fetal cells in the fluid can be grown in a laboratory and subjected to chromosomal and/or biochemical analysis.

Amoeboid movement *a-mee'-boyd* A means of cell locomotion characterized by the extension of pseudopodia, which pull the cell along as cytoplasm flows into them.

Amplitude *am'-pli-tude* Largeness; fullness or wideness of extent or range.

Ampulla *am-pul'-lah* A saclike dilation of a duct or tube.

Anabolism *an-ab'-o-li-zem* The energy-requiring building-up phase of metabolism in which simpler substances are synthesized into more complex substances.

Anaerobic *an-ay-er-o'-bick* Requiring no oxygen to live or grow.

Anaphase *an'-a-faze* The phase of mitosis in which chromosomes migrate to the poles.

Anastomosis *a-nas-to-mo'-sis* A union or joining of blood vessels or other tubular structures.

Androgen *an'-dro-jen* A hormone that controls male secondary sex characteristics.

Anemia *a-nee'-mee-ah* A decreased number of erythrocytes or decreased percentage of hemoglobin in the blood.

Aneurysm *an'-yu-riz-em* A localized dilation of an artery wall caused by weakening of the wall.

Angina pectoris *an-jee'-nah pek'-tor-is* A severe, suffocating chest pain caused by brief lack of oxygen to heart muscle.

Angiotensin *an-jee-o-ten'-sin* A vasoconstrictor substance found in the blood.

Anion *an'-eye-on* An ion carrying one or more negative charges and therefore attracted to a positive pole.

Anorexia *an-or-ek'-see-ah* Loss of appetite or desire for food.

Anorexia nervosa *ner-vo'-sah* A nervous condition

in which an extreme loss of appetite leads to emaciation, malnutrition, and possible worse consequences.

Anoxia *an-ok'-see-uh* A deficiency of oxygen.

Anterior The front of an organ or part; the ventral surface.

Antibody A specialized substance produced by the body that can provide immunity against a specific antigen.

Anticodon The three nucleotides by which a transfer RNA recognizes an RNA codon.

Antidiuretic hormone (ADH) *an-tee-dye-u-re'-tik* A pituitary gland hormone that controls the reabsorption of water by the kidney.

Antigen *an'-ti-jen* Any substance—including toxins, foreign proteins, or bacteria—which, when introduced to the body, causes antibody formation.

Antithrombin *an-ti-throm'-bin* A substance in blood plasma that neutralizes thrombin, thereby inhibiting coagulation.

Antrum *an'-trum* An open space or chamber; a cavity.

Anus *ay'-nus* The distal end of the digestive tract and the outlet of the rectum.

Aorta *ay-or'-tah* The major systemic artery; arises from the left ventricle of the heart.

Aortic arch *ay-or'-tik* The curved and most superior portion of the aorta.

Aortic body A receptor in the aortic arch sensitive to changing oxygen, carbon dioxide, and pH levels of the blood.

Aphasia *a-fay'-zhe-ah* A loss of the power of speech or a defect in speech.

Aplastic anemia *a-plas'-tic* Anemia caused by inadequate production of erythrocytes resulting from inhibition or destruction of the red bone marrow.

Appendix *a-pen'-dicks* A wormlike sac attached to the large intestine.

Aqueous humor *ay'-kwee-us hyu'-mer* The watery fluid of the anterior portion of the eye.

Arachnoid *a-rak'-noid* Weblike; specifically, the weblike, middle layer of the three meninges.

Areola *ah-ree'-o-lah* (1) Any minute opening or space in a tissue; (2) the circular, pigmented area surrounding the nipple.

Arteriole *ar-te'-ree-ole* A minute artery.

Arteriosclerosis *ar-te-ree-o-skle-ro'-sis* Any of a number of proliferative and degenerative changes in the arteries.

Artery A vessel that carries blood away from the heart.

Arthritis *ar-thright'-us* Inflammation of the joints.

Articulate *ar-tik'-u-late* To join together in such a way as to allow motion between the parts.

Ascites *as-sigh'-teez* Accumulation of serous fluid in the abdominal cavity.

Asphyxia *as-fik'-see-ah* Loss of consciousness resulting from a deficiency in the oxygen supply.

Asthma *az'-ma* A disease or allergic response characterized by bronchial spasms and difficult breathing.

Astigmatism *a-stig'-ma-tiz-em* A visual defect resulting from irregularity in the lens or cornea of the eye causing the image to be out of focus.

Ataxia *a-tak'-see-a* Lack of muscular coordination.

Atelectasis *at-e-lak'-ta-sis* Lung collapse or incomplete expansion of a lung.

Atherosclerosis *ath-er-o-skle-ro'-sis* Changes in the walls of large arteries consisting of lipid deposits on the artery walls.

Atom The smallest particle of an element; composed of electrons, protons, and neutrons; capable of existing individually or in combination with atoms of the same or another element.

Atomic number The number of protons in the nucleus of a chemical element.

Atomic weight The number of protons plus neutrons in the atomic nucleus of an element.

Atresia *a-tree'-zhee-ah* (1) The abnormal closure of a body canal or opening; (2) degeneration of the ovarian follicle and the ovum within it.

Atrioventricular bundle (AV bundle) *a-tree-o-ven-trik'-u-lar* The bundle of specialized fibers serving to conduct impulses from the AV node to the right and left ventricles; failure of the AV bundle results in heart block; also called bundle of His.

Atrioventricular node (AV node) A specialized compact mass of conducting cells located at the atrioventricular junction in the heart.

Atrium *ay'-tree-um* A chamber of the heart receiving blood from the veins.

Atrophy *a'-tro-fee* A reduction in size or wasting away of an organ or cell resulting from disease or lack of use.

Auditory *aw'-di-to-ree* Pertaining to the sense of hearing.

Auditory ossicles *aw'-di-to-ree oss'-i-kuls* The three tiny bones serving as transmitters of vibrations and located within the middle ear: the malleus, incus, and stapes.

Auditory (eustachian) tube *yoo-stay'shee-un* The connection between the middle ear and the pharynx.

Auricle *aw'-ri-kel* The external ear.

Auscultation *aws-kul-tay'-shun* The act of examination by listening to body sounds.

Autoimmune response *aw-to-im-myoon'* The production of antibodies or effector T cells that attack a person's own tissue.

Automaticity *aw-to-ma-ti'-si-tee* The ability of a structure, organ, or system to initiate its own activity.

Autonomic *aw-to-nom'-ik* Self-directed; self-regulating; independent.

Autonomic nervous system The division of the nervous system that functions involuntarily and is responsible for innervating cardiac muscle, smooth muscle, and glands.

Autosome *aw'-to-some* A chromosome that is not a sex chromosome.

Avogadro's number *av-o-gad'-roze* The number of molecules in one mole of a substance ($=6.06 \times 10^{23}$).

Axilla *aks-il'-ah* The armpit.

Axon *aks'-on* The process of a nerve cell by which impulses are carried away from the cell; efferent process; the conducting portion of a nerve cell.

Bacteria Any of a large group of microorganisms, usually non-spore-producing and generally one-celled; found in humans and other animals, plants, soil, air, and water; have a broad range of functions.

Baroreceptor *bar-o-ree-sep'-tor* A receptor that is stimulated by pressure changes.

Basal ganglia *bay'-zel gan'-glee-ah* Gray matter structures deep inside each of the cerebral hemispheres.

Basal metabolic rate *met-ah-bol'-ik* The rate at which energy is expended (heat produced) by the body per unit time under controlled (basal) conditions: 12 hours after a meal, at rest.

Base A substance that accepts hydrogen ions; capable of uniting with an acid to form salt and water.

Basement membrane A thin layer of substance to which epithelial cells are attached in mucous surfaces.

Basophil *bay'-so-fil* White blood cells that have relatively large cytoplasmic granules that appear reddish-purple to blue-black when stained with Wright's stain.

B cells Lymphocytes that are committed to differentiate into the antibody-producing plasma cells involved in humoral immunity.

Benign *bee-nine'* Not malignant.

Biscuspid *bigh-kus'-pid* Having two points or cusps.

Bifurcation *bi-fur-ka'-shun* Division into two branches.

Bile A greenish-yellow or brownish fluid produced in and secreted by the liver, stored in the gallbladder, and released into the small intestine.

Bilirubin *bil-i-roo'-bin* The red pigment of bile.

Biliverdin *bil-i-ver'-din* The green pigment of bile.

Biofeedback An area of research in which normally subconscious activity is raised to the conscious level.

Biopsy *bigh'-op-see* The removal and examination of live tissue.

Bipolar cells of the retina Neural elements of the retina with which the rods and cones connect.

Blastocyst *blas'-to-sist* A stage of early embryonic development.

Blood-brain barrier A mechanism that inhibits passage of materials from the blood into brain tissues and cerebrospinal fluid.

Blood group systems Sets of antigenic determinants found on the surfaces of human red blood cells.

Bolus *bo'-lus* (1) A rounded mass of food prepared by the mouth for swallowing; (2) a concentrated mass of a pharmaceutical preparation, usually given intravenously.

Bowman's capsule *bo'-manz* The double-walled cup at the beginning of a nephron.

Brachial *bray'-ki-al* Pertaining to the arm.

Bradycardia *brad-i-kar'-dee-ah* Slowness of the heart rate; below 60 beats per minute.

Brain stem The portion of the brain consisting of the medulla, pons, and midbrain.

Broca's area A portion of the left frontal lobe of the brain particularly developed on the left side of the brain in right-handed persons; damage to this area results in impaired speech.

Bronchitis *brong-kigh'-tis* Inflammation of the bronchi.

Bronchus *brong'-kus* One of the two large branches of the trachea leading to the lungs.

Buccal *bu'-kal* Pertaining to the cheek.

Buffer A substance or substances that tend to stablize the pH of a solution.

Bundle branch block A blocking of heart action resulting from a local lesion of the bundle of His; delayed contraction of one ventricle.

Bursa *bur'-sah* A small sac or cavity filled with fluid and located at friction points, especially joints.

Calcitonin *kal-si-to'-nin* A hormone released by the thyroid that lowers calcium and phosphate levels of the blood.

Calculus *kal'-ku-lus* A stone formed within various body parts.

Callus *kal'-us* (1) A bonelike material that protrudes between ends of a fractured bone; (2) a thickening of the skin caused by rubbing or friction.

Calorie *kal'-or-ee* A unit of heat; the amount of heat required to raise the temperature of 1 g of water from 15°C to 16°C.

Calyx *kay'-liks* A cuplike division of pelvis of the kidney.

Canal A duct or passageway; a tubular structure.

Canaliculi *kan-al-ik'-u-lee* Extremely small tubular passages or channels.

Cancer A malignant, invasive cellular tumor that has the capability of spreading throughout the body or body parts.

Capacitation *ka-pass-i-tay'-shun* The process in which sperm undergo changes in the female reproductive tract making them capable of fertilization.

Capillary *kap'-il-lar-ee* A minute blood vessel connecting arterioles with venules.

Carbohydrates *kar-bo-high'-drates* Organic compounds composed of carbon, hydrogen, and oxygen with the hydrogen and oxygen frequently present in a 2:1 ratio; include starches, sugars, cellulose.

Carbonic anhydrase *kar-bon'-ik an-high'-drase* An enzyme that facilitates the combination of carbon dioxide with water to form carbonic acid.

Carcinogen *kar-sin'-o-jen* Cancer-causing agent.

Carcinoma *kar-sin-o'-ma* Cancer; a malignant growth.

Cardiac *kar'-dee-ak* Pertaining to the heart.

Cardiac muscle Specialized muscle of the heart.

Cardiac output The blood volume (in liters) ejected per minute by the left ventricle.

Carotid artery *ka-rot'-id* A major artery in the neck.

Carotid body A receptor at the bifurcations of the common carotid arteries sensitive to changing oxygen, carbon dioxide, and pH levels of the blood.

Carotid sinus *sigh'-nus* A dilation of a common carotid artery at its bifurcation; involved in regulation of systemic blood pressure.

Cartilage *kar'-ti-lej* White, semiopaque, fibrous connective tissue.

Catabolism *ka-tab'-a-liz-em* The process in which living cells break down complex substances into simpler substances; destructive metabolism.

Catalyst *kat'-a-list* A substance that affects the rate of a chemical reaction.

Cataract *kat'-a-rakt* Partial or complete loss of transparency of the crystalline lens of the eye or its capsule.

Catecholamines *kat-e-kole'-uh-meens* Substances derived from the simple chemical catechol, including epinephrine and norepinephrine.

Catheter *ka'-the-ter* A narrow, hollow tube that can be inserted into a body cavity for withdrawal of fluids.

Cation *kat'-eye-on* An ion with a positive charge and therefore attracted toward a negative pole (cathode) in electrolytic cells.

Cecum *see'-kum* The blind-end pouch at the beginning of the large intestine.

Cell The basic biological unit of living organisms (except viruses), containing a nucleus and a variety of organelles.

Cell mediated immunity Immunity mediated by specialized proteins called antibodies.

Cell membrane The selectively permeable membrane forming the outer boundary of body cells; also called plasma membrane or unit membrane.

Cellulose *sel'-u-lose* A fibrous carbohydrate that is the main structural component of plant tissues.

Central nervous system (CNS) The brain and the spinal cord.

Centriole *sen'-tree-ole* A minute body found in the nucleus of the cell; active in cell division.

Cerebellum *ser-e-bel'-lum* Part of the hindbrain; controls movement coordination; consists of two hemispheres and a central portion (vermis).

Cerebrospinal fluid *ser-e-bro-spy'-nal* The fluid produced in the cerebral ventricles; fills the ventricles and surrounds the central nervous system.

Cerebrum *ser'-i-brum or se-ree'-brum* The largest part of the brain; consists of right and left cerebral hemispheres.

Cerumen *se-roo'-men* Earwax.

Cervical *ser'-vi-kal* Refers to the neck or the necklike portion of an organ or structure.

Cervix *ser'-vix* (1) The cylindrical, inferior portion of the uterus leading to the vagina; (2) any necklike structure.

Chemoreceptor *kem-o-ree-sep'-tor* Receptors sensitive to various chemical stimulations and changes.

Chemotaxis *kem-o-tak'-sis* Movement of a cell, organism, or part of an organism toward or away from a chemical substance.

Chiasma *kee-az'-ma* A crossing or intersection of two structures, such as the optic nerves.

Cholecystectomy *kol-e-sis-tek'-tom-ee* Removal of the gallbladder.

Cholecystokinin *kol-e-sis-to'-kin-in* An intestinal hormone that stimulates gallbladder contraction and pancreatic enzyme release, but inhibits gastric acid secretion when the stomach is actively secreting.

Cholesterol *ko-les'-ter-ol* An organic alcohol found in animal fats and oil as well as in most body tissues, especially bile.

Cholinergic fibers *ko-lin-ur'-jick* Nerve endings that, upon stimulation, release acetylcholine at their terminations.

Chondroblast *kon'-dro-blast* A cell that forms the fibers and matrix of cartilage.

Chondrocyte *kon'-dro-site* A mature cartilage cell.

Chorion *kor'-ee-on* The outermost fetal membrane; forms the placenta.

Choroid *ko'-roid* (1) Skinlike; (2) a portion of the vascular tunic.

Chromatid *kro'-ma-tid* One of the two strands of a duplicated chromosome.

Chromatin *kro'-ma-tin* The substance of chromosomes.

Chromosome *kro'-mo-some* The structures in the nucleus that carry the hereditary factors (genes).

Chyme *kime* The semifluid contents of the stomach consisting of partially digested food and gastric secretions.

Chymotrypsin *kime-oh-trip'-sin* An enzyme secreted into the gastrointestinal tract by the pancreas that acts to break down proteins into amino acids.

Cilia *sil'-lee-ah* Tiny, hairlike projections on cell surfaces that move in a wavelike manner.

Circadian *sir-kay-dee'-an* Daily; on a daily cycle.

Circumcision *sur-kum-si'-zhun* Removal of the foreskin of the penis.

Cirrhosis *sir-o'-sis* A chronic disease, particularly of the liver, characterized by an overgrowth of connective tissue.

Citric acid cycle A set of reactions serving to convert organic compounds to CO_2, releasing electrons.

Clitoris *kli'-to-ris* A structure in the female that is homologous to the penis in the male.

Clone One or more offspring of like genetic constitution produced by asexual reproduction.

Cochlea *koak'-lee-ah* A cavity of the inner ear resembling a snail shell.

Codon *ko'-don* A group of three nucleotides in DNA or mRNA that specify the insertion of an amino acid into protein.

Coenzyme *ko-en'zime* A nonprotein substance associated with and activating an enzyme.

Coitus *ko'-i-tus* Sexual intercourse.

Collagen *kol'-ah-jen* The main structural protein of tendons, bones, and cartilage.

Colloid *kol'-loid* Solute particles dispersed in a medium; particles do not settle out readily and do not pass through many natural membranes.

Colloidal osmotic pressure (COP) *kol-loi'-dal os-mot'-ick* The pressure exerted on a membrane by the particles in a colloid; usually refers to the osmotic pressure of blood plasma and body fluids resulting from the presence of protein.

Colostrum *kol-os'-trum* The first milk secreted after pregnancy.

Coma *ko'-ma* Unconsciousness from which a person cannot be aroused.

Competitive inhibitor An enzyme inhibitor that competes with a substrate for the active site.

Complement system A system consisting of a series of plasma proteins that normally circulate in the blood in inactive forms. The activation of the system's initial components initiates a cascading series of reactions that ultimately produce active complement molecules, which can act as chemotactic agents, enhance phagocytosis, and exert other effects.

Concave *kon'-kave* Having a curved, depressed surface.

Condom *kon'-dom* A sheath worn over the penis during sexual intercourse to prevent conception and/or infection.

Conduction *kon-duk'-shun* The transfer of energy from molecule to molecule by direct contact.

Cones One of the two types of photosensitive cells in the retina of the eye.

Congenital *kon-jen'-i-tal* Existing at birth.

Conjunctiva *kon-junk-ti'-va* The thin protective membrane on the insides of the eyelids and the anterior surface of the eye itself.

Conjunctivitis *kon-junk-ti-vigh'-tis* An inflammation of the conjunctiva of the eye.

Connective tissue A primary type of tissue; form varies extensively, as does function, which includes support and storage.

Contraception *kon-tra-sep'-shun* The prevention of conception; birth control.

Contractility *kon-trak-til'-i-tee* A substance's ability to shorten or develop tension upon the application of a stimulus.

Contralateral *kon-tra-lat'-er-al* Opposite; acting in unison with a similar part on the opposite side of the body.

Convection *kon-vek'-shun* The constant transfer of heat by movement of heated particles.

Convergence *kon-verj'-ence* Turning toward or approaching a common point from different directions.

Convoluted *kon'-vo-lu-ted* Rolled, coiled, or twisted.

Copulation *kop-u-lay'-shun* Sexual intercourse.

Cornea *kor'-nee-ah* The transparent anterior portion of the eyeball.

Corona radiata *ko-ro'-nah ray-dee-aw'-tah* (1) The arrangement of elongated follicle cells around a mature ovum; (2) the crownlike arrangement of nerve fibers radiating from the inner capsule of the brain to every part of the cerebral cortex.

Corpus *kor'-pus* Body; the major portion of an organ.

Corpus luteum *lew-tee-um* The cells that remain of the Graafian follicle after the ovum has been released; it produces estrogens and progesterone.

Cortex *kor'-teks* The outer surface layer of an organ.

Corticoid *kor'-ti-koid* A substance whose function and properties are similar to those of corticosteroids.

Corticosteroids *kor'-ti-ko-ste'-roidz* The steroid hormones released by the adrenal cortex.

Cortisol *kor'-ti-sol* A glucocorticoid produced by the adrenal cortex.

Costal *kos'-tal* Pertaining to the ribs.

Covalent bond A bond in which the electrons of an atom are shared by another atom.

Cramp A painful, involuntary contraction of a muscle.

Cranial *kray'-nee-al* Pertaining to the skull.

Cranial nerves Twelve pairs of nerves that arise directly from the brain.

Creatine phosphate *kree'-a-ten fos'-fate* A compound that serves as an alternative energy source for muscle tissue.

Crenation *kre-nay'-shun* The shriveling of an erythrocyte resulting from withdrawal of water.

Cretinism *kree'-tin-izm* A severe thyroid deficiency in the young that leads to stunted physical and mental growth.

Crista *kris'-ta* A crest or ridge.

Cross bridges Projections that extend at regular intervals from the thick filaments in myofibrils.

Crossing over The process of exchange between chromosomes.

Cryptorchidism *kript-or'-kid-izm* A developmental defect in which the testes fail to descend into the scrotum.

Cupula *ku'-pu-la* A domelike structure.

Cushing's syndrome *ku'-shingz sin'-drome* A disease produced by excess secretion of adrenocortical hormones; characterized by adipose tissue accumulation, weight gain, and osteoporosis, for example.

Cutaneous *ku-tay'-nee-us* Pertaining to the skin.

Cyanosis *sigh-a-no'-sis* A bluish coloration of the mucous membranes and skin caused by a significant concentration of reduced hemoglobin in the arterial blood.

Cyst *sist* A sac with a distinct wall, containing fluid or other material; may be pathological or normal.

Cystitis *sis-tigh'-tis* Inflammation of the urinary bladder.

Cytochromes *sigh'-to-kromes* A set of proteins functioning in electron transport.

Cytokinesis *sigh-to-ki-nee'-sis* The changes in the cytoplasm during cell division.

Cytology *sigh-tol'-o-jee* The science concerned with the study of cells.

Cytoplasm *sigh'-to-plaz-um* The protoplasm of a cell other than that of the nucleus.

Cytosine *sigh'-to-seen* One of the nucleic acid bases; pairs with guanine.

Deamination *dee-am-i-nay'-shun* The removal of an amino group from an organic compound by reduction, hydrolysis, or oxidation.

Defecation *def-e-kay'-shun* The elimination of the contents of the bowels (feces).

Deglutition *dee-gloo-tish'-un* The act of swallowing.

Dehydration *dee-high-dray'-shun* A condition resulting from excessive loss of water.

Dehydrogenation *dee-high-dro-je-nay'-shun* Removal of hydrogen atoms (and electrons).

Denaturation The unfolding of macromolecules due to opening of weak bonds.

Dendrite *den'-drite* Branching; the branching processes that are part of the receptive portion of a nerve cell.

Dental caries *den'-tal kar'-eez* Tooth cavity.

Deoxyribonucleic acid (DNA) *dee-ox-i-rye-bo-*

nu-kle'ik A nucleic acid found in all living cells: carries the organism's hereditary information.

Depolarization *dee-po-lar-i-zay'-shun* The neutralization to a state of nonpolarity; the loss of a negative charge inside the cell.

Depressor *dee-pres'-sor* Any substance that causes slowing, reduction of activity, or inhibition of another structure, organ, or substance.

Dermatitis *der-ma-tigh'-tis* An inflammation of the skin; nonspecific skin allergies.

Dermis *der'-mis* The deep layer of dense, irregular connective tissue of the skin; also called corium.

Desmosome *des'-mo-some* Small, apposed, ellipsoidal plates in membranes of adjacent cells.

Diabetes insipidus *dye-uh-bee'-teez in-sip'-i-dus* A disease characterized by passage of a large quantity of urine of low specific gravity plus intense thirst and dehydration; a hypothalamic disorder is the cause.

Diabetes mellitus *mel'-li-tus* A disease caused by a relative insulin deficiency, leading to failure of the body tissue to oxidize carbohydrates at a normal rate.

Dialysis *dye-al'-i-sis* The separation of substances from one another in a solution by taking advantage of their differing diffusibilities through porous membranes.

Diapedesis *dye-a-pe-dee'-sis* The passage of blood cells through unruptured vessel walls into the tissues.

Diaphragm *dye'-a-fram* (1) Any partition or wall separating one area from another; (2) a muscle that separates the thoracic cavity from the abdominopelvic cavity.

Diastole *dye-as'-to-lee* A period (between contractions) of relaxation and dilation of the heart during which it fills with blood.

Differentiation Production of different cell types during development.

Diffusion *di-fu'-zhun* The spreading of particles in a gas or solution with a movement toward uniform distribution of particles.

Digestion *di-jest'-yun* The bodily process of breaking down foods chemically and mechanically into compounds capable of being absorbed by body cells.

Dilate *dye'-late* To stretch; to open; to expand.

Diploid A cell or organism with two chromosomal sets.

Distal *diss'-tal* Farthest from the center or midpoint of a limb structure.

Diverticulum *dye-ver-tik'-u-lum* A pouch or sac in the walls of a hollow organ or structure.

Dizygotic (fraternal) twins Twins derived from two separate fertilized eggs.

DNA polymerase An enzyme that synthesizes DNA fibers.

Dominant An allele, or the corresponding trait, that is expressed in all heterozygous cells or organisms.

Dorsal *dor'-sal* Pertaining to the back; posterior.

Double helix The structure of DNA molecules.

Downs syndrome *dounz sin'-drome* An abnormality resulting from the presence of an extra copy of the genetic material contained in chromosome number 21. Characteristics include mental retardation and altered physical appearance.

Duct A canal or passageway.

Duodenum *doo-o-dee'-num* The first part of the small intestine.

Dura mater *doo'-rah may'-ter* The outermost and toughest of the three membranes (meninges) covering the brain and spinal cord.

Dysfunction *dis-funk'-shun* Lack of normal function; disorder.

Dysmenorrhea *dis-men-or-ree'-ah* Difficult, painful menstruation.

Dyspnea *disp'-nee-ah* Labored, difficult breathing.

Dystrophy *dis'tro-fee* A disorder caused by defective or abnormal development; degeneration.

Ectopic *ek-to'-pik* Not in the normal place; for example, in an ectopic pregnancy the fertilized ovum implants at a place other than the uterus.

Edema *e-dee'-mah* An abnormal accumulation of fluid in body parts or tissues; causes swelling.

Effector *ef-fek'-tor* A structure supplied by a motor nerve ending; a muscle or secretory cell.

Efferent *ef'-er-ent* Carrying away or away from, especially a nerve fiber that carries impulses away from the central nervous system.

Ejaculation *e-jak-u-lay'-shun* The sudden ejection of a fluid from a duct, especially semen from the penis.

Elastin *e-las'-tin* The main protein in elastic fibers of connective tissues.

Electrocardiogram (ECG) *ee-lek-tro-kar'-dee-o-gram* A graphic record of the electric current associated with heartbeats.

Electroencephalogram (EEG) *ee-lek-tro-en-sef'-a-lo-gram* A graphic record of the activity of nerve cells in the brain.

Electrolyte *ee-lek'-tro-lite* A substance that, when in solution, is capable of conducting an electric current.

Electron   *ee-lek'-tron*   A negatively charged particle in motion around the nucleus of an atom.

Electron acceptor   A substance that receives electrons from an electron donor, becoming reduced.

Electron donor   A substance that transfers electrons to an acceptor, thereby becoming oxidized.

Electron transport chain   A series of reactions that transfer electrons from various donors to oxygen.

Embolism   *em'-bo-liz-em*   The obstruction of a blood vessel by a clot floating in the blood; may also be a bubble of air in the vessel (air embolism).

Embryo   *em'-bree-oh*   An organism in its early stages of development; in humans, the first two months after conception.

Emesis   *em'-e-sis*   Vomiting.

Emphysema   *em-fi-see'-muh*   A condition caused by overdistension of the pulmonary alveoli or abnormal presence of air or gas in body tissues.

Encephalitis   *en-sef-a-ligh'-tis*   An inflammation of the brain.

Endocardial   *en-do-kar'-di-al*   Pertaining to the inner lining of the heart.

Endocarditis   *en-do-kar-di'-tis*   An inflammation of the inner lining of the heart.

Endocardium   *en-do-kar'-di-um*   The membrane lining the interior of the heart; endothelium and connective tissue.

Endocrine glands   *en'-do-krin*   Ductless glands that secrete their products into the extracellular space around the gland cells and from which the products enter the bloodstream.

Endoderm   *en'-do-derm*   Tissue forming the digestive tube and its associated structures.

Endometrium   *en-do-me'-tree-um*   The mucous membrane lining of the uterus.

Endoplasmic reticulum   *en-do-plaz'-mik re-tik'-u-lum*   A membranous network of tubular or saclike channels through the cytoplasm of a cell.

Endorphins   *en-dor'-fins*   Opiatelike substances found in the brain the amio acid sequence of which includes the amino acid sequence of an enkephalin at one end.

Endothelium   *en-do-thee'-lee-um*   The single layer of epithelial cells that line the walls of the heart and the vessels that carry blood and lymph.

Energy   The capacity to do work.

Enkephalins   *en-keff'-a-lins*   Opiatelike substances found in the brain.

Enzyme   *en'-zime*   A substance formed by living cells that acts as a catalyst in bodily chemical reactions.

Eosinophil   *ee-o-sin'-o-fil*   A granular white blood cell whose granules appear reddish-orange when stained with Wright's stain.

Epidermis   *e-pi-der'-mis*   The outer layer of cells of the skin.

Epididymis   *e-pi-did'-i-mis*   That portion of the seminal duct in which sperm mature and are transported from testes to body exterior.

Epiglottis   *e-pi-glot'-tis*   The elastic membrane-covered cartilage at the back of the throat; guards the glottis during swallowing.

Epinephrine   *e-pi-nef'-rine*   The chief hormone of the adrenal medulla.

Epithelium   *e-pi-thee'-lee-um*   One of the primary tissues; covers the surface of the body and lines the body cavities, ducts, and vessels.

Equilibrium   *ee-kwi-lib'-ri-um*   Balance; a state when opposite reactions or forces counteract each other exactly.

Erythrocyte   *e-ree'-throw-site*   Red blood cell.

Erythropoiesis   *e-rith-ro-poi-ee'-sis*   The formation process of erythrocytes.

Erythropoietin   *e-rith-ro-poi-ee'-tin*   A hormone, produced mainly by the kidneys, that stimulates production of red blood cells.

Estrogen   *es'-tro-jen*   Any substance that stimulates female secondary sex characteristics, female sex hormones.

Eupnea   *yoop-nee'-ah*   Easy, normal breathing.

Excretion   *eks-kree'-shun*   The elimination of waste products from the body.

Exocrine glands   *eks'-o-krin*   Glands that typically secrete their products by way of ducts onto body surfaces.

Exogenous   *eks-og'-en-us*   Developing or originating outside the body, organ, or part.

Expiration   *ex-pi-ray'-shun*   The act of expelling air from the lungs.

Extracellular   *eks-tra-sel'-u-lar*   Outside a cell.

Extracellular fluid   Fluid within the body but outside the cells.

Extrinsic   *eks-trin'-zik*   Originating from outside an organ or part.

Exudate   *eks'-yu-date*   Material including fluid, pus, or cells that has escaped from blood vessels and has been deposited in tissues.

Fallopian tube   *fal-low'-pee-an*   The oviduct or uterine tube; the tube through which the ovum is transported to the uterus.

Fascia   *fash'-ee-ah*   The layers of fibrous tissue under the skin or covering and separating muscles.

Fasciculus *fas-ik'-u-lus* A bundle of nerve, muscle, or tendon fibers separated by connective tissues.

Fatty acid An organic acid with a more or less long chain of carbon atoms.

Feces *fee'-seez* Material discharged from the bowel composed of food residue, secretions, and bacteria.

Feedback A system of complex causality in which the product of a process influences the process that generated it in return.

Fenestrated *fen'-es-tray-ted* Pierced with one or more small openings.

Fertilization The joining of sperm and oocyte to produce a zygote.

Fetus *fee'-tus* The unborn young; in humans, from the third month in the uterus until birth.

Fibrillation *fib-ri-lay'-shun* Irregular, uncoordinated contraction of muscle fibers.

Fibrin *figh'-brin* The fibrous insoluble protein formed during the clotting of blood.

Fibrinogen *figh-brin'-o-gen* A protein that is converted to fibrin during blood-clotting.

Filtration *fil-tray'-shun* The passage or straining of a solvent and dissolved substances through a membrane or filter.

Fissure *fis'-sure* (1) A groove or cleft; (2) the deepest linear depressions on the brain.

Fistula *fis'-tu-lah* An abnormal passage between organs or between a body cavity and the outside.

Flaccid *flak'-sid* Soft; flabby; relaxed.

Flagella *fla-jel'-ah* Long whiplike extensions of the cell membrane of some bacteria; serve as agents for locomotion.

Flavin adenine dinucleotide (FAD) A protein-associated electron carrier.

Flexion *flek'-shun* Bending; the movement that decreases the angle between bones.

Focal length *fo'-kal* The distance from the lens to the focal point.

Focal point The point at which light rays converge behind the lens.

Follicle *fol'-i-kal* A small sac or gland.

Follicle-stimulating hormone (FSH) A hormone produced by the anterior pituitary that stimulates ovarian follicle production in females and sperm production in males.

Formed elements Red blood cells, platelets, and white blood cells.

Fovea *fo-vee'-ah* A small depression or pit.

Fulcrum *ful'-krum* The pivot point of a lever.

Fundus *fun'-dus* The base of an organ; that part farthest from the opening of the organ.

Galactosemia *gal-ak-to-see'-me-ah* A genetic anomaly causing intolerance to milk and milk sugars.

Gallbladder The sac beneath the right lobe of the liver used for bile storage.

Gallstones Particles composed primarily of cholesterol that are occasionally formed in gallbladder and bile ducts.

Gamete *gam'-eet* Male or female reproductive cell.

Gametogenesis *gam-e-to-jen'-e-sis* The origin and formation of gametes.

Ganglion *gan'-glee-on* A group of nerve-cell bodies, usually located in the peripheral nervous system.

Gap junction Intercellular specialization with the cell membranes of adjacent cells only 20 Å apart.

Gastrin A hormone that stimulates gastric secretion, especially hydrochloric acid release.

Gastroenteritis *gas-tro-en-ter-i'-tis* An inflammation of mucosa of stomach and intestine.

Gastroesophageal sphincter *gas-tro-e-soff-a-jee'-al sfink'-ter* Narrowing between the esophagus and the stomach.

Gastrointestinal tract *gas-tro-in-tes'-tin-al* A compartmentalized hollow tube that begins with the mouth and ends with the anus.

Gene *jeen* One of the biological units of heredity located on chromosomes; transmits hereditary message.

Genetic counseling *jen-e'-tik* The counseling of parents about the chances of their children having particular genetic abnormalities.

Genetics *jen-e'-tiks* The science of heredity.

Genitalia *jen-i-tay'-lee-ah* The external sex organs.

Genotype *jen'-o-tipe* The genetic composition of an individual.

Gestation *jes-tay'-shun* The period of pregnancy; about 280 days for humans.

Gland An organ specialized to secrete or excrete substances for further use in the body or for elimination.

Glaucoma *glaw-ko'-mah* An abnormal elevation of the pressure within the eye.

Globin *glow'-bin* The protein component of hemoglobin.

Glomerulus *glom-er'-u-lus* The capillaries surrounded by Bowman's capsule in the kidney.

Glottis *glot'-tis* The opening between the vocal cords; entrance to the larynx.

Glucagon *gloo'-ka-gon* A hormone formed by islets of Langerhans in the pancreas; raises the glucose level of blood.

Glucocorticoids *gloo-ko-kor'-ti-koidz* The adrenal cortex hormones that affect metabolism of fats and carbohydrates.

Glucose *gloo'-kose* The principal sugar in the blood.

Glycerol *gliss-e-rol* An important alcoholic component of fat.

Glycogen *gligh-ko-jen* An animal starch; the main carbohydrate stored in animal cells.

Glycogenesis *gleye-ko-jen'-e-sis* The body's formation of glycogen from other carbohydrates.

Glycogenolysis *gleye-ko-jen-ol'-i-sis* The body's breakdown of glycogen to glucose.

Glycolysis *gligh-kol'-i-sis* The body's breakdown of glucose into simpler compounds, especially lactic acid.

Glyconeogenesis *gleye-ko-nee-o-jen'-e-sis* The formation of carbohydrates from substances that are not carbohydrates, such as protein or fat.

Goblet cells The individual cells of the respiratory and digestive tracts that function as glands.

Goiter *goi'-ter* An enlargement of the thyroid gland.

Gonad *go'-nad* A gland or organ producing gametes; an ovary or testis.

Gonadotropins *go-nad-o-tro'-pinz* The gonad-stimulating hormones produced by the anterior pituitary.

Graafian follicle *graf'-ee-an fol'-i-kal* A mature ovarian follicle.

Graded response A response whose magnitude varies directly with the strength of the stimulus.

Granulocytes *gran'-you-lo-sites* White blood cells that contain large numbers of granules in their cytoplasm.

Gray matter The gray area of the central nervous system; contains neurons.

Groin The junction of the thigh and the trunk; the inguinal area.

Growth hormone A hormone that stimulates growth in general; produced in the anterior pituitary; also called somatotropin.

Guanine *gwah'-neen* One of the nucleic acid bases; pairs with cytosine.

Gustation *gus-tay'-shun* Taste.

Gyrus *jigh'-rus* A convolution on the surface of the cerebral cortex.

Hair cells Receptors that have one or more cilialike processes that project from one side of a cell; they are involved in detection of many different kinds of mechanical forces.

Haploid A cell or organism with a single chromosomal set.

Hay fever An acute allergic reaction of conjunctiva and upper air passages due to pollen sensitivity.

Heart block A defective transmission of impulses from atrium to ventricle.

Heart murmur An abnormal heart sound.

Heat exhaustion The collapse of an individual after heat exposure without failure of the body's heat-regulating mechanism.

Heat stroke The failure of the heat-regulating ability of an individual under heat stress.

Hematocrit *hem-a'-to-krit* The percentage of erythrocytes to total blood volume.

Heme *heem* The iron-containing pigment that is essential to oxygen transport by hemoglobin.

Hemiplegia *hem-i-plee'-jee-ah* Paralysis of one side of the body.

Hemocytoblasts *hee-mo-sigh'-to-blasts* Stem cells that give rise to all the formed elements of the blood.

Hemoglobin *hee'-mo-glo-bin* The oxygen-transporting component of erythrocytes composed of heme and globin.

Hemolysis *hee-mol'-i-sis* The destruction of erythrocytes.

Hemophilia *hee-mo-phil'-i-a* A clotting defect caused by an inherited genetic absence of a blood-clotting factor.

Hemopoiesis *hem-o-poi'-ee-sis* The formation of blood.

Hemorrhage *hem'-o-ridj* The escape of blood from the vessels by flow through ruptured walls; bleeding.

Heparin *hep'-a-rin* A substance that prevents clotting found in many tissues, especially the liver.

Hepatitis *hep-at-eye'-tis* An inflammation and/or infection of the liver.

Hernia *her'-nee-ah* The abnormal protrusion of an organ or a body part through the containing wall of its cavity.

Herpes simplex *her'-peez sim'-pleks* A virus that can cause a fever blister or cold sore; also a venereal disease.

Hertz A measure of sound frequency; a Hertz is one cycle per second.

Heterosexuality *he-ter-o-seks-u-al'-i-tee* Sexual interest in or desire for persons of the opposite sex.

Heterozygous *het-er-o-zigh'-gus* A situation in which two different alleles occur at a given locus on homologous chromosomes.

High-energy bonds   Chemical bonds whose hydrolysis releases a substantial amount of energy.

Histamine   *his'-ta-meen*   A substance present in many cells which causes vasodilation and increased vascular permeability.

Homeostasis   *hom-ee-o-stas'-sis*   The existence of a relatively constant internal environment within the body; the state when the body organs function together to maintain a stable internal environment for the general well-being of the entire body.

Homosexuality   *ho-mo-seks-u-al'-i-tee*   Sexual interest in or desire for persons of the same sex.

Homozygous   *ho-mo-zigh'-gus*   A situation in which the same allele occurs at a given locus on homologous chromosomes.

Hormones   *hor'-mones*   The secretions of endocrine glands; responsible for specific regulatory effects on certain parts or organs.

Human chorionic gonadotropin (HCG)   *kor-ee-on'-ik go-nad-oh-tro'-pin*   A hormone produced by cells of the placenta.

Humoral immunity   *hyoo'-mer-al im-myoo'-ni-tee*   Immunity mediated by certain populations of lymphoid cells called T cells.

Hydrocarbon   *high-dro-kar'-bon*   A molecule composed of only carbon and hydrogen.

Hydrochloric acid   *high-dro-klo'-rik*   HCl; facilitates protein digestion in the stomach; produced by parietal cells.

Hydrogen ion   *high'-dro-jen eye'-on*   A hydrogen atom minus an electron (=a proton).

Hydrolysis   *high-drol'-i-sis*   The process in which a chemical compound unites with water and is then split into smaller molecules.

Hydrophilic   *high-dro-fil'-ik*   (Water-loving). Having chemical attraction to water.

Hydrophobic bond   *high-dro-fo'-bik*   The tendency of hydrophobic chemical groups such as —CH₃ to pull together away from water.

Hydroxyapatite   *high-drok-see-ap'-a-tite*   The mineral that imparts rigidity to bones.

Hydroxyl ion   *high-drok'-sil*   An OH⁻ group.

Hypermetropia   *high-per-me-tro'-pee-uh*   Farsighted-ness.

Hypertension   *high-per-ten'-shun*   High blood pressure.

Hypertonic   *high-per-ton'-ik*   Excessive, above normal tone or tension.

Hypertrophy   *high-per'-tro-fee*   An increase in the size of a tissue or organ independent of the body's general growth.

Hypodermis   *high-po-der'-mis*   The subcutaneous connective tissue; also called superficial fascia.

Hypothalamus   *high-po-thal'-a-mus*   The region of the diencephalon forming the floor of the third ventricle of the brain.

Hypothermia   *high-po-ther'-mi-ah*   Subnormal body temperature.

Hypotonic   *high-po-ton'-ik*   Below normal tone or tension.

Hypoxemic hypoxia   *high-pock-see'-mik high-pock'-see-uh*   A condition in which decreased oxygen is available to tissues because of decreased Po₂ in arterial blood.

Hypoxia   *hip-ox'-ee-a*   A condition in which a physiologically inadequate amount of oxygen is available to tissues.

Identical twins   Twins formed by the splitting of a fertilized egg or early embryo.

Ileum   *il'-ee-um*   The lower part of the small intestine between the jejunum and the cecum of the large intestine.

Immune surveillance   *im-myoon' sir-vail'-lantz*   The body's immune response to cancer.

Immunity   *im-myoon'-i-tee*   The body's ability to resist many organisms and chemicals that can damage tissues.

Immunoglobulin   *im-myoon-o-glob'-you-lin*   An antibody.

Impetigo   *im-pe-tee'-go*   A highly contagious skin infection common in children.

Impotence   *im'-po-tense*   (1) Lack of power, inability; (2) a male's inability to have sexual intercourse or maintain an erection.

In vitro   *in vee'-tro*   In a test tube, glass, or artificial environment.

In vivo   *in vee'-vo*   In the living body.

Infarct   *in-farkt'*   A region of dead, deteriorating tissue resulting from blood flow interference.

Inferior (caudal)   Pertaining to a position near the tail end of the long axis of the body.

Inflammation   *in-flam-may'-shun*   A physiological response of the body to tissue injury; includes dilation of blood vessels and an increase in vessel permeability.

Infundibulum   *in-fun-di'-bu-lum*   A funnel-shaped body part or passageway.

Inner cell mass   An accumulation of cells in the blastocyst from which the embryo develops.

Innervation   *in-ner-vay'-shun*   The supply of distribution of nerves or nerve stimuli to a part.

Inspiration   *in-spi-ray'-shun*   The drawing of air into the lungs.

Insulin   *in'-su-lin*   The hormone produced in the

pancreas affecting carbohydrate and fat metabolism, blood glucose levels, and other metabolic processes.

Integumentary system   *in-teg-u-men'-tar-ee*   The skin and its accessory structures.

Intercellular   *in-ter-sel'-u-lar*   Between the cells of the body or part.

Intercellular matrix   *may'-triks*   The material between adjoining cells.

Interferon   *in-ter-fer'-on*   A chemical that is able to provide some protection against virus invasion of the body; inhibits viral growth.

Interkinesis   *in-ter-ke-nee'-sis*   The period between two successive cell divisions.

Internal environment   The environment within the body made up of the fluid portion of the blood and the interstitial fluid that continually bathes the cells.

Internal respiration   The exchange of gases between blood and tissue fluid and between tissue fluid and cells.

Interphase   The period between two mitoses in the cell cycle.

Interstitial fluid   *in-ter-stish'-al*   The fluid between the cells or body parts.

Intracellular   *in-tra-sel'-u-lar*   Within a cell.

Intracellular fluid   Fluid within a cell.

Intrafusal muscle fibers   *in-tra-fyoo'-zal*   Specialized muscle cells found in muscle spindles.

Intramural pressure   *in-tra-mu'-ral*   The pressure built up in the walls of an organ.

Intrinsic factor   *in-trin'-sik*   A glycoprotein substance required for vitamin B_{12} absorption.

Intrinsic muscle   A muscle that has both its origin and its insertion in an organ.

Invert   To turn inward.

Involuntary muscle   A muscle not under control of the will.

Ion   *eye'-on*   An atom with a positive or negative electric charge.

Ionic bond   *eye-on'-ik*   A bond in which oppositely charged ions are held together by the attraction between opposite charges.

Ipsilateral   *ip-si-lat'-er-al*   Situated on the same side.

Iris   *eye'-ris*   The pigmented, circular diaphragm in front of the eye's lens.

Ischemia   *is-kee'-mee-uh*   A local decrease in blood supply resulting from obstruction or decrease of arterial inflow.

Islets of Langerhans   *eye'-lets lang'-er-hanz*   Clumps of endocrine cells in the pancreas.

Isometric   *i-so-met'-rik*   Of the same length.

Isotonic   *i-so-ton'-ik*   Having uniform tension under pressure or stimulation.

Isotope   *i'-so-tope*   A different form of a given element; isotopes have the same atomic number but different mass numbers.

Jaundice   *jawn'-dis*   A yellow color of the skin due to an accumulation of bile pigments; hyperbilirubinemia.

Jejunum   *je-joo'-num*   The part of the small intenstine between the duodenum and the ileum.

Joint   The junction of two or more bones; an articulation.

Juxtamedullary   *jux'-ta-med'-u-lair-y*   Referring to the inner portion of the cortex of the kidney, adjacent to the medulla.

Karyokinesis   *kar-ee-o-ken-ee'-sis*   See *mitosis*.

Keratin   *ker'-a-tin*   A fibrous insoluble protein present in the epidermis of the skin.

Ketosis   *kee-to'-sis*   An abnormal condition during which an excess of ketone bodies are produced.

Kilocalorie   One thousand calories.

Kinase   An enzyme that transfers phosphate groups to its substrate.

Kinetic energy   *ki-net'-ik*   The energy of motion.

Kinesthesia   *ki-nez-thee'-zhah*   The ability to perceive muscle movement.

Kinins   *kigh'-ninz*   Group of polypeptides that dilate arterioles, increase vascular permeability, act as powerful chemotactic agents, and induce pain.

Krebs cycle   The citric acid cycle; a series of reactions during which energy is liberated from metabolism of carbohydrates, fats, and amino acids.

Labor   *la'-bor*   The period characterized by strong, rhythmic uterine contractions that precedes the birth of a baby.

Lacrimal   *la'-kri-mal*   Pertaining to tears.

Lactation   *lak-tay'-shun*   The production and secretion of milk.

Lacteal   *lak'-tee-al*   The special lymphatic capillaries of the small intestine that take up chyle.

Lactic acid   *lak'-tik*   The product of anaerobic glycolysis, especially in muscle.

Language   A symbol system in which spoken words represent objects or concepts.

Laryngitis   *lar-en-jight'-us*   An inflammation of the larynx.

Larynx   *lar'-inks*   The organ of the voice; located between the trachea and the base of the tongue.

Lateral   *lat'-ur-al*   Away from the midline of the body.

Lateral sacs   The reticular sites in muscle from which calcium is released.

Learning   The ability to modify behavior in response to experience.

Lens   The elastic, doubly convex structure behind the pupil of the eye; focuses the light entering the eye on the retina.

Lesion   *lee'-zhun*   A tissue injury or wound.

Leukemia   *loo-kee'-mee-ah*   A cancerous condition in which there is an excessive production of leukocytes.

Leukocyte   *loo'-ko-site*   A white blood cell.

Ligament   *lig'-a-ment*   A band or sheetlike fibrous tissue connecting bones or parts.

Lipases   Enzymes that digest lipids.

Lipid   *lip'-id*   A substance that is almost insoluble in water but soluble in fat solvents; fatty acids and fats.

Lobe   A curved, rounded structure or projection.

Locus   *low'-kus*   The position that a gene occupies on a chromosome.

Lumbar puncture   A procedure involving insertion of a needle between the third and fourth lumbar vertebrae and into the subarachnoid space to sample cerebrospinal fluid.

Lumen   *loo'-men*   The space inside a tube, blood vessel, or intestine.

Luteinizing hormone   *lu'-tee-in-eye-zing*   An anterior pituitary hormone that stimulates maturation of cells in the ovary and acts on interstitial cells of the male testis.

Lymph   *limf*   The watery fluid in the lymph vessels collected from the tissue fluids.

Lymph node   A mass of lymphatic tissue.

Lymphatic system   *lim-fat'-ik*   A system of vessels carrying lymph closely related anatomically and functionally to the circulatory system.

Lymphocyte   *lim'-fo-site*   An agranular white blood cell.

Lymphokines   *lim'-fo-kines*   Substances involved in cell mediated immune responses that enhance the basic inflammatory response and subsequent phagocytosis.

Lysosomes   *ligh'-so-somes*   Tiny organelles that originate from the Golgi apparatus and contain strong digestive enzymes.

Lysozyme   *ligh'-so-zime*   An enzyme capable of destroying certain kinds of bacteria.

Macrophages   *mak'-ro-fajes*   Phagocytic cells found throughout the connective tissues of the body; they are involved in removing debris such as dead cells and foreign matter.

Macula lutea   *mak'-u-la lew'-tee-uh*   The slightly yellow region lateral to the optic disc.

Mammary glands   *mam'-mar-ee*   Milk-producing glands of the breasts.

Malignant   *ma-lig'-nant*   Life threatening; pertains to diseases that spread and lead to death, such as cancer.

Margination   *mar-jin-ay'-shun*   The adhesion of white blood cells to the walls of capillaries in the early stage of inflammation.

Mass number   The combined number of protons and neutrons in the nucleus of an atom.

Mast cells   Cells found throughout the body that contain histamine and other substances that stimulate the flow of mucus and the outpouring of plasma from capillaries.

Mastication   *mas-ti-kay'-shun*   The act of chewing.

Meatus   *mee-ay'-tus*   The external opening of a canal.

Mechanoreceptor   *mek-an-o-ree-sep'-tor*   A receptor sensitive to mechanical pressures such as touch, sound, or contractions.

Medial   *mee'-dee-al*   Toward the midline of the body.

Mediated transport   *mee'-dee-ay-ted*   Transport involving protein carrier molecules.

Medulla   *med-u'-la*   The central portion of certain organs.

Megakaryocytes   *meg-a-kar'-ee-o-sites*   Large cells in the bone marrow whose fingerlike processes break off to form platelets.

Meiosis   *my-o'-sis*   The last two cell divisions in gamete formation producing nuclei with half the full number of chromosomes (haploid).

Melanin   *mel'-a-nin*   The dark pigment responsible for skin color.

Melanocyte   *mel-an'-o-site*   A cell that produces melanin.

Menarche   *me'-nar-kee*   The onset of menstruation.

Meninges   *men-in'-jeez*   The membranes that cover the brain and spinal cord.

Meningitis   *men-in-ji'-tis*   An inflammation of the meninges covering the brain and spinal cord.

Menopause   *men'-o-pawz*   The physiological termination of menstrual cycles.

Menses   *men'-seez*   The recurrent monthly discharge of menstruation.

Menstruation   *men-stroo-ay'-shun*   The periodic, cyclic discharge of blood, secretions, tissue, and mucus from the mature female genital canal in the absence of pregnancy.

Mesenteries *mes'-en-ter-eez* The doubled-layered membranes of the peritoneum that support most of the organs in the abdominal cavity.

Messenger RNA The type of RNA molecules that direct the incorporation of amino acids into proteins.

Metabolic rate *met-a-bol'-ik* The energy expended by the body per unit time.

Metabolism *me-tab'-o-liz-em* The sum total of the chemical reactions that occur in the body.

Metabolize *me-tab'-o-lize* To transform substances into energy or materials the body can use or store by means of anabolism or catabolism.

Metaphase The phase of mitosis or meiosis in which chromosomes contract and come together prior to migration to the cell poles.

Metastasize *me-tas'-ta-size* The spread of disease from one body part or organ into another not directly connected to it.

Microbodies *my-kro-bod'-eez* Membrane-bounded cytoplasmic structures containing oxidative enzymes; also called peroxisomes.

Microfilament *my-kro-fil'-a-ment* Filaments associated with contractile activities of the cell and developmental modifications of cell and organ shape.

Microtubule *my-kro-too'-bule* Cytoplasmic structures not bound in membranes having a support function and associated with movement processes.

Microvilli *my-kro-vil'-lee* The tiny protoplasmic projections formed on the free surfaces of some epithelial cells; appear brushlike when viewed through a microscope.

Mineralocorticoid *min-er-al-o-kor'-ti-koid* An adrenal cortical steroid hormone that regulates mineral metabolism and fluid balance.

Minerals The inorganic chemical compounds found in nature.

Mitochondria *my-to-kon'-dree-ah* The cytoplasmic organelles in the form of granules, rods, filaments responsible for generation of metabolic energy for cellular activities.

Mitosis *my-to'-sis* The division of the cell nucleus into two new nuclei that each contain the same genetic information as the original nucleus; often followed by division of the cytoplasm of a cell; also called karyokinesis.

Molar *mo'-lar* A solution concentration determined by mass of solute—one liter of a one-molar solution contains an amount of solute equal to the solute's molecular weight in grams.

Molecular weight *mo-lek'-you-lar* The sum of the atomic weights of all atoms of a molecule.

Molecule *mol'-e-kewl* A complex of two or more atoms held together by covalent bonds.

Moles Elevations of the skin that are pigmented.

Monocyte *mon'-o-site* A large, single-nucleus white blood cell.

Monomers *mon'-o-merz* The units that combine to form polymeric molecules.

Monozygotic twins *See* identical twins.

Monosomy *mo-no'so-me'* The condition in which one of two homologous chromosomes is missing.

Motor end plate The region of a muscle cell membrane that contains receptors for the neurotransmitter acetylcholine.

Motor nerve cells The nerves that carry impulses leaving the brain and spinal cord.

Motor neurons Neurons that carry information from the central nervous system to muscles or glands.

Motor unit A neuron and the muscle cells it supplies.

mRNA *See* messenger RNA.

Mucus *mew-kus* A sticky, thick fluid secreted by mucous glands and mucous membranes that keeps the free surface of membranes moist.

Mucous membranes The membranes that form the linings of the digestive, respiratory, urinary, and reproductive tracts.

Multiple sclerosis *skler-o'-sis* A chronic condition characterized by destruction of the myelin sheaths of neurons in the spinal cord and the brain.

Multipolar neurons *mul-ti-pol'-ar* Neurons that have one long axon and numerous dendrites.

Muscle fibers Muscle cells.

Muscle spindles The complex encapsulated structures found in skeletal muscles that are sensitive to stretching.

Muscle twitch A single rapid contraction of a muscle followed by relaxation.

Muscular dystrophy *dis'-tro-fee* A progressive disorder marked by atrophy and stiffness of the muscles.

Mutation *my-tay'-shun* An alteration in the genetic material.

Myelin *my'-e-lin* The white, lipid substance forming a sheath around some nerves.

Myelinated fibers *my'-e-li-nay-ted* Axons (projections of a nerve cell) covered with myelin.

Myelitis *my-a-light'-us* An inflammation of the spinal cord.

Myocardial infarction *my-o-kar'-dee-al in-fark'-shun* A condition characterized by dead tissue areas in the myocardium caused by interruption of blood supply to the area.

Myocardium *my-o-kar'-di-um* The cardiac muscle layer of the wall of the heart.

Myofibril *my-o-figh'-bril* A fibril found in the cytoplasm of the muscle.

Myofilament *my-o-fil'-a-ment* The filamentous structures making up a sarcomere consisting of thick and thin types.

Myogenic *my-o-jen'-ik* Originating in muscle; of muscular origin.

Myoglobin *my-o-glo'-bin* An oxygen binding substance found in muscle.

Myometrium *my-o-me'-tree-um* The thick uterine musculature.

Myopia *my-o'-pee-ah* Nearsightedness.

Myosin *my'-o-sin* One of the principal proteins found in muscle.

Necrosis *ne-kro'-sis* The death or disintegration of a cell or tissues caused by disease or injury.

Negative feedback Feedback that tends to cause the level of a variable to change in a direction opposite to that of an initial change.

Nephron *nef'-ron* The functional part of the kidney.

Nerve fiber Axon (nerve cell projection) together with certain sheaths or coverings.

Neuralgia *noo-ral'-jee-ah* A severe paroxysmal (spasm-producing) pain along the course of a nerve.

Neuritis *noo-righ'-tis* An inflammatory or degenerative condition of the nerves.

Neurohypophysis *noo-ro-high-pof'-i-sis* The portion of the pituitary gland derived from the brain.

Neuromuscular junction *noo-ro-mus'-ku-lar* The junction where a motor neuron approaches the membrane of a skeletal muscle cell.

Neurons *noo'-ronz* The nerve cells that transmit messages throughout the body.

Neutron *noo'-tron* An uncharged particle located in the nucleus of an atom.

Neutrophil *noo'tro-fil* The most abundant of the white blood cells.

Nitrogen balance The state in a normal adult in which the nitrogen excreted equals the nitrogen intake in the form of food.

Nodes of Ranvier *rahn-vee-ay* Gaps between adjacent Schwann cells wrapped around an axon.

Nondisjunction *non-dis-jungk'-shun* Failure of separation of paired chromosomes during cell division.

Nucleic acid One of the major classes of cellular macromolecules. Includes DNA and RNA.

Nucleoli *noo-klee'-o-lee* Small spherical bodies in the cell nucleus.

Nucleotide *noo'-klee-o-tide* A component of DNA and RNA consisting of a sugar, a nitrogenous base, and a phosphate group.

Nucleus *noo'-klee-us* (1) A dense central body in most cells containing the genetic material of the cell; (2) the core of an atom.

Nutrient *noo'-tree-ent* Any externally supplied substance used in metabolism.

Nutrition The process of obtaining and utilizing nutrients.

Nystagmus *nis-tag'-mus* An oscillatory movement of the eyeballs.

Obesity *o-bee'-sit-ee* A condition of a person being overweight; often leads to other complications.

Occipital *ok-sip'-i-tal* Pertaining to the back of the head area.

Occlusion *o-kloo'-zhun* Closure or obstruction.

Olfaction *ol-fak'-shun* Smell.

Oncotic pressure *on-kot-ik* The osmotic force exerted by plasma proteins.

Oogenesis *o-o-jen'-e-sis* The process of origin, growth, and formation of the ovum.

Ophthalmic *of-thal'-mik* Pertaining to the eye.

Optic *op'-tik* Pertaining to the eye.

Optic chiasma *op'-tik kee-az'-ma* The meeting of the optic nerves after entering the cranium.

Opsonins *op'-se-nins* Proteins that coat specific foreign particles and thereby render them more susceptible to phagocytosis.

Oral Relating to the mouth.

Organ A part of the body combining two or more tissues to perform a specialized function.

Organelle *or-gan-el'* A specialized structure or part of a cell having a definite function to perform.

Organic molecules A class of molecules consisting of hydrocarbons and hydrocarbon derivatives.

Orgasm *or'-gaz-um* The intense emotional and physical climax associated with sexual stimulation.

Osmoreceptor *os-mo-ree-sep'-tor* A structure sensitive to osmotic pressure.

Osmosis *os-mo'-sis* The passage (diffusion) of a solvent through a membrane in response to a concentration difference across the membrane.

Ossicles *os'-si-kalz* The three bones of the middle

Osteoblasts *os'-tee-o-blasts* The bone-forming cells.

Osteoclasts *os'-tee-o-klasts* The large cells that break down or erode bone substance.

Osteomalacia *os-tee-o-mal-a'-she-ah* The softening of bone resulting from vitamin D deficiency in the adult.

Osteomyelitis *os-tee-o-my-a-light'-us* A disease in which the periosteum, the contents of the marrow cavity, and the bone tissue become inflamed.

Osteoporosis *os-tee-o-por'-o-sis* An increased softening of the bone resulting from a gradual reduction in the rate of bone formation while the rate of bone breakdown remains normal; a common condition in older people.

Otitis media *o-tigh'-tis mee'-dee-ah* Middle-ear infection.

Otolith *o'to-lith* One of the small calcareous masses in the utricle and saccule of the inner ear.

Ovarian cycle *o-va'-ree-an* The monthly cycle of follicle development, ovulation, and corpus luteum formation in an ovary.

Ovary *o'-va-ree* The female sex organ in which ova (eggs) are produced.

Ovulation *ov-u-lay'-shun* The maturation and release of an ovum from an ovary.

Ovum *o'vum* The female gamete (germ cell); an egg cell.

Oxidation *oks-i-day'-shun* The loss of electrons by a molecule; the process of substances combining with oxygen.

Oxidative phosphorylation *oks'-i-day-tiv fos-for-i-lay'-shun* The synthesis of ATP coupled with electron transport.

Oxygenated *oks'-i-je-nay-ted* The condition in which a substance is saturated with oxygen.

Oxygen debt *oks'-i-jen* The volume of oxygen required after exercise in excess of oxygen consumption in the resting state for an equivalent length of time.

Oxyhemoglobin *oks-i-hee'-mo-glo-bin* Hemoglobin that is combined with oxygen.

Oxytocin *oks-i-to'-sin* A hormone released by the posterior pituitary that stimulates contractility of smooth muscles of the uterus and myoepithelial cells surrounding the alveoli of the mammary glands.

Pacemaker *See* sinoatrial node.

Palate *pal'-et* The roof of the mouth.

Palpation *pal-pay'-shun* Examination by touch.

Pancreas *pan'-kree-as* The gland located behind the stomach, between the spleen and the duodenum, producing both endocrine and exocrine secretions.

Pancreatic juice *pan-kree-at'-ik* A secetion of the pancreas containing enzymes for digestion of carbohydrates, proteins, and lipids.

Pancreatitis *pan-kree-a-tigh'-tis* An inflammation of the pancreas.

Papilla *pa-pil'-lah* A small elevation, nipple-shaped or cone-shaped.

Papillary muscles *pa'-pil-lar-ee* Cone-shaped muscles such as those that project into the cardiac ventricular lumen.

Paralysis *pa-ral'-i-sis* The loss of muscle function or of sensation.

Paraplegia *par-a-plee'-gee-ah* Paralysis of the lower limbs.

Parasympathetic division *par-a-sim-pa-thet'-ik* A division of the autonomic nervous system; also referred to as the craniosacral division.

Parathyroid glands *par-a-thigh'-roid* The several small endocrine glands posterior to the capsule of the thyroid gland.

Parathyroid hormone The hormone released by the parathyroid glands that regulates blood calcium level.

Parenteral *par-en'-ter-al* Occurring through some route other than the alimentary canal, such as intravenous.

Parietal *par-eye'-i-tal* Pertaining to the walls of a cavity.

Parotid *par-ot'-id* Located near the ear.

Partial pressure The fraction of the total pressure of a gas mixture that is contributed by a particular gas.

Parturition *par-toor-i'-shun* The act of giving birth.

Pathogenesis *path-o-jen'-e-sis* The development of a disease.

Pelvis A basin-shaped structure, especially the lower portion of the body's trunk.

Penis *pee'-nis* The male organ of copulation and urinary excretion.

Pepsin An enzyme capable of digesting proteins in an acid pH.

Pepsinogen *pep-sin'-o-jen* An inactive precursor of pepsin.

Peptide bond *pep'-tide* A bond joining the amino group of one amino acid to the acid carboxyl group of a second amino acid with the loss of a water molecule.

Pericardium *per-i-kar'-dee-um* The closed membranous sac enveloping the heart.

Peripheral nervous system (PNS) *per-if'-er-al* A

system of nerves that connect the outlying parts of the body and their receptors with the central nervous system.

Peripheral resistance   The impedance to blood flow offered by the systemic blood vessels.

Peristalsis   *per-is-tal'-sis*   The progressive wave of contraction seen in tubes.

Peritoneum   *per-i-ton-ee'-um*   The membrane lining the interior of the abdominal cavity and surrounding the viscera within the cavity.

Peritonitis   *per-i-ton-eye'-tis*   An inflammation of the peritoneum.

Permeability   *per-mee-a-bil'-i-tee*   That property of membranes that permits transit of molecules and ions.

pH   The symbol for hydrogen ion concentration; a measure of the relative acid or base level of a substance or solution.

Phagocyte   *fag'-o-site*   A cell capable of engulfing and digesting particles or cells harmful to the body.

Phagocytosis   *fag-o-sigh-to'-sis*   The ingestion of foreign solids by cells.

Phantom pain   A phenomenon whereby a person who has undergone amputation continues to feel pain from the amputated body part.

Pharynx   *far'-inks*   The muscular, membranous tube extending from the base of the skull to the esophagus.

Phenotype   *fee'-no-tipe*   The observable characteristics of an individual as determined both genetically and environmentally.

Phlebitis   *fle-by'-tis*   An inflammation of a vein.

Phospholipid   *fos-fo-lip'-id*   The type of lipid found in the membranes of cells.

Photoreceptor   *fo-to-ree-sep'-tor*   The specialized receptor cells that can convert light energy into a nerve impulse.

Physiology   The branch of biology that deals with the functions of living organisms or their parts.

Pigment   A colored substance selectively absorbing light of certain wavelengths.

Pinna   *pin'-nah*   The irregularly shaped elastic cartilage covered with skin forming the most prominent portion of the outer ear.

Pinocytosis   *pi-no-sigh-to'-sis*   The engulfing of liquid by cells.

Pituitary gland   The gland located beneath the brain that serves a variety of functions including regulation of the gonads, thyroid, and adrenal cortex.

Placenta   *pla-sen'-tah*   The organ to which the embryo attaches by the umbilical cord for nourishment and waste removal; has an endocrine function as well.

Plasma   *plaz'-ma*   The fluid portion of the blood or lymph.

Plasma cells   Cells that produce antibodies.

Plasma membrane   The membrane that surrounds the cell cytoplasm.

Plasmin   *plaz'-min*   A protein that breaks down the fibrin meshwork of a blood clot.

Platelet   *plate'-let*   One of the disc-shaped components of blood; involved in clotting.

Platelet plug   A group of platelets that adhere to the walls of damaged blood vessels.

Pleura   *ploor'-ah*   The membrane covering the lungs and lining the thoracic cavity.

Pleurisy   *plur'-i-see*   Prolonged inflammation of the visceral and parietal pleuras, making breathing painful.

Plexus   *plek'-sus*   A network of interlacing nerves or anastomosing blood vessels or lymphatics.

Plica   *ply'-ka*   A fold.

Pneumothorax   *noo-mo-tho'-raks*   The presence of air or gas in a pleural cavity.

Podocyte   *pod'-o-site*   An epithelial cell located on the basement membrane of the glomerulus, spreading thin cytoplasmic projections over the membrane.

Polar body   A minute cell produced during oogenesis.

Polarized   *po'-lar-ized*   The state of an unstimulated neuron in which the inside of the cell is relatively negative in comparison to the outside.

Poliomyelitis   *po'-lee-o-my-a-light'-us*   A viral disease that may destroy the cell bodies of efferent neurons to skeletal muscles, especially in the cervical and lumbar areas of the spinal cord.

Polycythemia   *pol-ee-sigh-theem'-ee-a*   A condition in which there is a net increase in the total circulating erythroyte mass of the body.

Polymer   *pol'-ee-mer*   A molecule consisting of a sequence of similar units or monomers.

Polynucleotide   A chain of nucleotides, as in nucleic acids.

Polypeptide   *pol-ee-pep'-tide*   A small chain of amino acids.

Polyribosome   *pol-ee-ribe'-o-some*   A multiple structure composed of several ribosomes held together by a molecule of messenger RNA; also called polysome.

Polysome   *pol'-ee-some*   See polyribosome.

Pons (1) Any bridgelike structure or part; (2) the structure connecting the cerebellum with the brain stem providing linkage between upper and lower levels of the central nervous system.

Positive feedback Feedback that tends to cause the level of a variable to change in the same direction as an initial change.

Postganglionic (postsynaptic) neuron *post-gang-lee-on'-ik* A neuron of the autonomic nervous system having its cell body in a ganglion with the axon extending to an organ or tissue.

Postsynaptic potential An alteration in the polarity of a postsynaptic membrane as a result of the influence of a neurotransmitter.

Pregagnlionic (presynaptic) neuron *pree-gang-lee-on'ik* A neuron of the autonomic nervous system having its cell body in the brain or spinal cord and its axon terminating in a ganglion.

Pressoreceptor *pre-so-ree-sep'-tor* A nerve ending in the wall of the carotid sinus and aortic arch sensitive to vessel stretching.

Proctologic *prok-to-loj'-ik* Pertaining to the rectum or anus.

Progesterone *pro-jes'-ter-own* A hormone responsible for preparing the uterus for the fertilized ovum.

Prone Refers to a body lying horizontally with the face downward.

Pronucleus *pro-noo'-clee-us* One of two nuclear bodies (one male and one female) of a newly fertilized ovum, the fusion of which results in formation of a cleavage nucleus.

Prophase *pro'-faze* The first phase of mitosis.

Prostaglandins *pros-ta-glan'-dinz* A specialized group of fatty acids that serve as chemical messengers.

Protein *pro'-teen* A complex nitrogenous substance found in various forms in animals and plants as the principal component of protoplasm.

Proteinuria *pro-te-in-oo'-ree-ah* The passage of albumin or other protein in the urine.

Prothrombin *pro-throm'-bin* A circulating protein that is converted to the enzyme thrombin in the presence of prothrombin activator.

Proton *pro'-ton* A particle carrying a positive charge; located in the nucleus of an atom.

Protrude *pro-trood'* To project or assume an abnormally prominent position.

Proximal *proks'-i-mal* Toward the attached end of a limb or the origin of a structure.

Psoriasis *so-rye'-uh-sis* A chronic inflammatory skin disease characterized by development of red patches covered with silvery, overlapping scales.

Ptyalin *tie'-a-lin* A starch-splitting enzyme contained in saliva.

Puberty *pu'-ber-tee* The period at which reproductive organs become functional.

Pulmonary *pul'-mon-ar-ee* Pertaining to the lungs.

Pulmonary circuit The circulatory vessels of the lungs.

Pulmonary edema *e-dee'-mah* An effusion of fluid into the air sacs and interstitial tissue of the lungs.

Pulse The rhythmic expansion in arteries resulting from heart contraction; can be felt from outside the body.

Pupil An opening in the center of the iris through which light enters the eye.

Purkinje fibers *pur-kin'-jee* The modified cardiac muscle fibers forming the final portion of the conduction system of the heart.

Pus The fluid product of inflammation composed of white blood cells, the debris of dead cells, and a thin fluid.

Pyelonephritis *pie-el-o-nef-rye'-tis* An inflammation of the kidney pelvis and surrounding kidney tissues.

Pyloric glands *pie-lor'-ik* The glands located in the region of the pylorus of the stomach.

Pyloric region The final portion of the stomach preceding the duodenum.

Pyrogen *pie'-ro-jen* An agent that induces fever.

Quadriplegia *quad-ri-plee'-jee-ah* The paralysis of all four limbs.

Radiate *ray'-dee-ate* Diverging from a central point.

Receptor *ree-sep'-tor* A structure that converts information about conditions in the body's internal or external environments into neural signals.

Recessive *ree-seh'-siv* An allele, or the corresponding trait, that is expressed only in homozygous cells or organisms.

Reduction The gain of electrons by a molecule.

Reflex Automatic, stereotyped reactions to stimuli.

Refracted Bent.

Refractory period *ree-frak'to-ree* The period of resistance to stimulation immediately after responding.

Relative refractory period The period following stimulation during which a stronger than usual stimulation can evoke an action potential.

Renal *ree'-nal* Pertaining to the kidney.

Renal calculus *ree'-nal kal'-ku-lus* A kidney stone.

Renal circulation That part of the circulatory system that routes blood through the kidneys.

Renin *ren'-in* A substance produced by the kidneys which converts angiotensinogen to angiotensin.

Respiration *res-per-ay-shun* Utilization of oxygen as electron acceptor. Also, the exchange of air between the lungs and the atmosphere.

Resting potential The electrical potential (about −70 mV) across a neuronal membrane in the resting state.

Reticulocyte *re-tik'-u-lo-site* A nonnucleated, young erythrocyte.

Reticulum *re-tik'-u-lum* A fine network.

Retract To draw back, shorten, contract.

Rhodopsin *ro-dop'-sin* The photopigment contained in the rods of the retina.

Rh system A group of surface antigens on red blood cells.

Ribonucleic acid (RNA) *rye-bo-nu-kle'-ik* The nucleic acid that contains ribose.

Ribosomal RNA The RNA molecules that form the backbones of ribosomes.

Ribosomes *rye'-bo-somes* The cytoplasmic structures at which proteins are synthesized.

Rickets *rik'-ets* A disease occurring in infants and young children characterized by softening of the bone caused by demineralization from malnutrition.

Rods One of the two types of photosensitive cells in the retina.

Roentgenogram *rent'-ge-ne-gram* An x-ray film.

Saliva *sa-ligh'-vah* The combined secretions of salivary and mucous glands of the mouth.

Salt A compound that, when dissolved in water, dissociates into cations other than hydrogen ions and anions other than hydroxide ions.

Sarcomere *sar'-ko-meer* A unit within a myofibril that extends from Z line to Z line.

Sarcoplasmic reticulum *sar-ko-plas'-mik re-tik'-u-lum* The membranous network running through skeletal muscle cells.

Sclera *skle'-rah* The firm, fibrous outer layer of the eyeball; functions for protection and maintenance of eyeball shape.

Scrotum *skro'-tum* The two-layered sac enclosing the testes.

Sebaceous glands *se-bay'-shus* Glands that develop from and empty their sebum secretion into hair follicles.

Sebum *see'-bum* The secretion of sebaceous glands; oily substance rich in lipids.

Secretion *see-kree'-shun* The passage of material formed by a cell from the inside to the outside of its plasma membrane.

Semen *see'-men* The fluid produced by male reproductive structures; contains sperm.

Semilunar valves *sem-i-loo'-nar* Valves that prevent blood return to the ventricle after contraction.

Seminiferous tubules *sem-i-nif'-er-us* Highly convoluted tubes within the testes containing spermatocytes.

Sensory nerve A nerve that contains only the processes of sensory neurons and carries nerve impulses to the central nervous system.

Sensory nerve cell An initiator of nerve impulses following receptor activity.

Serous fluid *ser'-us* A thin, watery fluid.

Serum *se'-rum* The amber-colored fluid that exudes from clotted blood as the clot shrinks and then no longer contains fibrinogen.

Sex chromosome *kro'-mo-some* Chromosome that determines sex of the fertilized egg; X and Y chromosomes.

Sex-linked genes Genes located in sex chromosomes.

Sinoatrial node *sigh-no-ay'-tree-al* The mass of specialized myocardial cells in the wall of the right atrium.

Skeletal muscle Muscle that is primarily attached to bones and that is responsible for the voluntary movements of parts of the skeleton.

Smooth (nonstriated) muscle Muscle consisting of spindle-shaped, unstriped (nonstriated) muscle cells; involuntary muscles.

Sodium-potassium pump A transport system that transfers Na^+ ions out of cells and K^+ ions into cells, using energy.

Solute *sol'-yoot* The dissolved substance in a solution.

Solution A homogenous mixture of two or more components.

Somatic nervous system *so-ma'-tik* A division of the peripheral nervous system; also called the voluntary nervous system.

Sperm The mature male germ cell; a spermatozoon.

Spermatogenesis *sper-ma-to-jen'-e-sis* The process of meiosis (cell division) in the male to produce mature male germ cells.

Sphincter *sfink'-ter* A muscle surrounding and enclosing an orifice.

Sprain The wrenching of a joint producing stretching or laceration of the ligaments.

Squamous *skway'-mus* Pertaining to flat and thin

cells that form the free surface of epithelial tissue.

Stagnant hypoxia *stag'-nant high-pock'-see-uh* A condition marked by reduced available oxygen caused by slowed blood circulation.

Stasis *stay'-sis* (1) A decrease or stoppage of flow; (2) a state of equilibrium.

Static balance *stat'-ic* Balance concerned with changes in the position of the head.

Stenosis *sten-o'-sis* Constriction or narrowing.

Steroids *ster'-oidz* A specific group of chemical substances including certain endocrine secretions and cholesterol.

Stimulus *stim'-u-lus* An excitant or irritant; an alteration in the environment of a living thing producing a response.

Strabismus *stra-bis'-mus* A squint; that abnormality of the eyes in which the visual axes do not meet at the desired objective point.

Stratum *stray'-tum* A layer.

Stress Any stimulus that directly or indirectly causes neurons of the hypothalamus of the brain to release corticotropin releasing factor (CRF).

Striated muscle *stry'-ay-ted* Muscle consisting of cross-striated (cross-striped) muscle fibers.

Stricture *strik'-chur* A contraction or inward pinching of a canal or duct.

Stroke A condition, sudden in onset, in which a cerebral blood vessel is blocked or ruptured.

Stroke volume A volume of blood ejected by the left ventricle during a systole.

Stroma *stro'-mah* The supporting framework of an organ including connective tissue, vessels, and nerves.

Sty An inflammation of the connective tissue of the eyelid, near a hair follicle.

Subcutaneous *sub-ku-tay'-nee-us* Beneath the skin.

Sublingual *sub-ling'-wal* Located beneath the tongue.

Substrate A substance that is altered in the course of an enzyme-catalyzed reaction.

Substrate level phosphorylation Synthesis of ATP coupled to reactions of glycolysis or of other nonrespiratory pathways.

Sulcus *sul'-kus* A furrow or linear groove; on the brain, a less deep depression than a fissure.

Summation *sum-may'-shun* The accumulation of effects, especially those of muscular, sensory, or mental stimuli.

Superficial (external) Located close to or on the body surface.

Superior Refers to the head or upper; higher.

Supine Refers to a body lying with the face upward.

Surface tension The contractile surface of a liquid or structure by which it tends to present the least possible surface; liquid meniscus formation.

Surfactant *sur-fact'-ant* A substance on pulmonary alveoli walls which reduces surface tension thus preventing collapse.

Suspension *sus-pen'-shun* A dispersing of particles throughout a body of liquid.

Suspensory ligament of eye *sus-pen'-so-ree lig'-a-ment* Fibrous ligament that holds the lens in place in the eye.

Sweat glands The glands that secrete a watery solution of sodium chloride (salt water).

Sympathetic division A division of the autonomic nervous system; opposes parasympathetic functions.

Synapse *sin'-apse* A region of communication between neurons.

Synaptic cleft *sin-ap'-tik* The space at a chemical synapse between neurons.

Synaptic delay *sin-ap'tik* The time required for an impulse to cross a synapse between two neurons.

Synovial fluid *sin-o'-vi-al* A fluid secreted by the synovial membrane; lubricates joint surfaces and nourishes articular cartilages.

System A group of organs that function cooperatively to accomplish a common purpose.

Systemic *sis-tem'-ik* General; pertaining to the whole body.

Systemic circuit The blood vessels of the body with the exception of the pulmonary vessels.

Systemic edema *e-dee'-ma* An accumulation of fluid in body organs or tissues.

Systole *sis'-to-lee* The contraction phase of a cardiac cycle.

Systolic pressure *sis-tol'-ik* The pressure generated by the left ventricle during systole.

T cells A heterogeneous group of lymphocytes that include those cells committed to participating in cell-mediated immune responses.

Tachycardia *tak-ee-kar'-dee-ah* The abnormal, excessive rapidity of heart action; over 100 beats per minute.

Taste buds The receptors for taste on the tongue, roof of mouth, pharynx, and larynx.

Temporal summation *tem'-po-ral sum'-may-shun* The arrival of many nerve impulses at a synapse within a short period causing release of sufficient transmitter to initiate a nerve impulse in the postsynaptic cell.

Tendon *ten'-don* A band of dense fibrous tissue

forming the termination of a muscle and attaching the muscle to a bone.

Testis *tes'-tis* The male primary sex organ that produces sperm.

Testosterone *tes-tos'-te-rone* A steroid hormone produced in the testes.

Tetanus *tet'-a-nus* (1) The tense, contracted state of a muscle; (2) an infectious disease.

Thalamus *thal'-a-mus* The mass of gray matter at the base of the brain.

Theca *thee'-ka* A sheath.

Thermoreceptor *ther-mo-ree-sep'-tor* A receptor sensitive to temperature changes.

Thoracic *tho-ras'-ik* Refers to the chest.

Thorax *tho'-raks* That portion of the trunk above the diaphragm and below the neck.

Threshold The lower limit of stimulus capable of producing an impression on consciousness or evoking a response in an irritable tissue.

Thrombin *throm'-bin* An enzyme that induces clotting by converting fibrinogen to fibrin.

Thrombocytopenia *throm-bo-sigh-to-pee'-nee-ah* A condition in which there is a decrease in the number of blood platelets below normal.

Thrombophlebitis *throm-bo-fle-by'-tis* An inflammation of a vein associated with blood-clot formation

Thrombus *throm'-bus* A clot that is fixed or stuck to a vessel wall.

Thymine *thigh'-meen* A nucleic acid base found in DNA; pairs with adenine.

Thymus gland *thigh'-mus* A potential source of hormonal material; active in immune response.

Thyroid gland *thigh'-roid* One of the largest of the body's endocrine glands.

Thyroid-stimulating hormone (TSH) A pituitary hormone that accelerates growth of the thyroid gland and stimulates production of thyroxine.

Thyrotropin *See* thyroid-stimulating hormone (TSH).

Tissue A group of similar cells and fibers forming a distinct structure.

Tolerance *tol'-er-antz* The failure of the body to mount a specific immune response against a particular antigen.

Tomography *toe-mog'-ra-fee* X-ray photography of a specific structure in a certain layer of tissue in the body, in which images of structures in other layers are eliminated.

Tonic *ton'-ik* Refers to the state of continuous muscular or neuron activity.

Toxemia *tok-see'-mee-ah* A condition in which blood contains poisonous products.

Toxic *tok'-sik* Poisonous.

Trachea *tray'-kee-uh* The windpipe; the cartilaginous and membranous tube extending from larynx to bronchi.

Tract A collection of nerve fibers having the same origin, termination, and function.

Trait *trate* A characteristic or quality of an individual.

Transcription the synthesis of RNA using DNA as a template.

Transducer *trans-doo'ser* An agent that converts energy forms; for example, receptors of the nervous system transfer light into a nerve impulse.

Transfer RNA The class of molecules that bring the amino acids to the site where they are incorporated into proteins.

Translation the use of RNA as a template in the synthesis of protein.

Transmutation *trans-mu-tay'-shun* The change of one element into another.

Trauma *traw'-mah* An injury, wound, or shock; usually produced by external forces.

Triacylglycerol *trigh-a-sil-gli'-ser-ol* A substance in which three fatty acids are attachecd to one glycerol molecule (as in oil, butter fat . . .).

Tricarboxylic acid cycle *trigh-car-bok-sil'-ik* *See* Krebs cycle.

Trisomy *trigh'-so-mee* The condition in which one chromosome is represented three times rather than the normal two.

tRNA *See* transfer RNA.

Tropic *tro'-pik* Pertains to nutrition.

Trophoblast *trof'-o-blast* Outer sphere of cells of the blastocyst.

Trypsin *trip'-sin* An active enzyme that splits proteins.

Tubal ligation *too-bal li-gay'-shun* A form of surgical sterilization in which the uterine tubes are tied and severed.

Tubal pregnancy An ecotopic pregnancy that occurs within a uterine tube.

Tumor An abnormal growth of cells; a swelling; cancerous at times.

Tunica *too'-ni-kah* A covering or tissue coat; membrane layer.

Twitch A brief contraction of muscle in response to a stimulus.

Tympanic membrane *tim-pan'-ik* The eardrum.

Ulcer *ul'-ser* A lesion or erosion of an epithelial

surface, such as gastric ulcer of stomach.

Umbilical cord *um-bil'-i-kal* A structure bearing arteries and veins connecting the placenta and the fetus.

Umbilicus *um-bil'-i-kus* The navel; marks site that gave passage to umbilical vessels in the fetal stage.

Unmyelinated fibers *un-my'e-li-nay-ted* Nerve axons that are not covered with myelin.

Uracil *yoo'-ra-sil* One of the bases in RNA; pairs with adenine.

Urea *yoo-ree'-ah* The main nitrogen-containing waste excreted in the urine.

Uremia *yoo-ree'-mee-ah* A toxic accumulation in the blood of substances normally excreted by the kidneys.

Ureter *yoo-ree'-ter* The tube that carries urine from kidney to bladder.

Urethra *yoo-ree'-thrah* The canal through which urine passes from the bladder to the outside of the body.

Vacuole *vak'-yoo-ole* A clear space in a cell.

Valvular insufficiency *val'-vu-lar* A condition in which the cusps of the cardiac valves do not close tightly.

Varicose vein *var-uh-kos'* A dilated, knotted, tortuous vein.

Vas A duct; vessel.

Vasa recta *va'-sa rek'-ta* Capillary branches that supply loops of Henle and collecting ducts.

Vascular *vas'-ku-lar* Pertaining to channels or vessels.

Vasoconstriction *vaz-o-kon-strik'-shun* The narrowing of blood vessels.

Vasodilatation *vaz-o-die-lay-tay'-shun* The relaxation of the smooth muscles of the vascular system producing dilated vessels.

Vasomotion *vaz-o-mo'-shun* An increase or decrease in caliber of a blood vessel.

Vasomotor center *vaz-o-mo'tor* An area of brain concerned with regulation of blood vessel resistance.

Vasomotor nerve fibers *vaz-o-mo'-tor* The motor nerve fibers that regulate the constriction or dilation of blood vessels.

Vasopressin *vaz-o-pres'-sin* Another name for antidiuretic hormone.

Vein A vessel carrying blood away from the tissues toward the heart.

Ventral *ven'-tral* Anterior or front.

Ventricle *ven'-tri-kal* (1) A small cavity or pouch; (2) the blood propulsion chamber of the heart.

Ventricles of the brain *ven'-tri-kalz* The cavities in the interior of the brain.

Venule *ven'-ule* A small vein.

Vertigo *ver'-tee-go* Dizziness; the feeling of movement such as a sensation that the external environment is revolving.

Vesicle *ves'-i-kal* A small liquid-filled sac or bladder.

Villi Fingerlike projections on the inner surface of the small intestine.

Viscera *vis'-ser-ah* The internal organs.

Visceral *vis'-ser-al* Pertaining to the internal part of a structure or the internal organs.

Viscosity *vis-kos'-i-tee* The state of being sticky or thick.

Visual acuity *a-kyoo'-i-ty* The ability of the eye to distinguish detail.

Vital capacity The volume of air that can be expelled from the lungs by forcible expiration after the deepest inspiration.

Vitamins The organic compounds required by the body in minute amounts for physiological maintenance and growth.

Voluntary muscle Muscle under control of the will; skeletal muscle.

Wernicke's area *ver'-ne-kez* A portion of the left temporal lobe of the brain; damage to this area results in a loss of comprehension of spoken language.

White blood cell A blood cell that does not contain hemoglobin. Also known as a leukocyte.

White matter The white substance of the central nervous system; the nerve fibers.

Zygote *zy'-goat* The fertilized ovum before splitting (cleavage); produced by union of two gametes.

PHOTO AND ART ACKNOWLEDGMENTS

Fig. 2.19   courtesy of Dr. Irving Geis

3.7a   B. F. King, School of Medicine, University of California, Davis/BPS

3.7b   B. F. King, School of Medicine, University of California, Davis/BPS

3.7c   B. F. King, School of Medicine, University of California, Davis/BPS

3.12   courtesy of Dr. Barry Batzing

3.13   courtesy of Dr. Barry Batzing

3.8   Kimball, J. W., BIOLOGY, 4/e; Addison-Wesley Pub. Co., 1978

3.14   Kimball, J. W., BIOLOGY, 4/e; Addison-Wesley Pub. Co., 1978

3.17b   courtesy of E. de Harven, Rockefeller University

3.17c   courtesy of E. de Harven, Rockefeller University

5.8   Vander, Sherman, Luciano, HUMAN PHYSIOLOGY, 3/e; McGraw-Hill, 1980

6.3   courtesy of C. Franzini-Armstrong, University of Pennsylvania

8.13   Guyton, TEXTBOOK OF MEDICAL PHYSIOLOGY, W. B. Saunders Pub. Co., 1981

8.21   The Division of Ophthalmology, Stanford Medical Center; Ivan Pieper, Ophthalmic Photographer

9.2   © William Thompson, 1982

9.4   Dr. J. A. Hobson and Hoffman-La Roche, Inc. from Dreamstage catalog

10.17   Bloom & Fawcett, A TEXTBOOK OF HISTOLOGY, 10/e, W. B. Saunders, 1975

11.2a   courtesy of Dr. Irving Geis

11.7   Kage/Peter Arnold, Inc.

12.7   Vander, Sherman, Luciano, HUMAN PHYSIOLOGY, 3/e, McGraw-Hill, 1980

13.6   ©William Thompson, 1982

14.9   Andrejs Liepins, Sloan-Kettering Institute for Cancer Research

16.20   Purtillo, A SURVEY OF HUMAN DISEASES, 1978, Addison-Wesley Pub. Co.

16.21   Purtillo, A SURVEY OF HUMAN DISEASES, 1978, Addison-Wesley Pub. Co.

17.13   ©William Thompson, 1982

Note: Illustrative or tabular material is indicated in *italics*. Pages containing illustrations are not specified separately when they occur within an inclusive reference. Italicized page references may also contain textual references to a subject.